Model Predictive Control

Model Predictive Control

Fundamentals and Practice

Jay H. Lee
University of Southern California

Niket S. Kaisare
Indian Institute of Technology Madras

Carlos E. García
Shell Oil Co. (Retired)

WILEY

Published by John Wiley & Sons, Inc., Hoboken, New Jersey.

For general information on our other products and services or for technical support, please contact our Customer Care Department within the United States at (800) 762-2974, outside the United States at (317) 572-3993 or fax (317) 572-4002.

Wiley also publishes its books in a variety of electronic formats. Some content that appears in print may not be available in electronic formats. For more information about Wiley products, visit our web site at www.wiley.com.

Library of Congress Cataloging-in-Publication Data has been applied for:

Hardback ISBN: 9781394333295
ePdf ISBN: 9781394333318
ePub ISBN: 9781394333301
oBook ISBN: 9781394333325

Cover Design: Wiley
Cover Images: Westend61/Getty Images, Bim/Getty Images,
Westend61/Getty Images, Antro_Nero/stock.adobe.com

Contents

Foreword

Model Predictive Control (MPC) occupies a unique position in control engineering, distinguished by its seamless integration of rigorous theory and industrial efficacy. Originally developed to address the complexities of multivariable systems and operational constraints, MPC has evolved into a cornerstone of advanced control across the process industries, energy sectors, and autonomous systems.

Throughout my career in both academic research and industrial application, I have observed a persistent disconnect between theoretical instruction and practical implementation. While academia often prioritizes mathematical elegance, industry demands reliability and measurable performance—objectives that can sometimes obscure the underlying methodology. This book effectively bridges that gap.

Model Predictive Control: Fundamentals and Practice frames MPC as an evolution of classical and optimal control rather than an isolated technique. By elucidating the relationship between linear MPC and Linear-Quadratic (LQ) control, as well as moving horizon estimation (MHE) and Kalman filtering, the authors provide a unified framework that enhances both conceptual clarity and design precision.

A primary strength of this text is its unwavering commitment to practicality. Essential components—such as system identification, estimation, and constraint management—are treated as core pillars of control design rather than peripheral topics. This approach directly addresses the real-world challenges of modeling inaccuracies and process limitations. Furthermore, the transition into advanced subjects like Nonlinear MPC and Repetitive MPC is handled with a focus on problem-solving utility rather than mere abstraction.

Whether used as a graduate-level textbook or a reference for practicing engineers, this book empowers readers to navigate the space between theory and application. I am confident it will remain an enduring resource for anyone seeking a comprehensive, practical mastery of MPC.

Seville, Spain

Eduardo F. Camacho

Foreword

A Korean, an Indian, and a Boricua walk into a bar. Many years of hard work later, a book hits the stores. Hallelujah! It is about MPC.

Control engineering is a challenging subject for the uninitiated, among whom we include students studying the area—through choice or requirement—process engineers needing to manage control systems, and industry leaders managing the engineers. Largely, this difficulty is attributable to the dynamic nature of process systems—their memory reflecting the system's past as well as the current inputs—and the alteration of these dynamics that occurs with closing the feedback loop, which, after all, is why we use control. The machinery of control system design concentrates on this open-loop to closed-loop dynamic transformation, which is indeed a serious intellectual challenge. For those in the know, the cognoscenti, this has meant job security and legendary capabilities. As the subject further develops, those barriers to entry usually rise.

Luckily, however, there are others who seek to update, clarify, explain, and educate. And luckier still, this includes excellent communicators Jay Lee, Niket Kaisare, and Carlos García, who possess extraordinary skills with the theory and virtuoso experience in applications together with a commitment to simplifying the path to expertise. The driver here is modernity; people need to acquire control design skills, availing themselves of all the modern machinery, notably software packages, and to do this without an interminable apprenticeship. The vehicle is to use examples from the process industries to substantiate and instantiate the concepts. This is where commercial experience plays such an important part; abstract notions become concrete.

MPC has been a game-changer for feedback control applications. The measure of success of an engineering technology or methodology is whether it has changed the way that things are done. And the effectiveness of that change is most irrefutably validated in commercial adoption and quantifiable demonstrations of improvement in achieved performance or time to design, redesign, and implementation. This becomes further apparent with the ensuing appearance of multiple competitive software packages capable of speeding realization and operation. MPC has ticked each of these hallmarks of success. This book by three major figures in the area seeks to ease the entrée into MPC without sidelining the details, motivations, and problem domains. Sufficiently comprehensive to provide (eventually) an expert resource, the book preserves its focus on context and accessibility.

What MPC allows us to achieve is quite direct feedback control system design based on a range of possible dynamic system models. It does this for multi-input multi-output (MIMO) systems, and it allows us to include operating constraints and design objectives up front into its formulation. And it performs this while preserving understanding of the process. Constraints in process control are the design equivalent of the medical maxim primum non nocere or first do no harm. That is, avoid breaking things and then proceed to improve. This is a natural human approach to design sensibly ordered with subsequent optimization. MPC matches its design approach with this ethic, placing optimal methods subsidiary to the constraints. Further, MPC exploits modern computational power while preserving the human factors of design. It keeps sophisticated optimization machinery under the hood.

MPC's conceptual trick to relegate the closed-loop difficulty is to rely on open-loop models to proceed to controller design via the ingredients of constraints, objective functions, and dynamic optimization. To complete this in practice requires firstly, a dynamic model, then a mechanism to capture or summarize the historical system memory encapsulated within that model in a state estimate, and then the MPC design based on these elements plus constraints and objective. The authors cover the gamut of design from relevant modeling from experimental data (system identification) to application of these models in state estimation to MPC design and implementation using the model and the state estimate.

The grandeur and finesse in this book are to compose the overarching control design problem by addressing in an approachable fashion the constituent parts within a carefully developed historical and industrial context. It provides the reader with a rapid but reasoned entrée to MPC in the process industries via a judicious selection of the core elements undergirding the methods. Practical applications are used alongside the development, and associated theoretical explanations clarify the mathematical basis. The commercial success of these applications adds credibility and heft while motivating the reader. Yet the book will remain a resource for the practitioner and researcher.

Welcome to the world of Model Predictive Control. Let's take a look inside ...

San Diego, CA, USA *Robert Bitmead*

Foreword

Control and automation are essential components of virtually every dynamic system. While feedback control traces its origins back to antiquity and applies to even the simplest systems, a general, integrated framework is critical for stability and performance guarantees, especially as dynamic systems grow in size and complexity. Over the past century this task has been conducted by SISO PID controllers analyzed in the frequency domain and tuned with heuristic approaches. Further limitations of this approach occur due to ad hoc strategies for MIMO systems and in handling constraints.

In contrast, MPC has key advantages in working directly with plant data to develop predictive dynamic models, systematically handling input and output constraints, and serving as a general-purpose MIMO controller. Over the past five decades, MPC has grown into a highly successful control strategy, evolving from industrial practice and validation and aided by rigorous control theory and advances in optimization algorithms. Currently, MPC is the modern control paradigm for process and manufacturing, with emerging applications in robotics and logistics as well. MPC elements comprise problem formulations and solution strategies that include predictive modeling, constructing suitable objective and constraint functions, and real-time correction via moving horizons, both for identification and control. Subsequent research has refined and validated these elements through rigorous analysis, development of computational tools, and practical real-world applications.

Divided into three parts, this book explores and traces MPC's industrial past, its current state of the art with rigorous control concepts and its future development to address open challenges. It begins with a practical, industrial evolution, starting with data-driven step response models and dynamic matrix control, as exemplified with an included Heavy Oil Fractionator (HOF) case study from Shell R&D. The second part traces the parallel academic research of linear process control systems, influenced by industry but also strengthened and systematized with rigorous analysis. The third part emphasizes advanced, ongoing MPC developments, along with open questions that guide future research.

Part I provides the initial industrial background, starting with data-driven modeling and identification of open-loop stable linear time-invariant models, leading to derivation of Dynamic Matrix Control with Finite Step Response models and input/output constraints. Compactly formulated as a quadratic programming problem that is solved over a prediction horizon and implemented online with a moving horizon, this framework also extends to modified controller objective and constraint functions, along with data filters that handle disturbances and modeling errors.

To complement MPC's industrial development with fundamental control theory, Part II develops stability and robustness guarantees for setpoint tracking and disturbance rejection and a systematic framework to develop predictive models. The equivalence between transfer functions in $z-$ and $s-$domains, finite impulse response models and state space discrete models, as well as input delays and stability concepts are discussed, with a focus on linear state space models with related property definitions for control and estimation (i.e., stability, controllability, reachability, stabilizability, observability, and detectability).

Augmented state-space models are constructed to accommodate disturbance models, leading to direct consideration of state estimation strategies (e.g., pole placement, classical Kalman filters, and moving horizon estimation [MHE]) and unconstrained MPC strategies (with pole placement, open-loop least squares solutions, and recursive dynamic programming approaches). Important principles in this derivation are (i) relations between *infinite horizon* and *finite horizon* optimization and (ii) the *separation principle*, which allows independently designed MPC and the Kalman estimators to provide stable, offset-free performance, and (iii) extensions to constrained systems with the necessary cautions addressed through *Lyapunov stability*, but further complicated as the separation principle no longer applies.

These issues naturally lead to an overview of the broad subject of system identification. Considered here are parametric identification methods and structures (e.g., ARX, ARMAX, ARIMA, Box Jenkins), identifiability with persistent excitation, and classes of identification methods including predictive error methods, maximum likelihood (MLE) and Bayesian estimation, nonparametric identification techniques that yield empirical transfer function estimates, and subspace identification to obtain the best system order. This overview also highlights key practical considerations from a user's perspective, featuring experimental design, PRBS signals, and data preprocessing.

Part III deals with current research in advanced MPC and addresses open questions and challenges for larger, more complex systems. Concepts from Part II are further extended to include inferential control with unmeasured outputs, multi-rate estimation, and sequential linearization approaches to address nonlinear models. The latter topic segues into the emerging field of nonlinear MPC, where solution strategies are presented along with a framework for Lyapunov stability. Extensions for output disturbance models and MHE for state estimation are developed and demonstrated within the MATLAB framework. Finally, repetitive MPC and Iterative Learning MPC are developed for periodic and run-to-run batch processes, which operate in repeated cycles and never go to a steady state. Here, lifted model formulations lead to equivalent MPC approaches with appropriate terminal constraints and augmented systems with output errors and disturbances.

This volume presents the broad development of MPC in a clear, accessible manner; it provides a comprehensive treatment of MPC concepts and theory, reinforced with examples and case studies using the MATLAB MPC Toolbox. Rigorous analyses are outlined, and detailed derivations and proofs are deferred to appendices for the interested reader. Moreover, further exposition on related computational modeling and numerical optimization is cited, but the focus remains with the elements of MPC. Consequently, the presentation is suitable for a broad audience: from students to researchers to practicing control engineers. This material is suitable as a textbook (with exercises) as well as a comprehensive reference.

Moreover, this book serves as a bridge between academic research and industrial development and presents an exciting story of a major milestone in control and automation. With ongoing advances in computer science and applied mathematics, MPC continues to evolve with more complex applications: from laboratory experiments to enterprise-wide optimization. This volume is an essential guide to this evolution, as well as a valuable resource for students, industrial researchers, and academic researchers alike.

Pittsburgh, PA, USA

Lorenz T. Biegler

Foreword

MPC has developed over several decades into a well-established framework for the control of constrained dynamical systems. Its defining feature—the repeated solution of a finite-horizon optimal control problem using an explicit model—was initially motivated by practical considerations, particularly in industrial process control. Over time, the widespread adoption of this idea prompted a systematic examination of its theoretical properties and limitations, leading to a substantial body of work on stability, robustness, and performance.

The evolution of MPC has not followed a linear path. Rather, progress has come through refinements, reinterpretations, and extensions driven by shortcomings in earlier formulations. Issues of feasibility, closed-loop stability, robustness to uncertainty, and computational tractability have repeatedly shaped the field, and the resulting theory reflects an ongoing effort to adapt optimal control formulations to real-time implementation.

This book is situated within that historical development. Instead of surveying all approaches to MPC, it presents a personal account of one line of research and the ideas that motivated it. The material reflects the authors' long engagement with core problems in predictive control, and many of the results are closely connected to their own contributions. The perspective is therefore consistent and unified, shaped by their modeling assumptions, analytical tools, and design objectives.

I worked with the two senior authors during an earlier stage of their academic development, in my role as their doctoral supervisor. The work presented in this volume builds on themes that were already evident at that time: a focus on structural properties of predictive control schemes, an interest in rigorous analysis, and an attention to the implications of theoretical results for implementation. These themes provide continuity across the chapters and help to explain the organization of the material.

A central aspect of the book is its emphasis on how specific design elements in MPC—such as cost functions, constraints, and terminal conditions—emerged in response to particular analytical or practical concerns. By adopting a personal historical perspective, the authors clarify why certain formulations became standard and how alternative approaches arose. This perspective is useful not only for understanding existing methods but also for recognizing the assumptions on which they rest.

The exposition assumes prior familiarity with systems and control and is aimed primarily at graduate students and technically oriented practitioners. Mathematical arguments are developed where essential, while informal discussion is used to provide motivation and context.

MPC remains an active area of research, with ongoing advances in applications, computation, and theory. This book documents one trajectory within that broader landscape. By placing its results in historical context, it offers readers a basis for assessing both the field's achievements and the open questions that continue to shape it.

Zurich, Switzerland *Manfred Morari*

Preface

This textbook is intended to strengthen and broaden the foundation in feedback control that undergraduate students typically acquire. Traditionally, undergraduate control education has centered on feedback theory and the engineering of loop designs for regulating single-input single-output (SISO) linear systems. However, this scope often limits the ability to handle the complexity of real-world control challenges. This necessitates a broader skill set for effective control loop implementation. However, practical application of skills learned in academic environments is typically limited to relatively simple problems, which can leave graduates unprepared for the complexities they face as professional control engineers.

Hence, there is great value in teaching control fundamentals in a way that enables new engineers to apply them effectively from the outset. With this goal in mind, we seek to bridge the gap between academic training and industrial practice by integrating advances in modern control theory and methods. Central to this effort is a comprehensive treatment of Model Predictive Control (MPC), which has—over the past several decades—become the advanced control strategy of choice across the process industries and beyond. This popularity is largely because MPC allows engineers to tackle complex and economically significant control problems in a systematic and practical manner.

By merging theoretical concepts with practical applications, the textbook prepares students not only to understand but also to implement MPC and other advanced control techniques effectively. This preparation helps mitigate the steep learning curve often encountered in real application settings and accelerates the transition from novice to competent engineers capable of tackling sophisticated control challenges. This approach ultimately enriches the educational experience of students and enhances their readiness for professional practice in modern industrial environments.

Sample problems in the book are carefully selected and closely linked to industrial applications, ensuring that theoretical concepts are firmly grounded in practical scenarios. This approach enriches the learning experience while equipping readers with the skills and insight needed to address the challenges they are likely to encounter in their professional careers. The content is structured to facilitate easy application of basic and advanced concepts in model-based optimal control. After the introductory chapter (Chapter 1), the core material is structured into the following three parts:

Part I: Early Industrial MPC Algorithms: Introduces impulse-response and step-response model-based MPC algorithms that originated in the process industry. Although these early methods have inherent limitations (e.g., applicability restricted to open-loop stable systems), they are included not only for their historical significance, but also because they remain effective educational and practical tools for both beginners and experienced engineers.

Part II: Basics of Linear Systems, Optimal Control, and System Identification: Explores the fundamental concepts and techniques underlying linear systems and optimal control theory.

Part III: Advanced MPC Algorithms: Builds on the foundations established in the first two parts to present state-space and nonlinear MPC methods, with attention given to both the theoretical foundations and the practical aspects of implementing advanced MPC strategies.

In addition, Appendix provides a concise treatment of linear transformations and stochastic processes, as well as introductory guidance on using SIMULINK and the MATLAB MPC Toolbox to support practical implementation and simulation. The MATLAB codes used in this book are available on a public GitHub repository at https://github.com/Lee-MPC-Book.

Although this book is centered on MPC, it is designed to serve as a textbook for general graduate-level process control courses. For a course devoted specifically to MPC, Parts I and III provide the core material. Because Part III presumes prior knowledge of linear systems and optimal control/estimation, incorporating selected elements from Part II is recommended. If the course targets a multidisciplinary cohort, Part I may be omitted, as it is most relevant to applications in process industries. At the same time, Part I can serve as a valuable supplement to the undergraduate process control curriculum, which needs to go beyond its traditional emphasis on Laplace transforms and PID-based single-loop design.

With the expanded state-space formulation (covered in Part III), MPC has moved well beyond its traditional role in process control and is now applied across a wide range of engineering disciplines, including mechanical, aerospace, and electrical engineering. To reflect this versatility, we adopt a general approach in system descriptions and control formulations in Parts II and III, ensuring the material is relevant to a broad control engineering audience.

The origins of this book date back to the 1980s, beginning with the first author's PhD studies under Prof. Manfred Morari at Caltech, in collaboration with Dr. Carlos García at Shell. At that time, MPC was just gaining traction in refineries and petrochemical plants, drawing strong interest from both academia and industry. Like many emerging technologies, its development advanced rapidly through the 1990s and into the new millennium. This swift evolution rendered earlier drafts and contributions outdated or incomplete, causing repeated delays and nearly leading to the project's abandonment.

More than three decades after its inception, we are pleased to finally bring this project to completion. Over the years, partial drafts of the book have circulated widely and, according to Google Scholar, have been cited unofficially more than 200 times. Although the field of MPC has advanced greatly since its early days, many of its core theoretical foundations—such as stability guarantees—were established in the 1990s. Since then, numerous texts have appeared, often emphasizing either theory or practical applications. Yet there remains a clear need for a comprehensive textbook that bridges the gap between these perspectives, integrating MPC with traditional optimal control methods in a manner that is both rigorous and accessible. It is our hope that this book, despite its long delay, fulfills that need and serves as a valuable resource for both students and professionals in the field.

Los Angeles, California, USA
January 30, 2026

Jay H. Lee

Acknowledgments

We wish to express our sincere gratitude to the many collaborators and colleagues who have contributed, directly and indirectly, to the realization of this book.

Jay H. Lee and Carlos E. García are especially indebted to their former PhD advisor, Prof. Manfred Morari, who initiated this book project and involved us during its early stages. Although he has since chosen not to continue his participation after his retirement, we remain profoundly grateful for the insights, inspiration, and knowledge he has shared with us over the years. His pioneering contributions to model predictive control (MPC) have provided both the intellectual foundation and enduring motivation for this work.

We gratefully acknowledge the influence of our mentors and colleagues in the process systems engineering and control communities, whose foundational research established the theoretical and practical pillars of MPC. The vibrant exchanges at international conferences, workshops, and collaborative consortia have been a continuous source of inspiration and valuable perspective throughout the development of this text.

J.H.L. acknowledges the support and inspiration of students and colleagues at the many institutions he has been privileged to be part of. This includes not only Auburn University, Purdue University, Georgia Institute of Technology, Korea Advanced Institute of Science and Technology (KAIST), and now the University of Southern California, where he has held full-time positions, but also DuPont, Weyerhaeuser, and Seoul National University, where he spent his sabbatical leaves. He continues to draw energy and inspiration from the brilliant students with whom he has had the good fortune to work. It has been a long and rewarding journey—one filled with the joy of discovery, collaboration, and lasting friendship.

C.E.G. especially acknowledges the invaluable lessons gained during his 36-year career at Shell, which began when he was hired by Charlie Cutler to continue the development of DMC. He is deeply appreciative of the insights shared by Shell practitioners around the world through their experiences in implementing, developing, maintaining, and justifying MPC. This practical understanding—of both the value MPC delivers and the challenges associated with its implementation—has been an important asset in shaping this textbook.

N.S.K. gratefully acknowledges the support and encouragement of his students and colleagues, whose help made the challenging task of writing this book possible. His experience at ABB Corporate Research in Bangalore, India—working on control and optimization problems—provided valuable real-world insights that significantly influenced the approach presented in this book.

We also extend our appreciation to the institutions and funding agencies that have supported our research and teaching efforts, enabling sustained exploration of advanced topics in model predictive control and its integration with learning and optimization-based methods.

Finally, we are deeply thankful to our families for their patience, understanding, and unwavering support during the many years devoted to writing this book. Their encouragement has been a constant source of strength and motivation.

Jay H. Lee, Carlos E. García, and Niket S. Kaisare

1

Introduction

It took some time but its complete dominance in so many applications has continued to surprise me—though it is completely deserved, of course.

– Manfred Morari

In this chapter, the authors aim not only to provide an initial introduction to Model Predictive Control (MPC) but also to underscore its relevance in industrial settings, where control is routinely applied. The objective is to give readers a clearer understanding of how MPC differs from classical control methods and, in doing so, to develop an appreciation for its significance and a motivation to delve deeper into its study.

Although MPC is a general framework applicable to a wide range of control problems and is now employed across many industries, this introduction focuses primarily on the process industries—where MPC was first demonstrated in large-scale applications and remains most widely adopted. This emphasis aligns with the overall progression of the book. We begin with traditional step-response-based MPC (Part I), which is largely confined to process control applications due to its reliance on open-loop stable dynamics, and then proceed to more general state-space MPC formulations equipped with state estimators (Parts II and III). Accordingly, in the latter part of this chapter, we provide a brief introduction to MPC methods capable of handling more general dynamics and very fast sampling rates, which are particularly relevant to automotive systems, robotics, and other mechanical systems.

With this primary focus on process control, we aim to highlight the need to strengthen undergraduate education by integrating practical insights into its fundamentals. This is achieved by presenting a nuanced perspective on the challenges encountered in industrial practice. To this end, we examine a representative control problem that illustrates the level of complexity and breadth typically faced by control engineers in modern industrial settings.

1.1 What Is MPC?

MPC is an advanced optimal control strategy that operates on the following core principles (refer to Figure 1.1):

Dynamic Optimization: At each sampling instant, the control inputs (also called manipulated inputs or variables) are computed by solving a dynamic optimization problem in real time.

Use of a Dynamic Model: MPC relies on an explicit dynamic model that captures how manipulated variables and measured disturbances affect process outputs.

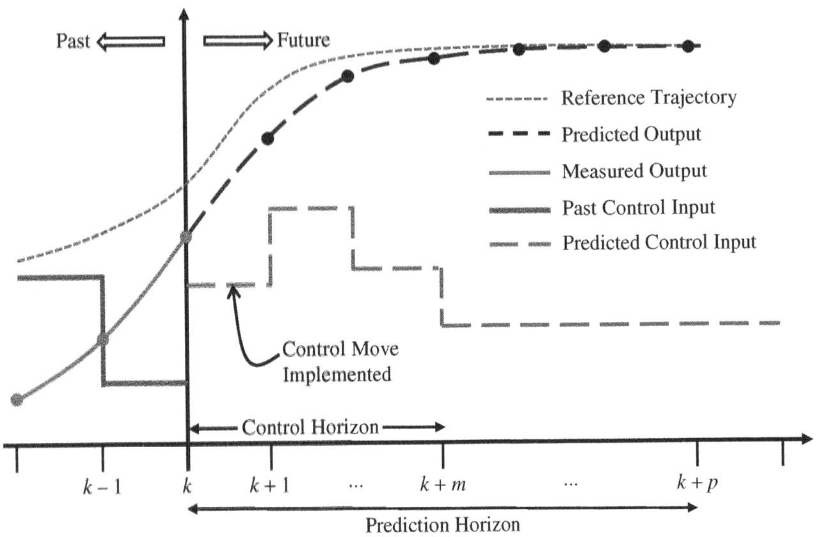

Figure 1.1 The basic idea of model predictive control (MPC).

Prediction and Correction: Future output trajectories are predicted with the model and continuously updated using actual process measurements.

Objective Function: Manipulated input actions are determined by minimizing an objective function—typically a weighted sum of output deviations from a reference trajectory, penalties on input variability, and/or process economic objectives—so as to dynamically guide the process toward, and/or maintain it close to, its operational goals.

Operational Constraints: Inequality constraints on both outputs and manipulated inputs are explicitly incorporated into the optimization problem to satisfy operational limits.

Receding Horizon: The optimization problem is re-solved at every sampling instant, with the prediction horizon shifting forward in time.

To better understand the principles of MPC, consider an example formulation for a simple first-order single-input single-output (SISO) process (the index k will be used throughout this book to denote the current sampling instant):

Optimization Problem Formulation: At each discrete sampling instant k, the optimization problem solves for the manipulated variable $u(k)$. In practice, most MPC algorithms compute the *change* in the manipulated variable, denoted as $\Delta u(k)$.

Dynamic Model Usage: The following first-order process model predicts the output y at the next time step $k + 1$:

$$y(k+1|k-1) = \alpha y(k|k-1) + \beta \Delta u(k) \tag{1.1}$$

resulting from a manipulated input change made at time k. Here, α and β are coefficients (analogous to the time constant and steady-state gain) that describe the system dynamics. The notation $y(i|j)$ indicates the prediction of y at sampling time (i) made using measurements up to time j.

Recursive Prediction: Eq. (1.1) can be applied recursively to extend the prediction horizon. For example, the two-step-ahead prediction is:

$$\begin{aligned} y(k+2|k-1) &= \alpha y(k+1|k-1) + \beta \Delta u(k+1) \\ &= \alpha^2 y(k|k-1) + \alpha\beta\Delta u(k) + \beta\Delta u(k+1). \end{aligned} \tag{1.2}$$

Continuing until the pth-step-ahead we get:

$$y(k+p|k-1) = \alpha^p y(k|k-1) + \sum_{i=1}^{p} \alpha^{p-i}\beta\Delta u(k+i-1). \tag{1.3}$$

Measurement Error Correction: If the model prediction $y(k|k-1)$ deviates from the measured output $y_m(k)$, a bias correction is applied:

$$\begin{aligned} &y(k+\ell|k) = y(k+\ell|k-1) + \text{BIAS}(k), \qquad \ell = 1, \dots, p \\ &\text{where, BIAS} = y_m(k) - y(k|k-1). \end{aligned} \tag{1.4}$$

This correction aligns the prediction with the measurement. In the absence of further information, the same bias is typically applied to all future predictions $y(k+\ell|k)$, illustrating a basic feedback mechanism in MPC. We note that this simple bias-correction approach, characteristic of early MPC algorithms used in the process industries, is limited to open-loop stable systems and is not optimal in general. For broader applicability, this idea can be extended by employing state estimation techniques, which yield optimal predictions under an assumed disturbance model.

Objective Function Formulation: The controller minimizes a cost function that balances output tracking and control effort:

$$\min_{\Delta u(k)} \sum_{\ell=1}^{p} [(y(k+\ell|k) - r(k+\ell)]^2 + \sum_{\ell=1}^{m} [\gamma_\ell(\Delta u(k+\ell-1)]^2, \tag{1.5}$$

where
p = prediction horizon (number of future steps considered),
m = control horizon (number of control moves optimized),
γ_ℓ = weighting factor penalizing input changes,
$r(k+\ell)$ = reference trajectory over the prediction horizon.

Constraints on Manipulated Inputs and Outputs: Inputs and outputs are typically bound to ensure feasibility and safety, for example:

$$y_{\text{low}} \le y(k+\ell|k) \le y_{\text{high}}, \qquad u_{\text{low}} \le u(k+\ell|k) \le u_{\text{high}}$$

when it is desired to enforce high and low limits that are constant over all future sampling times.

Receding Horizon: Although m future moves are computed, only $\Delta u(k)$ is implemented. At the next sampling time, the horizon shifts forward, new measurements are incorporated, and the optimization is solved again.

These core principles form the foundation of MPC and will be explored throughout this book in progressively more complex systems. In Section 1.2, we explain why MPC has become the preferred tool for addressing advanced control challenges.

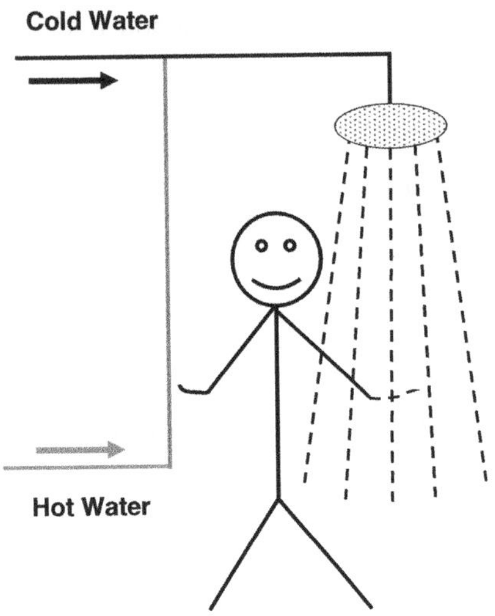

Figure 1.2 The shower temperature control system.

Example 1.1: *Shower Water Temperature Control.*[1]
To illustrate the principles of MPC, we use a simple time delay control example commonly taught in courses: a bather adjusting the hot water valve to control shower temperature (Figure 1.2). A disturbance might occur if upstream cold water is diverted, raising the shower temperature.

Because hot water pipes are long, temperature changes are sensed only after several seconds. A bather expecting immediate feedback often overadjusts, leading to oscillations. In MPC terms, the "mental model" of instantaneous response does not match the true delayed dynamics.

Assume that a 10% increase in valve opening raises the temperature by 5 °F, sensed after six seconds.

A simple discrete-time model for this pure-gain-with-dead-time process is:

$$T(k+1) = T(k) + 0.5\Delta V(k+1-\theta_d), \quad \theta_d = 6. \tag{1.6}$$

A pure integral controller computes valve moves as:

$$\Delta V(k) = K_I(T_r - T_m(k)). \tag{1.7}$$

For $\theta_d = 1$, a deadbeat response occurs at $K_I = 2.0$. With delay, however, a higher gain destabilizes the system, requiring detuning.

In MPC, predictions are made explicitly with delay and corrected using a bias term:

$$\begin{aligned} T(k+1|k) &= T(k|k-1) + 0.5\Delta V(k-5) + BIAS(k), \\ BIAS(k) &= T_m(k) - T(k|k-1). \end{aligned} \tag{1.8}$$

Expanding six steps ahead yields:

$$T(k+6|k) = T(k|k-1) + 0.5\sum_{\ell=1}^{5} \Delta V(k-\ell) + BIAS(k) + 0.5\Delta V(k). \tag{1.9}$$

1 We dedicate this example to the memory of the late Dr. Babatunde "Tunde" Ogunnaike, a dear friend to one of the authors since they were fellow students at the University of Wisconsin-Madison. Every one of his presentations on control of time delay systems started with the shower control problem and the "stick bather" cartoon.

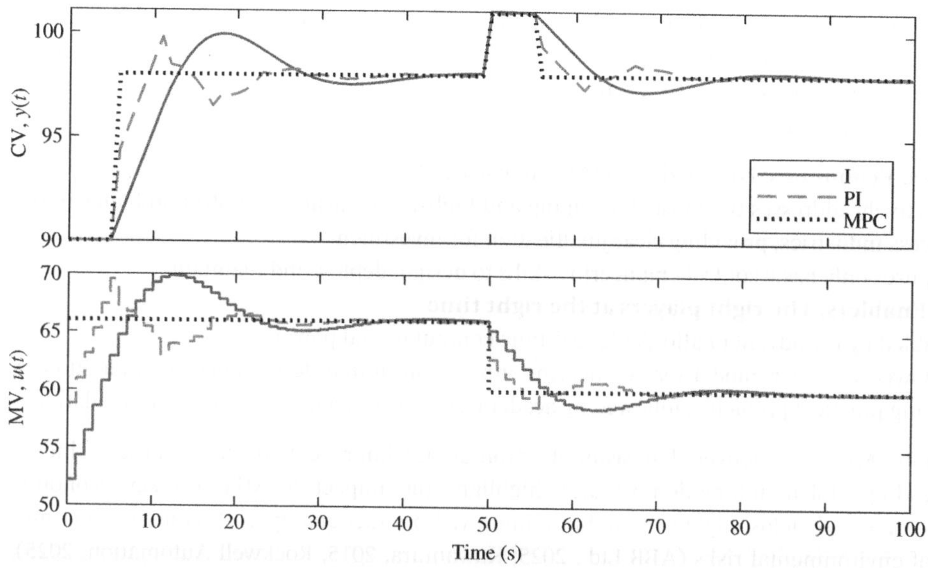

Figure 1.3 Shower temperature control using I, PI, and MPC controllers. Set point = 98 °F. Disturbance = +3 °F at 50 seconds.

Using a simplified objective of setting $T(k+6|k)$ to the setpoint T_r, the optimal law is:

$$\Delta V(k) = 2.0(T_r - T_m(k)) - \sum_{\ell=1}^{5} \Delta V(k-\ell), \tag{1.10}$$

combining integral action with correction for delayed moves.

Consider an initial steady state at 90 °F with the goal of reaching 98 °F, followed by a +3 °F step disturbance at 50 seconds. Figure 1.3 compares integral (I), proportional-integral (PI), and MPC controllers. Integral control is stable only at reduced gain ($K_I = 0.268$), its response is slow, and it shows long overshoot and sluggish disturbance rejection. PI control improves speed and reduces overshoot, though it is still limited. In contrast, MPC provides fastest possible rise to the set point and disturbance rejection by incorporating process gain information into the system.

This example also highlights sensitivity: A deadbeat-tuned MPC is robust to moderate gain mismatch (0.4–0.7) but unstable if the dead time is misspecified. Later chapters show how this can be mitigated with longer horizons and nonzero input weighting term γ_ℓ.

1.2 Why MPC?

MPC has become the tool of choice for advanced process control (APC) in the process industries, and its adoption in other industries and disciplines continues to grow. For the foreseeable future, a significant number of process control engineers will be designing, commissioning, and/or maintaining MPC loops within an industrial corporation or a control software vendor company. And as its adoption in other fields continues to grow, new areas of activity for researchers are likely to open up. In the following sections, we explore the factors behind MPC's success in the process industries. In doing so, we hope to educate future practitioners and theoreticians of MPC on the issues behind justifying its use in industry and future investments in research.

The authors have been actively involved in the evolution of MPC throughout much of its history, each bringing a unique perspective shaped by their backgrounds in industrial application, theoretical advancement, and education. We attribute the early success of MPC in refineries and petrochemical plants to two primary factors:

Attractive Economics: High value for the investment

- Effectively addresses the most challenging and high-value advanced control problems in the process industries, providing clear justification for investment
- Requires only basic control engineering skills to design, deploy, and maintain

Timely Enablers: The right players at the right time

- Enabled by the advent of affordable real-time computational power
- Followed a proven model for engineering innovation: initial development by practitioners solving practical problems, followed by academic efforts to establish theoretical foundations

In short, MPC has delivered measurable profitability improvements in all process industries. Leading global automation vendors highlight the impact of MPC on key economic drivers—increased yields, higher throughput, improved product quality, cost reduction, and mitigation of environmental risks (ABB Ltd., 2025; Kawamura, 2015; Rockwell Automation, 2025). Despite decades of successful implementation, significant opportunities remain. A 2025 market analysis report (Correa, 2025) projects an annual growth rate of 10.7% for the MPC segment from 2024 to 2033, reflecting the continued and expanding demand for its capabilities.

More broadly, MPC offers a flexible framework for specifying control objectives, naturally accommodating the integration of economic goals with control performance requirements. It is one of the few control strategies that can explicitly handle inequality constraints, making it appealing for applications in other domains, such as automotive systems, robotics, and mechatronics.

In the next section, we examine the economic drivers behind the adoption and sustained success of MPC in the process industries. The historical enablers—both industrial and academic—that supported its development will be discussed in subsequent sections. While the examples and discussion are framed within the context of the process industries, the insights are broadly applicable across a range of domains. Whether controlling a vehicle, managing a power distribution network, or optimizing a chemical reactor, the decision to deploy MPC is ultimately guided by economic value.

1.2.1 Economic Drivers of APC/MPC

Control engineers have access to a variety of tools to address industrial control problems. The decision to implement MPC—whether in an existing unit operation or a new plant—follows the same rationale as any capital investment: the anticipated benefits must outweigh the total cost. This includes not only the cost of the MPC software and associated hardware but also the full cost of deployment and long-term maintenance—collectively referred to as the total cost of ownership.

Automation functions in the process industries are typically organized into three hierarchical layers based on complexity and execution frequency, as illustrated in Figure 1.4. Historically, each function has been tied to specific hardware platforms. However, with the rapid advances in computational power and system integration, these boundaries have become increasingly blurred, allowing greater flexibility in deploying advanced control strategies such as MPC across different layers of automation.

- **Base Level Control Layer:** This layer includes sensors, actuators, and fast-acting control loops that respond to rapid disturbances. Controllers are often arranged hierarchically—for example, cascaded loops for valve positioning and flow control—executing in milliseconds to seconds. Operators or advanced algorithms adjust their set points to achieve economic and operational goals such as meeting product quality, maximizing throughput, minimizing energy use, and ensuring safety.

- **Advanced Control Layer:** This layer contains algorithms that adjust base-level set points to optimize unit performance, with execution intervals from seconds to minutes. The term "advanced control" has replaced the older "supervisory control" as the distributed control system (DCS) and programmable logic controller (PLC) technologies have matured. Today, DCS handle both supervisory and base control, enabling the deployment of more sophisticated tools like MPC. Operator buy-in remains crucial—poorly integrated loops are often disabled, losing their intended value.
- **Optimization Layer:** Real-time optimizers determine the most profitable process operating points, typically updated every few minutes to hours. These values are passed to advanced control layers, which maintain the operation at the optimal point. Optimizers rely on steady-state first-principles models with hundreds of equations to capture nonlinearities and constraints. They are most valuable when slow-changing disturbances (e.g., feedstock changes or fouling) or operational requirements shift the optimum.

Table 1.1 summarizes the relative benefits and capital costs associated with each automation layer. As with any investment, the return must justify the cost—both in deployment and long-term ownership.

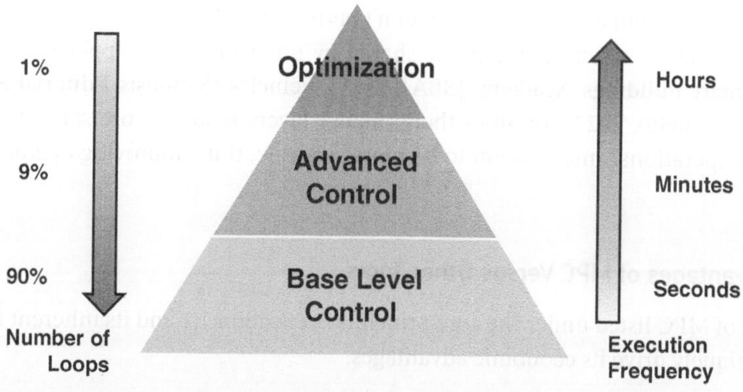

Figure 1.4 Layers of automation in the process industries.

Table 1.1 Project economics of automation layers.

Layer	Execution frequency	Number of loops (%)	Benefit potential (%)	Capital cost (%)
Base level control	ms to secs	90	15	65
Advanced control	secs to mins	9	65	15
Optimization	mins to hrs	1	20	20

Benefits: Automation systems help operators meet performance goals by continuously adjusting the process to handle disturbances and target changes. This allows operators to focus on exceptions rather than routine control.

The most important performance metric is loop uptime since automation provides no benefit when systems are offline. Higher-level loops depend on the uptime of lower layers—particularly the base level, which by itself delivers only about 15% of the benefits but enables the remaining 85% to be delivered by advanced control and optimization.

Costs: Costs include capital costs and maintenance costs.

Capital costs include design, installation, and commissioning of control infrastructure, such as control rooms, equipment, and engineering services. Base layer systems, while having the highest capital cost, provide foundational stability and reliable operation and are therefore essential for enabling benefits from higher layers. Projects targeting advanced control and optimization often require upgrades to sensors and controllers at the base level, particularly process analyzers. Their capital cost is primarily driven by the engineering expertise needed for model development, configuration, and tuning.

Maintenance costs are incurred because ongoing engineering support for monitoring, troubleshooting, and updating is critical to sustaining performance. Real-time optimizers require dedicated engineers with deep knowledge of optimization, process operations, and control systems, as well as strong communication skills. Despite such high support needs, they have been successfully deployed in many petrochemical applications (AVEVA Group Ltd., 2025). Advanced control loops also require expert support. Since most process engineers lack specialized control knowledge, companies often assign dedicated control engineers to each unit, sometimes with vendor support. High uptime is essential—operators will disable loops that don't perform reliably.

This hierarchical automation system paradigm is shared by other discipline applications, for example, buildings (Smart Buildings Academy [SBA], 2019), vehicles (Synopsis Editorial Staff, 2025), and aviation (Cesconetto, 2022). In all of them, higher layers require more sophisticated tools to enable human operational intervention to be more strategic, thus improving operational performance.

1.2.2 Economic Advantages of MPC Versus Other Tools

The technical features of MPC listed under the core principles in Section 1.1 and its inherent feedback loop structure uniquely drive its economic advantages.

1.2.2.1 Dynamic Model Prediction

The primary advantage of MPC lies in its explicit utilization of dynamic models that simultaneously relate multiple plant inputs to their outputs. This feature enables MPC to transparently manage a variety of dynamic characteristics, including dead times, integrators, and nonlinearities. Furthermore, these predictions of both outputs and inputs can be displayed to operators, enhancing their understanding of controller actions and contributing to improved loop uptime.

1.2.2.2 Objective Function

We have previously highlighted the enhanced capability of MPC to track output reference trajectories. The tuning parameters of the controller—horizon lengths and weights—have a clear and direct influence on how closely the output adheres to its target, as well as on the smoothness of changes in the control input. This clarity significantly reduces the effort required for troubleshooting and tuning by plant personnel.

A notable feature of MPC is its ability to manage "non-square" control problems, where there is a mismatch between the number of control inputs and controlled outputs. In scenarios where the number of controlled outputs exceeds the control inputs, it becomes generally impossible to maintain all outputs at their set points, thus necessitating the prioritization of control objectives. This prioritization is effectively handled through the strategic use of weight parameters in the cost minimization function and constraints, enabling a more customized and efficient control approach.

Furthermore, MPC facilitates the trade-off between economic and control performance objectives through the adjustment of weight parameters. Advanced control tools, even those based on linear models, can achieve many optimization goals. With modern MPC software supporting dynamic nonlinear models (see Amrit et al. (2015) as an example), the need to invest in expensive optimization systems can often be avoided.

1.2.2.3 Constraint Handling

Advanced control algorithms such as MPC are designed to drive and maintain the plant at its optimal operating point, typically situated at the intersection of various constraints. Achieving this with traditional SISO controllers can be challenging, particularly when the optimal set of intersections shifts frequently due to changing process conditions.

Moreover, disturbances and process changes can inadvertently push the process toward these limits. The predictive capabilities of MPC allow early corrective actions to be taken to prevent potential violations of these constraints. Without such preemptive measures, these impending violations would likely necessitate manual intervention by operators, including potentially shutting down the loop to address the issues directly.

1.2.2.4 Design and Commissioning

With an accurate model, designing an MPC controller becomes straightforward. After the model is established, the controller is configured by tuning the optimization parameters in (1.5) through offline simulations, with the goal of achieving satisfactory servo and regulatory performance.

This transparency in the design process can significantly reduce engineering and commissioning costs. However, it is important not to underestimate the effort and cost involved in developing the dynamic model. If the initial commissioning effort is dedicated to developing the most accurate model possible, the subsequent steps, including tuning the loops online, become considerably less labor-intensive.

1.2.2.5 Maintenance

Once commissioned, these tuning parameters can be adjusted in real time to "detune" the controller if instabilities arise or performance degrades due to model inaccuracies. This adjustment is feasible because the bias term $y_m - y(k|k-1)$ in (1.4) now provides feedback on the model error.

In the majority of cases, process dynamics do not change frequently or significantly enough for model error to become problematic, particularly if the initial model development was thorough. However, for applications where plant dynamics change frequently, some vendors like Yokogawa Corporation (2015) offer tools that provide early detection of performance degradation, which may necessitate updating the internal model.

Although the initial high cost of developing models during commissioning is often cited as a drawback, it is important to consider this expense in the context of overall lifecycle costs. The lower maintenance requirements of MPC, compared to traditional solutions like multiple SISO loops with selectors and limits, justify this upfront investment by potentially reducing long-term operational costs. One final note on the cost of ownership of APC/MPC tools: Standardizing on a single tool with consistent structure, tuning parameters, and design methodology across all applications within a corporation can lead to substantial savings in engineering training, operator acceptance, and the replication of applications. This standardization allows for the rotation of operating staff and control engineers across different units—and even across plant sites—without significant degradation of control system performance due to skill gaps.

In the next two sections, we explore the history of MPC deployments and their influence on control research and theory. We aim to demonstrate how timing has been crucial to its success thus far and how the collaboration between industry and academia has established a theoretical foundation that is likely to drive further expansion in the future.

1.3 Historical Overview

The development and adoption of MPC in the process industries was driven by necessity. Petrochemical processes in particular are inherently multivariable, with varying constraints and dynamic characteristics that include dead times, simultaneous slow and fast output responses, coexisting ramp and stable outputs, etc. In addition, energy integration and reduced inventories create significant dynamic interactions between units. Such characteristics have historically made control of these processes difficult, both for operators and for automatic control systems.

The need for an advanced control paradigm that could handle the complexities of physicochemical processes in a cost-effective and transparent manner had been understood for decades prior to the advent of MPC. This challenge was eloquently expressed by Lee and Weekman (1976) of Mobil Oil. They particularly pointed out that the existing advanced control theory developed to solve the minimum-time, minimum-fuel consumption problem of the aerospace industry was inadequate to address challenging industrial problems like the control of a fluidized catalytic cracker (FCC) unit. By then, control of an FCC had been accepted by the technical community as the benchmark problem for APC solutions.

Ideas for model-based solutions to complex control problems have been proposed since the 1950s. The emerging US space program drove the development of optimal estimation and control techniques, especially the Kalman filter for state estimation (Kalman, 1960). Small spaceborne computers executed offline-calculated optimal control laws using onboard PID controllers. Since aerospace vehicle dynamics are relatively better understood than those of other systems, guaranteeing stability was more manageable. Moreover, once developed for a particular vehicle, solutions could be replicated, thus achieving significant cost-effectiveness.

1.3.1 Early Computer Control

The breakthrough came with the advent of affordable real-time computers capable of executing supervisory control. By the 1950s, numerous computer-based supervisory control projects had been reported across the oil and petrochemical industries, including installations at Monsanto, B.F. Goodrich, Riverside Cement, and Union Carbide (Stout and Williams, 1995). A particularly notable effort was a collaboration between Standard Oil Co. of California and IBM, where data from an FCC unit at the El Segundo refinery was transmitted via teletype to an IBM 7090 mainframe in San Francisco. Optimal process settings were computed and relayed back every 15–20 minutes, after which operators manually implemented the adjustments. Building on this success, an on-site process computer was later installed, eliminating the need for telecommunication and enabling automated adjustments.

In a 1959 article, the Thompson, Ramo, and Wooldridge (TRW) Corporation reported commissioning what they described as the first successful real-time automatic computer control system in the process industries, implemented at the Texaco Port Arthur refinery (Anon, 1959). The article even documented the exact time and date of this milestone, marking a turning point in the history of process control:

> *Shortly before 11 a.m. on Mar. 12, a veteran Texas Co. process operator named Marvin Voight flipped the switch ... The action closed the loop in the first fully automatic, computer-controlled industrial process.*

TRW invested in this project to promote their RW-300 "desk-sized" 16KB computer. The project took two and a half years to complete because instruments that could transmit electronic signals had to be developed. It could handle 110 measurements, 16 control inputs, and had 7 basic loops executing every five minutes. No details are provided on the supervisory control schemes used though it is likely they mimicked their existing analog controller structure. Regardless, this was an amazing control engineering feat.

The article curiously mentions that implementing the same application on an FCC would involve "hundreds of complex interrelationships ... and outputs to the controls in the many dozens." *"It's not a job we can do tomorrow"* was the quote from a Texaco manager. This is evidence that the demand for solving more advanced control problems had been around for years, but the computational capability and advanced tools were lacking.

1.3.2 The Pioneers

Despite the potential benefits demonstrated by these early projects, computer-based control did not achieve widespread adoption in the process industries at the time, primarily due to the high levels of effort and expense associated with its implementation. In a 2010 interview, Karl Åström said that it was not until the 1970s that *"it became possible to ... use computer control for smaller projects and for smaller problems"* (Lundberg, 2010).

With the arrival and evolution of cheaper, more reliable, and more powerful microprocessors and associated distributed control systems, computer-based process control began to flourish. Around this time, several practitioners reported significant successes with the use of computer-model-based control in managing various refinery units (Richalet et al., 1978; Cutler and Ramaker, 1979; Cutler and Hawkins, 1988). A landmark publication in this field was by Charlie Cutler, who reported the use of a multivariable model-based control algorithm that handled constraints called dynamic matrix control (DMC) (Cutler and Ramaker, 1979, 1980). Concurrently, several similar algorithms were introduced by Richalet et al. (1978); Rouhani and Mehra (1982); and Martin (1984).

Cutler was a process engineer at the Shell Oil Co. New Orleans refinery. He implemented DMC for the first time in 1973. It was patented in 1982 (Prett et al., 1982), giving the control of an FCC as an example of their invention. In a 2020 interview (Larson, 2020), Cutler said:

> *But the computers of the time wouldn't let us solve real multivariable problems. We had to wait for computers with more horsepower. By the early 80s, we had computers that were reasonably good number crunchers, so I decided we could commercialize.*

By coincidence, Cutler delivered the first presentation of his seminal DMC paper on April 4, 1979—exactly 20 years to the day after TRW and Texaco reported the first successful computer control deployment.

The computational challenges in solving advanced control problems were not limited to the petrochemical industry. Though it is widely accepted that Cutler conceived the receding horizon idea on his own and was the first to deploy it full scale in the process industries, he was not the first to publish a receding horizon optimal control problem formulation. For example, in a 1964 article, Francis Kishi from TRW (Kishi, 1964) documented a quadratic optimization algorithm for "reentry aerospace vehicle control," which is virtually the same formulation as DMC using a finite impulse response (FIR) model. It explicitly says the optimization objective was "minimized periodically" at every time interval k over a fixed time "optimization interval" into the future. Yet there is no mention of any application, most likely because of insufficient onboard computational capability. Others who published MPC-like approaches during the same time period are listed under Subsection 1.4.1.

MPC thus parallels the trajectory of many other digital technologies: the need for a new approach was well recognized, and even early solutions existed, long before sufficient computing power made practical deployment feasible. Once that capability arrived, MPC adoption expanded rapidly.

1.3.3 Adoption Growth

Driven by early successes and a sharp decline in implementation barriers, model-based computer control rapidly gained traction across refineries and petrochemical plants in the Western world. Different companies developed proprietary approaches—often with unique features—under names such as MPHC, DMC, MAC, and ID-COM. These eventually became commercial software trademarks that remain in use today. Over time, the collective term Model Predictive Control (MPC) emerged and was widely adopted to describe this class of techniques (García et al., 1989).

The growth of MPC spurred the rise of specialized vendors offering commercial software and engineering services. Initially led by startups like DMCC, Setpoint, and Profimatics, the industry has since consolidated through mergers and acquisitions. Today, MPC technologies are embedded within the broader automation platforms of major companies such as AspenTech, Honeywell, Schneider Electric, ABB, and Emerson (Qin and Badgwell, 2003). This evolution has enhanced both the accessibility and scalability of MPC across a wide range of industrial applications.

MPC has become a standard feature in refineries and petrochemical plants around the world and has expanded into other sectors such as chemicals, pulp and paper, food processing, and mining. A survey by Qin and Badgwell (2003), based on input from five major MPC vendors, already showed substantial industrial adoption. However, their figures excluded in-house applications and implementations by smaller vendors, suggesting that actual usage may be significantly higher—possibly by an order of magnitude. Today, MPC is firmly established as a core component of modern process control systems.

Building on its success in refineries and the chemical process industry, MPC has expanded into fields such as automotive mechatronics, robotics, and aerospace—domains where constraint handling is equally critical. A key distinction in these newer applications is that the underlying dynamics are often unstable, necessitating more general model representations such as state-space models together with state observers. In addition, these applications require much faster sampling rates. While traditional process industries operate with sampling intervals on the order of minutes, these systems often require sampling in the range of seconds to milliseconds.

This introduces a major challenge: each optimal control problem must be solved in real time, within much tighter computational deadlines. To meet this requirement, the concept of fast MPC was developed. Fast MPC leverages specialized optimization algorithms that exploit the problem's structure and, in some cases, incorporates precomputed offline solutions to dramatically accelerate online computations.

Today, MPC is routinely implemented in applications with control loops running at frequencies of several hertz or higher, often on embedded platforms. Reported implementations include:

- Vehicle traction control (Borrelli et al., 2006).
- Active suspension systems (Giorgetti et al., 2005).
- Direct injection stratified charge engines (Giorgetti et al., 2006).
- Thrust-vectoring in ducted fan aircraft (Dunbar et al., 2002).
- Automotive powertrain control (Lagerberg and Egardt, 2005; Saerens et al., 2008).
- Magnetically actuated mass-spring-damper systems (Di Cairano et al., 2007).
- Power converters (Richter et al., 2010; Kouro et al., 2009).
- Multicore thermal management (Zanini et al., 2010).

These examples highlight MPC's growing versatility and its evolution into high-speed, embedded control environments.

Thus far, we have explored the key factors behind MPC's widespread adoption in the process industries—particularly in contrast to other optimal control techniques like Linear Quadratic Gaussian (LQG) control, which has seen limited industrial impact. One major enabler was the timely availability of affordable and reliable computing hardware. Additionally, MPC's intrinsic advantages—many of which are summarized under the core principles in Section 1.1—make it especially well-suited for process control applications.

In Section 1.4, we discuss how the continued collaboration between industry and academia has not only supported MPC's past success but also established a strong theoretical foundation for its future expansion into new domains beyond the process industries.

1.4 Impact of MPC on Control Research

Several papers from the optimal control literature of the 1950s and 1960s introduced foundational concepts that closely resemble what is now known as MPC. As previously noted, Kishi (1964) proposed a receding horizon optimal control strategy for reentry aerospace vehicles, while Propoi (1963) applied linear programming to manage constrained linear systems. Perhaps most notably, Lee and Markus (1967) captured the essence of MPC with remarkable foresight:

> *One technique for obtaining a feedback controller synthesis from knowledge of open-loop controllers is to measure the current control process state and then compute very rapidly the open-loop control function. The first portion of this function is then used during a short time interval, after which a new measurement of the process state is made, and a new open-loop control function is computed for this measurement. The procedure is then repeated.*

This early articulation foreshadowed the core principles of MPC—iterative state feedback, real-time optimization, and predictive control—highlighting that the conceptual foundations of MPC were recognized well before its widespread implementation became feasible.

1.4.1 Early Theoretical Developments

Early industrial implementations of MPC, which began appearing in the literature in the late 1970s and early 1980s, were initially met with skepticism by the academic community. These first-generation approaches relied heavily on finite impulse or step-response models and involved numerous tunable parameters, but lacked a rigorous theoretical framework to guarantee closed-loop stability or optimal performance. As a result, they were often regarded as heuristic and ad hoc. Yet, the underlying principles of MPC bore a strong resemblance to classical optimal control methods, such as linear quadratic (LQ) control—a connection that did not go unnoticed. This link had initially been obscured by the fact that classical optimal control traditionally focused on state regulation, whereas MPC was developed primarily for output set point tracking in an input–output framework.

As industrial success stories accumulated and interest grew, the process control research community began to re-examine MPC with renewed seriousness. This sparked a wave of theoretical advancements that greatly strengthened the analytical foundations of commercial MPC algorithms. The productive interplay between industrial practice and academic research was well captured by Richalet (1993), one of the pioneers of industrial MPC, who observed:

> *In the last 10 years, [...] the academic community devoted more attention to Model Based Predictive Control. Thus, it is to be noted with some satisfaction, that theory and practice supported each other quite efficiently in this methodology. Progress, which has consequently been steady, is now accelerating.*

One of the earliest and most influential examples of the synergy between industry and academia was the development of the Internal Model Control (IMC) framework. Originating from the graduate thesis work of one of the authors, IMC was directly inspired by the early industrial implementations of MPC. It offered new insights into the structure of MPC and clarified how key parameters—such as the prediction horizon—affect closed-loop stability (García and Morari, 1982).

Over time, IMC evolved beyond its initial connection to MPC and became recognized as a robust control strategy in its own right, particularly in the context of process control (Morari and Zafiriou, 1989). We will explore the core principles of IMC in more detail in Chapter 4 and Appendix E.

It is important to acknowledge the pivotal role that Shell Oil Company played in fostering the early synergy between theory and practice, as highlighted by Richalet. One of the foundational papers that inspired the development of IMC was Shell's seminal work on Dynamic Matrix Control (DMC) by Charlie Cutler, discussed earlier. Notably, Shell's control research group remained highly active throughout the following decade, regularly publishing their advancements (García, 1984; García and Morshedi, 1986; Prett and García, 1988) and collaborating with academic researchers cited throughout this book.

This level of openness was part of a deliberate strategy by Shell to stimulate academic innovation. Their rationale was twofold: to accelerate the adoption of new theoretical insights ahead of competitors and to cultivate a talent pipeline of graduates already familiar with their technologies. One of Shell's most impactful initiatives was the publication of a benchmark challenge problem and the organization of workshops to showcase proposed solutions (Prett and Morari, 1987; Prett et al., 1990). This benchmark, introduced in Section 1.4.2 and used throughout this book, became a unifying reference point for the MPC research community. Many of the academic pioneers behind MPC's major theoretical breakthroughs in the 1990s and beyond participated in these workshops early in their careers, underscoring Shell's influence in shaping the field.

In the 1980s, another branch of MPC, known as Generalized Predictive Control (GPC), emerged from the field of *adaptive control* (Clarke et al., 1987b). GPC was introduced as a more robust alternative to minimum variance control, which had been widely criticized for its sensitivity to model inaccuracies. Consistent with adaptive control strategies of the time, GPC relied on input/output (transfer function) models and incorporated stochastic elements from the outset. By contrast, early industrial MPC implementations were based on step-response models and were largely deterministic, with no explicit treatment of disturbances. This distinction made GPC less practical—if not cumbersome—for multivariable constrained systems, typical of oil and chemical industry applications, where more complex dynamics had to be addressed than in most adaptive control problems.

Although GPC did not gain significant traction within the process industries, it played an important conceptual role. Its early contributions to prediction-based control helped broaden the theoretical foundation of MPC and influenced subsequent developments. As such, GPC remains a valuable part of the historical evolution of MPC, especially in its integration into a wider range of industrial applications.

1.4.2 State-space Model Formulation and Stability Results

Early industrial MPC algorithms, such as DMC and ID-COM, were built primarily on FIR or step-response models. At the time, many practitioners believed that MPC's robustness stemmed from the use of these highly overparameterized models. However, by the late 1980s, researchers began to challenge this view, showing that FIR models were not essential to MPC's effectiveness. The observed robustness could, in fact, be achieved through other model structures.

This realization led to increased interest in state-space formulations, which offered several distinct advantages (Lee et al., 1994; Muske and Rawlings, 1993), including more compact representations and the ability to model unstable dynamics. By the early 1990s, the process control community had begun to reinterpret impulse and step-response models within a state-space framework and to adopt standard state-space system representations (Morari and Lee, 1991; Ricker, 1991; Lee et al., 1994; Muske and Rawlings, 1993).

This shift opened the door to applying well-established tools from estimation and control theory. For instance, integrating the Kalman filter into MPC clarified the role of disturbance modeling—something that early methods like DMC had addressed only implicitly and in a limited way—and enabled the systematic treatment of stochastic disturbances. It also highlighted deep connections between MPC and earlier theoretical developments, including receding horizon control (Propoi, 1963) and LQ optimal control (Anderson and More, 1996; Scokaert and Rawlings, 1998). This convergence of theory and practice enhanced MPC's robustness and flexibility, extending its applicability to a wider range of control problems across diverse industries.

The issue of closed-loop stability in MPC remained largely unresolved until the early 1990s, when pivotal work by Rawlings and coworkers (Rawlings and Muske, 1993), followed by others (Chmielewski and Manousiouthakis, 1996; Mayne et al., 2000), addressed this critical gap. These researchers drew attention to a body of literature—previously underappreciated by the process control community—on the stability of receding horizon control (Kwon and Pearson, 1977; Kleinman, 1974; Keerthi and Gilbert, 1988). Building on these earlier insights, they proposed an infinite-horizon MPC formulation that could be implemented in a finite-horizon setting while still guaranteeing closed-loop stability.

This approach not only resolved key theoretical concerns but also clarified the connection between unconstrained MPC and classical optimal control techniques, such as the Linear

Quadratic Regulator (LQR) and LQG frameworks. By aligning MPC more closely with these established methods, the field gained a stronger theoretical foundation, bolstering its credibility and accelerating its adoption across a wider range of engineering applications.

1.4.3 Other Theoretical Developments

MPC has been widely embraced by the nonlinear control community as a practical and powerful alternative for handling constrained nonlinear systems—scenarios where traditional nonlinear control techniques often become difficult or inapplicable. While integrating a nonlinear system model into MPC is conceptually straightforward, it introduces the challenge of solving a nonconvex optimization problem online—a task that is computationally demanding (Mayne and Michalska, 1990; Michalska and Mayne, 1993; Biegler, 2000; Lee and Ricker, 1994).

As with linear MPC, several formulations have been developed to ensure closed-loop stability for deterministic nonlinear systems. Early results relied on the assumption that the nonlinear optimization problem could be solved to global optimality in real time—a major practical limitation. Thus, the main barrier to nonlinear MPC was initially viewed as computational, specifically the need to efficiently and reliably solve large-scale nonlinear programs online. Later developments relaxed this requirement, shifting the focus to ensuring that each successive control input yields an improvement over the previous one, rather than achieving a globally optimal solution. This shift enabled more practical real-time implementations of nonlinear MPC.

Beyond nonlinear MPC, several additional research directions have shaped the field:

- *Robust MPC* incorporates model uncertainty directly into the optimization problem (Campo and Morari, 1987). A key challenge is that the standard MPC strategy of solving repeated *open-loop* optimizations with closed-loop updates often yields overly conservative and insufficiently robust results, since it ignores the effect of future measurement updates—a factor irrelevant only when model uncertainty is absent. Approaches based on dynamic programming (Lee and Yu, 1997) and linear matrix inequalities (LMI) (Kothare et al., 1996) have been proposed, but their computational complexity has hindered widespread practical use.
- *Mixed Logic MPC* extends the framework to hybrid systems by modeling both continuous dynamics and discrete events using Mixed Logic Dynamical Systems (Bemporad and Morari, 1999). This allows for the inclusion of binary inputs, states, and logic constraints to represent switching behavior and prioritized objectives.
- *Explicit MPC* eliminates online optimization by precomputing the control law offline using multiparametric programming (Bemporad et al., 2002). The resulting control action can be implemented via a simple table lookup. However, this approach suffers from exponential growth in complexity with respect to system dimension and prediction horizon, making it suitable only for low-dimensional systems.

1.4.4 Lessons Learned Along the MPC Journey

Those familiar with optimal control will recognize that the challenges addressed by MPC closely resemble classical optimal control problems studied in the 1950s and 1960s. Traditional methods sought to derive an explicit optimal feedback law offline, typically through the Hamilton–Jacobi–Bellman (HJB) equation. However, the HJB equation is notoriously difficult to solve, with analytical solutions available only in limited cases—such as the LQR problem. Consequently, classical methods were impractical for many real-world problems, particularly those involving inequality constraints.

MPC introduced a transformative shift by avoiding the need to solve the HJB equation explicitly. Instead, it leverages modern computational power to solve the open-loop optimal control problem online, at each sampling instant. This receding-horizon strategy allows MPC to dynamically apply optimal control actions without deriving a closed-form feedback law. As a result, MPC can handle more complex and constrained problems that arise in industrial applications. Thus, MPC represents a major leap forward, since it directly computes and applies optimal actions in real time. This capability marks a significant advancement in control engineering, bridging theory and industrial practice.

We have discussed how MPC's success within the process industries stems largely from its technological advantages—enabling high-performance control with reduced cost of ownership. Its rapid adoption can also be attributed to the timely availability of real-time computational power and the productive synergy between industrial practitioners and academic researchers. This success was not merely serendipitous. It offers important lessons that can—and should—guide the development of any new control methodology:

- Tackle unsolved problems with broad and significant impact.
- Develop tools that are intuitive and practical for stakeholders to use.
- Always consider the total cost of ownership.
- Don't be discouraged by computational or hardware limitations; innovate around them or help eliminate them.
- Recognize that practical solutions often emerge before formal theory.

In Section 1.5, we present a typical industrial control problem to illustrate the kinds of challenges that MPC was designed to address.

1.5 A Typical Industrial Control Problem

To appreciate the value of advanced control methods such as MPC, consider the heavy oil fractionator shown in Figure 1.5, a key refinery unit. Crude oil is heated in a furnace and fed to the fractionator, where it is separated into four product streams—lights, naphtha, light gas oil, and short residue—each routed to downstream units.

Refineries are among the most energy-intensive facilities, so engineers employ heat integration strategies to improve recovery. In the fractionator, pump-around circuits (Q1–Q3) serve as heat sources for other units, recovering energy at higher temperatures than the overhead condenser. The stripper section further improves product recovery. Plant-wide optimization is used to set operating conditions based on crude/product prices, feed properties, and equipment constraints. Methods range from heuristic rules to advanced linear or nonlinear programming. Even small efficiency gains translate into large profits, underscoring the importance of precise operation.

Material balance is maintained by four primary loops: reflux drum level (via naphtha reflux), stripper bottom level (via side stream draw), column bottom level (via short residue draw), and column pressure (via lights draw). Additional flow controllers regulate naphtha and light gas oil withdrawals. Six temperature measurements and boiling-point analyzers provide quality monitoring, though analyzers are prone to outages. Key performance specifications are naphtha and light gas oil compositions, most sensitive to Q2 and Q3 reflux duties. A secondary objective is maximizing Q1 duty to recover steam.

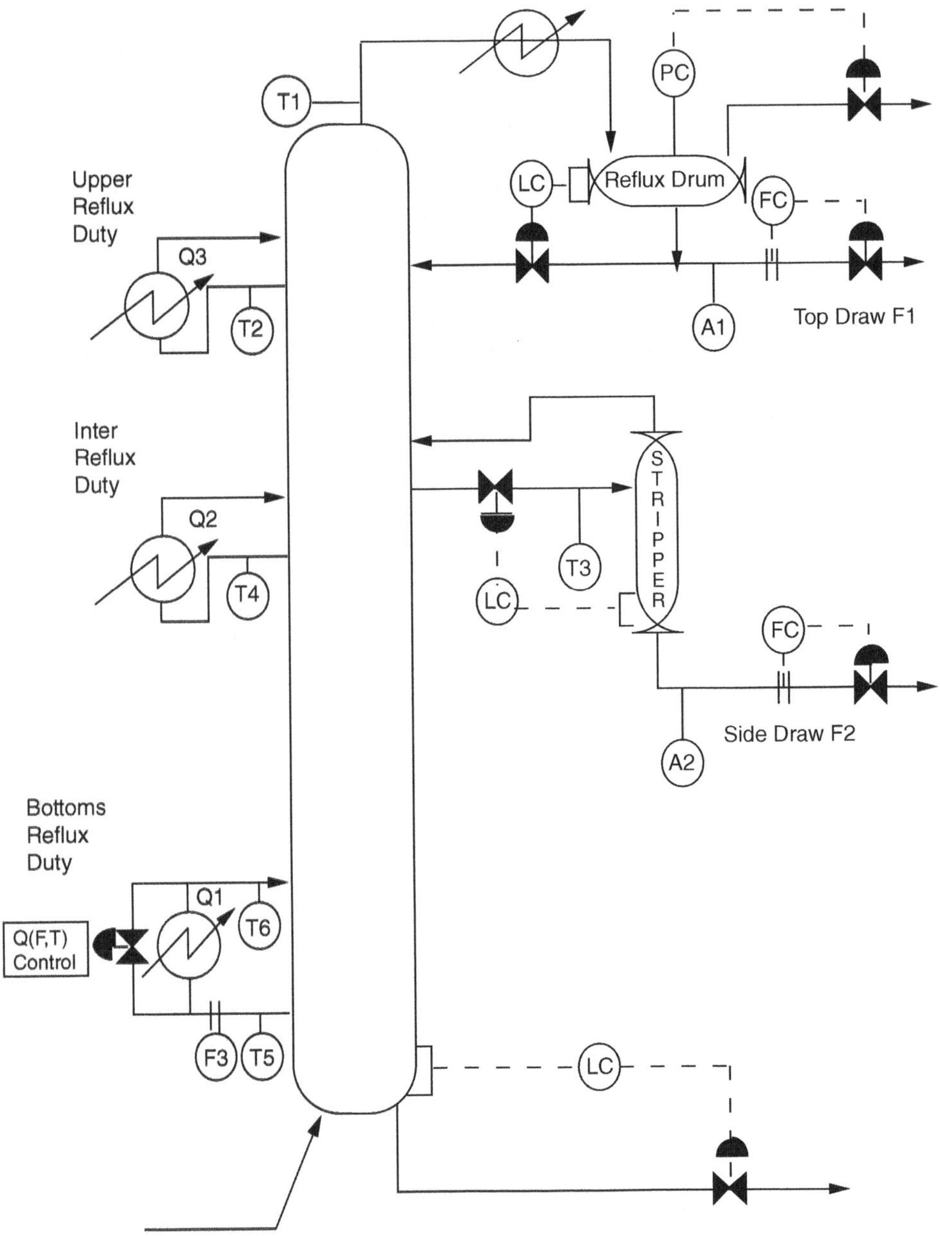

Figure 1.5 A heavy oil fractionator with heat integration and material balance loops.

Constraints include valve position limits, rate-of-change restrictions, minimum bottom temperature (T5), and product composition bounds. Robust control is required to handle these constraints, especially when analyzers fail or disturbances occur.

To structure the problem, Figure 1.6 distinguishes inputs and outputs. Inputs are either manipulated variables (e.g., Q1, F1, F2) or disturbances (e.g., Q2, Q3). Outputs include controlled variables (A1, A2 compositions) and associated variables (T5).

Several challenges arise:

Multivariable System: The system involves two controlled variables (A1 and A2) and one associated variable (T5). Due to the strong interactions between these variables, adjusting one input affects all outputs, as shown in Figure 1.7. Although one might consider pairing F1 with A1 and F2 with A2 (as illustrated in Figure 1.8), such a decoupled control structure proves ineffective, causing so-called controller fighting. A better approach is to implement a multivariable controller (Figure 1.9) that adjusts F1 and F2 simultaneously and in a coordinated manner, mitigating interactions and improving overall control performance.

Non-square Systems: Two outputs (A1, A2) must be controlled with three inputs (F1, F2, Q1). The extra degree of freedom can be used for economics or as a backup when other inputs are constrained.

Constraints: When inputs saturate (e.g., F1 at max flow), independent regulation of A1 and A2 fails. Options include controlling an average composition, prioritizing one variable, or reallocating Q1. Anticipating such events is essential.

Optimizing Control: Bottom reflux duty (Q1) is typically tuned online, balancing heat recovery with composition control. This requires tight coordination between optimization and regulatory control.

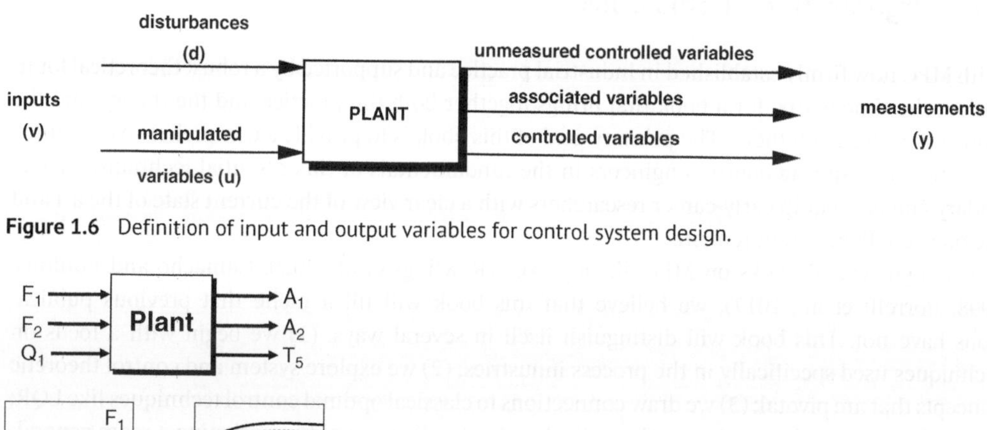

Figure 1.6 Definition of input and output variables for control system design.

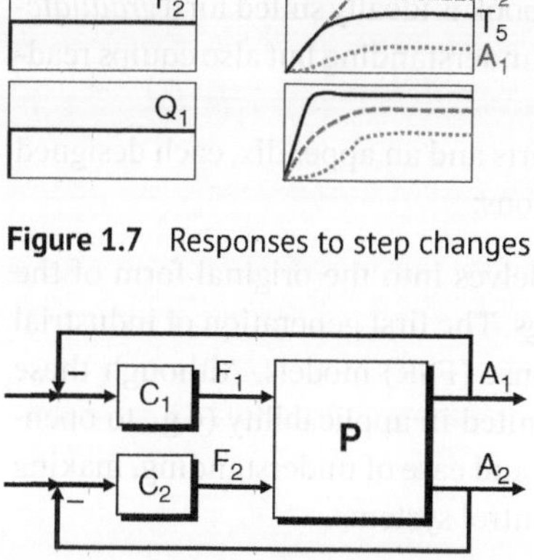

Figure 1.7 Responses to step changes in the three manipulated inputs of the heavy oil fractionator.

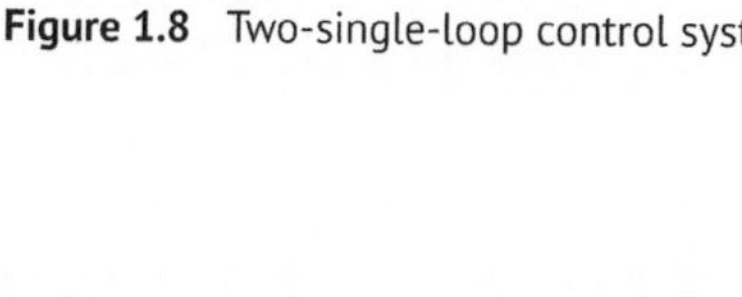

Figure 1.8 Two-single-loop control system.

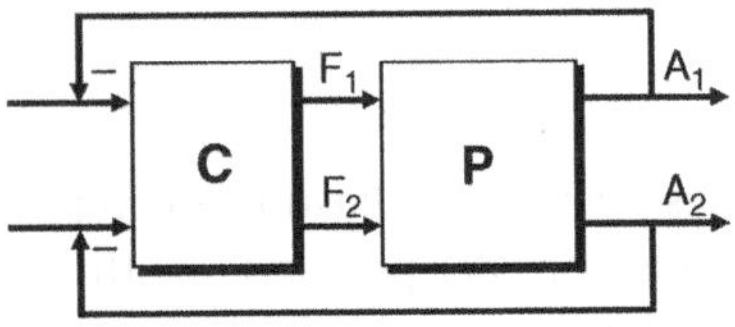

Figure 1.9 Multivariable control system.

Fault Tolerance: The system must maintain acceptable performance despite analyzer outages or actuator faults, without manual restructuring.

Individually, these challenges can be handled by traditional techniques: multivariable controllers, anti-windup schemes, selectors, or simple optimizing controllers. But in modern refineries, multiple issues arise simultaneously, making ad hoc fixes costly and complex to integrate.

This motivates a unified methodology: MPC. MPC can handle multivariable interactions, constraints, optimization, and robustness within a single framework, offering a scalable and practical solution unmatched by traditional methods.

1.6 Organization of This Book

With MPC now firmly established in industrial practice and supported by a robust theoretical foundation, the time is ripe for a book that brings together both the practice and the theory, and the connections between them. The primary goal of this book is to provide a comprehensive resource for educating aspiring control engineers in the fundamentals of this essential technique. A secondary aim is to equip early-career researchers with a clear view of the current state of the art and the open challenges that remain.

Although several books on MPC already exist (Rawlings et al., 2020; Camacho and Bordons, 1998; Borrelli et al., 2017), we believe that this book will fill a niche that previous publications have not. This book will distinguish itself in several ways: (1) we begin with a focus on techniques used specifically in the process industries; (2) we explore system and control theoretic concepts that are pivotal; (3) we draw connections to classical optimal control techniques like LQR; (4) we demonstrate how a deeper theoretical understanding can aid in designing a more generalized and extended form of MPC. Additionally, by covering related system and control theoretical concepts, such as linear systems and system identification, this book is ideally suited for a *graduate-level course in process control*. This approach not only enhances understanding but also equips readers to apply MPC effectively across various practical scenarios.

The remainder of the book is structured into three distinct parts and an appendix, each designed to build upon the knowledge established in the preceding sections:

Part I: Early Industrial MPC Algorithms: The first part delves into the original form of the MPC algorithm as it was first conceived in industrial settings. The first generation of industrial MPC techniques focused on the FIR and finite step response (FSR) models. Although these model types and their associated MPC formulations are limited in applicability (e.g., to open-loop stable systems), they are notable for their intuitiveness and ease of understanding, making them accessible even to those without formal training in control systems.

Part II: Basics of Linear Systems, Optimal Control, and System Identification: The second part introduces system and control-theoretic concepts essential for aspiring control engineers and researchers. It begins with basic modeling concepts before moving on to optimal state estimation techniques, such as Kalman filtering, and control strategies like LQR and LQG controls, presented in both unconstrained and constrained forms. Such techniques are largely based on state-space system representations, which are more compact and general than FIR or FSR models, and thus provide the foundation for the more general and versatile MPC techniques covered in Part III. Finally, this part covers the fundamental methods and theories of system identification, which are crucial for building system models from data.

Part III: Advanced MPC: The third part builds on the foundations laid in the first two parts to explore more sophisticated forms of MPC. This segment expands MPC into more general forms to broaden its applicability and enhance performance. Topics include state-space MPC, nonlinear MPC, repetitive MPC, and the computational challenges associated with these advanced techniques. By the end of this part, readers are expected to have a robust understanding sufficient to design new applications, and support and enhance existing MPC implementations.

Appendix: Related Topics and Supplementary Materials: The Appendix provides a concise review of linear algebra and stochastic processes—foundational topics necessary for understanding the advanced material presented in Parts II and III. Although most students have encountered matrices and vectors, fewer have studied linear transformations, which are essential for analyzing systems with multiple inputs and outputs. The Appendix also includes brief overviews of model reduction techniques and IMC (a robust control strategy closely related to MPC), extended derivations related to stochastic state estimation and control (which are placed here to preserve the flow of the main text), and introductory material on SIMULINK and the MATLAB MPC Toolbox.

Note that sections marked with {∗} at the end of the title contain optional material, which can be skipped without loss of continuity. Through this structured approach, the book aims to provide readers with a comprehensive understanding of both the foundational and advanced aspects of MPC, preparing them to address a wide range of practical control challenges in modern academic research and industrial environments.

Astute readers will notice that certain themes in system modeling and optimal control recur throughout the book. This repetition is intentional: these topics, often underemphasized in traditional control texts, are essential for practical applications.

One such theme is *disturbance modeling*. Traditional MPC methods, such as DMC, use simple feedback correction to remove steady-state offsets in controlled variables (CVs). However, explicit disturbance modeling can significantly improve performance and extend the applicability of MPC to more challenging cases, particularly when CVs are unmeasured or measured infrequently with long delays.

Another recurring theme is *integral action*. While traditional MPC inherently ensures offset-free tracking, this is not the case for general optimal estimation and control methods such as Kalman filtering and LQ control. In these frameworks, integral action must be deliberately incorporated—typically through state augmentation—to guarantee zero steady-state error.

Finally, we emphasize the conceptual distinction between open-loop optimal feedback control (OLOFC) and feedback optimal control in the context of stochastic systems. MPC follows the OLOFC paradigm: an open-loop optimal control problem is solved over a finite-horizon, with feedback introduced by re-solving at each sampling instant. In deterministic systems, the difference between OLOFC and true feedback optimal control is often negligible. In stochastic settings, however, explicitly designing feedback policies to account for uncertainties can be critical, as they greatly enhance robustness and overall performance.

As a final note, a few comments on notation are in order. Maintaining consistent notation across 13 chapters and 6 appendices has been a considerable challenge—particularly since different sections were drafted at widely different times over a writing period spanning more than 30 years. Although we have made every effort to ensure consistency, readers may still encounter occasional discrepancies. We have made every effort to define new symbols as they are introduced and have repeated key definitions to enhance readability, even at the risk of some redundancy.

Exercises

1. In Example 1.1, a bather used a hot water valve to control the water temperature.
- Use the shower process model (1.6) for $\theta_d = 1$ and the integral control law (1.7) to show that $K_I = 2.0$ produces a deadbeat response in the output. Use them to also show that the onset of instability is at $K_I = 4.0$.
- Do the same for $\theta_d = 6$ to show that the integral controller is unstable for $K_I = 2.0$.
- Derive the MPC control law (1.10) from the prediction model and the BIAS term in (1.8).
- Use the shower process model (1.6) for $\theta_d = 6$ and the MPC output deadbeat control law (1.10) to show that the closed-loop system response is unstable for a time delay of 4 in the prediction model. (Hint: In the prediction model summation term, only add the last four moves, or $\Delta V(k - \ell)$ for $\ell = 1$ to 4.)

2. Develop a constrained control problem of your choice, formulate the corresponding process model, and derive the associated MPC control law.

Part I

Early Industrial MPC Algorithms

2

Step Response Modeling and Identification

To find out what happens when you change something, it is necessary to change it.

– Box, Hunter, and Hunter

Computer control algorithms use mathematical model descriptions of the system to be controlled. The purpose of these dynamical models is to determine the required manipulated input moves in real time. Early industrial practitioners of MPC developed such models by performing input changes on a running plant. The model parameters are then computed so that the model output closely matches the controlled output response data. In this chapter we introduce the models and the techniques for testing and parameter fitting used by those pioneers.

Figure 2.1 depicts a computer-controlled system. The controlled output signal $y(t)$ is sampled at intervals t_s, resulting in a sequence of measurements $\{y(0), y(1), \dots, y(k), \dots\}$. Arguments $0, 1, \dots, k, \dots$ denote the time measured in multiples of the sampling interval t_s. Based on this sequence, the control algorithm determines a sequence of inputs $\{u(0), u(1), \dots, u(k), \dots\}$. The input time function $u(t)$ is obtained by holding the input constant over a sampling interval:

$$u(t) = u(k) \quad kt_s < t \leq (k+1)t_s.$$

Throughout the book, we will consider that the control task is carried out by such a *sampled-data control system.*

The appropriate control action depends on the dynamic characteristics of the plant to be controlled. Therefore, the control algorithm is usually designed based on a plant's *model,* which describes its dynamic characteristics (at least approximately). In the context of Figure 2.1 we are looking for a plant model that allows us to determine the sequence of outputs $\{y(0), y(1), \dots, y(k), \dots\}$ resulting from a sequence of inputs $\{u(0), u(1), \dots, u(k), \dots\}$.

In general, we are interested not only in the response of the plant to the manipulated variable (MV) u but also in its response to other inputs, such as a disturbance variable (DV) d. We will use the symbol v to denote a "general" input, which can be manipulated (u) or influenced by external perturbations (d). Thus, the general plant model will relate a sequence of inputs $\{v(0), v(1), \dots, v(k), \dots\}$ to a sequence of outputs $\{y(0), y(1), \dots, y(k), \dots\}$.

Initially, we will assume for our derivations that the system has only a single input v and a single output y (Single-Input Single-Output [SISO] system). Later we will generalize the modeling concepts to describe Multi-Input Multi-Output (MIMO) systems.

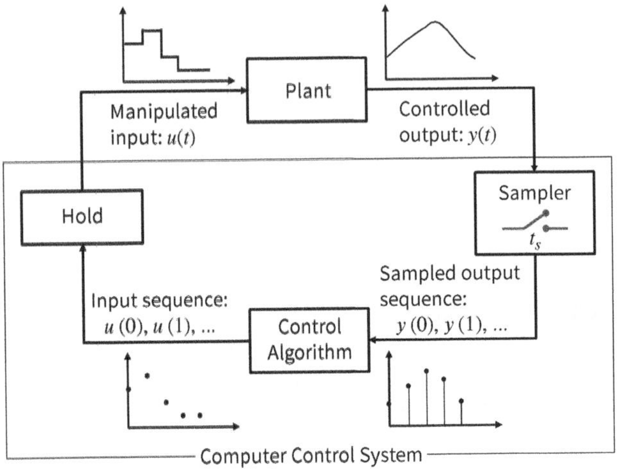

Figure 2.1 Sampled-data control system.

In this chapter, we introduce the basic modeling assumptions underlying *convolution models* and show how they can be used to predict output responses to input sequences. We also provide guidance on how such models can be identified from experimental data.

2.1 Linear Time-invariant Systems

Let us consider two specific input sequences

$$\mathbf{v}^1(0) = \{v^1(0), v^1(1), \dots, v^1(k), \dots\}$$
$$\mathbf{v}^2(0) = \{v^2(0), v^2(1), \dots, v^2(k), \dots\}.$$

Throughout this book, we will adopt the convention that the input sequences are zero for negative time, i.e., $v(k) = 0,\ \text{for}\ k < 0$.

For example, $\mathbf{v}^1(0)$ might represent a step

$$\mathbf{v}^1(0) = \{1, 1, \dots, 1, \dots\},$$

and $\mathbf{v}^2(0)$ a pulse.

$$\mathbf{v}^2(0) = \{1, 0, \dots, 0, \dots\}.$$

Let us assume that for a particular system P, these two input sequences give rise to the output sequences

$$\mathbf{y}^1(0) = \{y^1(0), y^1(1), \dots, y^1(k), \dots\}$$
$$\mathbf{y}^2(0) = \{y^2(0), y^2(1), \dots, y^2(k), \dots\},$$

respectively. Symbolically, we can also write

$$\mathbf{v}^1(0) \xrightarrow{P} \mathbf{y}^1(0) \qquad \text{and} \qquad \mathbf{v}^2(0) \xrightarrow{P} \mathbf{y}^2(0).$$

Let us consider two other input sequences:

$$\mathbf{v}^3(0) \triangleq \alpha\mathbf{v}^1(0) = \{\alpha v^1(0), \alpha v^1(1), \ldots, \alpha v^1(k), \ldots\},$$

where $-\infty < \alpha < \infty$ is a real constant, and

$$\mathbf{v}^4(0) \triangleq \mathbf{v}^1(0) + \mathbf{v}^2(0) = \{v^1(0) + v^2(0), v^1(1) + v^2(1), \ldots, v^1(k) + v^2(k), \ldots\}.$$

If changing the input magnitude by a factor α changes the output magnitude by the same factor

$$\alpha\mathbf{v}^1(0) = \mathbf{v}^3(0) \xrightarrow{P} \mathbf{y}^3(0) = \alpha\mathbf{y}^1(0),$$

and if the response to a sum of two input sequences is the sum of the two output sequences

$$\mathbf{v}^1(0) + \mathbf{v}^2(0) = \mathbf{v}^4(0) \xrightarrow{P} \mathbf{y}^4(0) = \mathbf{y}^1(0) + \mathbf{y}^2(0),$$

for arbitrary choices of $\mathbf{v}^1(0)$, $\mathbf{v}^2(0)$ and α, then the system P is called *linear*.

Linearity is a powerful property with the following implication: if the input sequences $\mathbf{v}^1(0), \mathbf{v}^2(0), \mathbf{v}^3(0), \ldots$ produce the output sequences $\mathbf{y}^1(0), \mathbf{y}^2(0), \mathbf{y}^3(0), \ldots$, respectively, then the response to any linear combination of the inputs is simply the same linear combination of the outputs. In other words, the process output can be determined by the principle of *superposition*.

$$\alpha_1\mathbf{v}^1(0) + \alpha_2\mathbf{v}^2(0) + \cdots + \alpha_k\mathbf{v}^k(0) + \cdots \xrightarrow{P}$$
$$\alpha_1\mathbf{y}^1(0) + \alpha_2\mathbf{y}^2(0) + \cdots + \alpha_k\mathbf{y}^k(0) + \cdots \quad \text{where} \ -\infty < \alpha_i < \infty.$$

In practice, there are no truly linear systems. Linearity is a mathematical idealization that allows us to utilize a wealth of very powerful theories. In a small neighborhood around a fixed operating point, the behavior of many systems is approximately linear. Because the purpose of control is usually to keep the system operation close to a fixed point, the set point, linearity is often a very good assumption.

Let us now define the shifted input sequence

$$\mathbf{v}^1(-\ell) = \left\{\underbrace{0, 0, \ldots, 0}_{\ell}, v^1(0), v^1(1), v^1(2), \ldots, v^1(k), \ldots\right\},$$

where the first ℓ elements are zero. Here we have simply moved the signal $\mathbf{v}^1(0)$ to the right by ℓ time steps. If the resulting system output sequence is shifted in the same manner

$$\mathbf{y}^1(-\ell) = \left\{\underbrace{0, 0, \ldots, 0}_{\ell}, y^1(0), y^1(1), y^1(2), \ldots, y^1(k), \ldots\right\},$$

and if the system property:

$$\mathbf{v}^1(-\ell) \xrightarrow{P} \mathbf{y}^1(-\ell),$$

holds for arbitrary input sequences $\mathbf{v}^1(0)$ and arbitrary integers ℓ, then the system is referred to as *time-invariant*. Time invariance implies, for example, that if tomorrow the same input sequence is applied to the plant as today, then the same output sequence will result. Throughout this part of the book we will assume that *linear time-invariant* models are adequate for the systems we are trying to describe and control.

2.2 Impulse/Step Response Models

Finite Impulse Response (FIR) and Finite Step Response (FSR) models are simple and intuitive linear models derived directly from observed process responses. They are typically applied to stable systems. Before introducing these models in detail, we first provide a brief discussion of system stability.

Let us perturb the input v in a stepwise manner and observe the corresponding output y. Figure 2.2 illustrates three broad types of system behavior. A *stable system* settles to a steady state following the input perturbation. An *unstable system* does not settle, with its output continuing to diverge over time. The output of an *integrating system* increases asymptotically in a ramp-like fashion, changing approximately linearly with time. An *exponentially unstable system* exhibits output growth in an exponential manner.

Most processes encountered in the process industries are *stable* (Figure 2.3a). An example of an *integrating process* is shown in Figure 2.3b, where a controller installed at the tank outlet delivers a constant flow. If a step change occurs in the inflow, the level will increase in a ramp-like manner. In the case of a pulse change, the level shifts to a different permanent value. Some chemical reactors, when left uncontrolled, can be *exponentially unstable*. For safety reasons, however, reactors are usually designed to be stable, ensuring that a runaway will not occur if the control system fails. The control concepts discussed in this part of the book are therefore not applicable to exponentially unstable systems.

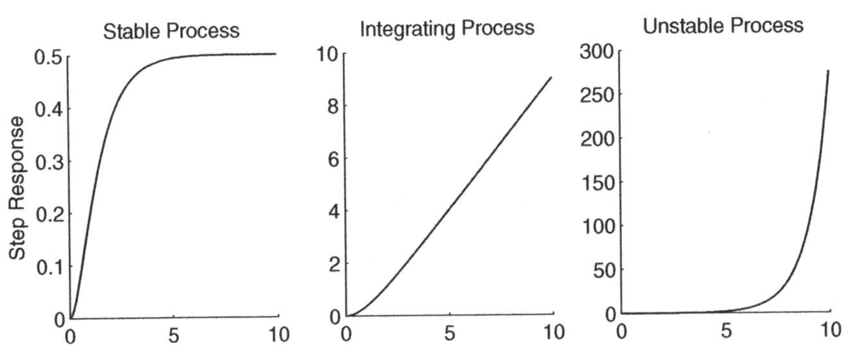

Figure 2.2 Output responses to a step input change for stable, integrating, and unstable systems.

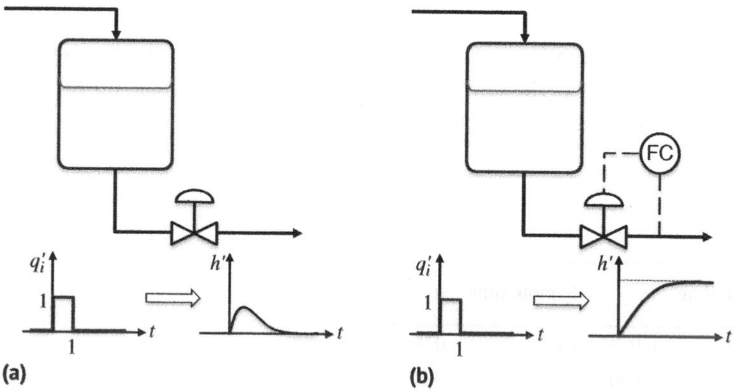

Figure 2.3 Water tank example: (a) stable system and (b) integrating system.

2.2.1 Impulse Response Models

Let us inject a unit pulse

$$\mathbf{v}(0) = \{1, 0, \ldots, 0, \ldots\},$$

into a system at rest and observe the response

$$\mathbf{y}(0) = \{h_0, h_1, \ldots, h_n, h_{n+1} \cdots\}.$$

We will assume that the system has the following characteristics:

- $h_0 = 0$, i.e., the system does not react immediately to the input
- $h_k = 0$ for $k > n$, i.e., the system settles after n time steps.

Such a system is referred to as an FIR system, with the corresponding matrix of FIR coefficients given by

$$H = [h_1, h_2, \ldots, h_n]^T. \tag{2.1}$$

An FIR model provides a reasonable approximation for many practical systems. For example, in Figure 2.3a, the outflow through a fixed opening depends on the liquid height in the tank. Following a pulse inflow perturbation, the level eventually returns to its original value, and the system can be well approximated by an FIR model. In contrast, when the outflow is maintained constant by a flow controller, as in Figure 2.3b, the pulse adds liquid volume that permanently changes the tank level. This *integrating system* cannot be represented by an FIR model.

In this book we use Laplace transforms to model linear continuous processes. We assume our readers have basic working knowledge of such transforms.

Example 2.1: *FIR Model for a Continuous-time System.*
Consider the continuous-time system

$$G(s) = \frac{1}{5s + 1},$$

sampled once every second ($t_s = 1$). The impulse response of the system is the output of the system to a unit impulse in the discrete-time: $\mathbf{v}(0) = \{1, 0, 0, 0, \dots\}$. This impulse in discrete-time is equivalent to a unit pulse of duration equal to the sampling time (see Figure 2.1). Thus, the impulse response model is the response of the system to the input

$$v(s) = \frac{1}{s}(1 - e^{-s}) :$$

Sample time	h_i	Sample time	h_i	Sample time	h_i
1	0.1813	11	0.0245	21	0.0033
2	0.1484	12	0.0201	22	0.0027
3	0.1215	13	0.0164	23	0.0022
4	0.0995	14	0.0135	24	0.0018
5	0.0814	15	0.011	25	0.0015
6	0.0667	16	0.009	26	0.0012
7	0.0546	17	0.0074	27	0.001
8	0.0447	18	0.006	28	0.0008
9	0.0366	19	0.005	29	0.0007
10	0.03	20	0.0041	30	0.0005

The dynamic behavior of this system can be effectively captured by retaining $n = 30$ coefficients.

Example 2.2: *FIR Model of a Discrete-time System.*
Consider the following discrete-time system with sampling time $t_s = 1$:

$$y(k) = 0.5y(k-1) + v(k-1).$$

Starting with $y(0) = 0$, response of this system to impulse $\mathbf{v}(0) = \{1, 0, 0, 0, \dots\}$ can be obtained by recursively solving the above equation. We used MATLAB™ function `impulse` to generate the impulse response:

Sample time	h_i	Sample time	h_i	Sample time	h_i
1	1	5	0.0625	9	0.0039
2	0.5	6	0.0313	10	0.002
3	0.25	7	0.0156	11	0.001
4	0.125	8	0.0078	12	0.0005

The dynamic behavior of this system can be effectively captured by retaining $n = 12$ coefficients.

For linear time-invariant systems, shifting the input pulse

$$\mathbf{v}(0) = \{0, 1, 0, \dots, 0, \dots\},$$

will simply shift the impulse response

$$\mathbf{y}(0) = \{0, h_0, h_1, \dots, h_n, h_{n+1} \cdots\}.$$

The dynamical behavior of an FIR system is completely characterized by the set of FIR coefficients. The system response can be computed by superposition as follows. Any arbitrary input

$$\mathbf{v}(0) = \{v(0), v(1), v(2)\dots\},$$

can be represented as a sum of impulses

$$\begin{aligned}\mathbf{v}(0) = &\{1, 0, 0, \dots\}\, v(0)\\ &+ \{0, 1, 0, \dots\}\, v(1)\\ &+ \{0, 0, 1, 0\dots\}\, v(2)\\ &+ \dots\end{aligned}$$

The system output is obtained by summing the impulse responses weighted by their respective impulse strength $v(i)$

$$\begin{aligned}\mathbf{y}(0) = &\{0, h_1, h_2\dots, h_n, 0, 0, \dots\}\, v(0)\\ &+ \{0, 0, h_1, h_2, \dots, h_n, 0, 0, \dots\}\cdot v(1)\\ &+ \{0, 0, 0, h_1, h_2, \dots, h_n, 0, 0, \dots\}\cdot v(2)\\ &+ \dots\\ = &\{0,\ h_1 v(0),\ h_2 v(0) + h_1 v(1),\ h_3 v(0) + h_2 v(1) + h_1 v(2), \dots\}.\end{aligned}$$

By direct inspection of the derivation, we see that at any time the output is given by

$$y(k) = \sum_{i=1}^{n} h_i v(k-i). \tag{2.2}$$

The coefficient h_i represents the effect of an input applied i sampling intervals earlier on the current output $y(k)$. To compute this output, the last n input values, $v(k-1), v(k-2), \dots, v(k-n)$, must therefore be stored in memory.

2.2.2 Step Response Models

For a unit input step

$$\mathbf{v}(0) = \{1, 1, \dots, 1, \dots\},$$

injected into the system at rest, the output will be

$$\mathbf{y}(0) = \{0, h_1, h_1 + h_2, h_1 + h_2 + h_3, \dots\}, \tag{2.3}$$

$$\triangleq \{0, s_1, s_2, s_3, \dots\}. \tag{2.4}$$

Here $s_1, s_2, s_3, \dots$ are called the step response coefficients.

For linear time-invariant systems, the shifted step

$$\mathbf{v}(0) = \{0, 1, 1, \dots, 1, \dots\},$$

will give rise to a shifted step response

$$\mathbf{y}(0) = \{0, 0, s_1, s_2, s_3, \dots\}.$$

The matrix of step response coefficients

$$S = [s_1, s_2, \dots, s_n]^T, \tag{2.5}$$

is a *complete* description of how a particular input affects the system output when the system is at rest. Any arbitrary input

$$\mathbf{v}(0) = \{v(0), v(1), v(2), \dots\},$$

can be represented as a sum of steps

$$\begin{aligned}\mathbf{v}(0) = &\{1, 1, 1, 1, \dots\}v(0)\\ &+ \{0, 1, 1, 1, \dots\}(v(1) - v(0))\\ &+ \{0, 0, 1, 1, \dots\}(v(2) - v(1))\\ &+ \dots,\end{aligned}$$

where we will define $\Delta v(i) \triangleq v(i) - v(i-1)$ to denote the input changes from one time step to the next. The system output is obtained by summing the step responses weighted by their respective step heights $\Delta v(i)$:

$$\begin{aligned}\mathbf{y}(0) = &\{0, s_1, s_2, s_3, \dots, s_n, s_n, s_n, \dots\}\, v(0)\\ &+ \{0, 0, s_1, s_2, \dots, s_n, s_n, s_n, \dots\}\, \Delta v(1)\\ &+ \{0, 0, 0, s_1, s_2, \dots, s_n, s_n, \dots\}\, \Delta v(2)\\ &+ \dots,\end{aligned}$$

$$\begin{aligned}\text{Thus, } \mathbf{y}(0) = \{&0,\\ &s_1 v(0),\\ &s_2 v(0) + s_1 \Delta v(1),\\ &\dots\\ &s_n v(0) + s_{n-1}\Delta v(1) + s_{n-2}\Delta v(2) + \dots + s_1 \Delta v(n-1),\\ &s_n \underbrace{(v(0) + \Delta v(1))}_{v(1)} + s_{n-1}\Delta v(2) + s_{n-2}\Delta v(3) + \dots + s_1 \Delta v(n)\}.\end{aligned}$$

We see that at a particular time, the output is given by

$$y(k) = \sum_{i=1}^{n-1} s_i \Delta v(k-i) + s_n v(k-n). \tag{2.6}$$

The coefficient s_i expresses the effect of an input change that occurred i sample intervals in the past on the present output $y(k)$. In order to compute this output, we need to keep the last n inputs in memory.

The step response coefficients are directly calculable from the impulse response coefficients and vice versa.

$$s_k = \sum_{i=1}^{k} h_i, \tag{2.7}$$

$$h_k = s_k - s_{k-1}. \tag{2.8}$$

Example 2.3: *Step and Impulse Response Models for Discrete-time System.*
Let us revisit the system in Example 2.2: In this example, we will

- Determine the step response coefficients
- Verify Eq. (2.8), and
- Verify the superposition principle for $\mathbf{v}(0) = \{0.5, 1.5, 1, 0, 0, \dots, 0, \dots\}$.

The step response coefficients can be obtained from the difference equation

$$y(k) = 0.5y(k-1) + v(k-1),$$

with $\mathbf{v}(0) = \{1, 1, 1, \dots\}$, or by using the MATLAB™ function `step`.

The table below summarizes the step response coefficients s_i along with the differences $s_i - s_{i-1}$.

Sample time	s_i	$s_i - s_{i-1}$	Sample time	s_i	$s_i - s_{i-1}$
1	1	1	7	1.9844	0.0156
2	1.5	0.5	8	1.9922	0.0078
3	1.75	0.25	9	1.9961	0.0039
4	1.875	0.125	10	1.998	0.0019
5	1.9375	0.0625	11	1.999	0.001
6	1.9688	0.0313	12	1.9995	0.0005

The reader can verify from results in Example 2.2 that Eq. (2.8) is indeed satisfied.

The response of the system to the given input signal is computed recursively. The response is plotted in Figure 2.4 and is shown below:

Time	0	1	2	3	4	5	6	7	8	...
$y = \{$	0	0.5	1.75	1.875	0.9375	0.4688	0.2344	0.1172	0.0586	...}.

Computing the response of the system using impulse response model:

0.5×	{0.0	1.0	0.5	0.25	0.125	0.0625	0.0313	0.0156	0.0078	...}+
1.5×		{0.0	1.0	0.5	0.25	0.125	0.0625	0.0313	0.0156	...}+
1.0×			{0.0	1.0	0.5	0.25	0.125	0.0625	0.0313	...}
$y = \{$	0	0.5	1.75	1.875	0.9375	0.4688	0.2344	0.1173	0.0586	...}

Computing using the step response model also yields the same result:

0.5×	{0.0	1.0	1.5	1.75	1.8750	1.9375	1.9688	1.9844	1.9922	...}+
1.0×		{0.0	1.0	1.5	1.750	1.8750	1.9375	1.9688	1.9844	...}+
−0.5×			{0.0	1.0	1.5	1.750	1.8750	1.9375	1.9688	...}+
−1.0×				{0.0	1.0	1.5	1.750	1.8750	1.9375	...}
$y = \{$	0	0.5	1.75	1.875	0.9375	0.4688	0.2344	0.1173	0.0586	...}

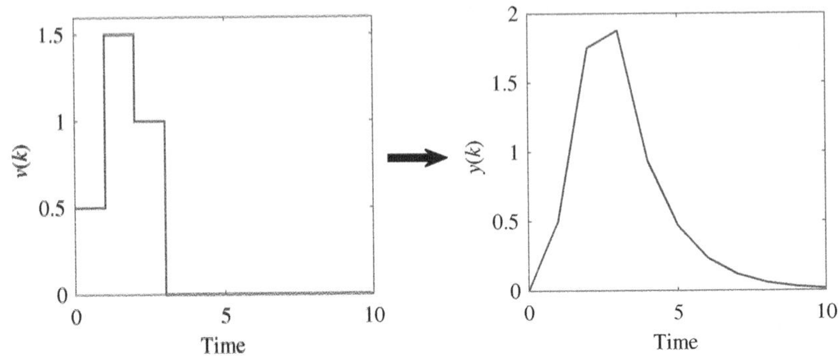

Figure 2.4 Input profile (left) and response (right) of the discrete-time linear system.

2.3 Multi-step Prediction

The objective of a controller is to determine control actions that produce desirable output behavior in the future. Achieving this requires the ability to efficiently predict the system's future output. For a dynamic system, this behavior depends both on past inputs applied to the process and on the inputs to be applied in the future. To separate these effects, all information from past inputs is captured in the *dynamic state* of the system. Thus, the future output behavior is determined by the current system state in conjunction with the present and future inputs.

For an FIR system, we may choose the past n inputs as the state x

$$x(k) = [v(k-1),\ v(k-2), \ldots,\ v(k-n)]^T. \tag{2.9}$$

This state compactly summarizes all relevant past input information for the FIR system and enables computation of its future evolution, given the present input $v(k)$ and the future inputs $v(k+1), v(k+2), \ldots$.

The representation of the state is usually not unique, and alternative choices are possible. For example, instead of using the n past inputs, we could define the state as the *effect* of those past inputs on the future outputs over the next n steps. In other words, the state may be defined as the present output together with the next $n-1$ future outputs—assuming that the present and future inputs are zero. These two state representations are equivalent, as they both have dimension n and are uniquely related by a linear map.

The latter choice of state is more convenient for predictive control computations. It makes explicit how the system evolves in the absence of control action, thereby simplifying the determination of control actions required to achieve a desired future output behavior. In the following sections, we discuss multistep prediction for FIR and step-response models based on this state representation.

2.3.1 Recursive Multi-step Prediction Based on FIR Model

Let us define the state at time k as

$$\bar{Y}(k) = \left[\bar{y}_0(k),\ \bar{y}_1(k), \ldots,\ \bar{y}_{n-1}(k)\right]^T, \tag{2.10}$$

where

$$\bar{y}_i(k) \stackrel{\Delta}{=} y(k+i) \text{ for } v(k+j) = 0;\ j \geq 0, \tag{2.11}$$

Thus, we define the ith system state $\bar{y}_i(k)$ as the system output at time $k+i$, under the assumption that the system inputs are zero beyond time k ($\upsilon(k+j)=0;\ j\geq 0$). This state captures the effect of past inputs and fully characterizes the evolution of the system output under the assumption of zero present and future inputs. To determine the actual future output, we simply add to the state the contributions of the present and future inputs using Eq. (2.2):

$$y(k+1)=\bar{y}_1(k)+h_1\upsilon(k), \tag{2.12}$$

$$y(k+2)=\bar{y}_2(k)+h_1\upsilon(k+1)+h_2\upsilon(k), \tag{2.13}$$

$$y(k+3)=\bar{y}_3(k)+h_1\upsilon(k+2)+h_2\upsilon(k+1)+h_3\upsilon(k), \tag{2.14}$$

and so on. We can express these equations in matrix form:

$$\begin{bmatrix} y(k+1) \\ y(k+2) \\ \vdots \\ \vdots \\ y(k+p) \end{bmatrix} = \underbrace{\begin{bmatrix} \bar{y}_1(k) \\ \bar{y}_2(k) \\ \vdots \\ \vdots \\ y_p(k) \end{bmatrix}}_{\text{effect of past inputs, stored in } \bar{Y}(k)} + \underbrace{\begin{bmatrix} h_1 \\ h_2 \\ \vdots \\ \vdots \\ h_p \end{bmatrix} \upsilon(k) + \begin{bmatrix} 0 \\ h_1 \\ \vdots \\ \vdots \\ h_{p-1} \end{bmatrix} \upsilon(k+1) + \ldots + \begin{bmatrix} 0 \\ 0 \\ \vdots \\ \vdots \\ h_1 \end{bmatrix} \upsilon(k+p-1)}_{\text{effect of future inputs (yet to be determined)}}, \tag{2.15}$$

The first term represents the effect of past inputs, summarized as part of the state. The subsequent terms capture the influence of the hypothesized future inputs, corresponding to the system's response to impulses applied at future time steps.

The dynamic state of the system $\bar{Y}(k)$ captures all relevant information at time k. To obtain the state at $k+1$, defined as

$$\bar{Y}(k+1)=\left[\bar{y}_0(k+1),\ \ \bar{y}_1(k+1),\ldots,\ \ \bar{y}_{n-1}(k+1)\right]^T \quad \text{with} \tag{2.16}$$

$$\bar{y}_i(k+1)\stackrel{\Delta}{=}y(k+1+i) \text{ for } \upsilon(k+1+j)=0; j\geq 0, \tag{2.17}$$

we add the effect of the input $\upsilon(k)$ at time k to the state $\bar{Y}(k)$:

$$\bar{y}_0(k+1)=\bar{y}_1(k)+h_1\upsilon(k), \tag{2.18}$$

$$\bar{y}_1(k+1)=\bar{y}_2(k)+h_2\upsilon(k), \tag{2.19}$$

$$\vdots$$

$$\bar{y}_{n-1}(k+1)=\bar{y}_n(k)+h_n\upsilon(k). \tag{2.20}$$

Note that $\tilde{y}_n(k)$ is not part of the state at time k, but under the FIR assumption it is known to be zero. By defining the matrix

$$M_i = \left.\begin{bmatrix} 0 & 1 & 0 & \dots & \dots & 0 & 0 \\ 0 & 0 & 1 & 0 & \dots & 0 & 0 \\ \vdots & & & & & & \vdots \\ 0 & 0 & \dots & \dots & \dots & 0 & 1 \\ 0 & 0 & \dots & \dots & \dots & 0 & 0 \end{bmatrix}\right\} n, \tag{2.21}$$

the state update can be written compactly as

$$\tilde{Y}(k+1) = M_i \cdot \tilde{Y}(k) + Hv(k). \tag{2.22}$$

This is the state update equation for finite impulse response models. It uses the state at time k, $\tilde{Y}(k)$, and the current input $v(k)$ to obtain the next state $\tilde{Y}(k+1)$. Multiplication by the matrix M_i represents the simple operation of shifting the vector $\tilde{Y}(k)$ forward and setting the last element to zero.

2.3.2 Recursive Multi-step Prediction Based on Step-response Model

Let us define the state at time k as

$$\tilde{Y}(k) = \left[\tilde{y}_0(k),\ \tilde{y}_1(k), \dots,\ \tilde{y}_{n-1}(k)\right]^T, \tag{2.23}$$

where

$$\tilde{y}_i(k) \stackrel{\Delta}{=} y(k+i) \text{ for } \Delta v(k+j) = 0;\ \ j \geq 0. \tag{2.24}$$

In this case, the state $\tilde{y}_i(k)$ is defined as the system output at time $k+i$ under the assumption that *input changes* are zero from time k onward ($\Delta v(k+j) = 0;\ j \geq 0$). Because of the FIR assumption, the step response settles after n steps, i.e., $\tilde{y}_{k+n-1}(k) = \tilde{y}_{k+n}(k) = \dots = \tilde{y}_{\infty}(k)$. Hence, the state $\tilde{Y}(k)$ completely characterizes the system output evolution when present and future input changes are zero.

To determine the actual future output, we add to the state the contributions of the present and future input changes. From Eq. (2.6) we obtain:

$$y(k+1) = \sum_{i=1}^{n-1} s_{i+1}\Delta v(k-i) + s_n v(k-n) + s_1 \Delta v(k), \tag{2.25}$$

$$= \tilde{y}_1(k) + s_1 \Delta v(k). \tag{2.26}$$

Continuing for $k+2, k+3, \dots$:

$$y(k+2) = \tilde{y}_2(k) + s_1\Delta v(k+1) + s_2\Delta v(k), \tag{2.27}$$

$$y(k+3) = \tilde{y}_3(k) + s_1\Delta v(k+2) + s_2\Delta v(k+1) + s_3\Delta v(k), \tag{2.28}$$

and so on. These equations can be expressed in matrix form as

$$\begin{bmatrix} y(k+1) \\ y(k+2) \\ \vdots \\ \vdots \\ y(k+p) \end{bmatrix} = \underbrace{\begin{bmatrix} \tilde{y}_1(k) \\ \tilde{y}_2(k) \\ \vdots \\ \vdots \\ \tilde{y}_p(k) \end{bmatrix}}_{\text{effect of past inputs and current bias, stored in } \tilde{Y}(k)} \tag{2.29}$$

$$+ \underbrace{\begin{bmatrix} s_1 \\ s_2 \\ \vdots \\ \vdots \\ s_p \end{bmatrix} \Delta v(k) + \begin{bmatrix} 0 \\ s_1 \\ \vdots \\ \vdots \\ s_{p-1} \end{bmatrix} \Delta v(k+1) + \ldots\ldots + \begin{bmatrix} 0 \\ 0 \\ \vdots \\ \vdots \\ s_1 \end{bmatrix} \Delta v(k+p-1).}_{\text{effect of future input changes (yet to be determined)}}$$

The first term captures the effect of past inputs and the bias held at the last value of $v(k-1)$, while the subsequent terms represent the contributions of future input changes.

To obtain the state at $k+1$, defined as

$$\tilde{Y}(k+1) = \left[\tilde{y}_0(k+1),\ \ \tilde{y}_1(k+1),\ \ldots,\ \ \tilde{y}_{n-1}(k+1)\right]^T \quad \text{with} \tag{2.30}$$

$$\tilde{y}_i(k+1) \stackrel{\Delta}{=} y(k+1+i) \text{ for } \Delta v(k+1+j) = 0;\ j \geq 0, \tag{2.31}$$

we add the effect of the input change $\Delta v(k)$ at time k to the state $\tilde{Y}(k)$:

$$\tilde{y}_0(k+1) = \tilde{y}_1(k) + s_1 \Delta v(k), \tag{2.32}$$

$$\tilde{y}_1(k+1) = \tilde{y}_2(k) + s_2 \Delta v(k), \tag{2.33}$$

$$\vdots$$

$$\tilde{y}_{n-1}(k+1) = \tilde{y}_n(k) + s_n \Delta v(k). \tag{2.34}$$

Note that $\tilde{y}_n(k) = \tilde{y}_{n-1}(k)$ due to the FIR assumption. By defining

$$M = \left.\begin{bmatrix} 0 & 1 & 0 & \ldots & \ldots & 0 & 0 \\ 0 & 0 & 1 & 0 & \ldots & 0 & 0 \\ \vdots & & & & & & \vdots \\ 0 & 0 & \ldots & \ldots & \ldots & 0 & 1 \\ 0 & 0 & \ldots & \ldots & \ldots & 0 & 1 \end{bmatrix}\right\} n, \tag{2.35}$$

we can compactly express the state update as

$$\tilde{Y}(k+1) = M \cdot \tilde{Y}(k) + S\Delta v(k). \tag{2.36}$$

Here, multiplication by the matrix M simply shifts the vector $\tilde{Y}(k)$ forward while repeating the last element. In actual computer implementations, there is no need to explicitly construct this matrix or perform the multiplication. The recursive relation Eq. (2.36) is known as the *step response model* of the system.

As seen from the derivation, the FIR and step response models are closely related, differing mainly in their state definitions. In the FIR model, future inputs are assumed to be zero, whereas in the step response model, future input *changes* are set to zero. The input representation also differs: the FIR model represents future inputs as pulses, while the step response model represents them as steps. Because the step response model expresses future inputs in terms of changes Δv, it is especially convenient for incorporating integral action in the controller, as will be shown.

2.3.3 Multivariable Generalization

The model equation (2.36) generalizes naturally to the case when the system has n_y outputs $y_\ell; \ell = 1, \dots, n_y$ and n_v inputs $v_j; j = 1, \dots, n_v$. We define the output *vector*

$$y(k-1) = \left[y_1(k-1), \dots, y_{n_y}(k-1)\right]^T,$$

the input *vector*

$$\Delta v(k-1) = \left[\Delta v_1(k-1), \dots, \Delta v_{n_v}(k-1)\right]^T,$$

and step response coefficient matrix

$$S_i = \begin{bmatrix} s_{1,1,i} & s_{1,2,i} & \cdots & s_{1,n_v,i} \\ s_{2,1,i} & & & \\ \vdots & & & \\ s_{n_y,1,i} & s_{n_y,2,i} & \cdots & s_{n_y,n_v,i} \end{bmatrix},$$

where $s_{\ell,m,i}$ is the ith step response coefficient relating the mth input to the lth output. The overall $(n \cdot n_y) \times n_v$ step response matrix is obtained by stacking the individual step response coefficient matrices:

$$S = \begin{bmatrix} S_1 \\ S_2 \\ \vdots \\ S_n \end{bmatrix}.$$

The state of the multiple output system at time k is defined as

$$\tilde{Y}(k) = \begin{bmatrix} \tilde{y}_0(k) \\ \tilde{y}_1(k) \\ \vdots \\ \tilde{y}_{n-1}(k) \end{bmatrix} \quad \text{with} \tag{2.37}$$

$\tilde{y}_i(k) \triangleq y(k+i)$ for $\Delta v(k+j) = 0; j \geq 0$.

With these definitions, the updated equation in the multivariable case becomes

$$\tilde{Y}(k+1) = M \cdot \tilde{Y}(k) + S\Delta v(k), \tag{2.38}$$

$$\text{where} \quad M = \left.\begin{bmatrix} 0 & I & 0 & \dots & \dots & 0 & 0 \\ 0 & 0 & I & 0 & \dots & 0 & 0 \\ \vdots & & & & & & \vdots \\ 0 & 0 & \dots & \dots & \dots & 0 & I \\ 0 & 0 & \dots & \dots & \dots & 0 & I \end{bmatrix}\right\} n,$$

where I is the identity matrix of dimension $n_y \times n_y$.

The recursive relation Eq. (2.38) is known as the *step response model* of the multivariable system. It was originally introduced in the first formulation of Dynamic Matrix Control (DMC).

2.4 Examples

The following examples illustrate different types of step response models.

Example 2.4: *SISO Process with Dead Time and Disturbance.*
Consider the following model, taken from the heavy oil fractionator control problem to be introduced later in Chapter 3:

$$y(s) = \underbrace{\frac{5.72}{60s+1}e^{-14s}}_{G_{sd}}u(s) + \underbrace{\frac{1.52}{25s+1}e^{-15s}}_{G_{ird}}d(s). \tag{2.39}$$

The step responses in Figure 2.5 illustrate the effects of the "side draw" (u, manipulated variable) and the "intermediate reflux duty" (d, disturbance) on the "side endpoint." The sampling time is 7 minutes ($t_s = 7$), and 35-step response coefficients ($n = 35$) were generated. Both responses exhibit dead times of approximately 14 minutes.

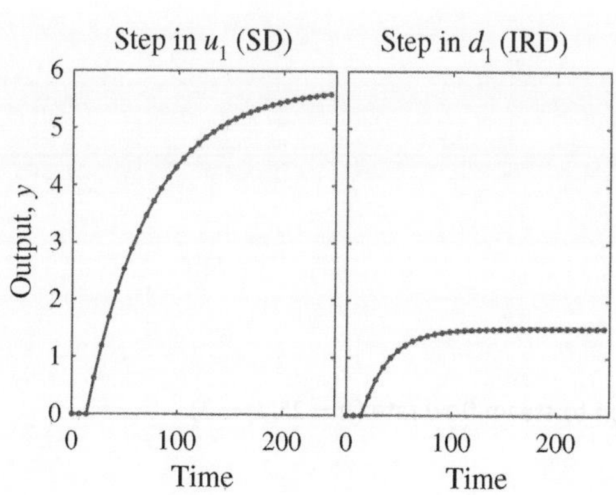

Figure 2.5 Responses of side endpoint for step change in side draw (left) and intermediate reflux duty (right). ($n = 35$, $t_s = 7$).

Example 2.5: *SISO Process with Inverse Response.*
Consider:

$$G = \frac{2.69(-6s+1)}{(20s+1)(5s+1)}e^{-1.5s}. \tag{2.40}$$

The step response in Figure 2.6 shows the experimentally observed effect of a change in steam flow rate on product concentration in a multieffect evaporator system. In the long term, an increase in steam rate enhances evaporation and thereby increases concentration. However, the accompanying rise in diluting flow initially reduces concentration. This behavior is a classical example of a system with an inverse response. Controlling systems with an inverse response poses a special challenge: the controller must not be misled by the initial inverse effect and must act with caution.

Example 2.6: *MIMO Process.*
Consider the model of a high-purity distillation column:

$$G = \frac{1}{75s+1}\begin{bmatrix} 0.878 & -0.864 \\ 1.082 & -1.096 \end{bmatrix}, \tag{2.41}$$

with overhead (y_1) and bottom (y_2) compositions as the two outputs and reflux (u_1) and boil up (u_2) as the two inputs. Figure 2.7 shows the response of the system to unit step changes in each input individually.

Note that all responses are very similar, differing only in sign, which will lead to control difficulties, as will be demonstrated in Chapter 3. Next, let us perturb both manipulated variables *simultaneously*. The resulting responses differ significantly in magnitude depending on the *direction* of the input vector u. As shown in Figure 2.8, when perturbations are applied in the $[1, -1]^T$ direction (i.e., u_1 increases while u_2 decreases, or vice versa), the "gain" of this MIMO system is large. In contrast, in other *directions* (e.g., $[1,\ 1]^T$), the gain is much smaller.

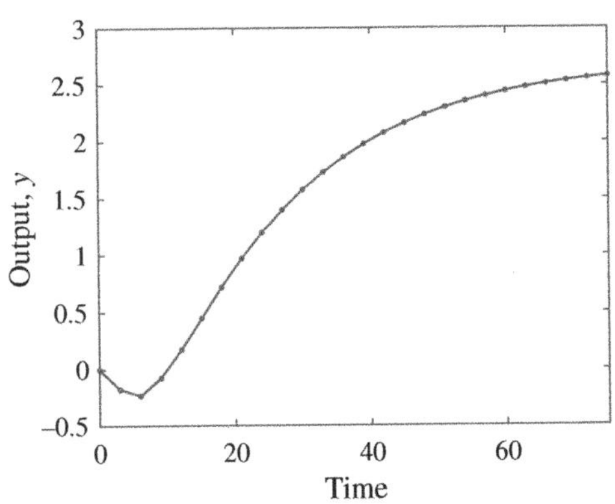

Figure 2.6 Response of evaporator product concentration to steam flow rate ($n = 25$, $t_s = 3$).

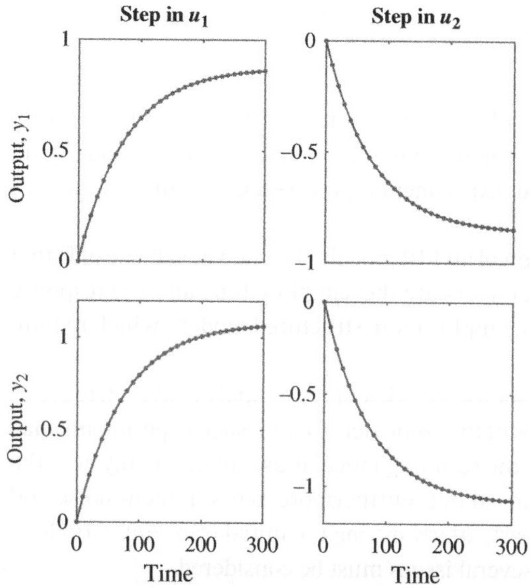

Figure 2.7 Response of bottom and top composition of a high-purity distillation column to step changes in reflux and boil up ($n = 30, t_s = 10$).

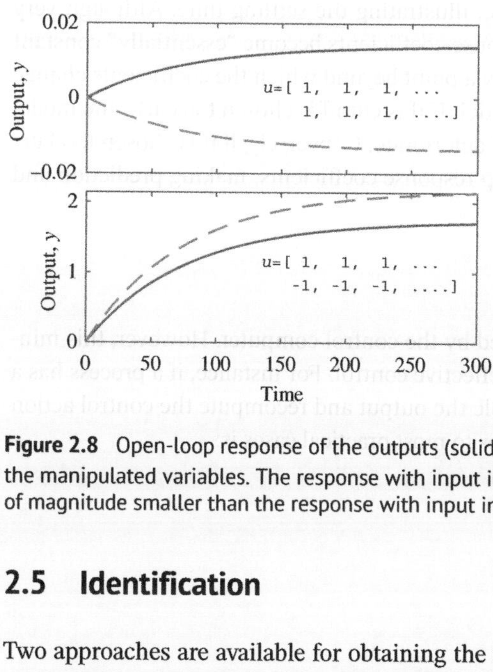

Figure 2.8 Open-loop response of the outputs (solid lines: y_1, dashed lines: y_2) for two different changes in the manipulated variables. The response with input in $[1,\ 1]^T$ direction shown in the top panel is an order of magnitude smaller than the response with input in $[1, -1]^T$ shown in the bottom panel.

2.5 Identification

Two approaches are available for obtaining the models required for prediction and control move computation. The first is to derive the differential equations representing material, energy, and momentum balances and solve them numerically. In this way, the system's response to a step input can be simulated and a step response model obtained. This "fundamental" modeling approach is often complex and requires substantial engineering effort, since the physicochemical phenomena must be understood and all process parameters either known or estimated experimentally. Its main advantage is that the resulting differential equation models are typically valid over a wide range of operating conditions.

The second approach is to perturb the inputs of the real process and record the output responses. By directly relating inputs to outputs, a process model can then be derived. This approach is referred to as *process identification*. In practice, especially when the process is complex or the underlying fundamental phenomena are not well understood, experimental process identification is the preferred modeling approach in industry.

In this section, we discuss the direct identification of an FIR model. The main advantage of fitting an FIR model is that the only user-specified parameters are the sampling time and the response length n. This contrasts with other techniques that employ more structured models, which require an additional structure-identification step.

One of the simplest ways to obtain the step response coefficients is to apply a step increase or decrease to one of the inputs. The main limitations of this approach are that such experiments may be difficult to perform, and even when successful, the resulting model is usually valid only near the operating region in which the experiment was carried out. Furthermore, measurement noise and disturbances can significantly affect the results, particularly during the initial transient periods.

For the experimental identification approach, several issues must be considered.

2.5.1 Settling Time

Figure 2.9 provides a schematic of a step response, illustrating the settling time. Although very few physical processes are truly FIR, the step response coefficients become "essentially" constant after a certain time. By inspection, we must identify a point beyond which the coefficients change negligibly and define a *truncated* step response model. If this cutoff is chosen too early, the model will be inaccurate and controller performance will deteriorate. Conversely, if it is chosen too late, the model will include an excessive number of step response coefficients, making prediction and control computations unnecessarily cumbersome.

2.5.2 Sampling Time

Usually, the minimum sampling interval is dictated by the control computer. However, this minimum interval is often shorter than necessary for effective control. For instance, if a process has a dead time of 10 minutes, it is unnecessary to sample the output and recompute the control action every 5 seconds. A useful rule of thumb that applies to most practical cases is:

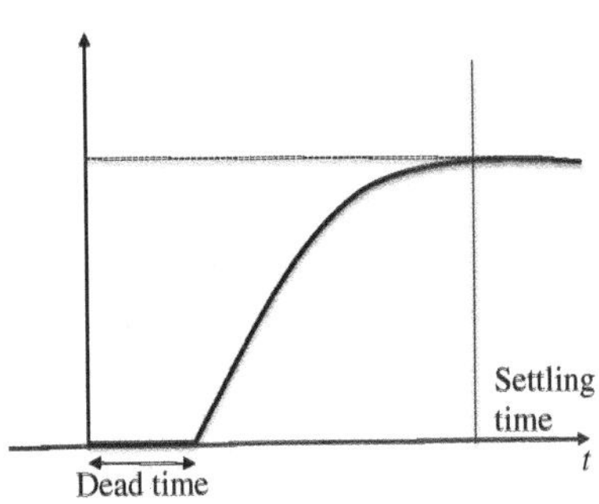

Figure 2.9 A typical step response.

Sampling time = max {0.03 × settling time, 0.3 × dead time}.

For processes with large dead times, the sampling time is determined primarily by the dead time, whereas for processes with short dead times, it is governed by the settling time.

In practice, step response models with about 30 coefficients are typically used (i.e., Settling Time/Sampling Time ≈ 30). If significantly fewer or more coefficients are required to adequately describe the response, the sampling time should be adjusted, e.g., the sampling time may be decreased if more coefficients are required to adequately capture the system dynamics.

2.5.3 Choice of the Input Signal for Experimental Identification

The simplest approach is to step up or step down the inputs to obtain the step response coefficients. However, attempting to determine the impulse response directly from step experiments can present several experimental challenges (Figure 2.10).

First, the plant must initially be at *steady state*, which is often difficult to achieve due to uncontrollable disturbances. A typical response starting from an unsteady initial condition is shown in Figure 2.10b. Disturbances that occur during the experiment can also distort the response, as illustrated in Figure 2.10c.

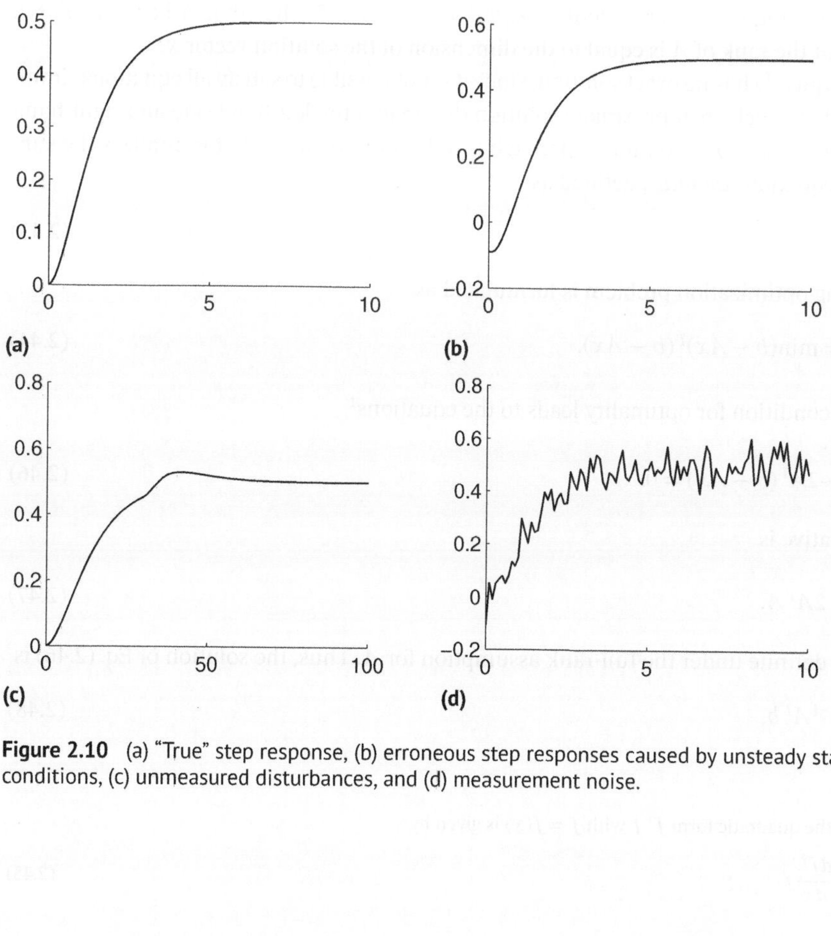

Figure 2.10 (a) "True" step response, (b) erroneous step responses caused by unsteady state initial conditions, (c) unmeasured disturbances, and (d) measurement noise.

Finally, excessive measurement noise may obscure the step response, as in Figure 2.10d. In such cases, it is often necessary to increase the magnitude of the input step to improve the signal-to-noise ratio of the output signal. Large steps, however, are usually discouraged by operating personnel because they can lead to undesirable consequences, such as off-spec products when operating conditions are perturbed too strongly.

In the presence of significant noise, disturbances, or nonsteady state initial conditions, it is preferable to use input signals that are more "random" in nature. The notion of signal randomness will be made precise in a later, advanced section of the book. Such signals can help mitigate the adverse effects of noise, disturbances, and initial condition offsets. Nevertheless, we recommend that engineers attempt a step test first, as it remains the simplest and most straightforward experiment.

2.5.4 The Linear Least Squares Problem

To explain the basic technique for identifying an FIR model from arbitrary input signals, we first introduce some fundamental concepts from linear algebra. Consider the following system of linear equations:

$$b = Ax, \tag{2.42}$$

where the matrix A has more rows than columns. In other words, there are more equations than unknowns, thus making the linear system *overspecified*. Assume further that A has *full column rank*, meaning that the rank of A is equal to the dimension of the solution vector x.

Such a system typically has no exact solution, since it is not possible to satisfy all equations simultaneously. Instead, we seek an approximate solution that makes the left-hand side and right-hand side "as close as possible." One common approach is to find the vector x that minimizes the sum of squares of the equation *residuals*, defined as

$$\rho = b - Ax. \tag{2.43}$$

The corresponding optimization problem is formulated as

$$\min_x \rho^T \rho = \min_x (b - Ax)^T (b - Ax), \tag{2.44}$$

whose necessary condition for optimality leads to the equations[1]

$$\frac{d(\rho^T \rho)}{dx} = -2A^T (b - Ax) = 0. \tag{2.46}$$

The second derivative is

$$\frac{d^2(\rho^T \rho)}{dx^2} = 2A^T A, \tag{2.47}$$

which is positive definite under the full-rank assumption for A. Thus, the solution of Eq. (2.46) is

$$x = (A^T A)^{-1} A^T b, \tag{2.48}$$

1 The derivative of the quadratic form $f^T f$ with $f = f(x)$ is given by

$$\frac{d(f^T f)}{dx} = 2\frac{df^T}{dx} f. \tag{2.45}$$

which minimizes the objective function and is known as the *least squares* solution of the overspecified system. The matrix $(A^T A)^{-1} A^T$ is called the *generalized inverse* of A. When A is square and nonsingular, the generalized inverse reduces to the standard inverse A^{-1}.

In some cases, it may be desirable to assign different weights to the residual components. In this case, the solution that minimizes the 2-norm of the weighted residual vector $\Lambda\rho$ is

$$x = (A^T \Lambda^T \Lambda A)^{-1} A^T \Lambda^T \Lambda b \tag{2.49}$$

where Λ is a diagonal matrix of weighting coefficients.

Although Eqs. (2.48) and (2.49) are algebraically correct, direct matrix inversion is rarely used in numerical computations. In practice, most software packages rely on more numerically robust methods, such as the QR algorithm, to obtain least squares solutions. For further details, the reader is referred to the numerical analysis literature.

2.5.5 Linear Least Squares Identification

The standard algorithm for parameter fitting in identification is the least squares method or one of its variants. It is used to fit models that relate a single output variable to multiple process inputs, and the same algorithm can be applied to all outputs using the same test data. Directly fitting an FIR model offers several advantages, particularly for beginners, since it requires no *a priori* process knowledge beyond an estimate of the settling time. In what follows, we provide a brief derivation of least squares identification and introduce some modifications to address the drawbacks of directly fitting an FIR model.

Consider a process with a single output y and a set of inputs υ_j, ; $j = 1, \dots, n_\upsilon$, described by an impulse response model:

$$y(k) = \sum_{j=1}^{n_\upsilon} \sum_{i=1}^{n} h_{j,i} \upsilon_j(k-i) + \nu(k-1), \tag{2.50}$$

where $\nu(k-1)$ captures effects on y not represented by the model. The horizon n is chosen sufficiently large so that ν accounts only for unmeasured effects such as noise, disturbances, and steady-state bias.

In practice, steady states between inputs and outputs are difficult to define because processes vary over time due to disturbances. For this reason, the model is commonly expressed in terms of incremental changes in inputs and outputs between successive sample times:

$$\Delta y(k) = \sum_{j=1}^{n_\upsilon} \sum_{i=1}^{n} h_{j,i} \Delta\upsilon_j(k-i) + \Delta\nu(k-1). \tag{2.51}$$

This form is obtained directly by differencing Eq. (2.50) at times k and $k-1$. Here, $\Delta\nu$ is assumed to be uncorrelated with $\Delta\upsilon$.

We estimate the impulse response coefficients by collecting input and output data over $N+n$ time intervals from an online experiment. Writing Eq. (2.51) for the past N intervals up to the current time $N+n$ gives:

$$\underbrace{\begin{bmatrix} \Delta y(n+1) \\ \Delta y(n+2) \\ \vdots \\ \Delta y(n+N) \end{bmatrix}}_{Y_N} = \underbrace{\begin{bmatrix} \Delta v_1(n) & \cdots & \Delta v_1(1) & \Delta v_2(n) & \cdots & \Delta v_{n_v}(1) \\ \Delta v_1(n+1) & \cdots & \Delta v_1(2) & \Delta v_2(n+1) & \cdots & \Delta v_{n_v}(2) \\ & \vdots & & & \vdots & \\ \Delta v_1(n+N-1) & \cdots & \Delta v_1(N) & \Delta v_2(n+N-1) & \cdots & \Delta v_{n_v}(N) \end{bmatrix}}_{\Phi_N} \times \underbrace{\begin{bmatrix} h_{1,1} \\ \vdots \\ h_{1,n} \\ h_{2,1} \\ \vdots \\ h_{n_v,n} \end{bmatrix}}_{\theta} + \underbrace{\begin{bmatrix} \Delta v(n+1) \\ \Delta v(n+2) \\ \vdots \\ \Delta v(n+N) \end{bmatrix}}_{V_N}. \tag{2.52}$$

A natural objective in selecting θ is to minimize the squared norm of the residual vector V_N. With this objective, the parameter estimation problem reduces to the least squares formulation introduced in Section 2.5.4:

$$Y_N = \Phi_N \theta, \tag{2.53}$$

where Y_N is the vector of past output measurements, Φ_N is the regressor matrix containing past input data, and θ is the parameter vector to be estimated. If the number of data points N exceeds the number of parameters in θ ($n_v \cdot n$), the least squares estimate of θ—that is, the choice of θ that minimizes $(Y_N - \Phi_N\theta)^T(Y_N - \Phi_N\theta)$—is given by Eq. (2.48):

$$\hat{\theta}_N = [{\Phi_N}^T \Phi_N]^{-1} {\Phi_N}^T Y_N(k). \tag{2.54}$$

Example 2.7: *Least Squares Identification for SISO System.*
Consider applying the least squares identification scheme to the linear system:

$$\frac{0.5}{19.36s^2 + 7.04s + 1}$$

with sampling time of 1.0, and compare it with step tests. Assume the output data is corrupted by measurement noise.

Figure 2.11 shows the typical measured output response of the system to a uniformly distributed input signal. The experiment duration is approximately 500 minutes, and a model with $n = 40$ FIR coefficients is sought. Since the system does not suffer from unmeasured disturbances, nonsteady initial conditions, or sensor/actuator bias, the data can be used directly without differencing.

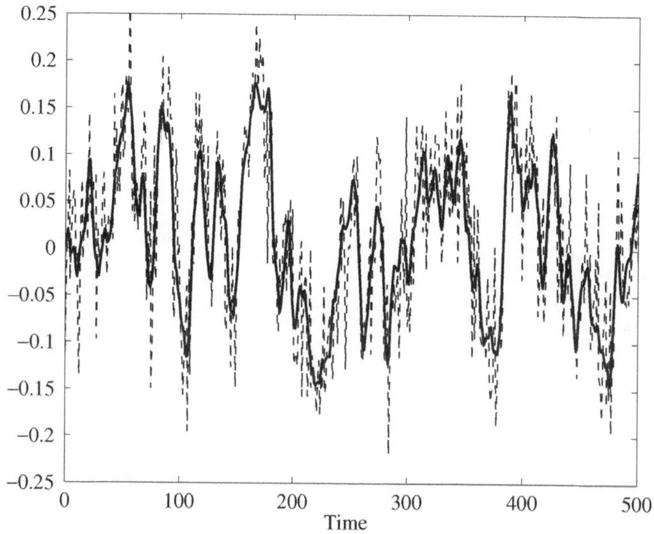

Figure 2.11 Measured output response (dashed) and "true" response in absence of measurement noise (solid).

Full MATLAB Code
The complete MATLAB code for least squares identification of the FIR model is available for download on GitHub:
https://github.com/Lee-MPC-Book

MATLAB LiveScript
A MATLAB LiveScript implementation is also available on GitHub, which walks through the individual sections of the code with explanations of their purpose.

Figure 2.12 shows the FIR coefficients obtained using least squares fitting (circles connected by dashed lines). FIR coefficients were also obtained using impulse tests for the same realization of measurement noise. A single impulse test yields very noisy FIR coefficients. To provide a fairer comparison, the coefficients shown in Figure 2.12 (squares) represent an average of 12 impulse tests (480 data points).

This example is revisited in greater detail in Chapter 10, which provides an in-depth discussion of system identification techniques.

In practical identification experiments, the variance of the error $\nu(k)$ may vary across data points. For example, the engineer may know that data collected during certain intervals is corrupted by more severe noise than others. In such cases, it is reasonable to assign lower weight to these less reliable data points when estimating θ. This can be achieved by minimizing a weighted sum of squared residuals, with higher weights assigned to data points believed to be more accurate. The corresponding least squares problem is formulated as:

$$(Y_N - \Phi_N\theta)^T\Lambda_\nu^T\Lambda_\nu(Y_N - \Phi_N\theta),$$

which yields the solution:

$$\hat{\theta}_N = [\Phi_N{}^T\Lambda_\nu^T\Lambda_\nu\Phi_N]^{-1}\Phi_N{}^T\Lambda_\nu^T\Lambda_\nu Y_N(k). \tag{2.55}$$

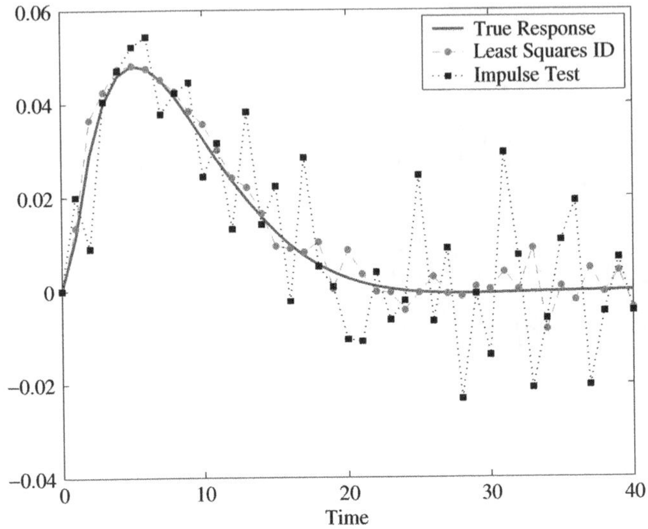

Figure 2.12 Comparison of least squares identification (circles) and impulse testing (squares) with true FIR coefficients (solid line).

It can be shown that if the underlying system is truly linear and n is chosen sufficiently large so that Δv and Δv are uncorrelated, the resulting estimate is *unbiased* (i.e., correct on average). Moreover, under mild conditions, the estimate converges to the true value as the number of data points N increases. Reliable software implementations are widely available to compute such weighted least squares parameter estimates.

A major drawback of this approach is the need to collect a large number of data points due to the many parameters that must be fitted. In the presence of noise, the parameter variance can become so large that the fit is essentially useless. As a result, the estimated step response often appears non-smooth, with many sharp peaks.

A simple way to mitigate this problem is to add a penalty term on the magnitude of the *changes* in the step response coefficients, i.e., the impulse response coefficients. Specifically, θ is chosen to minimize

$$(Y_N - \Phi_N\theta)^T \Lambda_\nu^T \Lambda_\nu (Y_N - \Phi_N\theta) + \theta^T \Lambda_\theta^T \Lambda_\theta \theta,$$

where Λ_θ is a weighting matrix that penalizes large variations in the impulse response coefficients. In effect, the penalty discourages sharp fluctuations in the step response coefficients. This technique is known as *regularization* in data regression.

As before, the problem can be written in least-squares form:

$$\begin{bmatrix} \Lambda_\nu Y_N \\ 0 \end{bmatrix} = \begin{bmatrix} \Lambda_\nu \Phi_N \\ \Lambda_\theta \end{bmatrix} \theta, \tag{2.56}$$

yielding the solution

$$\hat{\theta}_N = [\Phi_N^T \Lambda_\nu^T \Lambda_\nu \Phi_N + \Lambda_\theta^T \Lambda_\theta]^{-1} \Phi_N^T \Lambda_\nu^T \Lambda_\nu Y_N. \tag{2.57}$$

This modification to the standard least-squares identification algorithm generally produces smoother step responses. The main drawback is that the optimal choice of Λ_θ is often unclear. If Λ_θ is chosen too large, the estimates may become severely biased even with large data sets. Conversely, if it is chosen too small, the step response may not be smoothed sufficiently. More advanced statistical methods for reducing parameter variance will be discussed in the later chapters of this book.

The procedure described above can also be applied to fit measured disturbance models for use in feedforward compensation within MPC. To minimize difficulties when fitting such models, the manipulated inputs should be designed to remain uncorrelated with the disturbance. Naturally, the measured disturbance must also exhibit sufficient excitation, which may be difficult to guarantee in practice.

It is worth emphasizing that process identification is typically the most time-consuming step in implementing MPC. In this section, we have provided only a rudimentary discussion of the most basic identification technique. A more rigorous treatment of this method, together with a presentation of more advanced identification algorithms, will be given in Chapter 10 in the advanced part of this book.

Exercises

1. A unit-magnitude pulse input was applied uniformly over a duration of two minutes to a linear process in order to obtain its step response. The following output data were recorded during the pulse test:

Time (min)	y	Time (min)	y	Time (min)	y
0.5	0.0363	7.5	0.3051	14.5	0.0404
1	0.1269	8	0.2692	15	0.0345
1.5	0.2503	8.5	0.2364	15.5	0.0295
2	0.391	9	0.2068	16	0.0251
2.5	0.5021	9.5	0.1802	16.5	0.0214
3	0.5583	10	0.1565	17	0.0183
3.5	0.5764	10.5	0.1356	17.5	0.0156
4	0.5687	11	0.1173	18	0.0132
4.5	0.5444	11.5	0.1012	18.5	0.0113
5	0.5098	12	0.0871	19	0.0096
5.5	0.4698	12.5	0.0749	19.5	0.0081
6	0.4274	13	0.0643	20	0.0069
6.5	0.3849	13.5	0.0551		
7	0.3439	14	0.0472		

Using these data, obtain the FIR model of the system assuming a sampling period of 1 minute ($t_s = 1$).

2. For an FIR system, the state can be defined as a sequence of the n most recent past inputs, as shown in Eq. (2.9).

- Using this state definition, derive the p-step ahead output prediction equation in the form of Eq. (2.15).

- Also derive the state update equation of the form

$$x(k+1) = M_1 x(k) + M_2 v(k),$$

where M_1 and M_2 are appropriate system matrices.

3. Consider the following MIMO system:

$$G = \begin{bmatrix} \dfrac{1.7}{60.48s^2 + 15.6s + 1} & \dfrac{0.5}{19.36s^2 + 7.04s + 1} \\ \dfrac{0.3}{10.89s^2 + 4.62s + 1} & \dfrac{1.0}{36s^2 + 12s + 1} \end{bmatrix}, \tag{2.58}$$

- Simulate or compute the step responses for each element of the transfer matrix G.
- Based on the step responses, construct the matrix of step response coefficients that define the FSR model for this MIMO system. Assume a sampling time of $t_s = 1$.

4. Verify that Eq. (2.57) is the solution to the regularized least squares problem given in Eq. (2.56).

5. Repeat Example 2.7, this time assuming a significantly larger measurement noise variance. Since the noise affects all measurements equally, set $\Lambda_\nu = I$ in Eq. (2.57). Investigate the influence of the regularization matrix Λ_θ on the quality of the identified model:

- Let $\Lambda_\theta = \alpha I$, where α is a scalar and I is the identity matrix. Vary α from 0.01 to 10 and observe the resulting changes in model quality.
- Let Λ_θ be a diagonal matrix with entries that start at zero and increase with the index n. Analyze how this nonuniform regularization impacts the model.

6. Extend the least squares identification method to the MIMO system given in Eq. (2.58). Using a series of 10-step changes occurring between 1, 0, and -1 with a duration of 20 time intervals each, estimate the FIR model using the least squares approach. Assume the outputs are corrupted by additive white Gaussian noise with covariance $(0.04)^2$. You may use the following MATLAB code to generate the measurement noise:

```
nu = randn(2, N) * 0.04; % nu is a 2XN matrix for
 N (=200) time steps
```

7. Repeat the identification above, but in the presence of a constant output bias in addition to the white noise: a bias of 1.0 in the first output and 0.5 in the second. Demonstrate that direct application of the least squares method leads to biased estimates, and show that output differencing is required to obtain an accurate model in the presence of constant output bias.

3

Dynamic Matrix Control: The Basic Algorithm

My goal was to make a generic controller that could be configured for each application. It would be easier to install and maintain.

– Charlie Cutler

Dynamic Matrix Control (DMC) was one of the earliest commercial implementations of the Model Predictive Control (MPC) framework. This chapter introduces the fundamental principles of the DMC algorithm.

3.1 The Concept of Moving Horizon Control

Consider the schematic shown in Figure 3.1. At the current time step k, a dynamic process model is used to predict the system behavior over a horizon of length p. Based on this prediction, future trajectories of the process output in response to candidate inputs are computed. An optimization problem is then solved to determine the current and future values of the manipulated variables such that the predicted outputs exhibit desirable or optimal behavior.

A common objective is to minimize the sum of squared deviations between the predicted output and a specified reference trajectory (setpoint) over the prediction horizon. This optimization can also incorporate constraints on both the manipulated variables and the outputs, reflecting physical, safety, or operational limitations.

Although the concept is appealing, it would not work well in practice if the entire sequence of the computed control moves was applied blindly over the full prediction horizon. In reality, disturbances and modeling errors can cause significant deviations between the predicted and actual system behavior, making the precomputed control actions suboptimal as time progresses.

To address this, only the first control move is implemented at time k. At the next time step, $k + 1$, new measurements are obtained, the prediction horizon is shifted forward by one sample time step, and the optimization problem is resolved using the most recent system information. This strategy is known as *moving horizon control* or *receding horizon control.*

A similar strategy is employed in many non-technical contexts. An example is in chess, where one selects a move by evaluating all possible outcomes over a specified "depth" (analogous to the horizon). At the next turn, the evaluation is repeated based on the current board position. This reevaluation is necessary because the opponent's actual move makes many of the previously considered branches irrelevant. Another example is investment planning: A five-year strategy may

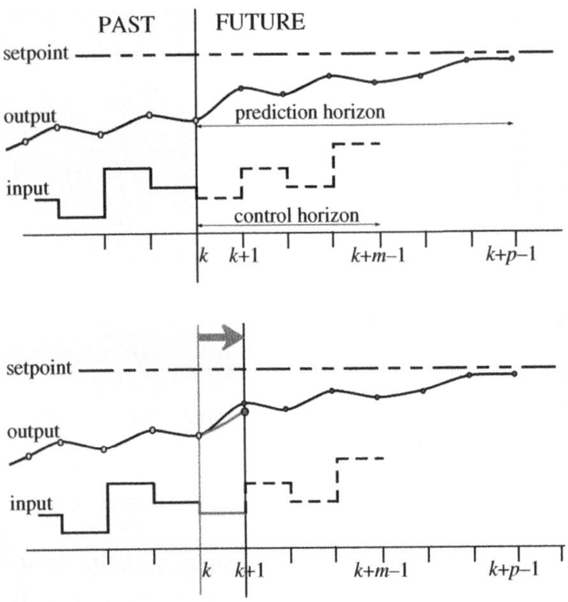

Figure 3.1 Schematic of moving horizon control.

be formulated to maximize returns, but as economic conditions evolve, a new five-year plan is periodically developed over a shifted horizon to reflect the latest market information.

The DMC algorithm relies on a prediction equation that expresses future system outputs as functions of the manipulated inputs and disturbances. This prediction model is essential for computing the optimal future control actions via an optimization procedure, as introduced in Chapter 2. In this chapter, we will first revisit the prediction equation, then present the objective function and formulate the resulting optimization problem. We will also provide comments on its solution. Finally, we will discuss the various tuning parameters available to the user for shaping the controller's performance.

3.2 Multi-step Prediction

Consider the block diagram shown in Figure 3.2, where three types of external inputs influence the system. The first is the manipulated variable (MV) u, whose effect on the output—typically a controlled variable (CV)—is represented by the transfer function P_u. The second is the measured disturbance variable (DV) d_m, with its effect described by P_{d_m}. The subscript m is used to emphasize that only measured disturbances are included in the model. The third is the unmeasured and unmodeled disturbance w_y, which enters the system additively at the output. The overall system can thus be described by

$$y = \begin{bmatrix} P_u & P_{d_m} \end{bmatrix} \begin{bmatrix} u \\ d_m \end{bmatrix} + w_y. \tag{3.1}$$

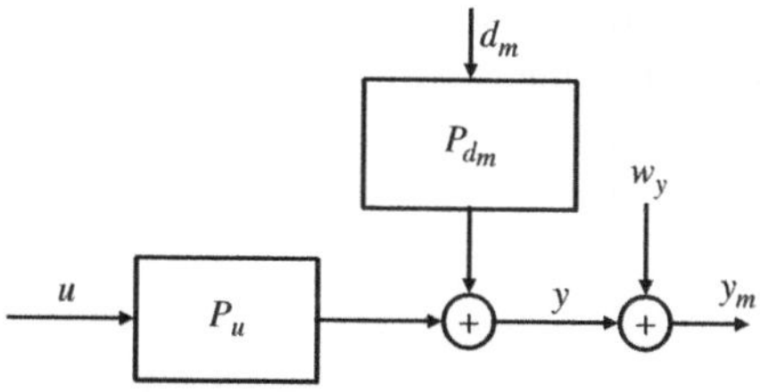

Figure 3.2 Basic problem setup. The output disturbance w_y captures the effect of unmeasured / unmodeled disturbances as well as plant-model mismatch.

Assume that step response matrices S^u and S^{d_m} are available for the system dynamics corresponding to P_u and P_{d_m}, respectively. We now show how to use these matrices to create a recursive prediction model, as in Eq. (2.38) where

$$S = \begin{bmatrix} S^u & S^{d_m} \end{bmatrix}, \tag{3.2}$$

driven by the known overall input

$$\Delta v(k) = \begin{bmatrix} \Delta u(k) \\ \Delta d_m(k) \end{bmatrix}. \tag{3.3}$$

Let us adopt (2.37) as the system state

$$\tilde{Y}(k) = \begin{bmatrix} \tilde{y}_0(k) \\ \tilde{y}_1(k) \\ \vdots \\ \tilde{y}_{n-1}(k) \end{bmatrix}. \tag{3.4}$$

By definition, the ith state, $\tilde{y}_i(k)$, is the output of the system at $(k+i)$ predicted at time k assuming that the system inputs remain constant at their most recent values. In other words

$$\begin{aligned} &\tilde{y}_i(k) \triangleq y(k+i) && i \in [0, n-1] \\ \text{assuming} \quad &\Delta u(k+j) = \Delta d_m(k+j) = 0 && j \geq 0. \end{aligned} \tag{3.5}$$

Furthermore, the state excludes any unmeasured disturbance information. Accordingly, it is assumed in the definition that

$$w_y(k) = w_y(k+1) = \cdots = 0. \tag{3.6}$$

Recall from (2.38) that, at each time k, $\tilde{Y}(k)$ is updated using

$$\begin{aligned} \tilde{Y}(k) &= M \cdot \tilde{Y}(k-1) + S^u \Delta u(k-1) + S^{d_m} \Delta d_m(k-1) \\ &= M \cdot \tilde{Y}(k-1) + S \Delta v(k-1). \end{aligned} \tag{3.7}$$

This equation reflects the effect of the input change $\Delta v(k-1)$ on the future evolution of the system, assuming *no further input changes*. The influence of this input change appears through the step response matrix S, which captures the dynamic behavior of the system.

The predicted output for the next p time steps can be expressed as:

$$\begin{bmatrix} y(k+1|k) \\ y(k+2|k) \\ y(k+3|k) \\ \vdots \\ y(k+p|k) \end{bmatrix} = \begin{bmatrix} \tilde{y}_1(k) \\ \tilde{y}_2(k) \\ \tilde{y}_2(k) \\ \vdots \\ \tilde{y}_p(k) \end{bmatrix} + \begin{bmatrix} w_y(k+1|k) \\ w_y(k+2|k) \\ w_y(k+3|k) \\ \vdots \\ w_y(k+p|k) \end{bmatrix}$$

$$+ \begin{bmatrix} S_1^{d_m} \\ S_2^{d_m} \\ S_3^{d_m} \\ \vdots \\ S_p^{d_m} \end{bmatrix} \Delta d_m(k|k) + \begin{bmatrix} 0 \\ S_1^{d_m} \\ S_2^{d_m} \\ \vdots \\ S_{p-1}^{d_m} \end{bmatrix} \Delta d_m(k+1|k) + \cdots + \begin{bmatrix} 0 \\ 0 \\ 0 \\ \vdots \\ S_1^{d_m} \end{bmatrix} \Delta d_m(k+p-1|k) \tag{3.8}$$

$$+ \begin{bmatrix} S_1^{u} \\ S_2^{u} \\ S_3^{u} \\ \vdots \\ S_p^{u} \end{bmatrix} \Delta u(k|k) + \begin{bmatrix} 0 \\ S_1^{u} \\ S_2^{u} \\ \vdots \\ S_{p-1}^{u} \end{bmatrix} \Delta u(k+1|k) + \cdots + \begin{bmatrix} 0 \\ 0 \\ 0 \\ \vdots \\ S_1^{u} \end{bmatrix} \Delta u(k+p-1|k).$$

The first term on the right-hand side—the first p elements of the state—represents the predicted future evolution of the system under the assumption of no further input changes. The subsequent terms account for the influence of the unmeasured or unmodeled disturbances $w_y(k+i|k)$, the measured disturbances $\Delta d_m(k+i|k)$, and present and future changes in the manipulated inputs $\Delta u(k+i|k)$, which are yet to be determined. The notation $y(k+i|k)$ denotes the prediction of $y(k+i)$ based on information available at time k, and the same convention applies to Δd_m and w_y.

Values of the disturbance variables are not available for times greater than current time k and must be *predicted* in a rational manner. From the measurement of $d_m(k)$ at time k, $\Delta d_m(k|k) = d_m(k) - d_m(k-1)$ is known. Unless additional process information or "upstream" measurements are available to infer future disturbance behavior, it is assumed that the disturbances remain unchanged going forward, i.e.,

$$\Delta d_m(k+1|k) = \Delta d_m(k+2|k) = \cdots = \Delta d_m(k+p-1|k) = 0. \tag{3.9}$$

This assumption is reasonable when disturbances vary infrequently. Similarly, future unmodeled disturbances $w_y(k+i|k)$ are assumed to remain constant

$$w_y(k|k) = w_y(k+1|k) = w_y(k+2|k) = \cdots = w_y(k+p|k), \tag{3.10}$$

$$\text{where} \quad w_y(k|k) \approx y_m(k) - \tilde{y}_0(k). \tag{3.11}$$

Here, $y_m(k)$ denotes the measured plant output at time k, and $\tilde{y}_0(k)$—the first component of the state vector $\tilde{Y}(k)$—is the model-predicted output at time k. The difference between the measured and predicted outputs provides a practical estimate of the overall effect of unmodeled disturbances.

Often, the control horizon m is chosen to be shorter than the prediction horizon p. The manipulated inputs are assumed to be adjusted only over the next m steps ($m \leq p$). Beyond this horizon, the input changes are assumed to be zero

$$\Delta u(k+m|k) = \Delta u(k+m+1|k) = \cdots = \Delta u(k+p-1|k) = 0. \tag{3.12}$$

With these assumptions, Eq. (3.8) simplifies to

$$\underbrace{\begin{bmatrix} y(k+1|k) \\ y(k+2|k) \\ y(k+3|k) \\ \vdots \\ y(k+m|k) \\ \vdots \\ y(k+p|k) \end{bmatrix}}_{\mathcal{Y}(k+1|k)} = \underbrace{\begin{bmatrix} \tilde{y}_1(k) \\ \tilde{y}_2(k) \\ \tilde{y}_3(k) \\ \vdots \\ \tilde{y}_m(k) \\ \vdots \\ \tilde{y}_p(k) \end{bmatrix}}_{\substack{\mathcal{M}\tilde{Y}(k) \\ \text{stored state vector}}} + \underbrace{\begin{bmatrix} S_1^{dm} \\ S_2^{dm} \\ S_3^{dm} \\ \vdots \\ S_m^{dm} \\ \vdots \\ S_p^{dm} \end{bmatrix} \Delta d_m(k)}_{\substack{\mathcal{S}^{dm}\Delta d_m(k) \\ \text{feedforward term}}} + \underbrace{\begin{bmatrix} y_m(k) - \tilde{y}_0(k) \\ y_m(k) - \tilde{y}_0(k) \\ y_m(k) - \tilde{y}_0(k) \\ \vdots \\ y_m(k) - \tilde{y}_0(k) \\ \vdots \\ y_m(k) - \tilde{y}_0(k) \end{bmatrix}}_{\substack{\mathcal{I}_p(y_m(k) - \tilde{y}_0(k)) \\ \text{feedback term}}}$$

$$+ \underbrace{\begin{bmatrix} S_1^u & 0 & 0 & \cdots & 0 \\ S_2^u & S_1^u & 0 & \cdots & 0 \\ S_3^u & S_2^u & S_1^u & \cdots & 0 \\ \vdots & \vdots & \ddots & \ddots & \vdots \\ S_m^u & S_{m-1}^u & S_{m-2}^u & \cdots & S_1^u \\ \vdots & \vdots & \ddots & \ddots & \vdots \\ S_p^u & S_{p-1}^u & S_{p-2}^u & \cdots & S_{p-m+1}^u \end{bmatrix}}_{\substack{\mathcal{S}^u \\ \text{dynamic matrix}}} \underbrace{\begin{bmatrix} \Delta u(k|k) \\ \Delta u(k+1|k) \\ \Delta u(k+2|k) \\ \vdots \\ \Delta u(k+m-1|k) \end{bmatrix}}_{\substack{\Delta\mathcal{U}(k) \\ \text{future input moves}}}. \tag{3.13}$$

This prediction equation can be readily modified if alternative assumptions are made regarding the future behavior of the measured and unmeasured disturbances. For example, if the measured disturbances are expected to follow a *ramp-like* trend, one can assume:

$$\Delta d_m(k) = \Delta d_m(k+1|k) = \cdots = \Delta d_m(k+p-1|k). \tag{3.14}$$

3.3 Objective Function

The operational requirements of the plant define the performance objectives of the control system. These objectives must be formulated mathematically so that a control law can be derived in algorithmic form. In DMC, a *quadratic* objective function is employed, which can be expressed in its simplest form as[1]

$$\min_{\Delta u(k|k)\ldots\Delta u(k+m-1|k)} \sum_{\ell=1}^{p} |y(k+\ell|k) - r(k+\ell|k)|^2. \tag{3.15}$$

This criterion minimizes the sum of squared deviations between the predicted CV values and a time-varying reference trajectory or setpoint $r(k+\ell|k)$ over the p-step prediction horizon. The quadratic form penalizes larger deviations more heavily than smaller ones, thereby keeping the output close to the desired trajectory on average and avoiding large excursions.

The manipulated inputs are assumed to remain constant beyond m intervals, as shown in Eq. (3.12), where $m \leq p$ always holds. The values of m and p influence the closed-loop performance and should be selected by the user based on the control objectives and system characteristics. In particular, m, which defines the number of degrees of freedom in optimization, has a significant impact on the computational burden.

Due to inherent process interactions, it may not be possible to keep all outputs simultaneously close to their respective reference trajectories. In practice, this means that only a subset of the outputs can be tightly controlled, often at the expense of allowing larger deviations in others. This trade-off can be explicitly managed by introducing weighting matrices into the objective function:

$$\min_{\Delta u(k|k)\ldots\Delta u(k+m-1|k)} \sum_{\ell=1}^{p} |\Gamma_\ell^y [y(k+\ell|k) - r(k+\ell|k)]|^2. \tag{3.16}$$

For example, consider a system with two outputs, y_1 and y_2, and constant diagonal weighting matrices of the form

$$\Gamma_\ell^y = \begin{bmatrix} \gamma_1 & 0 \\ 0 & \gamma_2 \end{bmatrix}, \quad \forall \ell. \tag{3.17}$$

In this case, the objective becomes

$$\min_{\Delta u(k|k)\ldots\Delta u(k+m-1|k)} \left\{ \gamma_1^2 \sum_{\ell=1}^{p} [y_1(k+\ell|k) - r_1(k+\ell|k)]^2 + \gamma_2^2 \sum_{\ell=1}^{p} [y_2(k+\ell|k) - r_2(k+\ell|k)]^2 \right\}. \tag{3.18}$$

The larger the weight assigned to a particular output, the greater its influence on the overall objective function. Consequently, the controller will place more emphasis on keeping that output closer to its reference trajectory.

Finally, the manipulated variable moves required to track a given output trajectory may be too aggressive to be acceptable in practice. This issue can be addressed by incorporating a penalty term on the manipulated variable moves into the objective function:

$$\min_{\Delta u(k|k)\ldots\Delta u(k+m-1|k)} \sum_{\ell=1}^{p} |\Gamma_\ell^y [y(k+\ell|k) - r(k+\ell|k)]|^2 + \sum_{\ell=1}^{m} |\Gamma_\ell^u [\Delta u(k+\ell-1)]|^2. \tag{3.19}$$

1 $||x||$denotes the norm $(x^T x)^{\frac{1}{2}}$ of vector x.

Note that larger elements in the matrix Γ_ℓ^u lead to smaller resulting input moves. As a consequence, the output trajectory will be followed more slowly or less precisely. Therefore, the relative magnitudes of Γ_ℓ^y and Γ_ℓ^u govern the trade-off between aggressive trajectory tracking and moderation of control effort.

Of course, not all practical performance criteria are fully captured by this quadratic objective. Nonetheless, many control problems can be effectively framed as trajectory tracking tasks, making this formulation highly useful. Most importantly, it results in an optimization problem that can be solved efficiently using well-established techniques.

3.4 Constraints

In many control applications, the desired performance cannot be captured solely by a trajectory tracking objective. Instead, various practical requirements are more naturally formulated as constraints on process variables.

Three common types of process constraints are:

Manipulated Variable Constraints: Hard bounds on the input $u(k)$, typically reflecting physical limitations such as valve saturation or actuator range limits.

Manipulated Variable Rate Constraints: Hard limits on the change in manipulated variables, $\Delta u(k)$, which are used to restrict the rate of change and avoid excessive control actions.

Output Variable Constraints: These may be either hard or soft limits imposed on system outputs to prevent undesirable behavior such as overshoot or undershoot. Output constraints fall into two categories:

- *Controlled Variables:* These outputs with setpoints may also be subject to constraint bounds. In the objective function, deviations from their setpoints are minimized, while the limits are simultaneously enforced to ensure feasibility and safe operation.
- *Associated Variables:* These outputs do not have setpoints but must remain within specified bounds. That is, they are constrained but not directly penalized in the objective function—corresponding rows of Γ_ℓ^y are set to zero in Eq. (3.19) to exclude them from the tracking cost.

The three types of constraints in DMC are incorporated into the optimization problem by expressing them as linear inequality constraints. In the following, we explicitly formulate these inequalities.

3.4.1 Manipulated Variable Constraints

The solution vector in DMC includes not only the current move to be implemented but also the future moves over the next $m - 1$ intervals. While it is possible to avoid constraint violations by restricting only the current move, enforcing constraints on future moves allows the controller to anticipate and prevent potential violations in advance, thereby yielding improved overall performance. The value of the manipulated variable at a future time $k + \ell$ is constrained by

$$u_{\text{low}}(\ell) \leq \left[u(k-1) + \sum_{j=0}^{\ell} \Delta u(k+j|k) \right] \leq u_{\text{high}}(\ell), \quad \forall \ell.$$

The limits $u_{\text{low}}(\ell)$ and $u_{\text{high}}(\ell)$, although constant in most practical cases, may in general vary over the horizon. These constraints can be compactly expressed in matrix form as

$$\begin{bmatrix} I_L \\ -I_L \end{bmatrix} \Delta\mathcal{U}(k) \leq \begin{bmatrix} u_{\text{high}}(0) - u(k-1) \\ \vdots \\ u_{\text{high}}(m-1) - u(k-1) \\ -u_{\text{low}}(0) + u(k-1) \\ \vdots \\ -u_{\text{low}}(m-1) + u(k-1) \end{bmatrix}, \tag{3.20}$$

where

$$I_L = \begin{bmatrix} I & 0 & \cdots & 0 \\ I & I & \cdots & 0 \\ \vdots & \vdots & \ddots & \vdots \\ I & I & \cdots & I \end{bmatrix}. \tag{3.21}$$

3.4.2 Manipulated Variable Rate Constraints

Often, MPC is applied in supervisory mode where limitations exist on how quickly lower-level controller setpoints can be adjusted. These limitations are handled by imposing constraints on the sizes of the manipulated variable moves. Specifically, the constraints are expressed as

$$\begin{bmatrix} I \\ -I \end{bmatrix} \Delta\mathcal{U}(k) \leq \begin{bmatrix} \Delta u_{\text{max}}(0) \\ \vdots \\ \Delta u_{\text{max}}(m-1) \\ \Delta u_{\text{max}}(0) \\ \vdots \\ \Delta u_{\text{max}}(m-1) \end{bmatrix}, \tag{3.22}$$

where $\Delta u_{max}(\ell) > 0$ specifies the maximum allowable magnitude of each move. These bounds may vary over time to provide additional tuning flexibility.

3.4.3 Output Variable Constraints

The algorithm utilizes the output predictions from Eq. (3.13) to anticipate and prevent future constraint violations:

$$\mathcal{Y}_{\text{low}} \leq \mathcal{Y}(k+1|k) \leq \mathcal{Y}_{\text{high}}. \tag{3.23}$$

Substituting the expression for $\mathcal{Y}(k+1|k)$ from Eq. (3.13), we obtain the corresponding constraints on $\Delta\mathcal{U}(k)$:

$$\begin{bmatrix} \mathcal{S}^u \\ -\mathcal{S}^u \end{bmatrix} \Delta\mathcal{U}(k) \leq \begin{bmatrix} -(\mathcal{M}\tilde{Y}(k) + S^{dm}\Delta d_m(k) + \mathcal{I}_p(y_m(k) - \tilde{y}_0(k))) + \mathcal{Y}_{\text{high}} \\ \mathcal{M}\tilde{Y}(k) + S^{dm}\Delta d_m(k) + \mathcal{I}_p(y_m(k) - \tilde{y}_0(k)) - \mathcal{Y}_{\text{low}} \end{bmatrix}, \tag{3.24}$$

where

$$\mathcal{Y}_{low} = \begin{bmatrix} y_{\text{low}}(1) \\ y_{\text{low}}(2) \\ \vdots \\ y_{\text{low}}(p) \end{bmatrix}; \qquad \mathcal{Y}_{\text{high}} = \begin{bmatrix} y_{\text{high}}(1) \\ y_{\text{high}}(2) \\ \vdots \\ y_{\text{high}}(p) \end{bmatrix}$$

are the lower and upper bounds on the output trajectories over the prediction horizon p.

3.4.4 Combined Constraints

The manipulated variable constraints Eq. (3.20), manipulated variable rate constraints Eq. (3.22), and output variable constraints Eq. (3.24) can be compactly expressed using a single inequality:

$$\mathcal{C}^u \Delta \mathcal{U}(k) \leq \mathcal{D}(k+1|k) \tag{3.25}$$

Here, $\mathcal{C}^u$ is the composite matrix that aggregates all constraint matrices from the left-hand sides of the previous inequalities:

$$\mathcal{C}^u = \begin{bmatrix} I_L \\ -I_L \\ I \\ -I \\ \mathcal{S}^u \\ -\mathcal{S}^u \end{bmatrix}. \tag{3.26}$$

The corresponding right-hand side vector $\mathcal{D}(k+1|k)$ collects the bounds and error terms from each constraint category:

$$\mathcal{D}(k+1|k) = \begin{bmatrix} u_{\text{high}}(0) - u(k-1) \\ \vdots \\ u_{\text{high}}(m-1) - u(k-1) \\ -u_{\text{low}}(0) + u(k-1) \\ \vdots \\ -u_{\text{low}}(m-1) + u(k-1) \\ \Delta u_{\text{max}}(0) \\ \vdots \\ \Delta u_{\text{max}}(m-1) \\ \Delta u_{\text{max}}(0) \\ \vdots \\ \Delta u_{\text{max}}(m-1) \\ -(\mathcal{M}\tilde{Y}(k) + S^{d_m}\Delta d_m(k) + \mathcal{I}_p(y_m(k) - \tilde{y}_0(k))) + \mathcal{Y}_{\text{high}} \\ \mathcal{M}\tilde{Y}(k) + S^{d_m}\Delta d_m(k) + \mathcal{I}_p(y_m(k) - \tilde{y}_0(k)) - \mathcal{Y}_{\text{low}} \end{bmatrix}. \tag{3.27}$$

3.5 Quadratic Programming Solution of the Control Problem

3.5.1 Quadratic Programs

Before formulating the DMC optimization problem, we briefly review key concepts from nonlinear programming. In particular, we consider the following standard form of a QP

$$\begin{aligned} \min_{x} \quad & x^T H x + 2g^T x \\ \text{s.t.} \quad & Cx \leq \mathrm{d}. \end{aligned} \tag{3.28}$$

Here, H is a symmetric matrix referred to as the *Hessian* matrix; g is the *gradient* vector; C is the matrix that represents the inequality constraint equations; and d is the corresponding inequality constraint vector.

This problem involves minimizing a quadratic objective function in the decision variables x, subject to a set of linear inequality constraints. In the absence of constraints, the solution can be obtained analytically by applying the necessary conditions for optimality

$$\frac{d}{dx}\left(x^T H x + 2g^T x\right) = 2Hx + 2g = 0. \tag{3.29}$$

The unique unconstrained solution is given by:

$$x = -H^{-1} g. \tag{3.30}$$

The second derivative is

$$\frac{d^2(x^T H x + 2g^T x)}{dx^2} = 2H,$$

indicating that a sufficient condition for a minimum in the unconstrained case is that the Hessian matrix H is positive definite.

When inequality constraints are introduced, strict positive definiteness of H is no longer required for the optimization problem to be well-posed. For example, if $H = 0$, the problem reduces to

$$\begin{aligned} \min_{x} \quad & g^T x \\ \text{s.t.} \quad & Cx \leq \mathrm{d}, \end{aligned} \tag{3.31}$$

which is a *Linear Programming* (LP) problem. The solution to an LP always lies at a constraint intersection—that is, a vertex of the feasible polyhedron. This property does not generally hold for QP solutions.

Although positive definiteness of H is not strictly necessary, it enables the use of more efficient QP algorithms. In particular, QP algorithms can leverage the pre-inverted Hessian matrix when H is positive definite and constant (i.e., when the model and weight parameters remain unchanged), thereby reducing the computational burden significantly.

3.5.2 Formulation as a Quadratic Program

We utilize the prediction Eq. (3.13) to reformulate the objective in Eq. (3.19) as:

$$\min_{\Delta \mathcal{U}(k)} \left\{ |\Gamma^y \left[\mathcal{Y}(k+1|k) - \mathcal{R}(k+1|k)\right]|^2 + |\Gamma^u \Delta \mathcal{U}(k)|^2 \right\}. \tag{3.32}$$

Incorporating the constraints from Eq. (3.25), the resulting optimization problem becomes

$$\begin{aligned}&\min_{\Delta\mathcal{U}(k)} \{|\Gamma^y\,[\mathcal{Y}(k+1|k)-\mathcal{R}(k+1|k)]\,|^2+|\Gamma^u\Delta\mathcal{U}(k)|^2\};\\ &\text{s.t. } \mathcal{Y}(k+1|k)=\mathcal{M}\tilde{Y}(k)+\mathcal{S}^{d_m}\Delta d_m(k)+\mathcal{I}_p(y_m(k)-\tilde{y}_0(k))+\mathcal{S}^u\Delta\mathcal{U}(k)\\ &\qquad C^u\Delta\mathcal{U}(k)\le\mathcal{D}(k+1|k),\end{aligned} \tag{3.33}$$

where

$$\Gamma^u=\text{diag}\{\Gamma_1^u,\cdots,\Gamma_m^u\} \tag{3.34}$$

and

$$\Gamma^y=\text{diag}\{\Gamma_1^y,\cdots,\Gamma_p^y\} \tag{3.35}$$

are block-diagonal weight matrices, and

$$\mathcal{R}(k+1|k)=\begin{bmatrix} r(k+1|k)\\ r(k+2|k)\\ \vdots\\ r(k+p|k)\end{bmatrix} \tag{3.36}$$

is the reference trajectory vector over the prediction horizon.

Substituting the prediction equation into the objective function, we obtain

$$\begin{aligned}&\|\,\Gamma^y\,[\mathcal{Y}(k+1|k)-\mathcal{R}(k+1|k)]\,\|^2+\|\,\Gamma^u\Delta\mathcal{U}(k)\|^2\\ &=|\Gamma^y\,[\mathcal{S}^u\Delta\mathcal{U}(k)-E_p(k+1|k)]\,|^2+|\Gamma^u\Delta\mathcal{U}(k)|^2 &(3.37)\\ &=\Delta\mathcal{U}^T(k)\left(\mathcal{S}^{u^T}\Gamma^{y^T}\Gamma^y\mathcal{S}^u+\Gamma^{u^T}\Gamma^u\right)\Delta\mathcal{U}(k) &(3.38)\\ &\quad-2E_p(k+1|k)^T\Gamma^{y^T}\Gamma^y\mathcal{S}^u\Delta\mathcal{U}(k)+E_p^T(k+1|k)\Gamma^{y^T}\Gamma^yE_p(k+1|k).\end{aligned}$$

Here, we define

$$\begin{aligned}E_p(k+1|k)&=\begin{bmatrix} e(k+1|k)\\ e(k+2|k)\\ \vdots\\ e(k+p|k)\end{bmatrix}\\ &\triangleq\mathcal{R}(k+1)-[\mathcal{M}\tilde{Y}(k)+\mathcal{S}^{d_m}\Delta d_m(k)+\mathcal{I}_p\,(y_m(k)-\tilde{y}_0(k))],\end{aligned} \tag{3.39}$$

which is the measurement-corrected vector of predicted output deviations from the reference trajectory (i.e., tracking errors), assuming zero future control moves. Note that this expression incorporates the influence of measurable disturbances through the term $\mathcal{S}^{d_m}\Delta d_m(k)$.

The optimization problem with a quadratic objective and linear inequality constraints, as formulated above, is a standard QP. By expressing the DMC problem in the standard QP form,[2] we obtain

$$\begin{aligned}&\min_{\Delta\mathcal{U}(k)}\ \Delta\mathcal{U}(k)^T\mathcal{H}\Delta\mathcal{U}(k)+2\mathcal{G}^T\Delta\mathcal{U}(k)\\ &\quad\text{s.t. } C^u\Delta\mathcal{U}(k)\le\mathcal{D}(k+1|k),\end{aligned} \tag{3.40}$$

2 The term $E_p^T(k+1|k)E_p(k+1|k)$ is independent of $\Delta\mathcal{U}(k)$ and can therefore be omitted from the objective function.

where the Hessian and gradient are defined as

$$\mathcal{H} = \mathcal{S}^{u^T}\Gamma^{y^T}\Gamma^{y}\mathcal{S}^{u} + \Gamma^{u^T}\Gamma^{u} \tag{3.41}$$

$$\mathcal{G} = -\mathcal{S}^{u^T}\Gamma^{y^T}\Gamma^{y}E_p(k+1|k). \tag{3.42}$$

3.6 Implementation

As explained in the introduction to this chapter, DMC is implemented in a *moving horizon* fashion. This means that the QP derived above is solved at every time step. Such an approach provides flexibility to reconfigure the algorithm online, allowing it to adapt to unexpected situations. For instance, if an actuator fails during operation, the upper and lower constraint limits on the corresponding manipulated variable can be set equal, effectively removing it from the optimization. The MPC problem is then solved using the remaining manipulated variables. Likewise, the weighting parameters in the objective function can be modified online, enabling the user to tune the controller behavior as needed. In this section, we discuss various implementation topics associated with DMC.

3.6.1 Moving Horizon Algorithm

The constrained MPC algorithm is implemented online as follows.

1. *Pre-computation of the Hessian and Constraint Matrices.* Calculate the Hessian matrix $\mathcal{H}$ and the constraint matrix C^u. These matrices remain constant throughout unless the DMC controller is retuned, e.g., by changing the weight parameters or horizon sizes.
2. *Preparation.* Keep the manipulated variables constant for at least n time intervals, i.e., $(\Delta u(-1) = \Delta u(-2) = \ldots = \Delta u(-n) = 0)$, and assume zero measured disturbances during the same period, i.e., $(\Delta d_m(-1) = \Delta d_m(-2) = \ldots = \Delta d_m(-n) = 0)$. Under these conditions, the system will be at rest at time $k = 0$.
3. *Initialization* ($k = 0$). Measure the current output $y(0)$ and initialize the model prediction vector as follows:[3]

$$\tilde{Y}(k) = \Bigg[\underbrace{y_m(0)^T,\ y_m(0)^T,\ \ldots,\ y_m(0)^T}_{n}\Bigg]^T. \tag{3.43}$$

The following steps are taken at each time step k:

4. *State Update:* Update the state $\tilde{Y}(k)$ according to

$$\tilde{Y}(k) = M \cdot \tilde{Y}(k-1) + S^u \Delta u(k-1) + S^{d_m} \Delta d_m(k-1). \tag{3.44}$$

5. *Measurements:* Record the measurements of the output $(y_m(k))$ and the disturbance $(\Delta d_m(k))$.

3 If Eq. (3.43) is used for initialization and changes in the past n inputs actually occurred, the algorithm may not operate smoothly during its initial steps. In particular, the transition from *manual* to *automatic* mode may introduce a disturbance—i.e., it will not be "bumpless."

6. *Computation of the Error Vector:* First compute $w_y(k) = y_m(k) - \tilde{y}_0(k)$, where the first element of $\tilde{Y}(k)$ is the model prediction of the output. Then, compute the reference trajectory error vector:

$$E_p(k+1|k) = \mathcal{R}(k+1|k) - \left[\mathcal{M}\tilde{Y}(k) + S^{d_m}\Delta d_m(k) + \mathcal{I}_p w_y(k)\right]. \tag{3.45}$$

7. *Computation of the QP Gradient Vector:* Compute the gradient vector for the QP:

$$\mathcal{G} = -\mathcal{S}^{u^T}\Gamma^{y^T}\Gamma^y E_p(k+1|k). \tag{3.46}$$

8. *Computation of the QP Constraint Vector:* Compute the constraint vector for the QP:

$$\mathrm{d} = \begin{bmatrix} u_{\text{high}}(0) - u(k-1) \\ \vdots \\ u_{\text{high}}(m-1) - u(k-1) \\ -u_{\text{low}}(0) + u(k-1) \\ \vdots \\ -u_{\text{low}}(m-1) + u(k-1) \\ \Delta u_{\max}(0) \\ \vdots \\ \Delta u_{\max}(m-1) \\ \Delta u_{\max}(0) \\ \vdots \\ \Delta u_{\max}(m-1) \\ E_p(k+1|k) - \mathcal{R}(k+1|k) + \mathcal{Y}_{\text{high}} \\ -E_p(k+1|k) + \mathcal{R}(k+1|k) - \mathcal{Y}_{\text{low}} \end{bmatrix}. \tag{3.47}$$

9. *Solve the QP:* Solve using a QP solver:

$$\begin{aligned} &\min_x \frac{1}{2}x^T\mathcal{H}x + \mathcal{G}^T x \\ \text{such that} \quad &\mathcal{C}^u x \leq \mathrm{d} \end{aligned}. \tag{3.48}$$

10. *Input Implementation:* Apply the control input $u(k) \leftarrow u(k-1) + \begin{bmatrix} I_{n_u} & 0 & \cdots & 0 \end{bmatrix} x^*$ to the plant, where x^* is the optimal solution of the QP formulated above.
11. Increment $k \leftarrow (k+1)$ and return to step 3.

Note that the sequence of moves implemented in the moving horizon QP will generally differ from the planned sequence $\Delta\mathcal{U}(k)$.

3.6.2 DMC Examples

We now present a few examples to illustrate the application of the DMC algorithm. In particular, we consider the control of the SISO process introduced in Example 2.4 of Chapter 2. The open-loop response of this system was discussed in Figure 2.5.

Example 3.1: *Set-point Tracking for the SISO process of Example 2.4 using DMC.* Consider the process from Example 2.4. Neglecting the disturbance, the process model is given by:

$$G = \frac{5.72}{60s + 1} e^{-14s}. \tag{3.49}$$

The objective is to drive the system from rest to a setpoint value of $r = 1$, subject to the input constraints $-0.2 \leq u \leq 0.2$ and $|\Delta u| \leq 0.04$. The sampling interval is 7 minutes (i.e., $t_s = 7$), and the prediction and control horizons are chosen as $p = 10$ and $m = 4$, respectively. The weighting matrices are set to $\gamma^y = 1$ and $\gamma^u = 0$.

Figure 3.3 shows the closed-loop response of the system under constrained DMC control. A complete MATLAB™ script of the DMC algorithm for this example (no disturbance case) is provided in GitHub repository at: https://github.com/Lee-MPC-Book

The following example illustrates the importance of incorporating feedforward control when measured disturbances are available. We consider disturbance rejection for the same system presented in Example 2.4 from Chapter 2.

Example 3.2: *Disturbance rejection using DMC for the unconstrained system in Example 2.4.* Consider the process from Example 2.4:

$$y(s) = \frac{5.72e^{-14s}}{60s + 1} u(s) + \frac{1.52e^{-15s}}{25s + 1} d_m(s). \tag{3.50}$$

We design a DMC controller to reject a unit step disturbance and regulate the output to the origin. To clearly demonstrate the disturbance rejection behavior, we assume the system is unconstrained. Two scenarios are considered: (i) the disturbance is measured and (ii) the disturbance becomes unmeasured.

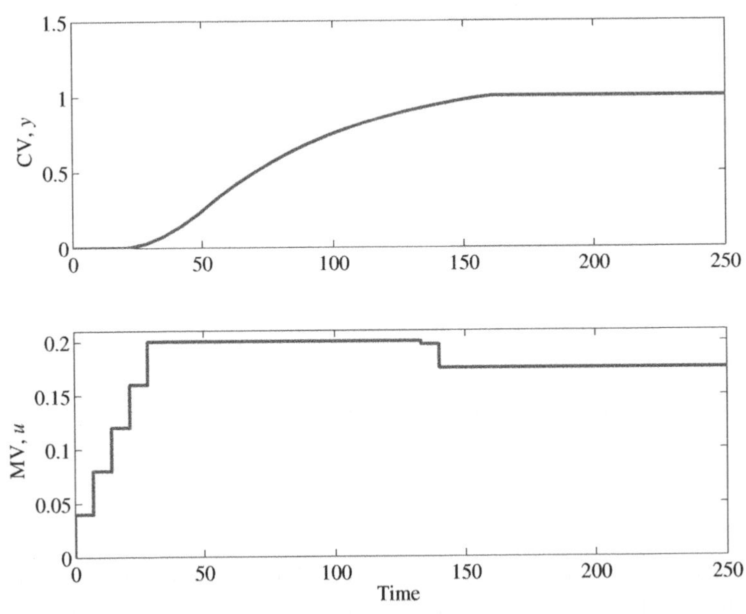

Figure 3.3 Application of DMC control for setpoint tracking in a SISO system with input constraints ($p = 12$, $m = 4$).

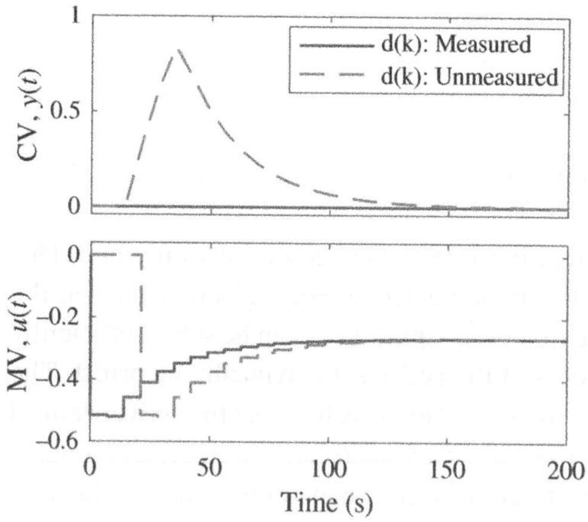

Figure 3.4 Application of DMC control for disturbance rejection. Solid lines: measured disturbance; dashed lines: unmeasured disturbance.

(i) Measured Disturbance: When the disturbance $d_m(k)$ is measured, the model anticipates its effect, allowing the DMC controller to take corrective action immediately—before the impact of the disturbance in the output can be detected.

The prediction and control horizons are set to $p = 10$ and $m = 4$, respectively. The weighting matrices are chosen as $\gamma^y = 1$ and $\gamma^u = 0$. Figure 3.4 illustrates the response of the system using the DMC controller. As shown by the solid line, the controller is able to perfectly reject the disturbance under the assumption of a perfect model.

However, the assumption of a perfect model is often unrealistic. As we will discuss in Section 3.7, using $\gamma^u = 0$ can lead to excessively aggressive and potentially unstable control actions. This is exacerbated when there is model error. Therefore, in practical applications, some penalty on the manipulated variable is generally advisable.

(ii) Unmeasured Disturbance: Now consider the case where the disturbance becomes unmeasured—e.g., due to a sensor failure. The underlying plant remains unchanged, but the disturbance information is no longer available to the DMC model. As a result, the controller can no longer anticipate the disturbance effect.

The closed-loop response under the same controller settings is shown as the dashed line in Figure 3.4. In this case, the DMC controller reacts only after the disturbance manifests in the measured output, through the output error $w_y(k) = y_m(k) - \tilde{y}_0(k)$. Once the disturbance is detected via this output deviation, the controller begins to take corrective action.

3.6.3 Efficient Solutions to the QP

In a moving-horizon framework, the QP in Eq. (3.48) is solved at every sampling instant using the updated prediction $\mathcal{Y}(k+1|k)$. Among the quantities involved, only the vectors $E_p(k+1|k)$ and $\mathcal{D}(k+1|k)$ vary with time, while the Hessian matrix $\mathcal{H}$ and the constraint matrix C^u remain constant across time steps. This constancy enables the use of parametric QP algorithms that exploit the pre-inverted Hessian to significantly reduce online computation. Naturally, if any of Γ^y, Γ^u, or the step response coefficients change, the Hessian must be recomputed and inverted—ideally in the background to avoid burdening the online computation.

Quadratic programming is a *convex* optimization problem and is therefore fundamentally tractable, meaning that a globally optimal solution within a given tolerance is guaranteed. QPs are well studied, and efficient, reliable algorithms are readily available. General-purpose solvers such as *QPSOL* can be used, but tailored algorithms that exploit the problem structure—such as sparsity in the Hessian and constraint matrices—are often more efficient and scalable for real-time implementation.

The conventional method for solving quadratic programs is the so-called *Active Set* method. This approach begins by assuming a subset of the constraints to be active. For a given active set, the resulting problem—where the active constraints are treated as equalities—can be solved efficiently using Lagrange multipliers. However, the initial guess of the active set is typically incorrect. The Karush–Kuhn–Tucker (KKT) conditions[4] are used to assess and iteratively adjust the active set until the correct one is identified. Most active set algorithms follow a *feasible path* strategy, meaning all iterates must satisfy the constraints. As a result, the total number of constraints can significantly affect computational performance, especially in real-time or embedded implementations.

An alternative to the active set approach is the *Interior Point (IP)* method, also known as the *Barrier* method. The key idea behind IP methods is to confine the solution within the feasible region by augmenting the objective function with a so-called barrier term that penalizes proximity to the constraint boundaries. Newton's method is then applied to the modified objective to iteratively approach the optimum. Various versions of IP algorithms have been developed, many of which leverage the dual problem formulation for improved computational efficiency. Although originally proposed for linear programs (LPs), IP methods have been successfully extended to QPs and other classes of constrained optimization problems. While no formal proof guarantees the number of steps, it is commonly observed in practice that Newton-based IP methods converge within 5–50 iterations. Considerable research has been devoted to applying IP methods to QPs arising in MPC, though a detailed discussion of these algorithms is beyond the scope of this book.

In the QP formulation presented earlier, the prediction equations were substituted directly into the objective function and constraints, thereby eliminating the equality constraints. This leads to a formulation in which the optimization variables are limited to $\Delta\mathcal{U}$, known as the *condensed QP* formulation. While this approach reduces the number of decision variables, it results in a dense Hessian matrix lacking block-diagonal structure, which limits the applicability of sparse matrix techniques. An alternative is the *sparse QP* formulation, in which the prediction equations are retained as explicit equality constraints. In this case, the optimization variables include both $\Delta\mathcal{U}$ and the predicted outputs $\mathcal{Y}$, increasing the problem dimensionality. However, the resulting Hessian preserves its sparse structure, enabling efficient exploitation of sparsity. The relative computational performance of *condensed* versus *sparse* QP formulations depends on the problem size and the solver used.

The computational performance of QPs depends on both the problem characteristics and the algorithm employed. As the number of constraints increases, the number of iterations typically grows, leading to longer solution times. This, in turn, may limit the achievable minimum control execution time. Additionally, the size of the QP—determined by the number of degrees of freedom, $m \cdot n_u$—has a direct impact on the computational load, with execution time generally scaling proportionally.

Storage requirements are also influenced by the problem dimensions. Specifically, the size of the Hessian matrix grows quadratically with the number of degrees of freedom. Moreover, due to the

4 The KKT conditions are necessary for optimality in general constrained optimization problems, and for QPs, they are both necessary and sufficient.

prediction mechanism, the state vector $\tilde{Y}(k)$ must be retained for the subsequent control cycle, as the update of the state vector at the next time step will require it.

3.6.4 Proper Constraint Formulation

Many engineering control problems naturally involve constraints. These are formulated as linear inequalities and incorporated directly into the QP problem. However, one must exercise caution against overusing constraints, as they can significantly complicate the prediction and analysis of control behavior under real-world operating conditions. Constraints should be introduced when essential and applied with careful consideration.

Importantly, while unconstrained MPC operates as a linear feedback controller, introducing constraints transforms it into a *nonlinear* control law. As a result, the controller's behavior can vary dramatically between small and large deviations, often in unexpected and undesirable ways. These effects are typically difficult to anticipate through analysis alone. The following example illustrates this point.

Example 3.3: *Nonlinearity of the constrained MPC law.* Consider again the system from Example 3.1, and compare the behavior of constrained and unconstrained MPC for three different setpoints: $r = 0.25$, $r = 0.5$, and $r = 1.0$. The input rate constraints are removed, but the input limits remain as $-0.2 \leq u \leq 0.2$. The tuning parameters are $m = 4$, $p = 10$, $\gamma^y = 1$, and $\gamma^u = 5$.

Figure 3.5 shows the system response under constrained DMC for the three setpoint changes. For $r = 0.5$, no constraints are active throughout the response, and the behavior is linear: the response for $r = 0.5$ is exactly twice that of $r = 0.25$, with both settling around 100 seconds. However, for $r = 1.0$, the response becomes noticeably slower, taking nearly 150 seconds to approach the setpoint. This is due to saturation of the manipulated variable at its upper limit, $u_{max} = 0.2$.

If the input constraint is relaxed and the system is controlled with an unconstrained DMC, the response for $r = 1.0$ becomes exactly twice that of $r = 0.5$. This clearly demonstrates that unconstrained DMC behaves as a linear control law, while the presence of constraints introduces nonlinearity.

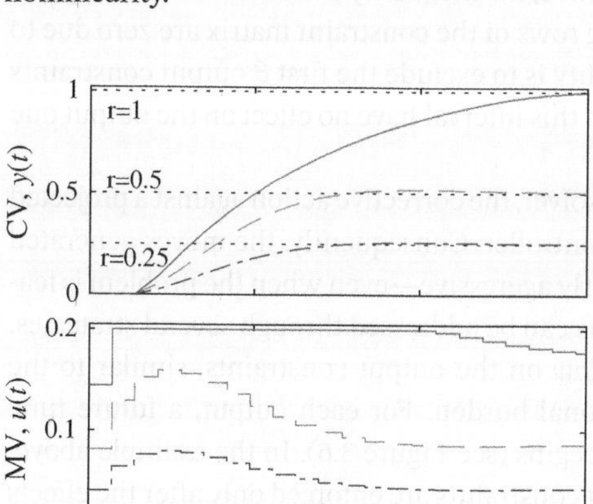

Figure 3.5 Input constraints result in nonlinear DMC behavior for a setpoint of $r = 1.0$. The response is not a linear scaling of the response for $r = 0.5$ due to actuator saturation.

Secondly, constraints can significantly increase the computational effort required to solve the QP. Therefore, they should be introduced judiciously. For instance, rather than imposing output constraints across the entire prediction horizon, one may enforce them only at *selected* time steps. Unless the system exhibits strong oscillatory behavior, placing constraints at the early time steps and at the end of the horizon is often sufficient to maintain the output within acceptable bounds throughout the horizon.

It is important to note that a predicted constraint violation does not necessarily imply an actual violation in closed-loop operation, and vice versa, due to the receding-horizon nature of MPC. The role of future constraints is primarily to prevent the controller from making myopic decisions in the current move.

In some cases, output constraints can lead to *infeasibility*. A QP is deemed infeasible when no choice of $\Delta\mathcal{U}(k)$ exists that satisfies all constraints, regardless of the objective function value. Physically, infeasibility may occur when output constraints are active but the manipulated variables lack sufficient authority to meet them—either due to hard limits or inherent system delays (e.g., dead time). To ensure robust operation, the MPC algorithm must include provisions to detect and manage infeasibility, such as constraint relaxation or prioritized constraint handling.

A simple example of infeasibility arises in the presence of dead time in the system response. To illustrate, consider a SISO system with a dead time of θ sampling intervals. The output constraint equations for this system take the form:

$$\begin{bmatrix} 0 & 0 & \cdots & 0 \\ \vdots & & & \vdots \\ 0 & 0 & \cdots & 0 \\ -S^u_{\theta+1} & 0 & \cdots & 0 \\ -S^u_{\theta+2} & -S^u_{\theta+1} & \cdots & 0 \\ \vdots & & & \vdots \end{bmatrix} \Delta\mathcal{U}(k) \geq \begin{bmatrix} c(k+1|k) \\ \vdots \\ c(k+\theta|k) \\ c(k+\theta+1|k) \\ c(k+\theta+2|k) \\ \vdots \end{bmatrix}$$

If any of the values $c(k+1|k), \ldots, c(k+\theta|k)$ are positive, the inequality cannot be satisfied regardless of the value of $\Delta\mathcal{U}(k)$, because the corresponding rows of the constraint matrix are zero due to the dead time. A practical way to avoid this infeasibility is to exclude the first θ output constraints from the QP formulation, since control actions within this interval have no effect on the output due to the delay.

Because inequalities are handled *exactly* by the QP solver, the corrective action against a projected violation can resemble that of a very tightly tuned controller. Consequently, the moves generated by the QP to prevent constraint violations may be overly aggressive—even when the problem is feasible. Both infeasibilities and excessive control actions can be addressed through several strategies.

One such strategy is the use of a *constraint window* on the output constraints, similar to the approach suggested earlier for reducing computational burden. For each output, a future time $k + H_c$ is selected at which constraint enforcement begins (see Figure 3.6). In the example above, H_c should be chosen to be at least $\theta + 1$ to ensure that constraints are enforced only after the effects of dead time or inverse response have subsided. For a given scenario, there exists a minimum value of H_c required to ensure feasibility. If H_c is chosen too small, feasibility may not be achievable; if chosen too large, constraint violations may be tolerated for an extended period.

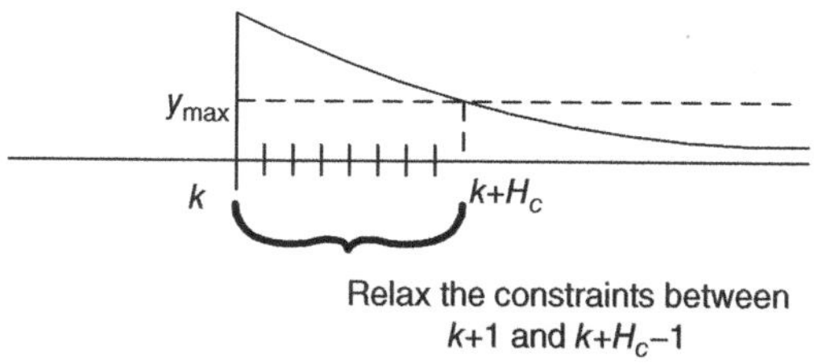

Figure 3.6 Relaxing the constraints.

In many cases, increasing H_c reduces the severity of the constraint violation but allows it to persist longer. This introduces a trade-off between the magnitude and duration of potential constraint violations.

In general, selecting an appropriate value of H_c for each constrained output to achieve a suitable compromise is challenging. In multivariable systems, constraints may need to be relaxed based on the relative priority of each constrained variable. This process becomes even more complex due to the time-varying nature of disturbances and the potential occurrence of actuator or sensor faults, which affect both the required amount and timing of constraint relaxation.

As a result, a more flexible and systematic approach is to *soften* the constraints by introducing a slack variable ϵ and penalizing constraint violations via an additional term in the objective function:

$$\min_{\epsilon,\Delta\mathcal{U}(k)} [\text{Usual Objective}] + \lambda\epsilon^2$$

$$y_{\min} - \epsilon \leq y(k+\ell|k) \leq y_{\max} + \epsilon \quad \text{plus other constraints.}$$

This formulation enables the optimization to balance between minimizing the original performance objective and minimizing the extent of constraint violation, as measured by ϵ^2. The parameter λ controls the trade-off between these competing goals. For finer control, a separate slack variable ϵ and penalty parameter λ can be introduced for each output and time step, allowing the degree and distribution of constraint violations to be adjusted with greater granularity.

3.6.5 Choice of Horizon Length

On the one hand, the prediction horizon p and control horizon m should be kept short to reduce computational burden; on the other hand, they must be long enough to prevent short-sighted control behavior. A short control horizon m relative to p is typically conservative, as it imposes artificial constraints—namely, forcing the control to remain constant beyond m steps—that are not actually present under the moving-horizon policy. Consequently, a small m relative to p often results in cautious control actions.

Conversely, selecting a small prediction horizon p (relative to m) leads to short-sightedness and typically results in aggressive control. For instance, if constraints are enforced only over a short horizon p, the controller may drive the system into a "dead alley" from which recovery is difficult—requiring large manipulated variable moves or constraint violations.

In the idealized case where p and m are infinite, and there are no disturbances or unmeasured inputs, the control sequence computed at time k is identical to the sequence actually implemented under the moving-horizon policy. Thus, the control action is truly optimal. However, as the horizons p and m are shortened, the computed and implemented control sequences increasingly diverge. Consequently, the short-term objective being optimized becomes less representative of the actual long-term performance under receding horizon control, which is generally undesirable.

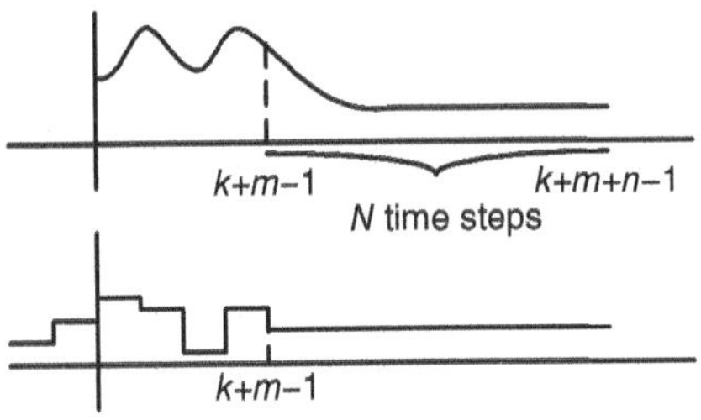

Figure 3.7 Choosing the horizon.

In general, it is advisable to select a small control horizon m to reduce computational effort, while ensuring it is large enough to provide adequate degrees of freedom for optimization. The prediction horizon p should be chosen as large as practical—ideally as large as necessary to capture the full impact of the optimized control actions. In fact, setting $p = \infty$ would fully account for the future effects of the control sequence.

This objective can be realized in several practical ways. For finite impulse response (FIR) models, since the last input change considered occurs at $k + m - 1$, the system settles by $k + m + n - 1$. This makes $p = m + n - 1$ a reasonable and effective choice for the prediction horizon (see Figure 3.7). Additionally, enforcing the terminal output to match the setpoint ensures that no residual error remains beyond the prediction horizon. This heuristic is widely adopted in commercial MPC implementations.

Alternatively, an infinite-horizon objective can be approximated by adding a large terminal output penalty at the end of the prediction horizon. This encourages the predicted output to settle to zero (or to the reference trajectory) by the final time step, thereby extending the effective influence of the prediction horizon.

3.6.6 Input Blocking

While a large control horizon is generally preferred for better controller performance, the available computational resource may limit its size. One way to relax this limit is through a procedure called *Blocking*, which allows the user to "block out" the input moves at selected locations from the calculation by setting them to zero *a priori*. This reduces the number of input moves that are computed through the optimization, hopefully without a significant sacrifice in the solution quality. Obviously, judicious selection of blocking locations is critical for achieving the intended effect. The selection is done mostly on an *ad hoc* basis, though there are some qualitative rules like blocking less of the immediate moves and more of the distant ones.

At a general level, blocking can be expressed as follows:

$$\Delta \mathcal{U} = \mathbf{B} \Delta \mathcal{U}^b \tag{3.51}$$

where $\Delta \mathcal{U}^b$ represents the reduced input parameters to be calculated through the optimization. $\mathbf{B}$ is the blocking matrix that needs to be designed for good performance. Typically, the rows of $\mathbf{B}$ corresponding to the blocked moves would contain all zeros. In general, columns of $\mathbf{B}$ can be designed to represent different basis in the input space. Note that the dimension of $\mathcal{U}^b$, which is less than that of $\mathcal{U}$, must also be determined in the design.

3.6.7 Filtering of the Feedback Signal

In practice, feedback measurements often contain significant noise and other fast-varying components. Because DMC models unmeasured disturbances as a constant bias in the prediction, the high-frequency components of the feedback signal must be filtered out to obtain a meaningful long-term forecast. To this end, the feedback signal is typically passed through a low-pass filter—commonly of first or second order—before being used in the prediction equation.

A more systematic approach involves state estimation, which is discussed in a later chapter. This technique enables modeling of the statistical characteristics of both disturbances and noise, allowing for optimal filtering based on a principled framework.

It is also worth noting that filtering the feedback signal can serve as an additional tuning mechanism for the DMC controller. Because the effect of input move penalties on closed-loop behavior varies with system dynamics, some strategies—such as Internal Model Control (IMC)—advocate setting these weights to zero (or very small values) and instead rely on feedback filtering to shape the closed-loop response. This can provide a desirable trade-off between responsiveness and robustness. These ideas will be discussed further in the next chapter, which covers potential enhancements to the basic DMC formulation.

3.7 Examples: Analysis and Guidelines

In this section, we analyze the effects of various tuning parameters on closed-loop performance. Through a series of examples, we aim to illustrate key behavioral patterns and derive general tuning guidelines. Unless otherwise noted, the system responses shown in this section are step responses, allowing for consistent comparison across different tuning scenarios.

3.7.1 Unconstrained SISO Systems

We begin by analyzing the influence of DMC tuning parameters on controller performance for single-input single-output (SISO) systems. Two representative processes are considered: a first-order plus time delay system, whose step response is shown in Figure 2.5, and a system exhibiting inverse response behavior, shown in Figure 2.6. The analysis presented in this section is limited to unconstrained systems. The effects of constraints are discussed separately in Section 3.7.2.

Example 3.4: *Setpoint change response for SISO system with dead time.* In this example, we examine how the tuning parameters—p, m, and γ^u—affect the performance of DMC for the unconstrained system described in Example 3.1.

Figure 3.8 shows the impact of the input move penalty γ^u, with the prediction horizon fixed at $p = 20$ and the control horizon at $m = 5$. When $\gamma^u = 0$, the DMC controller applies highly aggressive control actions, causing the system to reach the setpoint ($r = 1$) immediately after the dead time and maintain it thereafter. An interesting manipulated variable (MV) behavior is observed around $t = 260$ (solid line): although the MV initially settles, it exhibits a small increase for one sampling interval. This behavior results from model inaccuracies due to truncation of the step response matrix. As γ^u increases, the control response becomes increasingly conservative, leading to more gradual convergence to the setpoint. The corresponding MV trajectories also become significantly smoother.

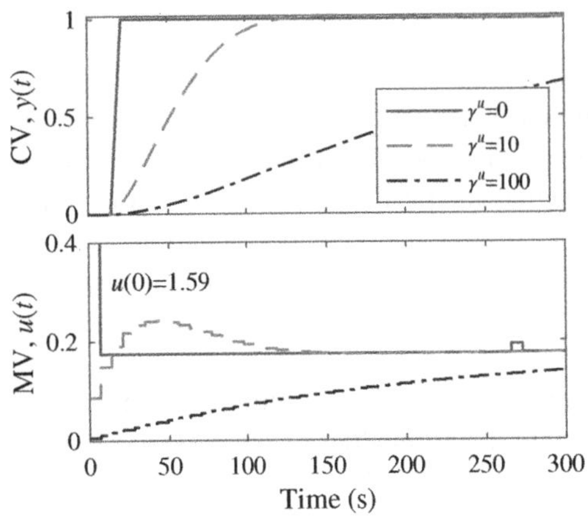

Figure 3.8 Effect of γ^u for DMC control of an unconstrained first-order with dead time SISO system with no disturbances ($p = 20$, $m = 5$, $\gamma^y = 1$, $r = 1$).

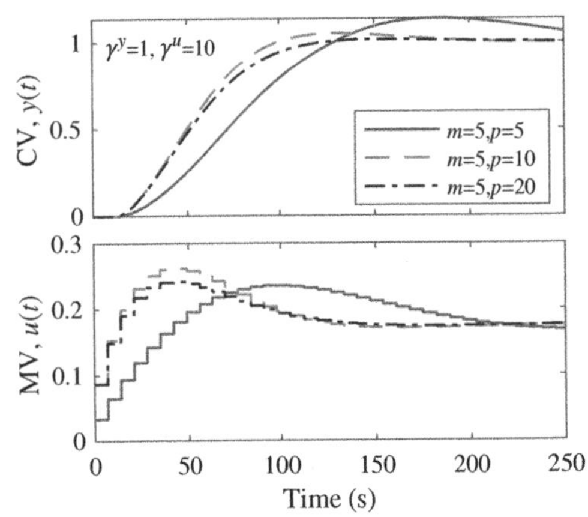

Figure 3.9 Effect of varying the prediction horizon p for DMC control of an unconstrained first-order with dead time SISO system with no disturbances ($\gamma^y = 1$, $\gamma^u = 10$, $r = 1$).

Figure 3.9 illustrates the effect of varying the prediction horizon while keeping the control horizon fixed at $m = 5$. The prediction horizon is increased starting from $p = 5$. As p increases, the closed-loop response becomes less oscillatory. When $p = m = 5$, the controller fails to account for the long-term effects of the optimized input moves, resulting in undesirable overshoot and slow settling. Beyond $p = 20$, further increases in the prediction horizon have negligible impact on performance. As subsequent examples will demonstrate, it is generally advisable to select the prediction horizon such that $p > m$.

Example 3.5: *Disturbance rejection for SISO systems with dead time.* Consider the system of Example 3.2. We investigate the effect of tuning parameters on the rejection of a unit step disturbance d_m—both measured and when the measurement is lost—for an unconstrained system.

Figure 3.10 illustrates the influence of the input move weight γ^u, Figure 3.11 shows the effect of changing the prediction horizon p, and Figure 3.12 highlights the impact of control horizon m.

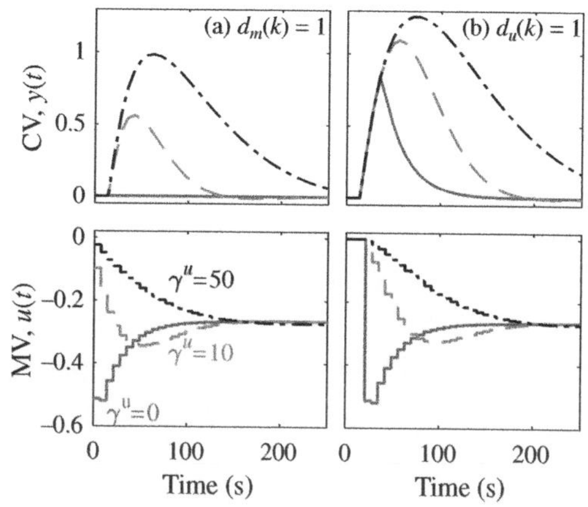

Figure 3.10 Effect of γ^u on rejection of (left) measured disturbance $d_m(k)$ and (right) unmeasured disturbance $d_u(k)$. DMC control is applied to an unconstrained first-order plus dead time SISO system ($\gamma^y = 1$, $p = 20$, $m = 5$).

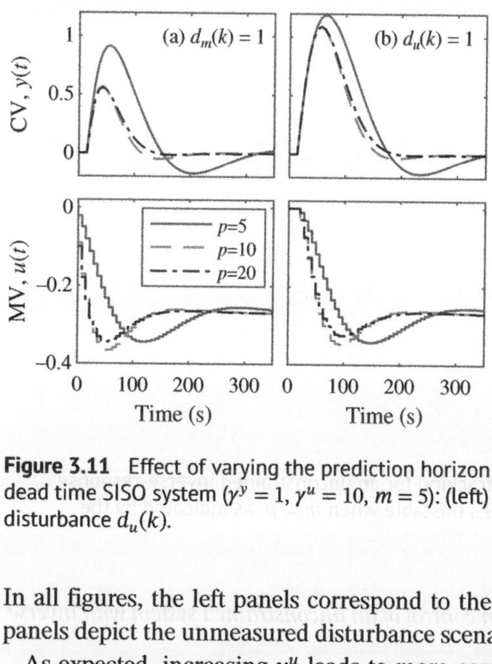

Figure 3.11 Effect of varying the prediction horizon p for DMC control of an unconstrained first-order plus dead time SISO system ($\gamma^y = 1$, $\gamma^u = 10$, $m = 5$): (left) measured disturbance $d_m(k)$ and (right) unmeasured disturbance $d_u(k)$.

In all figures, the left panels correspond to the case of a measured disturbance, while the right panels depict the unmeasured disturbance scenario.

As expected, increasing γ^u leads to more conservative control actions and slower disturbance rejection. The advantage of a measured disturbance is clearly evident: the controller responds proactively before the disturbance affects the output. For unmeasured disturbances, rejection is necessarily reactive and thus slower. In general, a longer prediction horizon improves performance, although diminishing returns are observed beyond a certain point. Similarly, increasing the control horizon provides additional flexibility but may increase computational burden.

Figure 2.6 shows the open-loop response of a system exhibiting inverse response behavior. Such systems present unique challenges for DMC, particularly due to the initial direction of the output being opposite to the steady-state change. The following examples illustrate several important characteristics of DMC performance when applied to systems with inverse response.

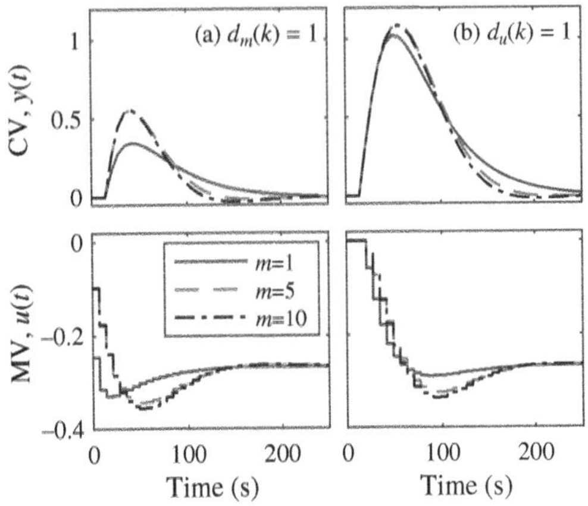

Figure 3.12 Effect of varying the control horizon m on rejection of (left) measured disturbance $d_m(k)$ and (right) unmeasured disturbance $d_u(k)$. DMC control is applied to an unconstrained first-order plus dead time SISO system ($\gamma^y = 1, \gamma^u = 10, p = 20$).

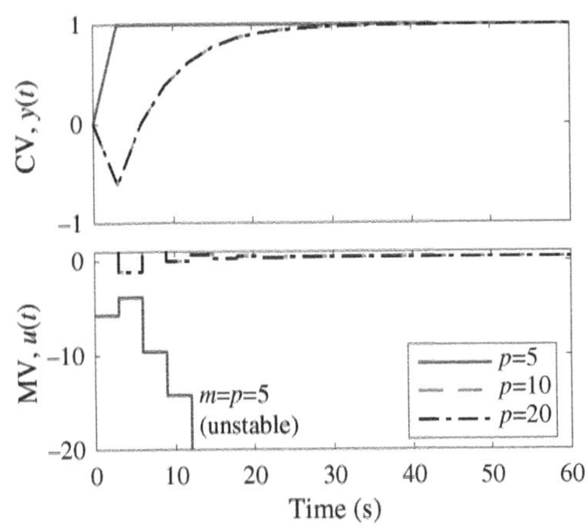

Figure 3.13 Effect of prediction horizon p on setpoint tracking for an unconstrained inverse-response system ($\gamma^y = 1, \gamma^u = 0, m = 5, r = 1.$). The system becomes unstable when $m = p$, as indicated by the diverging MV trajectory.

Example 3.6: *Effect of DMC tuning parameters on control of an unconstrained system with inverse response.*

Consider the system with inverse response introduced in Example 2.5:

$$g = \frac{2.69(-6s + 1)}{(20s + 1)(5s + 1)} e^{-1.5s}.$$

The open-loop step response of this system can be found in Figure 2.6. This example will highlight the risks associated with aggressive tuning, even when the model is known exactly.

Figures 3.13–3.15 illustrate how DMC tuning affects closed-loop performance. While the general trends are consistent with previous examples, inverse-response systems exhibit a greater tendency toward instability when tuned too aggressively. Specifically, when $\gamma^u = 0$, the system becomes unstable for the choice of $p = m$. Instability is evidenced by the unbounded growth of the manipulated variable u, even though the output y remains near the setpoint—at least initially. This occurs

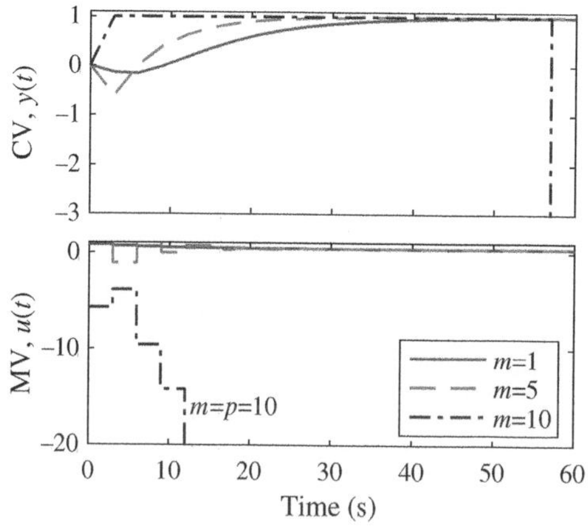

Figure 3.14 Effect of control horizon m on setpoint tracking for an unconstrained inverse-response system ($\gamma^y = 1, \gamma^u = 0, p = 10, r = 1$). Instability arises when $m = p$.

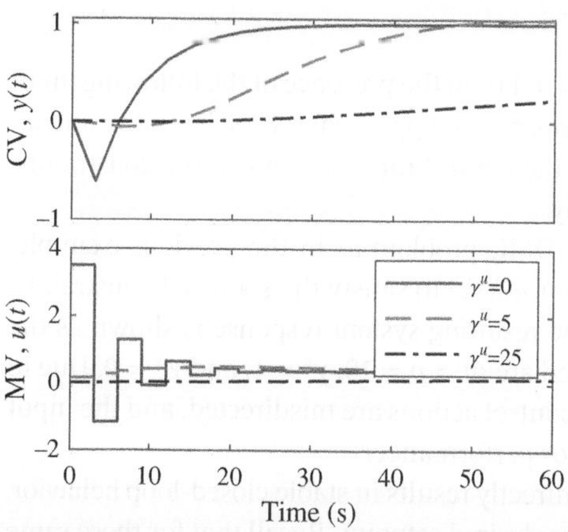

Figure 3.15 Effect of control weight γ^u on the setpoint response for an unconstrained inverse-response system ($p = 10, m = 5, r = 1$).

because the inverse response can mislead the optimizer into overcompensating with increasingly large inputs. As seen in Figure 3.13 ($p = 5$) and Figure 3.14 ($m = 10$), the input magnitude increases without bound, despite steady-state behavior in the output. In practice, this would eventually lead to actuator saturation and large deviations in the output.

3.7.2 Constrained SISO Systems

Constraints are an essential aspect of most chemical process operations. This section explores how various constraints influence the performance of DMC.

Example 3.7: *Effect of input constraints on a system with inverse response.*

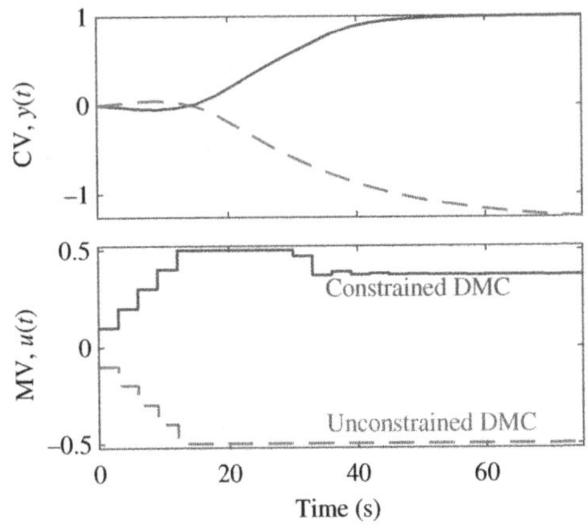

Figure 3.16 DMC performance in the presence of input constraints for a system with inverse response. Responses from constrained DMC and unconstrained DMC (with input clipping) are shown as solid and dashed lines, respectively.

Consider the system of Example 3.6 controlled at $r = 1$, but in the presence of the following input constraints, $-0.5 \leq u(k) \leq 0.5$ and $|\Delta u| \leq 0.1$. We compare two approaches: (i) solving the unconstrained DMC and applying a post hoc "clipping" of the control input to enforce the constraints, and (ii) solving the constrained DMC problem directly.

In the first approach, we solve the unconstrained DMC problem as in the previous example, and clip the first element of the computed input vector $u(k|k)$ to satisfy the specified constraints. This clipped value is then applied to the process. The resulting system response is shown as the dashed lines in Figure 3.16. The DMC parameters used are $m = p = 10$, $\gamma^y = 1$, and $\gamma^u = 0$. Due to the inverse response nature of the system, the initial control actions are misdirected, and the input eventually saturates at its lower bound, leading to poor performance.

In contrast, solving the constrained DMC problem directly results in stable closed-loop behavior, and the controller successfully drives the system to the desired setpoint. Recall that for these same tuning parameters, the unconstrained DMC yielded instability (see Figure 3.14, dashed-dotted line). This example highlights not only the stabilizing effect of properly handled constraints but also the importance of solving the constrained optimization problem rather than relying on ad hoc input clipping.

Example 3.8: *Effect of output constraints on a system with inverse response.*
Consider the system:

$$G = \frac{-s+1}{(s+1)(2s+1)} \qquad \text{sampling time, } t_s = 0.3$$

subject to output constraints $-0.3 \leq y \leq 0.3$ and no input constraints. The open-loop behavior exhibits an inverse response. At $t = 0$, an unmeasured output step disturbance of 0.3 is introduced, causing the output constraint to become active. The controller's objective is to reject this disturbance and regulate the output to zero.

Figure 3.17 compares the responses of constrained and unconstrained DMC. To reject the disturbance, a negative control move Δu is required (since the process gain is positive). However, due to the system's inverse response, any immediate negative input causes the output to initially rise, leading to a violation of the output constraint. Consequently, the constrained DMC problem becomes

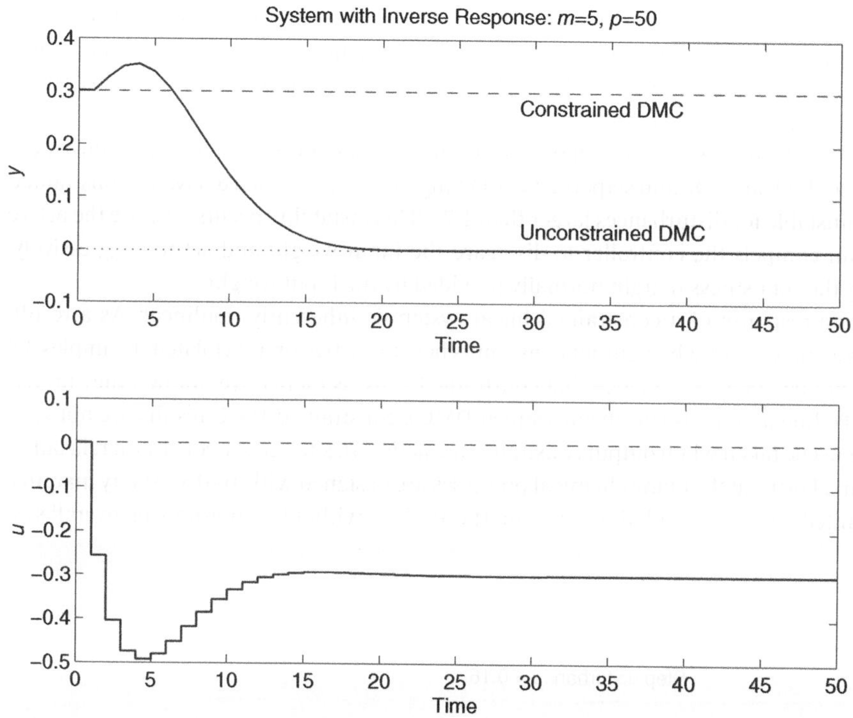

Figure 3.17 Response of a system with inverse dynamics under output constraints. The constrained DMC without a constraint relaxation strategy fails to respond due to infeasibility, whereas the unconstrained DMC achieves disturbance rejection but with temporary constraint violations.

infeasible, and the controller is unable to take any corrective action. In contrast, the unconstrained DMC proceeds despite the temporary constraint violation, effectively rejecting the disturbance and regulating the output. This example highlights the importance of implementing a constraint relaxation strategy that can appropriately soften output constraints when strict feasibility cannot be maintained.

An alternative approach is to apply constraint relaxation, as illustrated in Figure 3.6. For instance, setting $H_c = 5$ enables the controller to regulate the system effectively. In this formulation, output constraints are enforced only from time $k + H_c + 1$ to $k + p$. Since no other constraints are active in this example, the closed-loop response under constraint relaxation becomes equivalent to that of the unconstrained DMC. This is because, once the output $y(k)$ returns to the feasible region, the behavior of constrained and unconstrained DMC becomes identical.

Example 3.9: *A system with unmodeled dead time.*

Consider a true system (plant) described by:

$$G = \frac{e^{-0.15s}}{s+1}$$

with a sampling time of $t_s = 0.1$ and output constraints $-1 \leq y \leq 1$. The model used for DMC computations is $G_{\text{model}} = 1/(s+1)$, neglecting the delay term $e^{-0.15s}$, thus introducing *model-plant mismatch*. The weighting parameters are $\gamma^y = 1$ and $\gamma^u = 0.4$, with horizons $m = p = 1$. Unconstrained DMC ensures stable performance for $\gamma^u \geq 0.2$.

Figure 3.18 shows the output response of the constrained DMC to unmeasured step disturbances of magnitudes 1.6, 1.72, and 1.8, and setpoint $r = 0$. Owing to the unmodeled delay, the constrained DMC becomes unstable for disturbances larger than 1.72. This instability occurs because the active output constraint compels the controller to disregard the input weight and act too aggressively, thereby eroding the robustness margin normally provided by the input weight.

The closed-loop response of a constrained linear system is inherently nonlinear. As a result, unusual behaviors may occur when output constraints become active or are violated. Examples 3.8 and 3.9 should not be taken as evidence that constrained DMC is ineffective. Rather, they underscore that, unlike linear systems (or unconstrained DMC), constrained DMC results are not scalable, and care must be taken when output constraints are active. In such cases, constraint relaxation may be necessary. Fortunately, many chemical processes are designed with 10–15% safety margins, allowing for limited constraint violations over short durations without compromising overall system integrity.

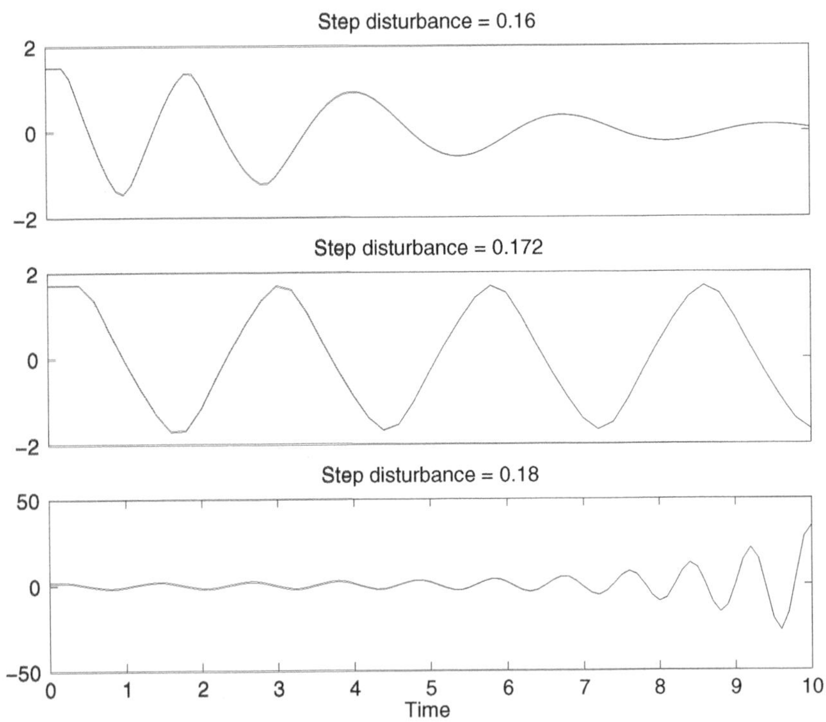

Figure 3.18 Limit of stability for a constrained DMC with plant model mismatch ($p = m = 1, r = 0$).

3.7.3 MIMO System with Strong Gain Directionality

The general tuning guidelines and closed-loop response characteristics discussed for SISO systems extend to many MIMO systems as well. These MIMO systems typically do not introduce significantly different control challenges from those encountered in the SISO case, aside from the added complexity of managing multiple variables. However, certain MIMO systems with pronounced gain directionality, such as the high-purity distillation column introduced in Example 2.6, present a different level of difficulty and can be much more challenging to control.

Recall that the responses of both outputs to each manipulated input are quite similar (see Figure 2.7 in Chapter 2). In contrast, when the two manipulated inputs are perturbed simultaneously, the system exhibits strikingly different behavior depending on whether the input combination aligns with the high-gain or low-gain direction, as illustrated in Figure 2.8. The consequences of this directional sensitivity for DMC control performance will be explored in the following sections.

Example 3.10: *Control of a highly correlated MIMO process. Effect of tuning weights.* Consider the high-purity distillation column in Eq. (2.41):

$$G = \frac{1}{75s+1}\begin{bmatrix} 0.878 & -0.864 \\ 1.082 & -1.096 \end{bmatrix}$$

The sample time is chosen as 10 minutes ($t_s = 10$). We use DMC to drive the system to the setpoint $r = [1,\ 0]^T$. The horizons are set to $m = 5$, $p = 20$, with the output control weight $\gamma^y = 10I$. The effect of varying the input weight γ^u is shown in Figure 3.19. When $\gamma^u = 0$, the outputs reach their setpoints of $r = [1,\ 0]^T$ immediately, though with very large values of the initial input move. With $\gamma^u = I$, DMC is still able to track the setpoints quickly. If the input weights were further increased to $\gamma^u = 10I$ (i.e., equal to γ^y), the system would take 15000 minutes to reach the setpoint (results not shown). As γ^u increases, the control inputs become more conservative, resulting in a slower response.

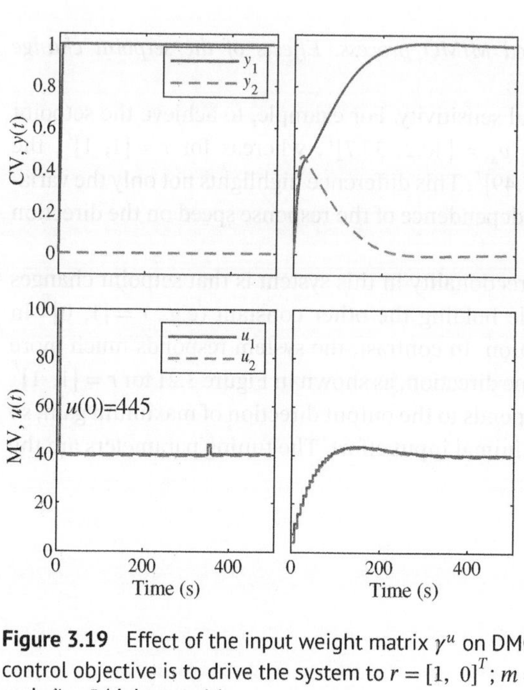

Figure 3.19 Effect of the input weight matrix γ^u on DMC performance for the high-purity distillation. The control objective is to drive the system to $r = [1,\ 0]^T$; $m = 5$, $p = 20$, $\gamma^y = 10I$, and with $\gamma^u = 0$ (left panels) and $\gamma^u = I$ (right panels).

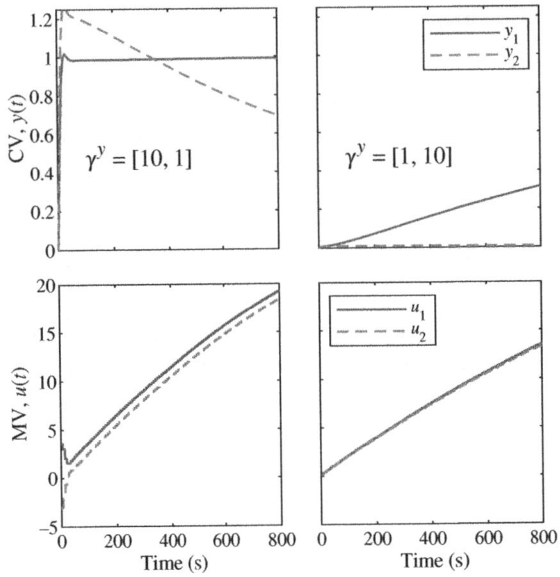

Figure 3.20 Effect of the output weighting matrix γ^y on DMC performance for the high-purity distillation example with $r = [1,\ 0]^T$. Parameters: $m = 5$, $p = 20$, $\gamma^u = I$. Solid lines: u_1 and y_1; dashed lines: u_2 and y_2.

The output weight matrix γ^y can also be adjusted to prioritize one output over the other. For example, increasing the penalty on y_2 causes the controller to minimize its deviation more strictly, even if y_1 takes longer to reach the setpoint. In contrast, emphasizing y_1 leads to a faster response in that variable, while allowing more deviation in y_2. Figure 3.20 compares two cases: $\gamma^y = \text{diag}([10,\ 1])$ and $\gamma^y = \text{diag}([1,\ 10])$. As expected, the controller adjusts its action according to the relative output penalties. With $\gamma^y = [10,\ 1]^T$, the output y_1 is controlled at its setpoint quickly, whereas with $\gamma^y = [1,\ 10]^T$, the output y_2 remains close to 0 for the entire duration.

Example 3.11: *Control of a highly correlated MIMO process. Effects of the setpoint change direction.*

This MIMO system exhibits strong directional sensitivity. For example, to achieve the setpoint $r = [1,\ 0]^T$, the steady-state input required is $u_{ss} = [40.2,\ 39.7]^T$, whereas for $r = [1,\ 1]^T$, the required input is much smaller: $u_{ss} = [8.51,\ \ -7.49]^T$. This difference highlights not only the variation in input magnitude but also a pronounced dependence of the response speed on the direction of the setpoint change.

An important consequence of strong gain directionality in this system is that setpoint changes requiring adjustment in only one output while holding the other constant (e.g., $r = [1,\ 0]^T$ in Figure 3.19) result in slow and difficult regulation. In contrast, the system responds much more rapidly when both outputs are moved in the same direction, as shown in Figure 3.21 for $r = [1,\ 1]^T$ or $r = [1,\ 1.25]^T$. The case $r = [1,\ 1.25]^T$ corresponds to the output direction of maximum gain, so the setpoint change is achieved quickly with minimal input effort. The tuning parameters are the same as those used in Figure 3.19(b).

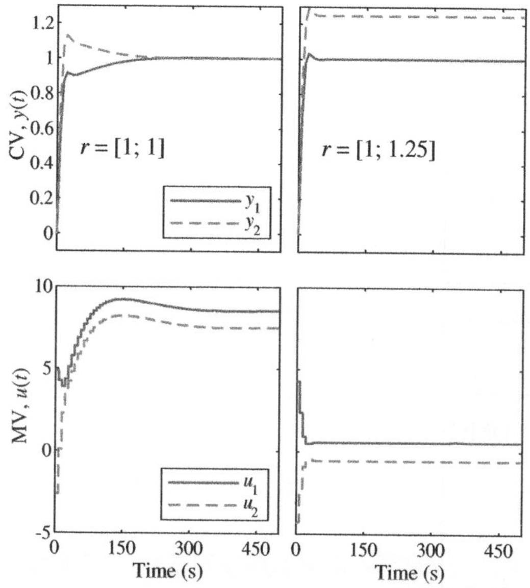

Figure 3.21 Response of the high-purity distillation system for setpoint changes in high-gain directions. DMC tuning parameters: $m = 5$, $p = 20$, $\gamma^u = I$, $\gamma^y = 10I$. Solid lines: u_1 and y_1; dashed lines: u_2 and y_2.

3.7.4 Constrained MIMO Systems

The interesting dynamics that arise due to constraints in SISO systems were discussed in Section 3.7.2. In this section, we examine the effects of various input and output constraints on the performance of MIMO systems.

Example 3.12: *High purity distillation column. Comparison between unconstrained and constrained DMC on a plant with constraints:*

Consider the system from Example 3.10, where setpoint tracking was demonstrated using DMC for an unconstrained system. We now analyze DMC performance in the presence of the following input constraints: $-10 \leq u_i \leq 10$ and $|\Delta u_i| \leq 1$.

We compare the performance of unconstrained and constrained DMC for a step change in the setpoint to $r = [0 \quad 0.4]^T$. In the unconstrained case, DMC computes the input moves without considering constraints, and the computed $\Delta u(k)$ is "clipped" to satisfy them. In the constrained case, the optimization explicitly incorporates the constraints. The controller tuning parameters are $m = 5$, $p = 20$, $\gamma^y = I$, and $\gamma^u = 0.1I$.

Based on the steady-state gain matrix, the input values required to drive the system to the setpoint are $[-12.21 \quad -12.41]^T$. As a result, when unconstrained DMC is applied with input saturation, performance deteriorates significantly once the inputs reach their lower bound, leading to excessive deviation in the outputs, as seen in Figure 3.22a. In contrast, the constrained DMC yields much better performance (Figure 3.22b). Although the output does not reach the setpoint due to actuator limits, it eventually settles at more reasonable values. Moreover, the controller weights can be adjusted to prioritize one output over the other, if desired.

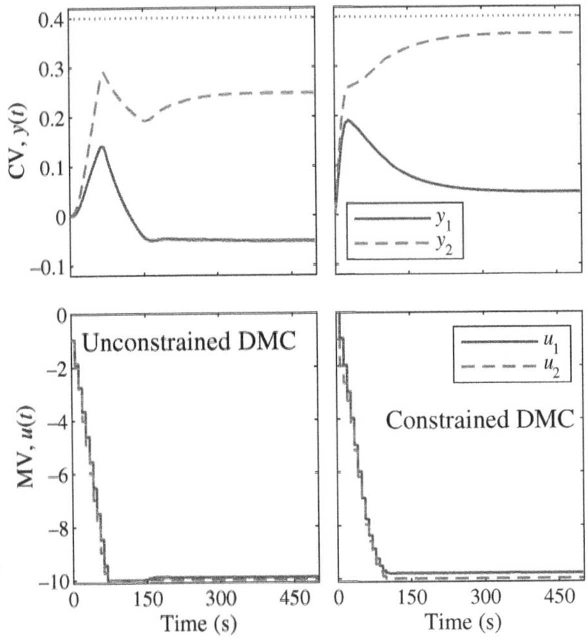

Figure 3.22 Comparison of constrained and unconstrained DMC performance for the high-purity distillation column example ($m = 5$, $p = 20$, $\gamma^u = 0.1$). (left) Response with unconstrained DMC followed by input clipping; (right) Response with constrained DMC using quadratic programming.

3.7.5 Conclusions and General Tuning Guidelines

The following general observations can be drawn from the analysis presented in this chapter. For both SISO and MIMO systems, the control becomes more aggressive and the system response faster when:

- both the prediction horizon p control horizon m are small—more specifically, when the gap $p - m$ is small, leading to "short-sighted" control actions, and
- the control weight matrix Γ^u is small.

However, such aggressive tuning reduces the robustness of the closed-loop system to model-plant mismatch. As a result, there are practical limits on the allowable aggressiveness, beyond which the closed-loop system may become unstable. In general, this implies:

- a lower bound on the horizon p
- an upper bound on the control horizon m,
- a lower bound on the control weight Γ^u

Although in rare special cases these bounds may not be necessary, they are almost always required in practice when model-plant mismatch is present. The effects of m and p are most pronounced when $\Gamma^u = 0$, while overall performance tends to be most sensitive to variations in Γ^u.

In practice, good closed-loop performance is often obtained when $p = m + n$, meaning the prediction horizon is chosen so that the effects of all considered input moves have fully settled. Additionally, one can impose a terminal constraint that forces the output to reach the setpoint at the end of the prediction horizon. This effectively makes the prediction horizon infinite—a condition known to promote stable and predictable closed-loop behavior.

In MIMO systems, the output weighting matrix Γ^y offers additional flexibility. By adjusting individual weights, one can prioritize control of specific outputs. A larger weight enforces tighter control of the corresponding output variable.

The tuning principles established for unconstrained systems remain broadly applicable in the presence of constraints. However, constraints render the control law nonlinear and non-scalable. Interestingly, input constraints may have a stabilizing effect by moderating aggressive control actions. Nonetheless, the presence of constraints increases the computational burden due to the need to solve a QP at each step. In some cases, constraint softening may be necessary to maintain QP feasibility.

Finally, while it is possible to make the weighting matrices time-varying, this is rarely done in practice due to the lack of straightforward tuning guidelines.

3.8 Case Study: Control of the "Shell Heavy Oil Fractionator" Using DMC

This case study demonstrates the application of the MATLAB MPC Toolbox (see Appendix F for an overview) to design a model predictive controller for a complex, practically motivated control problem. The system under consideration is a heavy-oil fractionation column (introduced in Chapter 1), which presents challenges typical of industrial applications, including multivariable interactions, constraint handling, and optimization.

The objective is to show how the toolbox can be used to develop and test a suitable MPC strategy through simulation. Interested readers are encouraged to attempt solving the problem using the standard step response-based DMC technique introduced earlier in this chapter. A reference solution will be provided at the end of the section.

3.8.1 Heavy Oil Fractionator: Background

Typically, a heavy oil fractionator (Figure 3.23) is used to separate crude oil into multiple product streams for further downstream processing. The vaporization of the feed stream is highly

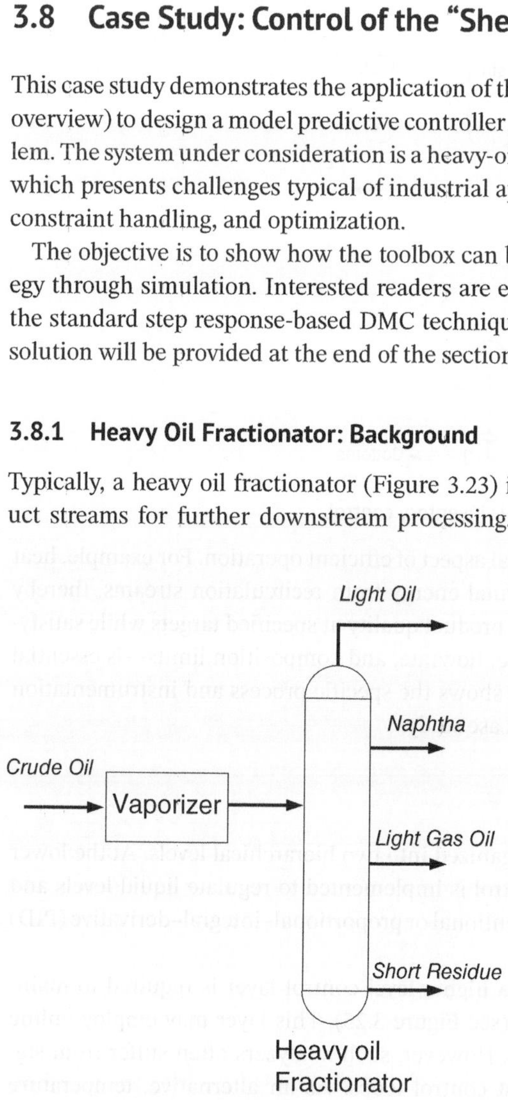

Figure 3.23 Heavy oil fractionator.

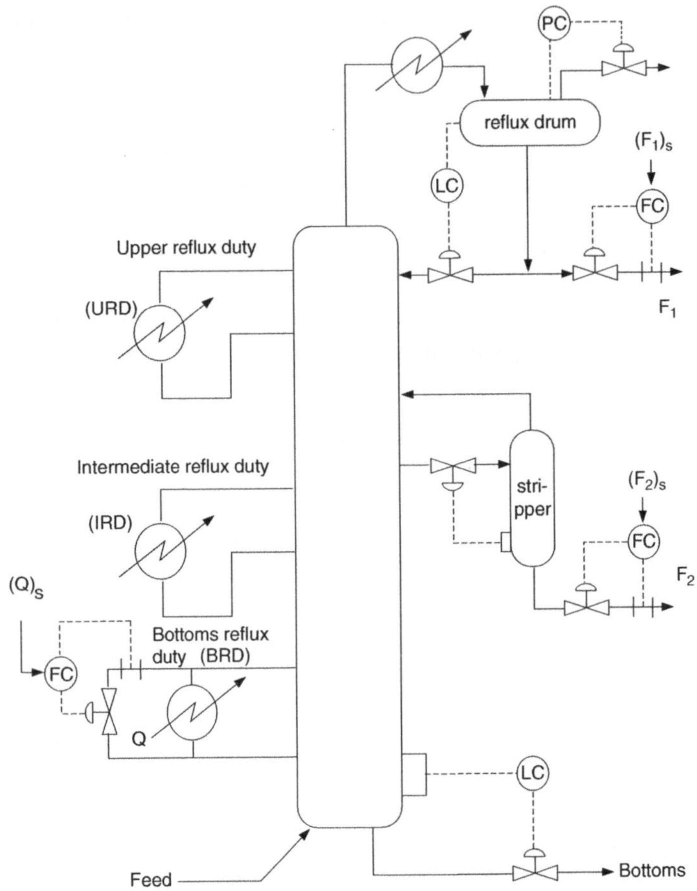

Figure 3.24 Control of heavy oil fractionator – level 1: inventory control.

energy-intensive, making heat integration a critical aspect of efficient operation. For example, heat exchangers are usually designed to recover thermal energy from recirculation streams, thereby improving overall energy efficiency. Maintaining product quality at specified targets while satisfying operational constraints—such as temperature, flowrate, and composition limits—is essential for safe and economical operation. Figure (3.24) shows the specific process and instrumentation equipment of the particular column used in this case study.

3.8.2 Control Structure Description

The control architecture of the fractionator is organized into two hierarchical levels. At the lower level—illustrated in Figure 3.24—inventory control is implemented to regulate liquid levels and flow rates. This layer does not require MPC; conventional or proportional–integral–derivative (PID) controllers are usually sufficient.

Beyond basic inventory and flow regulation, a higher-level control layer is required to maintain the desired quality of the product streams (see Figure 3.25). This layer may employ online composition analyzers for direct measurements. However, such analyzers often suffer from significant delays, limiting their usefulness in fast control loops. As an alternative, temperature measurements—which are faster and easier to obtain—can serve as proxies for composition control.

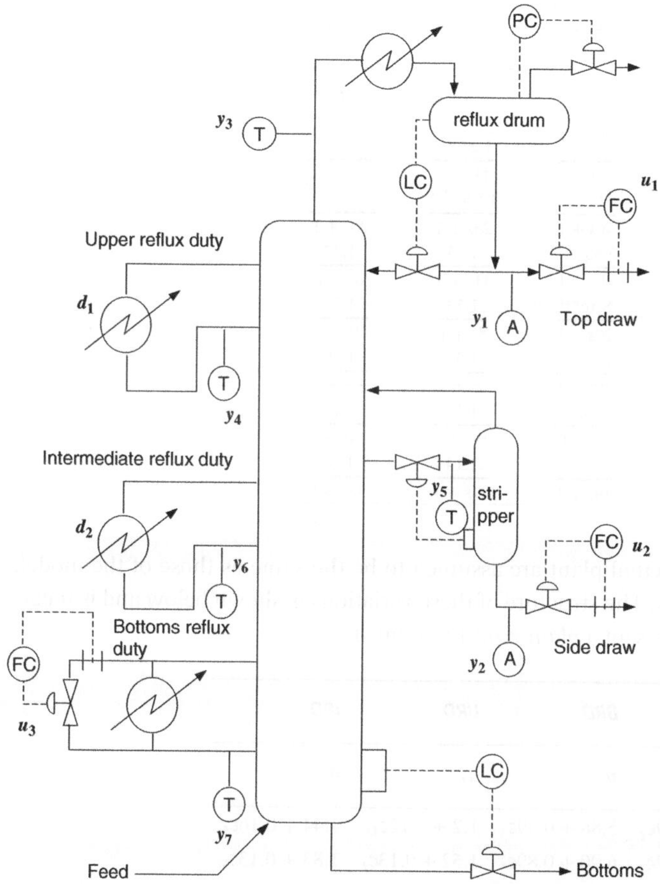

Figure 3.25 Control of heavy oil fractionator – level 2: quality control.

In addition to regulating product quality, the upper control layer enforces operational constraints and pursues optimization objectives that may conflict with basic control goals. At this level, MPC offers significant advantages through its ability to manage multivariable interactions, enforce constraints, and perform predictive optimization.

At the MPC level, the control objectives are prioritized as follows:

1. Ensure that output variable y_7 remains above its minimum threshold of −0.5
2. Maintain output variables y_1 and y_2 at their respective setpoints to ensure product quality.
3. Minimize the manipulated variable u_3, which controls the bypass valve opening, in order to maximize heat recovery through the BRD (Bottoms Reflux Duty).

These objectives reflect a balance between safety constraints, product quality targets, and energy efficiency considerations.

3.8.2.1 Input Output Transfer Functions

The dynamic behavior of the heavy oil fractionator is described by a set of linear transfer functions that relate the manipulated variables (MVs) and disturbances to the controlled variables (CVs) and other outputs. The table below summarizes the first-order plus time delay models for each input–output pair.

	TD	***SD***	***BRD***	***URD***	***IRD***
	u_1	u_2	u_3	d_1	d_2
TEP y_1	$\frac{4.05e^{-27s}}{50s+1}$	$\frac{1.77e^{-28s}}{60s+1}$	$\frac{5.88e^{-27s}}{50s+1}$	$\frac{1.2e^{-27s}}{45s+1}$	$\frac{1.44e^{-27s}}{40s+1}$
SEP y_2	$\frac{5.39e^{-18s}}{50s+1}$	$\frac{5.72e^{-14s}}{60s+1}$	$\frac{6.9e^{-15s}}{40s+1}$	$\frac{1.52e^{-15s}}{25s+1}$	$\frac{1.83e^{-15s}}{20s+1}$
TT y_3	$\frac{3.66e^{-2s}}{9s+1}$	$\frac{1.65e^{-20s}}{30s+1}$	$\frac{5.53e^{-2s}}{40s+1}$	$\frac{1.16}{11s+1}$	$\frac{1.27}{6s+1}$
URT y_4	$\frac{5.92e^{-11s}}{12s+1}$	$\frac{2.54e^{-12s}}{27s+1}$	$\frac{8.1e^{-2s}}{20s+1}$	$\frac{1.73}{5s+1}$	$\frac{1.79}{19s+1}$
SDT y_5	$\frac{4.13e^{-5s}}{8s+1}$	$\frac{2.38e^{-7s}}{19s+1}$	$\frac{6.23e^{-2s}}{10s+1}$	$\frac{1.31}{2s+1}$	$\frac{1.26}{22s+1}$
IRT y_6	$\frac{4.06e^{-8s}}{13s+1}$	$\frac{4.18e^{-4s}}{33s+1}$	$\frac{6.53e^{-1s}}{9s+1}$	$\frac{1.19}{19s+1}$	$\frac{1.17}{24s+1}$
BRT y_7	$\frac{4.38e^{-20s}}{33s+1}$	$\frac{4.42e^{-22s}}{44s+1}$	$\frac{7.2}{19s+1}$	$\frac{1.14}{27s+1}$	$\frac{1.26}{32s+1}$

The transfer functions for the actual plant are assumed to be the same as those of the model, except for potential gain variations. The structure of these variations is shown below and is incorporated into the simulation to represent a *plant-model mismatch*.

	TD	***SD***	***BRD***	***URD***	***IRD***
	u_1	u_2	u_3	d_1	d_2
TEP y_1	$4.05 + 2.11\epsilon_1$	$1.77 + 0.39\epsilon_2$	$5.88 + 0.59\epsilon_3$	$1.2 + 0.12\epsilon_4$	$1.44 + 0.16\epsilon_5$
SEP y_2	$5.39 + 3.29\epsilon_1$	$5.72 + 0.57\epsilon_2$	$6.90 + 0.89\epsilon_3$	$1.52 + 0.13\epsilon_4$	$1.83 + 0.13\epsilon_5$
TT y_3	$3.66 + 2.29\epsilon_1$	$1.65 + 0.35\epsilon_2$	$5.53 + 0.67\epsilon_3$	$1.16 + 0.08\epsilon_4$	$1.27 + 0.08\epsilon_5$
URT y_4	$5.92 + 2.34\epsilon_1$	$2.54 + 0.24\epsilon_2$	$8.10 + 0.32\epsilon_3$	$1.73 + 0.02\epsilon_4$	$1.79 + 0.04\epsilon_5$
SDT y_5	$4.13 + 1.71\epsilon_1$	$2.38 + 0.93\epsilon_2$	$6.23 + 0.30\epsilon_3$	$1.31 + 0.03\epsilon_4$	$1.26 + 0.02\epsilon_5$
IRT y_6	$4.06 + 2.39\epsilon_1$	$4.18 + 0.35\epsilon_5$	$6.53 + 0.72\epsilon_3$	$1.19 + 0.08\epsilon_4$	$1.17 + 0.01\epsilon_5$
BRT y_7	$4.38 + 3.11\epsilon_1$	$4.42 + 0.73\epsilon_2$	$7.2 + 1.33\epsilon_3$	$1.14 + 0.18\epsilon_4$	$1.26 + 0.18\epsilon_5$

3.8.2.2 Problem Statement

The following table defines five simulation cases, each representing different combinations of disturbances and gain variations.

The system must operate subject to the following constraints:

	d_1	d_2	ϵ_1	ϵ_2	ϵ_3	ϵ_4	ϵ_5
Case I:	0.5	0.5	0	0	0	0	0
Case II:	−0.5	−0.5	−1	−1	−1	1	1
Case III:	−0.5	−0.5	1	−1	1	1	1
Case IV:	0.5	−0.5	1	1	1	1	1
Case V:	−0.5	−0.5	−1	1	0	0	0

- Input bounds: $-0.5 \leq u_1, u_2, u_3 \leq 0.5$
- Input move limits: $|\Delta u_1|, |\Delta u_2|, |\Delta u_3| \leq 0.05$
- Output bounds: $-0.5 \leq y_1 \leq 0.5, y_7 \geq -0.5$
- Minimum sampling time: ≥ 1 min

The input constraints are treated as *hard constraints*, meaning they must be strictly enforced. The output constraints are *soft constraints*, which should be satisfied if feasible. In cases of infeasibility, the constraint on y_1 should be relaxed before relaxing the constraint on y_7.

The primary control objective is to regulate y_1 and y_2—the product stream compositions—at their respective setpoints, while satisfying all constraints under each of the five disturbance and uncertainty cases. As a secondary objective, the controller should minimize u_3 to maximize energy recovery through reduced bypass flow.

In addition to the five standard simulation cases, several failure scenarios will be tested. Specifically, for Case I, simulate the situation in which:

- One actuator (u_1 or u_2) becomes "stuck" (i.e., loses controllability).
- One of the composition sensors (measuring y_1 or y_2) goes offline.

Note that d_1 and d_2 are *unmeasured disturbances* in this case study. Therefore, the MPC controller does not include models for these disturbances—they are present only in the "plant" model used during closed-loop simulation to evaluate controller performance under real-world conditions.

3.8.2.3 Our Solution Strategy

We will use the MPC Toolbox in MATLAB to address this simulation problem. Other MATLAB toolboxes such as SIMULINK and the Graphical User Interface (GUI) tools may also be used for visualization and implementation, but our primary focus will remain on the MPC Toolbox.

The block diagram of the control scenario is shown in Figure 3.26. The system involves both manipulated inputs and measured outputs. Among the measured outputs, the primary control objectives are to regulate y_1 and y_2, which represent composition variables and are typically subject to significant measurement delays. While it is common in industrial practice to employ *inferential measurements* (e.g., temperature) as proxies for composition, this alternative is not considered in the present strategy, as it can be more effectively addressed using MPC with state estimation, introduced in later chapters.

The output variables of interest can be categorized as follows:

Controlled Variables: Maintain y_1 and y_2 at their respective setpoints.
Associated Variables: Ensure that y_7 remains above its specified minimum threshold.
Optimized Variables: Minimize in u_3, which indirectly serves as a performance output related to energy recovery. This is accomplished by configuring as a controlled variable with setpoint at its minimum limit. Therefore, it is both a manipulated input and a performance output in this problem.

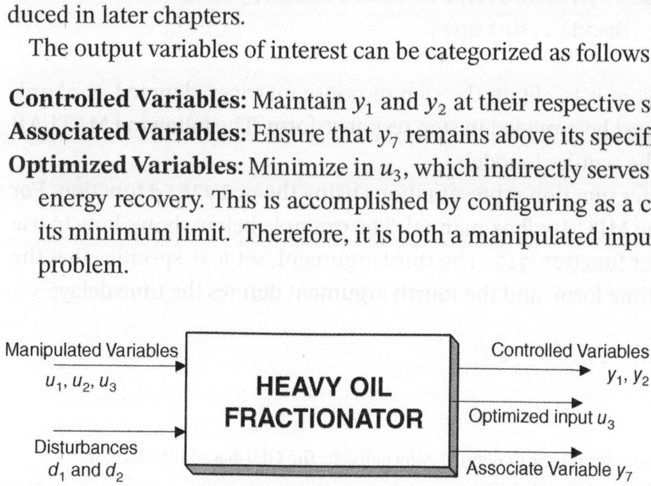

Figure 3.26 Control scenario block diagram.

Translating the above control objectives into mathematical terms, we next formulate the **model** that will be used for MPC design and simulation. Outputs and manipulated variables are related by:

$$\begin{bmatrix} y_1 \\ y_2 \\ y_7 \\ u_3 \end{bmatrix} = \tilde{G}^u \begin{bmatrix} u_1 \\ u_2 \\ u_3 \end{bmatrix} \tag{3.52}$$

where,

$$\tilde{G}^u = \begin{bmatrix} g_{11} & g_{12} & g_{13} \\ g_{21} & g_{22} & g_{23} \\ g_{71} & g_{72} & g_{73} \\ 0 & 0 & 1 \end{bmatrix}. \tag{3.53}$$

The model for the actual **plant** is given by

$$\begin{bmatrix} y_1 \\ y_2 \\ y_7 \\ u_3 \end{bmatrix} = G^u \begin{bmatrix} u_1 \\ u_2 \\ u_3 \end{bmatrix} + G^d \begin{bmatrix} d_1 \\ d_2 \end{bmatrix}. \tag{3.54}$$

Here, the model matrix $\tilde{G}^u$ may be intentionally defined to differ from the true plant matrix G^u in order to simulate a model–plant mismatch scenario. This reflects realistic conditions where the predictive model used by the controller does not perfectly match the actual process dynamics.

To solve the control problem, we use the `cmpc` algorithm provided in the MATLAB MPC Toolbox.[5] The `cmpc` function is specifically designed to solve constrained MPC problems.

The syntax for using the `cmpc` algorithm is as follows:

```
[y,u,ym] = cmpc(plant,model,ywt,uwt,M,P,tend,r,ulim,ylim,...
           tfilter,dplant,dmodel,dstep);
```

The first step in solving the problem is to obtain the step response matrices, denoted as `plant` and `model`. Both of these models must be provided in step response form. The following MATLAB commands[6] are used to construct the required models.

We begin by generating the transfer function representations using the `poly2tfd` function. For additional details, please refer to the MPC Toolbox manual. The example below shows how to use this function to generate the transfer function `g11`. The third argument, set to 0, specifies that the transfer function is in continuous-time form, and the fourth argument defines the time delay:

```
g11 = poly2tfd(4.05, [50 1], 0, 27);
```

5 Please refer to the MPC Toolbox documentation for full details. Alternatively, the GUI-based interface or SIMULINK can be used for implementation and visualization.

6 The commands described here are either standard MATLAB or MPC Toolbox functions. This example assumes that the MPC Toolbox is installed. If it is not, some of the commands may not be available.

This corresponds to the transfer function:

$$\frac{4.05\,e^{-27s}}{50s+1}.$$

After all relevant transfer functions are defined (see Eq. 3.53), we convert them into the required step response form using the `tfd2step` function. It is important to note that the transfer functions should be organized column-wise, not row-wise, when constructing the full matrix. The scalar variable `Ny` specifies the number of outputs (i.e., the number of rows in the transfer function matrix).

```
delt = 5;                                      % Interval size
tfin = 500;                 % Truncation time of step response
Ny = 4;                   % Number of outputs (y1, y2, y7, u3)

model = tfd2step(tfin,delt,Ny,g11,g21,g71,gu31,...
    g12,g22,g72,gu32,g13,g23,g73,gu33);
```

To verify the validity of the `tfd2step` command, you may consider computing the step response matrix manually. One approach is to take the inverse Laplace transform of each transfer function to obtain the corresponding time-domain step response.

Recall that the step response of a system is obtained by taking the inverse Laplace transform of the transfer function multiplied by $1/s$, the Laplace transform of a unit step input. Thus, to compute the step response manually, one multiplies the transfer function by $1/s$ and then applies the inverse Laplace transform.

For reference, the first few elements of the `model` step response matrix are listed below:

$$\gg model = \begin{bmatrix} \begin{bmatrix} 0.0000 & 0.0000 & 0.0000 \\ 0.0000 & 0.0000 & 0.0000 \\ 0.0000 & 0.0000 & 1.6659 \\ 0.0000 & 0.0000 & 1.0000 \end{bmatrix} \\ \begin{bmatrix} 0.0000 & 0.0000 & 0.0000 \\ 0.0000 & 0.0000 & 0.0000 \\ 0.0000 & 0.0000 & 2.9464 \\ 0.0000 & 0.0000 & 1.0000 \end{bmatrix} \\ \begin{bmatrix} 0.0000 & 0.0000 & 0.0000 \\ 0.0000 & 0.0945 & 0.0000 \\ 0.0000 & 0.0000 & 3.9306 \\ 0.0000 & 0.0000 & 1.0000 \end{bmatrix} \\ \begin{bmatrix} 0.0000 & 0.0000 & 0.0000 \\ 0.2113 & 0.5443 & 0.8108 \\ \vdots & \vdots & \vdots \end{bmatrix} \end{bmatrix}$$

We need two step response matrices because there can be a mismatch between the actual `plant` and our `model`. Moreover, the disturbance transfer functions are not to be incorporated into the model because the disturbances are not measured. Thus, we introduce the disturbances only in the plant and refer to them as `dplant`, while `dmodel` is left as an empty matrix.

The next step is to specify the weighting matrices (Γ^y and Γ^u). This is done by defining the vectors `ywt` and `uwt`, respectively. Constraint limits are specified using the vectors `ulim` and `ylim`. With these values in place, we can call the function `cmpc` to perform the control computation.

Below is a sample program for Case II. The setpoints for y_1 and y_2 are both set to 0. The lowest value of u_3 that satisfies the constraints is approximately $u_3 = 0.01$, so the setpoint for u_3 was chosen as 0.

```
% Project - Control of Heavy Oil Fractionator

% ------------------ OBTAIN "MODEL" ------------------
% To Obtain Modelled Step Response Coefficient
%    Model is assumed unaffected by variations in gain
E1=0; E2=0; E3=0;

c11=4.05+2.11*E1; c12=1.77+0.39*E2; c13=5.88+0.59*E3;
c21=5.39+3.29*E1; c22=5.72+0.57*E2; c23=6.90+0.89*E3;
c71=4.38+3.11*E1; c72=4.42+0.73*E2; c73=7.20+1.33*E3;

% To obtain the Transfer Functions
g11=poly2tfd(c11, [50 1], 0, 27);
g12=poly2tfd(c12, [60 1], 0, 28);
g13=poly2tfd(c13, [50 1], 0, 27);

g21=poly2tfd(c21, [50 1], 0, 18);
g22=poly2tfd(c22, [60 1], 0, 14);
g23=poly2tfd(c23, [40 1], 0, 15);

g71=poly2tfd(c71, [33 1], 0, 20);
g72=poly2tfd(c72, [44 1], 0, 22);
g73=poly2tfd(c73, [19 1]);

gu31=poly2tfd(0, [1]);
gu32=poly2tfd(0, [1]);
gu33=poly2tfd(1, [1]);

% --------- OBTAIN STEP RESPONSE MATRIX -------------
% Not all the above measured variables are controlled.
% y1 and y2 are controlled
% y7 is associated variable
% u3 is also to be optimized.
delt = 5;                                % Interval size
tfin = 500;                % Truncation time of step response
Ny = 4;                 % Number of outputs (y1, y2, y7, u3)

model = tfd2step(tfin,delt,Ny,g11,g21,g71,gu31,...
    g12,g22,g72,gu32,g13,g23,g73,gu33);
```

```
% --------------- SIMULATION SCENARIOS ---------------
% Case 2
E1=-1; E2=-1; E3=-1; E4=1; E5=1;
d1=-0.5; d2=-0.5;
y7min=-0.5;                         % minimum value of y7

% ------------------ OBTAIN "PLANT" ------------------
c11=4.05+2.11*E1; c12=1.77+0.39*E2; c13=5.88+0.59*E3;
c21=5.39+3.29*E1; c22=5.72+0.57*E2; c23=6.90+0.89*E3;
c71=4.38+3.11*E1; c72=4.42+0.73*E2; c73=7.20+1.33*E3;

% To obtain the Transfer Functions
g11=poly2tfd(c11, [50 1], 0, 27);
g12=poly2tfd(c12, [60 1], 0, 28);
g13=poly2tfd(c13, [50 1], 0, 27);

g21=poly2tfd(c21, [50 1], 0, 18);
g22=poly2tfd(c22, [60 1], 0, 14);
g23=poly2tfd(c23, [40 1], 0, 15);

g71=poly2tfd(c71, [33 1], 0, 20);
g72=poly2tfd(c72, [44 1], 0, 22);
g73=poly2tfd(c73, [19 1]);

% ------- OBTAIN STEP RESPONSE MATRIX (PLANT)-----------
% Note that u3 remains the same for plant as well
% Using same value of delt
plant = tfd2step(tfin,delt,Ny,g11,g21,g71,gu31,...
    g12,g22,g72,gu32,g13,g23,g73,gu33);

% -------------- OBTAIN DISTURBANCE MODEL --------------
dc11=1.20+0.12*E4; dc12=1.44+0.16*E5;
dc21=1.52+0.13*E4; dc22=1.83+0.13*E5;
dc71=1.14+0.18*E4; dc72=1.26+0.18*E5;

d11=poly2tfd(dc11, [45 1], 0, 27);
d12=poly2tfd(dc12, [40 1], 0, 27);

d21=poly2tfd(dc21, [25 1], 0, 15);
d22=poly2tfd(dc22, [20 1], 0, 15);

d71=poly2tfd(dc71, [27 1]);
d72=poly2tfd(dc72, [32 1]);
```

```
du1=poly2tfd(0,  [1]);
du2=poly2tfd(0,  [1]);      % Disturbances don't affect u3

dplant = tfd2step(tfin,delt,Ny,d11,d21,d71,du1,...
    d12,d22,d72,du2);
dmodel=[];                         % No measured Disturbances

% ---------------- CONTROLLER PARAMETERS ---------------
ywt=[10 10 0 10];          % Penalty-weights control outputs
uwt=[1 1 1];               % Penalty-weights for input steps
M = 10;
P = 20;                    % Control and Prediction Horizons
tend = 300;                                        % End Time
r = [0 0 y7min 0];                               % Set Points

% Constraints on Inputs |u(i)|<=0.5 and |del_U(i)|<=0.05
ulim = [-0.5 -0.5 -0.5 0.5 0.5 0.5 0.05 0.05 0.05];

% Constraints on Outputs      -0.5<=y1<=0.5 and y7min<=y7
ylim = [-0.5 -inf y7min -0.5 0.5 inf inf 0.5];

tfilter=[];                    % No data filtering required
dstep=[d1 d2];

[yp,u,ym] = cmpc(plant,model,ywt,uwt,M,P,tend,r,ulim,...
    ylim,tfilter,dplant,dmodel,dstep);

% ---------------- PLOTTING THE RESULTS ----------------
figure (3); plotall (yp,u,delt);
subplot(211); legend('y1','y2','y7','u3',0);
subplot(212); legend('u1','u2','u3',0);
```

Responses to the five cases are shown in Figures 3.27–3.34. Modifying the above program for the other cases is straightforward. Only the values of E1 … E5, D1, and D2 need to be updated accordingly. The cases involving a stuck actuator at u_1 and the loss of feedback for output y_1 are left as exercises. Below are a few helpful hints:

- For a stuck actuator, constrain the corresponding input to a very narrow range of operation to simulate its immobility.
- If the feedback filtering time delay is set to ∞, it effectively results in no feedback.

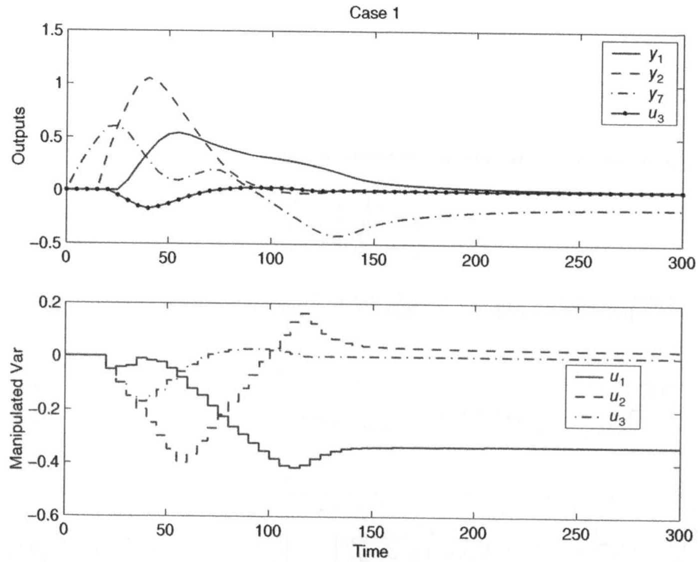

Figure 3.27 Response for Case 1. All constraints are satisfied, and u_3 is maintained at its setpoint of 0. Although a lower value of u_3 would be feasible in this case, the setpoint was intentionally chosen as 0 (see Case 2 in Figure 3.28).

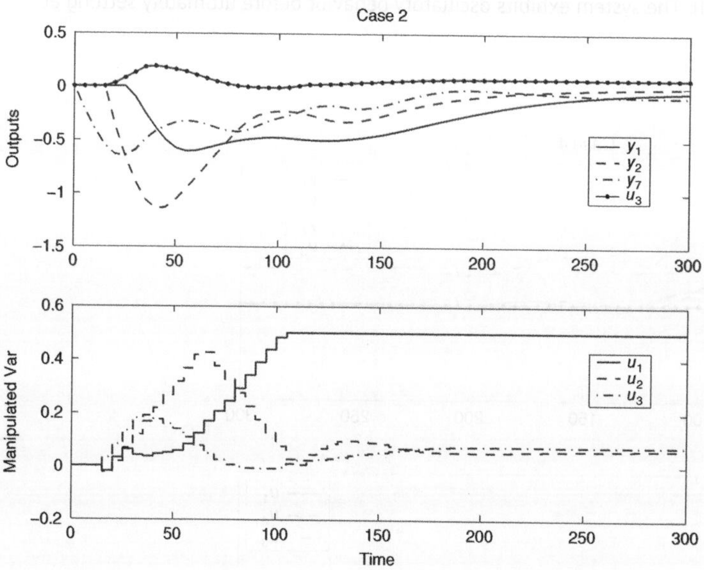

Figure 3.28 Response for Case 2. u_3 exhibits a slight offset from its setpoint of 0.

The control objectives are to maintain y_1 and y_2 at their respective setpoints while minimizing the value of u_3. In Case II, all constraints are satisfied with u_3 set to 0 or higher. To prevent infeasibility, the setpoint for u_3 was selected as 0. However, in Case I, it is possible to achieve even lower values of u_3 without violating constraints. This suggests the potential for a bi-level optimization approach, where the setpoint for u_3 is not fixed *a priori*, but instead determined through a linear optimization problem integrated within the MPC framework.

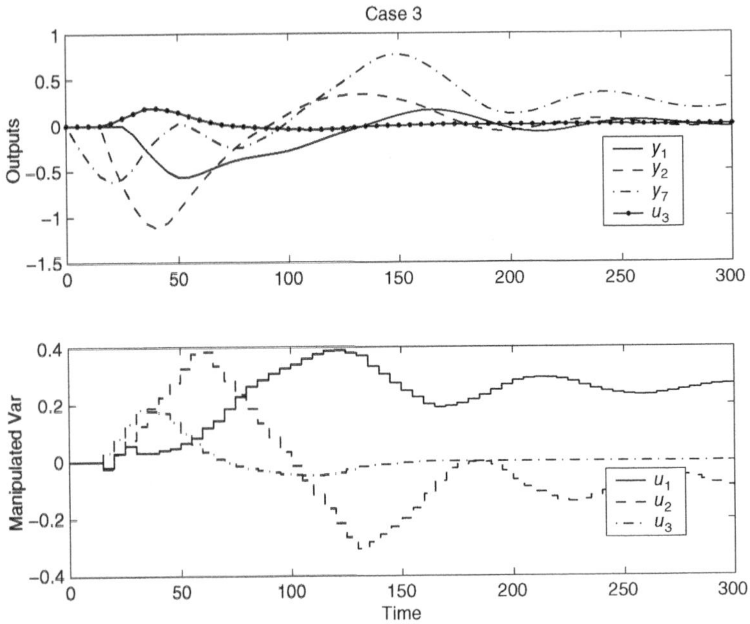

Figure 3.29 Response for Case III. The system exhibits oscillatory behavior before ultimately settling at the desired setpoints.

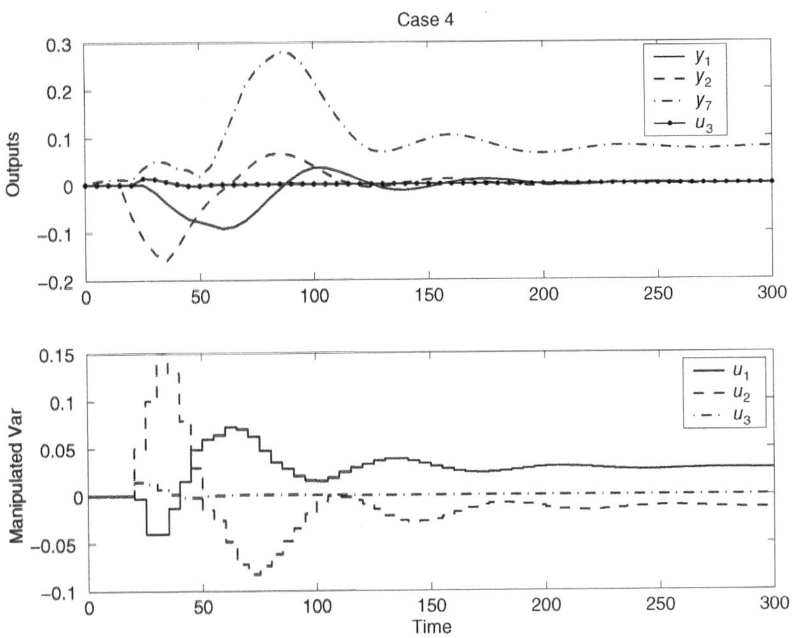

Figure 3.30 Response for Case IV. Disturbances d_1 and d_2 have opposite signs, partially canceling each other's effects. As a result, the system reaches steady state faster than in other cases.

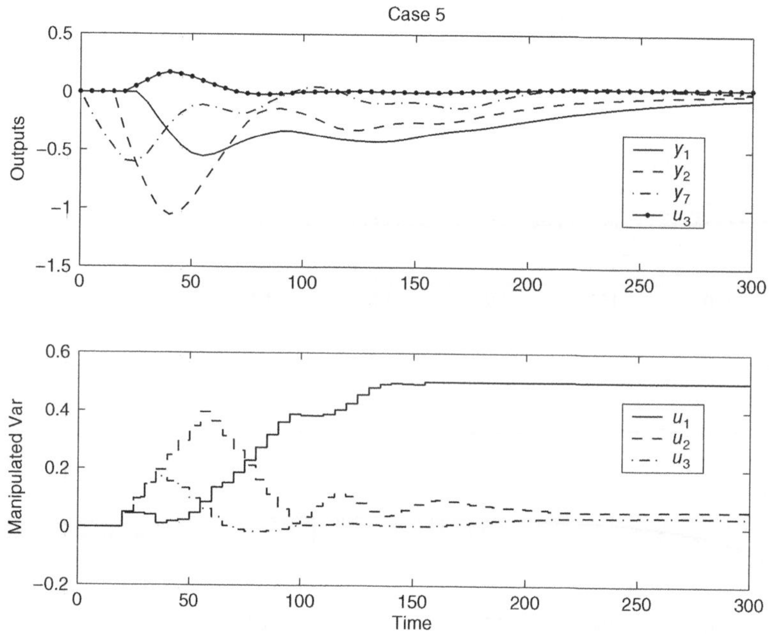

Figure 3.31 Response for Case V. Behavior is similar to that observed in Case II.

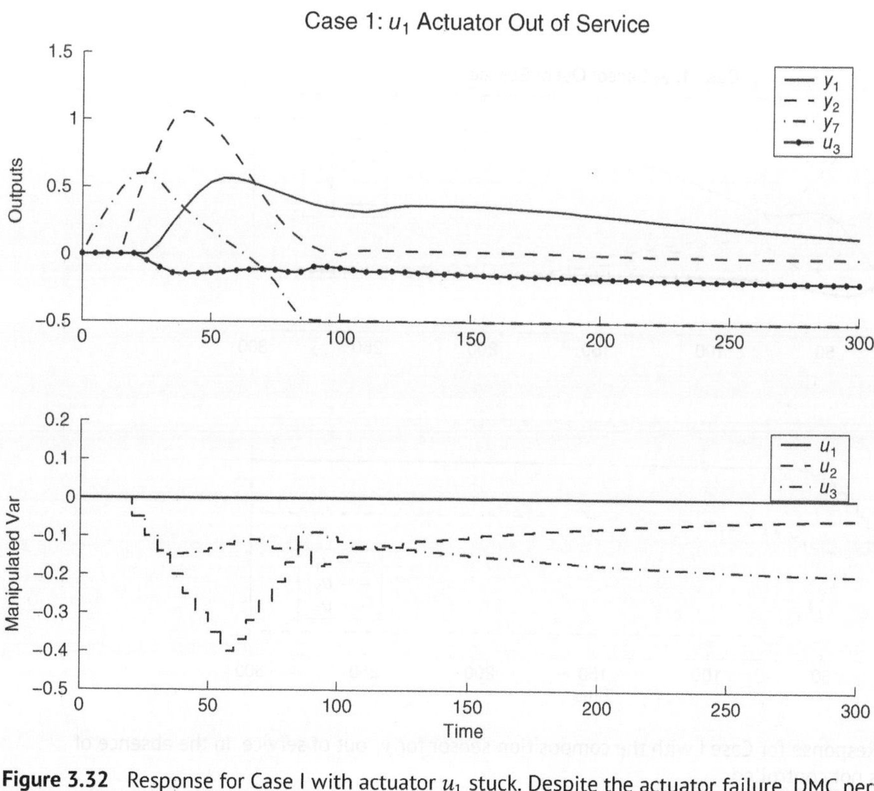

Figure 3.32 Response for Case I with actuator u_1 stuck. Despite the actuator failure, DMC performs effectively and maintains control.

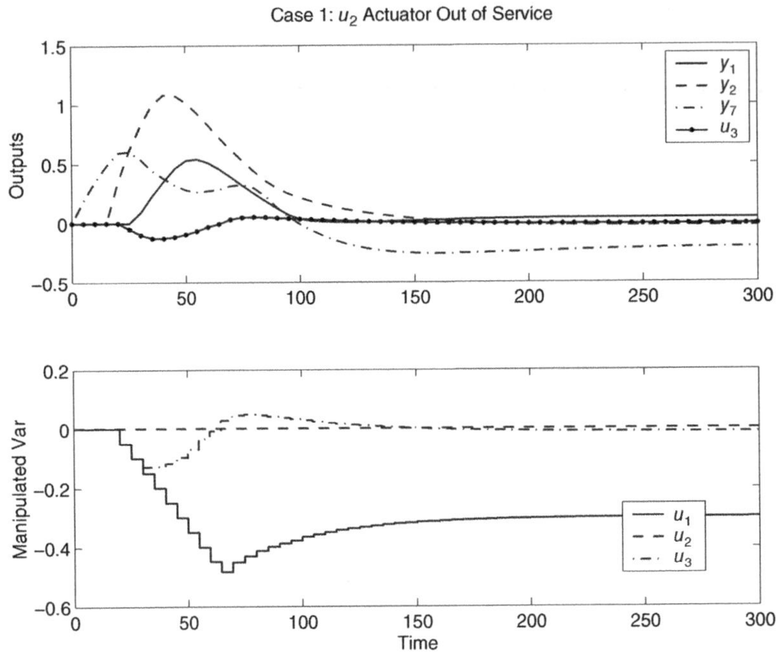

Figure 3.33 Response for Case I with actuator u_2 stuck. This scenario illustrates the advantage of applying MPC to a MIMO system.

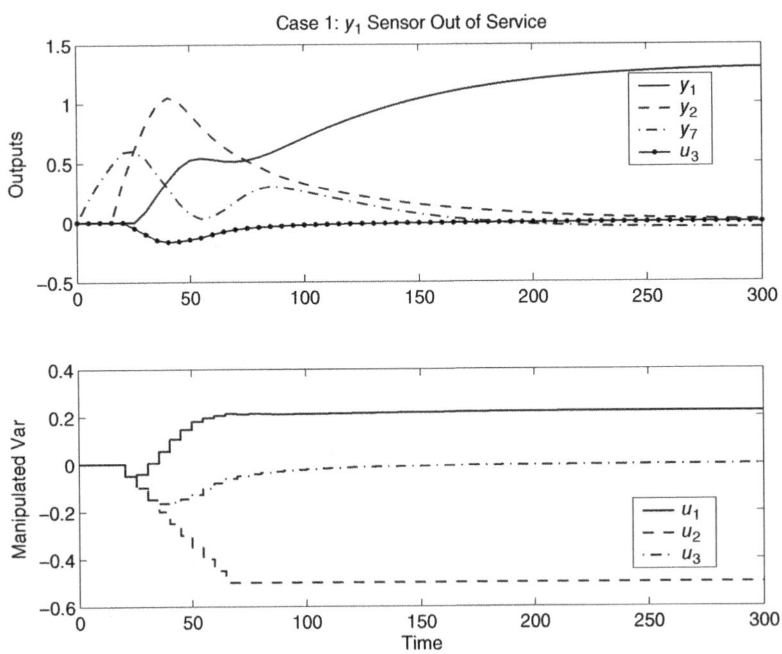

Figure 3.34 Response for Case I with the composition sensor for y_1 out of service. In the absence of feedback, y_1 is not controlled.

When the sensor for y_1 is out of service, composition feedback becomes unavailable. The controller operates under the assumption that y_1 remains unchanged at its initial value. However, in reality, y_1 violates its constraint due to the absence of corrective action. This issue may be mitigated through the use of redundant sensors or by implementing inferential control strategies that utilize secondary measurements—such as the temperature readings y_3 through y_6—as proxies for composition.

In addition to providing robustness against sensor failures, inferential measurements can significantly enhance controller performance. Composition measurements typically involve substantial dead time, during which the controller is unable to react to process disturbances. By contrast, temperature measurements are more readily available and can facilitate more timely control actions.

A detailed treatment of inferential control is beyond the scope of this chapter, but will be discussed further in Part III of this book.

Exercises

1. Extend the SISO DMC program developed in Example 3.1 to incorporate feedforward and feedback correction, accounting for both disturbance effects and model-plant mismatch. Implement this extended controller on a SISO system assuming the DMC model is:

$$G_{u,model} = \frac{6e^{-14s}}{54s+1}, \quad G_{d,model} = \frac{1.6e^{-15s}}{27s+1}.$$

Evaluate the controller performance for the following scenarios:

(i) Setpoint tracking with $r = 1.0$,
(ii) Rejection of a measured disturbance of $d = 0.5$,
(iii) Rejection of an unmeasured disturbance of $d = 0.5$.

2. Repeat the previous example for setpoint tracking ($r = 1.0$) while simultaneously rejecting a measured disturbance of $d = 0.5$, assuming the same model-plant mismatch as described above. Repeat for an unmeasured disturbance.

3. Extend the DMC controller developed in Example 3.1 to the multivariable system described by the Wood and Berry distillation column model:

$$\begin{bmatrix} Y_D \\ Y_B \end{bmatrix} = \begin{bmatrix} \dfrac{12.8e^{-s}}{16.7s+1} & \dfrac{-18.6e^{-3s}}{21s+1} \\ \dfrac{6.6e^{-7s}}{10.9s+1} & \dfrac{-19.4e^{-3s}}{14.4s+1} \end{bmatrix} \begin{bmatrix} L \\ V \end{bmatrix} \tag{3.55}$$

Use the sample time of 3 minutes ($t_s = 3$).

4. The step response model in the previous example was obtained by directly converting the transfer function model. In Chapter 2, an alternative approach based on least-squares identification was introduced for estimating step response models from input-output data.

Perform DMC simulations for setpoint tracking using the system defined in Exercise 3 of Chapter 2. Compare the controller performance under the following three modeling scenarios:

- FIR model obtained using the `step` command in MATLAB. This case assumes perfect model knowledge with no plant-model mismatch.

- FIR model obtained by applying step tests to the system and collecting noisy output data. Recall that the measurement noise with $\sigma_v^2 = 0.05^2$ was generated using: `randn(N,1)*0.05`.
- FIR model obtained through least-squares identification using the same noisy data as in the previous case.

5. In the high-purity distillation 2×2 control example, assume that the gains associated with y_1 are over-predicted by 20%, while those associated with y_2 are under-predicted by 20%. The model used for DMC calculations is therefore

$$G_{model} = \frac{1}{75s+1}\begin{bmatrix} 1.054 & -1.037 \\ 0.866 & -0.877 \end{bmatrix}$$

Simulate the closed-loop response of a DMC controller designed with this model under the model–plant mismatch for a setpoint change of $r = [1,\ 0]^T$. Use the tuning parameters $m = 5$, $p = 20$, $\gamma^y = 10I$, and $\gamma^u = I$.

4

Dynamic Matrix Control–Extensions and Variations

Discovery consists of seeing what everybody has seen and thinking what nobody has thought.

– Albert Szent-Györgyi

In Chapter 3, we examined the Dynamic Matrix Control (DMC) algorithm, one of the earliest and most influential algorithms in the development of Model Predictive Control (MPC). In this chapter, we explore alternative formulations used in other industrial MPC implementations, along with potential enhancements that can broaden the applicability of DMC and improve its performance.

4.1 Features Found in Other Industrial Algorithms

DMC was one of the first MPC algorithms to achieve successful application in industrial settings. In its original form, DMC did not use quadratic programming (QP) to enforce constraints. Instead, it incorporated an additional output term in the prediction model to steer the control input back into the feasible region whenever a predicted future input neared a constraint. While effective in practice, this approach was somewhat *ad hoc* and lacked a formal theoretical foundation. It wasn't until the 1980s that engineers at Shell Oil introduced a more systematic approach using QP to explicitly and rigorously manage input and output constraints. This enhanced version became known as QDMC. Today, the original DMC framework remains in use through the commercial software package DMC-PLUS, marketed by Aspen Technology.

In addition to DMC and QDMC, several other MPC algorithms have seen widespread industrial adoption. These include Model Predictive Heuristic Control (MPHC), which led to commercial products such as IDCOM and SMC-IDCOM developed by Setpoint (now part of Aspen Technology and Emerson); Hierarchical Constraint Control (HIECON) and Predictive Functional Control (PFC), both marketed by Adersa; Predictive Control Technology (PCT) developed by Profimatics (now part of Honeywell); and Robust Model Predictive Control Technology (RMPCT), currently marketed by Honeywell. Although these algorithms share the same fundamental principles, they differ in implementation details. Rather than delving into the specifics of each algorithm, we will focus on highlighting some commonly used features that are not present in the basic DMC/QDMC framework.

4.1.1 Reference Trajectories

In DMC, the optimization penalizes deviations of the output from a fixed setpoint (Figure 4.1a). In some cases, however, tight regulation to an exact setpoint may not be necessary; instead, the goal may be to maintain the controlled output within a specified zone (Figure 4.1b). By contrast, algorithms such as IDCOM, HIECON, and PFC allow the user to define not only the final target but also the desired path by which the output approaches it. This is achieved by introducing a *reference trajectory* for each CV at every sampling instant, typically formulated as a first-order path from the current output value to the setpoint. The time constant of this path serves as a tuning parameter that shapes the speed of the closed-loop response. This idea is depicted in Figure 4.1c.

Reference trajectories offer an intuitive mechanism for adjusting the aggressiveness of the controller. While DMC modulates aggressiveness primarily through the weighting matrix on the control move (input increment) penalty, the use of reference trajectories allows tuning via the output response speed, which is often more transparent and easier to interpret. Unlike input weighting parameters, whose effect on closed-loop dynamics can be highly system-dependent, reference trajectories directly shape the output behavior in a user-specified manner.

4.1.2 Coincidence Points

Some commercial algorithms, such as IDCOM and PFC, offer the option of penalizing output error only at a limited number of specific time points within the prediction horizon—referred to as coincidence points. This approach is primarily motivated by the desire to reduce computational load.

When the number of control moves must be kept small—often to maintain computational efficiency—a long prediction horizon may still be necessary, particularly in systems with significant inverse response or long time constants. In such cases, applying penalties at every step in the horizon can lead to overly conservative or sluggish control behavior. This issue can be mitigated by penalizing output deviations only at a few strategically chosen points. In the extreme case, requiring the output to match the reference trajectory at a single time point can be achieved with just

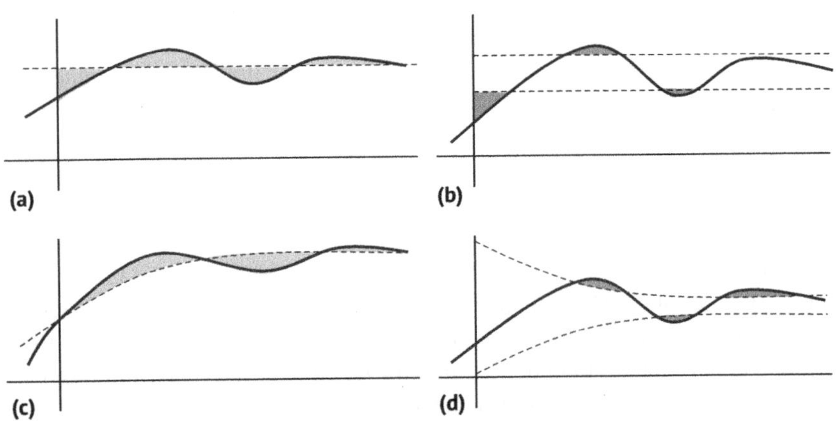

Figure 4.1 Illustration of output penalties in various MPC formulations. Quadratic penalties are applied based on deviation of the output from: (a) a fixed setpoint, (b) deviation outside a designated no-penalty zone, (c) a reference trajectory, and (d) a time-varying funnel. Shaded regions indicate areas where output violations incur a quadratic penalty.

one control move. This simplified formulation was implemented in IDCOM-M, a derivative of the original IDCOM algorithm, marketed by Setpoint.

The selection of coincidence points is clearly critical for performance, especially when only a few points are used. While some heuristic guidelines exist, there is no systematic method for choosing them. Since the response dynamics can vary significantly across outputs, coincidence points are typically defined separately for each controlled variable.

4.1.3 The Funnel Approach

The Robust Model Predictive Control Technology (RMPCT) algorithm distinguishes itself from other MPC formulations by aiming to keep each controlled output within a user-defined region known as a funnel, rather than tracking a fixed reference trajectory. A typical funnel shape is illustrated in Figure 4.1d. The funnel is specified by the user through upper and lower bounds, as well as a performance ratio, which is the desired time for the output to return to the acceptable zone divided by the open-loop response time. The funnel can be narrowed for exact setpoint control or widened for range control applications.

At each control step, the algorithm solves a quadratic program of the form

$$\min_{y^r,\Delta u} \sum_{i=1}^{p} \|y(k+i|k) - y^r(k+i|k)\|_Q^2 + \sum_{j=0}^{m-1} \|\Delta u(k+j|k)\|_R^2 \tag{4.1}$$

or alternatively,

$$\min_{y^r,u} \sum_{i=1}^{p} \|y(k+i|k) - y^r(k+i|k)\|_Q^2 + \sum_{j=0}^{m-1} \|u(k+j|k) - u^r\|_R^2 \tag{4.2}$$

subject to standard input and output constraints, along with the funnel constraint

$$y^f_{\min}(k+i|k) \le y^r(k+i|k) \le y^f_{\max}(k+i|k), 1 \le i \le p. \tag{4.3}$$

Here, $y^f_{\min}(k+i|k)$ and $y^f_{\min}(k+i|k)$ define the funnel bounds at time $k+i$, and u^r denotes the desired steady-state input. The reference trajectory y^r is treated as a free optimization variable constrained to lie within the funnel.

Typically, a high output weight $Q \gg R$ is used to prioritize keeping the outputs within the funnel. In this setup, the problem can be viewed as a multi-objective optimization: the primary goal is to minimize violation of the funnel constraints, and the secondary goal is to minimize input movement or deviation from u^r. When the output can be maintained within the funnel, the first term becomes zero. In such cases, the problem becomes a *degenerate QP*, often with infinitely many solutions. The algorithm selects the minimum-norm solution, yielding the smallest required input adjustment—hence the term *Robust MPC*.

If no feasible input trajectory exists to maintain the outputs within the funnel, the optimizer prioritizes minimizing the violation magnitude. This approach is particularly useful in multivariable systems, where specifying desirable trajectories for individual outputs is challenging due to interaction. Instead of prescribing exact trajectories, the user defines acceptable dynamic zones, and the controller determines the minimal input action needed to remain within—or minimally violate—those zones.

Table 4.1 Optimization resulting from different spatial and temporal norms.

		Temporal Norm		
Spatial Norm	–	1	2	∞
	1	LP	QP	LP
	2	QP	QP	QP
	∞	LP	QP	QP

4.1.4 Use of Other Norms

While the 2-norm is most commonly used in defining the MPC objective function, alternative norms are certainly possible. In particular, the use of the 1-norm (sum of absolute values) has been extensively explored, and the infinity norm has also been investigated—typically with the objective of minimizing the worst-case deviation over the prediction horizon.

Both the 1-norm and infinity norm lead to a linear programming (LP) formulation. This is advantageous from a computational perspective, as LP is a well-established class of problems with extensive theoretical foundations and mature solver support, particularly due to its prominence in economics and operations research. However, one major challenge in using these formulations is tuning. LP solutions tend to lie at the intersection of active constraints (vertices of the feasible polytope), and small changes in tuning parameters—such as input weights—can cause the optimal solution to shift abruptly from one vertex to another. This can result in discontinuous or unpredictable controller behavior. In contrast, QP formulations yield smoother, more continuous solutions with respect to tuning parameters, making them generally more robust and preferable for control applications.

Table 4.1 summarizes the optimization problem structures that result from different combinations of spatial and temporal norms. Spatial norms are norms applied to the input and output vectors at each sampling instant, whereas temporal norms are those applied across the time steps over the prediction horizon.

4.1.5 Input Parameterization

In some algorithms, such as Predictive Functional Control (PFC), the input trajectory is parameterized using continuous basis functions, such as polynomials. This approach is particularly useful when the control objective involves accurately tracking smooth setpoint trajectories—as is common in mechanical servo applications—especially when the sampling interval is too large to achieve this with piecewise constant inputs.

In contrast, other commercial algorithms like HIECON and IDCOM-M compute only a single control move at each step, corresponding to $m = 1$ in DMC. This greatly simplifies the optimization problem. However, using $m = 1$ in a standard DMC framework generally limits closed-loop performance. These algorithms overcome this limitation by employing a single coincidence point, where the output is required to exactly match the reference value, thereby compensating for the reduced control horizon.

4.1.6 Model Conditioning

In multivariable systems, it is common for two or more outputs to respond in a very similar way to all inputs. This phenomenon is known as *ill-conditioning* and is typically characterized by a nearly singular process gain matrix. A key implication of ill-conditioning is the difficulty in independently controlling the affected outputs—it may require excessively large input movements to steer the outputs in certain directions. Using an ill-conditioned model in control calculations is problematic: it can lead to numerical issues (e.g., inversion of nearly singular matrices), unnecessary actuator effort, and even closed-loop instability.

While model conditioning is typically evaluated during the controller design phase, real-time changes in the control structure—due to constraints, actuator saturation, or sensor/actuator failures—can introduce ill-conditioning during operation. Therefore, it is important to ensure at every execution step that the control algorithm does not attempt to directly invert an ill-conditioned model.

In DMC, this issue can be mitigated by including a substantial input move penalty in the objective function. This regularization increases the effective diagonal dominance of the matrix being inverted in the least-squares solution, thereby improving numerical stability.

Other algorithms that do not include an explicit input move penalty must explicitly check for ill-conditioning at runtime. For instance, RMPCT addresses this through a method known as *singular value thresholding*. In this approach, the process gain matrix undergoes a singular value decomposition (SVD) to identify CV directions with insufficient gain. Directions corresponding to singular values below a user-defined threshold are excluded from control, and only the remaining "high-gain" directions are actively regulated. Similarly, SMC-IDCOM handles ill-conditioning based on user-defined priority rankings of CVs. When ill-conditioning is detected, CVs are progressively dropped from the control problem in order of increasing priority until the condition number of the gain matrix reaches an acceptable level. In cases where two CVs behave similarly, the less critical one can be assigned a lower rank, allowing the controller to drop it if needed. Even when a CV is dropped, the expectation is that it may still be regulated indirectly due to its dynamic similarity with the retained, higher-priority CVs.

4.1.7 Prioritization of CVs and MVs

In most practical control problems, it is not possible to satisfy all constraints while simultaneously driving every input and output to its desired target. As a result, priorities must be established to reflect the relative importance of competing objectives. In DMC, these priorities are encoded through weighting parameters in the quadratic penalty terms of the objective function. However, for large and complex systems, determining appropriate weights that yield the desired control behavior can be a challenging and time-consuming task. Moreover, the resulting weights may vary widely in magnitude, potentially leading to numerical conditioning issues.

To address these challenges, algorithms such as HIECON and SMC-IDCOM allow users to rank objectives explicitly in order of importance. For instance, constraint satisfaction may be given top priority, taking precedence over other objectives such as setpoint tracking or economic optimization. Similarly, driving CVs to their setpoints may be deemed more critical than steering MVs to their economically optimal values.

These algorithms solve a series of optimizations in hierarchical fashion: the most important objective is addressed first, and any remaining degrees of freedom are then allocated to subsequent objectives in descending order of priority. They also support ranking individual CVs and MVs. This allows, for example, controlled constraint softening—where constraints are relaxed according to a specified priority—or prioritized setpoint tracking, where higher-ranked CVs are regulated more tightly than those with lower importance.

4.2 Connection with Internal Model Control

Early MPC schemes published in the late 1970s enjoyed the structure depicted in Figure 4.2 called the Internal Model Control (IMC) structure (García and Morari, 1982). It is derived by adding and subtracting the modeled effect of the manipulated variables within a conventional feedback loop. Any controller can be represented using either structure.

As discussed earlier, DMC compares the measured output to the step response model prediction, treating the difference as a bias term added to the prediction. This bias, combined with an optimization step to compute the control move, categorizes DMC as an IMC controller.

We now illustrate the benefits of IMC for open-loop stable, linear SISO processes. The three basic components of IMC are:

- *Process* (P_u): The output y_m results from both the manipulated input u and disturbance effects w_y.
- *Model* ($\tilde{P}_u$): The input u is fed to a process model whose output is subtracted from the measured process output to yield the IMC feedback signal $\hat{w}_y$. The notation means that it is an estimate of the unmeasured disturbance effects w_y.
- *Controller* (C_{IMC}): The IMC feedback signal is subtracted from the reference r, and the resulting error is used to compute the next control move.

Notably, $\hat{w}_y$ is the same bias term described in Eq. (1.4), representing the discrepancy between the measured and modeled outputs. In MPC, the IMC controller corresponds to the prediction equation and the optimization routine used to generate the control input.

When the model perfectly matches the process ($P_u = \tilde{P}_u$), the IMC feedback signal $\hat{w}_y$ reflects only the effect of disturbances—that is, $\hat{w}_y = w_y$. In this case, the controller acts as an open-loop feedforward controller, since no control action is being fed back. However, when there is model mismatch, $\hat{w}_y$ also captures the influence of the manipulated input on that mismatch, enabling the controller to compensate and maintain stability.

A key advantage of the IMC structure is that it naturally decouples the influence of controller tuning parameters on closed-loop performance from their influence on stability under model errors. In practice, with an accurate model, controller design reduces to designing a feedforward controller, with C_{IMC} defined as a stable inverse of the process model. This insight led to the development of IMC as a robust control strategy, where a low-pass filter is systematically applied to the feedback signal (i.e., measurement minus model output) to "detune" the controller for improved robustness to model errors (Morari and Zafiriou, 1989).

As discussed in Section 1.4, the IMC framework provided an early theoretical foundation for understanding MPC and its distinctions from traditional advanced control schemes. Although IMC has since evolved into an independent methodology for robust control, its structure continues to illuminate the underlying strengths of MPC.

Implementing a controller within an IMC framework offers greater transparency in controller design and tuning compared to the conventional feedback structure. However, because the model operates in an open-loop fashion—without direct measurement correction—this approach cannot handle open-loop unstable dynamics. Moreover, in multivariable systems subject to disturbance and measurement noise, making optimal and systematic use of measurements becomes

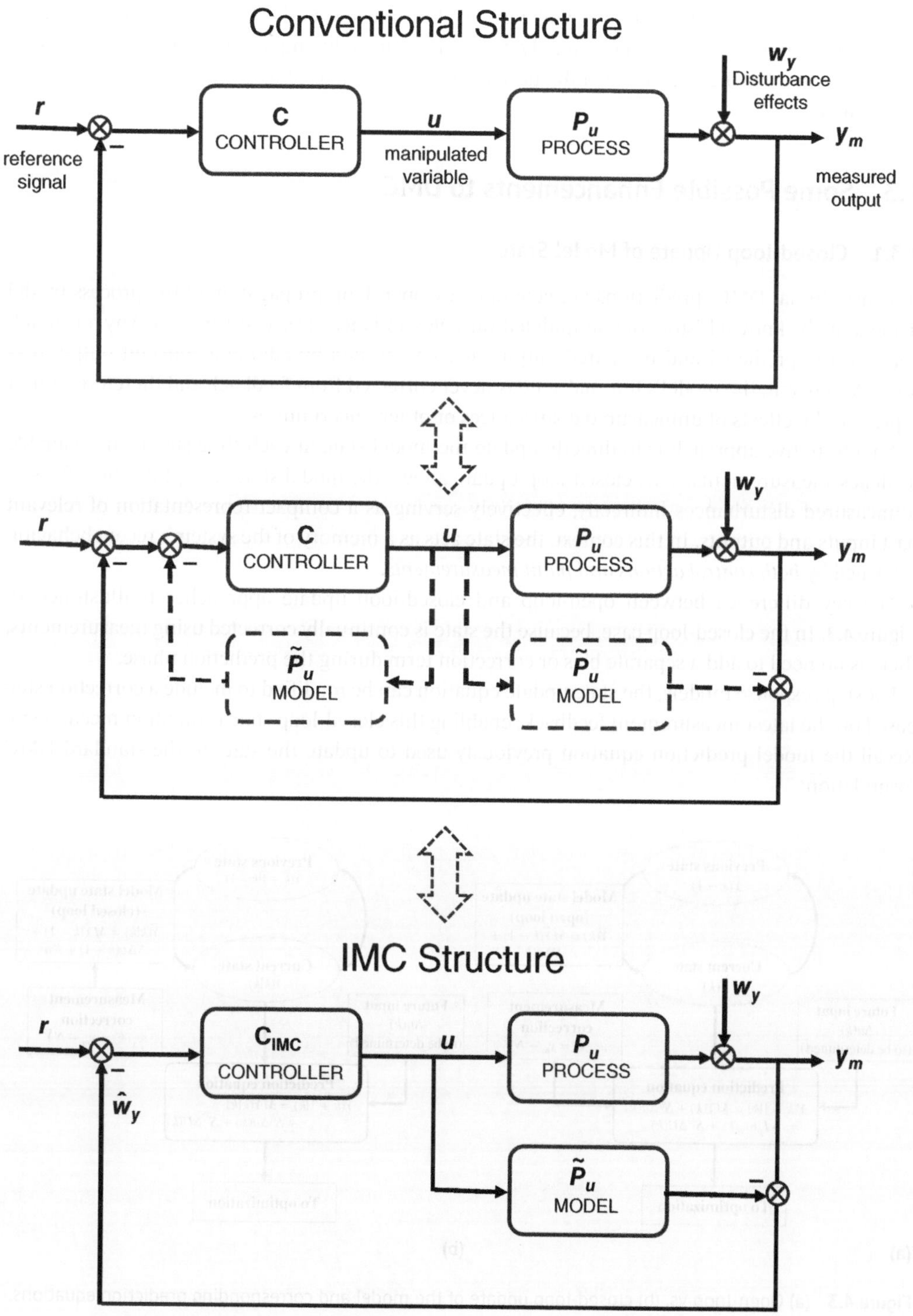

Figure 4.2 Comparison of conventional feedback control and Internal Model Control (IMC) loop structures.

challenging. For these reasons, we will next explore an alternative approach that uses measurements to update model states directly, rather than introducing a bias term into the prediction equations. Further details on IMC design for linear, stable SISO processes are provided in Appendix E.

4.3 Some Possible Enhancements to DMC

4.3.1 Closed-loop Update of Model State

In conventional DMC, predictions are generated by open-loop propagation of the process model using only the known history of manipulated variables and measured disturbances. Any mismatch between the predicted and measured outputs is compensated by adding a constant output-bias term. As a result, the model's internal state is never corrected from feedback and therefore cannot represent the effects of unmeasured disturbances or other uncertainties.

An alternative approach is to directly update the model state at each time step using available feedback measurements. This closed-loop update allows the model state to capture the effects of unmeasured disturbances indirectly, effectively serving as a compact representation of relevant past inputs and outputs. In this context, the state acts as a memory of the system's recent behavior, informed by *both control actions and plant measurements.*

The key difference between open-loop and closed-loop update approaches is illustrated in Figure 4.3. In the closed-loop case, because the state is continually corrected using measurements, there is no need to add a separate bias or correction term during the prediction phase.

For step response models, the state update equation can be modified to include a correction step based on the latest measurement feedback, enabling this closed-loop state estimation mechanism. Recall the model prediction equation previously used to update the state in the standard DMC formulation:

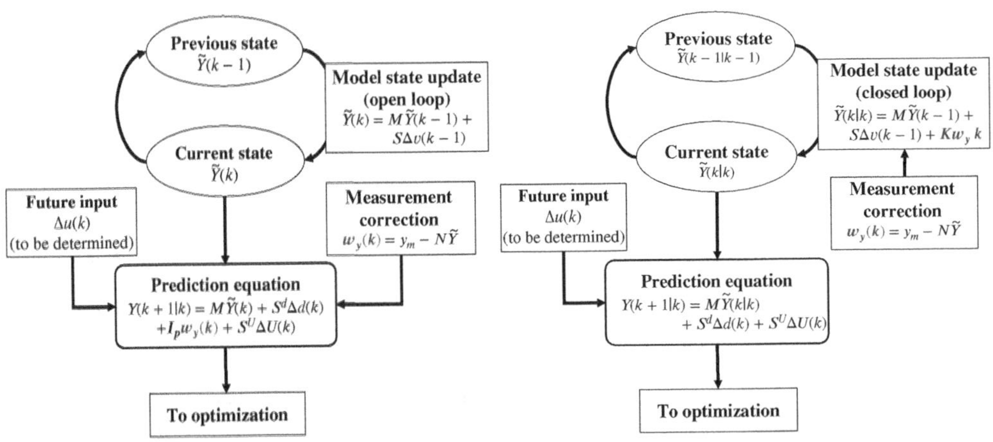

(a) (b)

Figure 4.3 (a) Open-loop vs. (b) closed-loop update of the model and corresponding prediction equations.

Model Prediction:

$$\tilde{Y}(k|k-1) = M \cdot \tilde{Y}(k-1|k-1) + S\Delta v(k-1). \tag{4.4}$$

Here, $\tilde{Y}(\cdot|k-1)$ denotes the estimate of the output trajectory $\tilde{Y}(\cdot)$ obtained at $k-1$, incorporating all measurement information available up to that point. In the conventional DMC formulation, this was the sole update applied to the model state, which is propagated using known control inputs alone.

We now introduce a *correction* step to adjust the model prediction $\tilde{Y}(k|k-1)$ using the discrepancy between the actual measurement $y_m(k)$ at time k and the predicted first-step output $\tilde{y}_0(k|k-1)$, which corresponds to the first element of the predicted output trajectory $\tilde{Y}(k|k-1)$.

Feedback Correction:

$$Y(k|k) = Y(k|k-1) + \mathbf{K}(y_m(k) - \tilde{y}_0(k|k-1)). \tag{4.5}$$

The matrix $\mathbf{K}$ is known as the observer gain. To recover the effect of the conventional DMC correction—where a constant bias equal to the current output error $(y_m(k) - \tilde{y}_0(k|k-1))$ is added to all future predictions—we choose:

$$\mathbf{K} = \mathbf{I} \quad = \left.\begin{bmatrix} I \\ I \\ \vdots \\ I \end{bmatrix}\right\} n. \tag{4.6}$$

This choice distributes the same correction uniformly across the entire prediction horizon, mimicking the constant bias correction strategy traditionally used in DMC. Note that, element by element, the correction equation becomes:

$$\begin{aligned}
\tilde{y}_0(k|k) &= \tilde{y}_0(k|k-1) + [y_m(k) - \tilde{y}_0(k|k-1)] \\
\tilde{y}_1(k|k) &= \tilde{y}_1(k|k-1) + [y_m(k) - \tilde{y}_0(k|k-1)] \\
&\vdots \quad \vdots \quad \vdots \\
\tilde{y}_{n-1}(k|k) &= \tilde{y}_{n-1}(k|k-1) + [y_m(k) - \tilde{y}_0(k|k-1)].
\end{aligned}$$

where the same correction term, $[y_m(k) - \tilde{y}_0(k|k-1)]$, is added uniformly to all future predicted outputs $\tilde{y}_1(k|k-1), \ldots, \tilde{y}_{n-1}(k|k-1)$. This is equivalent to applying a constant bias correction based on the current output error, which is assumed to persist over the prediction horizon.

Substituting Eqs. (4.4) and (4.5), we obtain the *state estimator*:

$$\begin{aligned}
\tilde{Y}(k|k) = M \cdot \tilde{Y}(k-1|k-1) + S\Delta v(k-1) + \\
\mathbf{I}\,[y_m(k) - N\,(M \cdot \tilde{Y}(k-1|k-1) + S\Delta v(k-1))],
\end{aligned} \tag{4.7}$$

where $N = \underbrace{\begin{bmatrix} I & 0 & 0 & \ldots & 0 \end{bmatrix}}_{n}.$

This expression allows us to compute the current state estimate $\tilde{Y}(k|k)$ based on the previous estimate $\tilde{Y}(k-1|k-1)$, the previous input move $\Delta v(k-1)$ and the current measurement $y_m(k)$. The matrix I is the *observer gain*.

One of the main advantages of the closed-loop formulation is that it opens the door to using well-established estimation theory to design K optimally, especially when statistical information about disturbances and measurement noise is available. Another important benefit is its ability to handle systems with unstable dynamics. Running an unstable model in open-loop mode would typically result in numerical overflow or divergence, but a properly designed closed-loop estimator can maintain numerical stability.

In a later section of this chapter, we will provide a simplified discussion of noise filtering. A more rigorous and general treatment of observer design, including the derivation and properties of the optimal K, will be presented in Chapter 7 in Part II of the book.

4.3.2 Integrating Dynamics

Integrating dynamics are common in chemical processes. In such systems, the step response does not settle to a finite steady-state value. As a result, the standard DMC formulation—based on a finite step response model—cannot be applied directly.

For systems with integrating dynamics, an equivalent assumption to finite settling is that, after n time steps, all stable dynamics have decayed, and the remaining response consists of a pure ramp, as illustrated in Figure 2.2b. The step response of such a system can be represented as:

$$\{S_1, S_2, \ldots, \ldots, S_n, S_n + (S_n - S_{n-1}), S_n + 2(S_n - S_{n-1}), \ldots, \ldots\}. \tag{4.8}$$

We define the system state, as before, using:

$$\tilde{Y}(k) = \left[\tilde{y}_0(k)^T,\ \tilde{y}_1(k)^T,\ \ldots, \tilde{y}_{n-1}(k)^T\right]^T. \tag{4.9}$$

The state evolves over time according to:

$$\tilde{Y}(k+1) = M^I \tilde{Y}(k) + S\Delta v(k), \tag{4.10}$$

where $S = \begin{bmatrix} S_1 & \cdots & S_n \end{bmatrix}^T$, and the modified shift matrix M^I is:

$$M^I = \left.\begin{bmatrix} 0 & 1 & 0 & \ldots & \ldots & 0 & 0 \\ 0 & 0 & I & 0 & \ldots & 0 & 0 \\ \vdots & & & & & & \vdots \\ 0 & 0 & \ldots & \ldots & \ldots & 0 & I \\ 0 & 0 & \ldots & \ldots & \ldots & -I & 2I \end{bmatrix}\right\} n. \tag{4.11}$$

The assumption of a pure ramp leads to the update equation for the last element:

$$\tilde{y}_{n-1}(k+1) = \tilde{y}_{n-1}(k) + (\tilde{y}_{n-1}(k) - \tilde{y}_{n-2}(k)) + S_n\Delta v(k). \tag{4.12}$$

For integrating systems, the model-prediction error is better modeled as a ramp, not as a constant bias. The future output prediction can thus be expressed as:

$$\underbrace{\begin{bmatrix} y(k+1|k) \\ y(k+2|k) \\ \vdots \\ y(k+p|k) \end{bmatrix}}_{\mathcal{Y}(k+1|k)} = \underbrace{\begin{bmatrix} \tilde{y}_1(k) \\ \tilde{y}_2(k) \\ \vdots \\ \tilde{y}_p(k) \end{bmatrix}}_{\mathcal{M}^I \tilde{Y}(k),\ \text{from memory}} + \underbrace{\begin{bmatrix} S_1^{d_m} \\ S_2^{d_m} \\ \vdots \\ S_p^{d_m} \end{bmatrix} \Delta d_m(k)}_{\substack{\mathcal{S}^{d_m}\Delta d_m(k) \\ \text{feedforward term}}} + \underbrace{\begin{bmatrix} w_y(k|k) + (w_y(k|k) - w_y(k-1|k-1)) \\ w_y(k|k) + 2\,(w_y(k|k) - w_y(k-1|k-1)) \\ \vdots \\ w_y(k|k) + p\,(w_y(k|k) - w_y(k-1|k-1)) \end{bmatrix}}_{\text{feedback term}} + \underbrace{\begin{bmatrix} S_1^u & 0 & \cdots & \cdots & 0 \\ S_2^u & S_1^u & 0 & \cdots & 0 \\ \vdots & \vdots & \ddots & \ddots & \vdots \\ S_m^u & S_{m-1}^u & \cdots & \cdots & S_1^u \\ \vdots & \ddots & \ddots & \ddots & \vdots \\ S_p^u & S_{p-1}^u & \cdots & \cdots & S_{p-m+1}^u \end{bmatrix}}_{\mathcal{S}^u,\ \text{dynamic matrix}} \underbrace{\begin{bmatrix} \Delta u(k|k) \\ \Delta u(k+1|k) \\ \vdots \\ \Delta u(k+m-1|k) \end{bmatrix}}_{\substack{\Delta\mathcal{U}(k) \\ \text{future input moves}}} \tag{4.13}$$

where $w_y(k|k) = y_m(k) - \tilde{y}_0(k)$ is the model prediction error. The matrix $\mathcal{M}^I$ is extended to project ramp behavior for $p > n$:

$$\mathcal{M}^I = \left.\begin{bmatrix} 0 & I & 0 & \dots & \dots & 0 & 0 \\ 0 & 0 & 1 & 0 & \dots & 0 & 0 \\ \vdots & & & & & & \vdots \\ 0 & 0 & \dots & \dots & \dots & 0 & I \\ 0 & 0 & \dots & \dots & \dots & -I & 2I \\ 0 & 0 & \dots & \dots & \dots & -2I & 3I \\ \vdots & \vdots & \vdots & \vdots & \vdots & \vdots & \vdots \end{bmatrix}\right\} p. \tag{4.14}$$

However, the above open-loop formulation presents a practical challenge: the model-predicted outputs (e.g., terms like $\tilde{y}_i(k)$) will grow unbounded over time due to the presence of integrators. This may eventually lead to numerical overflow in the control system, not because the physical process is unstable, but because the model states are not corrected to account for unmeasured disturbances or numerical inaccuracies.

To prevent this issue, it is essential to apply a closed-loop update to the model state using measurement feedback, as previously discussed for stable systems. For integrating systems, we propose a similar closed-loop update using the same structure as in Eqs. (4.4) and (4.5).

Model Prediction:

$$\tilde{Y}(k|k-1) = M^I \cdot \tilde{Y}(k-1|k-1) + S\Delta v(k-1). \tag{4.15}$$

Feedback Correction:

$$\tilde{Y}(k|k) = \tilde{Y}(k|k-1) + (\mathbf{I} + \mathbf{I}')(y_m(k) - \tilde{y}_0(k|k-1)), \tag{4.16}$$

where

$$\mathrm{I}' = \begin{bmatrix} 0 \\ I \\ 2I \\ \vdots \\ (n-1)I \end{bmatrix} \tag{4.17}$$

and hence

$$\mathbf{I} + \mathbf{I}' = \begin{bmatrix} I \\ 2I \\ 3I \\ \vdots \\ nI \end{bmatrix}. \tag{4.18}$$

The matrix $\mathbf{I} + \mathbf{I}'$ serves as the observer gain for integrating systems. The motivation for this form can be understood by considering the special scenario, where an unknown step disturbance enters the integrator. In such cases, the prediction error behaves like a ramp. Using the correction law in Eqs. (4.15) and (4.16), the predicted future outputs $\tilde{y}_\ell(k|k)$ for $\ell \geq k$ will match the actual process outputs $y_m(k+\ell)$, effectively canceling the disturbance's impact on prediction.

In general, when the system contains both stable and integrating outputs, the observer gain must be selected accordingly: use Eq. (4.6) for stable outputs and Eq. (4.18) for integrating ones.

4.3.3 Noise Filter

Let us revisit the correction steps for both stable and integrating systems:

- For stable systems, the correction takes the form:

$$\tilde{Y}(k|k) = \tilde{Y}(k|k-1) + \mathbf{I}(y_m(k) - \tilde{y}_0(k|k-1)). \tag{4.19}$$

- For integrating systems, it is given by:

$$\tilde{Y}(k|k) = \tilde{Y}(k|k-1) + (\mathbf{I} + \mathbf{I}')(y_m(k) - \tilde{y}_0(k|k-1)). \tag{4.20}$$

In both cases, the model prediction is corrected by the difference between the actual measurement $y_m(k)$ and the model's predicted output $\tilde{y}_0(k|k-1)$. This correction is applied either as a constant bias (for stable systems) or as a ramp bias (for integrating systems) to the future output trajectory $\tilde{Y}(k|k-1)$. This correction is appropriate when the prediction error results primarily from persistent disturbances or model mismatch. However, if the error is due to measurement noise, applying the full correction may be unwarranted. In most practical scenarios, the prediction error is a combination of disturbances, model inaccuracies, and measurement noise. Therefore, a more conservative correction is often desirable.

To account for this, the correction terms can be filtered. For stable systems, this yields:

$$\tilde{Y}(k|k) = \tilde{Y}(k|k-1) + \mathbf{I}F[y_m(k) - \tilde{y}_0(k|k-1)], \tag{4.21}$$

where F is a diagonal matrix defined as:

$$F = \text{diag}\left\{f_1, f_2, \ldots, f_{n_y}\right\}, \quad 0 < f_\ell \leq 1. \tag{4.22}$$

Rather than applying the full prediction error, this filtered correction reflects only a fraction f_ℓ of the error for each output y_ℓ. The value f_ℓ should be chosen based on the noise characteristics of the corresponding output: the higher the expected measurement noise, the smaller the corresponding f_ℓ should be.

To better understand the effect of the noise filter on control performance, consider a scenario in which the output suddenly jumps to a constant value due to an external disturbance (i.e., an "output disturbance"), so that $y_m(0) = \bar{y}$, and assume that neither the disturbance nor the manipulated variables change thereafter—i.e., $\Delta d(k) = 0$, $\Delta u(k) = 0$ for all k. From the model update step (Eq. 4.4), the prediction evolves as:

$$\tilde{Y}(k|k-1) = M\tilde{Y}(k-1|k-1) \tag{4.23}$$

Applying the filtered correction from Eq. (4.21), we get:

$$\begin{aligned} \tilde{Y}(k|k) &= \tilde{Y}(k|k-1) + \mathbf{I}F(y_m(k) - y(k|k-1)) \\ &= M\tilde{Y}(k-1|k-1) + \mathbf{I}Fy_m(k) - \mathbf{I}FNM\tilde{Y}(k-1|k-1) \\ &= (I - \mathbf{I}FN)M\tilde{Y}(k-1|k-1) + \mathbf{I}Fy_m(k). \end{aligned} \tag{4.24}$$

Here, N selects the first predicted output from the state:

$$N = \Big[\underbrace{I \;\; 0 \;\; \cdots \;\; 0}_{n}\Big]. \tag{4.25}$$

This expression shows that the correction acts like a first-order low-pass filter on the measurement residual $y_m(k) - N\tilde{Y}(k)$, similar to the filtering strategy used in IMC. The convergence of the predicted output to the true measurement follows an exponential trajectory with an effective time constant given by:

$$\text{Stable system time constant: } -\frac{T}{\ln(1 - f_\ell)}, \tag{4.26}$$

where T is the sampling period and f_ℓ is the filter gain for output y_ℓ.

In principle, one might consider applying a similar filtering strategy to integrating systems by scaling the observer gain $\mathrm{I} + \mathrm{I}'$ from Eq. (4.18) with the filter matrix F. However, this often leads to highly oscillatory behavior and is generally undesirable. According to optimal filtering theory (introduced later in Part II), a more effective filtering approach for integrating systems is to structure the correction term as follows:

$$\tilde{Y}(k|k) = \tilde{Y}(k|k-1) + (\mathbf{I} \cdot F + \mathbf{I}' \cdot F')(y_m(k) - \tilde{y}_0(k|k-1)). \tag{4.27}$$

Here, F is the same diagonal matrix as before (Eq. 4.22), and F' is defined as:

$$F' = \text{diag}\left\{f_1', f_2', \ldots, f_{n_y}'\right\} \tag{4.28}$$

with each element given by:

$$f_i' = \frac{f_i^2}{2 - f_i}. \tag{4.29}$$

This formulation ensures that both stable and integrating systems use the same underlying tuning parameter $f_\ell \in (0, 1)$ for each output. The amount of filtering increases as f_ℓ decreases. When $f_\ell = 1$, there is no filtering, and the original unfiltered correction rules (Eqs. (4.21) and 4.20) are recovered.

In practice, most systems will contain a mix of stable and integrating outputs. For each output, the appropriate form of filtering—either Eq. (4.19) for stable outputs or Eq. (4.27) for integrating ones—should be applied. This leads to a unified correction expression using a generalized filter gain matrix K_F:

$$\tilde{Y}(k|k) = \tilde{Y}(k|k-1) + K_F(y_m(k) - \tilde{y}_0(k|k-1)). \tag{4.30}$$

The matrix K_F incorporates the appropriate gain structure for each output, based on its dynamic characteristics and noise sensitivity.

4.3.4 Bi-level Optimization

The MPC calculation can be decomposed into two sequential stages to provide added flexibility and functionality. The first stage involves a local *steady-state optimization*, which determines target values for the MVs and CVs. The second stage performs a *dynamic optimization*, which computes the optimal control moves to drive the system toward these targets over the prediction horizon.

While MPC may include a steady-state optimizer with an economic objective, it does not replace the higher-level, slower plant-wide nonlinear optimizer. The upper layer sets optimal operating targets/envelopes for the prevailing conditions, while the MPC's local steady-state optimization—run more frequently over a tighter operating region—adjusts those targets to accommodate disturbances and control errors. This local optimization is typically based on a linear steady-state model, which may be obtained by linearizing a nonlinear model or by using the steady-state version of the step response model employed in the dynamic optimization.

There are several motivations for incorporating steady-state optimization into MPC. For example, it may be used to perform economic optimization at a higher frequency, allowing the controller to respond more quickly to local disturbances and changes in operating conditions. Even in the absence of an explicit economic objective, steady-state optimization can help identify the best feasible targets for the CVs along with the corresponding MV settling values, ensuring that the dynamic optimizer operates within achievable and desirable bounds.

The two-stage optimization can be formulated as below:

- Step 1: *Steady-state Optimization*
  The steady-state behavior of the system can be described using a linear prediction model of the form:

$$y(\infty|k) = K_s\underbrace{(u(\infty|k) - u(k-1))}_{\Delta u_s(k)} + b(k), \tag{4.31}$$

where $y(\infty|k)$ and $u(\infty|k)$ represent the predicted steady-state output and input at time k and $\Delta u_s(k)$ is the corresponding cumulative input change.

When only m future input moves are considered:

$$\Delta u_s(k) = \Delta u(k) + \Delta u(k+1) + \dots\dots + \Delta u(k+m-1). \tag{4.32}$$

In the context of step response models:

$$y(\infty|k) = y(k+m+n-1|k) \tag{4.33}$$

and the gain matrix is simply $K_s = S_n$. The bias term $b(k)$ accounts for model memory and disturbances:

$$b(k) = \tilde{y}_{n-1}(k) + S_n^d \Delta d_m(k) + (y_m(k) - \tilde{y}_0(k)). \tag{4.34}$$

This steady-state model can be used to solve an economic optimization problem of the form:

$$\min_{\Delta u_s(k)} \ell(u(\infty|k), y(\infty|k)) \tag{4.35}$$

subject to steady-state input and output constraints. If $\ell(\cdot)$ is linear and the model is linear, this results in a linear programming (LP) problem.

Alternatively, other formulations are also possible, such as:

- **Minimum input movement formulation:**

$$\min_{\Delta u_s(k)} \|\Delta u_s(k)\| \tag{4.36}$$

- **Target tracking formulation:**

$$\min_{\Delta u_s(k)} \|r - y(\infty|k))\|_Q. \tag{4.37}$$

In the first case, the goal is to identify the smallest input movement that satisfies the constraints. In the second, the goal is to bring the output as close as possible to the setpoint within feasible operating limits. The resulting solution defines the target steady-state values $u^*(\infty|k)$ and $y^*(\infty|k)$ to be used in the dynamic stage.

- Step 2: *Dynamic Optimization*
The dynamic optimization is performed using the same prediction model as standard MPC. A quadratic cost function is minimized over the prediction horizon to determine the optimal trajectory to the steady-state targets:

$$\left[\sum_{i=1}^{m+n-2} (y(k+i|k) - y^*(\infty|k))^T Q (y(k+i|k) - y^*(\infty|k)) + \sum_{j=0}^{m-1} \Delta u^T(k+j|k) R \Delta u(k+j|k)\right] \tag{4.38}$$

subject to dynamic model constraints and input/output constraints.

To ensure consistency between the steady-state and dynamic stages, an additional constraint may be imposed:

$$\Delta u(k|k) + \Delta u(k+1|k) + \dots\dots + \Delta u(k+m-1|k) = \Delta u_s^*(k). \tag{4.39}$$

This enforces that the final predicted outputs at the end of the horizon match the steady-state targets, i.e., $y(k+m+n-1|k) = y^*(\infty|k)$. Together with $p = m+n-2$, this ensures that all residual dynamics have decayed, making the formulation equivalent to setting $p = \infty$.

The above equations apply to the case where the steady-state optimization is performed at the same frequency as the dynamic optimization (i.e., at every sampling instant). In that case, care must be taken to filter measurement noise and other high-frequency fluctuations in the feedback signal. Without such filtering—particularly in LP-based formulations—the optimal solution can vary substantially between successive time steps, potentially causing instability or undesirable chattering in the control actions. In case longer steady-state optimization execution times are desired, need to ensure the plant measurements and model predictions align with the correct time interval.

4.3.5 Product Property Estimation

Many CVs, such as product compositions and other quality-related properties, cannot be measured with the frequency, speed, accuracy, or reliability necessary for effective direct feedback control. Despite this, control of such variables is often critical to process performance. In these cases, the typical approach is to estimate these unmeasurable variables based on other, more readily measured process variables. This technique is known as inferential estimation. Since all process variables are influenced by a common set of disturbances and manipulated inputs, they are generally correlated. These correlations can be exploited to construct an estimator that predicts the unmeasured property variable from measurements of secondary variables.

For example, in a distillation column, the goal is to control the product compositions in the distillate and bottoms. Composition analyzers (e.g., gas chromatographs (GCs)) are often slow and sampled infrequently, introducing significant delays. By contrast, tray temperatures—which correlate with composition through vapor-liquid-equilibrium (VLE) relations and column dynamics—can be measured rapidly and reliably. Consequently, temperatures at carefully chosen trays are used as surrogate controlled variables or within inferential (soft-sensor) models to estimate product compositions and close the loop.

A common approach is to use linear regression techniques—such as least squares or partial least squares (PLS)—to develop a model of the form:

$$\hat{y}^p(k) = \mathbf{L}\begin{bmatrix} y_1^s(k-\delta_1) \\ \vdots \\ y_\ell^s(k-\delta_\ell) \end{bmatrix} \tag{4.40}$$

where $\hat{y}^p$ is the estimated product property, and y_i^s are secondary variables, and δ_i are the corresponding time delays. These delays account for different dynamic responses of the secondary variables to inputs or disturbances.

When direct measurements of the product property are occasionally available, they can be used to improve the accuracy of the inferential estimate by correcting its bias. Suppose the direct measurement of the property y^p becomes available after a delay of θ_d sample steps. A corrected estimate at time k can then be computed as:

$$\hat{y}^{p*}(k) = \hat{y}^p(k) + \left(y_m^p(k-\theta_d) - \hat{y}^p(k-\theta_d)\right) \tag{4.41}$$

where y_m^p is the delayed measured value and $\hat{y}^{p*}(k)$ is the bias-corrected inferential estimate.

For processes with strong nonlinearities or wide operating ranges, linear regression may be inadequate. In such cases, nonlinear regression techniques, such as artificial neural networks (ANNs), can be used to construct more accurate inferential models.

Exercises

1. In the servo control problem of Example 3.1 in Chapter 3, a step change in the setpoint is introduced at $t = 0$. As an alternative, consider tracking a reference trajectory defined by

$$r(s) = \frac{1}{\lambda s + 1}.$$

   Modify the DMC controller to follow this reference trajectory, and evaluate its performance for different values of $\lambda \in 10, 20, 40$.

2. Using the same system as in the previous exercise, implement closed-loop model updating with measurement correction. Test different values of the noise filter parameter f, and compare the resulting closed-loop responses with those obtained using the reference trajectory approach. Discuss how the choice of f influences the closed-loop time constant, referencing the relationship provided in Eq. (4.26).

3. In Chapter 3, a DMC controller was designed for the Wood and Berry column system:

$$\begin{bmatrix} Y_D \\ Y_B \end{bmatrix} = \begin{bmatrix} \dfrac{12.8e^{-s}}{16.7s+1} & \dfrac{-18.6e^{-3s}}{21s+1} \\ \dfrac{6.6e^{-7s}}{10.9s+1} & \dfrac{-19.4e^{-3s}}{14.4s+1} \end{bmatrix} \begin{bmatrix} L \\ V \end{bmatrix}.$$

   Repeat the DMC simulation for this system, but now using the closed-loop update formulation. Instead of tuning via input weights, tune the controller by adjusting the filter parameters f_ℓ, with the input weight Γ^u set to zero. Evaluate the closed-loop response and comment on the effectiveness of this tuning method. Which tuning approach did you find more intuitive or effective: adjusting input weights or adjusting the filter parameters f_ℓ? Use the sample time of 3 minutes ($t_s = 3$) as before.

Part II

Basics of Linear Systems, Optimal Control, and System Identification

5

Linear Time Invariant System Models

The fact that the model is an approximation does not necessarily detract from its usefulness because models are approximations. All models are wrong, but some are useful.

– George E. P. Box

In Part I of this book, we developed discrete-time control using the DMC algorithm without formally introducing the connection to sampling from continuous-time systems. In this chapter, we present different model types for linear time-invariant (LTI) systems and review key concepts in discrete-time linear systems. Although our emphasis will be on discrete-time models, we begin with their continuous-time counterparts to highlight important similarities and differences.

This chapter offers a brief introduction to digital control systems and outlines various model representations. We discuss two main representations: transfer functions and state-space models. Transfer functions remain popular due to their familiarity and ease of system identification, whereas state-space models provide a compact and flexible framework, particularly well-suited for multivariable systems. The finite impulse response (FIR) models used in DMC are viewed as a special class of models, called "convolution models."

Since multivariable systems are central to MPC, we will adopt the state-space framework as the foundation for developing and analyzing advanced MPC algorithms in later chapters. Chapter 6 is devoted to a detailed treatment of state-space models. While our focus will be on state-space formulations, this choice is not restrictive—linear systems can be readily transformed between transfer function and state-space forms, as we demonstrate here.

In many practical situations, a purely deterministic description of the system is inadequate for designing high-performance controllers. For example, when feedback measurements are sparse or unavailable, it becomes necessary to account for disturbances and their correlation structures. To address this, we briefly introduce stochastic disturbance modeling and show how it can be integrated with deterministic system models.

Our presentation is intended to be concise and focused. Readers seeking more comprehensive coverage of linear systems theory and stochastic modeling are referred to standard texts, such as Kwakernaak and Sivan (1972); Åström and Wittenmark (1990).

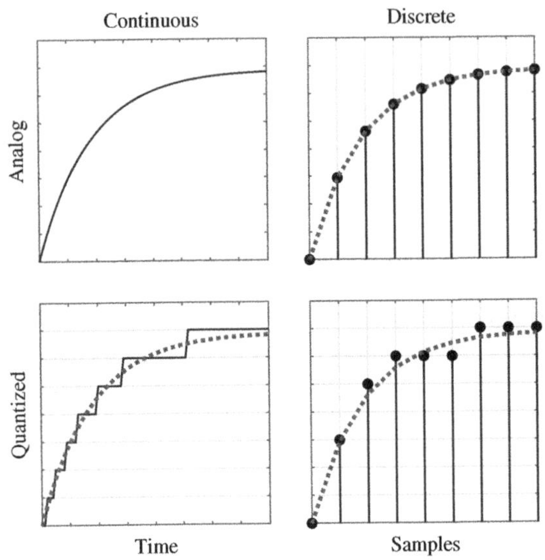

Figure 5.1 Classification of signals based on quantization and discretization.

5.1 Sampling and Reconstruction

5.1.1 Introduction to Digital Control

A schematic of a computer-controlled system was shown in Figure 2.1 in Chapter 2. The system consists of an analog-to-digital (A/D) converter, a discrete-time controller, and a digital-to-analog (D/A) converter. The process output is a *continuous-time* signal—that is, a signal defined over a continuous-time domain. In contrast, a *sampled* signal is defined only at discrete time instants, typically spaced at regular intervals called the *sampling interval.*

Figure 5.1 classifies signals according to their discretization and quantization properties: signals in the left column are continuous-time, while those in the right column are their discrete-time (sampled) counterparts. Signals may also be distinguished by whether their values are continuous or quantized. An *analog* signal can assume any real value within a specified range, whereas a *digital* signal—because of the finite precision of digital computers—can take only a finite set of distinct values within that range. Such signals are therefore said to be *quantized.*

Strictly speaking, *digital control* refers to the use of quantized, discrete-time signals for control purposes (corresponding to the bottom-right quadrant of Figure 5.1). However, given the fine resolution of modern digital instruments, we will not distinguish between discrete-time and digital signals and will use these terms interchangeably.

5.1.2 Sampling

Sampling is the process of converting an analog (continuous-time) signal into a digital (discrete-time) signal. Figure 5.2 shows the basic schematic of a sampler, and an example is illustrated in Figure 5.3. The continuous-time signal (dotted line) is sampled at uniform intervals of length t_s. Throughout this book, we assume *periodic sampling*, meaning that samples are taken at regular intervals.

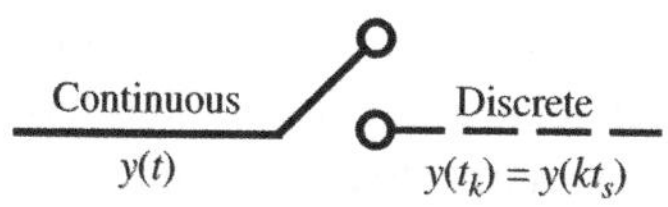

Figure 5.2 Schematic of a sampler.

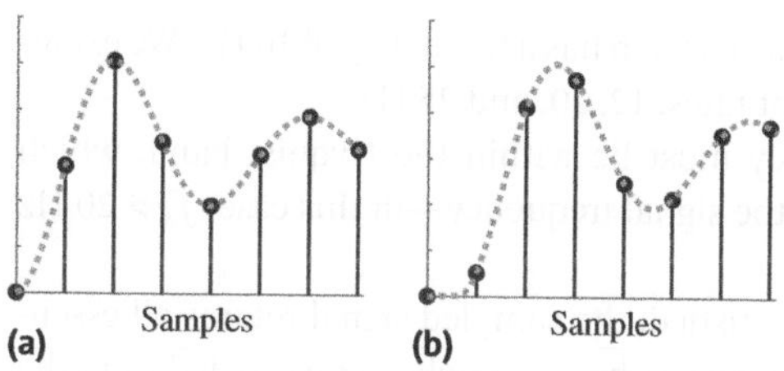

Figure 5.3 Sampling is a time-dependent operation.

The sampling frequency is defined as $f_s = 1/t_s$ (in Hz) or equivalently $\omega_s = 2\pi/t_s$ (in radians per unit time). The sampled signal $y(t_k)$ can be represented as a sequence of impulses, each with a size of $y(kt_s)$.

Figure 5.3 also highlights an important characteristic of sampling—it is a time-varying operation. If the continuous-time signal is delayed by θ, the corresponding sampled signal is not simply delayed by θ/t_s time steps unless θ is an exact multiple of t_s. This illustrates the sensitivity of discrete-time representations to the alignment of the sampling clock.

Finally, sampling inherently leads to a loss of high-frequency information. The question of how to choose an appropriate sampling rate to avoid aliasing and retain essential dynamics is addressed in the following section.

5.1.3 Aliasing

Shannon's Sampling Theorem states that, to reconstruct a sinusoidal signal exactly from its samples, the sampling rate must exceed twice the highest frequency component present in the signal. The frequency range $(-\omega_s/2, \omega_s/2)$, within which exact reconstruction is theoretically possible, is called the *Nyquist band*. In practical terms, it is impossible to recover any frequency component greater than π/t_s radians/time from a signal sampled at interval t_s.

When sampling a signal that contains high-frequency components beyond the Nyquist band, these components can become indistinguishable from lower-frequency components—a phenomenon known as the *aliasing effect*. That is, high-frequency content "folds back" into the Nyquist band and distorts the low-frequency information.

For instance, for $\omega_1 \in (-\omega_s/2, \omega_s/2)$, we have the identity:

$$\sin(\omega_1 kt_s) = \sin\left(\left(\omega_1 + \frac{2\pi}{t_s}m\right) kt_s\right),$$

which implies that frequency components at $\omega_1 + m\omega_s$ (for $m = \pm1, \pm2, \ldots$) are aliased to ω_1.

Example 5.1 illustrates the effects of sampling a sinusoidal signal at an insufficient rate, leading to visible distortion. Example 5.2 further demonstrates how high-frequency components not only get lost but also actively contaminate the reconstructed signal within the Nyquist band. Since noise typically manifests as high-frequency content, this aliasing can significantly degrade signal quality.

To prevent aliasing, it is necessary to filter out frequency components above the Nyquist limit *before* sampling. This is accomplished using an *anti-aliasing filter*. An ideal anti-aliasing filter has a gain (i.e., amplitude ratio) of one within the Nyquist band and drops sharply to zero outside it.

Example 5.1: *Sampling and Nyquist Band.*
Consider a sinusoidal signal defined by $y(t) = \sin(20\pi t)$, which has a frequency of 10 Hz. We examine the effect of sampling this signal at three different rates: 12, 20, and 25 Hz.

To avoid loss of information, the signal frequency must lie within the Nyquist band, which requires the sampling frequency f_s to exceed twice the signal frequency—in this case, $f_s > 20$ Hz (i.e., $\omega_s > 40\pi$ radians/s).

When sampled at 25 Hz, the Nyquist criterion is satisfied, the sampled signal retains all essential information, and the original continuous-time signal can be accurately reconstructed from the samples, with no aliasing effects.

Figure 5.4 illustrates the aliasing phenomenon for cases where the sampling rate is insufficient. The solid line represents the original continuous-time signal, and the dots indicate the sampled data. To help visualize the aliasing, the samples are connected with dotted lines.

At the critical sampling frequency $\omega_s = 40\pi$ (i.e., $f_s = 20$ Hz), the sinusoid appears as a constant in the sampled data. Thus, sampling exactly at the Nyquist rate leads to information loss. Therefore, the sampling frequency should be strictly greater than twice the highest frequency component.

At the lower sampling frequency of 12 Hz ($\omega_s = 24\pi$), the signal frequency lies outside the Nyquist band, and *frequency folding* occurs. Due to aliasing, the sampled signal appears to have a much lower frequency. This happens because it becomes indistinguishable from a sinusoid at $\omega = \omega_1 + n\omega_s$, where n is an integer. In this example, the alias appears at $\omega = \mp 4\pi$.

Example 5.2: *Aliasing/Frequency Folding.*
Consider a signal composed of two sinusoids:

$$y(t) = \sin(6\pi t) + 0.2\sin(20\pi t),$$

sampled at a rate of 12 samples/time. This corresponds to a sampling frequency of $\omega_s = 24\pi$ radians/time.

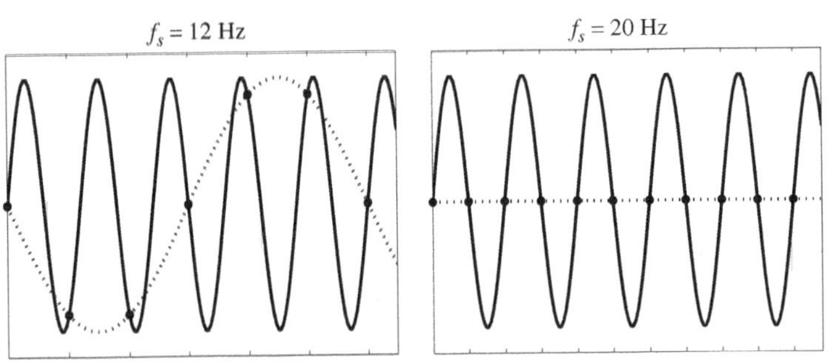

Figure 5.4 Aliasing effect due to low sampling frequency. The original signal (solid line) is a sinusoid with frequency 10 Hz, sampled at 12 and 20 samples/time. Dotted lines visually connect sampled points to show the aliasing behavior.

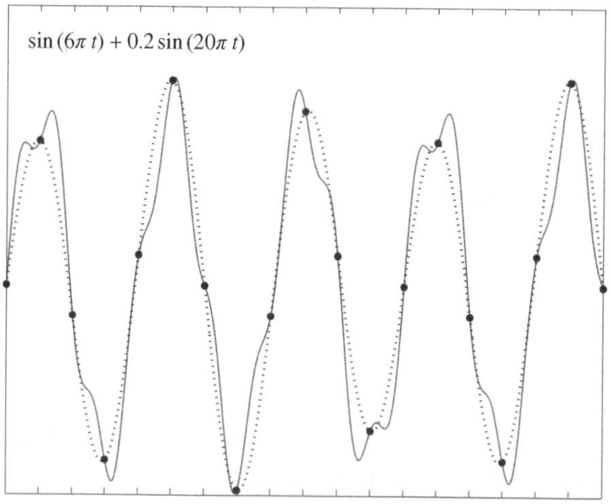

Figure 5.5 Aliasing effect of a signal with a frequency component beyond the Nyquist band. The true signal is shown as a solid line. The alias of the high-frequency sinusoid corrupts the sampled signal.

The first sinusoid has a frequency of 3 Hz ($\omega = 6\pi$), which lies well within the Nyquist band and is therefore sampled correctly. However, the second sinusoid has a frequency of 10 Hz ($\omega = 20\pi$), which exceeds the Nyquist limit ($\omega_s/2 = 12\pi$).

As a result, the higher-frequency component undergoes aliasing: its frequency is effectively "folded" into the Nyquist band, producing an alias at $\omega = 4\pi$. This phenomenon is illustrated in Figure 5.5. Although the original high-frequency component is not directly visible, its alias appears in the sampled signal, introducing distortion. In this way, the alias falls within the frequency range of interest and corrupts the lower-frequency content, rather than simply being lost.

5.1.4 Reconstruction

The process of retrieving a continuous-time signal from its discrete-time samples is known as *reconstruction*. According to Shannon's sampling theorem, the highest frequency that can be accurately reconstructed from sampled data is $\omega_s/2$, where ω_s is the sampling frequency. This defines the limit of recoverable information—the Nyquist frequency. Common reconstruction methods are discussed below.

5.1.4.1 Shannon's Reconstruction

$$y(t) = \sum_{k=-\infty}^{\infty} y(t_k) \frac{\sin\left(\omega_s(t - kt_s)/2\right)}{\left(\omega_s(t - kt_s)/2\right)}. \tag{5.1}$$

Equation (5.1) provides an exact formula for reconstructing a continuous-time signal from its discrete-time samples, under the assumption that the original signal contains no frequency components beyond the Nyquist band.

This method, known as Shannon's reconstruction, is theoretically ideal but *acausal*, as it requires knowledge of future samples to reconstruct the signal at a given time. As a result, it is not practical for real-time digital control applications, where signal reconstruction must be performed causally as data arrives.

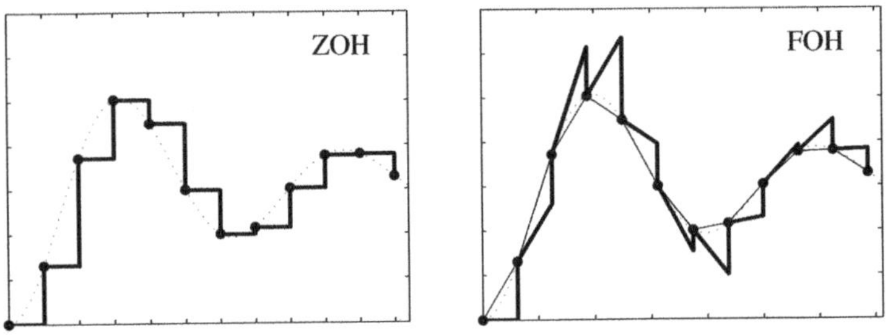

Figure 5.6 Zero and first-order hold.

5.1.4.2 Zero and Higher Order Hold

A widely used causal reconstruction method is the *zero-order hold* (ZOH), where the signal is held constant over the entire sampling interval $[kt_s, (k+1)t_s)$ at the value $y(t_k)$:

$$y(t) = y(t_k) \quad kt_s \leq t < (k+1)t_s. \tag{5.2}$$

An alternative is the *first-order hold* (FOH), which linearly extrapolates from the current and previous sample values:

$$y(t) = y(t_k) + \frac{t - kt_s}{t_s}(y(t_k) - y(t_{k-1})) \quad kt_s \leq t < (k+1)t_s. \tag{5.3}$$

Examples of signal reconstruction using ZOH and FOH are shown in Figure 5.6, where the dotted line denotes the original continuous-time signal and the dots indicate its sampled values.

Although higher-order holds can be defined in principle, they are seldom used in practice for constructing control inputs. For most control applications, including DMC, ZOH provides sufficiently accurate and effective signal reconstruction.

5.2 Introduction to *z*-transform

In the design and analysis of discrete-time systems, the *z*-transform plays a role analogous to the Laplace transform for continuous-time systems. The *z*-transform of a sampled signal $y(t_k)$ is defined as

$$Y(z) = \mathcal{Z}(y(t_k)) = \sum_{k=0}^{\infty} y(kt_s)z^{-k}, \tag{5.4}$$

The discrete signal $y(t_k)$ consists of the sequence:

$$y(0),\ y(t_s),\ y(2t_s),\ \ldots,\ y(kt_s),\ \ldots,$$

and can be represented as a train of impulses:

$$y(t) = \sum_{k=0}^{\infty} y(t_k)\delta(t - kt_s) \quad \xrightarrow{\mathcal{L}} \quad \hat{Y}(s) = \sum_{k=0}^{\infty} y(t_k)e^{-kt_s s}. \tag{5.5}$$

Comparing (5.4) and (5.5), we see that $z^{-1} \stackrel{\Delta}{=} e^{-sh}$. Therefore, z^{-1} corresponds to a delay of one sampling interval.

Equation (5.4) can be expanded as

$$Y(z) = y(0) + z^{-1}y(t_s) + z^{-2}y(2t_s) + \cdots + z^{-k}y(kt_s) + \cdots \tag{5.6}$$

This expansion is useful for deriving the *z*-transforms of standard signals. For example, consider a unit step input, where $y(t_k) = 1$ for all $k \geq 0$. Substituting, the *z*-transform of a unit step is

$$Y(z) = 1 + z^{-1} + z^{-2} + \cdots = \frac{1}{1 - z^{-1}}.$$

Forward Shift Operator: We introduce the forward shift operator, q, which is a discrete-time analog of the differential operator d/dt in continuous-time systems. It is defined in the discrete-time domain as

$$q\{y(t_k)\} = y(t_{k+1}).$$

Taking the *z*-transform, we get:

$$\mathcal{Z}\{q\{y(t_k)\}\} = zY(z).$$

The treatment of the z-transform in this book is intentionally concise and focused, sufficient to support the topics in the subsequent chapters. Readers interested in a deeper understanding are encouraged to consult specialized textbooks on discrete-time systems, such as Åström and Wittenmark (1990).

5.3 Transfer Function Models

Transfer functions provide a convenient and familiar means of representing system dynamics by establishing an algebraic relationship between the Laplace or z-transforms of the inputs and outputs. While transfer functions describe only the output response to changes in the input, this limitation is generally not problematic—most controller designs are primarily concerned with the input–output behavior.

5.3.1 Continuous Time

A continuous-time transfer function is a widely used model form for LTI systems. It relates the Laplace transforms of the input and output as follows

$$G_c(s) = \frac{y(s)}{u(s)} = \frac{\hat{\beta}_m s^m + \hat{\beta}_{m-1}s^{m-1} + \cdots + \hat{\beta}_1 s + \hat{\beta}_0}{s^n + \hat{\alpha}_{n-1}s^{n-1} + \cdots + \hat{\alpha}_1 s + \hat{\alpha}_0}, \qquad m < n. \tag{5.7}$$

Here, $y(s)$ and $u(s)$ denote the Laplace transforms of the output and input, respectively. This transfer function corresponds to the following time-domain differential equation:

$$\frac{d^n y}{dt^n} + \hat{\alpha}_{n-1}\frac{d^{n-1}y}{dt^{n-1}} + \cdots + \hat{\alpha}_0 y = \hat{\beta}_m \frac{d^m u}{dt^m} + \hat{\beta}_{m-1}\frac{d^{m-1}u}{dt^{m-1}} + \cdots + \hat{\beta}_0 u. \tag{5.8}$$

Taking the Laplace transform of Eq. (5.8) under the assumption of zero initial conditions yields the transfer function in Eq. (5.7). The parameter n is referred to as the *order* of the transfer function. Physical systems are generally *strictly proper*, meaning $m < n$, because real-world outputs do not react instantaneously to changes in inputs.

Time delays can be incorporated easily within the transfer function framework. For example, if there is a delay of θ in the input–output relationship, the transfer function becomes:

$$\frac{y(s)}{u(s)} = \frac{\hat{\beta}_m s^m + \hat{\beta}_{m-1} s^{m-1} + \cdots + \hat{\beta}_1 s + \hat{\beta}_0}{s^n + \hat{\alpha}_{n-1} s^{n-1} + \cdots + \hat{\alpha}_1 s + \hat{\alpha}_0} \cdot e^{-\theta s}. \tag{5.9}$$

This exponential term, $e^{-\theta s}$, captures the pure time delay in the system.

5.3.2 Discrete Time

In discrete time, system dynamics are represented using a difference equation, which serves as the discrete-time analogue of the continuous-time differential equation in Eq. (5.8):

$$y(t_k) = -\alpha_1 y(t_{k-1}) - \cdots - \alpha_n y(t_{k-n}) + \beta_1 u(t_{k-1}) + \beta_2 u(t_{k-2}) + \cdots + \beta_m u(t_{k-m}). \tag{5.10}$$

This form arises naturally by approximating derivatives with finite difference formulas. It explicitly includes delayed input and output terms to capture the memory effects inherent in dynamic systems.

The *delay operator* q^{-1}, defined by $q^{-1}y(t_k) = y(t_{k-1})$, provides a more compact representation of the same equation:

$$\left(1 + \alpha_1 q^{-1} + \cdots + \alpha_n q^{-n}\right) y(t_k) = \left(\beta_1 q^{-1} + \beta_2 q^{-2} + \cdots + \beta_m q^{-m}\right) u(t_k). \tag{5.11}$$

With slight abuse of notation, we can write this relationship as a rational function of the delay operator:

$$y(t_k) = \frac{\left(\beta_1 q^{-1} + \beta_2 q^{-2} + \cdots + \beta_n q^{-m}\right)}{\left(1 + \alpha_1 q^{-1} + \cdots + \alpha_n q^{-n}\right)} u(t_k). \tag{5.12}$$

Applying the z-transform to the difference equation yields the following transfer function:

$$\frac{Y(z)}{U(z)} = \frac{\left(\beta_1 z^{-1} + \beta_2 z^{-2} + \cdots + \beta_m z^{-m}\right)}{\left(1 + \alpha_1 z^{-1} + \cdots + \alpha_n z^{-n}\right)}. \tag{5.13}$$

This is the discrete-time analogue of the Laplace-domain transfer function and is often referred to as the pulse transfer function, as it characterizes the output response to a sequence of input pulses. The order of the system is $\max(m, n)$.

To express the transfer function in terms of positive powers of z, we multiply both numerator and denominator by $z^{\max(m,n)}$. Assuming $n > m$, we obtain:

$$G(z) \triangleq \frac{N(z)}{D(z)} = \frac{\left(\beta_1 z^{n-1} + \beta_2 z^{n-2} + \cdots + \beta_m z^{n-m}\right)}{\left(z^n + \alpha_1 z^{n-1} + \cdots + \alpha_n\right)}. \tag{5.14}$$

The *poles* of the system are the roots of the denominator polynomial $D(z) = 0$, and the *zeros* are the roots of the numerator polynomial $N(z) = 0$. Poles and zeros play a central role in determining the system's dynamic behavior, including stability and response characteristics.

While the transfer function form in Eq. (5.14) is convenient for analysis, the difference equation forms in Eqs. (5.10) or (5.11) are preferred for simulation, as they directly relate the current output to past outputs and past inputs through a recursive computation.

Comments:

- For notational convenience, the relationship $y(z) = G(z)u(z)$ will often be expressed in the time domain as $y(t_k) = G(q)u(t_k)$, or simply $y(k) = G(q)u(k)$ when it is clear from the context that the system is discrete time. This notation corresponds to the difference equation in (5.12).

- In the special case where all $\alpha_i = 0$, the system reduces to

$$y(k) = \beta_1 u(k-1) + \beta_2 u(k-2) + \cdots + \beta_m u(k-m). \tag{5.15}$$

If m is chosen to span the system's settling time, this expression defines a FIR model, as used in Part I of this book.

5.3.3 Transfer Matrix

For systems with n_u inputs and n_y outputs, the dynamic behavior is described by specifying $n_y \times n_u$ individual transfer functions. These transfer functions are typically organized into a "*transfer matrix,*" as shown below

$$G(z) = \begin{bmatrix} G_{1,1}(z) & G_{1,2}(z) & \cdots & G_{1,n_u}(z) \\ G_{2,1}(z) & \cdots & \cdots & G_{2,n_u}(z) \\ \vdots & \ddots & \ddots & \vdots \\ G_{n_y,1}(z) & \cdots & \cdots & G_{n_y,n_u}(z) \end{bmatrix}. \tag{5.16}$$

Here, $G_{i,j}(z)$ represents the transfer function from the jth input to the ith output.

5.3.4 Converting Continuous Transfer Function to Discrete Transfer Function

A discrete-time system $G(z)$ receives discrete input signals $U(z)$ and produces discrete output signals $Y(z)$. To interface with a continuous-time system $G_c(s)$, the discrete input signal is first converted into a piecewise-constant continuous signal using a ZOH. The resulting continuous output is then sampled to obtain the discrete output signal $Y(z)$.

The discrete-time transfer function corresponding to a continuous-time system with a ZOH can be derived using the following formula:

$$G(z) = \mathcal{Z}\left\{\underbrace{\mathcal{L}^{-1}\left\{G_c(s)\frac{1-e^{-t_s s}}{s}\right\}}_{y_i(t)}\right\}. \tag{5.17}$$

Here, $\mathcal{L}^{-1}$ denotes the inverse Laplace transform, and $\mathcal{Z}$ denotes the z-transform. The multiplication of $G_c(s)$ by $(1-e^{-t_s s})/s$ accounts for the ZOH behavior prior to discretization.

The intermediate function

$$y_i(t) \triangleq \mathcal{L}^{-1}\left\{G_c(s)\frac{1-e^{-t_s s}}{s}\right\}, \tag{5.18}$$

represents the system's continuous-time response to a discrete unit impulse input—i.e., a unit pulse starting at $t = 0$ and lasting for one sampling interval. This is consistent with the fact that the z-transform of a discrete unit impulse is 1. Since $Y(z) = G(z)U(z)$, the transfer function $G(z)$ is thus the z-transform of the output response to a unit-size pulse.

Conversion tables for commonly encountered transfer functions are available in many standard textbooks. However, these tables are not directly applicable when the system includes fractional delays—that is, delays that are not integer multiples of the sampling interval.

Example 5.3: *Discrete Transfer Function for an Integrator and Double Integrator.*

The Laplace transforms of the integrator and double integrator systems are given by $G_c(s) = 1/s$ and $G_c(s) = 1/s^2$, respectively. The corresponding discrete-time transfer functions can be derived using Eqs. (5.17) and (5.18).

For an integrator,

$$y_i(t) = \mathcal{L}^{-1}\left\{\frac{1 - e^{-t_s s}}{s^2}\right\} = tS(t) - (t - t_s)S(t - t_s),$$

where $S(t)$ and $S(t - t_s)$ are unit step functions starting at time 0 and t_s, respectively. Using the z-transform definition in Eq. (5.4), we compute the z-transform of $tS(t)$:

$$\begin{aligned}\sum_{k=0}^{\infty}(kt_s)z^{-k} &= t_s(z^{-1} + 2z^{-2} + \cdots + kz^{-k} + \cdots)\\ &= \frac{t_s z^{-1}}{(1 - z^{-1})^2} = \frac{t_s z}{(z - 1)^2}.\end{aligned}$$

Since $(t - t_s)S(t - t_s)$ is a one-sample delayed version of $tS(t)$, its z-transform is $z^{-1}\dfrac{t_s z}{(z-1)^2}$.

Thus, the discrete-time transfer function becomes:

$$G(z) = \frac{t_s z}{(z - 1)^2} - z^{-1}\left(\frac{t_s z}{(z - 1)^2}\right) = \frac{t_s}{z - 1}.$$

For the double integrator, we use:

$$\mathcal{L}^{-1}\left\{\frac{1}{s^3}\right\} = \frac{t^2}{2}, \text{ and } \mathcal{Z}\left\{t_s^2\right\} = \frac{t_s^2 z(1 + z)}{(z - 1)^3}.$$

Again, since $(t - t_s)^2 S(t - t_s)$ is a delayed version of $t^2 S(t)$, its z-transform is $z^{-1} \cdot \dfrac{t_s^2 z(1+z)}{2(z-1)^3}$.

Therefore, the transfer function is

$$\begin{aligned}G(z) &= \frac{t_s^2 z(1 + z)}{2(z - 1)^3} - \frac{t_s^2(1 + z)}{2(z - 1)^3}\\ &= \frac{t_s^2(1 + z)}{2(z - 1)^2}.\end{aligned}$$

5.3.5 Stability and Implications of Poles

Definition 5.1: A transfer function of the form (5.14) is said to be *stable* if the output y remains bounded in response to any bounded input u.

Consider a system with a single pole described by the difference equation:

$$y(t_k) = \alpha y(t_{k-1}) + u(t_{k-1}).$$

Assuming an arbitrary initial condition $y(t_0)$, the step response of the system is given by:

$$y(t_k) = \alpha^k y(t_0) + \left(\alpha^{k-1}u(t_0) + \alpha^{k-2}u(t_1) + \cdots + u(t_{k-1})\right).$$

Clearly, for the response to remain bounded, we must have $|\alpha| < 1$. Furthermore, negative values of α lead to oscillatory behavior. The effect of the pole location (i.e., the value of α) on the system's

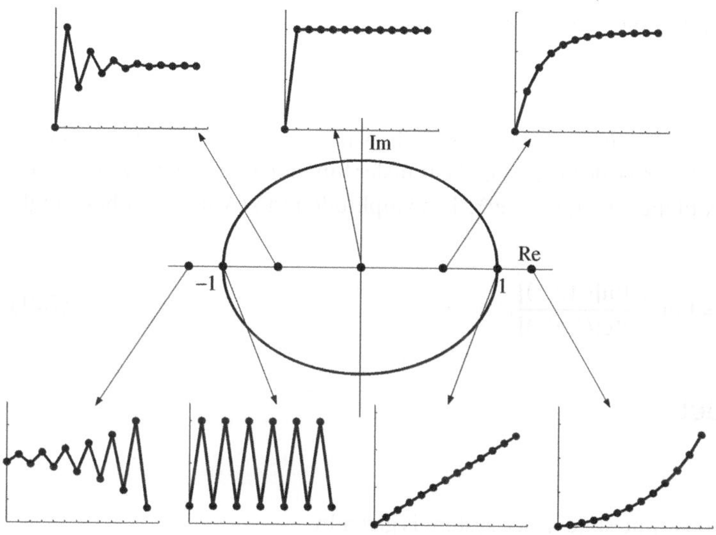

Figure 5.7 Effect of pole locations on the step response of a discrete-time system.

response is illustrated in Figure 5.7. The closer the pole is to the unit circle, the more sluggish is the transient response of the system.

Generalizing this result, the stability of a transfer function can be assessed by examining the locations of its poles. A transfer function is stable if and only if all its poles lie strictly inside the unit disk in the complex z-plane. For multivariable systems, a transfer matrix is considered stable if each of its individual transfer function elements is stable.

Beyond stability, poles and zeros of a discrete-time transfer function offer insight into the system's dynamic characteristics. These include the presence and nature of oscillations, degree of damping, and rate of decay. Recall from Section 5.2 that $z = e^{st_s}$. This relationship connects discrete-time behavior to that of continuous-time systems.

For instance, a continuous-time pole at $s = \gamma + i\omega$ maps to a discrete-time pole at $z = e^{\gamma t_s} e^{i\omega t_s}$. In continuous time, poles located in the left half of the complex plane (i.e., $\gamma < 0$) indicate stability. Since $|e^{\gamma t_s}| < 1$ when $\gamma < 0$, these poles are mapped to locations strictly within the unit disk in the discrete domain—consistent with the earlier discrete-time stability condition.

Attenuation rates in continuous-time systems depend on γ, with vertical lines of constant γ (i.e., constant attenuation) mapping to circles of constant radius in the z-plane. Similarly, lines of constant damping ratio in the s-plane become spiral curves in the z-domain, and horizontal lines of constant frequency ω (i.e., ω = constant) map to rays with angle ωt_s in the z-plane.

Finally, it is worth noting that the presence of a zero outside the unit disk leads to an *inverse response*, such as an initial undershoot in the step response.

5.3.6 Gain, Frequency Response

$$K = y(\infty)/u(\infty),$$

is defined as the *gain* of the system. For *stable* systems, the gain can be computed directly from the transfer function using:

$$K = \lim_{z \to 1} G(z). \tag{5.19}$$

This result follows from the final value theorem, which states:

$$\lim_{t \to \infty} y(t) = \lim_{z \to 1} z\, \hat{y}(z). \tag{5.20}$$

The frequency response of a discrete-time system describes how the system responds to sinusoidal inputs at different frequencies. It can be derived from the transfer function by evaluating $G(z)$ along the unit circle in the complex plane, i.e., for $z = e^{j\omega}$. The amplitude ratio (A.R.) and phase angle (ϕ) are then given by

$$\text{A.R.} = \left|G(e^{j\omega})\right| \qquad \phi = \tan^{-1} \frac{\text{Im}[G(e^{j\omega})]}{\text{Re}[G(e^{j\omega})]}. \tag{5.21}$$

5.4 State-space Model

5.4.1 Continuous Time

A standard form of a continuous-time state-space system is given by

$$\begin{aligned} \dot{x} &= A^c x + B^c u \\ y &= Cx + Du. \end{aligned} \tag{5.22}$$

Here, x denotes the state vector, u the input vector (manipulated and/or disturbance variables), and y the output vector, typically formed as linear combinations of the states (and possibly the inputs). For most physical systems, the output does not respond instantaneously to the input, so we adopt a strictly proper model and set $D = 0$ henceforth. The state vector x may comprise physical variables (from first-principles models) or auxiliary nonphysical variables introduced to retain the memory of past inputs needed for predicting future outputs.

The solution to the above differential equation with initial condition $x(0) = x_0$ is

$$x(t) = e^{A^c \cdot t} x_0 + \int_0^t e^{A^c \cdot (t-\tau)} B^c u(\tau) d\tau, \tag{5.23}$$

The matrix exponential in this expression is defined by the infinite power series:

$$e^{A_c t} = \sum_{n=0}^{\infty} \frac{(A^c t)^n}{n!}. \tag{5.24}$$

This series can be evaluated efficiently using the Cayley–Hamilton Theorem, as discussed in Appendix A.

5.4.2 Discrete Time

A discrete-time analogue of the continuous-time model in (5.22) is

$$\begin{aligned} x(t_{k+1}) &= Ax(t_k) + Bu(t_k) \\ y(t_k) &= Cx(t_k). \end{aligned} \tag{5.25}$$

When a ZOH is used, the system input $u(t)$ changes only at discrete, equally spaced sampling times $\ldots,\ t_k,\ t_{k+1},\ \ldots$. This discrete-time model may either be derived from a continuous-time system or obtained directly via system identification, which is addressed in Chapter 10.

For notational convenience, we rewrite the system as

$$\begin{aligned} x(k+1) &= Ax(k) + Bu(k) \\ y(k) &= Cx(k). \end{aligned} \tag{5.26}$$

Here, $x(k)$ denotes the state at the kth sampling instant. Allowing direct feedthrough gives $y(k) = Cx(k) + Du(k)$. In the above, with $D = 0$, the model is strictly proper, so $y(k)$ does not respond to a change in $u(k)$ within the same sample—the earliest effect appears at $k + 1$ (i.e., at least a one-sample delay).

Representing Time Delays

Time delays can be represented conveniently in the discrete-time setting. Consider the case where the input is delayed by n_θ sampling intervals (with $n_\theta \geq 1$). The system can be described as

$$\begin{aligned} x(k+1) &= Ax(k) + Bu(k - n_\theta) \\ y(k) &= Cx(k). \end{aligned} \tag{5.27}$$

This system can be transformed into the standard form of (5.26) by augmenting the state vector to include the delayed inputs:

$$\begin{aligned} \begin{bmatrix} x(k+1) \\ u(k-n_\theta+1) \\ \vdots \\ u(k-1) \\ u(k) \end{bmatrix} &= \begin{bmatrix} A & B & 0 & \cdots & 0 \\ 0 & 0 & I & \cdots & 0 \\ \vdots & \ddots & \ddots & \ddots & \vdots \\ 0 & 0 & 0 & \cdots & I \\ 0 & 0 & \cdots & \cdots & 0 \end{bmatrix} \begin{bmatrix} x(k) \\ u(k-n_\theta) \\ \vdots \\ u(k-2) \\ u(k-1) \end{bmatrix} + \begin{bmatrix} 0 \\ 0 \\ \vdots \\ 0 \\ I \end{bmatrix} u(k) \\ y(k) &= \begin{bmatrix} C & 0 & \cdots & 0 \end{bmatrix} \begin{bmatrix} x(k) \\ u(k-n_\theta) \\ \vdots \\ u(k-1) \end{bmatrix}. \end{aligned} \tag{5.28}$$

Delays of different lengths in individual input channels can be handled by extending this augmentation strategy accordingly. Delays in output channels can also be treated in a similar manner.

Example 5.4: *Consider the Discrete-time System.*

$$y(k) = 0.5y(k-1) + u(k-3). \tag{5.29}$$

This system exhibits an input delay of three sampling intervals ($3 \cdot t_s$), including the inherent one-sample delay between input u and output y.

The equivalent state-space representation is

$$\begin{aligned} \begin{bmatrix} x_1(k+1) \\ x_2(k+1) \\ x_3(k+1) \end{bmatrix} &= \begin{bmatrix} 0.5 & 1 & 0 \\ 0 & 0 & 1 \\ 0 & 0 & 0 \end{bmatrix} \begin{bmatrix} x_1(k) \\ x_2(k) \\ x_3(k) \end{bmatrix} + \begin{bmatrix} 0 \\ 0 \\ 1 \end{bmatrix} u(k) \\ y(k) &= \begin{bmatrix} 1 & 0 & 0 \end{bmatrix} x(k). \end{aligned} \tag{5.30}$$

5.4.3 Converting Continuous- to Discrete-time System

5.4.3.1 System Without Time Delay

Recall that the solution to the continuous-time state-space system (5.22) with initial condition $x(0) = x_0$ is given by

$$x(t) = e^{A^c \cdot t} x_0 + \int_0^t e^{A^c \cdot (t-\tau)} B^c u(\tau) d\tau. \tag{5.31}$$

Let $x(t_k)$ denote the state at sampling time t_k. By integrating (5.22) over the interval $[t_k, t_{k+1}]$, and assuming the input remains constant during this interval (i.e., $u(t) = u(k)$ for $t \in [t_k, t_{k+1}]$), we obtain:

$$\begin{aligned} x(t_{k+1}) &= e^{A^c \cdot (t_{k+1} - t_k)} x(t_k) + \int_{t_k}^{t_{k+1}} e^{A^c \cdot (t_{k+1} - \tau')} B^c u(k) d\tau', \\ &= e^{A^c \cdot t_s} x(t_k) + \left(\int_0^{t_s} e^{A^c \tau} B^c d\tau \right) u(k). \end{aligned} \tag{5.32}$$

Thus, the discrete-time system matrices are given by

$$\begin{aligned} x(k+1) &= Ax(k) + Bu(k) \\ y(k) &= Cx(k), \end{aligned} \tag{5.33}$$

$$A = e^{A^c \cdot t_s} \qquad B = \int_0^{t_s} e^{A^c \cdot \tau} d\tau \cdot B^c. \tag{5.34}$$

Equation (5.34) provides the standard formulas for converting continuous-time system matrices into their discrete-time counterparts under a ZOH assumption.

5.4.3.2 System with Input Delay

When the continuous-time system includes an input delay, the discretization procedure becomes more involved. Consider the continuous-time state-space model in (5.22) with an input delay of θ. First, consider $0 < \theta \leq t_s$. In this case, the delayed input $u(t - \theta)$ takes the following values over the interval $[t_k, t_{k+1})$:

$$u(t - \theta) = \begin{cases} u(k-1) & \text{for } t_k \leq t < t_k + \theta \\ u(k) & \text{for } t_k + \theta \leq t < t_{k+1}. \end{cases} \tag{5.35}$$

Integrating (5.22) from t_k to t_{k+1} with this piecewise-constant input yields:

$$x(k+1) = Ax(k) + B_1 u(k-1) + B_0 u(k), \tag{5.36}$$

$$\begin{aligned} B_1 &= e^{A^c \cdot (h-\theta)} \textstyle\int_0^{\theta} e^{A^c \cdot \tau} d\tau \cdot B^c, \\ B_0 &= \textstyle\int_0^{h-\theta} e^{A^c \cdot \tau} d\tau \cdot B^c. \end{aligned} \tag{5.37}$$

Augmenting the state vector to include $u(k-1)$:

$$\begin{bmatrix} x(k+1) \\ u(k) \end{bmatrix} = \begin{bmatrix} A & B_1 \\ 0 & 0 \end{bmatrix} \begin{bmatrix} x(k) \\ u(k-1) \end{bmatrix} + \begin{bmatrix} B_0 \\ I \end{bmatrix} u(k). \tag{5.38}$$

Here, the augmented state includes $u(k-1)$ because its influence is not fully captured by $x(k)$ due to the delay.

Generalizing to the case where $\theta = (n_\theta - 1)t_s + \theta'$, where $n_\theta = \texttt{ceil}(\theta/t_s)$ is an integer and $0 < \theta' \le t_s$. Then, the delayed input $u(t-\theta)$ is given by:

$$u(t-\theta) = \begin{cases} u(k - n_\theta) & t_k \le t < t_k + \theta' \\ u(k - n_\theta + 1) & t_k + \theta' \le t < t_{k+1}. \end{cases} \tag{5.39}$$

Integrating over $[t_k, t_{k+1})$ gives:

$$x(k+1) = Ax(k) + B_1 u(k - n_\theta) + B_0 u(k - n_\theta + 1), \tag{5.40}$$

where B_1 and B_0 are as in (5.37), with θ replaced by θ'.

The resulting standard discrete-time state-space model can be written by augmenting the state with delayed inputs:

$$\begin{bmatrix} x(k+1) \\ u(k-n_\theta+1) \\ u(k-n_\theta) \\ \vdots \\ u(k) \end{bmatrix} = \begin{bmatrix} A & B_1 & B_0 & 0 & \cdots & 0 \\ 0 & 0 & I & 0 & \cdots & 0 \\ \vdots & \ddots & \ddots & \ddots & \ddots & \vdots \\ \vdots & \ddots & \ddots & \ddots & \ddots & I \\ 0 & \cdots & \cdots & \cdots & \cdots & 0 \end{bmatrix} \begin{bmatrix} x(k) \\ u(k-n_\theta) \\ u(k-n_\theta+1) \\ \vdots \\ u(k-1) \end{bmatrix} + \begin{bmatrix} 0 \\ 0 \\ \vdots \\ 0 \\ I \end{bmatrix} u(k). \tag{5.41}$$

For systems with varying delays across different input and output channels, the discretization process can become quite complex. A practical approach is to discretize each input–output pair separately and then combine them into a single model. This typically results in redundant or non-minimal state representations, since some states can be merged across channels.

A minimal realization can be obtained from a non-minimal model by applying model reduction (Appendix C) to eliminate redundant or negligible states from the aggregated model.

Example 5.5: *Discretization of Double Integrator.*

Consider the double integrator system $\frac{y(s)}{u(s)} = \frac{1}{s^2}$. The corresponding time domain equation is

$$\frac{d^2x}{dt^2} = u.$$

This second-order differential equation can be written in state-space form as

$$\begin{bmatrix} \frac{dx_1}{dt} \\ \frac{dx_2}{dt} \end{bmatrix} = \begin{bmatrix} 0 & 1 \\ 0 & 0 \end{bmatrix} \begin{bmatrix} x_1 \\ x_2 \end{bmatrix} + \begin{bmatrix} 0 \\ 1 \end{bmatrix} u$$

$$y = \begin{bmatrix} 1 & 0 \end{bmatrix} \begin{bmatrix} x_1 \\ x_2 \end{bmatrix}.$$

We now discretize the continuous-time state-space model using the formulas in (5.34) to obtain a discrete-time system of the form (5.26).

The matrix exponential in the state transition matrix is computed as

$$A = e^{A^c t_s} = I + \begin{bmatrix} 0 & 1 \\ 0 & 0 \end{bmatrix} t_s = \begin{bmatrix} 1 & t_s \\ 0 & 1 \end{bmatrix},$$

since $(A^c)^2 = 0$ for this particular A^c. The input matrix is obtained as

$$B = \int_0^{t_s} e^{A^c s} ds \cdot B^c = \begin{bmatrix} t_s & t_s^2/2 \\ 0 & t_s \end{bmatrix} \begin{bmatrix} 0 \\ 1 \end{bmatrix} = \begin{bmatrix} \frac{t_s^2}{2} \\ t_s \end{bmatrix}.$$

Therefore, the discrete-time state-space model is

$$x(k+1) = \begin{bmatrix} 1 & t_s \\ 0 & 1 \end{bmatrix} x(k) + \begin{bmatrix} \frac{t_s^2}{2} \\ t_s \end{bmatrix} u(k), \tag{5.42}$$

$$y(k) = \begin{bmatrix} 1 & 0 \end{bmatrix} x(k). \tag{5.43}$$

State-space models will be used extensively throughout the remainder of the text; a dedicated chapter will examine their stability and other fundamental properties.

5.5 Conversion Between Discrete-time Models

This section presents systematic procedures for converting among the discrete-time model forms introduced so far and clarifies the assumptions, equivalences, and limitations associated with each conversion.

Figure 5.8 provides a summary diagram of available conversion procedures between model forms. As shown, there are often multiple possible pathways to convert one model type into another. While the discretization of continuous-time models has already been addressed earlier in this chapter, the discussion below focuses on interconversion between discrete-time model structures.

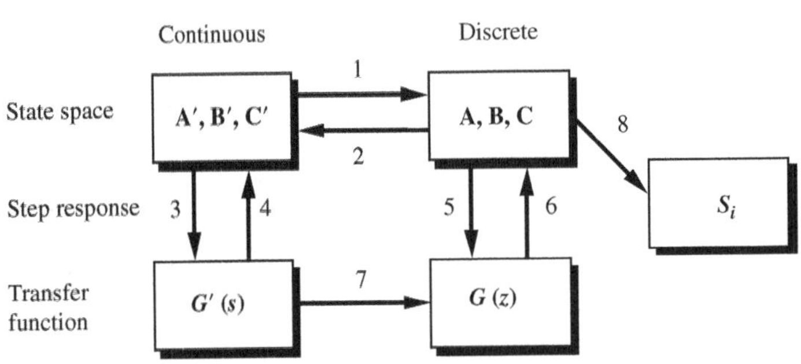

Figure 5.8 Summary of conversions among different LTI model structures. The numbered arrows indicate the conversion routes as referenced in the text.

5.5.1 Representing State-space System as Transfer Function

The pulse transfer matrix corresponding to the state-space model (5.26) can be obtained by taking the z-transform under the assumption that $x(0) = 0$:

$$\begin{aligned} zx(z) &= Ax(z) + Bu(z) \\ y(z) &= Cx(z). \end{aligned} \tag{5.44}$$

Solving for $x(z)$ from the first equation and substituting into the second gives:

$$\begin{aligned} x(z) &= (zI - A)^{-1}Bu(z) \\ y(z) &= Cx(z) = C(zI - A)^{-1}Bu(z). \end{aligned} \tag{5.45}$$

Hence, the corresponding transfer function is given by

$$G(z) = C(zI - A)^{-1}B. \tag{5.46}$$

Example 5.6: *Discretization of the Double Integrator from Example 5.3.*

Consider the double integrator system with transfer function $\frac{y(s)}{u(s)} = \frac{1}{s^2}$. In Example 5.5, this system was converted to state-space form and discretized with a sampling time $t_s = 1$, resulting in

$$\begin{aligned} x(k+1) &= \begin{bmatrix} 1 & t_s \\ 0 & 1 \end{bmatrix} x(k) + \begin{bmatrix} t_s^2/2 \\ t_s \end{bmatrix} u(k) \\ y(k) &= \begin{bmatrix} 1 & 0 \end{bmatrix} x(k). \end{aligned}$$

To obtain the corresponding pulse transfer function, we apply the z-transform as described in this section:

$$\begin{aligned} G(z) = C(zI - A)^{-1}B &= \begin{bmatrix} 1 & 0 \end{bmatrix} \begin{bmatrix} z-1 & -t_s \\ 0 & z-1 \end{bmatrix}^{-1} \begin{bmatrix} t_s^2/2 \\ t_s \end{bmatrix} \\ &= \frac{1}{(z-1)^2} \begin{bmatrix} 1 & 0 \end{bmatrix} \begin{bmatrix} z-1 & t_s \\ 0 & z-1 \end{bmatrix} \begin{bmatrix} t_s^2/2 \\ t_s \end{bmatrix} \\ &= \frac{t_s^2(z+1)}{2(z-1)^2}. \end{aligned}$$

This result matches the one obtained in Example 5.3 using the ZOH-based conversion route. Therefore, the two paths—direct ZOH discretization of the transfer function (route 7), and conversion through state-space form followed by z-transform (routes 4–1–5)—yield consistent results.

5.5.2 Realization of Transfer Function as State-space System

Any transfer function matrix can be realized in the form of a state-space system. While the transfer matrix corresponding to a given state-space model is unique, the reverse is not: a given transfer matrix can be realized by infinitely many different state-space representations.

Consider the single-input, single-output (SISO) transfer function:

$$\frac{y(z)}{u(z)} = \frac{b_1 z^{-1} + \cdots + b_n z^{-n}}{1 + a_1 z^{-1} + \cdots + a_n z^{-n}}. \tag{5.47}$$

This transfer function corresponds to the following input–output difference equation:

$$y(k) = -a_1 y(k-1) - \cdots - a_n y(k-n) + b_1 u(k-1) + \cdots + b_n u(k-n). \tag{5.48}$$

Two possible state-space realizations of the transfer function above are:

- **Controllable Canonical Form:**

$$\begin{aligned} x(k+1) &= \begin{bmatrix} -a_1 & -a_2 & \cdots & \cdots & -a_n \\ 1 & 0 & \cdots & \cdots & 0 \\ 0 & 1 & \cdots & \cdots & 0 \\ \vdots & & & & \vdots \\ 0 & 0 & \cdots & 1 & 0 \end{bmatrix} x(k) + \begin{bmatrix} 1 \\ 0 \\ 0 \\ \vdots \\ 0 \end{bmatrix} u(k) \\ y(k) &= \begin{bmatrix} b_1 & b_2 & \cdots & \cdots & b_n \end{bmatrix} x(k). \end{aligned} \tag{5.49}$$

- **Observable Canonical Form:**

$$\begin{aligned} x(k+1) &= \begin{bmatrix} -a_1 & 1 & 0 & \cdots & 0 \\ -a_2 & 0 & 1 & \cdots & 0 \\ \vdots & & & & \vdots \\ -a_{n-1} & 0 & 0 & \cdots & 1 \\ -a_n & 0 & 0 & \cdots & 0 \end{bmatrix} x(k) + \begin{bmatrix} b_1 \\ b_2 \\ \vdots \\ b_{n-1} \\ b_n \end{bmatrix} u(k) \\ y(k) &= \begin{bmatrix} 1 & 0 & \cdots & \cdots & 0 \end{bmatrix} x(k). \end{aligned} \tag{5.50}$$

For multi-input, multi-output (MIMO) systems, realizing a transfer matrix requires more care. If the transfer matrix is provided in matrix polynomial form, it can often be directly realized in a canonical form. For example, consider a transfer matrix given by:

$$(I + A_1 z^{-1} + \cdots \cdots + A_n z^{-n}) y(z) = (B_1 z^{-1} + \cdots \cdots + B_n z^{-n}) u(z), \tag{5.51}$$

This can be realized in observable canonical form as

$$\begin{aligned} x(k+1) &= \begin{bmatrix} -A_1 & I & \cdots & \cdots & 0 \\ -A_2 & 0 & I & \cdots & 0 \\ \vdots & & & \vdots & \vdots \\ -A_{n-1} & 0 & \cdots & \cdots & I \\ -A_n & 0 & \cdots & 0 & 0 \end{bmatrix} x(k) + \begin{bmatrix} B_1 \\ B_2 \\ \vdots \\ \vdots \\ B_n \end{bmatrix} u(k) \\ y(k) &= \begin{bmatrix} I & 0 & \cdots & \cdots & I \end{bmatrix} x(k). \end{aligned} \tag{5.52}$$

This realization, however, is not necessarily minimal. Redundant or unreachable states may be present and should be removed through model reduction techniques, which are discussed in Appendix C.

If the transfer matrix is not given in matrix polynomial form, it is often more practical to realize each element (i.e., each scalar transfer function) separately and then pack them into a single state-space model. This packed model is likely to be non-minimal and should again be reduced in order.

Example 5.7: *Consider the Continuous-time Transfer Function.*

$$g'(s) = \frac{e^{-2s}}{s+1}.$$

With a sampling time of $t_s = 1$, the corresponding pulse transfer function is obtained using:

$$\mathcal{Z}\left\{\mathcal{L}^{-1}\left\{\frac{1}{s+1}\frac{1-e^{-s}}{s}\right\}\right\} = \frac{0.6321}{z-0.3679}.$$

Accounting for the delay e^{-2s}, we introduce a two-step delay into the discrete-time domain, yielding the pulse transfer function:

$$g(z) = z^{-2}\frac{0.6321}{z-0.3679}.$$

This corresponds to the time-domain difference equation:

$$y(k) = 0.3679y(k-1) + 0.6321u(k-3).$$

A state-space realization in controllable canonical form is given by

$$A = \begin{bmatrix} 0.3679 & 0 & 0 \\ 1.0 & 0 & 0 \\ 0 & 1.0 & 0 \end{bmatrix}; \quad B = \begin{bmatrix} 1 \\ 0 \\ 0 \end{bmatrix}; \quad C = \begin{bmatrix} 0 & 0 & 0.6321 \end{bmatrix}.$$

5.5.3 Impulse and Step Responses of State-space System

The FIR and finite step response (FSR) models, used in Part I of this book, are examples of convolution models—models in which outputs are expressed as convolutions of input signals and model parameters:

$$\text{Finite Impulse Response:} colon\ y(k) = \sum_{i=1}^{n} H_i u(k-i), \tag{5.53}$$

$$\text{Finite Step Response:} colon\ y(k) = \sum_{i=1}^{n-1} S_i \Delta u(k-i) + S_n u(k-n). \tag{5.54}$$

For the discrete-time state-space model defined by A, B, C, let

$$B = \begin{bmatrix} b_1, b_2, \ldots, b_{n_u} \end{bmatrix} \text{ and } C = \begin{bmatrix} c_1^T, c_2^T, \ldots, c_{n_y}^T \end{bmatrix}^T,$$

where b_i and c_i denote the ith column and row of B and C, respectively.

The discrete-time unit impulse beginning at the jth sample is defined as:

$$\delta_j(t) = \begin{cases} 1 & \text{for } jt_s \le t < (j+1)t_s \\ 0 & \text{elsewhere.} \end{cases}$$

Assuming zero initial state, the response to an impulse in the ith input channel at $t = 0$, i.e.,

$$u_i(t) = \delta_0(t)$$
$$u_m(t) = 0 \qquad m \neq i,$$

yields the following output sequence:

$$\begin{aligned}
y^i_{imp}(0) &= 0\\
y^i_{imp}(1) &= Cb_i\\
y^i_{imp}(2) &= CAb_i\\
&\vdots\\
y^i_{imp}(n) &= CA^{n-1}b_i.
\end{aligned}$$

Here, $y^i_{imp}(k)$ denotes the output vector at time k resulting from a unit impulse applied to input u_i at time zero. The impulse response coefficient matrix at time k (for $k > 0$) is defined as

$$H_k \triangleq \begin{bmatrix} h_{1,1,k} & h_{1,2,k} & \dots & h_{1,n_u,k}\\ h_{2,1,k} & h_{2,2,k} & \dots & h_{2,n_{u}1,k}\\ \vdots & \vdots & & \vdots\\ h_{n_y,1,k} & h_{n_y,2,k} & \dots & h_{n_y,n_u,k} \end{bmatrix}, \tag{5.55}$$

$$\triangleq \begin{bmatrix} c_1A^{k-1}b_1 & c_1A^{k-1}b_2 & \dots & c_1A^{k-1}b_{n_u}\\ c_2A^{k-1}b_1 & c_2A^{k-1}b_2 & \dots & c_2A^{k-1}b_{nu}\\ \vdots & \vdots & & \vdots\\ c_{n_y}A^{k-1}b_1 & c_{n_y}A^{k-1}b_2 & \dots & c_{n_y}A^{k-1}b_{n_u} \end{bmatrix} = CA^{k-1}B, \tag{5.56}$$

where $h_{\ell,m,k}$ represents the response of output y_ℓ to a unit impulse in input u_m at time k.

The step response coefficient matrices, which capture the output due to a unit step change in input, are given by

$$S_k = \sum_{i=1}^{k} H_i. \tag{5.57}$$

5.5.4 Derivation of Transfer Matrix from Impulse Response

In Chapter 2, we introduced convolution models using heuristic arguments to derive the FIR representation. Here, we present a more rigorous derivation and establish its connection to the transfer function model.

Recall that the impulse response matrix is given by (5.55). Let $[H_k]i$ denote the ith column of the matrix H_k. Then, $[H_{k-j}]i$ represents the system's response at time k to a unit impulse $\delta_j(t)$ applied to the ith input channel. More generally, the term $[H_{k-j}]_i u_i(j)$ corresponds to the response at time k resulting from an impulse of magnitude $u_i(j)$ applied at time j to the ith input.

We can represent any signal generated by a ZOH as a sequence of discrete impulses:

$$u(t) = \begin{bmatrix} \sum_j u_1(j)\delta_j(t)\\ \sum_j u_2(j)\delta_j(t)\\ \vdots\\ \sum_j u_{n_u}(j)\delta_j(t) \end{bmatrix}.$$

By the principle of superposition—i.e., by summing the responses to individual impulses applied across different input channels and at different times—the system response at time k is given by

$$y(k) = \sum_{j=-\infty}^{k-1} \sum_{i=1}^{n_u} [H_{k-j}]_i u_i(j) = \sum_{j=-\infty}^{k-1} H_{k-j} u(j), \tag{5.58}$$

$$= \sum_{\ell=1}^{\infty} H_\ell u(k-\ell). \tag{5.59}$$

By summing only up to $j = k - 1$, we are implicitly assuming that the system is *causal*—that is, the output at time k depends only on inputs applied at times strictly prior to k. Equation (5.59) shows that the system output is expressed as a convolution: a weighted sum of past inputs, where the weights are given by the impulse response coefficients.

Taking the z-transform of (5.59), and assuming $u(k) = 0$ for $k < 0$, we obtain:

$$y(z) = H(z) \cdot u(z), \tag{5.60}$$

where $H(z)$ denotes the z-transform of the sequence H_k. Although the derivation is somewhat tedious, the result is intuitive: the pulse transfer matrix is simply the z-transform of the system's impulse response. This is consistent with the fact that the z-transform of a discrete-time unit impulse—i.e., a unit pulse occurring at $t = 0$ and lasting for one sample interval—is equal to 1.

For stable systems, a commonly used approximation of (5.59) is the FIR model:

$$y(k) = \sum_{i=1}^{n} H_i u(k-i). \tag{5.61}$$

Here, n is chosen large enough so that the contributions of the truncated terms are negligible. The corresponding z-domain representation is

$$\frac{y(z)}{u(z)} = H_1 \frac{1}{z} + H_2 \frac{1}{z^2} + \cdots + H_n \frac{1}{z^n}. \tag{5.62}$$

This shows that the transfer matrix of an FIR model has all its poles at the origin, reflecting the pure delay structure of the response.

5.5.5 From Impulse/Step Response to State-space Model

A state-space realization of the FIR model in controllable canonical form corresponding to (5.61) is given by

$$\begin{aligned} x(k+1) &= Ax(k) + Bu(k) \\ y(k) &= Cx(k), \end{aligned} \tag{5.63}$$

where

$$A=\left.\begin{bmatrix} 0 & 0 & 0 & \dots & \dots & 0 & 0 \\ I & 0 & 0 & 0 & \dots & 0 & 0 \\ 0 & I & 0 & 0 & \dots & 0 & 0 \\ \vdots & & \ddots & \ddots & & \vdots & \vdots \\ 0 & 0 & \dots & \dots & \dots & 0 & 0 \\ 0 & 0 & \dots & \dots & \dots & I & 0 \end{bmatrix}\right\} n\times n_u \qquad B=\begin{bmatrix} I \\ 0 \\ \vdots \\ 0 \end{bmatrix} \tag{5.64}$$

$$C=\begin{bmatrix} H_1 & H_2 & \cdots & H_n \end{bmatrix}.$$

In this realization, the state vector stores the past n input values:

$$x(k)=\begin{bmatrix} u^T(k-1) & u^T(k-2) & \cdots\cdots & u^T(k-n) \end{bmatrix}^T. \tag{5.65}$$

All eigenvalues of A are located at the origin.

The observable canonical form realization is given by

$$A=\left.\begin{bmatrix} 0 & I & 0 & \dots & \dots & 0 & 0 \\ 0 & 0 & I & 0 & \dots & 0 & 0 \\ \vdots & & & & & & \vdots \\ 0 & 0 & \dots & \dots & \dots & 0 & I \\ 0 & 0 & \dots & \dots & \dots & 0 & 0 \end{bmatrix}\right\} n\times n_y \qquad B=\begin{bmatrix} H_1 \\ H_2 \\ \vdots \\ H_n \end{bmatrix} \tag{5.66}$$

$$C=\begin{bmatrix} I & 0 & \cdots & 0 \end{bmatrix}.$$

It is important to note that state-space realizations derived from impulse or step responses are generally over-parameterized and non-minimal. In practice, the system order can often be reduced significantly through model reduction techniques. One such technique is balanced truncation, which is introduced in Appendix C. It is especially well suited for simplifying high-order realizations and offers favorable numerical properties.

We can similarly derive the state-space representation from a step response model using arguments presented in Section 2.3 in the earlier discussion of industrial MPC. The model will have a form similar to (2.36):

$$A=\left.\begin{bmatrix} 0 & I & 0 & \dots & \dots & 0 & 0 \\ 0 & 0 & I & 0 & \dots & 0 & 0 \\ \vdots & & & & & & \vdots \\ 0 & 0 & \dots & \dots & \dots & 0 & I \\ 0 & 0 & \dots & \dots & \dots & 0 & I \end{bmatrix}\right\} n\times n_y \qquad B=\begin{bmatrix} S_1 \\ S_2 \\ \vdots \\ S_n \end{bmatrix}, \tag{5.67}$$

$$C=\begin{bmatrix} I & 0 & \cdots & 0 \end{bmatrix}. \tag{5.68}$$

Comparing the state-space realizations in (5.66) and (5.67), we note that the last row of the matrix A differs between the two forms. In contrast to the impulse-response-based realizations, the above system incorporates the input in the form of Δu rather than u. This model introduces n_y integrators, reflecting the cumulative effect of Δu on the output y. Hence, this formulation is particularly useful for designing MPC controllers that inherently provide integral action.

Example 5.8: *Conversion to FIR Models*

Discretize the system of Figure 2.6 and convert it into various discrete-time LTI models. The transfer function of the system is given by

$$g = \frac{2.69(-6s+1)}{(20s+1)(5s+1)}e^{-1.5s}, \tag{5.69}$$

and the sampling interval is $t_s = 3$.

Impulse Response Model: The unit step response of this system was shown earlier. The impulse response coefficients can be obtained by differencing the step response data from Figure 2.6, resulting in

$$H = \begin{bmatrix} -0.1744 & -0.0568 & 0.1567 & 0.2478 & 0.2752 & 0.2709 & \cdots \end{bmatrix}^T.$$

A total of $n = 40$ coefficients were retained in the FIR model with a sampling interval of $t_s = 3$.

State-space Model: The continuous-time transfer function can be converted to state space using the MATLAB command `g1 = ss(g)`, followed by `c2d(g1, 3)` to obtain:

$$x(k+1) = \begin{bmatrix} 0.4448 & -0.08317 & 0.4166 \\ 0.5198 & 0.9647 & 0.3172 \\ 0 & 0 & 0 \end{bmatrix} x(k) + \begin{bmatrix} 0.6231 \\ 0.1244 \\ 1 \end{bmatrix} u(k).$$

$$y(k) = \begin{bmatrix} -0.3228 & 0.2152 & 0 \end{bmatrix} x(k)$$

State-space → Impulse Response: Using Eq. (5.56), we obtain:

$$H_1 = CB = -0.1744 \quad H_2 = CAB = -0.0568 \quad H_3 = CA^2B = 0.1567$$

$$H_4 = CA^3B = 0.2478 \quad H_5 = CA^4B = 0.2752 \quad H_6 = CA^5B = 0.2709 \quad \ldots$$

which matches the earlier impulse response model.

Impulse Response → State-space and Transfer Function: Conversion from the impulse response to state-space and transfer function models is straightforward. Substituting the FIR coefficients into (5.66) and (5.62), respectively, yields:

$$x(k+1) = \begin{bmatrix} 0 & 1 & 0 & 0 & \ldots & 0 \\ 0 & 0 & 1 & 0 & \ldots & 0 \\ \vdots & \vdots & & \ddots & \ddots & \vdots \\ 0 & 0 & 0 & \ldots & \ldots & 1 \\ 0 & 0 & 0 & \ldots & \ldots & 1 \end{bmatrix} x(k) + \begin{bmatrix} -0.1744 \\ -0.0568 \\ 0.1567 \\ \vdots \\ 0.0020 \end{bmatrix} u(k)$$

$$y(k) = \begin{bmatrix} 1 & 0 & 0 & \ldots & 0 \end{bmatrix} x(k),$$

$$g(z) = \frac{-0.1744z^{39} - 0.0568z^{38} + 0.1567z^{37} + \cdots + 0.0020}{z^{40}}.$$

One can verify the equivalence of these models by comparing their step responses with that of the original system (5.69). The responses are nearly indistinguishable, confirming that all three capture essentially the same dynamics.

Exercises

1. Derive the final value theorem for the z-transform, and use it to show that the steady-state gain of a stable system can be obtained by evaluating the transfer function at $z = 1$.
2. Starting from the convolution expression for an LTI system, derive the equation $y(z) = H(z)u(z)$, where $H(z)$ is the z-transform of the sequence of impulse response coefficient matrices H_k. This requires taking the z-transform of the convolution sum.
3. Consider the system:

$$\frac{dx}{dt} = \begin{bmatrix} -0.625 & -0.25 \\ 0.25 & 0 \end{bmatrix} x + \begin{bmatrix} 1 \\ 0 \end{bmatrix} u$$
$$y = \begin{bmatrix} 0 & 0.3 \end{bmatrix} x,$$

   - What are the eigenvalues of A^c?
   - Convert the above to a transfer function and compute its poles.
   - Discretize the state-space model with a sampling interval of $t_s = 1$.
   - What are the eigenvalues of the resulting A matrix?
4. Figure 5.7 illustrates the effect of pole locations on the response of a discrete-time transfer function model. Consider a continuous-time state-space model:

$$x' = A^c x + B^c u.$$

   Ignore the effect of input and consider only the autonomous system. Let λ^c denote the eigenvalues of A^c. As noted in Eq. (5.34), the corresponding discrete-time system matrix is given by $A = \exp(A^c t_s)$.
   - Under what conditions on λ^c is the continuous-time system stable?
   - What are the corresponding eigenvalues in the matrix A?
   - Show that a positive eigenvalue of A^c that leads to an unstable response corresponds to an eigenvalue of A outside the unit circle.
   - When does the continuous-time system produce an oscillatory response? What are the corresponding eigenvalues of A?
   - Relate these results to the behavior depicted in Figure 5.7.
5. Show that z-transform of the observer canonical form and the controller canonical form of state-space realization both lead to the same transfer function.
6. Consider the system:

$$y(k) = 0.5y(k-1) + 0.25y(k-2) + 0.125y(k-3) + u(k-1),$$

   - At steady state, $y(k) = y(k-i) = y_\infty$ and $u(k-1) = u_\infty$. Use this to compute y_∞ for a unit step input.
   - Take the z-transform of the equation to obtain the discrete-time transfer function.
   - Use the final value theorem to compute the system gain.
   - Convert the transfer function to a state-space model in the controllable canonical form.
   - Convert the transfer function to a state-space model in the observable canonical form.
7. For the same system as above, define the state vector as

$$x(k) \triangleq \begin{bmatrix} y(k) & y(k-1) & y(k-2) \end{bmatrix}^T,$$

   - Derive the corresponding state-space model in this basis.

- Using Eq. (5.46), convert the state-space model into a transfer function.
- Verify that this transfer function matches the result obtained in the previous problem.

8. For the same system,

$$y(k) = 0.5y(k-1) + 0.25y(k-2) + 0.125y(k-3) + u(k-1),$$

derive the step response model manually by assuming $y(-k) = 0$ and $u(-k) = 0$ for all $k > 0$. Show the evolution of $y(k)$ step by step in response to a unit step input.

9. The transfer function model $2.5/(2s+1)$ corresponds to the state-space system:

$$\frac{dx}{dt} = -0.5x + 1.25u \qquad y = 1 \cdot x,$$

Use hand calculations to answer the following:

- Discretize the above state-space model using a sampling interval $t_s = 0.5$.
- Determine how the discrete-time model changes for $2.5/(2s+1) \cdot e^{-2s}$, i.e., including a time delay of 2 seconds. Use the same $t_s = 0.5$.
- Discuss how the model changes if the delay is increased to 2.1 seconds, i.e., for $2.5/(2s+1) \cdot e^{-2.1s}$.

10. Consider the following MIMO system:

$$\begin{bmatrix} \dfrac{120s+2}{(100s+1)(20s+1)} & \dfrac{80s}{(100s+1)(20s+1)} \\[2ex] \dfrac{-80s}{(100s+1)(20s+1)} & \dfrac{-(120s+2)}{(100s+1)(20s+1)} \end{bmatrix},$$

- Discretize each individual transfer function and obtain a corresponding state-space realization.
- Combine the four SISO realizations into a single multivariable (MIMO) system. This can be done using block-diagonal matrices as follows:

$$x(k+1) = \begin{bmatrix} A_{11} & 0 & 0 & 0 \\ 0 & A_{12} & 0 & 0 \\ 0 & 0 & A_{21} & 0 \\ 0 & 0 & 0 & A_{22} \end{bmatrix} x(k) + \begin{bmatrix} B_{11} & 0 \\ 0 & B_{12} \\ B_{21} & 0 \\ 0 & B_{22} \end{bmatrix} u(k)$$

$$y(k) = \begin{bmatrix} C_{11} & C_{12} & 0 & 0 \\ 0 & 0 & C_{21} & C_{22} \end{bmatrix} x(k),$$

where (A_{ij}, B_{ij}, C_{ij}) denotes the state-space realization of the (i, j)th element of the original transfer matrix $G(s)$.

- What is the total order (i.e., total number of states) of the full MIMO system?

11. Apply balanced truncation to the full state-space model above to eliminate redundant states. What is the reduced-order model, and what is the minimum order required to preserve the input–output behavior within a reasonable tolerance?

Balanced truncation can be carried out in MATLAB by creating a reduction specification with `reducespec` (method `"balanced"`) and then applying `reduce`. Details are provided in Appendix C.

6

Discrete-time State Space Models

Mathematics is the art of reducing any problem to linear algebra.

- William Stein

In Chapter 5, we introduced the concepts of sampling and reconstruction and examined various forms of linear time-invariant system models. We demonstrated that these different model representations are equivalent in terms of their input–output behavior and can be readily transformed from one form to another.

From this point forward, we will primarily focus on state-space models, as they offer a compact and versatile framework for the design and analysis of MPC algorithms. The standard discrete-time state-space representation is given by

$$\begin{aligned} x(k+1) &= Ax(k) + Bu(k) \\ y(k) &= Cx(k). \end{aligned} \tag{6.1}$$

6.1 State-coordinate Transformation

In describing an input-output system using a state-space model, the choice of state variables is not unique—there are *infinitely many* ways to store the same internal system information. As seen earlier, a given transfer function can be realized in different canonical forms, such as the controllable or observable canonical form.

More generally, we can redefine the state coordinates without altering the system's input-output behavior. A new coordinate system can be introduced via the linear transformation $\tilde{x} = Tx$, where T is any nonsingular (invertible) square matrix. Applying this transformation to the state-space system Eq. (6.1), we obtain

$$\begin{aligned} T^{-1}\tilde{x}(k+1) &= AT^{-1}\tilde{x}(k) + Bu(k) \\ y(k) &= CT^{-1}\tilde{x}(k). \end{aligned} \tag{6.2}$$

Rearranging the terms yields the state-space system in the new coordinates:

$$\begin{aligned}\tilde{x}(k+1) &= \underbrace{TAT^{-1}}_{\tilde{A}}\tilde{x}(k) + \underbrace{TB}_{\tilde{B}}u(k)\\ y(k) &= \underbrace{CT^{-1}}_{\tilde{C}}\tilde{x}(k),\end{aligned} \tag{6.3}$$

Although state coordinate transformations do not change the input-output relationship between $u(k)$ and $y(k)$, certain choices of state variables may simplify analysis or design. We will soon illustrate how appropriate coordinate transformations can provide such advantages.

Example 6.1: *Conversion Between Two Canonical Forms.*
Consider the system from Example 5.7, where the transfer function was realized in the controllable canonical form as follows

$$\begin{aligned}x(k+1) &= \begin{bmatrix} 0.3679 & 0 & 0\\ 1.0 & 0 & 0\\ 0 & 1.0 & 0\end{bmatrix} x(k) + \begin{bmatrix} 1.0\\ 0\\ 0\end{bmatrix} u(k)\\ y(k) &= \begin{bmatrix} 0 & 0 & 0.6321\end{bmatrix} x(k).\end{aligned}$$

Now consider the coordinate transformation defined by

$$T = \begin{bmatrix} 0 & 0 & 0.6321\\ 0 & 0.6321 & -0.2325\\ 0.6321 & -0.2325 & 0\end{bmatrix}.$$

Note that elements such as $T_{2,3} = T_{3,2} = a_1 b_3$ reflect structure-specific terms arising from the original transfer function parameters.

Applying the transformation using Eq. (6.2), the state-space system in the new coordinates becomes

$$\begin{aligned}x(k+1) &= \begin{bmatrix} 0.3679 & 1.0 & 0\\ 0 & 0 & 1.0\\ 0 & 0 & 0\end{bmatrix} x(k) + \begin{bmatrix} 0\\ 0\\ 0.6321\end{bmatrix} u(k)\\ y(k) &= \begin{bmatrix} 1.0 & 0 & 0\end{bmatrix} x(k).\end{aligned}$$

This is the observable canonical form corresponding to the same transfer function. As expected, the input-output behavior remains unchanged despite the transformation of the internal state coordinates.

6.2 Stability

6.2.1 System Poles and Characteristic Equation

For the state-space system defined in Eq. (6.1), the eigenvalues of the state transition matrix A are referred to as the system's *poles*. These poles are obtained by solving the characteristic equation:

$$\det(A - \lambda I) = 0. \tag{6.4}$$

The solutions λ of this equation determine key dynamic properties of the system, including its stability. The location of the poles in the complex plane directly affects the system's transient and steady-state behavior.

Before exploring how pole locations influence dynamics, we first define the concept of stability in the context of state-space systems.

6.2.2 Stability

Definition 6.1: **Asymptotic Stability**: The state-space system described by Eq. (6.1) is said to be asymptotically stable if the state converges to the origin from any initial condition in the absence of input excitation; that is,

$$\lim_{k\to\infty} x(k) = 0 \qquad \forall x(0) \in \mathbb{R}^n \text{ with } u = 0.$$

A necessary and sufficient condition for asymptotic stability of the system Eq. (6.1) is that all eigenvalues of the matrix A lie strictly inside the unit circle in the complex plane. If any eigenvalue lies on or outside the unit circle, the system is unstable.

This condition follows directly from the modal decomposition of the system. Suppose, for simplicity, that A is diagonalizable with a complete set of eigenvectors $\{e_1, \dots, e_n\}$. We can choose these eigenvectors as the new basis set. Performing the coordinate transformation:

$$\tilde{x} = \begin{bmatrix} e_1 & \cdots & \cdots & e_n \end{bmatrix}^{-1} x, \tag{6.5}$$

transforms the state system Eq. (6.1) into

$$\begin{aligned} \tilde{x}(k+1) &= \begin{bmatrix} e_1 & e_2 & \cdots & e_n \end{bmatrix}^{-1} A \begin{bmatrix} e_1 & e_2 & \cdots & e_n \end{bmatrix} \tilde{x} \\ &= \begin{bmatrix} \lambda_1 & 0 & \cdots & 0 \\ 0 & \lambda_2 & \cdots & 0 \\ \vdots & \vdots & \ddots & \vdots \\ 0 & \cdots & 0 & \lambda_n \end{bmatrix} \tilde{x}. \end{aligned} \tag{6.6}$$

In the transformed coordinates, each state evolves independently and can be analyzed separately. From the equations

$$\tilde{x}_i(k+1) = \lambda_i \tilde{x}_i(k), \quad i = 1, \dots, n,$$

where x_i represents the i_{th} state, it is clear that all λ_is must have magnitudes less than one for asymptotic stability.

Definition 6.2: **BIBO Stability**: A state-space system described by Eq. (6.1) is said to be "bounded input / bounded output (BIBO) stable" if y remains bounded in response to any bounded u.

BIBO stability is a weaker condition than asymptotic stability; i.e., asymptotic stability implies BIBO stability, but the converse does not necessarily hold. This is because a system may contain simple integrators that are not influenced by the input u, yet still produce a bounded output y for any bounded input, thereby satisfying the BIBO stability condition despite being asymptotically unstable.

Example 6.2: Consider the following system with an integrating state:

$$\begin{bmatrix} x_1(k+1) \\ x_2(k+1) \end{bmatrix} = \begin{bmatrix} 0.2 & 0.5 \\ 0.0 & 1.0 \end{bmatrix} \begin{bmatrix} x_1(k) \\ x_2(k) \end{bmatrix} + \begin{bmatrix} 1 \\ 0 \end{bmatrix} u(k)$$
$$y(k) = \begin{bmatrix} 1 & 1 \end{bmatrix} \begin{bmatrix} x_1(k) \\ x_2(k) \end{bmatrix}. \tag{6.7}$$

The system is BIBO stable because the integrating state x_2 is not affected by the input $u(k)$ and thus does not grow unbounded in response to a bounded input. However, the system is not asymptotically stable, since x_2 remains constant at its initial value and does not converge to zero as $k \to \infty$.

6.2.3 Lyapunov Equation

We now introduce a fundamental concept related to the stability of dynamical systems, along with a matrix equation that will appear frequently throughout this book. These arise from the so-called *Lyapunov's Second Method*, a general framework for proving the stability of nonlinear systems.

In the present context, the idea can be paraphrased as follows: we seek a convex function $V(x) \geq 0$ such that $V(0) = 0$ and $V(x)$ strictly decreases as the state x evolves over time. If such a function exists, the system can be shown to be asymptotically stable.

For linear systems, a natural candidate is the quadratic form $V(x) = x^T P x$, where P is a symmetric matrix to be determined. To ensure that $V(x)$ qualifies as a Lyapunov function, the key requirement is that it decreases over time. Specifically, we require

$$\begin{aligned} V(x(k+1)) - V(x(k)) &= x^T(k+1)Px(k+1) - x^T(k)Px(k) \\ &= x^T(k)\left(A^T P A - P\right)x(k). \end{aligned} \tag{6.8}$$

Thus, for $x^T P x$ to serve as a valid Lyapunov function, we require

$$A^T P A - P = -Q \qquad \text{for some } Q > 0. \tag{6.9}$$

Equation (6.9) is known as the *Lyapunov equation* for discrete-time systems. It can be shown that, for any given $Q > 0$, there exists a unique positive definite solution $P > 0$ if and only if all the eigenvalues of A lie strictly inside the unit circle. The proof is left as an exercise.

6.3 Controllability, Reachability, and Stabilizability

For a state-space system intended for control, it is critical to determine whether the system is controllable, reachable, or stabilizable. While these concepts are closely related, they are not equivalent and differ in subtle but important ways.

- *Controllability* asks whether it is possible to drive the state of the system from an arbitrary initial condition to the origin in finite time using a suitable input sequence.
- *Reachability* is a related concept that concerns whether the system can be driven to an arbitrary final state (not necessarily the origin) in finite time.
- *Stabilizability* is a weaker notion: it requires only that the system can be driven to the origin asymptotically (as $k \to \infty$) from any initial state using a suitable input sequence, without the restriction of finite time.

This section defines these terms precisely and presents mathematical criteria for verifying them.

6.3.1 Definitions

Definition 6.3: A system described by Eq. (6.1) is said to be "controllable" if it is possible to find an input sequence that moves the system from an arbitrary initial state to the origin in *finite* time.

Definition 6.4: A system described by Eq. (6.1) is said to be "reachable" if it is possible to find an input sequence that moves the system from an arbitrary initial state to any desired state in finite time.

Definition 6.5: A system described by Eq. (6.1) is said to be "stabilizable" if it is possible to find an input sequence that returns the state to the origin asymptotically (as $k \to \infty$) from an arbitrary initial condition.

Note that a reachable system is always controllable, but a controllable system is not necessarily reachable. For instance, if $A^n x(0) = 0$, the system is controllable regardless of whether it is reachable.

While reachability or controllability is desirable, these properties may be too stringent for practical systems. In many cases, the minimum requirement for effective control is stabilizability. By definition, any stable system is automatically stabilizable.

6.3.2 Conditions for Reachability

For a system represented by Eq. (6.1) with n states, the following are some mathematical conditions for reachability.

6.3.2.1 Rank Condition

The most well-known test for controllability is the following rank condition:

$$\text{rank}\,(\mathcal{W}_n^c) = n, \tag{6.10}$$

$$\mathcal{W}_n^c = \begin{bmatrix} B & AB & \cdots & A^{n-1}B \end{bmatrix}. \tag{6.11}$$

Proof: This condition can be established using the Cayley–Hamilton theorem (see Appendix A). First, observe that the state evolution can be written as

$$x(k) = A^k x(0) + \underbrace{\begin{bmatrix} B & AB & \cdots & A^{k-1}B \end{bmatrix}}_{\mathcal{W}_k^c} \begin{bmatrix} u(k-1) \\ u(k-2) \\ \vdots \\ u(0) \end{bmatrix}. \tag{6.12}$$

Reachability requires that, for any initial state $x(0) \in \mathbb{R}^n$ and target state $x_t \in \mathbb{R}^n$, there exist a finite-horizon k and an input sequence $(u(0), \ldots, u(k-1))$ such that $x(k) = x_t$. For linear systems, this holds if rank $(\mathcal{W}_k^c) = n$ (for some $k \geq n$ in the time-invariant case), i.e., $\mathcal{W}_k^c$ has full row rank.

By the Cayley–Hamilton theorem, the controllability matrix satisfies rank$(\mathcal{W}_k^c) = $ rank$(\mathcal{W}_n^c)$ for all $k \geq n$. Therefore, the reachability condition reduces to the rank test in Eq. (6.10).

The matrix $\mathcal{W}_n^c$ is known as the *controllability matrix*. From the above discussion, it is clear that the range space of the controllability matrix defines the *reachable subspace*—the subspace of the state space within which the state can be freely manipulated using the input.

6.3.2.2 Hautus Condition

Another useful test for reachability is the so-called Hautus condition, given by

$$\text{rank}\left(\left[A - \lambda I \mid B\right]\right) = n \qquad \forall \lambda \in \Lambda_A, \tag{6.13}$$

where Λ_A denotes the spectrum (i.e., the set of eigenvalues) of A.

Proof: To see how the above condition arises, we first prove:

$$\text{Reachability} \quad \rightarrow \quad \text{rank}\left[A - \lambda I \mid B\right] = n \quad \forall \lambda \in \Lambda_A.$$

For this, we can prove the contrapositive instead

$$\text{Not reachable} \quad \leftarrow \quad \text{rank}\left[A - \lambda I \mid B\right] \neq n \quad \text{for some } \lambda \in \Lambda_A.$$

If rank $\left[A - \lambda I \mid B\right] \neq n$ for some $\lambda \in \Lambda_A$, then there exists a row vector x^T such that

$$x^T \left[A - \lambda I \mid B\right] = 0^T.$$

Since the above implies that $x^T(A - \lambda I) = 0^T$, x is a left eigenvector of A corresponding to λ, i.e.,

$$x^T A = \lambda x^T.$$

Since $x^T B = 0$, we also have

$$x^T \left[B \quad AB \quad \cdots \quad A^{n-1}B\right] = \left[x^T B \quad \lambda x^T B \quad \cdots \quad \lambda^{n-1} x^T B\right] = 0^T.$$

This implies that the controllability matrix does not have a full row rank and, therefore, system is not reachable.

We must also prove that

$$\text{Reachability} \quad \leftarrow \quad \text{rank}\left[A - \lambda I \mid B\right] = n \quad \forall \lambda \in \Lambda_A.$$

Again, we can prove the contrapositive instead

$$\text{Not reachable} \quad \rightarrow \quad \text{rank}\left[A - \lambda I \mid B\right] \neq n \quad \text{for some } \lambda \in \Lambda_A.$$

If a system is not reachable, the controllability matrix is rank-deficient, i.e.,

$$\exists\, x \text{ such that} \qquad x^T \left[B \quad AB \quad \cdots \quad A^{n-1}B\right] = 0^T$$

$$x^T B = 0, \quad x^T AB = 0, \quad \cdots \quad , \quad x^T A^{n-1} B = 0$$

$$x^T A^k B = 0 \quad \forall k \geq 0.$$

The above is a consequence of the Cayley–Hamilton theorem.

Since $x^T B = 0$, this implies that x is a left eigenvector of A. This further implies

$$\exists\, \lambda \in \Lambda_A \quad \text{such that} \quad x^T \left[A - \lambda I \mid B\right] = 0^T,$$

which means it is rank deficient. This completes the proof of the equivalence between reachability and the Hautus condition.

A related condition for testing stabilizability is given by

$$\text{rank}\left(\left[A - \lambda I \mid B\right]\right) = n \qquad \forall \lambda \in \Lambda_A^+, \tag{6.14}$$

where Λ_A^+ denotes the set of all eigenvalues of A that lie on or outside the unit disk. This condition ensures that all unstable modes (i.e., those not asymptotically decaying) are controllable.

The proof follows the same argument as that used for the reachability Hautus condition. The key difference is that we now restrict attention only to the unstable eigenvalues of A. If any such eigenvalue $\lambda \in \Lambda_A^+$ corresponds to a left eigenvector x satisfying $x^T B = 0$, then the mode associated with λ is uncontrollable and cannot be stabilized, violating the stabilizability condition.

6.3.3 Coordinate Transformation

There exists a coordinate transformation,

$$\begin{bmatrix} \tilde{x}_1 \\ \tilde{x}_2 \end{bmatrix} \begin{bmatrix} T_1^c & T_2^c \end{bmatrix}^{-1} x, \tag{6.15}$$

that transforms the state-space model into the form:

$$\begin{bmatrix} \tilde{x}_1(k+1) \\ \tilde{x}_2(k+1) \end{bmatrix} = \begin{bmatrix} \tilde{A}_{11} & \tilde{A}_{12} \\ 0 & \tilde{A}_{22} \end{bmatrix} \begin{bmatrix} \tilde{x}_1(k) \\ \tilde{x}_2(k) \end{bmatrix} + \begin{bmatrix} \tilde{B}_1 \\ 0 \end{bmatrix} u(k), \tag{6.16}$$

where $(\tilde{A}_{11}, \tilde{B}_1)$ is a reachable pair. This form is both intuitive and convenient: $\tilde{x}_1$ corresponds to the reachable portion of the state, while $\tilde{x}_2$ represents the unreachable portion, which cannot be influenced by the input.

To obtain this form, we construct T_1^c and T_2^c such that the columns of T_1^c span the range space of the controllability matrix $\mathcal{W}_n^c$, and the columns of T_2^c span its orthogonal complement. One way to determine these matrices is by performing a singular value decomposition (SVD) on the controllability matrix

$$\mathcal{W}_n^c = \begin{bmatrix} U_1 & U_2 \end{bmatrix} \begin{bmatrix} \Sigma_1 & 0 \\ 0 & 0 \end{bmatrix} \begin{bmatrix} V_1^T \\ V_2^T \end{bmatrix}, \tag{6.17}$$

and choosing

$$T_1^c = U_1, \quad T_2^c = U_2. \tag{6.18}$$

If all the eigenvalues of $\tilde{A}_{22}$ are zero, the system is controllable. If the eigenvalues of $\tilde{A}_{22}$ in Eq. (6.16) lie strictly inside the unit disk, the system is stabilizable. This means that the unreachable states evolve according to stable dynamics.

It should now be evident that reachability and stabilizability are intrinsic properties of the system. A lack of either cannot be addressed through control design alone—it requires modifying the system dynamics or redefining the set of inputs.

Example 6.3: *Understanding Controllability.* Consider the system

$$x(k+1) = \begin{bmatrix} 0.5 & -0.2 \\ 0.25 & 0 \end{bmatrix} x(k) + \begin{bmatrix} 0.5 \\ 0 \end{bmatrix} u(k),$$

and suppose we wish to design a control input that brings the system to the origin in one or two steps from an arbitrary initial condition.

Let the initial state be $[a_1 \;\; a_2]^T$. After applying an initial input $u(0) = u_0$, the state becomes

$$x(1) = \begin{bmatrix} 0.5a_1 - 0.2a_2 \\ 0.25a_1 \end{bmatrix} + \begin{bmatrix} 0.5u_0 \\ 0.0 \end{bmatrix}.$$

To drive the system to the origin in one step, we require $x(1) = \mathbf{0}$. This yields: $a_1 = 0$ and $u_0 = 0.4a_2$. Thus, any state lying on the x_2-axis can be driven to the origin in exactly one step.

Next, consider applying a second input $u(1)$ to reach the origin in two steps. The resulting state is

$$\begin{aligned} x(2) &= \begin{bmatrix} 0.5 & -0.2 \\ 0.25 & 0 \end{bmatrix} \begin{bmatrix} 0.5a_1 - 0.2a_2 + 0.5u_0 \\ 0.25a_1 \end{bmatrix} + \begin{bmatrix} 0.5u_1 \\ 0.0 \end{bmatrix} \\ &= \begin{bmatrix} 0.2a_1 - 0.1a_2 + 0.25u_0 + 0.5u_1 \\ 0.125a_1 - 0.05a_2 + 0.125u_0 \end{bmatrix}. \end{aligned}$$

Setting $x(2) = \mathbf{0}$ yields two equations in two unknowns (u_0 and u_1), leading to

$$u(0) = -a_1 + 0.4a_2 \qquad u(1) = 0.1a_1.$$

Hence, any initial state in $\mathbb{R}^2$ can be driven to the origin in two steps or fewer. The system is therefore controllable.

As a concrete example, consider the initial state $x(0) = [-0.2 \;\; 0.7]^T$. Applying the control inputs

$$u(0) = 0.48; \qquad x(1) = \begin{bmatrix} 0 \\ -0.05 \end{bmatrix} \qquad u(1) = -0.02; \qquad x(2) = \begin{bmatrix} 0 \\ 0 \end{bmatrix},$$

demonstrates that the system reaches the origin as expected.

Example 6.4: *Controllable Subspace.* Consider the following simple system, where a is an arbitrary scalar parameter:

$$\begin{bmatrix} x_1(k+1) \\ x_2(k+1) \end{bmatrix} = \begin{bmatrix} 1.5 + 0.5a & 0.5 - 0.5a \\ 1.5 - 0.5a & 0.5 + 0.5a \end{bmatrix} \begin{bmatrix} x_1(k) \\ x_2(k) \end{bmatrix} + \begin{bmatrix} 1 \\ 1 \end{bmatrix} u(k). \tag{6.19}$$

The controllability matrix is given by

$$\mathcal{W}_2^c = \begin{bmatrix} B & AB \end{bmatrix} = \left(\begin{bmatrix} 1 \\ 1 \end{bmatrix} \begin{bmatrix} 2 \\ 2 \end{bmatrix} \right).$$

Since the rank of $\mathcal{W}_2^c$ is 1, the system is not reachable.

To analyze the reachable subspace, we perform a SVD of $\mathcal{W}_2^c$:

$$\mathcal{W}_2^c = \begin{bmatrix} \frac{1}{\sqrt{2}} & \frac{1}{\sqrt{2}} \\ \frac{1}{\sqrt{2}} & -\frac{1}{\sqrt{2}} \end{bmatrix} \begin{bmatrix} \sigma_1 & 0 \\ 0 & 0 \end{bmatrix} \begin{bmatrix} \frac{1}{\sqrt{5}} & \frac{2}{\sqrt{5}} \\ \frac{2}{\sqrt{5}} & -\frac{1}{\sqrt{5}} \end{bmatrix}.$$

From this decomposition, we observe that the reachable subspace is spanned by the first left singular vector:

$$\begin{bmatrix} \frac{1}{\sqrt{2}} \\ \frac{1}{\sqrt{2}} \end{bmatrix}.$$

This subspace is illustrated in Figure 6.1.

To isolate the reachable dynamics, we perform a coordinate transformation:

$$\begin{bmatrix} \tilde{x}_1(k) \\ \tilde{x}_2(k) \end{bmatrix} = \begin{bmatrix} \frac{1}{\sqrt{2}} & \frac{1}{\sqrt{2}} \\ \frac{1}{\sqrt{2}} & -\frac{1}{\sqrt{2}} \end{bmatrix} \begin{bmatrix} x_1(k) \\ x_2(k) \end{bmatrix},$$

we obtain

$$\begin{bmatrix} \tilde{x}_1(k+1) \\ \tilde{x}_2(k+1) \end{bmatrix} = \begin{bmatrix} 2 & 1 \\ 0 & a \end{bmatrix} \begin{bmatrix} \tilde{x}_1(k) \\ \tilde{x}_2(k) \end{bmatrix} + \begin{bmatrix} \sqrt{2} \\ 0 \end{bmatrix} u(k).$$

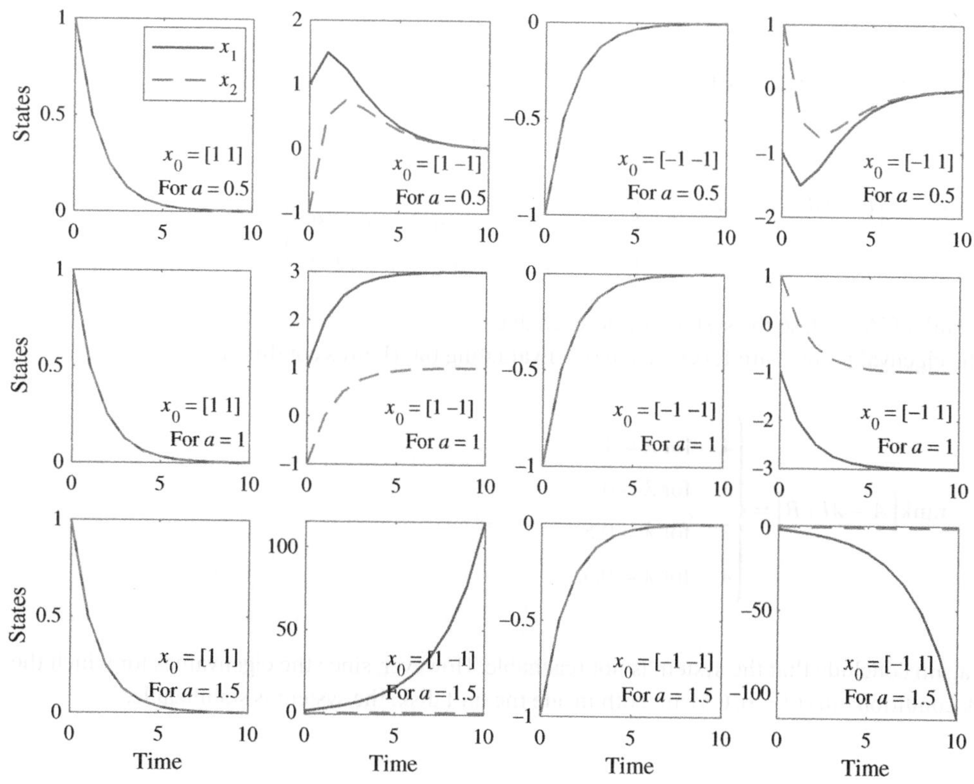

Figure 6.1 Transient responses of the system in Example 6.4 for various values of a. The system is stabilizable for $a = 0.5$ (top row), but not stabilizable for $a = 1$ (middle row) and $a = 1.5$ (bottom row). For $a = 1$ and $a = 1.5$, the response can be stable if the initial state lies in the *stabilizable subspace* defined by $x_1 = x_2$.

Here, $\tilde{x}_1 = (x_1 + x_2)/\sqrt{2}$ corresponds to the controllable portion of the state, while $\tilde{x}_2 = (x_1 - x_2)/\sqrt{2}$ evolves autonomously and cannot be influenced by the input. The stabilizability of the system depends on the parameter a: the system is stabilizable if and only if the autonomous dynamics are stable, i.e., if $|a| < 1$.

If $a \geq 1$, the system is not stabilizable. This behavior is illustrated in Figure 6.1, which shows the transient responses of the states $x_1(k)$ and $x_2(k)$ (starting at different $x(0)$) for various values of a. The bottom panels compare responses with $a = 1.5$, but with different initial conditions. When the initial state lies within the reachable subspace $x = [\alpha \quad \alpha]^T$, the system can still exhibit a stable closed-loop response—even when $a = 1.5$—highlighting the role of the controllable subspace in stabilization.

Example 6.5: Let us consider the following state-space system to further illustrate the concepts of reachability and stabilizability:

$$x(k+1) = \begin{bmatrix} 0.65 & 0.1 & 0 & -0.45 \\ 0.35 & 0.4 & 0.3 & -0.15 \\ 0 & 0.25 & 0.35 & -0.2 \\ -0.2 & -0.05 & -0.15 & 0.4 \end{bmatrix} x(k) + \begin{bmatrix} 0.5 \\ 0.5 \\ 0 \\ 0 \end{bmatrix} u(k). \tag{6.20}$$

$$y(k) = \begin{bmatrix} 2 & 0 & 0 & -2 \end{bmatrix} x(k)$$

The controllability matrix is

$$\mathcal{W}^c = \begin{bmatrix} B & AB & A^2B & A^3B \end{bmatrix} = \begin{bmatrix} 0.5 & 0.375 & 0.3375 & 0.3263 \\ 0.5 & 0.375 & 0.3375 & 0.3263 \\ 0 & 0.125 & 0.1625 & 0.1738 \\ 0 & -0.125 & -0.1625 & -0.1738 \end{bmatrix}.$$

The rank of $\mathcal{W}^c$ is 2, so the system is not reachable.

The eigenvalues of A are 1, 0, 0.5, and 0.3. Evaluating the Hautus condition:

$$\operatorname{rank}\begin{bmatrix} A - \lambda I \mid B \end{bmatrix} = \begin{cases} 4 & \text{for } \lambda = 1 \\ 3 & \text{for } \lambda = 0 \\ 3 & \text{for } \lambda = 0.5 \\ 4 & \text{for } \lambda = 0.3. \end{cases}$$

We again conclude that the system is not reachable. However, since the eigenvalues for which the rank condition fails ($\lambda = 0, 0.5$) are both inside the unit disk, the system is stabilizable.

Performing SVD on $\mathcal{W}^c$, we obtain

$$\mathcal{W}^c = \begin{bmatrix} 0.6823 & -0.1857 & 0.7071 & 0 \\ 0.6823 & -0.1857 & -0.7071 & 0 \\ 0.1857 & 0.6823 & 0 & 0.7071 \\ -0.1857 & -0.6823 & 0 & 0.7071 \end{bmatrix} \begin{bmatrix} 1.1438 & 0 & 0 & 0 \\ 0 & 0.2412 & 0 & 0 \\ 0 & 0 & 0 & 0 \\ 0 & 0 & 0 & 0 \end{bmatrix}$$

$$\times \begin{bmatrix} 0.5965 & -0.7699 & 0.1184 & -0.1932 \\ 0.4880 & 0.1296 & -0.5943 & 0.6260 \\ 0.4554 & 0.3995 & 0.7476 & 0.2721 \\ 0.4457 & 0.4804 & -0.2716 & -0.7048 \end{bmatrix}^T .$$

The first two left singular vectors u_1 and u_2 span the reachable subspace. The remaining singular vectors u_3 and u_4 correspond to the zero singular values and span the orthogonal complement. Since these vectors are not unique, any orthonormal basis spanning the same subspace will suffice. For instance, we could equivalently use

$$u_3 = \begin{bmatrix} 0.5 \\ -0.5 \\ -0.5 \\ -0.5 \end{bmatrix}, \qquad u_4 = \begin{bmatrix} 0.5 \\ -0.5 \\ 0.5 \\ 0.5 \end{bmatrix}.$$

Using the coordinate transformation $z = U^{-1}x = U^T x$, the state-space matrices in the new coordinates are

$$\tilde{A} = U^{-1}AU = \begin{bmatrix} 0.9136 & 0.3510 & 0.2281 & -0.1579 \\ 0.1510 & 0.3864 & -0.1139 & -0.0088 \\ 0 & 0 & 0.3 & -0.3 \\ 0 & 0 & -0.2 & 0.2 \end{bmatrix}$$

$$\tilde{B} = U^{-1}B = \begin{bmatrix} 0.6823 \\ -0.1857 \\ 0 \\ 0 \end{bmatrix}$$

$$\tilde{C} = CU = \begin{bmatrix} 1.7360 & 0.9931 & 1.4142 & -1.4142 \end{bmatrix}.$$

From these expressions, it is clear that the first two state components in the transformed coordinates are reachable, while the last two are not. The dynamics of the unreachable subsystem are governed by the matrix:

$$\begin{bmatrix} 0.3 & -0.3 \\ -0.2 & 0.2 \end{bmatrix},$$

whose eigenvalues are 0.5 and 0. Since both lie inside the unit disk, the unreachable modes are stable, and the system is therefore stabilizable.

6.4 Observability, Reconstructability, and Detectability

Another desirable property for output feedback control is *observability* (also referred to as *reconstructability*), which concerns whether it is possible to infer the entire state trajectory from the measured output sequence. A related but weaker property is *detectability*, which—analogous to stabilizability—requires that the unobservable modes be stable.

6.4.1 Definitions

Definition 6.6: System Eq. (6.1) is said to be *observable* if no two distinct initial states, under the same input sequence, produce identical output sequences for all time.

Definition 6.7: A system is called *detectable* if it is possible to construct, from the output measurements, a sequence of state estimates that converges asymptotically to the true state as $k \to \infty$, regardless of the initial estimate.

If a system is observable, then any arbitrary and unknown initial state $x(0)$ can be uniquely reconstructed from the input sequence $u(0), u(1), \dots, u(k-1)$ and the corresponding output sequence $y(0), y(1), \dots, y(k-1)$ in finite time k. For time-invariant systems, observability is thus equivalent to *reconstructability*, which addresses whether the current state can be determined from a finite history of past outputs and inputs.

6.4.2 Conditions for Observability

The following are some useful results for testing observability:

6.4.2.1 Rank Condition

The most widely used test for observability is the rank condition:

$$\text{rank}\,(\mathcal{W}_n^o) = n, \tag{6.21}$$

where

$$\mathcal{W}_n^o = \begin{bmatrix} C^T & (CA)^T & \cdots & (CA^{n-1})^T \end{bmatrix}^T. \tag{6.22}$$

This condition can be proven using an argument analogous to that used in the proof of reachability, via the Cayley–Hamilton theorem. The proof is left as an exercise.

The matrix $\mathcal{W}_n^o$ is referred to as the *observability matrix*. Its null space defines the *unobservable subspace*—that is, the set of initial states that produce zero output for all time. Any state trajectory starting entirely within the unobservable subspace cannot be distinguished from the zero state based on output measurements alone.

From the observability rank condition, we can deduce that observability of system Eq. (6.1) is equivalent to reachability of its *dual system*:

$$\begin{aligned} x(k+1) &= A^T x(k) + C^T u(k) \\ y(k) &= B^T x(k). \end{aligned} \tag{6.23}$$

6.4.2.2 Hautus Condition

Another widely used test for observability is the Hautus condition:

$$\operatorname{rank}\left(\left[A^T - \lambda I \mid C^T\right]\right) = n \qquad \forall \lambda \in \Lambda_A. \tag{6.24}$$

This condition follows directly from the duality principle: observability of Eq. (6.1) is equivalent to reachability of the dual system Eq. (6.23), and the Hautus condition for reachability applies accordingly.

For detectability, the Hautus condition becomes

$$\operatorname{rank}\left(\left[A^T - \lambda I \mid C^T\right]\right) = n \qquad \forall \lambda \in \Lambda_A^+. \tag{6.25}$$

6.4.3 Coordinate Transformation

As in the case of controllability, we can construct a coordinate transformation that brings the system Eq. (6.1) into a form that explicitly separates unobservable subspace from the rest. Define the transformation:

$$\begin{bmatrix} \tilde{x}_1 \\ \tilde{x}_2 \end{bmatrix} = \begin{bmatrix} T_1^o & T_2^o \end{bmatrix}^{-1} x. \tag{6.26}$$

which transforms the original system into

$$\begin{aligned} \begin{bmatrix} \tilde{x}_1(k+1) \\ \tilde{x}_2(k+1) \end{bmatrix} &= \begin{bmatrix} \tilde{A}_{11} & 0 \\ \tilde{A}_{21} & \tilde{A}_{22} \end{bmatrix} \begin{bmatrix} \tilde{x}_1(k) \\ \tilde{x}_2(k) \end{bmatrix} + \begin{bmatrix} \tilde{B}_1 \\ \tilde{B}_2 \end{bmatrix} u(k) \\ y(k) &= \begin{bmatrix} \tilde{C}_1 & 0 \end{bmatrix} \begin{bmatrix} \tilde{x}_1(k) \\ \tilde{x}_2(k) \end{bmatrix}. \end{aligned} \tag{6.27}$$

In this form, the pair $(\tilde{C}_1, \tilde{A}_{11})$ is observable, and the state $\tilde{x}_2$ is unobservable from the output $y(k)$. That is, the values of $\tilde{x}_2$ do not influence the current or future outputs and therefore cannot be reconstructed from measurements. The system Eq. (6.27) (and thus the original system Eq. [6.1]) is detectable if all eigenvalues of $\tilde{A}_{22}$ lie strictly inside the unit disk.

To construct such a transformation, choose T_2^o such that its columns span the null space of $\mathcal{W}_n^o$, and T_1^o as its orthogonal complement. A convenient way to obtain T_1^o and T_2^o is to apply the SVD to $\mathcal{W}_n^o$, as will be illustrated in the following examples.

Example 6.6: Let us consider the dual system of Eq. (6.19), which was analyzed in Example 6.4, to illustrate the concepts of observability and the unobservable subspace:

$$\begin{aligned} \begin{bmatrix} x_1(k+1) \\ x_2(k+1) \end{bmatrix} &= \begin{bmatrix} 1.5+0.5a & 1.5-0.5a \\ 0.5-0.5a & 0.5+0.5a \end{bmatrix} \begin{bmatrix} x_1(k) \\ x_2(k) \end{bmatrix} \\ y(k) &= \begin{bmatrix} 1 & 1 \end{bmatrix} \begin{bmatrix} x_1(k) \\ x_2(k) \end{bmatrix}. \end{aligned} \tag{6.28}$$

The observability matrix is

$$\mathcal{W}_2^o = \begin{bmatrix} C^T & A^T C^T \end{bmatrix}^T = \left(\begin{bmatrix} 1 \\ 1 \end{bmatrix} \begin{bmatrix} 2 \\ 2 \end{bmatrix} \right).$$

The rank of $\mathcal{W}_2^o$ is 1, so the system is not observable.

The SVD of $\mathcal{W}_2^o$ is

$$\mathcal{W}_2^o = \begin{bmatrix} \frac{1}{\sqrt{5}} & \frac{2}{\sqrt{5}} \\ \frac{2}{\sqrt{5}} & -\frac{1}{\sqrt{5}} \end{bmatrix} \begin{bmatrix} \sigma_1 & 0 \\ 0 & 0 \end{bmatrix} \begin{bmatrix} \frac{1}{\sqrt{2}} & \frac{1}{\sqrt{2}} \\ \frac{1}{\sqrt{2}} & -\frac{1}{\sqrt{2}} \end{bmatrix}.$$

The null space of $\mathcal{W}_2^o$ is spanned by the second right singular vector (i.e., $\left[\frac{1}{\sqrt{2}}, \frac{-1}{\sqrt{2}}\right]^T$), which defines the unobservable subspace of the system.

To separate the observable and unobservable components, we perform the coordinate transformation:

$$\begin{bmatrix} \tilde{x}_1(k) \\ \tilde{x}_2(k) \end{bmatrix} = \begin{bmatrix} \frac{1}{\sqrt{2}} & \frac{1}{\sqrt{2}} \\ \frac{1}{\sqrt{2}} & -\frac{1}{\sqrt{2}} \end{bmatrix} \begin{bmatrix} x_1(k) \\ x_2(k) \end{bmatrix},$$

we obtain this time

$$\begin{bmatrix} \tilde{x}_1(k+1) \\ \tilde{x}_2(k+1) \end{bmatrix} = \begin{bmatrix} 2 & 0 \\ 1 & a \end{bmatrix} \begin{bmatrix} \tilde{x}_1(k) \\ \tilde{x}_2(k) \end{bmatrix}$$

$$y(k) = \begin{bmatrix} \sqrt{2} & 0 \end{bmatrix} \begin{bmatrix} \tilde{x}_1(k) \\ \tilde{x}_2(k) \end{bmatrix}.$$

Here, $\tilde{x}_1$ represents the observable component of the state, while $\tilde{x}_2$ lies entirely in the unobservable subspace and does not influence the output. Since $\tilde{x}_2$ evolves autonomously, the detectability of the system depends on the parameter a. Specifically, if $a \geq 1$, the unobservable mode is unstable, and the system is not detectable.

Example 6.7: Let us revisit the system Eq. (6.20), previously used in Example 6.5 to demonstrate reachability and stabilizability. This time, we analyze the system for observability and detectability.

First, we compute the observability matrix:

$$\mathcal{W}^o = \begin{bmatrix} C^T & A^TC^T & (A^T)^2C^T & (A^T)^3C^T \end{bmatrix}^T = \begin{bmatrix} 2 & 1.7 & 1.55 & 1.475 \\ 0 & 0.3 & 0.45 & 0.525 \\ 0 & 0.3 & 0.45 & 0.525 \\ -2 & -1.7 & -1.55 & -1.475 \end{bmatrix}^T.$$

The rank of $\mathcal{W}^o$ is 2, so the system is not observable.

To apply the Hautus test, we evaluate the rank of the matrix $\begin{bmatrix} A^T - \lambda I & C^T \end{bmatrix}$ for each eigenvalue of A. The ranks are 4, 3, 4, 3 for $\lambda = 1, 0, 0.5, 0.3$, respectively. Hence, the system is not observable.

However, since the eigenvalues that result in a loss of rank ($\lambda = 0$ and 0.3) lie strictly inside the unit disk, the system is still detectable.

Next, we perform SVD of $\mathcal{W}^o$. The SVD of its transpose $(\mathcal{W}^o)^T$ is given by

$$(\mathcal{W}^o)^T = \begin{bmatrix} 0.6964 & -0.1227 & 0.7018 & -0.0867 \\ 0.1227 & 0.6964 & 0.0867 & 0.7018 \\ 0.1227 & 0.6964 & -0.0867 & -0.7018 \\ -0.6964 & 0.1227 & 0.7018 & -0.0867 \end{bmatrix} \times$$

$$\begin{bmatrix} 4.8615 & 0 & 0 & 0 \\ 0 & 0.6618 & 0 & 0 \\ 0 & 0 & 0 & 0 \\ 0 & 0 & 0 & 0 \end{bmatrix} \begin{bmatrix} 0.5730 & -0.7416 & 0.3478 & -0.0273 \\ 0.5022 & 0.001 & -0.7990 & 0.3308 \\ 0.4668 & 0.3724 & -0.0379 & -0.8013 \\ 0.4491 & 0.558 & 0.489 & 0.4978 \end{bmatrix}^T$$

The unobservable subspace is spanned by u_3 and u_4, the left singular vectors corresponding to the zero singular values of $(\mathcal{W}^o)^T$, which span the null space of $\mathcal{W}^o$. These vectors are not unique—any pair of orthonormal vectors spanning the same subspace will suffice. For instance, we could equivalently use:

$$u_3 = \begin{bmatrix} 0.5 \\ 0.5 \\ -0.5 \\ 0.5 \end{bmatrix}, \quad u_4 = \begin{bmatrix} 0.5 \\ -0.5 \\ 0.5 \\ 0.5 \end{bmatrix},$$

as they span the same subspace.

Using the coordinate transformation $z = U^{-1}x = U^T x$, we compute the transformed state-space matrices:

$$\tilde{A} = U^{-1}AU = \begin{bmatrix} 0.9294 & 0.1008 & 0 & 0 \\ 0.3008 & 0.5706 & 0 & 0 \\ 0.2549 & -0.0891 & 0.2350 & 0.0833 \\ 0.1261 & 0.0344 & 0.1833 & 0.065 \end{bmatrix}$$

$$\tilde{B} = U^{-1}B = \begin{bmatrix} 0.4095 \\ 0.2868 \\ 0.3942 \\ 0.3075 \end{bmatrix}$$

$$\tilde{C} = CU = \begin{bmatrix} 2.7855 & -0.4908 & 0 & 0 \end{bmatrix}.$$

In the transformed coordinates, the first two states are observable, and the remaining two are unobservable. The submatrix

$$\begin{bmatrix} 0.2350 & 0.0833 \\ 0.1833 & 0.065 \end{bmatrix},$$

governs the dynamics of the unobservable states. Since its eigenvalues are 0.3 and 0, which lie inside the unit disk, the system is detectable.

6.5 Kalman Decomposition and Minimal Realization

6.5.1 Kalman Decomposition

Kalman (1961) showed that a linear time-invariant state-space system can always be transformed via a suitable coordinate change into a structured form that reveals its internal reachability and observability properties. Specifically, there exists a coordinate transformation such that the system takes the form:

$$\begin{bmatrix} \tilde{x}_1(k+1) \\ \tilde{x}_2(k+1) \\ \tilde{x}_3(k+1) \\ \tilde{x}_4(k+1) \end{bmatrix} = \begin{bmatrix} \tilde{A}_{11} & \tilde{A}_{12} & 0 & 0 \\ 0 & \tilde{A}_{22} & 0 & 0 \\ \tilde{A}_{31} & \tilde{A}_{32} & \tilde{A}_{33} & \tilde{A}_{34} \\ 0 & \tilde{A}_{42} & 0 & \tilde{A}_{44} \end{bmatrix} \begin{bmatrix} \tilde{x}_1(k) \\ \tilde{x}_2(k) \\ \tilde{x}_3(k) \\ \tilde{x}_4(k) \end{bmatrix} + \begin{bmatrix} \tilde{B}_1 \\ 0 \\ \tilde{B}_3 \\ 0 \end{bmatrix} u(k)$$

$$y(k) = \begin{bmatrix} \tilde{C}_1 & \tilde{C}_2 & 0 & 0 \end{bmatrix} \begin{bmatrix} \tilde{x}_1(k) \\ \tilde{x}_2(k) \\ \tilde{x}_3(k) \\ \tilde{x}_4(k) \end{bmatrix}. \tag{6.29}$$

This decomposition separates the state into four components:

- $\tilde{x}_1$: states that are both reachable and observable,
- $\tilde{x}_2$: states that are observable but not reachable,
- $\tilde{x}_3$: states that are reachable but not observable,
- $\tilde{x}_4$: states that are neither reachable nor observable.

The transformation $\tilde{x} = \begin{bmatrix} T_1 & T_2 & T_3 & T_4 \end{bmatrix}^{-1} x$ achieves this structure, where each submatrix T_i is chosen to span the corresponding subspace:

- T_1: linearly independent vectors in the range of $\mathcal{W}_n^c$ but not in the null space of $\mathcal{W}_n^o$,
- T_2: linearly independent vectors in the orthogonal complement of the range of $\mathcal{W}_n^c$ but not in the null space of $\mathcal{W}_n^o$,
- T_3 and T_4: similarly defined to span the remaining disjoint subspaces.

This construction provides a canonical decomposition that explicitly reveals the structural properties of the system. An illustrative example is given below.

Example 6.8: *Kalman Decomposition of a Four-state System.*
We revisit the system from Eq. (6.20), previously analyzed in Examples 6.5 and 6.7 to illustrate reachability and observability. Here, we apply Kalman decomposition to separate the state space into subspaces with distinct reachability and observability properties.

From the preceding SVDs of the controllability and observability matrices, we extract the subspaces $\mathcal{R}_\perp$ (the orthogonal complement of the reachable subspace) and $\mathcal{S}$ (the unobservable subspace):

$$\mathcal{R}_\perp = \text{Span}\left\{\begin{bmatrix} 0.5 \\ -0.5 \\ -0.5 \\ -0.5 \end{bmatrix}, \begin{bmatrix} 0.5 \\ -0.5 \\ 0.5 \\ 0.5 \end{bmatrix}\right\}; \quad \mathcal{S} = \text{Span}\left\{\begin{bmatrix} 0.5 \\ 0.5 \\ -0.5 \\ 0.5 \end{bmatrix}, \begin{bmatrix} 0.5 \\ -0.5 \\ 0.5 \\ 0.5 \end{bmatrix}\right\}$$

$$\Downarrow$$

$$\Downarrow$$

$$\mathcal{R}_\perp \cap \mathcal{S} = \text{Span}\left\{\begin{bmatrix} 0.5 \\ -0.5 \\ 0.5 \\ 0.5 \end{bmatrix}\right\}; \quad \mathcal{R}_\perp \cup \mathcal{S} = \text{Span}\left\{\begin{bmatrix} 0.5 \\ -0.5 \\ -0.5 \\ -0.5 \end{bmatrix}, \begin{bmatrix} 0.5 \\ -0.5 \\ 0.5 \\ 0.5 \end{bmatrix}, \begin{bmatrix} 0.5 \\ 0.5 \\ -0.5 \\ 0.5 \end{bmatrix}\right\}$$

Based on this, we can classify the state directions as follows

- span$\left(\begin{bmatrix} 0.5 & -0.5 & 0.5 & 0.5 \end{bmatrix}\right)$ is neither reachable nor observable;
- span$\left(\begin{bmatrix} 0.5 & 0.5 & -0.5 & 0.5 \end{bmatrix}\right)$ is reachable but not observable;
- span$\left(\begin{bmatrix} 0.5 & -0.5 & -0.5 & -0.5 \end{bmatrix}\right)$ is observable but not reachable;
- span$\left(\begin{bmatrix} 0.5 & 0.5 & 0.5 & -0.5 \end{bmatrix}\right)$, the orthogonal complement of $\mathcal{R}_\perp \cup \mathcal{S}$, is both reachable and observable.

Using the coordinate transformation

$$z = \begin{bmatrix} 0.5 & 0.5 & 0.5 & 0.5 \\ 0.5 & -0.5 & 0.5 & -0.5 \\ 0.5 & -0.5 & -0.5 & 0.5 \\ -0.5 & -0.5 & 0.5 & 0.5 \end{bmatrix}^{-1} x,$$

we obtain the transformed system:

$$\tilde{A} = U^{-1}AU = \begin{bmatrix} 1 & 0.2 & 0 & 0 \\ 0 & 0.5 & 0 & 0 \\ 0.2 & 0.2 & 0.3 & 0.1 \\ 0 & 0.1 & 0 & 0 \end{bmatrix}$$

$$\tilde{B} = U^{-1}B = \begin{bmatrix} 0.5 \\ 0 \\ 0.5 \\ 0 \end{bmatrix}$$

$$\tilde{C} = CU = \begin{bmatrix} 2 & 2 & 0 & 0 \end{bmatrix}.$$

This form confirms Kalman decomposition: the transformed system reveals explicitly the reachable/observable structure of the original system.

6.5.2 Minimal Realization

A state-space system is said to be *minimal* if it is both reachable and observable. A *minimal realization* of a system refers to such a minimal system that is input-output equivalent to the original system but excludes any unreachable or unobservable state components. This terminology reflects the fact that, from an input-output perspective, unreachable or unobservable states do not contribute to the system behavior and can therefore be removed without affecting the input-output dynamics.

Applying Kalman decomposition to the system in Eq. (6.29), the minimal realization corresponds to the dynamics associated with $\tilde{x}_1$, the subspace that is both reachable and observable:

$$\begin{aligned} \tilde{x}_1(k+1) &= \tilde{A}_{11}\tilde{x}_1(k) + \tilde{B}u(k) \\ y(k) &= \tilde{C}_1\tilde{x}_1(k). \end{aligned} \tag{6.30}$$

Any system obtained by applying a state-space coordinate transformation to the above also qualifies as a minimal realization of the original system.

Example 6.9: *Minimal Realization of a 4-state System.*
From the Kalman decomposition of the system in Eq. (6.20), we identified a one-dimensional subspace that is both reachable and observable. A minimal realization, therefore, reduces the original system to

$$\begin{aligned} x(k+1) &= x(k) + 0.5u(k) \\ y(k) &= 2x(k). \end{aligned} \tag{6.31}$$

This corresponds to the scalar input-output relationship:

$$y(k) = y(k-1) + u(k-1), \tag{6.32}$$

which fully captures the input-output behavior of the original 4-state system.

In general, a high-order system can often be reduced to a lower-order model that retains the essential features of the input-output dynamics. This process is known as *model reduction*. An introduction to model reduction techniques is provided in Appendix C.

6.6 Disturbance Modeling

One of the primary objectives of control is to mitigate the impact of external disturbances on key process variables. While it is sometimes possible to eliminate disturbance sources through design changes, more commonly their effects must be counteracted by adjusting the manipulated variables. To design a controller that accomplishes this effectively, it is helpful to have a model that predicts how disturbances influence the outputs. Such a model may be developed from first principles or system identification. However, it is not necessary for the model to explicitly represent the physical origins of all disturbances. For the purposes of linear controller design, it is sufficient that the model captures the *aggregate effect* of disturbances on the output. This is particularly important because the physical sources of disturbances are often unknown or difficult to isolate.

In this section, we present stochastic models of disturbances and discuss how they can be integrated with deterministic system models for estimation and control.

6.6.1 Linear Stochastic System Model for Stationary Processes

Given the inherently uncertain nature of disturbances, a natural modeling framework is based on probability theory. Stochastic processes serve as the primary mathematical tool for representing such uncertainty. Basic concepts related to stochastic processes are reviewed in Appendix B.

In general, specifying a stochastic process requires defining the joint probability distribution over the entire time sequence—a task that is intractable in nearly all practical scenarios without simplifying assumptions. Two such assumptions that are commonly used are:

- *Stationarity*, which implies that the statistical properties (e.g., mean and covariance) of the process are time-invariant;
- *Normality*, which allows the process to be fully characterized by its first two moments—mean and covariance.

A widely adopted modeling approach for stationary stochastic disturbances is to represent them as the output of a linear system driven by an external white noise input. This is motivated by the fact that any stationary stochastic process (in terms of its first two moments) can be generated by a stable, stably invertible linear system excited by white noise. Within this framework, stochastic modeling reduces to identifying an appropriate linear system that reproduces the given mean and covariance structure. The system can be represented either in state-space form or as a transfer matrix.

6.6.1.1 Transfer Function

Spectral factorization theorem states that a given power spectral density (PSD) of a stationary random process can be factored into the product of a causal, stable filter and its Hermitian conjugate. This implies that any stationary stochastic process can be represented as the output of a linear system driven by white noise. Thus, the general form of such a model is

$$w(k) = H(q)\varepsilon(k). \tag{6.33}$$

Here, $w(k)$ denotes the so-called colored noise, $\varepsilon(k)$ is a zero-mean white noise sequence with covariance R_ε, and $H(q)$ is a stable, stably invertible transfer matrix. Without loss of generality, we impose $H(0) = I$, since any deviation from this condition can be absorbed into R_ε.

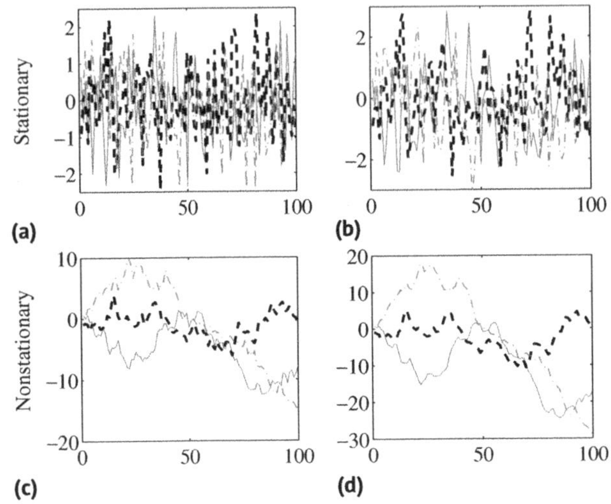

Figure 6.2 Stationary and nonstationary disturbances: (a) white noise, (b) colored stationary noise, (c) and (d) nonstationary disturbances. Three realizations are shown of each type of disturbance.

Figure 6.2a and b illustrate three sample realizations of a white noise sequence and a corresponding colored noise sequence, respectively. The colored noise is generated from the difference equation:

$$(1 - 0.5q^{-1})w(k) = q^{-1}\varepsilon(k).$$

The PSD of the process $w(k)$ can be derived by computing the autocovariance of $w(k)$ and taking its Fourier transform. The result is

$$\Phi_w(\omega) = H(e^{j\omega})R_\varepsilon H^T(e^{-j\omega}). \tag{6.34}$$

This expression provides a direct path for designing $H(q)$ to match a specified power spectrum (or equivalently, a target covariance function). Specifically, one can determine a spectral factor $\Phi_w^{1/2}(\omega)$ and then identify a stable, stably invertible transfer matrix whose frequency response matches this factor.

6.6.1.2 State-space Model

A stochastic process can also be modeled using the following stochastic difference equations in state-space form:

$$\begin{aligned} x_w(k+1) &= A_w x_w(k) + B_w \varepsilon(k) \\ w(k) &= C_w x_w(k) + \varepsilon(k), \end{aligned} \tag{6.35}$$

where $\varepsilon(k)$ is a zero-mean white noise sequence with covariance R_ε. All eigenvalues of A_w are assumed to lie inside the unit circle, ensuring stability.

We can now analyze the mean and covariance properties of the process. Taking expectations, we obtain

$$E\{x_w(k)\} = A_w E\{x_w(k-1)\} = A_w^k E\{x(0)\}, \tag{6.36}$$

$$E\{x_w(k)x_w^T(k)\} = A_w E\{x_w(k-1)x_w^T(k-1)\}A_w^T + B_w R_\varepsilon B_w^T, \tag{6.37}$$

$$E\{x_w(k+\tau)x_w^T(k)\} = A_w^\tau E\{x_w(k)x_w^T(k)\}. \tag{6.38}$$

From these relations, it follows that

$$\lim_{k\to\infty} E\{x(k)\} = 0$$
$$\lim_{k\to\infty} E\{x_w(k+\tau)x_w^T(k)\} = A_w^\tau \bar{P}_w, \tag{6.39}$$

where $\bar{P}_w$ is the steady-state covariance of $x_w(k)$, obtained as the unique positive semi-definite solution to the discrete-time Lyapunov equation:

$$\bar{P}_w = A_w \bar{P}_w A_w^T + B_w R_\varepsilon B_w^T. \tag{6.40}$$

Thus, the steady-state mean and covariance of the output $w(k)$ are given by

$$\bar{w} = \lim_{k\to\infty} E\{w(k)\} = 0$$
$$R_w(\tau) = \lim_{k\to\infty} E\{w(k+\tau)w^T(k)\} = C_w A_w^\tau \bar{P}_w C_w^T + R_\varepsilon. \tag{6.41}$$

Thus, the system defined by Eq. (6.35) becomes asymptotically stationary, with mean and autocovariance independent of the initial condition, as shown in Eq. (6.41).

Determining a suitable triple (A_w, B_w, C_w) to reproduce a given covariance function $R_w(\tau)$ is a nontrivial task, but several numerical methods are available. These will be discussed in a later chapter on system identification.

6.6.2 Stochastic System Models for Processes with Nonstationary Behavior

While the assumption of stationary disturbances is often a convenient simplification, it does not apply universally. Many real-world disturbances—particularly in chemical and biological processes—exhibit persistent or drifting behavior that is inconsistent with stationarity. For example, the bottom panels of Figure 6.2 show disturbance signals with sustained trends or abrupt shifts, which cannot be adequately captured by stationary stochastic models.

A general model of nonstationary noise would be difficult to specify and analyze due to its complexity. However, for many practical applications, nonstationary disturbances can be effectively represented using integrated white noise processes, which capture key features such as random jumps or Brownian-motion-like drifts.

An *integrated white noise process* is defined recursively as

$$\varepsilon_{\text{int}}(k+1) = \varepsilon_{\text{int}}(k) + \varepsilon(k), \tag{6.42}$$

where $\varepsilon(k)$ is a zero-mean white noise sequence with covariance R_ε. The covariance of $\varepsilon_{\text{int}}(k)$ grows unbounded with time:

$$E\{\varepsilon_{\text{int}}(k)\varepsilon_{\text{int}}^T(k)\} = E\{\varepsilon_{\text{int}}(k-1)\varepsilon_{\text{int}}^T(k-1)\} + R_\varepsilon, \tag{6.43}$$

indicating that $\varepsilon_{\text{int}}(k)$ is nonstationary.

More generally, a persistent or drifting disturbance can be modeled as a combination of a stationary component and an integrated white noise component. A typical state-space model for such a process is

$$x_w(k+1) = A_w x_w(k) + B_w \varepsilon_{\text{int}}(k)$$
$$w(k) = C_w x_w(k) + \varepsilon_{\text{int}}(k). \tag{6.44}$$

Since $\varepsilon_{\text{int}}(k)$ is nonstationary, the process $w(k)$ is also nonstationary. However, it is often advantageous to model the differenced process, defined as $\Delta w(k) \triangleq w(k) - w(k-1)$, which may be stationary even if $w(k)$ is not.

Differencing Eq. (6.44) over two time steps yields:

$$\begin{aligned} \Delta x_w(k+1) &= A_w \Delta x_w(k) + B_w \varepsilon(k) \\ \Delta w(k) &= C_w \Delta x_w(k) + \varepsilon(k). \end{aligned} \tag{6.45}$$

where $\varepsilon(k)$ is white noise. This formulation results in a zero-mean stationary model for $\Delta w(k)$, which is compatible with standard estimation and control frameworks. As before, any desired covariance function for $\Delta w(k)$ can be achieved by appropriately selecting (A_w, B_w, C_w).

Alternatively, a transfer function model can be used

$$w(k) = H(q)\varepsilon_{\text{int}}(k), \tag{6.46}$$

where $H(q)$ is a stable, stably invertible transfer function. Expressing the model in terms of the differenced variable yields:

$$\Delta w(k) = H(q)\varepsilon(k). \tag{6.47}$$

This leads to a stationary disturbance model suitable for control applications.

6.6.3 Models for Estimation and Control

The models used in the subsequent development of estimation and control theory combine both deterministic and stochastic components. The deterministic part captures the causal relationship between known inputs—such as manipulated variables—and the system outputs. The stochastic part models the statistical structure of the output variations due to unknown inputs, such as external disturbances or sensor noise.

The standard form of such a combined model is

$$\begin{aligned} x(k+1) &= Ax(k) + Bu(k) + \varepsilon_1(k) \\ y(k) &= Cx(k) + \varepsilon_2(k), \end{aligned} \tag{6.48}$$

where $\varepsilon_1(k)$ and $\varepsilon_2(k)$ are zero-mean Gaussian white noise sequences that represent process and measurement noise, respectively.

Such a model can be developed either from first principles (fundamental modeling) or through system identification techniques. However, it is important to emphasize that this is not merely a deterministic model with arbitrarily added white noise inputs. Simply appending white noise to a deterministic system does not, in general, yield an accurate statistical characterization of the residuals, and will typically result in poor predictive performance.

While the model Eq. (6.48) is often developed directly in this form, it can also be derived and interpreted as the superposition of two subsystems:

$$\begin{aligned} x_{\text{det}}(k+1) &= A_1 x_{\text{det}}(k) + B_1 u(k) \\ y_{\text{det}}(k) &= C_1 x_{\text{det}}(k), \end{aligned} \tag{6.49}$$

and

$$\begin{aligned} x_{\text{sto}}(k+1) &= A_2 x_{\text{sto}}(k) + \tilde{\varepsilon}_1(k) \\ y_{\text{sto}}(k) &= C_2 x_{\text{sto}}(k) + \tilde{\varepsilon}_2(k). \end{aligned} \tag{6.50}$$

Here, the total output is obtained as the sum of the deterministic and stochastic components:

$$y(k) = y_{\text{det}}(k) + y_{\text{sto}}(k). \tag{6.51}$$

In this sense, the two subsystems can be combined—by defining the augmented state as $x = \begin{bmatrix} x_{\text{det}}^T & x_{\text{sto}}^T \end{bmatrix}^T$—to obtain the overall system representation.

In this formulation:

- The deterministic subsystem models the predictable influence of the control input $u(k)$ on the output.
- The stochastic subsystem captures residual variations in the output using a statistical description (typically via a covariance function).
- This framework facilitates the development of optimal estimation and control strategies that account for both known inputs and random disturbances.

In the system identification section, we will discuss methods for constructing such combined models directly from input-output data, with attention to ensuring both structural consistency and statistical fidelity.

Example 6.10: *Disturbance Modeling in the Heavy Oil Fractionator.*
Consider the side-draw model from the heavy oil fractionator case study (Section 3.8), given by

$$y = G_{\text{sd}}u + G_{\text{ird}}d \quad = \frac{5.72}{(60s+1)}e^{-14s}u(s) + \frac{1.52}{(25s+1)}e^{-15s}d(s) \tag{6.52}$$

The step responses of these transfer functions were examined in Section 2.4. The disturbance in the reflux duty, $d(s)$, exhibiting drifting-type behavior, can be modeled as an integrated noise process. Using a sampling time of $t_s = 5$, the deterministic and stochastic continuous-time transfer functions are discretized into state-space form using MATLAB's `c2d` function and the approach discussed in Section 5.4.3. This yields

$$\begin{aligned} x_{\text{sd}}(k+1) &= \begin{bmatrix} 0.920 & 0.9514 & 0.2479 & 0 \\ 0 & 0 & 1.0 & 0 \\ 0 & 0 & 0 & 1.0 \\ 0 & 0 & 0 & 0 \end{bmatrix} x_{\text{sd}}(k) + \begin{bmatrix} 0 \\ 0 \\ 0 \\ 1.0 \end{bmatrix} u(k) \\ y_{\text{sd}}(k) &= \begin{bmatrix} 0.3813 & 0 & 0 & 0 \end{bmatrix} x_{\text{sd}}(k), \end{aligned} \tag{6.53}$$

and

$$\begin{aligned} x_{\text{ird}}(k+1) &= \begin{bmatrix} 0.8187 & 1.1329 & 0 & 0 \\ 0 & 0 & 1.0 & 0 \\ 0 & 0 & 0 & 1.0 \\ 0 & 0 & 0 & 0 \end{bmatrix} x_{\text{ird}}(k) + \begin{bmatrix} 0 \\ 0 \\ 0 \\ 1.0 \end{bmatrix} d(k) \\ y_{\text{ird}}(k) &= \begin{bmatrix} 0.2432 & 0 & 0 & 0 \end{bmatrix} x_{\text{ird}}(k). \end{aligned} \tag{6.54}$$

Since $d(k)$ in Eq. (6.54) is an integrated white noise, we first difference the model (see Section 6.6.2). The stochastic model may then be written as

$$\begin{bmatrix} \Delta x_{\text{ird}}(k+1) \\ y_{\text{ird}}(k+1) \end{bmatrix} = \begin{bmatrix} A_{\text{ird}} & 0 \\ C_{\text{ird}}A_{\text{ird}} & I \end{bmatrix} \begin{bmatrix} \Delta x_{\text{ird}}(k) \\ y_{\text{ird}}(k) \end{bmatrix} + \begin{bmatrix} B_{\text{ird}} \\ C_{\text{ird}}B_{\text{ird}} \end{bmatrix} \varepsilon(k) \tag{6.55}$$

Thereafter, we combine the two models by augmenting the state vector as

$$\tilde{x}(k) = \begin{bmatrix} x_{\text{sd}}^T(k) & \Delta x_{\text{ird}}^T(k) & y_{\text{ird}}^T(k) \end{bmatrix}^T,$$

to construct the composite state-space model in the form:

$$\begin{bmatrix} x_{\text{sd}}(k+1) \\ \Delta x_{\text{ird}}(k+1) \\ y_{\text{ird}}(k+1) \end{bmatrix} = \begin{bmatrix} A_{\text{sd}} & 0 & 0 \\ 0 & A_{\text{ird}} & 0 \\ 0 & C_{\text{ird}}A_{\text{ird}} & I \end{bmatrix} \begin{bmatrix} \Delta x_{\text{ird}}(k) \\ \Delta x_{\text{sd}}(k) \\ y_{\text{ird}}(k) \end{bmatrix} + \begin{bmatrix} B_{\text{sd}} \\ 0 \\ 0 \end{bmatrix} u(k) + \begin{bmatrix} 0 \\ B_{\text{ird}} \\ C_{\text{ird}}B_{\text{ird}} \end{bmatrix} \varepsilon(k)$$

$$y(k) = \begin{bmatrix} C_{\text{sd}} & 0 & I \end{bmatrix} \begin{bmatrix} x_{\text{sd}}(k) \\ \Delta x_{\text{ird}}(k) \\ y_{\text{ird}}(k) \end{bmatrix}, \tag{6.56}$$

where the matrices $(A_{\text{sd}}, B_{\text{sd}}, C_{\text{sd}})$ and $(A_{\text{ird}}, B_{\text{ird}}, C_{\text{ird}})$ are taken from Eqs. (6.53) and (6.54), respectively, and the white noise sequence $\varepsilon(k)$ represents the differenced disturbance variable $\Delta d(k)$.

Exercises

1. Consider the following discrete-time linear system:

$$x(k+1) = \begin{bmatrix} 0.5 & 0.3 \\ 0.2 & 0.6 \end{bmatrix} x(k) + \begin{bmatrix} 1 \\ 0 \end{bmatrix} u(k),$$

(a) Compute the eigenvalues of the state matrix A.

(b) Let the initial condition be $x(0) = \begin{bmatrix} 1 & 1 \end{bmatrix}^T$, and assume $u(k) = 0$ for all k (i.e., an autonomous system). Compute $x(10)$.

Repeat the above steps for the modified system:

$$x(k+1) = \begin{bmatrix} 0.5 & 0.3 \\ 0.2 & 1.2 \end{bmatrix} x(k) + \begin{bmatrix} 1 \\ 0 \end{bmatrix} u(k).$$

2. In Eq. (6.16), the state-space system is partitioned as

$$\begin{bmatrix} \tilde{x}_1(k+1) \\ \tilde{x}_2(k+1) \end{bmatrix} = \begin{bmatrix} \tilde{A}_{11} & \tilde{A}_{12} \\ 0 & \tilde{A}_{22} \end{bmatrix} \begin{bmatrix} \tilde{x}_1(k) \\ \tilde{x}_2(k) \end{bmatrix} + \begin{bmatrix} \tilde{B}_1 \\ 0 \end{bmatrix} u(k),$$

where the pair $(\tilde{A}_{11}, \tilde{B}_1)$ is controllable.

(a) Show that the overall system is not controllable.

(b) Show that, for any initial state $\tilde{x}_1(0)$, the system state can be driven to the origin in a finite number of steps.

(c) Show that the system is stabilizable if all eigenvalues of $\tilde{A}_{22}$ lie strictly inside the unit circle.

3. Consider the system:

$$x(k+1) = \begin{bmatrix} 0.4 & 0.6 \\ 0 & \alpha \end{bmatrix} x(k) + \begin{bmatrix} 1 \\ 0 \end{bmatrix} u(k)$$

(a) Show that the system is not controllable by computing the controllability matrix.

(b) Choose $\alpha < 1$ (e.g., $\alpha = 0.8$) and verify that the system is stabilizable.

(c) Repeat with $\alpha > 1$ (e.g., $\alpha = 1.2$) and demonstrate that the system is not stabilizable.

4. Consider the state-space system partitioned as

$$\begin{bmatrix} \tilde{x}_1(k+1) \\ \tilde{x}_2(k+1) \end{bmatrix} = \begin{bmatrix} \tilde{A}_{11} & 0 \\ \tilde{A}_{21} & \tilde{A}_{22} \end{bmatrix} \begin{bmatrix} \tilde{x}_1(k) \\ \tilde{x}_2(k) \end{bmatrix} + \begin{bmatrix} \tilde{B}_1 \\ \tilde{B}_2 \end{bmatrix} u(k)$$

$$y(k) = \begin{bmatrix} C_1 & 0 \end{bmatrix} \begin{bmatrix} \tilde{x}_1(k) \\ \tilde{x}_2(k) \end{bmatrix},$$

where the pair $(\tilde{C}_1, \tilde{A}_{11})$ is observable.

(a) Show that the full system is not observable.

(b) Demonstrate that $\tilde{x}_1(k)$ can be reconstructed from output measurements in finite steps.

(c) Show that the system is detectable if $\tilde{A}_{22}$ is stable.

5. Show that controllability of the system:

$$\begin{aligned} x(k+1) &= Ax(k) + Bu(k) \\ y(k) &= Cx(k), \end{aligned}$$

implies observability of its dual system:

$$\begin{aligned} x(k+1) &= A^T x(k) + C^T u(k) \\ y(k) &= B^T x(k). \end{aligned}$$

Use the structure of the controllability and observability matrices and the duality of their ranks to establish the result.

6. Prove that for any symmetric positive definite matrix $Q > 0$, there exists a unique symmetric positive definite solution $P > 0$ to the discrete-time Lyapunov equation

$$P = APA^T + Q, \tag{6.57}$$

if and only if the matrix A is stable (i.e., all eigenvalues of A lie strictly inside the unit circle).

7. Prove the rank condition for observability using the Cayley–Hamilton theorem. Explain in what sense the null space of the observability matrix $\mathcal{W}_n^o$ represents the unobservable subspace.

8. Prove the Hautus criterion for stabilizability: show that the pair (A, B) is stabilizable if and only if

$$\text{rank}\left(\left[A - \lambda I \mid B\right]\right) = n \qquad \forall \lambda \in \Lambda_A^+, \tag{6.58}$$

where Λ_A^+ denotes the set of all eigenvalues of A that lie on or outside the unit disk.

9. Derive the formula

$$\Phi_w(\omega) = H(e^{j\omega})R_\varepsilon H^T(e^{-j\omega}), \tag{6.59}$$

for the power spectral density (PSD) of the stochastic process defined by

$$w(k) = H(q)\varepsilon(k) \tag{6.60}$$

where $\varepsilon(k)$ is a zero-mean white noise sequence with covariance matrix R_ε.

10. Given a simple spectral density function $\Phi_w(\omega)$ (e.g., of the form $1 + 0.5\cos(\omega)$), perform the following:
(a) Compute a spectral factor $H(e^{j\omega})$ such that $\Phi_w(\omega) = H(e^{j\omega})H^*(e^{j\omega})$.
(b) Realize $H(q)$ as a stable and stably invertible transfer function.
(c) Generate a realization of white noise, pass it through the filter $H(q)$, and obtain the output sequence.
(d) Compute the empirical covariance function of the output and verify that its Fourier transform approximates the original spectral density.

11. Show that Eq. (6.56) is equivalent to

$$\begin{bmatrix} \Delta x_{\text{sd}}(k+1) \\ \Delta x_{\text{ird}}(k+1) \\ y(k+1) \end{bmatrix} = \begin{bmatrix} A_{\text{sd}} & 0 & 0 \\ 0 & A_{\text{ird}} & 0 \\ C_{\text{sd}}A_{\text{sd}} & C_{\text{ird}}A_{\text{ird}} & I \end{bmatrix} \begin{bmatrix} \Delta x_{\text{sd}}(k) \\ \Delta x_{\text{ird}}(k) \\ y(k) \end{bmatrix} + \begin{bmatrix} B_{\text{sd}} \\ 0 \\ CB_{\text{sd}} \end{bmatrix} \Delta u(k) + \begin{bmatrix} 0 \\ B_{\text{ird}} \\ C_{\text{ird}}B_{\text{ird}} \end{bmatrix} \varepsilon(k)$$

$$y(k) = \begin{bmatrix} 0 & 0 & I \end{bmatrix} \begin{bmatrix} \Delta x_{\text{sd}}(k) \\ \Delta x_{\text{ird}}(k) \\ y_{\text{ird}}(k) \end{bmatrix}.$$

This can be achieved by differencing both the deterministic system model in Eq. (6.53) and the stochastic system model in Eq. (6.54) before combining them through augmentation.

12. Revisit Problem 8 from Chapter 5, where a FIR model was developed. Perform balanced truncation on this FIR model and obtain a reduced-order realization that approximates the original system with lower complexity. Compare the time-domain responses of the original and reduced models.

7

State Estimation

I have been aware from the outset that the deep analysis of something which is now called Kalman filtering were of major importance. But even with this immodesty I did not quite anticipate all the reactions to this work.

– Rudolf Kalman

Estimating the values of state variables based on a process model and available measurements is a fundamental problem in control applications. In practice, it is rarely feasible to measure all disturbances and state variables directly. Therefore, one must estimate the system state using measured input and output data—a task known as *state estimation*.

We now revisit the standard state-space model introduced in Chapter 6:

$$\begin{aligned} x(k+1) &= Ax(k) + Bu(k) + \varepsilon_1(k) \\ y_m(k) &= Cx(k) + \varepsilon_2(k), \end{aligned} \tag{7.1}$$

where $\varepsilon_1(k)$ and $\varepsilon_2(k)$ are zero-mean white noise sequences with the following covariance structure:

$$E\left\{\begin{bmatrix} \varepsilon_1(k) \\ \varepsilon_2(k) \end{bmatrix}\begin{bmatrix} \varepsilon_1(k) \\ \varepsilon_2(k) \end{bmatrix}^T\right\} = \begin{bmatrix} R_1 & R_{1,2} \\ R_{1,2}^T & R_2 \end{bmatrix}. \tag{7.2}$$

The objective of state estimation is to determine $x(k+i)$ for $i \geq 0$, using the available inputs $u(j)$ and measured outputs $y_m(j)$ for $j \leq k$. Estimating $x(k)$ corresponds to *filtering*, while estimating $x(k+i)$ for $i > 0$ is referred to as *prediction*. In addition, one may be interested in estimating $x(k+i)$ for $i < 0$, a task known as *smoothing*. These three estimation problems are depicted schematically in Figure 7.1.

State estimation techniques range from simple open-loop estimators to statistically optimal methods such as the Kalman filter. In this chapter, we examine several widely used approaches. State estimation is a critical component of Model Predictive Control (MPC). A solid understanding of state estimation is essential for effective controller design. Moreover, it plays a central role in parameter estimation problems, such as those encountered in system identification, which will be discussed in a later chapter.

In the context of MPC, the primary role of state estimation is to realign the model state with the actual process state using available measurements, thereby enabling accurate multistep prediction of both measured and unmeasured outputs. While much of the literature on state estimation

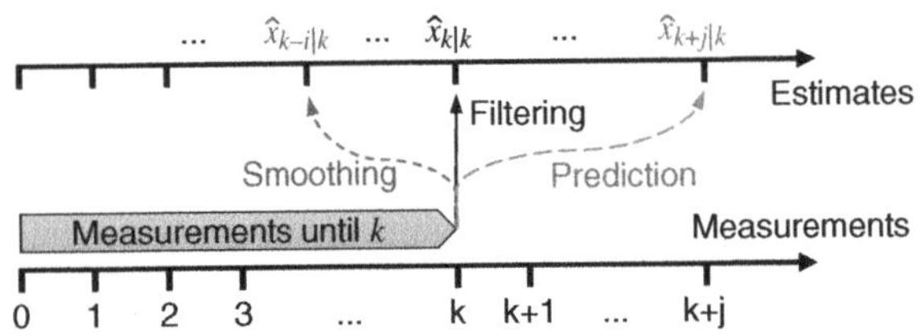

Figure 7.1 Comparison of prediction, filtering, and smoothing at time k. The lower timeline indicates the available measurements up to time k, while the upper timeline shows the corresponding state estimates computed for different time indices.

focuses on algorithmic techniques, the importance of appropriate disturbance modeling is often underemphasized. A common misconception is that satisfactory estimation can be achieved by simply adding white noise to the state and output equations of a deterministic model. In practice, this approach can yield poor results regardless of the estimation technique used. To obtain reliable estimates, disturbances (or their aggregate effects) must be modeled using suitable stationary or nonstationary stochastic processes. The system equations should then be augmented with the corresponding disturbance models, as illustrated in Chapter 6.

7.1 Linear Estimator Structure

We begin with a simple linear state estimator structure for the system defined in Eq. (7.1):

$$\hat{x}(k|k-1) = A\hat{x}(k-1|k-1) + Bu(k-1), \tag{7.3}$$

$$\hat{x}(k|k) = \hat{x}(k|k-1) + K_f(y_m(k) - C\hat{x}(k|k-1)). \tag{7.4}$$

Here, $\hat{x}(i|j)$ denotes an estimate of the state $x(i)$ based on measurements available up to time j. These two recursive equations allow us to compute the filtered state estimate $\hat{x}(k|k)$.

- Equation (7.3), called the *model propagation* or *model forwarding* step, projects the prior state estimate one step ahead based on the system model, without incorporating any new measurements.
- Equation (7.4), called the *measurement update*, corrects the predicted estimate using the observed discrepancy between the measured output $y_m(k)$ and its predicted value based on $\hat{x}(k|k-1)$. This discrepancy $(y_m(k) - C\hat{x}(k|k-1))$ is called the *innovation*.

In some applications—particularly when control inputs require a full sample period to compute—it is preferable to calculate the one-step-ahead prediction $\hat{x}(k+1|k)$ instead of the filtered estimate. This can be achieved by simply reversing the order of operations in (7.3) and (7.4), i.e., perform the measurement update first, followed by the model propagation:

$$\begin{aligned}\hat{x}(k|k) &= \hat{x}(k|k-1) + K_f\,(y_m(k) - C\hat{x}(k|k-1))\\ \hat{x}(k+1|k) &= A\hat{x}(k|k) + Bu(k),\end{aligned} \tag{7.5}$$

$$\Downarrow$$

$$\hat{x}(k+1|k) = A\hat{x}(k|k-1) + Bu(k) + \underbrace{AK_f}_{K}\{y_m(k) - C\hat{x}(k|k-1)\} \tag{7.6}$$

The free parameter in the above estimator is K_f (or equivalently, $K = AK_f$), commonly referred to as the *filter gain matrix* (or *predictor gain matrix*). In this chapter, we will use the terms filter and predictor interchangeably. When the context is clear, we may refer to K simply as the filter gain.

The choice of K is critical: It determines how the discrepancy between the measured output $y_m(k)$ and its predicted value is used to correct the state estimate. That is, K governs how measurement errors are translated into adjustments in the estimated state.

To evaluate the performance of the estimator, we define two types of estimation error:

$$x_e(k+1) \stackrel{\Delta}{=} x(k+1) - \hat{x}(k+1|k), \tag{7.7}$$

$$\tilde{x}_e(k) \stackrel{\Delta}{=} x(k) - \hat{x}(k|k). \tag{7.8}$$

In general, the gain matrix K should be selected to minimize the estimation error according to a suitable performance criterion, such as the expected value of the squared error.

The error dynamics equations can be derived directly from the estimator formulation. Substituting (7.6) into (7.7) and using (7.1), the equation for propagation of the one-step-ahead prediction error is obtained as

$$x_e(k+1) = (A - KC)x_e(k) + \varepsilon_1(k) + K\varepsilon_2(k). \tag{7.9}$$

Similarly, the error dynamics for the filtered estimate is given by

$$\tilde{x}_e(k+1) = (A - K_fCA)\tilde{x}_e(k) + (I - K_fC)\varepsilon_1(k) + K_f\varepsilon_2(k+1). \tag{7.10}$$

This deterministic linear estimator structure of (7.6) is commonly referred to as the Luenberger observer, named after David Luenberger, who introduced this method in his 1963 PhD thesis at Stanford University. The observer gain K serves as the primary design parameter and can be selected using various approaches, including pole placement (see Section 7.2) or optimal techniques such as the Kalman filter (see Section 7.3).

7.2 Observer Pole Placement

The theory of state observers is rooted in the behavior of the estimation error under deterministic conditions (i.e., with ε_1 and ε_2 set to zero). The primary concern is the reconstruction of the state sequence when the initial condition is unknown or only approximately known. From (7.9), it is clear that in the deterministic setting, the error dynamics are governed by the transition matrix $A - KC$. The stability of an observer is defined as follows:

Definition 7.1: An estimator of the form (7.6) for a system described by (7.1) is said to be *observer-stable* if $x_e(k) \to 0$ as $k \to \infty$ for any initial error $x_e(0)$, assuming $\varepsilon_1(i) = 0$ and $\varepsilon_2(i) = 0$ for all $i \geq 0$.

Under these assumptions, the error dynamics reduce to

$$x_e(k+1) = (A - KC)x_e(k). \tag{7.11}$$

It follows that observer stability requires all eigenvalues of $A - KC$ to lie strictly within the unit disk.

Moreover, note that the eigenvalues of $(A - K_f CA)$ are the same as those of $(A - \underbrace{AK_f}_{K} C)$, since XY and YX share the same eigenvalues when X and Y are square matrices. Therefore, stability of the one-step-ahead predictor (7.6) implies the stability of the filtered estimator (7.4), and vice versa.

The eigenvalues of $A - KC$ are referred to as the *observer poles*. Determining the gain K to place these poles at desired locations is known as *pole placement*. If the pair (C, A) is observable, it can be shown that the observer poles can be placed arbitrarily through an appropriate choice of K. The proof is left as an exercise.

Conventional pole placement techniques typically involve transforming the system model into a canonical form—such as the observer canonical form—so that the relationship between the desired pole locations and the entries of K becomes transparent. For further details on pole placement methods, see standard texts on linear control systems such as Åström and Wittenmark (1990); Kwakernaak and Sivan (1972).

Example 7.1: *Deadbeat Observer for a Two-state System.*
Consider the discrete-time state-space model obtained by sampling the continuous-time system $e^{-s}/(s+1)$ with a sampling interval $t_s = 1.5$

$$x(k+1) = \begin{bmatrix} 0.2231 & 0.3834 \\ 0 & 0 \end{bmatrix} x(k) + \begin{bmatrix} 0.3935 \\ 1 \end{bmatrix} u(k) + \varepsilon_1(k)$$
$$y_m(k) = \begin{bmatrix} 1 & 0 \end{bmatrix} x(k) + \varepsilon_2(k).$$

Let the observer gain be $K = \begin{bmatrix} k_1 & k_2 \end{bmatrix}^T$. Then,

$$A - KC = \begin{bmatrix} 0.2231 - k_1 & 0.3834 \\ -k_2 & 0 \end{bmatrix}.$$

To obtain a *deadbeat observer*, where all observer poles are at the origin, choose $K = \begin{bmatrix} 0.2231 & 0 \end{bmatrix}^T$. This results in eigenvalues at zero, ensuring that the estimation error converges to zero in finite time.

Example 7.2: *Observer Design for System in Canonical Form.*
Consider a general n^{th}-order system in observer canonical form:

$$x(k+1) = \begin{bmatrix} -a_1 & 1 & 0 & \cdots & 0 \\ -a_2 & 0 & 1 & \cdots & 0 \\ \vdots & \vdots & \ddots & \ddots & \vdots \\ -a_{n-1} & 0 & 0 & \cdots & 1 \\ -a_n & 0 & 0 & \cdots & 0 \end{bmatrix} x(k) + Bu(k) + \varepsilon_1(k)$$
$$y_m(k) = \begin{bmatrix} 1 & 0 & 0 & \cdots & 0 \end{bmatrix} x(k) + \varepsilon_2(k),$$

The observer error dynamics are governed by the matrix:

$$A - KC = \begin{bmatrix} -a_1 - k_1 & 1 & 0 & \cdots & 0 \\ -a_2 - k_2 & 0 & 1 & \cdots & 0 \\ \vdots & \vdots & \ddots & \ddots & \vdots \\ -a_{n-1} - k_{n-1} & 0 & 0 & \cdots & 1 \\ -a_n - k_n & 0 & 0 & \cdots & 0 \end{bmatrix}.$$

It is easy to verify that choosing

$$K = \begin{bmatrix} -a_1 & -a_2 & \cdots & -a_n \end{bmatrix}^T,$$

yields a deadbeat observer, since all eigenvalues of $A - KC$ are placed at the origin, ensuring finite-time convergence of the estimation error.

A major drawback of the pole placement approach to observer design is that there is generally no clear guidelines for selecting appropriate observer pole locations to achieve good estimation performance. While a basic trade-off exists between the speed of convergence and noise attenuation—faster observers tend to amplify measurement noise—this aspect can typically be handled through tuning the pole locations based on practical considerations.

A more subtle and challenging issue, however, is that the speed of error recovery is not determined solely by pole placement. For example, even a deadbeat observer—with all poles placed at the origin—may exhibit slow recovery from certain types of initial errors. This limitation becomes especially pronounced when the state dimension is large compared to the output dimension.

To illustrate this point, consider a scenario where a deadbeat observer yields the following error dynamics:

$$x_e(k+1) = \underbrace{\begin{bmatrix} e_1 & e_2 & \cdots & e_n \end{bmatrix} \begin{bmatrix} 0 & 1 & 0 & \cdots & 0 \\ 0 & 0 & 1 & \cdots & 0 \\ \vdots & & \ddots & & \vdots \\ \vdots & & \ddots & & 1 \\ 0 & 0 & \cdots & \cdots & 0 \end{bmatrix} \begin{bmatrix} e_1 & e_2 & \cdots & e_n \end{bmatrix}^{-1}}_{(A-KC)}. \tag{7.12}$$

Although all eigenvalues of $(A - KC)$ are zero, the convergence is not instantaneous in every direction. Specifically, an initial error aligned with the last eigenvector e_n will take n sampling steps to decay to zero. If this mode happens to be the one most susceptible to external disturbances or modeling errors, then the deadbeat observer—despite its theoretically fast poles—could yield poor practical performance. This issue is especially pronounced in systems where the state dimension is significantly larger than the measurement dimension.

This example underscores the fact that observer design requires more than just pole placement; attention must also be given to the structure of the eigenvectors and the system's sensitivity to various modes of estimation error.

Example 7.3: *Deadbeat Observer for an Unobservable System.*
Consider the system:

$$x(k+1) = \begin{bmatrix} 0 & 1 & 0 \\ 0 & 0 & 1 \\ 0 & 0 & 0 \end{bmatrix} x(k) + \begin{bmatrix} 0 \\ 0 \\ 1 \end{bmatrix} \varepsilon_1(k)$$
$$y_m(k) = \begin{bmatrix} 0 & 0 & 1 \end{bmatrix} x(k) + \varepsilon_2(k).$$

A straightforward pole placement design that yields a deadbeat observer sets the gain as $K = \begin{bmatrix} 0 & 0 & 0 \end{bmatrix}^T$. This, however, implies that the observer does not utilize any measurement correction and therefore performs no actual state estimation. In other words, although deadbeat dynamics are achieved in theory, the observer fails to account for the influence of ε_1 on x.

To improve estimation efficiency, it is often beneficial to incorporate statistical information about how disturbances influence the system dynamics. For example, although the state dimension may be large, only a few modes may be significantly affected by process noise. In such cases, estimation accuracy can be improved by focusing on reconstructing those dominant modes. These insights motivate the development of statistically optimal estimators, such as the Kalman filter, which we introduce in Section 7.3.

7.3 Kalman Filter

The filter gain matrix can also be selected to be statistically optimal. Specifically, it may be chosen to minimize the variance of the estimation error for a given stochastic system model. The resulting estimator is the celebrated *Kalman filter*, one of the most widely used techniques for state estimation. Originally introduced by Kalman in his seminal paper Kalman (1960), it has become the foundation of modern estimation theory. In what follows, we derive the filter for the simplified case where $R_{12} = 0$ and highlight its key properties.

To simplify the discussion, we assume that ε_1 and ε_2 are mutually uncorrelated ($R_{12} = 0$), with $R_1 \geq 0$ and $R_2 > 0$. The more general case involving correlated noise will be addressed later. Recall that the linear estimator in (7.5) can be expressed in the one-step-ahead predictor form as

$$\hat{x}(k+1|k) = A\hat{x}(k|k-1) + Bu(k) + K(k)\left[y_m(k) - C\hat{x}(k|k-1)\right]. \tag{7.13}$$

In this formulation, we allow the filter gain matrix $K(k)$ to vary with time for generality.

7.3.1 Derivation of the Optimal Filter Gain Matrix

Recall that the error dynamics for $x_e(k) = x(k) - \hat{x}(k|k-1)$ are given by

$$x_e(k+1) = \left[A - K(k)C\right]x_e(k) + \varepsilon_1(k) + K(k)\varepsilon_2(k). \tag{7.14}$$

Define

$$\begin{aligned} P(k) &\stackrel{\Delta}{=} \mathrm{Cov}\{x_e(k)\} \\ &= E\left\{(x_e(k) - E\{x_e(k)\})(x_e(k) - E\{x_e(k)\})^T\right\}. \end{aligned} \tag{7.15}$$

Assume that the initial estimate is unbiased, i.e., $E\{x_e(0)\} = 0$. Then, using the expectation rules outlined in Appendix B and applying them to (7.14), we find that $E\{x_e(k)\} = 0$ for all $k \geq 0$, and thus

$$\begin{aligned} P(k+1) &= E\left\{x_e(k+1)x_e^T(k+1)\right\} \\ &= (A - K(k)C)P(k)(A - K(k)C)^T + R_1 + K(k)R_2K^T(k). \end{aligned} \tag{7.16}$$

In this expression, we used the fact that $x_e(k)$, $\varepsilon_1(k)$, and $\varepsilon_2(k)$ in (7.14) are mutually independent.

Now let us choose $K(k)$ to minimize $\alpha^T P(k+1)\alpha$ for any arbitrary vector α. Through straightforward algebra, it can be shown that

$$\begin{aligned} \alpha^T P(k+1)\alpha = \alpha^T\big[&AP(k)A^T + R_1 - K(k)CP(k)A^T \\ &-AP(k)C^TK^T(k) + K(k)(R_2 + CP(k)C^T)K^T(k)\big]\alpha. \end{aligned}$$

Completing the square on the terms involving $K(k)$, we obtain

$$\begin{aligned} \alpha^T P(k+1)\alpha = \alpha^T\Big\{&\Big[K(k) - AP(k)C^T\left(R_2 + CP(k)C^T\right)^{-1}\Big] \\ &\times\left[R_2 + CP(k)C^T\right]\Big[K(k) - AP(k)C^T\left(R_2 + CP(k)C^T\right)^{-1}\Big]^T\Big\}\alpha \\ &+ \alpha^T\Big\{AP(k)A^T + R_1 - AP(k)C^T(R_2 + CP(k)C^T)^{-1}CP(k)A^T\Big\}\alpha. \end{aligned}$$

Since the first term is a nonnegative square term and the second term does not depend on $K(k)$, the choice of $K(k)$ that minimizes the expression is

$$K(k) = AP(k)C^T(R_2 + CP(k)C^T)^{-1}. \tag{7.17}$$

Note that this optimal solution is independent of the choice of α. Furthermore, substituting this value of $K(k)$ into the expression for $P(k+1)$ yields

$$P(k+1) = AP(k)A^T + R_1 - AP(k)C^T\left(R_2 + CP(k)C^T\right)^{-1}CP(k)A^T. \tag{7.18}$$

Given $\hat{x}(1|0)$ and $P(1)$, the pair of equations (7.17) and (7.18), together with the prediction equation (7.13), can be used recursively to compute $\hat{x}(k+1|k)$ and $P(k+1)$. These constitute the *time-varying Kalman filter* equations. The matrix recursion (7.18) is known as the *Riccati difference equation* (RDE).

Suppose $P(k)$ in the RDE converges to a steady-state solution P_∞ as $k \to \infty$. In that case, one may consider foregoing optimality during the transient period and instead implementing a filter with a constant gain matrix given by

$$K_\infty \triangleq AP_\infty C^T(R_2 + CP_\infty C^T)^{-1}, \tag{7.19}$$

where P_∞ is obtained either by iterating the RDE until convergence or by directly solving the following *algebraic Riccati equation* (ARE):

$$P_\infty = AP_\infty A^T + R_1 - AP_\infty C^T\left(R_2 + CP_\infty C^T\right)^{-1}CP_\infty A^T. \tag{7.20}$$

This formulation is referred to as the *steady-state* Kalman filter, in contrast to the time-varying Kalman filter. Properties of the ARE and the steady-state filter will be discussed later.

Remarks:

- Recall the relationship between the one-step-ahead predictor gain $K(k)$ and the filter gain $K_f(k)$, given by $K(k) = AK_f(k)$. Therefore, for optimal filtering, we can choose

$$K_f(k) = P(k)C^T(R_2 + CP(k)C^T)^{-1}. \tag{7.21}$$

  This gain matrix should be used to implement the optimal filter in the form of (7.4), which recursively computes $\hat{x}(k|k)$ rather than $\hat{x}(k+1|k)$.
- In the derivation above, we imposed a *linear estimator* structure and optimized the gain matrix with respect to a chosen cost criterion. Alternatively, it can be shown that the Kalman filter is the optimal estimator in the sense of minimizing the *conditional covariance* of the estimation error, provided that ε_1 and ε_2 are Gaussian (in addition to being white). A derivation based on probabilistic arguments is presented in Appendix D.1.
- The Kalman filter places the observer poles (i.e., the eigenvalues of $A - KC$) based on the relative uncertainty in the process model vs. the measurements. This behavior is illustrated in Example (7.4).

Example 7.4: *Kalman Filter for a Two-state System*
Consider the system from Example 7.1. We will construct a steady-state Kalman filter for different values of the noise covariances R_1 and R_2.

For the case $R_1 = I$ and $R_2 = 1$, we can solve the ARE using the MATLAB command `dare` to obtain:

$$P = \begin{bmatrix} 1.1739 & 0 \\ 0 & 1 \end{bmatrix} \qquad K = \begin{bmatrix} 0.1205 \\ 0 \end{bmatrix} \qquad \lambda(A - KC) = (0.1026, 0).$$

If we set $R_1 = 0$, the Kalman gain becomes $K = \begin{bmatrix} 0 \, 0 \end{bmatrix}$, and the observer poles are identical to the system poles. This corresponds to the case of perfect model certainty, where the estimator ignores measurement feedback entirely.

Conversely, when $R_2 = 0$, the solution yields:

$$P = \begin{bmatrix} 1.147 & 0 \\ 0 & 1 \end{bmatrix} \qquad K = \begin{bmatrix} 0.2231 \\ 0 \end{bmatrix} \qquad \lambda(A - KC) = (0, 0).$$

This configuration corresponds to a deadbeat observer, as all observer poles are placed at the origin.

Example 7.5: *Kalman Filter Design for the System in Example 7.3.*
We design a Kalman filter assuming $\text{cov}(\varepsilon_1) = 1$ and $\text{cov}(\varepsilon_2) = 0.2$. For these parameters, the Kalman gain and the steady-state error covariance are:

$$K = \begin{bmatrix} 0 \\ 0.8333 \\ 0 \end{bmatrix} \qquad P = \begin{bmatrix} 0.1667 & 0 & 0 \\ 0 & 0.1667 & 0 \\ 0 & 0 & 1 \end{bmatrix}.$$

A comparison of the true state values and their estimates is shown in Figure 7.2. The deadbeat observer discussed in Example 7.3 leaves the state uncorrected because of the zero observer gain. In contrast, the Kalman filter provides substantially improved predictions for the first and second

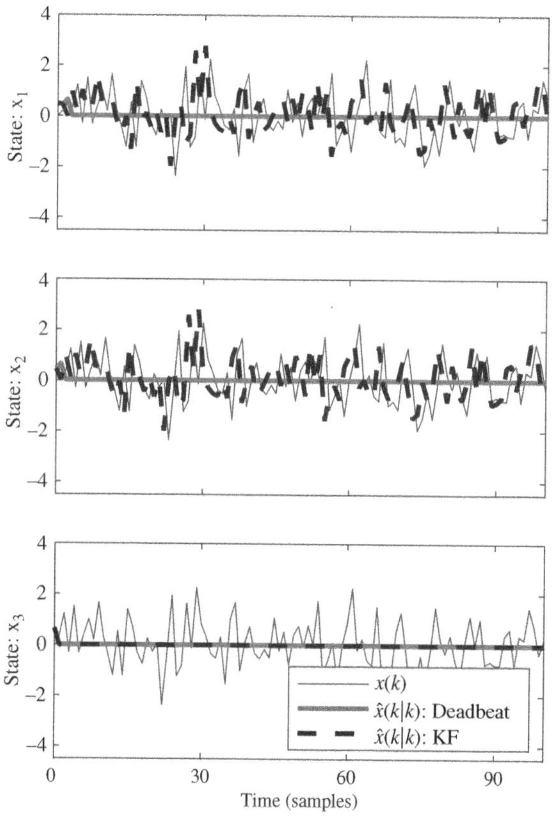

Figure 7.2 Comparison of actual state variables ($x(k)$: thin lines) with estimates from the deadbeat observer (solid lines) and the Kalman filter (dashed lines).

states. Interestingly, although the Kalman filter also places all observer poles at the origin, its estimation dynamics differ fundamentally from those of the deadbeat observer with zero gain, leading to markedly better overall performance.

Example 7.6: *Comparison of Pole Placement and Kalman Filter.*
Consider the following discrete-time state-space model:

$$x(k+1) = \begin{bmatrix} 0.6346 & 0.02865 & 0.02325 & 0.2662 \\ 0.9891 & 0.04599 & 0.03975 & 0.6882 \\ 0.7102 & 0.03516 & 0.03416 & 0.9322 \\ 0.1060 & 0.00792 & 0.01213 & 0.6866 \end{bmatrix} x(k) + \begin{bmatrix} 0.01448 \\ 0.1351 \\ 0.02001 \\ 0.002739 \end{bmatrix} \varepsilon(k),$$

$$y_m(k) = \begin{bmatrix} 0 & -28.1 & -27.3 & 0 \end{bmatrix} x(k) + \nu(k)$$

with $\text{cov}(\varepsilon) = 1$ and $\text{cov}(\nu) = 0.04$.

Since the system has four states, we need to place four estimator poles. However, it is not immediately clear where to place them optimally, other than ensuring they lie within the unit disk. Suppose we place the poles at $\lambda = 0.01, 0.1, 0.25, 0.4$. As shown in Figure 7.3, this design allows accurate estimation of the output $y(k) = \begin{bmatrix} 0 & -28.1 & -27.3 & 0 \end{bmatrix} x(k)$, but yields poor estimates for

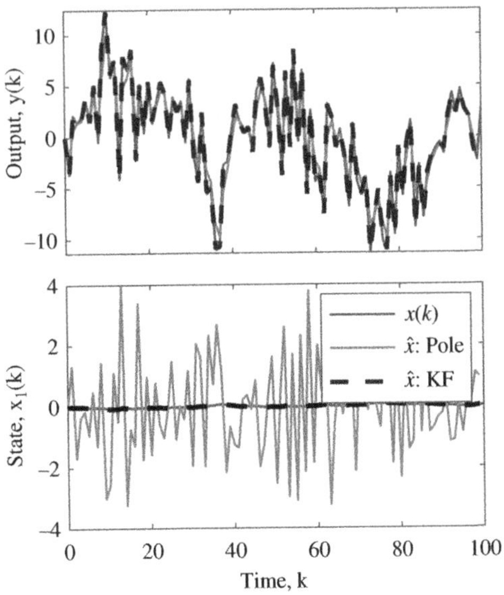

Figure 7.3 Estimates of the output variable $y(k)$ and the first state $x_1(k)$ using pole-placement design (solid lines) and the Kalman filter (dashed lines). True values are shown as thin lines.

the unmeasured state $x_1(k)$. The MATLAB codes for this example are uploaded on GitHub at: https://github.com/Lee-MPC-Book

In contrast, the Kalman filter is designed by solving the Riccati equation. The dashed lines in the figure represent the state estimates obtained using the Kalman filter. Unlike the pole-placement design, the Kalman filter provides significantly better estimates of $x_1(k)$, despite being unmeasured.

7.3.2 Correlated Noise Case

The assumption that $\varepsilon_1(k)$ and $\varepsilon_2(k)$ are independent may not always hold. The previous derivation for the independent case can be straightforwardly extended to handle the general case where these noise sequences are correlated. The resulting Kalman filter equations take the following form:

$$\hat{x}(k|k) = \hat{x}(k|k-1) + P(k)C^T(CP(k)C^T + R_2)^{-1}(y_m(k) - C\hat{x}(k|k-1)), \tag{7.22}$$

and

$$\hat{x}(k+1|k) = A\hat{x}(k|k) + R_{1,2}(CP(k)C^T + R_2)^{-1}(y_m(k) - C\hat{x}(k|k-1)), \tag{7.23}$$

where

$$\begin{aligned} P(k+1) = {} & AP(k)A^T + R_1 - (AP(k)C^T + R_{1,2}) \\ & \times (CP(k)C^T + R_2)^{-1}(CP(k)A^T + R_{1,2}^T). \end{aligned} \tag{7.24}$$

Note that, as indicated in Eq. (7.23), the measured output $y_m(k)$ appears explicitly in the one-step-ahead state prediction $\hat{x}(k+1|k)$—in contrast to the uncorrelated noise case.

These expressions can be derived by introducing a transformed system in which the correlated noise terms are rendered uncorrelated:

$$\begin{aligned} x(k+1) &= Ax(k) + \varepsilon_1(k) - R_{12}R_2^{-1}\left(Cx(k) + \varepsilon_2(k) - y_m(k)\right) \\ &= \underbrace{(A - R_{12}R_2^{-1}C)}_{A'} x(k) + R_{12}R_2^{-1}y_m(k) + \underbrace{\varepsilon_1(k) - R_{12}R_2^{-1}\varepsilon_2(k)}_{\varepsilon_1'(k)} \\ y_m(k) &= Cx(k) + \varepsilon_2(k). \end{aligned}$$

It is straightforward to verify that

$$E\left\{\begin{bmatrix}\varepsilon_1'(k)\\ \varepsilon_2(k)\end{bmatrix}\begin{bmatrix}\varepsilon_1'(k)\\ \varepsilon_2(k)\end{bmatrix}^T\right\} = \begin{bmatrix} R_1 - R_{12}R_2^{-1}R_{21} & 0 \\ 0 & R_2 \end{bmatrix}. \tag{7.25}$$

Since $\varepsilon_1'(k)$ and $\varepsilon_2(k)$ are now independent white-noise sequences, the standard Kalman filter formulas derived for the uncorrelated case can be directly applied to this transformed system. The reader is encouraged to verify that the resulting Kalman filter is same as the above. This is left as an exercise problem.

7.3.3 Stability of Kalman Filter

In Section 7.3.1, we derived the time-varying optimal filter gain matrix using the RDE, given in (7.18). If the sequence $P(k)$ converges, the limiting value P_∞ is a positive semi-definite solution to the ARE of (7.20). In many practical situations, one may implement the filter using this steady-state gain K_∞, accepting a possible loss of optimality during the transient phase. However, it is essential to verify that the resulting steady-state Kalman filter is stable—i.e., that the associated estimation error dynamics governed by $A - K_\infty C$ are stable.

To analyze this, we introduce two important definitions:

Definition 7.2: A positive semi-definite solution to the ARE (7.20) is called a *stabilizing solution* if all eigenvalues of the associated error transition matrix $A - K_\infty C$, with K_∞ defined by (7.19), lie strictly inside the unit circle.

Definition 7.3: A positive semi-definite solution to the ARE (7.20) is called a *strong solution* if all eigenvalues of $A - K_\infty C$ lie on or inside the unit circle.

These definitions naturally lead to the following three important questions regarding the behavior of the RDE:

1. Under what conditions does the sequence $P(k)$ converge to a steady-state solution P_∞?
2. If convergence occurs, when is P_∞ a stabilizing solution to the ARE (7.20)?
3. If P_∞ is stabilizing, under what conditions is it the unique positive semi-definite solution to the ARE?

The question of whether $P(k)$ converges is crucial, as it determines whether the ARE admits a meaningful (i.e., positive semi-definite) solution. The answer is:

The sequence $P(k)$ in the RDE (7.18) converges to P_∞ for any $P(0)$ if (C, A) is detectable.

When the system is detectable, the increase in covariance due to process disturbances and potentially unstable dynamics is eventually offset by the information gained from measurements. Notably, higher estimation uncertainty leads to greater correction via observation, yielding a natural balancing effect.

The second question—whether the steady-state solution is stabilizing—is equally important. Stability is a fundamental requirement, and optimality of the estimation error covariance does not by itself guarantee observer stability. Assuming that (C, A) is detectable, the following facts are useful:

- If $(A, R_1^{1/2})$ is stabilizable and $P(0) \geq 0$, then $P(k) \rightarrow P_\infty$ as $k \rightarrow \infty$, and P_∞ is a stabilizing solution of the ARE.
- If $(A, R_1^{1/2})$ has no uncontrollable modes on the unit circle and $P(0) > 0$, then $P(k) \rightarrow P_\infty$ as $k \rightarrow \infty$, and P_∞ is again a stabilizing solution.
- If $(A, R_1^{1/2})$ has uncontrollable modes on the unit circle, then P_∞ is a strong solution, but not a stabilizing solution of the ARE.

While these statements can be rigorously proven using algebraic techniques, we offer qualitative justification based on the expected asymptotic behavior of the estimation error covariance. The key insight is that, under detectability of (C, A), a stabilizing observer gain K always exists. However, the optimal filter may still fail to achieve stability if the estimation problem is ill-posed in this regard. For example, if both the process noise covariance and the initial error covariance are zero, the solution P_∞ is trivially the zero matrix, resulting in an open-loop observer—an unstable structure if the underlying system itself is unstable.

The stabilizability of $(A, R_1^{1/2})$ in the first statement implies that all unstable modes of A are directly influenced by the process noise. If the filter were not stabilizing, the covariance associated with these modes would grow without bound. Since the Kalman filter is designed to minimize estimation error covariance, it must counteract this unbounded growth. Therefore, for the filter to be optimal, it must necessarily be stabilizing.

The second statement considers the case where $(A, R_1^{1/2})$ may have unstable modes that are not excited by external disturbances. Even in such cases, observer stability can still be achieved. The stronger requirement $P(0) > 0$ ensures that any initial errors in these unforced modes can grow exponentially if uncorrected. To minimize the total estimation error, the filter must attenuate these errors, which forces the estimator to stabilize the associated modes. At steady state, this leads to a nonzero covariance that balances error growth and correction.

The issue with uncontrollable modes on the unit circle, as addressed in the third statement, is more subtle. Unlike exponentially unstable modes, initial errors in marginally stable modes do not grow unbounded over time. Consequently, the time-varying Kalman filter eventually suppresses these errors to zero. As a result, the corresponding steady-state covariance becomes zero, and the optimal filter ceases to apply any correction to these modes. Without correction, these modes remain unstabilized, even though the filter minimizes the error covariance in the optimal sense.

The question of whether the steady-state solution to the ARE is a unique positive semi-definite solution is important, especially when one attempts to solve the ARE directly rather than through iteration of the RDE. If multiple positive semi-definite solutions exist, solving the ARE directly may yield a non-stabilizing solution, even though a stabilizing solution also exists. The key result is:

> If $(A, R_1^{1/2})$ is stabilizable, then the stabilizing solution is the unique positive semi-definite solution to the ARE (7.20). If $(A, R_1^{1/2})$ has uncontrollable mode(s) outside the unit circle, then there exists at least one additional positive semi-definite solution besides the stabilizing one.

We offer here an intuitive explanation rather than a formal proof. When $(A, R_1^{1/2})$ has uncontrollable modes outside the unit circle, these modes cannot be influenced by the process noise. If the RDE is initialized with zero covariance in these directions, the corresponding entries in $P(k)$ remain zero during iteration. The resulting solution is then positive semi-definite but does not stabilize the corresponding modes. Hence, such a solution fails to be stabilizing, even though it satisfies the ARE.

Example 7.7: *Detectability and Convergence of Kalman Gain.*
Consider the system:

$$x(k+1) = \begin{bmatrix} \alpha & 0.5 \\ 0 & 0.2 \end{bmatrix} + \begin{bmatrix} 1 \\ 0 \end{bmatrix} \varepsilon_1(k)$$
$$y_m(k) = \begin{bmatrix} 0 & 1 \end{bmatrix} x(k) + \varepsilon_2(k),$$

with $\text{var}(\varepsilon_1(k)) = 1$ and $\text{var}(\varepsilon_2(k)) = 0.25$. We aim to determine the steady-state solution of the Riccati equation.

Note that the system is not observable, but it is detectable if and only if α lies strictly within the unit disk. To investigate the convergence of the RDE, we initialize with $P(0) = I$ and iterate the RDE for three different values of $\alpha \in 0.8, 1.0, 1.2$. The results are summarized in the following table:

α	$k = 1$	$k = 3$	$k = 8$	∞
0.8	$\begin{bmatrix} 1.765 & 0.05 \\ 0.05 & 0.02 \end{bmatrix}$	$\begin{bmatrix} 2.398 & 0.0016 \\ 0.0016 & 3\times10^{-5} \end{bmatrix}$	$\begin{bmatrix} 2.737 & 0 \\ 0 & 0 \end{bmatrix}$	$\begin{bmatrix} 2.778 & 0 \\ 0 & 0 \end{bmatrix}$
1.0	$\begin{bmatrix} 2.125 & 0.05 \\ 0.05 & 0.02 \end{bmatrix}$	$\begin{bmatrix} 4.183 & 0.0024 \\ 0.0024 & 3\times10^{-5} \end{bmatrix}$	$\begin{bmatrix} 9.19 & 0 \\ 0 & 0 \end{bmatrix}$	No convergence
1.2	$\begin{bmatrix} 2.565 & 0.05 \\ 0.05 & 0.02 \end{bmatrix}$	$\begin{bmatrix} 7.862 & 0.003 \\ 0.003 & 3\times10^{-5} \end{bmatrix}$	$\begin{bmatrix} 60.5 & 0 \\ 0 & 0 \end{bmatrix}$	No convergence

As seen from the results, the error covariance matrix converges to a steady-state solution P_∞ only when $\alpha < 1$, consistent with the detectability condition. When $\alpha \geq 1$, the system becomes undetectable and the Riccati recursion fails to converge.

Example (7.7) demonstrates that when (C, A) is detectable, the Riccati equation (7.18) converges to a steady-state solution P_∞. This convergence can be confirmed by experimenting with different choices of positive semi-definite $P(0)$. In particular, for $\alpha = 0.8$, the Riccati recursion reliably converges to a unique P_∞. In Example 7.8, we consider an observable system and illustrate the conditions under which the Kalman filter guarantees observer stability.

Example 7.8: *Stability of the Kalman Filter.*
Consider the system:

$$x(k+1) = \begin{bmatrix} 0.2 & 0.5 \\ 0 & \alpha \end{bmatrix} + \begin{bmatrix} 1 \\ 0 \end{bmatrix} \varepsilon_1(k)$$
$$y_m(k) = \begin{bmatrix} 1 & 0 \end{bmatrix} x(k) + \varepsilon_2(k),$$

with $\text{var}(\varepsilon_1(k)) = 1$ and $\text{var}(\varepsilon_2(k)) = 0.25$. Since the system is observable for all α, $P(k)$ converges to a steady-state solution P_∞, and the corresponding steady-state Kalman gain K_∞ can be computed.
Solving ARE:

Using the MATLAB command `dare`, the steady-state covariance and gain matrices are:

$$P_\infty = \begin{bmatrix} 1.008 & 0 \\ 0 & 0 \end{bmatrix} \quad K_\infty = \begin{bmatrix} 0.160 \\ 0 \end{bmatrix} \quad \lambda = (0.04, 0.8) \quad \text{for } \alpha = 0.8$$

$$P_\infty = [\ ] \quad K_\infty = [\ ] \quad \text{for } \alpha = 1.0$$

$$P_\infty = \begin{bmatrix} 1.562 & 1.285 \\ 1.285 & 2.981 \end{bmatrix} \quad K_\infty = \begin{bmatrix} 0.527 \\ 0.851 \end{bmatrix} \quad \lambda = (0.04, 0.833) \quad \text{for } \alpha = 1.2.$$

When $\alpha = 1.0$, MATLAB fails to find a stabilizing solution and returns empty matrices for P_∞ and K_∞.

Recursion of RDE:

Let us now compute P_∞ by iterating the RDE (7.18):

- For $\alpha = 0.8$, starting with $P(0) = I$, the RDE converges to the same P_∞ as above. Since $(A, R_1^{1/2})$ is stabilizable, the solution is unique for any positive semi-definite $P(0)$, including $P(0) = 0$.
- For $\alpha = 1.0$, the pair $(A, R_1^{1/2})$ has an uncontrollable mode on the unit circle. With $P(0) = I$, the RDE converges to:

$$P_\infty = \begin{bmatrix} 1.008 & 0 \\ 0 & 0 \end{bmatrix} \quad K_\infty = \begin{bmatrix} 0.160 \\ 0 \end{bmatrix} \quad \lambda = (0.04, 1.0) \quad \text{for } \alpha = 1.0.$$

  This is a non-stabilizing but *strong solution* of the ARE. The estimation error does not diverge, but it also does not vanish asymptotically.
- For $\alpha = 1.2$, the RDE with $P(0) = I$ converges to a stabilizing solution (shown above). However, the solution is not unique. If we instead start with $P(0) = 0$, the RDE converges to:

$$P_\infty = \begin{bmatrix} 1.008 & 0 \\ 0 & 0 \end{bmatrix} \quad K_\infty = \begin{bmatrix} 0.1603 \\ 0 \end{bmatrix}.$$

  This solution is not stabilizing, as it leaves an unstable mode uncorrected. Nevertheless, it remains a positive semi-definite solution of the ARE. Hence, when $(A, R_1^{1/2})$ is not stabilizable, the ARE may admit multiple positive semi-definite solutions, only a subset of which are stabilizing.

The preceding examples demonstrate the convergence behavior of the RDE under different conditions. When (C, A) is a detectable pair, the RDE converges. If, in addition, $\left(A, R_1^{1/2}\right)$ is stabilizable, the RDE converges to a unique positive semi-definite solution of the ARE, and the resulting Kalman filter is stable. If $(A, R_1^{1/2})$ has uncontrollable mode(s) on the unit circle, the RDE still converges to a unique solution, but this solution is only a strong solution—not stabilizing. When $(A, R_1^{1/2})$ is not stabilizable, the RDE admits multiple solutions: a stabilizing solution may exist, but it is no longer unique.

7.4 Extensions

7.4.1 Inferential Estimation

Inferential control refers to the situation in which one or more of the controlled variables is not directly measurable. In such cases, these variables must be *inferred* from available measurements using a state estimator. Consider Example 7.6; let us assume that the primary controlled variable

$$y_c(k) = C_c x(k), \tag{7.26}$$

was unmeasured. The Kalman filter can be used to estimate $\hat{y}_c(k) = C_c \hat{x}(k|k)$ from the measurements $y_m(j),\ j \le k$.

Such scenarios are common in process industries, where measured outputs typically consist of variables such as temperature, pressure, or pH that can be easily measured online. In contrast, key performance indicators (e.g., composition or product purity) are often difficult or costly to monitor continuously and may require offline laboratory analysis.

Referring to Figure 1.6 from Chapter 1, output variables $y(k)$ can include measured and unmeasured outputs, as well as auxiliary variables. If only some—and not all—output variables are measured, then $y_m(k)$ denotes the subset that is actually measured. The system model is then described by

$$\begin{aligned} x(k+1) &= Ax(k) + Bu(k) + B_d d(k) \\ y(k) &= Cx(k) \\ y_m(k) &= Hy(k) + v(k), \end{aligned} \tag{7.27}$$

where matrix H selects the measured outputs from $y(k)$. It has as many rows as measured outputs; each row contains a single entry of 1 at the corresponding location and zeros elsewhere. It is easy to see that when all outputs are measured, H is an identity matrix.

In Example 7.6, assume that $x_1(k)$ was the primary variable of interest. Thus:

$$y_c(k) = \begin{bmatrix} 1 & 0 & 0 & 0 \end{bmatrix} x(k). \tag{7.28}$$

We now proceed to define a general framework that will be applicable to a broader class of inferential control problems encountered in practice.

Example 7.9: *Inferential Estimation.* Consider the system of Example (7.6). Suppose that the primary controlled output of interest is x_1. Thus, the outputs include the measured variable (y_m) and the controlled variable y_c. We define the full output vector as

$$y \stackrel{\Delta}{=} \begin{bmatrix} 0 & -28.1 & -27.3 & 0 \\ 1 & 0 & 0 & 0 \end{bmatrix} x, \tag{7.29}$$

Since the output vector y contains both measured and unmeasured components, the measurement equation becomes:

$$y_m(k) = \underbrace{\begin{bmatrix} 1 & 0 \end{bmatrix}}_{H} y(k) + v(k).$$

7.4.2 Non-stationary (Integrating) Noise

The assumption that ε_1 and ε_2 are white noise sequences is often unrealistic in practical settings. In many cases, the noise processes are better represented as either stationary colored noise or non-stationary disturbances such as integrated white noise. A wide range of such scenarios can be effectively addressed using disturbance modeling, as introduced in Section 6.6.

In this section, we consider an example where the unmeasured disturbance follows an integrated white noise process. We demonstrate how to modify the Kalman filter to account for this non-stationary behavior through appropriate disturbance modeling. As we shall see in subsequent chapters, model augmentation is essential for achieving accurate state estimation in a number of practically important situations.

For illustration, we consider the following SISO system (a portion of which was previously introduced in Example 7.6):

$$\begin{aligned} x(k+1) &= \begin{bmatrix} 0.6346 & 0.02865 & 0.02325 & 0.2662 \\ 0.9891 & 0.04599 & 0.03975 & 0.6882 \\ 0.7102 & 0.03516 & 0.03416 & 0.9322 \\ 0.1060 & 0.00792 & 0.01213 & 0.6866 \end{bmatrix} x(k) \\ &\quad + \begin{bmatrix} 0.02949 \\ 0.1902 \\ 0.1670 \\ 0.009063 \end{bmatrix} u(k) + \begin{bmatrix} 0.01448 \\ 0.1351 \\ 0.02001 \\ 0.002739 \end{bmatrix} d(k) \\ y_m(k) &= \begin{bmatrix} 0 & -28.1 & -27.3 & 0 \end{bmatrix} x(k) + \nu(k). \end{aligned} \tag{7.30}$$

Consider that $d(k)$ represents an unmeasured disturbance, which shows a drifting-type behavior. Then, a typical disturbance model is given by

$$d(k+1) = d(k) + \varepsilon(k), \tag{7.31}$$

where $\varepsilon(k)$ is a white noise sequence with covariance R_ε.

To handle such non-stationary disturbances effectively, the standard Kalman filter must be extended to incorporate the disturbance dynamics. Before that, Example 7.10 considers the use of standard (unaugmented) Kalman filter for an integrated white noise disturbance.

Example 7.10: *Kalman filter: Comparison of Disturbance Modeled as White Noise and Random Walk.*

Consider the system defined by Eq. (7.30), where the *unmeasured disturbance* $d(k)$ evolves as an integrated white noise sequence given by Eq. (7.31), and the measurement noise $\nu(k)$ is a white noise sequence with covariance 0.25. We also introduce model-plant mismatch: while Eq. (7.30) defines the true plant, the estimation and control model uses:

$$B = \begin{bmatrix} 0.03277 \\ 0.2113 \\ 0.1856 \\ 0.01007 \end{bmatrix} \quad B_d = \begin{bmatrix} 0.01316 \\ 0.1228 \\ 0.01819 \\ 0.00249 \end{bmatrix}.$$

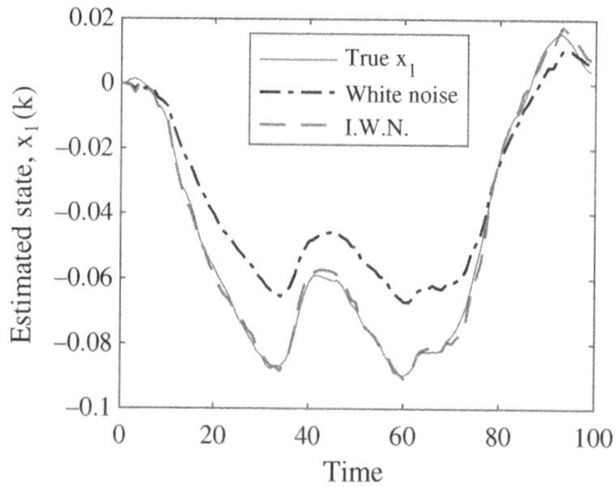

Figure 7.4 Comparison of unmeasured disturbance modeled as white noise vs. integrated white noise.

In this example, we apply the original Kalman filter (which assumes that $d(k)$ is white) to measurements generated using the non-stationary disturbance model. The results—dash-dot lines in Figure 7.4—show poor estimation performance for $x_1(k)$ due to the unmodeled non-stationarity of $d(k)$.

This illustrates the importance of disturbance modeling: accurately characterizing $d(k)$ as an integrating process is essential for robust and accurate state estimation. The benefits of this approach will be demonstrated in the following example.

As discussed in Section 6.6, we suitably augment the state to handle integrating dynamics and obtain the augmented model in a standard Kalman filter form:

$$\begin{aligned} \bar{x}(k+1) &= \bar{A}\bar{x}(k) + \bar{B}\bar{u}(k) + \varepsilon_1(k) \\ y_m(k) &= \bar{C}\bar{x}(k) + \varepsilon_2(k). \end{aligned} \tag{7.32}$$

We now discuss two options for augmenting the model to handle integrated white noise disturbance.

Option 1: Augmenting process model with disturbance states

A widely used approach for handling non-stationary unmeasured disturbances is to augment the system state with the disturbance $d(k)$ and apply the Kalman filter to estimate both the system state and the disturbance. In some applications, direct estimation of the disturbance itself may be beneficial. In this formulation, the augmented state is defined as

$$\bar{x}(k) \triangleq \begin{bmatrix} x(k) \\ d(k) \end{bmatrix}. \tag{7.33}$$

Since an integrated white noise disturbance evolves according to (7.31), the augmented model is written as

$$\underbrace{\begin{bmatrix} x(k+1) \\ d(k+1) \end{bmatrix}}_{\bar{x}(k+1)} = \underbrace{\begin{bmatrix} A & B_d \\ 0 & I \end{bmatrix}}_{\bar{A}} \underbrace{\begin{bmatrix} x(k) \\ d(k) \end{bmatrix}}_{\bar{x}(k)} + \underbrace{\begin{bmatrix} B_u \\ 0 \end{bmatrix}}_{\bar{B}} u(k) + \underbrace{\begin{bmatrix} 0 \\ \epsilon(k) \end{bmatrix}}_{\varepsilon_1(k)}$$
$$y_m(k) = \underbrace{\begin{bmatrix} HC & 0 \end{bmatrix}}_{\bar{C}} \underbrace{\begin{bmatrix} x(k) \\ d(k) \end{bmatrix}}_{\bar{x}(k)} + \underbrace{\nu(k)}_{\varepsilon_2(k)}. \tag{7.34}$$

Option 2: Differenced model augmented with outputs

Another approach for handling non-stationary disturbances is to difference the state update model and augment the state vector. Starting from the model (7.27), we take differences to obtain:

$$\begin{aligned} \Delta x(k+1) &= A\Delta x(k) + B\Delta u(k) + B_d\Delta d(k) \\ y(k+1) &= y(k) + C\Delta x(k+1) \\ &= y(k) + CA\Delta x(k) + CB\Delta u(k) + CB_d\Delta d(k). \end{aligned} \tag{7.35}$$

The second equation follows from the identity $y(k+1) = y(k) + \Delta y(k+1)$. We now define an augmented state:

$$\bar{x}(k) \triangleq \begin{bmatrix} \Delta x(k) \\ y(k) \end{bmatrix},$$

which yields the augmented model:

$$\underbrace{\begin{bmatrix} \Delta x(k+1) \\ y(k+1) \end{bmatrix}}_{\bar{x}(k+1)} = \underbrace{\begin{bmatrix} A & 0 \\ CA & I \end{bmatrix}}_{\bar{A}} \underbrace{\begin{bmatrix} \Delta x(k) \\ y(k) \end{bmatrix}}_{\bar{x}(k)} + \underbrace{\begin{bmatrix} B \\ CB \end{bmatrix}}_{\bar{B}} \Delta u(k) + \underbrace{\underbrace{\begin{bmatrix} B_d \\ CB_d \end{bmatrix}}_{\bar{B}_2} \varepsilon(k)}_{\varepsilon_1(k)}, \tag{7.36}$$

$$y_m(k) = \underbrace{\begin{bmatrix} 0 & H \end{bmatrix}}_{\bar{C}} \bar{x}(k) + \underbrace{\nu(k)}_{\varepsilon_2(k)}. \tag{7.37}$$

The corresponding process and measurement noise covariance matrices are:

$$\bar{R}_1 = \bar{B}_2 R_\varepsilon \bar{B}_2^T, \qquad \bar{R}_2 = R_\nu.$$

This formulation enables the Kalman filter to account for non-stationary behavior in the disturbance by applying differencing and augmenting the state.

Example 7.11: *Importance of Disturbance Modeling in Kalman Filter Design for Non-stationary Disturbances.*

We now demonstrate the use of state augmentation using Option 2 above. The first step is to construct the augmented state-space model. Starting from the system matrices $\{A, B, B_d, C, H\}$ defined in Eq. (7.30), we construct the augmented matrices $\{\bar{A}, \bar{B}, \bar{C}\}$ and the corresponding noise covariance matrices $\bar{R}_1$ and $\bar{R}_2$ in MATLAB as shown below:

```
% Augmenting the model
Aaug=[A zeros(4,2);
    C*A eye(2)];
Baug=[B; C*B];
Caug=[zeros(1,4) 1 0];
B2aug=[Bd; C*Bd];

% Error covariances
R1aug=B2aug*Reps*B2aug';
R2aug=Rnu;
Paug=eye(size(Aaug))*1e-8;
```

With the augmented model, filtered state estimates $\hat{x}(k|k)$ are updated as

```
xaug = xaug + Kaug*(yk-Caug*xaug);
```

where `Kaug` is the Kalman gain for the augmented system.

The dashed line in Figure 7.4 shows the predicted value $\hat{x}_1(k)$ using the differencing and augmenting approach. The filter performance improves significantly by appropriate modeling of the disturbance $d(k)$ as an integrated white noise sequence.

Remark 7.1: Soft Sensing: Example 7.11 illustrates that, with appropriate disturbance modeling, the Kalman filter serves as an effective tool for estimating unmeasured (but observable) variables. This approach is commonly referred to as inferential estimation or soft sensing.

Remark 7.2: Complete MATLAB Code: The full MATLAB script used to generate the results in Examples 7.10 and 7.11 is provided on GitHub: https://github.com/Lee-MPC-Book

7.4.3 Time-varying System

Consider the time-varying system:

$$\begin{aligned} x(k+1) &= A(k)x(k) + B(k)u(k) + \varepsilon_1(k) \\ y_m(k) &= C(k)x(k) + \varepsilon_2(k), \end{aligned} \tag{7.38}$$

where $\varepsilon_1(k)$ and $\varepsilon_2(k)$ are zero-mean white noise sequences with covariances $R_1(k)$ and $R_2(k)$, respectively. Such a system may, for instance, arise as a linear approximation to a nonlinear system along a trajectory.

Generalizing the Kalman filter to this time-varying case is straightforward. We have already derived the equations for the optimal time-varying estimator in the linear time-invariant case. The presence of time-varying system matrices does not fundamentally alter the estimator structure. The Kalman filter equations retain the same recursive form:

$$\hat{x}(k+1|k) = A(k)\hat{x}(k|k-1) + B(k)u(k) + K(k)\{y_m(k) - C(k)\hat{x}(k|k-1)\}, \tag{7.39}$$

where the Kalman gain and the error covariance update are given by

$$K(k) = A(k)P(k)C^T(k)(R_2(k) + C(k)P(k)C^T(k))^{-1}, \tag{7.40}$$

$$\begin{aligned} P(k+1) = {} & A(k)P(k)A^T(k) + R_1(k) - \\ & A(k)P(k)C^T(k)\left(R_2(k) + C(k)P(k)C^T(k)\right)^{-1}C(k)P(k)A^T(k). \end{aligned} \tag{7.41}$$

Unlike the time-invariant case, the covariance matrix $P(k)$ and gain matrix $K(k)$ do not, in general, converge to steady-state values.

7.4.4 Periodically Time-varying System

Let us consider the system:

$$\begin{aligned} x_k(t+1) &= A(t)x_k(t) + B(t)u_k(t) + \varepsilon_{1,k}(t) \\ y_{m,k}(t) &= C(t)x_k(t) + \varepsilon_{2,k}(t) \\ x_{k+1}(0) &= x_k(N), \end{aligned} \tag{7.42}$$

where, $t = 0, \cdots, N-1$. This represents a periodic linear system with period N. The variable k denotes the period index, while t, which ranges from 0 to $N-1$, is the time index within a single period.

Such periodic systems arise frequently in practice. Examples include:

- Batch processes that evolve along predefined trajectories.
- Cyclic operation in continuous processes, where system behavior repeats over time.
- Multirate sampled-data systems, where different variables are sampled at rates that are integer multiples of a base sampling interval. In such cases, only the C matrix may vary periodically, and the overall period is the least common multiple of the individual sampling intervals.

This structure enables the use of specialized estimation and control techniques that exploit the periodic nature of the dynamics.

The Kalman filter for the system above can be derived directly from the time-varying Kalman filter equations and is given by

$$\begin{aligned} \hat{x}_k(t+1|t) &= A(t)\hat{x}_k(t|t-1) + B(t)u_k(t) + \\ &\quad K_k(t)\left\{y_{m,k}(t) - C(t)\hat{x}_k(t|t-1)\right\} \quad t = 0, \cdots, N-1 \\ \hat{x}_{k+1}(0|-1) &= \hat{x}_k(N|N-1). \end{aligned} \tag{7.43}$$

$$K_k(t) = A(t)P_k(t)C^T(t)(R_2(t) + C(t)P_k(t)C^T(t))^{-1}. \tag{7.44}$$

$$\begin{aligned} P_k(t+1) &= A(t)P_k(t)A^T(t) + R_1(t) - A(t)P_k(t)C^T(t)\times \\ &\quad \left[R_2(t) + C(t)P_k(t)C^T(t)\right]^{-1} C(t)P_k(t)A^T(t), \quad t = 0, \cdots, N-1 \end{aligned} \tag{7.45}$$

$$P_{k+1}(0) = P_k(N). \tag{7.46}$$

Due to the periodic nature of the system, for detectable systems, the Kalman gain and covariance matrices converge to a *periodic steady-state* as $k \to \infty$:

$$K_k(t) \;\to K_\infty(t) \quad P_k(t) \;\to P_\infty(t).$$

As in the linear time-invariant case, if some transient performance loss is acceptable, the periodic solution can be used instead of the full time-varying implementation. This offers computational benefits, as the periodic gain sequence $K_\infty(t)$ can be precomputed and stored.

The periodic solutions $P_\infty(t)$, $t = 0, \cdots, N-1$ and $K_\infty(t)$, $t = 0, \cdots, N-1$ can be computed using two alternative approaches, discussed next.

- One approach is to iterate the RDE (7.45) until it converges to a periodic solution $P_\infty(t)$. Once obtained, the corresponding gains $K_\infty(t)$ can be computed using (7.44).

- Alternatively, the periodic system can be reformulated as an equivalent "lifted" system:

$$
\begin{aligned}
x_{k+1}(0) &= \Phi x_k(0) + \Gamma_1 \begin{bmatrix} u_k(0) \\ u_k(1) \\ \vdots \\ u_k(N-1) \end{bmatrix} + \Gamma_2 \begin{bmatrix} \varepsilon_{1,k}(0) \\ \varepsilon_{1,k}(1) \\ \vdots \\ \varepsilon_{1,k}(N-1) \end{bmatrix} \\
\begin{bmatrix} y_{m,k}(0) \\ y_{m,k}(1) \\ \vdots \\ y_{m,k}(N-1) \end{bmatrix} &= \Pi x_k(0) + \mathcal{D} \begin{bmatrix} u_k(0) \\ u_k(1) \\ \vdots \\ u_k(N-1) \end{bmatrix} + \begin{bmatrix} \varepsilon_{2,k}(0) \\ \varepsilon_{2,k}(1) \\ \vdots \\ \varepsilon_{2,k}(N-1) \end{bmatrix}.
\end{aligned}
\tag{7.47}
$$

Deriving the exact forms of Φ, Γ_1, Γ_2, Π, and $\mathcal{D}$ is left as an exercise. This lifted formulation allows the system to be treated as a linear time-invariant system (in terms of index k). One can then compute $P_\infty(0)$ by solving the corresponding ARE. With $P_\infty(0)$ in hand, the full periodic sequence $\{P_\infty(1), \cdots, P_\infty(N-1)\}$ can be obtained using the recursion in (7.44) and (7.45).

7.4.5 Measurement Delays

Measurement delays are common in process systems. For example, key variables such as concentration or product quality are often obtained via laboratory analysis and become available only after some delay. Let n_θ denote the measurement delay, so that the measurement taken at time $k - n_\theta$ becomes available at time k. This situation can be handled by augmenting the state to include delayed measurements. We define the augmented state as

$$
\bar{x}(k) \triangleq \begin{bmatrix} x(k) \\ y(k-1) \\ \vdots \\ y(k-n_\theta) \end{bmatrix}, \tag{7.48}
$$

leading to

$$
\bar{x}(k+1) = \bar{A}\bar{x}(k) + \bar{B}u(k) + \bar{\varepsilon}_1(k). \tag{7.49}
$$

$$
y_m(k) = \begin{bmatrix} 0 & 0 & \cdots & I \end{bmatrix} \bar{x}(k) + v(k). \tag{7.50}
$$

Augmenting the state-space model to account for both disturbance modeling and time delays was discussed in Chapter 6. The error covariance $P(k)$ and Kalman gain $K(k)$ for the delayed measurement case can then be computed using the Riccati equation (7.18) and gain formula (7.17), respectively, applied to the augmented model.

Remark 7.3: Fixed-lag Smoothing: In some applications, it is desirable to re-estimate past states (i.e., perform smoothing) when measurements arrive with a known delay. This can be handled by augmenting the state vector as

$$\bar{x}(k) \triangleq \begin{bmatrix} x(k) \\ x(k-1) \\ \vdots \\ x(k-n_\theta) \end{bmatrix}, \tag{7.51}$$

leading to

$$\bar{x}(k+1) = \bar{A}\bar{x}(k) + \bar{B}u(k) + \bar{\varepsilon}_1(k). \tag{7.52}$$

$$y_m(k) = \begin{bmatrix} 0 & 0 & 0 & \cdots & C \end{bmatrix} \bar{x}(k) + v(k). \tag{7.53}$$

This formulation enables smoothing of past states up to $x(k - n_\theta)$, hence the name "fixed-lag smoothing." However, since the state dimension is typically much larger than that of the output, this approach can significantly increase the size of the augmented state $\bar{x}(k)$, leading to higher computational cost.

Remark 7.4: With the increasing adoption of wireless communication and remote control in process plants, handling measurement delays has become more important due to communication latencies and packet loss. This issue has been extensively studied in the fields of communication and electrical engineering. Gopalakrishnan et al. (2011) provide a comparative review of various approaches for handling such delays. Their work includes several case studies, including a four-state binary distillation column example, and is recommended for readers interested in practical implementations.

Remark 7.5: Extensions to continuous-time systems are discussed in Kwakernaak and Sivan (1972). A comprehensive overview of various extensions and modifications of the Kalman filter—including the extended Kalman filter for nonlinear systems—is provided in Jazwinski (2007).

7.5 Least Squares Formulation of State Estimation

In this section, we present an alternative deterministic formulation of the optimal filtering problem. While this formulation reduces to the standard Kalman filter in its basic form, its optimization-based structure offers key advantages: it naturally accommodates constraints on the estimated variables and allows direct use of nonlinear models. We begin with a batch formulation that utilizes all available measurement data at once and then extend it to a recursive formulation using a moving horizon approach and the concept of *arrival cost.*

7.5.1 Batch Least Squares Formulation

Suppose at time step k one wishes to compute the best smoothed estimate of the state sequence $x(1), \cdots, x(k+1)$, given an initial state estimate $\hat{x}(1|0)$, the deterministic input sequence

$u(1), \cdots, u(k)$, and all collected measurements $y_m(1), \cdots, y_m(k)$. To this end, one can formulate and solve the following least-squares problem:

$$J_k(y_m(1), \cdots, y_m(k), u(1), \cdots, u(k), \hat{x}(1|0)) \triangleq \min_{z_1, \cdots, z_{k+1}} \left[z_1^{eT} Q_0 z_1^e + \sum_{i=1}^{k} \left(z_{i+1}^{eT} Q_1 z_{i+1}^e + y_i^{eT} Q_2 y_i^e \right) \right],$$

where

$$\begin{aligned} z_1^e &= z_1 - \hat{x}(1|0) \\ z_{i+1}^e &= z_{i+1} - (Az_i + Bu(i)), \\ y_i^e &= y_m(i) - Cz_i, \qquad i = 1, \cdots, k. \end{aligned} \tag{7.54}$$

The first term z_1^e represents the error in the initial state estimate $\hat{x}(1|0)$. The term $z_{i+1}^e, i = 1, \cdots, k$ corresponds to the error in the state transition predicted by $z_{i+1} = Az_i + Bu(i)$, and $y_i^e, i = 1, \cdots, k$ denotes the error in the output prediction given by $y_m(i) = Cz_i$. These error terms are penalized in the objective function, and the weighting matrices should reflect our relative confidence in the initial state estimate, the state transition model, and the output equation; higher weights indicate greater confidence. Although other norms could be used in the objective, the 2-norm formulation offers computational advantages and will serve as the most convenient choice for our subsequent discussion.

The notation J_k refers to both the least-squares problem and its optimal cost. The solution to the above least-squares problem, $z_1^*, \cdots, z_{k+1}^*$, yields *the estimate for the state sequence*, denoted as $\hat{x}(1|k), \cdots, \hat{x}(k+1|k)$.

The advantages of the optimization formulation are as follows:

- State constraints can be incorporated. For example, prior knowledge that the state must lie within a certain set can be enforced as

$$z_i \in \mathcal{X}. \tag{7.55}$$

- Constraints can also be imposed on the estimated errors. For instance,

$$z_i^e \in \mathcal{Z}_i \text{ and } y_i^e \in \mathcal{Y}_i. \tag{7.56}$$

  Incorporating such prior knowledge in the form of constraints can improve both the efficiency and robustness of the estimation process.
- The framework extends naturally to nonlinear systems. Consider the system

$$\begin{aligned} x(k+1) &= f(x(k), u(k), \varepsilon_1(k)) \\ y_m(k) &= g(x(k)) + \varepsilon_2(k). \end{aligned} \tag{7.57}$$

  This leads to a nonlinear least-squares problem, which retains the same conceptual structure but requires more computational effort to solve.

Although the method is formulated in a deterministic least-squares setting, the design remains nontrivial, as it still requires the selection of appropriate weighting matrices. This task is fundamentally analogous to specifying the covariance matrices in Kalman filter design. In fact, when we later establish the equivalence between the least-squares estimation approach and the Kalman filter, we will show that the weighting matrices can be explicitly related to the noise covariance matrices.

At first glance, the computational burden of solving the least-squares problem appears significantly higher than that of the Kalman filter, seemingly growing unbounded with time. However, for unconstrained linear systems, we will show that dynamic programming can be used to derive a recursive expression for $\hat{x}(k+1|k)$, effectively reducing the least-squares solution to the standard Kalman filter update. The main drawback of this approach is the loss of access to the full smoothed state sequence.

An alternative is to employ a *fixed-size moving window*, within which the least-squares problem is solved. By introducing an appropriate penalty on the initial state, the solution obtained in this limited window can match that of the full-batch problem exactly. The initial state estimate and its associated weighting matrix can be tuned to ensure consistency with the full problem.

Note that the dynamic programming approach is not applicable to constrained linear or nonlinear systems, where an analytical solution of the stage-wise least-squares problem is no longer available. In such cases, the constrained and/or nonlinear least-squares problem must be solved directly within a moving window using numerical optimization techniques. This significantly increases the computational burden, as mathematical programming is required. Furthermore, there is generally no systematic method for selecting the initial-state penalty to guarantee equivalence with the full-batch solution.

The use of a moving estimation window and related considerations will be discussed in greater detail in Section 7.5.3.

7.5.2 Recursive Solution and Equivalence with Kalman Filter

The dynamic programming approach can be applied to Eq. (7.54) to recursively compute the sequence $\hat{x}(k+1|k)$ using the following equations:

$$\begin{aligned}\hat{x}(k+1|k) = {} & A\hat{x}(k|k-1)- \\ & AQ_0^{-1}(k)C^T\left(CQ_0^{-1}(k)C^T+Q_2^{-1}\right)^{-1}C\,(y_m(k)-C\hat{x}(k|k-1)),\end{aligned} \tag{7.58}$$

and

$$\begin{aligned}Q_0^{-1}(k+1) = {} & AQ_0^{-1}(k)A^T+Q_1^{-1}- \\ & AQ_0^{-1}(k)C^T\left(CQ_0^{-1}(k)C^T+Q_2^{-1}\right)^{-1}CQ_0^{-1}(j)A^T,\end{aligned} \tag{7.59}$$

The derivation of these recursions is provided in Appendix D to avoid interrupting the flow of the main discussion. Since the development is based on dynamic programming, the associated algebra is somewhat involved. Additional details can be found in Section D.2.

These equations are equivalent to those of the Kalman filter if the weighting matrices are chosen as follows: $Q_0(k) = P^{-1}(k)$, $Q_1 = R_1^{-1}$, and $Q_2 = R_2^{-1}$. This equivalence implies that the estimate $\hat{x}(k+1|k)$ obtained from the least-squares formulation matches the Kalman filter estimate, provided the weighting matrices are set to the inverses of the corresponding covariance matrices: $P^{-1}(1)$, R_1^{-1}, and R_2^{-1}.

The equivalence between the unconstrained least-squares estimator and the Kalman filter is significant for several reasons:

- It provides a physically meaningful guideline for selecting the weighting matrices in the least-squares estimator.

- It offers insight into the conditions under which the least-squares approach performs well. In particular, it shows that the assumption of Gaussian noise is implicit in the formulation, just as in the Kalman filter.
- It highlights that minimizing the residual of a deterministic model without proper disturbance modeling is effectively equivalent to applying a Kalman filter with artificial white noise terms added. However, we've already argued that blindly adding white noise terms—without accounting for the actual nature of disturbances—can degrade estimation performance. As in Kalman filter design, accurate disturbance modeling remains critical in the least-squares approach.

7.5.3 Use of Moving Estimation Window

The full least-squares problem must be solved directly—rather than through the recursive formulation—when a smoothed estimate of the past state sequence is required, or when the problem involves constraints and/or a nonlinear model. To limit the computational burden, a common strategy is to use a *moving estimation window*, which restricts the optimization to a fixed number of recent time steps, i.e., time steps $k-m+1, \cdots, k$. Estimation performed in this manner is referred to as *moving horizon estimation* (MHE).

Formulation of MHE

Recall the full batch least-squares problem in Eq. (7.54). We define the following j-step arrival cost (see Section D.2 for details):

$$\Phi_j(\zeta) \stackrel{\Delta}{=} J_j \text{ with the constraint } z_{j+1} = \zeta.$$

This quantity captures the cumulative contribution of all measurements ($y_m(1), \cdots, y_m(j)$), knowledge on the initial state ($\hat{x}(1|0)$), and model equations up to time j, condensed into a cost function parameterized by the state z at time $j+1$. In forward dynamic programming, it serves a role analogous to the cost-to-go function in backward dynamic programming.

Consider an m-step estimation horizon. By applying forward dynamic programming arguments (see Section D.2), the least-squares problem for full batch estimation can be reformulated as the following m-step estimation problem posed at each time step k:

$$J_m\left(y_m(k-m+1), \cdots, y_m(k), u(k-m+1), \cdots, u(k), \right) \stackrel{\Delta}{=} \min_{z_{-m+1}, \cdots, z_1} \left[\Phi_{k-m}(z_{-m+1}) + \sum_{i=-m+1}^{0} z_{i+1}^{eT} Q_1 z_{i+1}^{e} + y_i^{eT} Q_2 y_i^{e} \right], \tag{7.60}$$

$$\text{where,} \quad z_{i+1}^e = z_{i+1} - (Az_i + Bu(k+i))$$
$$y_i^e = y_m(k+i) - Cz_i \qquad -m+1 \le i \le 0.$$

Here, $\Phi_{k-m}(z_{-m+1})$ denotes the $(k-m)$-step arrival cost defined earlier. It represents the optimal cost accumulated over the earlier time steps $1, \cdots, k-m$, subject to the constraint that the initial state of the moving window is fixed to z_{-m+1}. The solution to the above problem $z_{-m+1}^*, \cdots, z_1^*$ provides the estimated state sequence $x(k-m+1), \cdots, x(k+1)$, denoted as $\hat{x}(k-m+1|k), \cdots, \hat{x}(k+1|k)$.

Hence, in MHE, the least-squares estimation problem is posed over a fixed moving window of size m. To recover the full batch estimation solution exactly using the MHE strategy, one must be able to *construct a recursive analytical expression for the arrival cost* and incorporate it into Eq. (7.60). For unconstrained linear systems, a recursive formula for the arrival cost is derived in Section D.2 (see Eqs. D.33–D.35). These recursions are essentially equivalent to the Kalman filter update equations.

Constraints of the types given in Eqs. (7.55) and (7.56) can be incorporated into the MHE formulation. However, the presence of constraints means that the least-squares problem no longer admits an analytical solution. If the constraints are linear inequalities and the arrival cost $\Phi_{k-m}(z_{-m+1})$ is quadratic, the resulting optimization problem becomes a quadratic program (QP). If the system dynamics or output model are nonlinear, as in Eq. (7.57), the problem becomes a nonlinear program (NLP).

For constrained linear systems, as well as unconstrained or constrained nonlinear systems, it is generally not possible to compute the exact arrival cost recursively. In such cases, the only practical option is to approximate the arrival cost. It is preferable to use an approximation that underestimates the true cost. Specifically, in place of $\Phi_{k-m}(z_{-m+1})$ in Eq. (7.60), we use an approximate cost $\underline{\Phi}_{k-m}(z_{-m+1})$ satisfying

$$\underline{\Phi}_{k-m}(z_{-m+1}) \leq \Phi_{k-m}(z_{-m+1}) \quad \forall z_{-m+1} \in \mathcal{X}. \tag{7.61}$$

This approach ensures that the accuracy of the initial estimate z_{-m+1} is not overestimated. Statistically, this corresponds to avoiding the assumption of more information than is actually available. Although we do not pursue the mathematical analysis of this approximation in detail here, it is worth noting that a moving horizon estimator using an arrival cost that underestimates the true value leads to stability in the observer sense.

For a constrained linear problem, it can be shown that

$$\begin{aligned}\Phi_{k-m}(z_{-m+1}) \geq (z_{-m+1} &- \hat{x}(k-m+1|k-m))^T Q_0^{-1}(k-m+1) \\ &(z_{-m+1} - \hat{x}(k-m+1|k-m)) + \Phi(\hat{x}(k-m+1|k-m)),\end{aligned} \tag{7.62}$$

where $\hat{x}(k-m+1|k-m)$ is the estimate obtained from the constrained MHE at time $k-m$, and $Q_0(k-m+1)$ is computed using the recursion in Eq. (D.35).

This result implies that using a quadratic penalty on the initial state deviation—with the weighting matrix given by the Kalman filter update equation—yields a valid lower-bounding approximation of the arrival cost. The proof is somewhat subtle, as it relies on the fact that $\hat{x}(k-m+1|k-m)$ is computed from the constrained MHE rather than the unconstrained case (i.e., the Kalman filter). For a detailed treatment and formal proof, see the references cited at the end of this section.

For nonlinear problems, computing a tight lower bound for the arrival cost is significantly more challenging. The simplest admissible choice is to set $\underline{\Phi}_{k-m}(z_{-m+1}) = 0$, which trivially satisfies the lower-bound condition. However, this choice effectively discards all information prior to the estimation window; with a short horizon, it can result in poor estimation performance.

A more practical alternative is to use linearized statistics. In this approach, the quadratic initial penalty is computed as in the linear case, but based on a model obtained by linearizing the nonlinear system. However, due to the approximation inherent in linearization, the resulting penalty may not strictly lower-bound the true arrival cost.

To ensure a valid lower bound while still utilizing linearized information, one can combine

$$\begin{aligned}\underline{\Phi}_{k-m}(z_{-m+1}) = (z_{-m+1} &- \hat{x}(k-m+1|k-m))^T (Q_0^{-1}(k-m+1) + \gamma I)^{-1} \\ &(z_{-m+1} - \hat{x}(k-m+1|k-m)) + \Phi(\hat{x}(k-m+1|k-m)),\end{aligned} \tag{7.63}$$

where $\gamma \geq 0$. For sufficiently large γ, this expression serves as a valid lower-bounding approximation to $\Phi_{k-m}(z_{-m+1})$. As $\gamma \to \infty$, the approximation converges to the trivial case $\underline{\Phi}_{k-m}(z_{-m+1}) = 0$.

Example 7.12: *Moving Horizon Estimator for the System in Example 7.10.*
We develop an MHE with horizon $m = 10$ for the same system considered in Example 7.10, which includes both model-plant mismatch and an unmeasured integrated white noise disturbance. The structure of the implementation will follow that of the Kalman filter, with appropriate modifications to incorporate the MHE formulation. The MATLAB code is available at https://github.com/Lee-MPC-Book

The core of MHE is its optimization problem. The objective function defined in Eq. (7.60) is implemented as a standalone function, `J_MHE`, and passed to an optimization solver in the MATLAB code. In this implementation, we have taken m as reasonably large and ignored the arrival cost Φ_{k-m}.

Since the unmeasured disturbance is an integrated white noise sequence, we augment the system state with the disturbance, as in Eq. (7.34). The main MATLAB code for this example is similar to that in Example 7.11, with the MHE calculations replacing the Kalman filter update at each time step.

Figure 7.5 shows that the MHE provides estimation performance comparable to that of the Kalman filter presented earlier. The formulation of the objective function is central to the MHE framework. Extending the approach to nonlinear systems is conceptually straightforward: the linear model used in computing ε_1 is simply replaced with a nonlinear function.

A key advantage of MHE is its ability to explicitly incorporate constraints. For example, physical quantities such as species compositions or inlet flow rates are inherently nonnegative. These constraints may apply to states or disturbances. The following example considers a system in which the unmeasured disturbance is strictly positive and modeled as $|\varepsilon|$, where $\varepsilon \sim \mathcal{N}(0, Q_1)$ is Gaussian white noise.

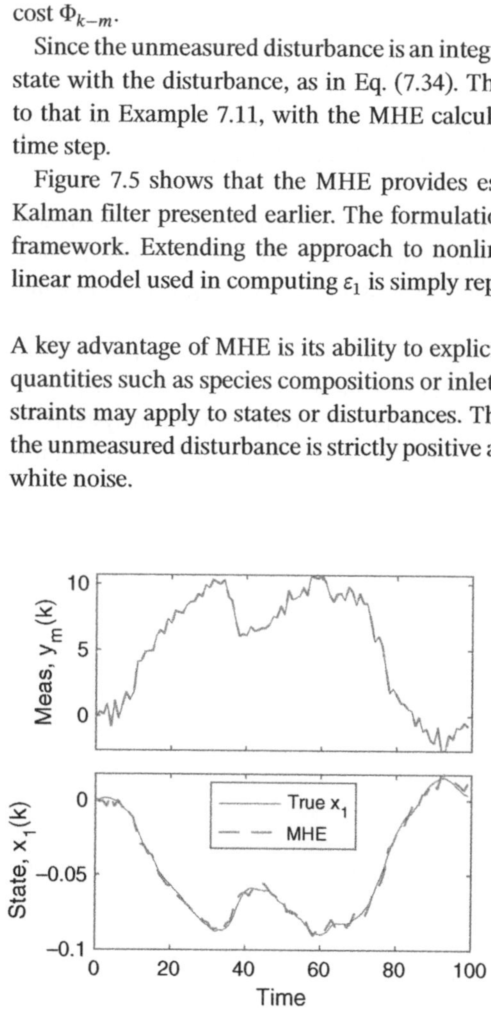

Figure 7.5 Comparison of the true values of $y_m(k)$ and $x_1(k)$ (thin lines) with the corresponding MHE estimates (dashed lines), under an unmeasured disturbance modeled as integrated white noise.

Example 7.13: *Handling Constraints in MHE.*
Consider the following system:

$$x(k+1) = \begin{bmatrix} 1 & 0.2 \\ -0.1 & 0.3 \end{bmatrix} x(k) + \begin{bmatrix} 0 \\ 1 \end{bmatrix} |\varepsilon(k)|$$

$$y_m(k) = \begin{bmatrix} 0.5 & -2 \end{bmatrix} x(k) + v(k).$$

Here, $v(k)$ is zero-mean white noise with variance 0.01, and $\varepsilon(k)$ is a zero-mean white noise disturbance with $\text{cov}(\varepsilon) = 1$.

We solve the state estimation problem using two formulations:

1. Unconstrained MHE, and
2. MHE with a nonnegativity constraint on $|\varepsilon(k)|$.

Figure 7.6 shows that the estimator performance is significantly better when the nonnegativity constraints are imposed when using the MHE.In this example, MHE was implemented with a horizon of $m = 5$. At each time step, the optimization variables include the initial state estimate $\hat{x}(k - m + 1|k - m)$ and the disturbance sequence $\{|\varepsilon(k - m + 1)|, \dots, |\varepsilon(k - 1)|\}$. In the constrained case, nonnegativity constraints are imposed on each $\varepsilon(i)$ in the horizon.

MHE offers a flexible framework for solving batch state estimation problems. The discussion and examples in this section illustrated its equivalence to the Kalman filter in the unconstrained linear case, as well as its ability to address more complex scenarios involving constraints and nonlinearities.

Some of the foundational contributions to MHE include the early works of Michalska and Mayne (1995) and Robertson et al. (1996). The stability of MHE in the presence of constraints was rigorously analyzed by Rao et al. (2001) for linear systems and extended to general nonlinear systems by Rao et al. (2003). Further, Robertson and Lee (2002) demonstrated how MHE can be used not only

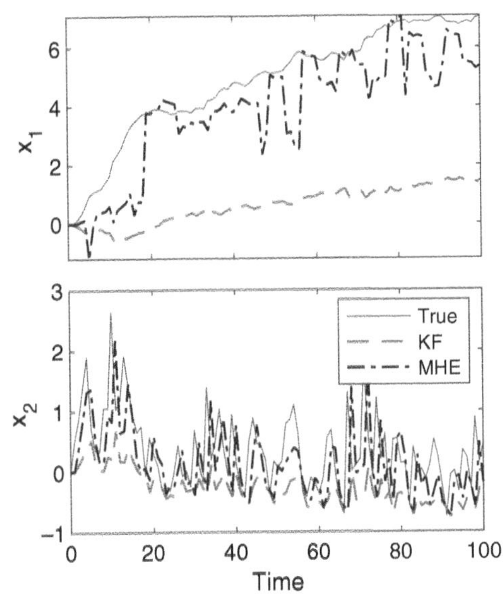

Figure 7.6 Comparison of Kalman filter and MHE with nonnegativity constraints on $d(i)$. The constrained MHE demonstrates improved tracking performance for the state variables.

for state estimation in constrained systems but also for constructing optimal estimators under non-Gaussian noise, including asymmetric distributions, by leveraging the constraint-handling capability of the framework.

Exercises

1. Derive the Kalman filter when the assumption of independence between the state and output noise does not hold, i.e., $R_{12} \neq 0$. Optimize the filter gain matrix as in Section 7.3.

2. Repeat the above derivation for the case $R_{12} \neq 0$ using a different approach. Rewrite the system so that the noise terms are independent. Note that system (7.1) can be transformed as

$$\begin{aligned} x(k+1) &= Ax(k) + Bu(k) + \varepsilon_1(k) - R_{12}R_2^{-1}\left(Cx(k) + \varepsilon_2(k) - y_m(k)\right) \\ &= \underbrace{(A - R_{12}R_2^{-1}C)}_{A'}x(k) + R_{12}R_2^{-1}y_m(k) + \underbrace{\varepsilon_1(k) - R_{12}R_2^{-1}\varepsilon_2(k)}_{\varepsilon_1'(k)} \\ y_m(k) &= Cx(k) + \varepsilon_2(k). \end{aligned}$$

Verify that

$$E\left\{\begin{bmatrix} \varepsilon_1'(k) \\ \varepsilon_2(k) \end{bmatrix}\begin{bmatrix} \varepsilon_1'(k) \\ \varepsilon_2(k) \end{bmatrix}^T\right\} = \begin{bmatrix} R_1 - R_{12}R_2^{-1}R_{21} & 0 \\ 0 & R_2 \end{bmatrix}. \tag{7.64}$$

Noting that the term $R_{12}R_2^{-1}y_m(k)$ does not affect the optimal gain computation, as it is known, use the standard formulae to complete the derivation of the Kalman filter.

3. **Derivation of the Kalman filter gain $K_f(k)$**: For many estimation and monitoring tasks, the filtered estimate $\hat{x}(k|k)$ and its error covariance $\bar{P}(k|k) \triangleq \mathrm{cov}\{\tilde{x}_e(k)\}$ are required. Recall:

$$x_e(k+1) \stackrel{\Delta}{=} x(k+1) - \hat{x}(k+1|k). \tag{7.7}$$

$$\tilde{x}_e(k) \stackrel{\Delta}{=} x(k) - \hat{x}(k|k). \tag{7.8}$$

Show that the following equations can be used to compute $\bar{P}(k|k)$:

$$\begin{aligned} P(k) &= A\bar{P}(k-1|k-1)A^T + R_1 \\ K_f(k) &= P(k)C^T\left(CP(k)C^T + R_2\right)^{-1} \\ \bar{P}(k|k) &= \left(I - K_fC\right)P(k). \end{aligned} \tag{7.65}$$

4. Show that for an observable system, the observer poles can be arbitrarily placed.

5. For a stochastic process:

$$\begin{aligned} x(k+1) &= \alpha x(k) + \varepsilon(k) \quad \mathrm{cov}(\varepsilon) = r_1 \\ y_m(k) &= x(k) + \nu(k) \quad \mathrm{cov}(\nu) = r_2, \end{aligned}$$

with $\mathrm{cov}(x(0)) = r_0$, compute the steady-state Kalman filter gain and the poles of the steady-state filter. Compare the filter poles to the system pole for $\alpha = 0.8$, 1.0, and 1.2.

6. Consider the system:

$$x(k+1) = \begin{bmatrix} 1 & 0.1 \\ -0.3 & 0.65 \end{bmatrix} + \begin{bmatrix} 0 \\ 1 \end{bmatrix} \varepsilon \quad R_1 = \text{cov}\{\varepsilon\} = 1$$

$$y_m(k) = \begin{bmatrix} 1 & 0 \end{bmatrix} x(k) + \nu(k) \qquad R_2 = \text{cov}\{\nu\} = 0.2.$$

- Design a deadbeat observer.
- Design an observer with poles p_1 and p_2 placed arbitrarily within the unit circle.
- For both designs, use the definition $P = E\left\{x_e x_e^T\right\}$ to compute the steady-state error covariance. Assume $P(k+1) - P(k) \to 0$ as $k \to \infty$.

7. For the system in Problem 6, design the steady-state Kalman filter. Obtain the steady-state gain K_∞, covariance matrix P_∞, and the poles of the resulting estimator.

8. Repeat the previous problem for several values of R_1 and R_2. Comment on the effect of these parameters on the estimator performance and filter poles.

9. **Simulations of the Observers**: Compare the performance of the pole-placement observer and the Kalman filter designed in Problems 6 and 7. Use `randn(100,1)` to generate one realization each for $\varepsilon(k)$ and $\nu(k)$. Initialize with $x(0) = \begin{bmatrix} 1 & 1 \end{bmatrix}$ and $\hat{x}(0) = \begin{bmatrix} 0 & 0 \end{bmatrix}$, and simulate for 100 time steps. Plot the true and estimated trajectories, and compute the estimation errors. Repeat the simulation for 50 Monte Carlo realizations to compare the performance of both estimators statistically.

10. Derive the structure of the system matrices for the lifted system (7.47) in terms of the time-varying matrices: Φ, Γ, Π, and $\mathcal{D}$.

11. Show that the batch least-squares problem yields the maximum a posteriori (MAP) estimate of the state sequence—i.e., the maximizer of the conditional density of the state sequence given the measurements.

8

Unconstrained Quadratic Optimal Control

In spite of its relatively long history, the (optimal control) problem has never been formulated rigorously from a mathematical point of view.

– Rudolf Kalman

Early Model Predictive Control (MPC) algorithms, such as Dynamic Matrix Control (DMC), share important conceptual and structural similarities with classical optimal control techniques—particularly linear quadratic (LQ) optimal control. These similarities are often overlooked because DMC is typically presented as a set point tracking method in an input–output framework, whereas LQ control is traditionally formulated as a state regulation problem to the origin. Addressing set point tracking within the LQ framework is possible but requires additional modeling or augmentation.

Nonetheless, both approaches share a quadratic objective function and incorporate feedback through measurement-based correction. Given this connection, it is instructive to revisit LQ optimal control—not only to highlight its similarities and differences with MPC, but also to explore how the classical framework can be extended to handle constraints, a defining feature of MPC. Moreover, the continued development of MPC can benefit from the theoretical rigor and foundational insights provided by classical optimal control.

In this chapter, we present the basic results of LQ optimal control. We derive both optimal and suboptimal control laws for finite-horizon and infinite-horizon problems, focusing on the unconstrained case. Extensions to constrained systems will be addressed in the following chapter.

The standard quadratic optimal control problem involves regulating the state to the origin—that is, driving the system state to zero from a given nonzero initial condition. For the linear system

$$x(k+1) = Ax(k) + Bu(k), \tag{8.1}$$

with initial condition $x(0)$, the goal is to determine an input sequence or feedback policy that minimizes the quadratic cost function

$$V_p \triangleq \sum_{i=0}^{p-1} \left[x^T(i)Qx(i) + u^T(i)Ru(i)\right] + x^T(p)Q_t x(p), \tag{8.2}$$

where Q is a positive semi-definite matrix, R is positive definite, and Q_t is a positive semi-definite matrix penalizing the terminal state. When the horizon length p is finite, the problem is referred to as a *finite-horizon* optimal control problem, whereas for $p = \infty$, it becomes the *infinite-horizon* problem.

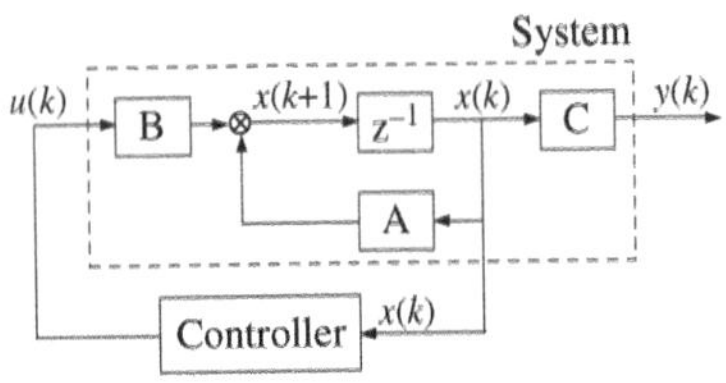

Figure 8.1 Block diagram of a linear state-space system with full state feedback.

Many practical control tasks—such as output regulation in the presence of constant disturbances or set point tracking—can be formulated as state regulation problems of this form.

8.1 Linear State Feedback Controller Design

Before introducing the LQ control problem, we first review the concept of linear state-feedback control for the system described by Eq. (8.1). In Part I, we observed that for an unconstrained control problem, the optimal control law often takes a linear form. When there are no bounds on the control input $u(k)$, a common choice for the controller is

$$u(k) = -Lx(k). \tag{8.3}$$

Figure 8.1 shows the block diagram of a system with state-feedback control. For the linear control law in Eq. (8.3), the "controller" block simply applies the feedback gain $-L(k)$ to the state vector. Substituting the feedback law (8.3) into the system model (8.1) yields the closed-loop dynamics:

$$x(k+1) = (A - BL)x(k). \tag{8.4}$$

The eigenvalues of the matrix $(A - BL)$ determine the closed-loop poles of the system. For the closed-loop system to be stable, these eigenvalues must lie strictly inside the unit circle in the complex plane.

As before, determining the controller gain L to assign specific locations to the closed-loop poles is referred to as pole placement controller design. Drawing from arguments similar to those used in pole placement observer design (see Section 7.2), the following points apply to linear state-feedback controllers:

- If (A, B) is a controllable pair, all closed-loop poles can be arbitrarily placed through appropriate choice of L.
- The controllable canonical form allows direct correspondence between the feedback gain L and the desired eigenvalues of $(A - BL)$.
- A *deadbeat controller* places all closed-loop poles at the origin.
- The deadbeat controller theoretically achieves perfect regulation in at most n sampling intervals. However, this often leads to overly aggressive control actions and poor robustness to modeling errors or disturbances.
- If (A, B) is not controllable and the controllability matrix has rank q $(q < n)$, at most q eigenvalues can be freely assigned. As illustrated in Example 6.5 and Section 6.5.1, an appropriate coordinate transformation can be used to express the system as

$$\begin{bmatrix} \tilde{x}_1(k+1) \\ \tilde{x}_2(k+1) \end{bmatrix} = \begin{bmatrix} \tilde{A}_{11} & \tilde{A}_{12} \\ 0 & \tilde{A}_{22} \end{bmatrix} \begin{bmatrix} \tilde{x}_1(k) \\ \tilde{x}_2(k) \end{bmatrix} + \begin{bmatrix} \tilde{B}_1 \\ 0 \end{bmatrix} u(k), \tag{8.5}$$

where $\tilde{x}_1(k)$ spans the q-dimensional controllable subspace. The feedback matrix L can influence the eigenvalues of $\tilde{A}_{11} - \tilde{B}_1\tilde{L}_1$, but not those of $\tilde{A}_{22}$.

- As with observers, pole placement design is of limited practical value, since the pole locations that yield satisfactory performance are not always evident. Additional performance or robustness criteria are therefore often required.

Example 8.1: *Pole Placement Design.*
Consider the linear system from Example 5.8:

$$g = \frac{2.69(-6s+1)}{(20s+1)(5s+1)}e^{-1.5s}.$$

With $t_s = 3$, the resulting discrete-time transfer function is

$$g_z = \frac{-0.1744z^2 + 0.189z + 0.1545}{z^3 - 1.41z^2 + 0.4724z}.$$

This transfer function can be converted into the following controllable canonical state-space form

$$x(k+1) = \begin{bmatrix} 1.41 & -0.4724 & 0 \\ 1 & 0 & 0 \\ 0 & 1 & 0 \end{bmatrix} x(k) + \begin{bmatrix} 1 \\ 0 \\ 0 \end{bmatrix} u(k),$$

$$y_m(k) = \begin{bmatrix} -0.1744 & 0.189 & 0.1545 \end{bmatrix} x(k).$$

We aim to design a linear state-feedback controller of the form

$$u(k) = -\begin{bmatrix} l_1 & l_2 & l_3 \end{bmatrix} x(k).$$

Using this control law, the closed-loop state matrix becomes

$$[A - BL] = \begin{bmatrix} 1.41 - l_1 & -0.4724 - l_2 & -l_3 \\ 1 & 0 & 0 \\ 0 & 1 & 0 \end{bmatrix}.$$

Deadbeat Controller: We choose the gain L such that all the closed-loop poles are placed at the origin. By observation, $l_1 = 1.41$, $l_2 = -0.4724$, $l_3 = 0$. This deadbeat controller drives the state to the origin in three steps. Figure 8.2 shows the system response when the controller is applied from the initial condition $x(0) = [1.0,\ 0.5,\ 1.0]^T$.

8.2 Finite-horizon Quadratic Optimal Control

Consider the finite-horizon quadratic optimal control problem:

$$J_p(x(0)) = \min_{u(0),\ldots,u(p-1)} \underbrace{\sum_{i=0}^{p-1} \left[x^T(i)Qx(i) + u^T(i)Ru(i)\right] + x^T(p)Q_t x(p)}_{V_p}, \tag{8.6}$$

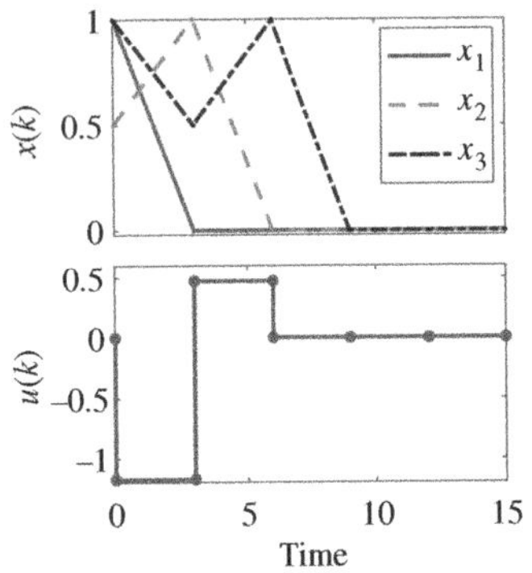

Figure 8.2 Application of deadbeat control.

subject to the deterministic system dynamics defined in Eq. (8.1) and the initial condition $x(0)$. We denote the optimal cost associated with this problem as $J_p(x(0))$, emphasizing its dependence on the initial condition $x(0)$.

There are two primary approaches to solving this problem. The conceptually simpler method involves explicitly deriving the relationship between the future state sequence and the future input sequence, and substituting it into the cost function to yield a least-squares optimization—similar to the DMC approach introduced in earlier chapters. A more advanced method is to apply dynamic programming to solve the problem recursively, stage by stage. We now review both approaches.

8.2.1 Open-loop Optimal Solution via Least Squares

The process model can be used to predict the future behavior of the system, starting from the initial state $x(0)$. Consider the ℓ-step state prediction:

$$\begin{aligned} x(\ell) &= Ax(\ell-1) + Bu(\ell-1) \\ &= A^2x(\ell-2) + ABu(\ell-2) + Bu(\ell-1) \\ &\vdots \\ &= A^\ell x(0) + \left[A^{\ell-1}Bu(0) + \cdots + ABu(\ell-2) + Bu(\ell-1)\right]. \end{aligned}$$

Thus, the entire state trajectory over the horizon can be expressed as a linear function of the initial condition and the future control inputs:

$$\begin{bmatrix} x(0) \\ x(1) \\ \vdots \\ \vdots \\ x(p) \end{bmatrix} = \begin{bmatrix} I \\ A \\ \vdots \\ \vdots \\ A^p \end{bmatrix} x(0) + \begin{bmatrix} 0 & \dots & \dots & 0 \\ B & 0 & \dots & 0 \\ AB & \ddots & \ddots & \vdots \\ \vdots & \ddots & \ddots & \vdots \\ A^{p-1}B & \dots & \dots & B \end{bmatrix} \begin{bmatrix} u(0) \\ \vdots \\ \vdots \\ u(p-1) \end{bmatrix}. \tag{8.7}$$

Denote the above as

$$\mathcal{X} = \mathrm{S}^x x(0) + \mathrm{S}^u \mathcal{U}, \tag{8.8}$$

where $\mathcal{X}$ is the stacked vector of future states, and $\mathcal{U}$ is the stacked vector of future control inputs. The quadratic cost function can now be written as

$$V_p(x(0)) = \mathcal{X}^T\bar{Q}\mathcal{X} + \mathcal{U}^T\bar{R}\mathcal{U}, \tag{8.9}$$

where $\bar{Q} = \text{blockdiag}\{Q, \ldots, Q, Q_t\}$ and $\bar{R} = \text{blockdiag}\{R, \ldots, R\}$.

Substituting Eqs.(8.8) into (8.9), we obtain:

$$\begin{aligned} V_p(x(0)) &= \left(S^x x(0) + S^u\mathcal{U}\right)^T \bar{Q}(S^x x(0) + S^u\mathcal{U}) + \mathcal{U}^T\bar{R}\mathcal{U} \\ &= \mathcal{U}^T\underbrace{(S^{uT}\bar{Q}S^u + \bar{R})}_{\mathcal{H}}\mathcal{U} - 2\underbrace{(-x(0)^T S^{xT}\bar{Q}S^u)}_{g^T}\mathcal{U} + x(0)^T S^{xT}S^x x(0) \end{aligned} \tag{8.10}$$

The above is a standard quadratic minimization problem, where $\mathcal{H}$ is positive definite due to the definiteness of R and $\bar{R}$, thereby guaranteeing a unique optimal solution.

By applying Lemma D.1 from Chapter 7, we find that the input sequence that minimizes V_p is given by:

$$\begin{aligned} \mathcal{U} &= \mathcal{H}^{-1}g \\ &= -\left(S^{uT}\bar{Q}S^u + \bar{R}\right)^{-1} S^{uT}\bar{Q}S^x x(0). \end{aligned} \tag{8.11}$$

Moreover, the corresponding optimal cost is

$$\begin{aligned} J_p(x(0)) &= -g^T\mathcal{H}g + x^T(0)S^{xT}S^x x^T(0) \\ &= x^T(0)\left[S^{xT}S^x - S^{xT}\bar{Q}S^u(S^{uT}\bar{Q}S^u + \bar{R})^{-1} S^{uT}\bar{Q}S^x\right] x(0). \end{aligned} \tag{8.12}$$

This result provides the *optimal open-loop input sequence* and the associated cost, derived through least-squares minimization using an explicit prediction model.

8.2.2 State Feedback Solution via Dynamic Programming

A more elegant and recursive approach to solving the finite-horizon optimal control problem is through dynamic programming. The idea is to decompose the problem into a sequence of one-step optimal control problems. The following definition formalizes this concept.

Definition 8.1: The optimal cost associated with the ℓ-step problem is defined as

$$J_\ell(x(p-\ell)) \triangleq \min_{u(p-\ell),\cdots,u(p-1)} \sum_{i=p-\ell}^{p-1} \left[x^T(i)Qx(i) + u^T(i)Ru(i)\right] + x^T(p)Q_t x(p), \tag{8.13}$$

subject to the system dynamics

$$x(i+1) = Ax(i) + Bu(i),$$

and initial condition $x(p-\ell)$. This quantity, denoted as the ℓ-step *cost to go*, is a function of the initial state $x(p-\ell)$ and represents the minimum achievable cost when solving the ℓ-step optimal control problem starting from that state.

Due to the structure of the system dynamics, the optimal control decision at time $p-1$ depends only on the current state $x(p-1)$—not on earlier states, such as $x(p-2)$, $x(p-3)$, etc. Based on this observation, we start by solving the one-step optimal control problem posed at time $p-1$:

$$J_1(x(p-1)) = \min_{u(p-1)} \left\{x^T(p-1)Qx(p-1) + u^T(p-1)Ru(p-1) + x^T(p)S(p)x(p)\right\}, \quad (8.14)$$

where $S(p) = Q_t$ is the terminal cost weight. Substituting the state equation $x(p) = Ax(p-1) + Bu(p-1)$ into the objective function, we obtain:

$$\begin{aligned} J_1(x(p-1)) = \min_{u(p-1)} &\{x^T(p-1)(A^TS(p)A+Q)x(p-1) + \\ &2x^T(p-1)A^TS(p)B \quad u(p-1) + u^T(p-1)(B^TS(p)B+R)u(p-1)\}. \end{aligned}$$

Applying Lemma D.1 from Chapter 7, the optimal control input at time $p-1$ is given by

$$u(p-1) = \underbrace{-(B^TS(p)B+R)^{-1}B^TS(p)A}_{L(p-1)}\ x(p-1), \quad (8.15)$$

and the corresponding one-step cost-to-go is

$$J_1\,(x(p-1)) = x^T(p-1)S(p-1)x(p-1), \quad (8.16)$$

where the cost-to-go matrix is defined as

$$S(p-1) = A^TS(p)A + Q - A^TS(p)B(B^TS(p)B+R)^{-1}B^TS(p)A. \quad (8.17)$$

We now move one step earlier to time $p-2$ and consider the two-step problem:

$$J_2(x(p-2)) = \min_{u(p-1),u(p-2)} \left\{\sum_{i=p-2}^{p-1} \left[x^T(i)Qx(i) + u^T(i)Ru(i)\right] + x^T(p)S(p)x(p)\right\}. \quad (8.18)$$

This is equivalent to a one-stage problem with cost-to-go inherited from the previous step:

$$J_2(x(p-2)) = \min_{u(p-2)} \{J_1(x(p-1)) + x^T(p-2)Qx(p-2) + u^T(p-2)Ru(p-2)\}, \quad (8.19)$$

$$= \min_{u(p-2)} \left\{x^T(p-1)S(p-1)x(p-1) + x^T(p-2)Qx(p-2) + u^T(p-2)Ru(p-2)\right\}. \quad (8.20)$$

Since the above has the same structure as the one-step problem, the optimal solution is

$$u(p-2) = \underbrace{-(B^TS(p-1)B+R)^{-1}B^TS(p-1)A}_{L(p-2)}x(p-2), \quad (8.21)$$

with the corresponding two-step cost-to-go:

$$J_2(x(p-2)) = x^T(p-2)S(p-2)x(p-2), \quad (8.22)$$

where

$$S(p-2) = A^TS(p-1)A + Q - A^TS(p-1)B\left(B^TS(p-1)B+R\right)^{-1}B^TS(p-1)A. \quad (8.23)$$

Continuing this backward recursion—i.e., successively solving $J_{p-k}(x(k))$ for $k = p-1, \dots, 0$—we arrive at the feedback-optimal solution for the original problem (8.6):

$$u(k) = -L(k)x(k) \qquad \text{for } k = p-1, \dots, 0, \tag{8.24}$$

where

$$L(k) = \left(B^T S(k+1)B + R\right)^{-1} B^T S(k+1)A, \tag{8.25}$$

$$S(k) = A^T S(k+1)A + Q - A^T S(k+1)B\left(B^T S(k+1)B + R\right)^{-1} B^T S(k+1)A. \tag{8.26}$$

The recursion in Eq. (8.26) is known as the *discrete-time Riccati equation* (also called the *Riccati difference equation*) and is initialized with $S(p) = Q_t$. It is solved backward in time to compute $S(k)$ for $k = p-1, \dots, 0$.

The optimal cost for the finite-horizon problem is given by

$$J_p(x(0)) = x(0)^T S(0)x(0). \tag{8.27}$$

8.2.3 Comparison of the Two Approaches

Dynamic programming decomposes a multistep optimal control problem into a sequence of single-step problems. The optimal feedback gain at each step, as given in Eq. (8.24), is computed recursively via the Riccati difference equation (RDE) in Eq. (8.26). This recursive structure involves only low-dimensional matrix operations and is computationally efficient.

In contrast, the explicit multistep prediction approach requires forming and inverting the Hessian matrix $\mathcal{H}$, whose dimension increases with the prediction horizon. For long horizons, this matrix becomes large, and special numerical techniques may be required to exploit its structure and reduce computational burden.

A key advantage of the dynamic programming approach is that it provides the optimal control in the form of an explicit *linear state-feedback law* at each time step. The closed-loop implementation of this feedback law enhances robustness to disturbances and model uncertainties, as the control action continuously adapts to the current state. In contrast, the explicit prediction-based solution yields an open-loop control sequence. To obtain an equivalent feedback representation, one must solve the following least-squares problem at each time step $k = 0, \dots, p-1$:

$$J_{p-k}(x(k)) = \min_{v_0,\cdots,v_{p-k-1}} \left\{ \sum_{i=0}^{p-k-1} \left[z_i^T Q z_i + v_i^T R v_i\right] + z_{p-k}^T Q_t z_{p-l} \right\} \tag{8.28}$$
$$\text{subject to system dynamics } z_{i+1} = Az_i + Bv_i.$$

Here, we initialize $z_0 = x(k)$, and z_i and v_i are temporary variables introduced solely for the optimization carried out at each time step. At time step k, they serve as surrogates for the predicted values of $x(k+i)$ and $u(k+i)$ based on information available at time k.

Among the computed input moves, only the first input v_0 is applied to the plant as $u(k)$, thereby recovering the same optimal feedback policy between $u(k)$ and $x(k)$.

However, the batch least-squares formulation provides a major advantage when constraints are imposed on the inputs and/or states. In such cases, dynamic programming becomes impractical because the convenient recursive structure no longer holds. In contrast, the least-squares approach—although lacking a closed-form solution—remains applicable by solving the constrained version of $J_{p-k}(x(k))$ *on-line* at each sampling instant using numerical optimization techniques. After computing the optimal input sequence, only the first control action $u(k)$ is applied to the system, while the remaining inputs are discarded. This procedure, known as *shrinking-horizon control*, is conceptually identical to the receding-horizon control strategy used in MPC.

8.3 Infinite-horizon Quadratic Optimal Control

For continuous processes operating over extended periods, it is natural to consider the infinite-horizon optimal control problem:

$$J_\infty(x(0)) = \min_{u(\cdot)} \left\{ V_\infty \triangleq \sum_{k=0}^{\infty} \left[x^T(k)Qx(k) + u^T(k)Ru(k) \right] \right\}, \tag{8.29}$$

with initial condition $x(0)$. In this section, we present both the optimal state feedback solution and a strategy based on a finite moving horizon. While the state feedback solution admits an analytical derivation, the moving horizon approach is introduced to motivate and facilitate the development of a practical framework for handling constraints—an issue that will be addressed in Chapter 9.

8.3.1 Optimal State Feedback Law: Asymptotic Solution of the Finite-horizon Problem

When the prediction horizon extends to infinity, solving for the optimal input sequence using the explicit least-squares approach becomes computationally intractable. In contrast, the dynamic programming framework remains applicable and yields a compact and elegant solution.

Assume that the system (A, B) is stabilizable and that $(Q^{1/2}, A)$ is detectable. Consider the infinite-horizon case, where the optimal control policy is given by

$$u(k) = -(B^T S(k+1)B + R)^{-1} B^T S(k+1)Ax(k), \qquad k = \infty, \dots, 0. \tag{8.30}$$

The corresponding RDE is

$$S(k) = A^T S(k+1)A + Q - A^T S(k+1)B\left(B^T S(k+1)B + R\right)^{-1} B^T S(k+1)A, \tag{8.31}$$

with terminal condition $S(\infty) = Q$. This recursion is solved backward in time. Under the stated assumptions, the sequence generated by (8.31) converges to a steady-state solution S_∞, which is the unique positive definite solution to the *algebraic Riccati equation* (ARE):

$$S_\infty = A^T S_\infty A + Q - A^T S_\infty B\left(B^T S_\infty B + R\right)^{-1} B^T S_\infty A. \tag{8.32}$$

The optimal state-feedback control law becomes:

$$u(k) = -\underbrace{(B^T S_\infty B + R)^{-1} B^T S_\infty A}_{L_\infty} x(k), \qquad k = 0, \dots, \infty. \tag{8.33}$$

The corresponding optimal cost is given by

$$J_\infty(x(0)) = x^T(0) S_\infty x(0). \tag{8.34}$$

This control policy is known as the asymptotic form of the *linear quadratic regulator* (LQR), or the infinite-horizon LQR. In Section 8.4, we will further examine the conditions under which the RDE (8.31) converges and explore key properties of the LQR solution.

8.3.2 Receding Horizon Implementation of the Finite-horizon Solution

As an alternative to the dynamic programming-based solution—which is not easily extendable to systems with constraints—we consider an approach based on the explicit multistep prediction of the future state trajectory. Since predicting the system response over an infinite horizon is impractical, we adopt a receding horizon control (RHC) strategy. In this framework, a finite-horizon optimal control problem is solved at each sampling instant using the current state as the initial condition. Only the first control input from the optimal sequence is implemented, and the process is repeated at the next time step with the updated state.

Specifically, at time step k ($k = 0, 1, \ldots$), given the current state $x(k)$, the following p-stage finite-horizon optimization problem is solved:

$$J_p(x(k)) = \min_{v_0,\ldots,v_{p-1}} \left\{ \sum_{i=0}^{p-1} \left[z_i^T Q z_i + v_i^T R v_i \right] + z_p^T Q_t z_p \right\} \tag{8.35}$$
$$\text{subject to system dynamics } z_{i+1} = Az_i + Bv_i.$$

As before, $z_0 = x(k)$, and z_i and v_i are temporary variables introduced solely for the optimization carried out at each time step. In the optimization performed at time step k, we use this compact notation in place of the more conventional $x(k+i|k)$ and $u(k+i|k)$ to streamline the exposition.

Thus, a finite-horizon open-loop optimal control problem is solved at each time step using the current plant state $x(k)$ as the initial condition (z_0). From the computed solution, only the first control input v_0 is applied as the plant input $u(k)$. At the next time step, the optimization problem $J_p(x(k+1))$ is solved again using the updated state feedback from the plant.

The finite-horizon solution has already been discussed in Section 8.2. One approach is to compute the control sequence $v_0, \ldots, v_{p-1}$ using the least-squares expression in (8.11) at each time step, and implement only the first control input $u(k)$.

An alternative approach is to solve the RDE (8.26) offline for a fixed horizon length p, and then apply the corresponding state-feedback law:

$$u(k) = \underbrace{-(B^T S_p B + R)^{-1} B^T S_p A}_{L_p} x(k). \tag{8.36}$$

Here, $S_p = S(0)$ is obtained by recursively solving the RDE:

$$S(i) = A^T S(i+1)A + Q - A^T S(i+1)B(B^T S(i+1)B + R)^{-1} B^T S(i+1)A, \tag{8.37}$$

with terminal condition $S(p) = Q_t$. This method yields the same optimal control law as the least-squares approach but avoids the need to solve a potentially large-scale least squares problem.

Note that the feedback control resulting from the above strategy is generally suboptimal. However, as we will see, by choosing the terminal weight matrix Q_t appropriately—specifically, to match the solution of the infinite-horizon Riccati equation—the control law derived from the finite-horizon problem can be made optimal for the infinite-horizon quadratic optimal control problem.

Example 8.2: *Finite- and Infinite-horizon LQ Control.*
Consider the system in Example 8.1, where the objective is to regulate the system to the origin, starting from the initial state $x(0) = [1.0\ 0.5\ 1.0]^T$, by minimizing the infinite-horizon cost (8.29) with $Q = I$ and $R = 100$. We compare the receding horizon implementation (with $p = 3$), the optimal infinite-horizon solution, and the deadbeat controller.

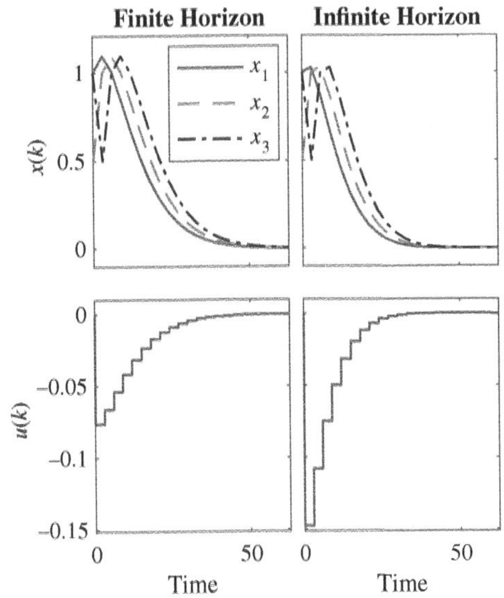

Figure 8.3 State and input trajectories for the three-state system under finite-horizon and infinite-horizon LQR control.

A large value of R is chosen to highlight the contrast between the three approaches. A large R penalizes control effort heavily, resulting in very conservative (cautious) actions.

Finite-horizon LQR: In the receding horizon case, we set $S(p) = I$ and solve the RDE (8.26) backward to obtain:

$$S_p = S(0) = \begin{bmatrix} 14.4312 & -4.5321 & 0 \\ -4.5321 & 3.8443 & 0 \\ 0 & 0 & 1 \end{bmatrix},$$

and the corresponding control law as

$$u(k) = \begin{bmatrix} 0.0959 & -0.0390 & 0 \end{bmatrix} x(k).$$

The infinite-horizon cost in the receding horizon implementation is computed using (8.2), with p chosen large enough so that x and u both return to the origin. The resulting cost is $V_\infty = 18.198$, and the corresponding input and state trajectories are shown in the left panels of Figure 8.3.

Infinite-horizon LQR: The optimal solution to the infinite-horizon problem is obtained by solving the ARE (8.32):

$$S_\infty = \begin{bmatrix} 23.2514 & -9.0804 & 0 \\ -9.0804 & 6.2094 & 0 \\ 0 & 0 & 1 \end{bmatrix},$$

and the corresponding control law as

$$u(k) = -\begin{bmatrix} 0.1922 & -0.0891 & 0 \end{bmatrix} x(k).$$

The infinite-horizon cost is $V_\infty = 16.723$, and the corresponding state and input trajectories are shown in the right panels of Figure 8.3. One can verify that the cost computed as $x^T(0)S_\infty x(0)$ also yields the same value.

Comparison with Deadbeat Control: For reference, the infinite-horizon cost of the deadbeat controller is 164.5, which is substantially higher. This outcome is expected, as the deadbeat controller exhibits highly aggressive behavior, and the large value of R penalizes such actions. Comparing Figure 8.2 with Figure 8.3 shows that the initial control move under deadbeat control is much more aggressive than that under LQR. Although similar cautious behavior can be obtained by placing the closed-loop poles farther from the origin, such manual tuning is generally cumbersome—particularly for high-dimensional systems.

Example 8.3: *Effect of Input Weight Matrix R.*
We revisit the previous problem with varying values of the input weight matrix R. The results are summarized in the following table:

R	S_∞	L_∞	λ_1, λ_2
10000	$\begin{bmatrix} 45.353 & -20.446 & 0 \\ -20.446 & 12.074 & 0 \\ 0 & 0 & 1 \end{bmatrix}$	$\begin{bmatrix} 0.0043 & -0.0021 & 0 \end{bmatrix}$	$[0.8556, 0.5496]$
100	$\begin{bmatrix} 23.251 & -9.0804 & 0 \\ -9.0804 & 6.2094 & 0 \\ 0 & 0 & 1 \end{bmatrix}$	$\begin{bmatrix} 0.1922 & -0.0891 & 0 \end{bmatrix}$	$0.6086 \pm 0.1132i$
1	$\begin{bmatrix} 4.5058 & -0.5018 & 0 \\ -0.5018 & 2.1826 & 0 \\ 0 & 0 & 1 \end{bmatrix}$	$\begin{bmatrix} 1.0624 & -0.3866 & 0 \end{bmatrix}$	$0.1736 \pm 0.2359i$
0.01	$\begin{bmatrix} 3.0219 & -0.0066 & 0 \\ -0.0066 & 2.0022 & 0 \\ 0 & 0 & 1 \end{bmatrix}$	$\begin{bmatrix} 1.4027 & -0.4708 & 0 \end{bmatrix}$	$0.0034 \pm 0.0393i$

As R decreases, the control gain L_∞ becomes more aggressive, and the closed-loop poles move closer to the origin. In the limit as $R \to 0$, the behavior of the infinite-horizon LQR approaches that of the deadbeat controller, as seen in Example 8.1.

8.3.3 Equivalence Between Finite- and Infinite-horizon Problems

As noted earlier, the feedback law resulting from the receding horizon implementation of a finite-horizon problem is, in general, only a suboptimal solution to the infinite-horizon quadratic optimal control problem. However, under specific assumptions and with particular choices of the terminal weighting matrix Q_t, the finite-horizon solution can be made equivalent to the infinite-horizon solution.

- We may choose Q_t such that the terminal cost $z_p^T Q_t z_p$ exactly matches the optimal cost from time $k + p$ to infinity, i.e.,

$$z_p^T Q_t z_p = \min_{v_{p+j}} \left\{ \sum_{j=0}^{\infty} z_{p+j}^T Q z_{p+j} + v_{p+j}^T R v_{p+j} \right\}. \tag{8.38}$$

From the discussion of the infinite-horizon optimal policy in Section 8.3.1, it is evident that such a Q_t can be obtained by solving the ARE:

$$Q_t = A^T Q_t A + Q - A^T Q_t B(B^T Q_t B + R)^{-1} B^T Q_t A. \tag{8.39}$$

With this choice of Q_t, the optimal solution to $J_p(z_0)$ coincides with that of $J_\infty(z_0)$.

- We may also choose Q_t such that the terminal cost equals the infinite-horizon cost from $k + p$ onward, under the assumption that no control action is applied beyond time $k + p$. That is, with $u_{p+i} = 0$ (for $i \geq 0$), the system evolves autonomously as

$$z_{i+1} = Az_i \qquad \text{for } i = p, \ldots, \infty$$

This assumption is meaningful only when the system is stable, since otherwise the cost diverges. Under this condition, we choose Q_t to satisfy:

$$z_p^T Q_t z_p = \sum_{i=p}^{\infty} z_i^T Q z_i. \tag{8.40}$$

It can be shown that a positive semi-definite Q_t satisfying Eq. (8.40) solves the Lyapunov equation:

$$AQ_t A^T + Q = Q_t. \tag{8.41}$$

A full proof is omitted. Solving $J_p(z_0)$ with this Q_t is thus equivalent to solving the infinite-horizon problem $J_{\infty,p}(z_0)$ with the assumption that $v_i = 0$ for $i \geq p$:

$$J_{\infty,p}(z_0) = \min_{u_0,\cdots,u_{p-1}} \left\{ \sum_{i=0}^{\infty} z_i^T Q z_i + \sum_{j=0}^{p-1} u_j^T R u_j \right\} \qquad u_j = 0, j \geq p. \tag{8.42}$$

- Now suppose again that no control is applied beyond horizon p, but the system is unstable. The previous formulation cannot be used because the Lyapunov equation (8.41) does not admit a positive semi-definite solution in this case. To ensure a finite cost, we must enforce that the unstable modes vanish at $k + p$. This can be done either by explicitly constraining them to zero or assigning infinite weight to them at the terminal time.

  To proceed, we apply a coordinate transformation that separates the stable and unstable modes:

$$\begin{bmatrix} \tilde{z}^s \\ \tilde{z}^u \end{bmatrix} = \begin{bmatrix} T_1 & T_2 \end{bmatrix}^{-1} z. \tag{8.43}$$

Here, the columns of T_1 and T_2 are the (generalized) eigenvectors corresponding to the stable and unstable eigenvalues, respectively. In these transformed coordinates, the system dynamics become

$$\begin{bmatrix} \tilde{z}^s_{i+1} \\ \tilde{z}^u_{i+1} \end{bmatrix} = \begin{bmatrix} \tilde{A}^s & 0 \\ 0 & \tilde{A}^u \end{bmatrix} \begin{bmatrix} \tilde{z}^s_i \\ \tilde{z}^u_i \end{bmatrix} + \begin{bmatrix} \tilde{B}^s \\ \tilde{B}^u \end{bmatrix} v_i, \tag{8.44}$$

where $\tilde{A}^u$ contains the unstable eigenvalues.

We then solve the Lyapunov equation for the stable subsystem:

$$\tilde{A}^s \tilde{P}_0 \tilde{A}^{sT} + G_1^T Q G_1 = \tilde{P}_0, \tag{8.45}$$

and define the terminal weighting matrix as

$$Q_t = T_1^T \tilde{P}_0 T_1 \tag{8.46}$$

Finally, we impose the terminal constraint:

$$\underbrace{\begin{bmatrix} 0 & I \end{bmatrix}\begin{bmatrix} T_1 & T_2 \end{bmatrix}^{-1} z_p}_{\tilde{z}_p^u} = 0, \tag{8.47}$$

which ensures that the unstable modes are zero at the end of the horizon, thereby rendering the infinite-horizon cost finite.

- Finally, we may consider a more restrictive terminal constraint that enforces the entire state to be zero at the end of the horizon, i.e., $z_p = 0$. Under this constraint, the infinite-horizon cost beyond time $k + p$ becomes zero, making the infinite-horizon problem effectively equivalent to the finite-horizon problem.

Note, however, that implementing any of the finite receding horizon control strategies discussed—except the first—generally yields a control law that differs slightly from the true solution to $J_\infty(x(k))$.

Naturally, when the state $x(k)$ is not directly measurable, one can employ a state estimator to compute $\hat{x}(k|k)$ and use it in place of $x(k)$ for computing the control input $u(k)$.

Example 8.4: *Comparison of Infinite-horizon Implementations.*
We revisit the finite-horizon implementation of Example 8.2, using the specific choices of Q_t discussed in this section.

With $Q = I$, $R = 100$, and $p = 3$, solving the ARE (8.32) yields:

$$S_\infty = \begin{bmatrix} 23.2514 & -9.0804 & 0 \\ -9.0804 & 6.2094 & 0 \\ 0 & 0 & 1 \end{bmatrix}.$$

It is straightforward to verify that the finite-horizon implementation in Example 8.2, with $Q_t = S_\infty$, yields the same control law and cost ($V_\infty = 16.723$) as the infinite-horizon LQ solution reported earlier.

Next, consider the second method, where Q_t is chosen as the solution to the Lyapunov equation (8.41). Using `dlyap` in MATLAB, we obtain:

$$S(3) = Q_t = \begin{bmatrix} 46.222 & -20.902 & 0 \\ -20.902 & 12.313 & 0 \\ 0 & 0 & 1 \end{bmatrix}.$$

Solving the RDE (8.26),

$$S_p = S(0) = \begin{bmatrix} 24.897 & -10.051 & 0 \\ -10.051 & 6.7829 & 0 \\ 0 & 0 & 1 \end{bmatrix}$$

$$u(k) = -\begin{bmatrix} 0.2128 & -0.1013 & 0 \end{bmatrix} x(k).$$

The resulting infinite horizon cost is $V_\infty = 16.763$, which is slightly higher than the optimal value obtained using $Q_t = S_\infty$.

8.4 Analysis

8.4.1 State Feedback Case

Implementing the linear state feedback law $u(k) = -Lx(k)$—obtained either from the infinite-horizon solution or from a receding horizon policy—on the system (8.1) results in the closed-loop dynamics:

$$x(k+1) = (A - BL)x(k). \tag{8.48}$$

Stability of the closed-loop system can be assessed by examining the eigenvalues of the matrix $A - BL$. However, it is desirable to establish stability conditions that do not require explicit eigenvalue computation, as this simplifies the controller design process significantly.

As we will show below, the infinite-horizon LQR guarantees closed-loop stability under mild and general assumptions for virtually any choice of design parameters. In Chapter 9, we will also demonstrate that receding horizon control policies—when designed to align with the infinite-horizon cost, either directly or indirectly through terminal constraints or weights—generally yield stable closed-loop behavior.

To analyze the infinite-horizon control problem, we consider the corresponding optimal estimation problem for the *dual system:*

$$x(k+1) = A^T x(k) + \varepsilon_1(k), \tag{8.49}$$

$$y_m(k) = B^T x(k) + \varepsilon_2(k), \tag{8.50}$$

where $\varepsilon_1(k)$ and $\varepsilon_2(k)$ are zero-mean white noise sequences with covariances Q and R, respectively. It can be verified that the estimator Riccati equation (7.18) for the dual system (8.50) is identical to the controller Riccati equation (8.31) for the original system (8.1). Note that the estimator Riccati equation evolves forward in time, while the controller Riccati equation runs backward; thus, setting $P(0) = S(\infty)$ aligns the two.

Furthermore, the asymptotic optimal estimator gain for the dual system, given by (7.19), is equal to the transpose of the asymptotic optimal controller gain for the original system, given in (8.33). The observer poles of the dual system are therefore identical to the poles of the original closed-loop system, since the eigenvalues of $(A^T - L_\infty^T B^T)$ match those of $(A - BL_\infty)$.

This duality allows us to infer important properties of the LQR and its associated Riccati equation from the Kalman filter analysis in Section 7.3.3. In particular, if the pair (B^T, A^T) is detectable, then the observer RDE (7.18) for the dual system converges asymptotically. By duality, this implies that the controller RDE (8.31) for the original system will also converge if the pair (A, B) is stabilizable.

From the earlier result, we know that, in addition to the detectability of the pair (B^T, A^T), if the pair $(A^T, Q_1^{1/2})$ is stabilizable, then the solution of the observer ARE (7.20) is unique, stabilizing, and positive semi-definite. By duality, this implies that if $(Q_1^{1/2}, A)$ is detectable, and the pair (A, B) is stabilizable, then the controller ARE (8.32) also admits a unique, stabilizing, positive semi-definite solution.

In summary, if the pair (A, B) is stabilizable and the pair $(Q^{1/2}, A)$ is detectable, then the controller RDE (8.31), initialized with $S(\infty) \geq 0$, converges to a unique positive semi-definite solution of the ARE (8.32). Moreover, all eigenvalues of $(A - BL_\infty)$ lie strictly inside the unit disk. The first condition is evidently necessary to ensure the existence of a stabilizing solution. The second condition requires that all unstable modes of the system be penalized in the cost function in a linearly independent manner. This guarantees that, if the infinite-horizon cost remains bounded, then $x \to 0$ and $u \to 0$ as $k \to \infty$.

8.4.2 Output Feedback Case

Consider the system

$$x(k+1) = Ax(k) + Bu(k), \tag{8.51}$$

$$y_m(k) = Cx(k), \tag{8.52}$$

together with an output feedback controller of the form:

$$\hat{x}(k+1|k+1) = A\hat{x}(k|k) + Bu(k) + K_f\,(y_m(k+1) - C(A\hat{x}(k|k) + Bu(k))), \tag{8.53}$$

$$u(k) = -L\hat{x}(k|k). \tag{8.54}$$

Defining the estimation error $x_e(k) = x(k) - \hat{x}(k|k)$ and combining the system and estimator equations, we obtain the augmented closed-loop dynamics:

$$\begin{bmatrix} x(k+1) \\ x_e(k+1) \end{bmatrix} = \begin{bmatrix} A - BL & -BL \\ 0 & A - K_f CA \end{bmatrix} \begin{bmatrix} x(k) \\ x_e(k) \end{bmatrix}. \tag{8.55}$$

Note that this system is one-way coupled: the evolution of $x(k)$ is influenced by $x_e(k)$, but not vice versa. Consequently, the eigenvalues of the overall closed-loop system matrix are simply the union of the eigenvalues of $(A - BL)$ and $(A - K_f CA)$. This structure implies that closed-loop stability of the observer-based output feedback controller can be achieved by designing the state feedback gain L and the observer gain K separately, each to ensure stability of their respective subsystems.

This property is known as the *separation principle*, and it holds for unconstrained linear systems: the control law and state estimator can be designed independently without compromising closed-loop stability.

8.4.3 Set Point Tracking and Disturbance Rejection

Several problems in process control are disturbance rejection or set point tracking problems. As seen in Part I, we typically formulate the output regulation problem as the minimization of the cost function:

$$\sum_{i=0}^{p} \underbrace{[y(i) - r]^T}_{e^T(i)} Q_e \underbrace{[y(i) - r]}_{e(i)} + \Delta u^T(i) R \Delta u(i), \tag{8.56}$$

where $e(i) = y(i) - r$ is the output tracking error.

To cast this into a standard LQ regulation framework, we augment the system with the error dynamics:

$$\underbrace{\begin{bmatrix} \Delta x(k+1) \\ e(k+1) \end{bmatrix}}_{\bar{x}(k+1)} = \underbrace{\begin{bmatrix} A & 0 \\ \mathrm{CA} & I \end{bmatrix}}_{\bar{A}} \underbrace{\begin{bmatrix} \Delta x(k) \\ e(k) \end{bmatrix}}_{\bar{x}(k)} + \underbrace{\begin{bmatrix} B \\ \mathrm{CB} \end{bmatrix}}_{\bar{B}} \Delta u. \tag{8.57}$$

Note that this effectively introduces integrators into the system model, which are necessary to achieve offset-free tracking of constant set points and rejection of constant disturbances.

With this augmented system written as

$$\bar{x}(i+1) = \bar{A}\bar{x}(i) + \bar{B}\Delta u(i), \tag{8.58}$$

one can solve the state regulation problem minimizing

$$\sum_{i=0}^{p} \left[\bar{x}^T(i)Q\bar{x}(i) + \Delta u^T(i)R\Delta u(i)\right] \quad \text{where} \qquad Q = \begin{bmatrix} 0 & 0 \\ 0 & Q_e \end{bmatrix}. \tag{8.59}$$

Note that the input penalty is applied to the incremental input $\Delta u(i)$ rather than to the input $u(i)$ itself. This formulation is crucial for achieving integral action, which ensures offset-free tracking of constant set points and rejection of constant disturbances. In contrast, applying a quadratic penalty directly to $u(i)$ can result in a steady-state offset.

8.5 Stochastic LQ Control

We now turn our attention to stochastic LQ control problems. This section summarizes the key results; detailed derivations and analysis are provided in Appendix D.

8.5.1 Finite-horizon Problem

Consider the stochastic linear system

$$x(k+1) = Ax(k) + Bu(k) + \varepsilon_1(k), \qquad \text{cov}\{\varepsilon_1\} = R_1,$$

with initial condition $x(0)$. The corresponding *open-loop optimal* quadratic control problem is

$$J_p^o(x(0)) = \min_{u(0),\cdots,u(p-1)} E\left\{\sum_{k=0}^{p-1} \left[x^T(k)Qx(k) + u^T(k)Ru(k)\right] + x^T(p)Q_t x(p)\right\}. \tag{8.60}$$

Using arguments analogous to those in the deterministic case, the batch optimization formulation yields the following *open-loop optimal* control input:

$$\mathcal{U} = -\left(\mathcal{S}^{uT}\bar{Q}\mathcal{S}^u + \bar{R}\right)^{-1}\mathcal{S}^{uT}\bar{Q}\mathcal{S}^x x(0), \tag{8.61}$$

with the corresponding optimal cost:

$$J_p^o(x(0)) = x^T(0)\left[\mathcal{S}^{xT}\mathcal{S}^x - \mathcal{S}^{xT}\bar{Q}\mathcal{S}^u\left(\mathcal{S}^{uT}\bar{Q}\mathcal{S}^u + \bar{R}\right)^{-1}\mathcal{S}^{uT}\bar{Q}\mathcal{S}^x\right]x(0) + \text{trace}\{\mathcal{S}^{\varepsilon T}\bar{Q}\mathcal{S}^{\varepsilon}\bar{R}_1\}, \tag{8.62}$$

where

$$\mathcal{S}^x = \begin{bmatrix} I \\ A \\ \vdots \\ A^p \end{bmatrix} \quad \mathcal{S}^u = \begin{bmatrix} 0 & 0 & \cdots & 0 \\ B & 0 & \cdots & 0 \\ AB & B & \cdots & 0 \\ \vdots & \vdots & \ddots & \vdots \\ A^{p-1}B & A^{p-1}B & \cdots & B \end{bmatrix} \quad \mathcal{S}^{\varepsilon} = \begin{bmatrix} 0 & 0 & \cdots & 0 \\ I & 0 & \cdots & 0 \\ A & I & \cdots & 0 \\ \vdots & \vdots & \ddots & \vdots \\ A^{p-1} & A^{p-2} & \cdots & I \end{bmatrix}.$$

Compared to the deterministic case, this cost expression includes an additional term—trace $\{S^{\varepsilon T}\bar{Q}S^{\varepsilon}\bar{R}_1\}$—which captures the contribution of the process noise. Readers interested in the detailed derivation are referred to Appendix D.

Example 8.5: *Open-loop Optimal Control via Least Squares.*
Consider the stochastic version of the system in Example 8.1:

$$x(k+1) = \begin{bmatrix} 1.41 & -0.4724 & 0 \\ 1 & 0 & 0 \\ 0 & 1 & 0 \end{bmatrix} x(k) + \begin{bmatrix} 1 \\ 0 \\ 0 \end{bmatrix} u(k) + \varepsilon_1(k),$$

with noise covariance $R_1 = 0.25I$, initial state $z_0 = \begin{bmatrix} 1 & 0.5 & 1 \end{bmatrix}^T$, and weight matrices $Q = Q_t = I$ and $R = 10$. Compute the LQ control actions using the least squares solution, and compare the cost for $p = 10$ with that obtained from Eq. (8.62).

To solve the problem, we construct the matrices $\mathcal{S}^x$, $\mathcal{S}^u$, and $\mathcal{S}^\varepsilon$. Then, using Eq. (8.62), we obtain $J_{10}^o = 95.7$.

A 10-step simulation is then carried out starting from z_0, using Eq. (8.61) to compute the control input at each step. The process noise is generated at every iteration using the `randn` command. The instantaneous cost $x(k)^T Qx(k) + u(k)^T Ru(k)$ is accumulated over the horizon, and the terminal cost $x(p)^T Qx(p)$ is added at the end. Averaging the resulting p-step costs over 100 Monte Carlo runs yields an empirical cost of 88.2. The slight discrepancy arises from the finite sample size used in estimating the average cost.

Remark 8.1: If we choose $R_1 = 0$ (i.e., no state noise), we recover the deterministic cost. It is easy to verify that the 10-step cost for the deterministic case is 8.91. Computing the cost using p-step simulation in absence of state noise yields the exact same cost of 8.91.

As before, the optimal *feedback policy* can be derived using dynamic programming. Starting with $S(p) = Q_t$, the following recursive equations apply:

$$u(k-1) = -\left(B^T S(k)B + R\right)^{-1} B^T S(k)A, \qquad k = p, \ldots, 1, \tag{8.63}$$

$$S(k-1) = A^T S(k)A + Q - A^T S(k)B\left(B^T S(k)B + R\right)^{-1} B^T S(k)A. \tag{8.64}$$

The resulting *closed-loop optimal* cost, derived in Appendix D, is

$$J_p^{cl}(z_0) = z_0^T S(0)z_0 + \sum_{j=1}^{p} \text{trace}\,[S(j)R_1]. \tag{8.65}$$

Example 8.6: *Feedback Optimal LQR via Dynamic Programming.*
Repeat Example 8.5 using the dynamic programming approach. Compare the closed-loop cost J^{cl} from Eq. (8.65) with the empirical cost averaged over 100 Monte Carlo simulations.

Using Eq. (8.65), the 10-stage cost is calculated as $J^{cl} = 38.3$, which is substantially lower than the open-loop optimal cost of 95.7.

Solving the DP recursion yields a closed-form feedback control policy:

$$u(k) = -L(k)x(k), k = p, \ldots, 0.$$

This control law was implemented, and the p-step cost, averaged over 100 Monte Carlo simulations, was found to be 38.9. As a reference, the corresponding deterministic problem (i.e., the same setup with $R_1 = 0$) produces the same cost as in Example 8.5, namely $J_p^{\text{det}} = 8.91$.

These examples highlight the key differences between the least squares and dynamic programming formulations. While both approaches yield the same result in the deterministic case, their interpretations diverge in the stochastic setting. The least squares method provides an *open-loop* optimal input sequence for a given initial condition z_0, whereas dynamic programming yields an *optimal feedback* law—an explicit policy that adapts to real-time state realizations.

8.5.2 Output Feedback LQ Control

We now turn our attention to the stochastic output feedback problem:

$$\begin{aligned} x(k+1) &= Ax(k) + Bu(k) + \varepsilon_1(k), \quad && \text{cov}\{\varepsilon_1\} = R_1 \\ y_m(k) &= Cx(k) + \varepsilon_2(k), && \text{cov}\{\varepsilon_2\} = R_2, \end{aligned}$$

with initial condition $x(0) = z_0$. Here, $\varepsilon_1(k)$ and $\varepsilon_2(k)$ are mutually independent, zero-mean Gaussian white noise sequences, and z_0 itself is a Gaussian random variable with mean $\hat{z}_0$ and covariance P_0.

The optimal quadratic control problem involves minimizing the expected value of the cost function given in (D.46). Derivation and analysis of the celebrated linear quadratic gaussian (LQG) result are provided in Appendix D. In brief, the separation principle holds for unconstrained linear systems with Gaussian noise: The optimal quadratic output feedback solution consists of a Kalman filter for state estimation and an LQR-based optimal state feedback regulator:

$$\hat{x}(k|k) = \hat{x}(k|k-1) + K_f(k)\,(y_m(k) - C\hat{x}(k|k-1)), \tag{8.66}$$

$$\text{with} \quad \hat{x}(k|k-1) = A\hat{x}(k-1|k-1) + Bu(k-1)$$

$$u(k) = -L(k)\hat{x}(k|k), \tag{8.67}$$

where $K_f(k)$ and $L(k)$ are the Kalman filter and LQR gains, respectively.

In the infinite-horizon case, the gains converge to steady-state values, denoted by $K_{f,\infty}$ and L_∞. The separation principle remains valid in this setting, meaning that the steady-state Kalman filter and the steady-state LQR can be designed independently and then combined to yield the optimal output feedback controller.

We conclude by addressing a problem of practical importance: Set point tracking in the presence of a random-walk disturbance, with the additional constraint that not all primary controlled variables are directly measured. This setup was introduced in Chapter 7 (see following Eq. 7.30) to handle integrating white noise disturbances within the Kalman filter framework:

$$x(k+1) = \begin{bmatrix} 0.6346 & 0.02865 & 0.02325 & 0.2662 \\ 0.9891 & 0.04599 & 0.03975 & 0.6882 \\ 0.7102 & 0.03516 & 0.03416 & 0.9322 \\ 0.1060 & 0.00792 & 0.01213 & 0.6866 \end{bmatrix} x(k)$$

$$+ \begin{bmatrix} 0.02949 \\ 0.1902 \\ 0.1670 \\ 0.009063 \end{bmatrix} u(k) + \begin{bmatrix} 0.01448 \\ 0.1351 \\ 0.02001 \\ 0.002739 \end{bmatrix} d(k)$$

$$y(k) = \begin{bmatrix} 0 & -28.1 & -27.3 & 0 \\ 1 & 0 & 0 & 0 \end{bmatrix} x(k)$$

$$y_m(k) = \underbrace{\begin{bmatrix} 1 & 0 \end{bmatrix}}_{H} y(k) + v(k).$$

Here, $d(k)$ represents an integrating white noise sequence, and $v(k)$ denotes Gaussian white measurement noise. Recall that in Example 7.11, the first output was measured.

To handle the integrating disturbance, the state was augmented as described in Eq. (7.32). However, since we are now also concerned with tracking a set point r, we must incorporate the deviation of the primary variable from the reference, defined as $e(k) \stackrel{\Delta}{=} y(k) - r$. We therefore define the augmented state:

$$\bar{x}(k) \stackrel{\Delta}{=} \begin{bmatrix} \Delta x(k) \\ y(k) - r \end{bmatrix},$$

which leads to the following augmented model

$$\bar{x}(k+1) = \bar{A}\bar{x}(k) + \bar{B}\Delta u(k) + \bar{B}_2 \varepsilon_1(k), \tag{8.68}$$

where

$$\bar{A} = \begin{bmatrix} A & 0 \\ \text{CA} & I \end{bmatrix} \quad \bar{B} = \begin{bmatrix} B \\ \text{CB} \end{bmatrix} \quad \bar{B}_2 = \begin{bmatrix} B_d \\ \text{CB}_d \end{bmatrix}.$$

Note that $y(k)$ represents all outputs. If only some of these outputs are measured, then the output equation becomes:

$$e_m(k) = \bar{C}\bar{x}(k) + \varepsilon_2(k), \qquad \bar{C} = \begin{bmatrix} 0 & H \end{bmatrix}. \tag{8.69}$$

The corresponding noise covariances are given by $\bar{R}_1 = \bar{B}_2 R_\varepsilon \bar{B}_2^T$ and $\bar{R}_2 = R_v$. Note that if all outputs are measured, then $H = I$, and the overall model takes a form similar to that of (8.57). The following example considers the case in which the primary controlled variable $y_2(k)$ is not directly measured but instead inferred from the measurement of $y_1(k)$.

Example 8.7: *Output Feedback "Inferential" Control with Unmeasured Random Walk Disturbance.*
Consider the set point-tracking problem for the system in Eq. 7.30, with a set point of $r = 0.5$, in the presence of an integrating white noise disturbance $d(k+1) = d(k) + \varepsilon_1(k)$, where $R_1 = 0.01$, and measurement noise with covariance $R_2 = 0.25$. The primary controlled variable $y_2(k) = x_1(k)$ is not directly measured but inferred from measurements of $y_m(k)$. A 10% model-plant mismatch is assumed in both the system and disturbance gains. The MATLAB code is provided on GitHub: https://github.com/Lee-MPC-Book

After defining the model and plant dynamics, we construct the augmented model to incorporate output error and disturbance:

```
%% Augmented Matrices
Aaug=[A zeros(4,2); C*A eye(2)];
Baug=[B; C*B];
Caug=[zeros(1,4) 1 0];
B2aug=[Bd; C*Bd];
R1aug=B2aug*R1*B2aug';
R2aug=R2;
```

Strong Solution of ARE: The augmented system possesses two eigenvalues on the unit circle due to the inclusion of output tracking errors for y_1 and y_2. Since only y_1 is measured while y_2 is not, the system is not detectable. Consequently, the steady-state Riccati equation does not admit a unique stabilizing solution. Nevertheless, the observer RDE, when iterated, converges to a strong solution, allowing the Kalman filter gain to be computed. This occurs because the model includes only one unmeasured disturbance, which can be inferred from the measurement of y_1.

Thereafter, the control logic proceeds as follows:

(i) Obtain measurement $y_m(k)$ and compute $e_m(k)$
(ii) Estimate state $\hat{\bar{x}}(k|k)$ using the Kalman filter.
(iii) Compute $\Delta u(k)$, update and apply control $u(k)$ to the plant.
(iv) Predict $\hat{\bar{x}}(k+1|k)$ for the next time step.
(v) Set $k \leftarrow (k+1)$ and repeat the steps.

Figure 8.4 illustrates that the output feedback LQG—with appropriate model augmentation—does a good job of set point tracking for the unmeasured $y_2(k)$ despite unmeasured disturbances and mismatched dynamics.

The same problem was addressed in Chapter 7 in Example 7.10, without control (i.e., $u(k) = 0$) to highlight the role of the Kalman filter and to compare performance with unaugmented and augmented models.

The theories of linear estimation and LQ control developed in this and the previous chapters form the essential foundation for state-space MPC, which will be addressed in the final part of this book. In particular, the detectability and stabilizability conditions—crucial for the design of stabilizing filters and controllers—constitute key theoretical results. Furthermore, the use of *state augmentation*,

$$\bar{x}(k+1) = \bar{A}\bar{x}(k) + \bar{B}\Delta u(k) + \varepsilon(k), \quad y_m(k) = \bar{C}\bar{x}(k) + \nu(k),$$

can be structured to ensure convergence and stability, facilitate effective feedback control, and extend the framework to address practically important control objectives such as offset-free set point tracking and disturbance rejection.

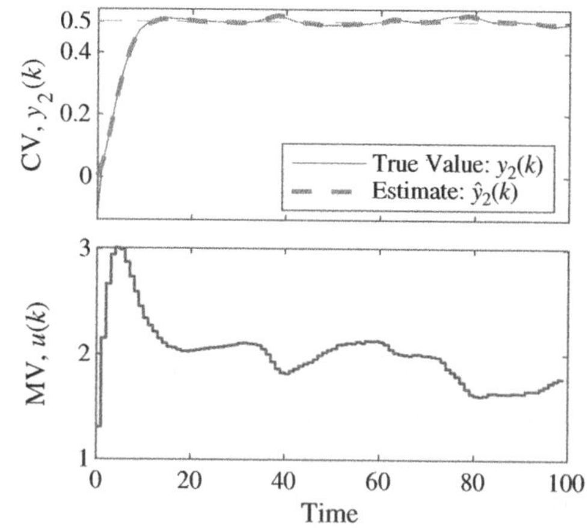

Figure 8.4 Performance of output feedback LQG in tracking set point $r = 0.5$ under unmeasured disturbance and model-plant mismatch. The primary controlled variable $y_2(k)$ is not directly measured but inferred through estimation.

Exercises

1 For the linear time-invariant deterministic system studied in this chapter, derive the quadratic optimal feedback controller for a given time-varying reference trajectory $\{r(k), k = 1, \ldots, p\}$ over a finite horizon using the dynamic programming approach.

2 Prove that the solution to Eq. (8.41) indeed provides the infinite-horizon cost as stated.

3 Consider the following system

$$x(k+1) = \begin{bmatrix} 1 & 0.1 \\ -0.3 & 0.65 \end{bmatrix} + \begin{bmatrix} 1 \\ 0.2 \end{bmatrix} u(k)$$
$$y_m(k) = \begin{bmatrix} 0.7 & 0.3 \end{bmatrix} x(k).$$

- Design a deadbeat controller for this system.
- Design a controller that puts closed-loop poles at l_1 and l_2

4 For the system above, design an LQR controller that minimizes the infinite-horizon cost $\sum x^T(i)Ix(i) + 0.1u^2(i)$. Compute the optimal control law $u(k) = -L_\infty x(k)$ and the closed-loop poles.

What is the LQ optimal control law if the infinite-horizon cost is instead $\sum_i y^2(i) + 0.1u^2(i)$?

5 Simulate the system above from the initial condition $x(0) = \begin{bmatrix} 1 & 0.5 \end{bmatrix}^T$ using each of the four controllers designed in the previous two problems.

6 Now consider the stochastic system

$$x(k+1) = \begin{bmatrix} 1 & 0.1 \\ -0.3 & 0.65 \end{bmatrix} + \begin{bmatrix} 1 \\ 0.2 \end{bmatrix} u(k) + \begin{bmatrix} 1 \\ 0 \end{bmatrix} \varepsilon(k)$$

$$y_m(k) = \begin{bmatrix} 0.7 & 0.3 \end{bmatrix} x(k) + \nu(k)$$

where, $R_1 = \text{cov}\{\varepsilon\} = 1$, $R_2 = \text{cov}\{\nu\} = 0.2$,

- Design an output feedback LQG controller.
- Simulate the closed-loop system with the initial condition $x(0) = \begin{bmatrix} 1 & 0.5 \end{bmatrix}^T$, assuming the initial state is unknown and using $\hat{x}(0) = 0$ as the estimator initialization. Generate realizations of the process and measurement noise using `randn(100,1)` with the specified covariances.
- Repeat the simulation 50 times using Monte Carlo sampling, generating new Gaussian noise realizations for each run.

7 For the infinite-horizon output feedback problem with the above stochastic system, compute the optimal cost for the LQG controller designed in the previous exercise:

- First, use the analytic expression you derived earlier for the infinite-horizon cost.
- Next, compute the same cost by solving the closed-loop Lyapunov equation (Eq. D.96).
- Finally, estimate the cost from the Monte Carlo simulations.

8 Repeat the previous problem assuming the state noise $\varepsilon(k)$ follows an integrating white noise model.

9

Constrained Quadratic Optimal Control

Engineers love to optimize problems. I ask: What's the goal? What are our constraints? What is the optimal, elegant way to get to that goal within those constraints?

– Ruchi Sanghvi

This chapter extends the linear quadratic optimal control framework from Chapter 8 to systems subject to constraints on inputs and states. The inclusion of constraints substantially increases the complexity of both control computation and closed-loop analysis. In particular, constraints lead to a nonlinear control law and, consequently, a nonlinear closed-loop system, which precludes the direct application of standard linear system analysis techniques.

9.1 Finite-horizon Problem

We now revisit the finite-horizon optimal control problem introduced in Section 8.2, but with the inclusion of constraints on states and inputs. The objective is to minimize the cost function:

$$J_p(x(0)) = \min_{u(0),\dots,u(p-1)} \left\{ \sum_{i=0}^{p-1} \left[x^T(i)Qx(k) + u^T(i)Ru(k)\right] + x^T(p)Q_t x(p) \right\}, \tag{9.1}$$

subject to the deterministic system dynamics:

$$x(k+1) = Ax(k) + Bu(k), \tag{9.2}$$

with initial condition $x(0)$. Furthermore, the state and control inputs are now constrained to remain within prescribed feasible sets, denoted by $\mathbb{X}$ and $\mathbb{U}$:

$$x(i) \in \mathbb{X}, \quad i = 1, \dots, p, \tag{9.3}$$

$$u(i) \in \mathbb{U}, \quad i = 0, \dots, p-1. \tag{9.4}$$

In the simplest case, these feasible sets may be defined by upper and lower bounds on the elements of x and u. More generally, the constraints may take the form of linear inequalities:

$$F \begin{bmatrix} x(i) \\ u(i) \end{bmatrix} \le g, \quad g > 0. \tag{9.5}$$

In general, we assume that the feasible sets are convex, compact, and contain the origin in their interior. These assumptions ensure that the origin is a feasible steady-state point of the system. We also assume that the constraint sets are time invariant. Under these conditions, the resulting problem becomes a constrained least-squares optimization, which typically does not admit a closed-form analytical solution and must be solved numerically.

For convenience, we refer to the above finite-horizon optimal control problem as $J_p(x(0))$. As before, the same symbol will be used to denote both the optimization problem and the resulting optimal cost.

In this constrained setting, the dynamic programming approach is no longer practically viable because the cost-to-go function cannot be expressed in closed form. Numerical implementation of dynamic programming requires discretizing the state space and recursively computing and storing the optimal cost at each discretized point, starting from the terminal stage and proceeding backward. However, the number of required grid points grows exponentially with the state dimension, making the approach computationally prohibitive. This phenomenon is commonly known as the "curse of dimensionality" in the dynamic programming literature.

A more practical alternative is to use the explicit multistep prediction-based least-squares formulation introduced in Section 8.2. By substituting the prediction equation from (8.8) into the objective function (8.9) and the state constraint (9.3), we arrive at the following *convex optimization* problem, which can be solved numerically for a given initial state $x(0)$:

$$\min_{\mathcal{U}} \left[\mathcal{U}^T(\mathsf{S}^{uT}\bar{Q}\mathsf{S}^u + \bar{R})\mathcal{U} + 2\mathcal{U}^T\mathsf{S}^{uT}\bar{Q}\mathsf{S}^x x(0)\right], \tag{9.6}$$

subject to

$$\mathsf{S}^x x(0) + \mathsf{S}^u\mathcal{U} \in \mathbb{X} \times \cdots \times \mathbb{X}, \tag{9.7}$$

$$\mathcal{U} \in \mathbb{U} \times \cdots \times \mathbb{U}. \tag{9.8}$$

When the state and input constraints can be expressed as linear inequalities, the resulting optimization problem is a quadratic program (QP), for which efficient off-the-shelf numerical solvers are readily available.

For a given $x(0)$, the optimization problem can be solved to obtain an *open-loop optimal* input trajectory. However, feedback control is often preferred to compensate for disturbances and model inaccuracies. One approach discussed in Chapter 8 is to resolve the optimization online at each time step with a "shrinking horizon." That is, at each k, solve the problem $J_{p-k}(x(k))$ using the current state $x(k)$, and apply only the first control input from the resulting optimal sequence.

9.2 Infinite-horizon Problem

We now consider the infinite-horizon optimal control problem:

$$\min_{u(0),\cdots,u(\infty)} \left\{\sum_{i=0}^{\infty} \left[x^T(i)Qx(i) + u^T(i)Ru(i)\right]\right\}, \tag{9.9}$$

subject to the system dynamics

$$x(k+1) = Ax(k) + Bu(k), \tag{9.10}$$

the initial condition $x(0)$ and constraints

$$x(k) \in \mathbb{X}, \qquad k = 1, \cdots, \infty, \tag{9.11}$$

$$u(k) \in \mathbb{U}, \qquad k = 0, \cdots, \infty. \tag{9.12}$$

This infinite-dimensional optimization problem cannot be solved directly via convex programming techniques. Instead, a practical alternative is the finite-horizon receding horizon control (RHC) strategy. In this approach, the finite-horizon problem $J_p(z_0)$—defined by (9.1–9.4)—is solved at *each time step* $k \in [0, \infty)$ using the current state $x(k)$ as the initial condition, i.e.,

$$\min_{v_0,\cdots,v_{p-1}} \left\{ \sum_{i=0}^{p-1} \left[z_i^T Q z_i + v_i^T R v_i \right] + z_p^T Q_t z_p \right\}, \tag{9.13}$$

with the initial condition $z_0 = x(k)$, and subject to

$$\text{the system model} \qquad z_{i+1} = A z_i + B v_i, \tag{9.14}$$

$$\text{and constraints} \qquad z_i \in \mathbb{X} \quad v_i \in \mathbb{U}. \tag{9.15}$$

Here, z_i and v_i are temporary variables introduced for the optimization carried out at each time step. In the optimization performed at time step k, they represent the predicted values of $x(k+i)$ and $u(k+i)$.

The control law is implemented in the feedback form:

$$u(k) = u^* \left(x(k) \right),$$

where $u^*(\cdot)$ denotes the (generally nonlinear) operator mapping the initial state z_0 to the first optimal input v_0^* computed from the above optimization. This requires solving a convex quadratic program online at each time step.

Example 9.1: *Finite-Horizon Constrained LQR on a Double Integrator.*
Consider a discrete-time double integrator system:

$$x(k+1) = \begin{bmatrix} 1 & t_s \\ 0 & 1 \end{bmatrix} x(k) + \begin{bmatrix} 0.5t_s^2 \\ t_s \end{bmatrix} u(k), \tag{9.16}$$

sampled at $t_s = 0.1$ s. A constrained LQR with horizon $p = 10$ will be implemented. The input constraints are

$$-0.5 \le u(k) \le 0.5.$$

The state and input weight matrices were chosen as

$$Q = Q_t = \text{diag}(1, 0.1), \qquad R = 0.1.$$

Starting with an initial condition of $\begin{bmatrix} 2 & -2 \end{bmatrix}^T$, the controller is simulated for $N = 200$ steps (i.e., 20*s*). At each time, the constrained quadratic programming (QP) problem (9.6) is solved using `quadprog`, with Hessian and gradient as defined in (8.10). Figure 9.1 shows the application of finite-horizon LQR.

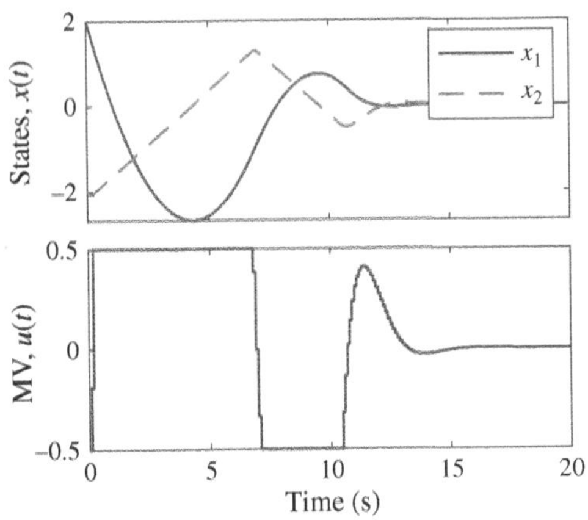

Figure 9.1 A constrained LQR for discrete-time double integrator with $p = 10$.

The choice of prediction horizon p and terminal weight Q_t is critical for ensuring both stability and satisfactory closed-loop performance. As shown in the unconstrained case, the infinite-horizon formulation provides strong theoretical guarantees. To approximate these properties in the constrained setting, one must carefully design the finite-horizon problem. In what follows, we extend these insights to the constrained case and discuss strategies for selecting terminal penalties and constraints that promote stability and performance. Several such strategies have been proposed in the literature, including notable contributions by Kwon and Pearson (1977), Keerthi and Gilbert (1988), Mayne and Mischalska (1990), and Rawlings and Muske (1993).

9.2.1 Options for Re-formulation as an Equivalent Finite-horizon Problem

We now consider strategies for formulating a finite-horizon optimal control problem that ensures feasibility, guarantees stability, and closely approximates the performance of the infinite-horizon solution. In line with the discussion in the unconstrained case (Section 8.2), the following approaches are presented in order of increasing restrictiveness.

9.2.1.1 Option 1: Sufficiently Large Prediction Horizon for the Terminal State to Be Within Unconstrained Domain

Since the origin lies in the interior of the feasible state and input sets, the optimal control will eventually become unconstrained as the system state approaches the origin. We can exploit this by selecting a *sufficiently long horizon* p so that the terminal state z_p lies in a region where the unconstrained linear quadratic regulator (LQR) control law is feasible. Specifically, we assume that p is chosen sufficiently large such that

$$\begin{aligned} v_{p+i} &= -L_\infty z_{p+i}, & v_{p+i} &\in \mathbb{U} \quad \forall i \geq 0, \\ z_p &= (A - BL_\infty)^i z_p, & z_p &\in \mathbb{X} \quad \forall i \geq 0, \end{aligned} \tag{9.17}$$

where L_∞ is the steady-state LQR gain obtained by solving the discrete algebraic Riccati equation (DARE):

$$A^T Q_t A + Q - A^T Q_t B (B^T Q_t B + R)^{-1} B^T Q_t A = Q_t. \tag{9.18}$$

If z_p satisfies the condition in (9.17), then the infinite-horizon constrained control problem can be solved exactly via the finite-horizon problem, provided that the terminal weight Q_t is chosen as the solution of (9.18). In that case, the terminal cost term satisfies

$$z_p^T Q_t z_p = \sum_{i=p}^{\infty} z_i^T Q z_i + v_i^T R v_i. \tag{9.19}$$

In summary, under Option 1:

- The prediction horizon p is chosen to be large enough that the terminal state enters a region where the unconstrained LQR controller is admissible;
- The terminal weighting matrix Q_t is set as the solution of the DARE (9.18);
- The finite-horizon optimal control problem (9.13–9.15) becomes equivalent to the infinite-horizon constrained problem.

In the literature, theoretical results have been established for determining an appropriate horizon length p as a function of the initial state norm $|x(k)|$, such that the resulting terminal state lies within the invariant set of the unconstrained LQR controller.

9.2.1.2 Option 2: Fixed Horizon with Terminal State Constraint for Unconstrained LQR Beyond Horizon to Remain Feasible

A major limitation of Option 1 is that the required horizon length p must be determined based on the current state $x(k)$, which is generally not possible *a priori*. Moreover, the required value of p may be quite large, potentially leading to numerical difficulties. To address this, we fix the horizon length p and impose a constraint on the terminal state:

$$z_p \in X_{\text{ma}}. \tag{9.20}$$

Here, X_{ma} denotes the *maximal output admissible set* associated with the unconstrained LQR controller. This set is defined as the largest set of states for which the LQR controller remains feasible and satisfies all system constraints indefinitely:

$$\begin{aligned} z_p \in X_{\text{ma}} \Rightarrow \quad & z_{p+i} \in \mathbb{X} \text{ and } \left(-L_\infty z_{p+i}\right) \in \mathbb{U} \text{ for } i \geq 0, \\ \text{where} \quad & z_{p+i} = (A - BL_\infty)^i z_p. \end{aligned}$$

This means that any state $z_p \in X_{\text{ma}}$ ensures that the unconstrained LQR controller will keep the system trajectory within admissible bounds for all future times $p + i$, $i \geq 0$, thereby satisfying the same condition as in (9.17). The concept of the maximal output admissible set and its application to RHC was first introduced by Gilbert and Tan (1991). A conceptual illustration of X_{ma} is shown in Figure 9.2.

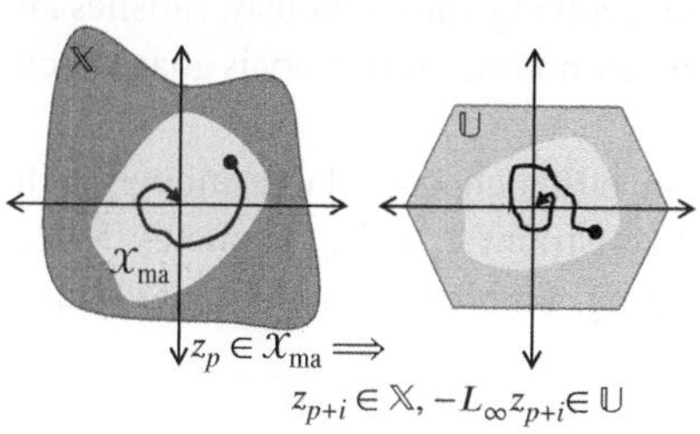

Figure 9.2 Conceptual illustration of the (maximal) output admissible set.

To implement this strategy:

- The horizon length p is fixed in advance;
- The terminal weight Q_t is chosen as the solution to the DARE (9.18);
- The finite-horizon optimization problem (9.13–9.15) is solved subject to the additional terminal constraint (9.20).

Remark 9.1: The essential difference between Options 1 and 2 lies in the treatment of the terminal condition. In Option 1, the horizon length p is dynamically adjusted so that $z_p \in X_{\text{ma}}$ is automatically satisfied, eliminating the need for an explicit terminal constraint. In contrast, Option 2 employs a fixed horizon and explicitly enforces $z_p \in X_{\text{ma}}$ as a terminal constraint. This approach can introduce conservatism or suboptimality when the constraint becomes active—that is, when p is not sufficiently large for the optimizer to reach X_{ma} without explicit enforcement. Full equivalence with the infinite-horizon solution is achieved only when this terminal constraint remains inactive at the optimum.

9.2.1.3 Option 3: Fixed Horizon with Terminal State Constraint for Zero Input Beyond Horizon to Remain Feasible

Rather than assuming an unconstrained LQR policy beyond the prediction horizon p, we may instead impose that no control action is taken for $k + p + i, i \geq 0$. That is,

$$\upsilon_{p+i} = 0 \qquad \forall i \geq 0.$$

This trivially satisfies the input constraints, $\upsilon_{p+i} \in \mathbb{U}$, but requires that the autonomous evolution of the system from z_p remains feasible:

$$z_{p+i} = A^i z_p, \qquad z_p \in \mathbb{X}. \tag{9.21}$$

Under this assumption, the terminal cost can be written as

$$z_p^T Q_t z_p = \sum_{i=p}^{\infty} z_i^T Q z_i,$$

where Q_t satisfies the discrete Lyapunov equation:

$$Q_t = A^T Q_t A + Q. \tag{9.22}$$

As in Option 2, we define a maximal admissible set, $\hat{X}_{\text{ma}}$, such that

$$z_p \in \hat{X}_{\text{ma}} \;\Rightarrow\; z_{p+i} \in \mathbb{X},\; i < 0$$

where $z_{p+i} = A^i z_p$,

$\hat{X}_{ma}$ represents the largest set of states from which the system, evolving autonomously, satisfies all state constraints indefinitely. If a terminal state $z_p \in \hat{X}_{\text{ma}}$, the autonomous trajectory is guaranteed to remain within $\mathbb{X}$ for all future time steps.

There are two main implementation paths under this option: either choose p large enough such that $z_p \in \hat{X}_{\text{ma}}$ is satisfied naturally, or enforce the terminal constraint $z_p \in \hat{X}_{\text{ma}}$ explicitly. Since computing $\hat{X}_{\text{ma}}$ may be restrictive or computationally demanding, a more conservative alternative is to define an additional constraint horizon h_c and enforce:

$$z_{p+i} \in \mathbb{X}, \quad \text{for } i = 1, \ldots, h_c.$$

If h_c is chosen sufficiently large, all future states will also remain admissible. Notably, increasing h_c does not significantly affect the optimization complexity since no additional control variables are introduced.

We summarize the implementation variants below:

Option 3A Choose p large enough so that $z_p \in \hat{X}_{\text{ma}}$ is satisfied without explicit enforcement. Solve the finite-horizon problem (9.13–9.15) using Q_t obtained from the Lyapunov equation (9.22).

Option 3B Fix p and impose the terminal constraint $x(k+p) \in \hat{X}_{\text{ma}}$. Alternatively, define an extended constraint horizon h_c and solve the following modified problem:

$$\min_{v_0 \cdots v_{p-1}} \left\{ \sum_{i=0}^{p-1} \left(z_i^T Q z_i + v_i^T R v_i \right) + z_p^T Q_t z_p \right\}, \tag{9.13}$$

$$\begin{aligned} & z_i \in \mathbb{X} \quad 1 \le i \le (p + h_c), \\ & v_i \in \mathbb{U} \quad 0 \le i \le p-1, \\ & Q_t = A^T Q_t A + Q. \end{aligned} \tag{9.23}$$

Option 3C If the system is unstable, the Lyapunov equation (9.22) has no positive semi-definite solution. In this case, we must eliminate the unstable components of the state at the terminal step. This can be done by decomposing the system via coordinate transformation into stable and unstable modes, $\tilde{z}^s$ and $\tilde{z}^u$. We then:

- Enforce $\tilde{z}_p^u = 0$ as a terminal constraint, and
- Compute the terminal weight Q_t for $\tilde{z}^s$ by solving the Lyapunov equation over the stable subspace, as discussed in Section 8.3.2.

Hence, for unstable systems, the terminal constraint $z_p \in \hat{X}_{\text{ma}}$ must be modified to additionally enforce that the unstable modes are driven to zero at the terminal step.

Remark 9.2: Unlike Option 2, Option 3 does not require a terminal constraint when no state constraints are present, as the assumption of zero input beyond the horizon inherently guarantees feasibility.

9.2.1.4 Option 4: Terminal Equality Constraint Enforcing All States to Be Zero

As discussed in Chapter 8 regarding the equivalence between finite and infinite-horizon LQR, the final option is to enforce the terminal condition $z_p = 0$. This is the most restrictive approach and is only applicable when the system is controllable. Alternatively, setting the terminal weight to a large value, i.e., $Q_t \to \infty I$, has the same effect as imposing a terminal equality constraint $z_p = 0$. When this constraint is applied in the infinite-horizon formulation, all terms beyond p in the cost function vanish, rendering the finite and infinite-horizon problems equivalent. However, it is important to note that implementing a terminal equality constraint is generally more difficult than enforcing inequality constraints.

Example 9.2: *Comparison of Infinite-Horizon Constrained and Unconstrained LQR for Double Integrator.*

Consider the double integrator of Example 9.1. In this example, we solve an infinite-horizon constrained LQR using Option 1 and compare it with an unconstrained LQR. Specifically, the following two controllers are compared:

1. *Unconstrained LQR (with clipping):* The infinite-horizon feedback gain is computed using the `dlqr` function. At each step, the control action is calculated using $u(k) = -Lx(k)$. The computed input is clipped to ensure that it satisfies the constraint $u_{\min} \leq u(k) \leq u_{\max}$. This results in a saturated LQR controller.
2. *Constrained LQR via QP:* A finite-horizon QP problem is solved, as before. To ensure that the horizon is long enough, we chose $p = 50$ and the terminal matrix Q_t is calculated from the DARE.

Figure 9.3 shows that the unconstrained LQR controller with input clipping drives the system to the origin, but the control signal saturates repeatedly. This makes the closed-loop behavior deviate from the ideal unconstrained LQR trajectory (c.f., Figure 9.4). On the other hand, the constrained LQR controller anticipates the input limits and generates smoother control trajectories.

The complete MATLAB code for this problem is available on GitHub at: https://github.com/Lee-MPC-Book

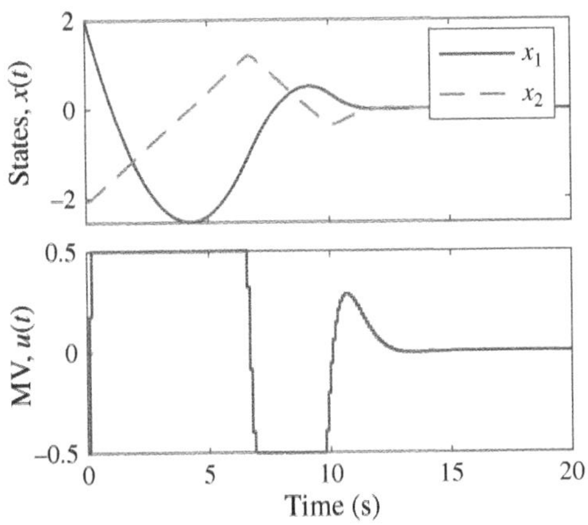

Figure 9.3 An infinite-horizon *unconstrained LQR* with input chopping for the double integrator system ($J = 250.6$).

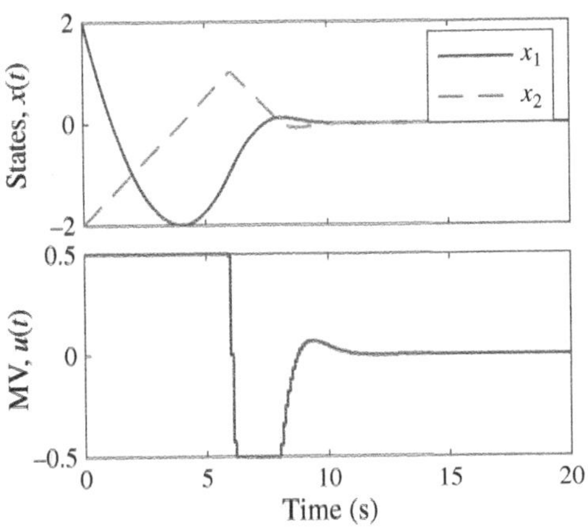

Figure 9.4 An infinite-horizon *constrained LQR (using Option 1)* for discrete-time double integrator with horizon $p = 10$ ($J = 146.9$).

9.2.2 Comparison of Various Options

Let us first compare the options presented here with those discussed in Section 8.2. In Options 1 and 2, the unconstrained infinite-horizon LQR is assumed beyond the horizon, with Q_t chosen as the solution to the DARE. In Options 3A through 3C, it is assumed that no further control action is taken beyond time $k + p$, and the system evolves autonomously; in this case, Q_t is determined from the Lyapunov equation to represent the tail cost. Option 4 requires the terminal state to return to the origin, in which case all constraints are trivially satisfied and no cost remains beyond the horizon.

Key observations comparing the approaches are summarized below:

- The options are listed in increasing order of restrictiveness, which generally correlates with increasing deviation from the optimal solution of the original constrained infinite-horizon problem.
- A key difficulty in Options 1 and 3A is the need to select a horizon length p such that the terminal constraint is always satisfied. In practice, this may require a large p, leading to significant computational burden.
- This issue is partially alleviated in Options 2 and 3B, where a fixed horizon p can be used while maintaining feasibility beyond the horizon through appropriate terminal constraints. Nevertheless, p must still be selected sufficiently large to ensure that these terminal constraints remain feasible.
- In Option 2, if p is sufficiently large such that the terminal constraint $z_p \in X_{\text{ma}}$ is inactive, then the finite-horizon solution exactly matches that of the infinite-horizon problem. However, the size of such a p may still pose computational challenges.
- In Option 2, the maximal output admissible set X_{ma}, which may have a complex geometric structure, may be approximated by a simpler polyhedral superset to reduce computational complexity while maintaining feasibility.
- In Option 3B, rather than imposing a terminal constraint $z_p \in \hat{X}_{\text{ma}}$, one can define a constraint horizon h_c and require that the state trajectory additionally satisfies the original constraints $z_{p+i} \in \mathbb{X}$ for $1 \leq i \leq h_c$ under zero input. Provided h_c is sufficiently large, feasibility in the open-loop trajectory is ensured. Since control inputs are only computed over the horizon of p, the added computational cost of checking the extended state trajectory remains modest.
- In Options 3C and 4, one or more state components must be driven to zero at the end of the horizon. This is often restrictive and can lead to infeasibility if the control horizon is too short. Moreover, terminal equality constraints are typically more challenging to handle numerically than inequality constraints.

9.3 Constraint Softening

In the presence of state constraints, the finite-horizon optimal control problem $J_p(z_0)$ may become infeasible. This can occur, for example, when a large disturbance enters the system, making it impossible to maintain the state within the feasible set over the prediction horizon. Clearly, a real-time control algorithm must not fail under such circumstances, and a relaxation mechanism is needed to ensure that a solution can always be computed.

In many applications, input constraints are considered "hard," meaning they must be strictly enforced. On the other hand, state constraints can often tolerate temporary or infrequent violations, though such violations may degrade performance. A common strategy is to soften the state constraints by introducing slack variables and penalizing their use in the objective function.

For example, consider the original state constraint:

$$Fz_i \leq g. \tag{9.24}$$

This can be relaxed to:

$$Fz_i \leq g + \epsilon_i, \tag{9.25}$$

where ϵ_i is a nonnegative slack variable vector representing constraint violations. To discourage excessive violation, a penalty term is added to the objective function:

$$\min_{v_0,\cdots,v_{p-1},\epsilon_0,\cdots,\epsilon_{p-1}} \left\{ \sum_{i=0}^{p-1} \left[z_i^T Q z_i + v_i^T R v_i + \epsilon_i^T Q_{\epsilon_i} \epsilon_i \right] + z_p^T Q_t z_p \right\}, \tag{9.26}$$

where Q_{ϵ_i} is a positive definite matrix that determines how heavily each component of the slack vector is penalized. This formulation provides flexibility, allowing different levels of penalization for different constraints and at different time steps. However, it also introduces a large number of tuning parameters, which can increase the design and computational burden.

To reduce complexity, several simplifications are possible:

- The matrices Q_{ϵ_i} are chosen to be identical for all i, ensuring that constraint violations are penalized uniformly across the prediction horizon.
- Each Q_{ϵ_i} is typically selected as a scalar multiple of the identity, i.e., $Q_{\epsilon_i} = q_{\epsilon_i} I$, so that violations of all state constraints are penalized equally.
- If the slack variable ϵ_i is held constant over time, only the worst violation across the horizon is penalized for each constraint.
- If a single scalar slack variable is shared among all constraints, only the maximum violation at each time step is penalized.

By selecting the structure of ϵ_i and the elements of Q_ϵ appropriately, the designer can balance computational cost against the severity and duration of constraint violations.

It is important to note that the inclusion of soft constraints creates a trade-off between performance and constraint satisfaction. Depending on the relative magnitudes of the weights Q, R, Q_t, and Q_ϵ, the optimizer may prefer to violate a constraint if doing so reduces the overall cost. As a result, even if a strictly feasible solution exists, it may not be selected unless constraint violations are penalized heavily. To address this, one can use the one-norm (instead of the two-norm) for the slack penalty to discourage distributed violations and promote sparsity.

9.4 Derivation of an Explicit Form of the Optimal Control Law via Multi-parametric Programming

The optimal state-feedback control law for the constrained LQR problem can be characterized explicitly through offline computation. The resulting control law is *piecewise affine*: within each polyhedral region of the state space—defined by a specific set of active constraints—the optimal control is given by an affine function of the state. The main challenge lies in identifying and characterizing these regions and the corresponding affine laws. In this section, we present the use of multi-parametric *QP* to derive an explicit solution to the constrained LQR problem. This approach was first introduced by Bemporad et al. (2002).

Recall the constrained finite-horizon optimal control problem described in Eqs. (9.6–9.8), which must be solved at each time step in MPC. If all constraints can be written as linear inequalities in the form of (9.5), then the problem reduces to a parametric QP of the form:

$$\min_{\mathcal{V}} \mathcal{V}^T \mathbf{H} \mathcal{V} + 2\mathcal{V}^T \mathbf{G} z_0, \tag{9.27}$$

subject to

$$\mathbf{C}\mathcal{V} \leq \mathbf{d} + \mathbf{D} z_0, \tag{9.28}$$

where $\mathcal{V}$ is the vector of optimization variables, and z_0 is the parameter (initial condition) of the problem.

To simplify the parametric dependence, we apply a change of variables:

$$\zeta = \mathcal{V} + \mathbf{H}^{-1}\mathbf{G} z_0.$$

Substituting this into the original problem yields:

$$\min_{\zeta} \zeta^T \mathbf{H} \zeta, \tag{9.29}$$

$$\mathbf{C}\zeta \leq \mathbf{d} + \mathbf{E} z_0, \tag{9.30}$$

where $\mathbf{E} = \mathbf{D} + \mathbf{C}\mathbf{H}^{-1}\mathbf{G}$.

The optimal solution ζ^* satisfies the *Karush–Kuhn–Tucker (KKT)* conditions:

$$\mathbf{H}\zeta^* + \mathbf{C}^T \lambda^* = 0, \tag{9.31}$$

$$(\lambda^*)^T (\mathbf{C}\zeta^* - \mathbf{d} - \mathbf{E} z_0) = 0, \tag{9.32}$$

$$\lambda^* \geq 0, \tag{9.33}$$

$$\mathbf{C}\zeta^* - \mathbf{d} - \mathbf{E} z_0 \leq 0. \tag{9.34}$$

Here, λ^* are the Lagrange multipliers associated with the inequality constraints. Each constraint can be classified as either:

- *Active*, when $\mathbf{C}\zeta^* - \mathbf{d} - \mathbf{E} z_0 = 0$, in which case the corresponding $\lambda_i^* \geq 0$;
- *Inactive*, when $\mathbf{C}\zeta^* - \mathbf{d} - \mathbf{E} z_0 < 0$, in which case $\lambda_i^* = 0$.

Once the active set is identified, the inequality constraints can be replaced by equality constraints for the active set, and the optimal solution can be computed by solving the resulting set of linear equations.

The first step in characterizing the polyhedral regions and their corresponding optimal feedback laws is to identify a feasible point z_0. This can be done, for example, by solving a linear program. From this solution, the active constraints—those for which $\lambda^* \neq 0$—can be identified. Let us denote the active constraints as

$$\hat{\mathbf{C}}\zeta - \hat{\mathbf{d}} - \hat{\mathbf{E}} z_0 = 0, \tag{9.35}$$

where the ^ notation denotes rows corresponding to the active set. We then solve the following set of equations:

$$\mathbf{H}\zeta^* + \hat{\mathbf{C}}^T \lambda^* = 0, \tag{9.36}$$

$$\left(\hat{\mathbf{C}}\zeta^* - \hat{\mathbf{d}} - \hat{\mathbf{E}} z_0\right) = 0. \tag{9.37}$$

This yields

$$\zeta^* = \mathbf{H}^{-1}\hat{\mathbf{C}}^T(\hat{\mathbf{C}}\mathbf{H}^{-1}\hat{\mathbf{C}}^T)^{-1}\left(\hat{\mathbf{d}} + \hat{\mathbf{E}}z_0\right), \tag{9.38}$$

$$\hat{\lambda}^* = -(\hat{\mathbf{C}}\mathbf{H}^{-1}\hat{\mathbf{C}}^T)^{-1}\left(\hat{\mathbf{d}} + \hat{\mathbf{E}}z_0\right). \tag{9.39}$$

Substituting these expressions into the KKT conditions (9.33–9.34), we obtain the critical region $\mathbf{CR_0}$:

$$\mathbf{CR_0} = \{z_0 : \mathbf{S}z_0 \leq \mathbf{r}\}, \tag{9.40}$$

where

$$\mathbf{S} = \begin{bmatrix} (\hat{\mathbf{C}}\mathbf{H}^{-1}\hat{\mathbf{C}}^T)^{-1}\hat{\mathbf{E}} \\ \mathbf{C}\left(\mathbf{L} + \mathbf{H}^{-1}\mathbf{G}\right) - \mathbf{E} \end{bmatrix}, \tag{9.41}$$

$$\mathbf{r} = \begin{bmatrix} -(\hat{\mathbf{C}}\mathbf{H}^{-1}\hat{\mathbf{C}}^T)^{-1}\hat{\mathbf{d}} \\ \mathbf{d} - \mathbf{C}\hat{\mathbf{p}} \end{bmatrix}. \tag{9.42}$$

This region $\mathbf{CR_0}$ consists of all z_0 for which the current active set remains valid.

The optimal control input is then given by:

$$\mathcal{V} = \mathbf{H}^{-1}\hat{\mathbf{C}}^T(\hat{\mathbf{C}}\mathbf{H}^{-1}\hat{\mathbf{C}}^T)^{-1}\left(\hat{\mathbf{d}} + \hat{\mathbf{E}}z_0\right) - \mathbf{H}^{-1}\mathbf{G}z_0, \tag{9.43}$$

$$= \mathbf{L}z_0 + \hat{\mathbf{p}}, \tag{9.44}$$

where

$$\mathbf{L} = \mathbf{H}^{-1}\hat{\mathbf{C}}^T(\hat{\mathbf{C}}\mathbf{H}^{-1}\hat{\mathbf{C}}^T)^{-1}\hat{\mathbf{E}} - \mathbf{H}^{-1}\mathbf{G}, \tag{9.45}$$

$$\hat{\mathbf{p}} = \mathbf{H}^{-1}\hat{C}^T(\hat{\mathbf{C}}\mathbf{H}^{-1}\hat{\mathbf{C}}^T)^{-1}\hat{\mathbf{d}}. \tag{9.46}$$

The control law $\mathcal{V} = \mathbf{L}z_0 + \hat{\mathbf{p}}$ is optimal within this region.

To characterize the entire state space, we can construct adjacent regions by modifying the active set. One approach is to flip one inequality at a time to generate candidate regions outside $\mathbf{CR_0}$, denoted by $\mathbf{R_1}$. For each new region, the same procedure is applied to identify the corresponding active set and compute its associated critical region. This recursive process continues until the relevant portion of the state space is fully partitioned.

An alternative approach is to systematically swap active constraints by removing one and adding a new one, effectively stepping from one region to a neighboring one. Through either method, the state space is eventually partitioned into a union of polyhedral regions, each with its own piecewise affine optimal control law. See Figure 9.5 for a schematic illustration of this recursive construction. In summary, this approach to characterizing the MPC control law shifts the computational burden from online to offline. The quadratic program is solved offline for all relevant regions of the state space. At runtime, the main task is to identify the polyhedral region to which the current state belongs and retrieve the corresponding linear feedback law through table lookup. This becomes a set-membership identification problem, which can be nontrivial when the number of regions is large.

The main determining factor for online computational load of the explicit MPC law is the number of regions. The number of regions depends on a number of factors, including the system size, prediction and control horizon size, and weight parameters. Move blocking can effectively reduce the complexity as well. In the literature, there exist useful methods to merge regions that contain

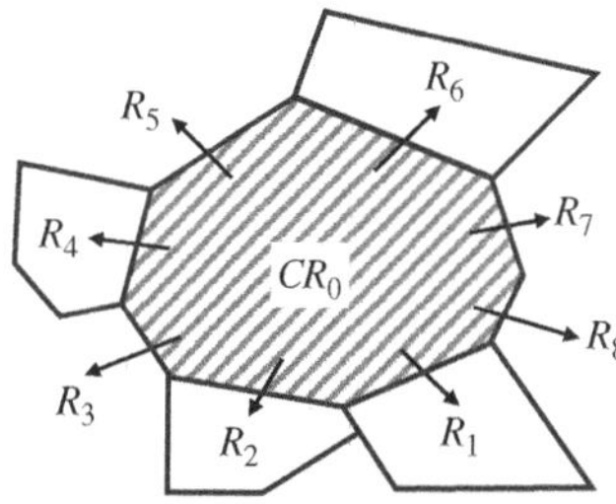

Recursive construction of critical regions in state space

Figure 9.5 Illustration of recursive partitioning of the state space into critical regions, each corresponding to a distinct set of active constraints.

the same control law, as the constructive algorithm can result in multiple regions with the same control law.

Several software packages that facilitate the calculation and implementation of explicit MPC control laws for constrained linear systems exist, including MATLAB's MPC Toolbox.

Example 9.3: *Constrained LQR via Parametric Programming.*
We now use the parametric programming approach to solve the constrained control problem for a double integrator from Example 9.2. We consider the problem of regulating this system to the origin using an MPC law that minimizes the following quadratic cost at each time step k:

$$\sum_{i=1}^{p} \{x_i^T Q x_i + u_i^T R u_i\}, \qquad Q = \text{diag}(1,\ 0.1),\ R = 0.1, \tag{9.47}$$

subject to the same input constraints

$$-0.5 \leq u_i \leq 0.5.$$

This can be solved using multi-parametric programming, as described in this section. It is available in the MATLAB MPC toolbox.

After specifying the model (9.16), constraints and LQR parameters, the controller object is generated using `generateExplicitMPC`. Figure 9.6 shows the polyhedral regions generated for this system.

Thereafter, `mpcmoveExplicit` function can be used to compute control moves, $u(k)$ at each time-step. Figure 9.7 shows the closed-loop control result. Comparing with those of Example 9.2, we observe that the performance closely matches that of the infinite-horizon constrained LQR.

9.5 Analysis

In this section, we analyze the stability of the finite-horizon RHC law, considering both state-feedback and output-feedback implementations. Stability results for various formulations of the constrained LQR problem have been established in the literature, including the works of Keerthi and Gilbert (1988), Mayne and Mischalska (1990), Rawlings and Muske (1993), and Chmielewski and Manousiouthakis (1996).

To ensure that the analysis is well-posed, we assume the system is *stabilizable*. Under constraints, stabilizability depends not only on the system dynamics but also on the initial condition.

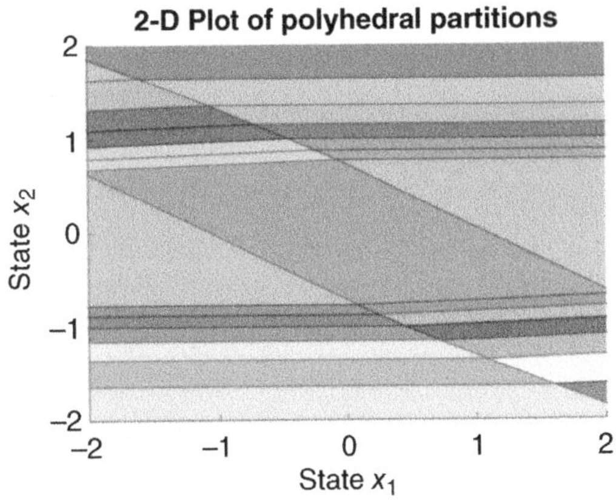

Figure 9.6 Various polyhedral regions generated for constrained LQR of a double integrator model.

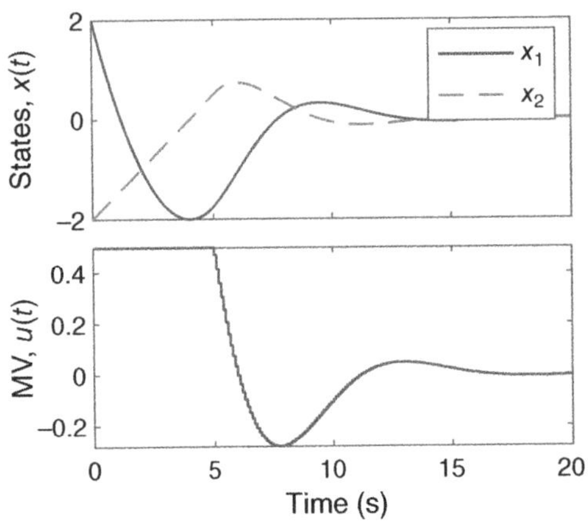

Figure 9.7 Constrained LQR using multi-parametric programming for discrete-time double integrator with horizon $p = 10$ ($J = 151.6$).

In the presence of unstable modes, global stabilizability cannot be guaranteed. In such cases, we restrict attention to initial conditions that lie within the domain of attraction of the receding horizon controller.

9.5.1 Stability Concepts and Lyapunov's Direct Method

While a single definition of stability is sufficient for linear systems, multiple notions are required for nonlinear systems. In this section, we consider Lyapunov stability (also known as stability in the sense of Lyapunov), asymptotic stability, and exponential stability. We also review Lyapunov's direct method, a powerful tool for analyzing the stability of nonlinear systems.

9.5.1.1 Stability Concepts

Definition 9.1: The equilibrium $x = 0$ is said to be *stable (in the sense of Lyapunov)* if, for any $\epsilon > 0$, there exists a corresponding $\delta > 0$ such that, if $|x(0)| < \delta$, then $|x(t)| < \epsilon$ for all $t \geq 0$. Otherwise, the equilibrium is said to be unstable.

Definition 9.2: The equilibrium $x = 0$ is said to be *asymptotically stable* if it is stable (in the sense of Lyapunov) and there exists some $\gamma > 0$ such that $|x(0)| < \gamma$ implies $x(t) \to 0$ as $t \to \infty$.

The second condition in the definition of asymptotic stability is referred to as *attractivity*. The set of all initial conditions from which the trajectories converge to the origin is known as the *domain of attraction*. Note that while attractivity and asymptotic stability are equivalent for linear systems, attractivity alone is not sufficient for asymptotic stability in the case of nonlinear systems.

These stability concepts are illustrated in Figure 9.8. For linear systems, a marginally stable system—one with some eigenvalues lying on the unit circle—would satisfy the definition of Lyapunov stability but not asymptotic stability.

We also introduce the notion of exponential stability.

Definition 9.3: The equilibrium $x = 0$ is said to be *exponentially stable* if it is stable (in the sense of Lyapunov) and there exists some $\gamma > 0$ such that, if $\| x(0) \| < \gamma$, then

$$\| x(t) \| \leq \alpha \| x(0) \| e^{-\lambda t}, \forall t > 0$$

for some $\alpha > 0$ and $\lambda > 0$.

Exponential stability is a stronger condition than asymptotic stability. For linear systems, however, the two notions are equivalent.

Finally, note that the above definitions of asymptotic and exponential stability refer to *local* stability. For *global* stability, the corresponding conditions must hold for any initial state, not just those within a neighborhood of the origin.

Definition 9.4: The equilibrium $x = 0$ is said to be *globally asymptotically* (or *globally exponentially*) *stable* if asymptotic (or exponential) stability holds for any initial condition.

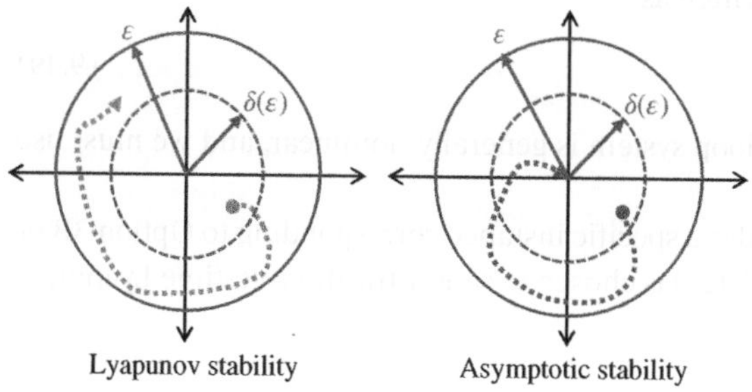

Figure 9.8 Concepts of stability for nonlinear systems.

9.5.1.2 Lyapunov's Direct Method

Lyapunov's direct method is based on the principle that if an energy—or energy-like—function of a system continuously decreases, then the system must eventually settle to an equilibrium. This method is among the most powerful and widely used tools for analyzing the stability of general nonlinear systems. We present it here in the context of a discrete-time system.

Consider an autonomous nonlinear system described by $x(k+1) = f(x(k))$. Let $V(x)$ be a scalar function defined in some neighborhood of the origin, specifically within the ball $B_\gamma = \{x : \| x \| < \gamma\}$. Suppose $V(x)$ satisfies the following conditions:

- $V(0) = 0$ and $V(x) > 0$ for all $x \neq 0$. Such a function is called locally *positive definite*.
- $V(x) < \infty$ for all $\| x \| < \infty$.
- Along trajectories of the system, $V(x(k+1)) - V(x(k)) \leq 0$ for all $k \geq 0$.

A function $V(x)$ that satisfies these properties is called a *Lyapunov function* for the system $x(k+1) = f(x(k))$.

The existence of such a Lyapunov function implies Lyapunov (i.e., marginal) stability. If the third condition is strengthened to

$$V(x(k+1)) - V(x(k)) < 0 \text{ for all } x \neq 0,$$

then $V(x)$ also establishes asymptotic stability.

To ensure *global* asymptotic stability, the above conditions must hold for all $x \in \mathbb{R}^n$, not just within some neighborhood. In addition, $V(x)$ must satisfy the condition of *radial unboundedness*:

$$V(x) \to \infty \text{ as } \| x \| \to \infty.$$

The primary challenge in applying Lyapunov's method lies in constructing a suitable Lyapunov function for a given system. In later sections, we will use this method to prove the asymptotic stability of suboptimal RHC schemes introduced earlier.

9.5.2 State Feedback Case

Consider the state feedback control law

$$u(k) = u^*(x(k)), \tag{9.48}$$

where u^* denotes the solution to the convex optimization problem $J_p(z_0)$ evaluated at $z_0 = x(k)$. The resulting closed-loop dynamics can be written as

$$x(k+1) = Ax(k) + Bu^*(x(k)). \tag{9.49}$$

Since u^* is a nonlinear mapping, the closed-loop system is generally nonlinear, and we must use nonlinear stability analysis tools.

To illustrate the stability analysis, we consider a specific instance corresponding to Option 3B or 3C. In this setting, the terminal weight Q_t in $J_p(z_0)$ is chosen to satisfy the discrete-time Lyapunov equation:

$$A^T Q_t A + Q = Q_t,$$

and a terminal constraint of the form $z_p \in \hat{X}_{\text{ma}}$ is imposed.

Assumptions: To establish asymptotic stability under the feedback law $u(k) = u^*(x(k))$, we make the following assumptions:

(i) The optimization problem $J_p(z_0)$ is feasible for the initial state $z_0 = x(0)$.
(ii) The corresponding optimal cost $J_p(z_0)$ is finite.

These conditions imply that:

- The initial state $x(0)$ lies within the stabilizable set.
- All imposed state constraints are satisfied over the prediction horizon.
- The terminal constraint $z_p \in \hat{X}_{\text{ma}}$ is feasible.

Since the origin ($x = 0$, $u = 0$) is assumed to lie in the interior of the feasible region, and the system is stabilizable, these assumptions hold in some neighborhood around the origin. As a result, there exists a positively invariant region in which the feedback law ensures feasibility and enables a meaningful stability analysis.

Proof of Convergence

We begin by establishing the asymptotic convergence of the state sequence to the origin. To do so, consider the cost function $J_p(x(k))$, which is equivalent to the following infinite-horizon cost with a finite control horizon:

$$J_{\infty,p}(x(k)) = \min_{v_0,\cdots,v_\infty} \left\{ \sum_{i=0}^{\infty} \left[z_i^T Q z_i + v_i^T R v_i \right] \right\}, \tag{9.50}$$

with $v_i = 0$, for $i \geq p$ and the initial condition $z_0 = x(k)$, and subject to the model dynamics $z_{i+1} = Az_i + Bv_i$ and constraints $z_i \in \mathbb{X}$, $v_i \in \mathbb{U}$.

We now show that

$$J_{\infty,p}(x(k)) \geq J_{\infty,p}(x(k+1)) + x^T(k)Qx(k) + u^T(k)Ru(k). \tag{9.51}$$

To derive this inequality, define the restriction operator $\mathcal{R}$ applied to the input sequence $\mathcal{V}$ as:

$$\mathcal{R}(\mathcal{U}) \triangleq \begin{bmatrix} 0 & I & 0 & \cdots & 0 \\ 0 & 0 & I & \cdots & 0 \\ & & \ddots & & \\ 0 & 0 & \cdots & \cdots & I \\ 0 & 0 & \cdots & \cdots & 0 \end{bmatrix} \mathcal{U}. \tag{9.52}$$

Then, $\mathcal{R}(\mathcal{V}^*(k))$, the restriction of the optimal input sequence for $J_{\infty,p}(x(k))$, is a feasible (but not necessarily optimal) input for $J_{\infty,p}(x(k+1))$. Let the corresponding suboptimal cost be denoted by $\hat{J}_{\infty,p}(x(k+1))$. It follows that:

$$J_{\infty,p}(x(k)) = \hat{J}_{\infty,p}(x(k+1)) + x^T(k)Qx(k) + u^T(k)Ru(k). \tag{9.53}$$

Since $\hat{J}_{\infty,p}(x(k+1)) \geq J_{\infty,p}(x(k+1))$, inequality (9.51) holds.

Now, inequality (9.51) can be rewritten as

$$J_{\infty,p}(x(i)) - J_{\infty,p}(x(i+1)) \geq x^T(i)Qx(i) + u^T(i)Ru(i). \tag{9.54}$$

By summing the above inequalities from $i = 0$ to $i = k - 1$, we obtain:

$$J_{\infty,p}(x(0)) - J_{\infty,p}(x(k)) \geq \sum_{i=0}^{k-1}[x^T(i)Qx(i) + u^T(i)Ru(i)]. \tag{9.55}$$

Since each term in the summation is nonnegative, the right-hand side is nonnegative, and the left-hand side is bounded below by zero. Given that $J_{\infty,p}(x(0)) < \infty$ and $J_{\infty,p}(x(k)) \geq 0$, the left-hand side is finite for all $k \geq 0$, and therefore the right-hand side must also remain finite for all k. Taking the limit as $k \to \infty$, we have:

$$\sum_{k=0}^{\infty}[x^T(k)Qx(k) + u^T(k)Ru(k)] < \infty. \tag{9.56}$$

Since both $Q > 0$ and $R > 0$, it follows that $x(k) \to 0$ and $u(k) \to 0$ as $k \to \infty$.

Proof of Lyapunov Stability

To complete the proof of asymptotic stability, we must establish Lyapunov stability in addition to convergence (attractivity). For this, define a ball around the origin:

$$\mathcal{X}_0 \stackrel{\Delta}{=} \{x : \| x \| \leq c\},$$

such that

$$J_{\infty,p}(x(0)) \leq b\|x(0)\|^2 \quad \forall x(0) \in \mathcal{X}_0 \text{ for some } b < \infty. \tag{9.57}$$

Such a ball always exists because, for states sufficiently close to the origin, the optimal control becomes linear, and the cost function is quadratic in x.

From the preceding argument, we have:

$$J_{\infty,p}(x(k)) \leq J_{\infty,p}(x(0)) \;\; \forall k > 0. \tag{9.58}$$

Moreover, since the objective function is quadratic with a positive-definite weighting matrix, it follows that:

$$\exists \; a > 0 \text{ such that } a \| x(k) \|^2 \leq J_{\infty,p}(x(k)). \tag{9.59}$$

Combining the two results above, we conclude that for all $x(0) \in \mathcal{X}_0$,

$$a \| x(k) \|^2 \leq b \| x(0) \|^2, \tag{9.60}$$

which implies

$$\| x(k) \| \leq \left(\frac{b}{a}\right)^{1/2} \| x(0) \|. \tag{9.61}$$

Thus, if the initial condition satisfies

$$\| x(0) \| \leq \min \left\{ \left(\frac{b}{a}\right)^{-1/2} \epsilon, c \right\}, \tag{9.62}$$

then, $\quad \| x(k) \| \leq \epsilon \quad \forall k \geq 0. \quad (9.63)$

Comments

- The assumption $Q > 0$ can be relaxed to $Q \geq 0$, provided that $(Q^{1/2}, A)$ is detectable. Detectability ensures that modes not directly penalized in the objective function—i.e., those evolving autonomously—are asymptotically stable. Therefore, the previous argument, applied only to the modes directly weighted by Q, is sufficient to establish asymptotic stability.
- Extending the proof to the case where state constraints are softened using slack variables, as in Eq. (9.26) with $p = \infty$, is straightforward if the same weighting matrix Q_ϵ is used over the horizon. The proof is left as an exercise. One potential complication in implementation is that $\hat{X}_{\text{ma}}$ in the terminal constraint now depends on the slack variable ϵ, making it impossible to compute $\hat{X}_{\text{ma}}$ *a priori*, thereby increasing the complexity of the numerical optimization.
- When state constraints are softened in the infinite-horizon formulation, Option 3B applied to stable systems can be shown to yield *global* asymptotic stability—indeed, even *exponential* stability. In contrast, Option 3C applied to unstable systems cannot ensure global stabilization under bounded inputs, since exponentially unstable systems require increasingly large control inputs as $|x(0)|$ grows. In such cases, only a finite domain of attraction exists.
- The stability proofs for Option 2 follow the same essential argument as above. One can show that the restricted control sequence obtained from the infinite-horizon solution at time k provides a feasible (though possibly suboptimal) solution at time $k + 1$.
- The stability proof for Option 4 relies on the relation:

$$\begin{aligned} J_p(x(k)) &= J_{p-1}(x(k+1)) + x^T(k)Qx(k) + u^T(k)Ru(k) \\ &\geq J_p(x(k+1)) + x^T(k)Qx(k) + u^T(k)Ru(k). \end{aligned} \tag{9.64}$$

  The inequality holds because the restriction of the solution to $J_p(x(k))$ is a feasible solution to $J_p(x(k+1))$, given that $z_{k+p} = 0$ implies $z_{k+p+1} = 0$. However, the assumption $J_p(x(0)) < \infty$ is more restrictive in this case.

Alternatively, we can apply Lyapunov's direct method discussed earlier. In this case, we consider the optimal cost function $J_p(x)$—defined as $J_p(z_0)$ with $z_0 = x(k)$—as a Lyapunov function candidate for the closed-loop system:

$$x(k+1) = Ax(k) + Bu^*(x(k)).$$

The function $J_p(x)$ is positive definite because $x = 0$ is an equilibrium point (hence $J_p(0) = 0$), and $Q > 0, R > 0$ ensure that $J_p(x) > 0$ for all $x \neq 0$. Moreover, $V(x(0)) < \infty$ by assumption. Therefore, to establish asymptotic stability using Lyapunov's method, we must verify that the cost difference $J_p(x(k+1)) - J_p(x(k))$ is negative definite.

Observe that: $J_p(x(k+1)) - J_p(x(k))$. Note that:

$$J_p(x(k+1)) - J_p(x(k)) = J_p(x(k+1)) - \left(J_{p-1}(x(k+1)) + x^T(k)Qx(k) + u^T(k)Ru(k)\right). \tag{9.65}$$

Whether this expression is guaranteed to be negative definite depends on the choice of p, Q_t, and the terminal constraint. The following two cases ensure the condition holds regardless of the choice of p:

- *Terminal constraint* $z_{k+p} = 0$: In this case, we have:

$$J_p(x(k+1)) \leq J_{p-1}(x(k+1)), \tag{9.66}$$

because the optimal solution to $J_{p-1}(x(k+1))$ is given by restricting the optimal solution of $J_p(x(k))$, which also serves as a feasible (though possibly suboptimal) solution to $J_p(x(k+1))$. Since this solution satisfies $z_{k+p} = 0$, it also results in $z_{k+p+1} = 0$, making the cost computed under $J_{p-1}(x(k+1))$ valid for the p-step problem.

As the term $-(x^T(k)Qx(k) + u^T(k)Ru(k))$ is negative definite, the negative semi-definiteness of $J_p(x(k+1)) - J_{p-1}(x(k+1))$ implies that $J_p(x(k+1)) - J_p(x(k)) < 0$.

- *Terminal weight Q_t from Lyapunov equation and terminal constraint $z_{k+p} \in \hat{X}_{ma}$:* As previously shown in Eq. (9.51),

$$J_{\infty,p}(x(k+1)) - J_{\infty,p}(x(k)) \leq -(x^T(k)Qx(k) + u^T(k)Ru(k)). \tag{9.67}$$

Since the right-hand side is a negative definite function the cost difference $J_{\infty,p}(x(k+1)) - J_{\infty,p}(x(k))$ is negative definite.

9.5.3 Output Feedback Case

Here we will show that, if the state feedback law $u(k) = u^*(x(k))$ that yields an asymptotically stable closed-loop system is coupled with a stable linear observer, then the resulting output feedback law $u(k) = u^*(\hat{x}(k|k))$ also yields an asymptotically stable closed-loop system. In other words, the stability of the state feedback controller and that of the observer can be checked separately for the stability of the combined system.

The key point in proving the above is that the stable MPC feedback laws we described earlier are nonlinear, but the nonlinearity is relatively well-behaved. Specifically, they are *Lipschitz continuous*, which means that

> there exists a fixed constant K such that $\| u^*(x+\epsilon) - u^*(x) \| \leq K \| \epsilon \|$ for all $\epsilon > 0$ and for all x.

For example, when the constraints are linear inequalities, we solve a QP for control calculation. We have already shown that the resulting feedback law $u^*(x)$ is a piecewise affine function of x, which is clearly Lipschitz continuous.

The closed-loop system under the output feedback controller can be written as follows:

$$\begin{aligned} x(k+1) &= Ax(k) + Bu^*(\hat{x}(k|k)) \\ &= Ax(k) + Bu^*(x(k)) + B[u^*(\hat{x}(k|k)) - u^*(x(k))], \end{aligned} \tag{9.68}$$

$x(k+1) = h(x(k)) \triangleq Ax(k) + Bu^*(x(k))$ is a nonlinear dynamic system, for which the origin is an asymptotically stable fixed point. In addition, $h(x(k))$ is Lipschitz continuous since $u^*(x(k))$ is Lipschitz continuous.

At this point, we draw upon a result in the literature that such "well-behaved" nonlinear systems are robust with respect to an exponentially decaying additive disturbance. That is,

> if the origin is an asymptotically stable equilibrium point for $x(k+1) = h(x(k))$ where $h(x)$ is a Lipschitz continuous function, then the origin remains an asymptotically stable fixed point for the perturbed system $x(k+1) = h(x(k)) + e(k)$ where $e(k)$ is an exponentially converging sequence.

The proof of the above result is omitted due to its complexity; interested readers are referred to relevant literature for details. Stability analysis of perturbed state-space systems with Lipschitz continuous nonlinearities is presented in the work of Halanay (1963). Applications of these results in the context of MPC can be found in Scokaert et al. (1997).

What remains to show for us to prove the asymptotic stability of the output feedback controller is that $B[u^*(\hat{x}(k|k)) - u^*(x(k))]$ is indeed an exponentially converging sequence. Since $u^*(x)$ is Lipschitz continuous,

$$\|u^*(\hat{x}(k|k)) - u^*(x(k)\| \leq K\|x(k) - \hat{x}(k|k)\| \text{ for some } K > 0. \tag{9.69}$$

Since $(x(k) - \hat{x}(k|k))$ is an exponentially converging sequence due to the assumption of a stable linear observer,

$$\| x(k) - \hat{x}(k|k) \| \leq \alpha e^{-\lambda k} \text{ for some } \alpha > 0 \text{ and } \lambda > 0, \tag{9.70}$$

and therefore

$$\| u^*(x(k)) - u^*(\hat{x}(k|k)) \| \leq K\alpha e^{-\lambda k}. \tag{9.71}$$

9.6 Stochastic Case (*)

Extending the optimal control formulations to stochastic systems is of obvious interest, as real processes experience disturbances and noises that are best described by random variables. However, we will see that there are a number of new complications, which have not been resolved in a satisfactory manner at this time.

First, in the stochastic case, the state constraints need to be formulated in a probabilistic manner. For instance, one may enforce

$$E\{x(k+i)\} \in \mathbb{X}, \quad i = 1, \cdots, p \tag{9.72}$$

However, the above does not guarantee the satisfaction of the constraint with much confidence. A better alternative is based on the *chance constraint*

$$\Pr\{x(k+i) \in \mathbb{X}, \quad i = 1, \cdots, p\} \geq \gamma, \tag{9.73}$$

where Pr denotes *probability* and γ is a parameter between 0 and 1. However, this constraint is much more difficult to handle computationally.

Second, no matter how the state constraints are formulated, derivation of the linear quadratic optimal *feedback* law for the constrained stochastic case is not possible because, unlike in the unconstrained case, the stochastic dynamic program for the constrained problem does not yield an analytical solution. An alternative is to adopt the strategy of *open-loop optimal feedback control*, which results by solving the open-loop optimal control problem at each time step. For instance, in an infinite-horizon problem with full state feedback, one can implement $u(k) = u^*(x(k))$ where $u^*(x(k))$ represents the optimal solution to the problem:

$$J_p(x(k)) = \min_{v_0,\cdots,v_{p-1}} E\left\{\sum_{i=0}^{p-1}[z_i^T Q z_i + v_i^T R v_i] + z^T(p)Q_t z(p)\right\}, \tag{9.74}$$

∗ This section can be skipped without losing continuity.

for a stochastic system

$$z_{i+1} = Az_i + Bv_i + \varepsilon_{1,i} \quad i = 0, \cdots, p-1, \tag{9.75}$$

initial condition $z_0 = x(k)$ and inequality constraints:

$$\Pr\{z_i \in \mathbb{X}\} > \gamma \quad i = 1, \cdots, p, \tag{9.76}$$

$$v_i \in \mathbb{U}, \quad i = 0, \cdots, p-1. \tag{9.77}$$

Aside from the computational challenges associated with enforcing the chance constraint, the control law $u(k) = u^*(x(k))$—obtained by solving the above optimization problem at each time step—does not, in general, represent the optimal feedback policy, even in the limit as $p \to \infty$. This contrasts with the unconstrained case, where the separation principle applies and the optimal feedback law reduces to a combination of the deterministic optimal state-feedback controller and an optimal state estimator. To obtain the true optimal feedback policy in the constrained case, one must directly optimize over the space of feedback policies.

If we replace the chance constraint with the less desirable alternative of $E\{x(k+i) \in \mathbb{X}\}$, the solution to the above open-loop optimal control problem turns out to be identical to the deterministic case. Hence, considering the stochastic nature of the system in the control calculation does not provide anything extra.

Despite these theoretical difficulties, the deterministic formulation has been applied successfully in many stochastic problems. It has been observed empirically that, since constrained linear systems represent only a minor departure from linear systems, especially when there are no state constraints, the combination of the optimal deterministic regulator (*i.e.*, the constrained LQR) with the optimal state estimator yields a satisfactory suboptimal solution in almost all cases.

Exercises

1. Prove the asymptotic stability of the finite-horizon MPC with Option 3 for the case of soft constraints with quadratic penalty.

2. Prove the asymptotic stability for the finite-horizon MPC with Option 4 (enforcing the terminal constraint $z_p = 0$).

3. Prove the quadratic state cost weighed by the solution to Eq. (9.22) represents the infinite-horizon cost-to-go under zero control.

4. Show that the MPC control laws defined by a quadratic program is Lipschitz continuous.

5. Consider the discrete-time, closed-loop LTI system:

$$\begin{aligned} x(k+1) &= A_{\text{cl}}\, x(k) \\ y(k) &= C\, x(k), \end{aligned} \tag{9.78}$$

with

$$A_{\text{cl}} = \begin{bmatrix} 0.8 & 0.1 \\ 0 & 0.9 \end{bmatrix}, \qquad C = \begin{bmatrix} 1 & 0 \\ 0 & 1 \end{bmatrix}. \tag{9.79}$$

The outputs are constrained to lie in the hyper-rectangle, i.e.,

$$\mathbb{X} = \left\{ x \in \mathbb{R}^2 : \begin{bmatrix} -1 \\ -1 \end{bmatrix} \leq Cx \leq \begin{bmatrix} 1 \\ 1 \end{bmatrix} \right\}. \tag{9.80}$$

The *maximal output admissible set* is

$$X_{\text{ma}} \triangleq \left\{ x \in \mathbb{R}^2 : \begin{bmatrix} -1 \\ -1 \end{bmatrix} \leq CA_{\text{cl}}^k x \leq \begin{bmatrix} 1 \\ 1 \end{bmatrix}, \ \forall k \in \mathbb{Z}_{\geq 0} \right\}. \tag{9.81}$$

(a) Show that A_{cl} is Schur (all eigenvalues strictly inside the unit circle). Explain why this implies X_{ma} is *nonempty, compact*, and *positively invariant* for the autonomous system under the constraint $\mathbb{X}$.

(b) Show that X_{ma} can be written as the infinite intersection of half-spaces:

$$X_{\text{ma}} = \bigcap_{k=0}^{\infty} \left\{ x : \begin{bmatrix} -1 \\ -1 \end{bmatrix} \leq CA_{\text{cl}}^k x \leq \begin{bmatrix} 1 \\ 1 \end{bmatrix} \right\}, \tag{9.82}$$

(c) Define the N-step approximation:

$$X_{\text{ma}}^N \triangleq \bigcap_{k=0}^{N} \left\{ x : \begin{bmatrix} -1 \\ -1 \end{bmatrix} \leq CA_{\text{cl}}^k x \leq \begin{bmatrix} 1 \\ 1 \end{bmatrix} \right\}. \tag{9.83}$$

- **(i)** Prove that $X_{\text{ma}}^{N+1} \subseteq X_{\text{ma}}^N$ (monotone nesting).
- **(ii)** Argue why $\{X_{\text{ma}}^N\}$ converges to X_{ma} as $N \to \infty$.

(d) Explicitly compute A_{cl}^k for $k = 0, 1, \ldots, 6$. Using these, write the linear inequalities defining X_{ma}^6 in the form

$$Gx \leq h, \tag{9.84}$$

and plot X_{ma}^6 in the (x_1, x_2)-plane. (A box $[-1, 1]^2$ superimposed for reference is helpful.)

(e) Use a norm bound to estimate an $N^\star$ for which X_{ma}^N is numerically indistinguishable from X_{ma} to a tolerance $\varepsilon = 10^{-3}$.

Hint: If $\| A_{\text{cl}}^N \|_\infty \leq \varepsilon$, then constraints for $k > N$ add at most $O(\varepsilon)$ tightening.

(f) Sample a set of initial conditions on the boundary of X_{ma}^6 and simulate $x(k+1) = A_{\text{cl}}x(k)$ for $k = 0, \ldots, 50$. Verify that $x(k) \in \mathbb{X}$ for all k. Then sample points just *outside* X_{ma}^6 and show that the constraint is eventually violated.

(g) Repeat items (c)-(f) with

$$C = \begin{bmatrix} 1 & 1 \\ 1 & -0.5 \end{bmatrix}, \qquad \mathbb{X} = \left\{ x : \begin{bmatrix} -1 \\ -0.8 \end{bmatrix} \leq Cx \leq \begin{bmatrix} 1 \\ 0.8 \end{bmatrix} \right\}. \tag{9.85}$$

Comment on how output mixing (nonidentity C) changes the geometry of X_{ma}.

6. Consider the discrete-time double integrator system:

$$x(k+1) = \begin{bmatrix} 1 & t_s \\ 0 & 1 \end{bmatrix} x(k) + \begin{bmatrix} \frac{t_s^2}{2} \\ t_s \end{bmatrix} u(k), \tag{9.86}$$

$$y(k) = \begin{bmatrix} 1 & 0 \end{bmatrix} x(k), \tag{9.87}$$

with sampling time $t_s = 0.5$ s, state vector

$$x(k) = \begin{bmatrix} \text{position} \\ \text{velocity} \end{bmatrix}, \tag{9.88}$$

and scalar control input $u(k)$ (acceleration).

The objective is to track a reference position r_y while satisfying the constraints:

$$-0.5\ \text{m/s}^2 \le u(k) \le 0.5\ \text{m/s}^2, \tag{9.89}$$

$$-2\ \text{m} \le y(k) \le 2\ \text{m}, \tag{9.90}$$

$$-1.0\ \text{m/s} \le \dot{y}(k) \le 1.0\ \text{m/s}. \tag{9.91}$$

The cost function to be minimized at each time step is:

$$J = \sum_{i=1}^{p} \left| y(k+i) - r_y \right|_Q^2 + \sum_{i=0}^{m-1} |u(k+i)|_R^2, \tag{9.92}$$

where p is the prediction horizon, m is the control horizon, $Q = 1$, and $R = 0.1$.

(a) Write the prediction model in condensed form:

$$\mathcal{Y} = \mathcal{S}^x z_0 + \mathcal{S}^u \mathcal{V}, \tag{9.93}$$

and identify $\mathcal{S}^x$ and $\mathcal{S}^u$. Note that $z_0 = x(k)$.

(b) Express the position, velocity, and input constraints as:

$$C\mathcal{V} \le d + D z_0. \tag{9.94}$$

(c) Set $p = 10$ and $m = 3$, and solve the resulting QP for a step change in

$$r_y : 0\ \text{m} \rightarrow 1\ \text{m}. \tag{9.95}$$

(d) Apply only the first control move $u(k)$ at each time step and resolve the QP. Plot $y(k)$, $\dot{y}(k)$, and $u(k)$ over 30 s.

(e) Compare the closed-loop response with an **unconstrained LQR** solution. Discuss how changing p, m, or R affects performance and constraint satisfaction.

7. Consider the discrete-time linear system:

$$x(k+1) = \begin{bmatrix} 1 & 1 \\ 0 & 1 \end{bmatrix} x(k) + \begin{bmatrix} 0.5 \\ 1 \end{bmatrix} u(k),$$

$$y(k) = \begin{bmatrix} 1 & 0 \end{bmatrix} x(k),$$

with the state vector $x(k) \in \mathbb{R}^2$ and scalar control input $u(k) \in \mathbb{R}$.

The MPC problem is defined with the following parameters:

- Prediction horizon: $p = 3$
- State weighting: $Q = I_2$
- Input weighting: $R = 0.1$
- Terminal weight Q_t obtained from the discrete-time LQR solution for (Q, R).

The enforced constraints are:

$$-1 \leq v_i \leq 1, \quad i = 0, 1, 2$$
$$-5 \leq (z_1)_i \leq 5, \quad i = 1, 2, 3$$

(a) Write the finite-horizon MPC problem in the standard condensed QP form

$$\min_{\mathcal{V}} \frac{1}{2}\mathcal{V}^T H \mathcal{V} + z_0^T G \mathcal{V},$$

where $\mathcal{V} = [v_0, v_1, v_2]^T$. Clearly define H and G in terms of the system matrices.

(b) Compute the unconstrained optimal control law

$$u(k) = Lx(k),$$

and determine the region of $x(k)$ for which this law satisfies all constraints.

(c) Determine the polyhedral regions in the state space corresponding to:
- Unconstrained control
- Upper input saturation
- Lower input saturation

Include the effect of state constraints on these regions.

(d) Starting from $x(0) = [4, 0]^T$, simulate the closed-loop system using the explicit MPC law and verify constraint satisfaction.

(e) Compare the explicit MPC solution with an online QP-based MPC solution for the same parameters. Discuss when they differ.

10

System Identification

The cardinal sin in control is to believe that the plant is given.

– Karl Åström

Identifying process dynamics is often the most time-consuming and resource-intensive step in implementing a model predictive controller. It typically demands significant user expertise and careful experimentation. This chapter provides an introduction to various identification methods and highlights key practical considerations. Given that system identification is a broad and deep subject, worthy of a dedicated volume, our aim here is to offer a high-level overview and a foundation for further study. Hence, the treatment of individual methods will be brief and informal. Additional references are provided in the text for readers interested in more comprehensive and rigorous discussions.

10.1 Problem Overview

The goal of system identification is to develop a model that predicts system behavior based on input–output data collected from the process, as illustrated in Figure 10.1. For practical reasons, the search is often restricted to linear models. Both known and unknown inputs can influence the outputs. Since not all inputs vary in a deterministic fashion, it is often beneficial to identify models that include both deterministic and stochastic components. A general model structure used in system identification is

$$y(k) = G(q)u(k) + H(q)\varepsilon(k), \tag{10.1}$$

where $\varepsilon(k)$ is a zero-mean white noise sequence. Various standard model structures for $G(q)$ and $H(q)$ will be introduced in Section 10.2.

System identification methods can be broadly categorized into two classes based on how input–output data are translated into mathematical models: parametric and nonparametric identification. Parametric identification starts with a predefined, typically compact, model structure characterized by a finite set of parameters, which are then estimated from the data. In nonparametric identification, little to no assumption is made about the underlying model structure. A typical example of the latter is finite impulse response (FIR) modeling, introduced in Chapter 2. Note that FIR modeling may also be interpreted as a special case of parametric identification when the impulse response is assumed to have finite length.

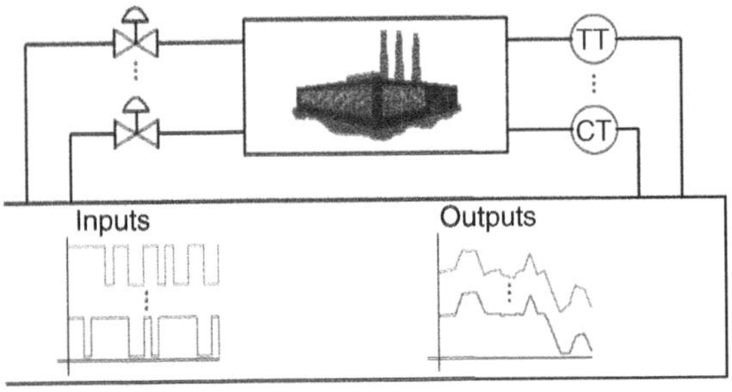

Figure 10.1 Conceptual illustration of system identification.

It is essential to keep in mind the ultimate purpose of the model—in our case, the design of a feedback control system. The accuracy of a model should not be evaluated in isolation, but in terms of how well it predicts output behavior when used within the intended closed-loop control framework. This objective should guide all phases of the identification process, including test input design, data preprocessing, model structure selection, and parameter estimation.

In this chapter, we will first introduce common model structures and parametric identification methods, including the prediction error method (PEM) and statistical approaches such as maximum likelihood estimation (MLE) and Bayesian estimation. We will also discuss two nonparametric identification techniques: frequency response identification and subspace identification.

10.2 Model Structures

10.2.1 Finite Impulse Response (FIR) Model

Recall the FIR model introduced earlier in Eq. (2.2) of Chapter 2:

$$y(k) = \sum_{i=1}^{n} h_i u(k-i) + e(k),$$

where $e(k)$ represents the aggregated model error.

The FIR model belongs to a broader class of nonparametric models known as *convolution models*, which take the form:

$$y(k) = \sum_{i=-\infty}^{\infty} h_i u(k-i). \tag{10.2}$$

For strictly causal systems, $h_i = 0$ for all $i \leq 0$. Moreover, if the system is stable, the impulse response coefficients decay over time. This allows us to approximate the convolution sum by truncating it after a finite number of terms, i.e., $h_i \approx 0$ for $i > N$, leading to the FIR model in (2.2).

10.2.2 Structures for Parametric Identification

Various discrete-time linear time-invariant (LTI) models were introduced in Chapter 5. Recall the example of difference equation of the form (5.10):

$$y(k) = \sum_{i=1}^{n} (-\alpha_i) y(k-i) + \sum_{i=1}^{m} \beta_i u(k-i).$$

Using the forward shift operator in the time domain, defined as $q\{y(k)\} = y(k+1)$, this equation can be expressed more compactly as

$$A(q)y(k) = B(q)u(k)$$
$$\text{i.e., } y(k) = G(q)u(k).$$

To account for modeling errors, disturbances, and measurement noise, an additive residual term is introduced:

$$y(k) = G(q)u(k) + w_y(k).$$

Thus, the general structure for parametric identification is given by

$$y(k) = G(q,\theta)u(k) + H(q,\theta)\varepsilon(k), \tag{10.3}$$

where θ is the vector of *model parameters.*

The *process model* $G(q,\theta)$ represents the causal relationship between the deterministic input $u(k)$ and the output $y(k)$. Although $u(k)$ typically denotes a manipulated input, it may also include measured disturbances. $\varepsilon(k)$ is a zero-mean white noise sequence. Together with the *noise model* $H(q,\theta)$, it characterizes the statistical properties—such as the autocorrelation and cross-correlation—of the residual sequence $y(k) - G(q,\theta)u(k)$.

For a stationary process, the noise model $H(q,\theta)$ is typically assumed—without loss of generality—to be a stable, stably invertible, and normalized transfer function (i.e., $H(\infty,\theta) = 1$). This assumption is justified by the *spectral factorization theorem*, which states that any power spectrum can be factorized into a stable and stably invertible component (Åström and Wittenmark, 1990). For processes that exhibit random mean shifts (nonstationary behavior), it becomes necessary for the noise model to include an integrator. An equivalent approach is to reformulate the identification problem using the differenced signals: replacing $y(k)$ and $u(k)$ with $\Delta y(k) = y(k) - y(k-1)$ and $\Delta u(k) = u(k) - u(k-1)$, respectively.

Within the general structure, various parameterizations are possible. We begin by discussing some of the most commonly used ones in the single-input, single-output (SISO) context.

10.2.2.1 ARX Model

If we model the output as a linear function of past inputs and outputs with an additive noise term, we obtain

$$y(k) + a_1 y(k-1) + \cdots + a_n y(k-n) = b_1 u(k-1) + \cdots + b_m u(k-m) + \varepsilon(k). \tag{10.4}$$

When the residual term $\varepsilon(k)$, called *equation error*, is modeled as a white noise sequence, the resulting structure is known as the ARX model—short for autoregressive with eXogenous input.

The ARX model corresponds to the following parameterization of the process and noise models:

$$\begin{aligned} G(q,\theta) &= \frac{B(q)}{A(q)} \triangleq \frac{b_1 q^{-1} + \cdots + b_m q^{-m}}{1 + a_1 q^{-1} + \cdots + a_n q^{-n}} \\ H(q,\theta) &= \frac{1}{A(q)} \triangleq \frac{1}{1 + a_1 q^{-1} + \cdots + a_n q^{-n}}, \end{aligned} \tag{10.5}$$

In general, an ARX model of relatively high order may be required to adequately capture both the process and noise dynamics. To understand why, recall the general structure (10.3) can be written as

$$H^{-1}(q,\theta)y(k) = H^{-1}(q,\theta)G(q,\theta)u(k) + \varepsilon(k). \tag{10.6}$$

Assuming $H^{-1}(q,\theta)$ is stable, and that $G(q,\theta)$ is also stable, we can approximate the left-hand side and the input-dependent term as

$$\begin{aligned} H^{-1}(q,\theta) &\approx 1 + a_1 q^{-1} + \cdots + a_n q^{-n} \\ H^{-1}(q,\theta)G(q,\theta) &\approx 1 + b_1 q^{-1} + \cdots + b_m q^{-m}, \end{aligned} \tag{10.7}$$

for sufficiently large n and m.

10.2.2.2 ARMAX Model

A natural extension of the ARX structure is to model the equation error as a moving average of white noise:

$$\begin{aligned} y(k) + a_1 y(k-1) + \cdots + a_n y(k-n) = b_1 u(k-1) + \cdots + b_m u(k-m) \\ + \varepsilon(k) + c_1 \varepsilon(k-1) + \cdots + c_\ell \varepsilon(k-\ell). \end{aligned} \tag{10.8}$$

This is known as the AutoRegressive Moving Average with eXogeneous inputs (ARMAX) model. Here, the noise transfer function is parameterized as

$$H(q,\theta) = \frac{C(q)}{A(q)} \triangleq \frac{1 + c_1 q^{-1} + \cdots + c_\ell q^{-\ell}}{1 + a_1 q^{-1} + \cdots + a_n q^{-n}} \tag{10.9}$$

Due to the added numerator in the noise model, an ARMAX model can represent both process and noise dynamics more compactly than an ARX model. In fact, any linear state-space system of order n (driven by deterministic and white noise inputs) admits an equivalent ARMAX representation of order n. However, this added flexibility comes at a cost: parameter estimation becomes more challenging, and overparameterization can lead to a *loss of identifiability*—that is, multiple parameter sets may produce identical input–output behavior, resulting in nonunique model representations.

10.2.2.3 Output Error Model

Both ARX and ARMAX models place common poles in G and H. In some cases, however, it may be more appropriate to model them separately. One such parametrization is the output error (OE) structure, defined as

$$\begin{aligned} \tilde{y}(k) + a_1 \tilde{y}(k-1) + \cdots + a_n \tilde{y}(k-n) &= b_1 u(k-1) + \cdots + b_m u(k-m) \\ y(k) &= \tilde{y}(k) + \varepsilon(k). \end{aligned} \tag{10.10}$$

Here, $\tilde{y}(k)$ represents the noise-free output. The disturbance $\varepsilon(k)$ is typically assumed to be white noise. Under this assumption, the OE structure corresponds to the transfer functions:

$$G(q,\theta) = \frac{B(q)}{A(q)} \quad \text{and} \quad H(q) = 1. \tag{10.11}$$

At first glance, this may seem limiting, as in many real-world processes, disturbances are autocorrelated and not adequately represented by white noise. However, the OE structure can be generalized. For example, when $H(q)$ is not equal to 1 but is known *a priori* (i.e., the residual sequence is a colored noise with a known spectrum), we can rewrite the model as

$$\underbrace{H^{-1}(q)y(k)}_{y_f(k)} = G(q,\theta)\underbrace{H^{-1}u(k)}_{u_f(k)} + \varepsilon(k). \tag{10.12}$$

This form fits the standard OE structure as in (10.10), after prefiltering the input and output to decorrelate the noise.

One complication of the OE model is that the terms $\tilde{y}(k)$ are not directly measurable and depend implicitly on the model parameters. As a result, parameter estimation becomes more complex than in the ARX model.

Remark 10.1: The FIR model may be considered a special case of the ARX or OE models with $A(q) = 1$. Consequently, FIR models can also be treated as parametric models.

10.2.2.4 Box–Jenkins Model

A natural generalization of the OE model is to allow the disturbance transfer function to be a rational function with unknown parameters. This leads to the Box–Jenkins (BJ) model, which has the form:

$$y(k) = \frac{B(q)}{A(q)}u(k) + \frac{C(q)}{D(q)}\varepsilon(k). \tag{10.13}$$

This structure is quite general, as it allows separate and flexible modeling of both the process and noise dynamics. However, the parameter estimation problem becomes nonlinear, and identifiability issues can arise.

10.2.2.5 General Model

The most general model structure takes the form:

$$A(q)y(k) = \frac{B(q)}{F(q)}u(k) + \frac{C(q)}{D(q)}\varepsilon(k). \tag{10.14}$$

Here, only a subset of the poles is shared between the process and noise models. This formulation encompasses all previously discussed models as special cases.

10.2.3 Key Issues in Parametric Models

10.2.3.1 Integrator

The models discussed above assume that ε is a white noise sequence. However, when persistent errors are present, a stable and stably invertible noise model $H(q)$ must include an integrator to capture the non-decaying behavior. All of the model structures can be generalized to accommodate this by including an integrator in $H(q,\theta)$.

For example, the ARMAX model can be extended as follows

$$y(k) = \frac{B(q)}{A(q)}u(k) + \frac{1}{1-q^{-1}}\frac{C(q)}{A(q)}\varepsilon(k), \tag{10.15}$$

which is referred to as the ARIMAX model (where "I" stands for "integrated"). From a parameter estimation standpoint, this formulation is equivalent to

$$\Delta y(k) = \frac{B(q)}{A(q)}\Delta u(k) + \frac{C(q)}{A(q)}\varepsilon(k), \tag{10.16}$$

where $\Delta y(k) = y(k) - y(k-1)$ and $\Delta u(k) = u(k) - u(k-1)$. The same approach applies to the other model structures as well.

10.2.3.2 Model Identifiability

Model identifiability refers to the ability of an identification procedure to determine a unique set of model parameters that accurately represent the true system. Loss of identifiability can occur for two main reasons:

Structural nonuniqueness: The model structure may inherently allow multiple parameter sets θ to yield identical input–output behavior. This is often a result of overparameterization or redundant parameterization.

Insufficient data excitation: The input–output data may lack the richness needed to distinguish between competing models. In such cases, even a well-structured model may not be uniquely identifiable due to lack of informative data.

The first issue is tied to the mathematical form of the model, while the second depends on the excitation level and information content of the collected data. Both must be addressed to ensure reliable identification.

Example 10.1: *Loss of Identifiability for an ARMAX System.*
Certain model structures can suffer from parameter nonuniqueness when overparameterized—a phenomenon known as "loss of identifiability."

Consider a system governed by the ARMAX model:

$$y(k) = \frac{\beta_1 q^{-1}}{1 + \alpha_1 q^{-1}} u(k) + \frac{1 + \zeta_1 q^{-1}}{1 + \alpha_1 q^{-1}} \varepsilon(k). \tag{10.17}$$

Suppose we attempt to fit an overparameterized ARMAX structure:

$$y(k) = \frac{b_1 q^{-1} + b_2 q^{-2}}{1 + a_1 q^{-1} + a_2 q^{-2}} u(k) + \frac{1 + c_1 q^{-1} + c_2 q^{-2}}{1 + a_1 q^{-1} + a_2 q^{-2}} \varepsilon(k). \tag{10.18}$$

This model can be reparameterized as

$$y(k) = \frac{\beta_1 q^{-1}(1 + \beta_2 q^{-1})}{(1 + \alpha_1 q^{-1})(1 + \alpha_2 q^{-1})} u(k) + \frac{(1 + \zeta_1 q^{-1})(1 + \zeta_2 q^{-1})}{(1 + \alpha_1 q^{-1})(1 + \alpha_2 q^{-1})} \varepsilon(k), \tag{10.19}$$

where the constraints $\beta_2 = \alpha_2$ and $\zeta_2 = \alpha_2$ ensure equivalence with the real system in Eq. (10.17). This condition implies that infinitely many parameter sets can reproduce the same input–output behavior as the real system, thereby resulting in a loss of identifiability.

Similar issues can arise in OE and BJ structures. In contrast, the ARX structure avoids such ambiguity because it lacks a noise model with poles that can cancel with those in the numerator.

10.2.3.3 Multivariable System

The model structures described earlier extend naturally to multivariable systems by generalizing $A(q), B(q), \cdots$, to matrix-valued functions. For example,

$$A(q) = I + A_1 q^{-1} + \cdots + A_n q^{-n}.$$

Extensions to multivariable systems are mostly straightforward but can introduce additional complexity. In this setting, $A(q)$, $B(q)$, etc., are matrix polynomials of appropriate dimensions. For example, the multiple-input, multiple-output (MIMO) ARX model takes the form:

$$y(k) = A^{-1}(q)B(q)u(k) + A^{-1}(q)\varepsilon(k), \tag{10.20}$$

where $A(q)$ and $B(q)$ are $n_y \times n_y$ and $n_y \times n_u$ matrix polynomials, respectively.

One approach is to fit each individual SISO model—using one of the model structures described above—by varying one input at a time. Alternatively, experiments can be performed by varying all inputs simultaneously, followed by fitting each output independently. This is called *multiple-input, single-output (MISO) identification*. In this case, the only modification is that $B(q)$ becomes a row vector consisting of n_u polynomials, where n_u is the number of inputs. The parameter estimation problem remains structurally the same, differing only in the number of parameters.

However, SISO or MISO identification cannot capture the shared dynamics and cross-correlations among outputs. To account for these interactions, MIMO identification must be employed, wherein all outputs are simultaneously fitted within a single multivariable model.

Parameterizing the coefficient matrices requires careful consideration. Treating all matrix entries as independent unknowns can lead to overparameterization, which can cause *loss of identifiability*. In general, obtaining a parsimonious model that yields a unique set of parameter values requires significant prior knowledge, such as the observability indices of the outputs. Starting with an identifiable model structure is especially important because many of the parametric model structures lead to nonlinear parameter estimation problems. Nonlinear estimation can suffer from numerical problems when the parameter values are not uniquely determined.

10.3 Parametric Identification Methods

In parametric identification, the model structure is selected before the parameter estimation process. The model parameters are then estimated using the available input–output data. If the identification dataset reflects the intended closed-loop conditions, this approach leads to models that are well-suited for control design. Although a specific structure is assumed during estimation, the model can be refined iteratively based on the quality of fit, through residual analysis or other validation techniques.

The key results are summarized in this section. For a more comprehensive treatment of parametric identification methods across different model structures, refer to Ljung (1998) and Soderstrom and Stoica (1989).

10.3.1 Prediction Error Method

Consider the system given in Eq. (10.3):

$$y(k) = G(q, \theta)u(k) + H(q, \theta)\varepsilon(k),$$

where $G(q, \theta)$ and $H(q, \theta)$ are rational transfer functions parameterized by θ, and $\varepsilon(k)$ is a white noise sequence. The optimal one-step-ahead predictor of $y(k)$ under this model is given by

$$\hat{y}(k|k-1) = G(q, \theta)u(k) + \left(I - H^{-1}(q, \theta)\right)(y(k) - G(q, \theta)u(k)), \tag{10.21}$$

which uses only past values of y and u, as $(I - H^{-1}(q, \theta))$ contains at least one delay (since $I - H^{-1}(\infty, \theta) = 0$). The prediction error is defined as

$$e_{\text{pred}}(k) \triangleq y(k) - \hat{y}(k|k-1). \tag{10.22}$$

By comparing Eqs. (10.3) and (10.21), we see that the prediction error simplifies to $\varepsilon(k)$, confirming that (10.21) is indeed the optimal one-step predictor under the assumed model structure.

Since the central role of the model is to predict future outputs, a natural identification objective is to select the parameter vector θ that minimizes the sum of squared prediction errors over the available data. The parameter estimation problem is then formulated as

$$\arg\min_{\theta} \sum_{k=1}^{N} \|e_{\text{pred}}(k, \theta)\|_2^2, \tag{10.23}$$

where $\|\cdot\|_2$ denotes the Euclidean norm. While other norms can be used, the 2-norm is most common due to its desirable statistical properties. Using Eq. (10.21), the prediction error can be explicitly written as

$$e_{\text{pred}}(k, \theta) = H^{-1}(q, \theta)\,(y(k) - G(q, \theta)u(k)). \tag{10.24}$$

This approach to parameter estimation is known as the *Prediction Error Method* (PEM). Developed by Lennart Ljung and collaborators, PEM is described in detail in Ljung (1998) and is implemented in MATLAB's System Identification Toolbox. The numerical complexity of PEM depends on the chosen model structure, which we describe below with examples.

10.3.1.1 Linear Least-squares Cases

For certain model structures, minimizing the 2-norm of the prediction error results in a linear least-squares (LS) problem. An example is the ARX structure, where $G(q, \theta) = \frac{B(q)}{A(q)}$ and $H(q, \theta) = \frac{1}{A(q)}$. The prediction error becomes

$$\begin{aligned} e_{\text{pred}}(k, \theta) &= A(q)y(k) - B(q)u(k) && (10.25)\\ &= y(k) + a_1 y(k-1) + \cdots + a_n y(k-n) \\ &\quad - b_1 u(k-1) - \cdots - b_m u(k-m). && (10.26) \end{aligned}$$

Since $e_{\text{pred}}(k, \theta)$ is linear in the unknown parameters, minimizing $\sum_k e^2_{\text{pred}}(k, \theta)$ is a standard linear LS problem.

Example 10.2: *Use the PEM to Identify the Parameters of an ARX Model.*
Consider the following ARX model:

$$y(k) - 0.54y(k-1) + 0.14y(k-2) = 0.2u(k-1) + 0.1u(k-2) + \varepsilon(k),$$

where $\varepsilon(k)$ is zero-mean white noise with variance 0.1^2. The prediction error (10.26) becomes

$$e_{\text{pred}}(k) = y(k) - \underbrace{\begin{bmatrix} -y(k-1) & -y(k-2) & u(k-1) & u(k-2) \end{bmatrix}}_{\phi^T(k)} \underbrace{\begin{bmatrix} a_1 \\ a_2 \\ b_1 \\ b_2 \end{bmatrix}}_{\theta}.$$

We generate the test input signal as a uniformly distributed random input signal. The true plant is simulated using the input signal, resulting in the dataset $\{u(1), \dots, u(N), y(1), \dots, y(N)\}$. The data can be stacked as Y_N and Φ_N, and the parameter vector θ can be obtained using linear LS.

MATLAB Code and Results: This linear LS is a non-iterative process. The MATLAB code implementing the PEM for an ARX model is available on GitHub at https://github.com/Lee-MPC-Book

The identified parameter values for one realization yield the following:

$$y(k) - 0.545y(k-1) + 0.168y(k-2) = 0.198u(k-1) + 0.102u(k-2) + \varepsilon(k).$$

Remark 10.2: In practical applications, the true process model is unknown. The identified model can be validated by comparing its predicted outputs with the measured outputs from an independent test dataset and by assessing the whiteness of the resulting residuals.

Some other model structures—such as FIR model with known disturbance characteristics—also result in a linear LS problem. When $G(q,\theta) = \sum_{i=1}^{n} h_i q^{-i}$ and $H(q)$ contains no unknown parameters, we can write:

$$e_{\text{pred}}(k,\theta) = y_f(k) - h_1 u_f(k-1) - \cdots - h_n u_f(k-n), \tag{10.27}$$

where $y_f(k) = H^{-1}(q)y(k)$ and $u_f(k) = H^{-1}(q)u(k)$. The dependence on θ is linear, and PEM reduces to a linear LS problem. The same formulation applies to Laguerre or other orthogonal basis expansions.

10.3.1.2 Nonlinear LS Cases

For other model structures, such as ARMAX or BJ, the PEM becomes a nonlinear LS problem. For example, for an *ARMAX model*, $G(q,\theta) = \dfrac{B(q)}{A(q)}$ and $H(q,\theta) = \dfrac{C(q)}{A(q)}$. The prediction error becomes

$$e_{\text{pred}}(k,\theta) = \frac{A(q)}{C(q)}y(k) - \frac{B(q)}{C(q)}u(k). \tag{10.28}$$

The resulting problem is a nonlinear regression problem. A *pseudo-linear regression* (PLR) approach can be used as an approximation. The optimal predictor can be written as

$$\hat{y}(k|k-1) = B(q)u(k) + [1 - A(q)]y(k) + [C(q) - 1][y(k) - \hat{y}(k|k-1)]. \tag{10.29}$$

The derivation is provided in an exercise at the end of the chapter. This recursive structure is known as the pseudo-linear regression form of the ARMAX model. As the prediction $\hat{y}(k|k-1)$ appears on both sides, this form is not truly linear. However, past predictions (based on previous parameter estimates) can be treated as fixed, and the parameters can be updated iteratively via linear LS. This results in an approximate, recursive estimation scheme.

Example 10.3: *Pseudo-Linear Regression for ARMAX Model.*

Consider a simple ARMAX system, as introduced in Example 10.1 of Eq. (10.17):

$$y(k) = \frac{\beta_1 q^{-1}}{1 + \alpha_1 q^{-1}}u(k) + \frac{1 + \zeta_1 q^{-1}}{1 + \alpha_1 q^{-1}}\varepsilon(k).$$

The one-step-ahead predictor for this system is given by

$$\hat{y}(k|k-1) = \frac{\zeta_1 - \alpha_1}{1 + \zeta_1 q^{-1}}q^{-1}y(k) + \frac{\beta_1}{1 + \zeta_1 q^{-1}}q^{-1}u(k). \tag{10.30}$$

This expression can be derived by substituting the ARMAX model (10.17) into the general prediction formula (10.21).

Rewriting this expression in an explicit regression form, we obtain:

$$\begin{aligned}\hat{y}(k|k-1) &= -\alpha_1 y(k-1) + \beta_1 u(k-1) + \zeta_1 e(k-1) \\ &= \begin{bmatrix} -y(k-1) & u(k-1) & e(k-1) \end{bmatrix} \begin{bmatrix} \alpha_1 \\ \beta_1 \\ \zeta_1 \end{bmatrix}.\end{aligned} \tag{10.31}$$

While this appears to be a linear regression model, it is not strictly linear in the parameters, because the term $e(k-1) = y(k-1) - \hat{y}(k-1|k-2)$ itself depends on the parameter vector $\theta = [\alpha_1, \beta_1, \zeta_1]^T$.

In pseudo-linear regression, the past prediction errors $e(k-1)$ are treated as known quantities, based on the parameter estimates obtained from the previous iteration. The parameters are then iteratively updated using linear LS estimation, effectively treating the recursive relationship as if it were exact.

10.3.2 Properties of Linear LS Identification

We have seen that for many model structures, the PEM can be formulated as a linear regression problem. The general form of such a problem is

$$y(k) = \phi^T(k)\theta + e(k), \tag{10.32}$$

where y is the observed (or filtered) output, ϕ is the regressor vector constructed from past inputs and/or outputs, θ is the unknown parameter vector to be estimated, and e is the residual error, which depends on the choice of θ. In LS identification, data is collected over N sample intervals, and the parameter estimate θ_N^{LS} is obtained by minimizing the sum of squared residuals:

$$\theta_N^{LS} = \arg\min_{\theta} \sum_{k=1}^{N} e^2(k). \tag{10.33}$$

For such problems, the linear regression model (10.32) over N samples can be expressed compactly as

$$\hat{Y}_N = \Phi_N \theta + E_N \tag{10.34}$$

$$\text{where,} \quad \hat{Y}_N = \begin{bmatrix} y(1) \\ y(2) \\ \vdots \\ y(N) \end{bmatrix} \quad \Phi_N = \begin{bmatrix} \phi^T(1) \\ \phi^T(2) \\ \vdots \\ \phi^T(N) \end{bmatrix} \quad E_N = \begin{bmatrix} e(1) \\ e(2) \\ \vdots \\ e(N) \end{bmatrix}. \tag{10.35}$$

The LS solution that minimizes the squared error is then given by

$$\hat{\theta}_N^{LS} = (\Phi_N^T \Phi_N)^{-1} \Phi_N^T Y_N \tag{10.36}$$

provided that $\Phi_N^T \Phi_N$ is nonsingular. This solution is the *best linear unbiased estimate* (BLUE) under the assumption that $e(k)$ is zero-mean, uncorrelated, and has constant variance.

Example 10.4: *Identification of a Linear ARMAX Model.*
We use MATLAB's `System Identification Toolbox` to identify an ARMAX model for the system in Example 2.7. The true system is

$$G(s) = \frac{0.5}{19.36s^2 + 7.04s + 1}. \tag{10.37}$$

The output measurements are corrupted with zero-mean Gaussian white noise of variance 0.04^2. For further details on system setup, refer to Example 2.7.

MATLAB Code and Results: The MATLAB code for ARMAX identification using the SysId toolbox is available on GitHub at https://github.com/Lee-MPC-Book

An ARMAX model with model order $n_a = 2, n_b = 2, n_c = 2$ was found to best fit the data. Figure 10.2 compares the impulse response of the original system with that of the identified model. Measurement noise (`e`) with variance 0.0016 was added, approximately one order of magnitude smaller than the output signal variance, to simulate a realistic noise setting.

A few key points in using the System Identification Toolbox are

- We use `idinput` to generate a pseudo-random binary sequence (PRBS) as the input excitation. PRBS is discussed in detail in Section 10.6.2.
- The input–output dataset—`allData`—is constructed using `iddata` command.
- The ARMAX model is identified using

```
sys=armax(allData,[2 2 2 1],opt);
```

- The command `armax` uses an *iterative* nonlinear optimization routine, with ARX identification used to obtain an initial model.
- The optimization typically converges in four to six iterations for this example. These iterations can be tracked using `opt.Display='on';` option.

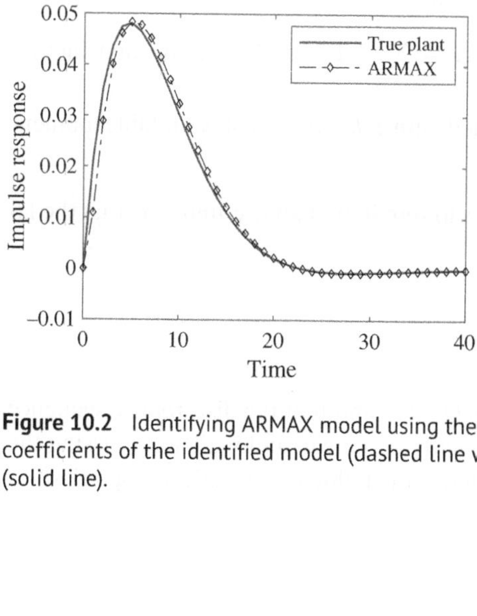

Figure 10.2 Identifying ARMAX model using the System Identification Toolbox. The impulse response coefficients of the identified model (dashed line with circles) closely match the true FIR coefficients (solid line).

Remark 10.3:

1. An ARX model can be identified in MATLAB using the `arx` function. ARX model estimation is a linear regression problem and does not require iterative optimization.
2. The SysID toolbox uses nonlinear optimization to estimate parameters of ARMAX, which are more efficient than pseudo-linear regression.
3. The identified model includes an estimate of the noise variance, along with statistical information such as parameter standard deviations. These aspects will be discussed in a later section.

10.3.2.1 Convergence

To discuss the convergence of the LS estimate, let us assume that the underlying system (from which the data are collected) is described by

$$y(k) = \phi^T(k)\theta_o + \varepsilon(k), \tag{10.38}$$

where θ_o is the true parameter vector and $\varepsilon(k)$ represents disturbances, noise, and other unmodeled effects.

We can gain insight into the behavior of the LS estimate by rewriting the solution in the following form:

$$\begin{aligned} \hat{\theta}_N^{\mathrm{LS}} &= \left[\frac{1}{N}\sum_{k=1}^{N}\phi(k)\phi^T(k)\right]^{-1}\frac{1}{N}\sum_{k=1}^{N}\phi(k)\left[\phi^T(k)\theta_o + \varepsilon(k)\right] \\ &= \theta_o + \left[\frac{1}{N}\sum_{k=1}^{N}\phi(k)\phi^T(k)\right]^{-1}\frac{1}{N}\sum_{k=1}^{N}\phi(k)\varepsilon(k). \end{aligned} \tag{10.39}$$

This expression shows that the estimation error arises from the second term:

$$\left[\frac{1}{N}\sum_{k=1}^{N}\phi(k)\phi^T(k)\right]^{-1}\frac{1}{N}\sum_{k=1}^{N}\phi(k)\varepsilon(k).$$

A key property of the LS estimator is that, under mild conditions, $\hat{\theta}_N^{\mathrm{LS}}$ converges to the true parameter vector θ_o as $N \to \infty$. This convergence holds provided:

- The regressors $\phi(k)$ are *persistently exciting*, so that $\frac{1}{N}\sum_{k=1}^{N}\phi(k)\phi^T(k)$ remains nonsingular and converges to a positive definite matrix.
- The noise sequence $\varepsilon(k)$ is zero-mean, uncorrelated with $\phi(k)$, and satisfies suitable moment conditions (e.g., bounded variance).

Under these assumptions, the second term converges to zero in probability, implying that the LS estimate is consistent:

$$\hat{\theta}_N^{\mathrm{LS}} \to \theta_o \quad \text{as} \quad N \to \infty.$$

10.3.2.2 Statistical Properties

Assume again that the true system is described by Eq. (10.38), and that the disturbance sequence $\varepsilon(k)$ is independent and identically distributed (i.i.d.) with zero mean and variance r_ε. Using the expression for the LS estimate in Eq. (10.39), we can derive the following statistical properties:

Unbiasedness:

Taking the expectation of the estimation error:

$$E\{\hat{\theta}_N^{\text{LS}} - \theta_0\} = E\left\{\frac{1}{N}\left(\sum_{k=1}^{N} \phi(k)\phi^T(k)\right)^{-1} \frac{1}{N}\sum_{k=1}^{N} \phi(k)\varepsilon(k)\right\} \\ = 0. \tag{10.40}$$

This confirms that the LS estimator is unbiased under the stated assumptions.

Covariance:

$$E\left\{(\hat{\theta}_N^{\text{LS}} - \theta_o)(\hat{\theta}_N^{\text{LS}} - \theta_o)^T\right\} \\ = \left(\frac{1}{N}\sum_{k=1}^{N} \phi(k)\phi^T(k)\right)^{-1} \left(\frac{1}{N^2}\sum_{k=1}^{N} \phi(k)r_\varepsilon\phi^T(k)\right)\left(\frac{1}{N}\sum_{k=1}^{N} \phi(k)\phi^T(k)\right)^{-1} \\ = \left(\frac{1}{N}\sum_{k=1}^{N} \phi(k)\phi^T(k)\right)^{-1} \frac{r_\varepsilon}{N} \\ = r_\varepsilon(\Phi_N^T\Phi_N)^{-1}. \tag{10.41}$$

This expression quantifies the uncertainty in the parameter estimate and can be used to construct confidence intervals.

Confidence Regions:

Assuming a normal distribution for the estimator, confidence regions for the parameter vector θ can be constructed as ellipsoids. For a given confidence level, the boundary of this ellipsoid is determined using the chi-squared (χ^2) distribution. This allows for hypothesis testing and model validation based on statistical significance.

10.3.3 Persistency of Excitation

In linear LS identification, convergence of the parameter estimates to their true values in the presence of noise requires that:

$$\text{rank}\left\{\lim_{N\to\infty} \frac{1}{N}\sum_{k=1}^{N} \phi(k)\phi^T(k)\right\} = \dim\{\phi\}. \tag{10.42}$$

This condition is closely related to the concept of *persistency of excitation*. A signal $u(k)$ is said to be *persistently exciting of order n* if:

$$\text{rank}\{C_u^n\} = n, \tag{10.43}$$

where

$$C_u^n = \lim_{N\to\infty} \frac{1}{N}\sum_{k=1}^{N} \begin{bmatrix} u(k-1)u(k-1) & u(k-1)u(k-2) & \cdots & u(k-1)u(k-n) \\ u(k-2)u(k-1) & u(k-2)u(k-2) & \cdots & u(k-2)u(k-n) \\ \vdots & \ddots & \ddots & \vdots \\ u(k-n)u(k-1) & u(k-n)u(k-2) & \cdots & u(k-n)u(k-n) \end{bmatrix}. \tag{10.44}$$

This is equivalent to requiring that the power spectrum of $u(k)$ be nonzero at n or more distinct frequency points between $-\pi$ and π.

Now, suppose $\phi(k)$ consists of past inputs and outputs. A necessary and sufficient condition for (10.42) to hold is

> The input is persistently exciting of order $\dim\{\phi\}$.

This condition follows trivially when $\phi(k)$ consists only of n past inputs, as in FIR models. In that case,

$$\lim_{N\to\infty} \frac{1}{N}\sum_{k=1}^{N} \phi(k)\phi^T(k) = C_u^n. \tag{10.45}$$

The condition also holds when $\phi(k)$ includes filtered past inputs $u_f(k-1), \ldots, u_f(k-n)$, where $u_f(k) = H^{-1}(q)u(k)$. Note that:

$$\Phi_{u_f}(\omega) = \frac{\Phi_u(\omega)}{|H(e^{j\omega})|^2}. \tag{10.46}$$

Thus, if $u(k)$ is persistently exciting of order n, then $u_f(k)$ is as well. It is less obvious, but still true, that this condition holds even when $\phi(k)$ includes output terms (as in the ARX model). A formal proof is left as an exercise at the end of this chapter.

An important conclusion is that to ensure convergence of parameter estimates to their true values, the input signal $u(k)$ must be designed to be persistently exciting of order $\dim\{\theta\}$. The following examples illustrate this idea:

- A pulse input is not persistently exciting, as the rank of C_u^1 is zero.
- A step input is persistently exciting of order 1.
- A single-step test is inadequate in the presence of significant noise, since only one parameter (the steady-state gain) can be identified without error.
- A sinusoidal input is persistently exciting of order 2, as its spectrum is nonzero at two distinct frequency points.
- A random signal can be persistently exciting of any order, provided its spectrum is nonzero over a frequency interval.
- A periodic signal with period n can be persistently exciting of at most order n.

Failure to satisfy the persistency of excitation condition does not imply that parameter estimation is impossible. Rather, it means that the estimates will not converge to the true parameter values, no matter how much data is collected.

Example 10.5: *Importance of Persistent Excitation: Step Tests Lead to Poor Model Performance.* We now repeat Example 10.4 using a step input signal instead of PRBS. The step input is applied at time 0 using:

```
v=ones(nSamp,1);
```

The rest of the code remains the same as that in Example 10.4. MATLAB gives the following warning when we try FIR identification:

```
Warning: Matrix is close to singular or badly scaled.
Results may be inaccurate.
```

Both FIR and ARMAX models identified using this step-response data exhibit very poor performance.

A user might naïvely assume that a step test, as sometimes suggested by MATLAB, could yield reasonable models. For instance, one could try

```
v=[zeros(100,1); ones(nSamp-100,1)];
```

With this input, MATLAB no longer produces the singularity warning.

We conducted 50 Monte Carlo simulations, identifying FIR and ARMAX models using data generated from this single-step input. Many of the resulting models performed arbitrarily poorly. Figure 10.3 shows one of the better-performing realizations for both FIR and ARMAX models.

Remark 10.4: The matrix C_u^n computed using Eq. (10.44) is ill-conditioned. Although step tests are inadequate for accurate dynamic modeling, they may still be useful for estimating time delays and steady-state gains.

10.3.4 Frequency-domain Bias Distribution Under PEM

The earlier discussion on parameter convergence assumed the existence of a "true" parameter vector. However, even when parameter estimates converge to their best possible values (in a LS sense),

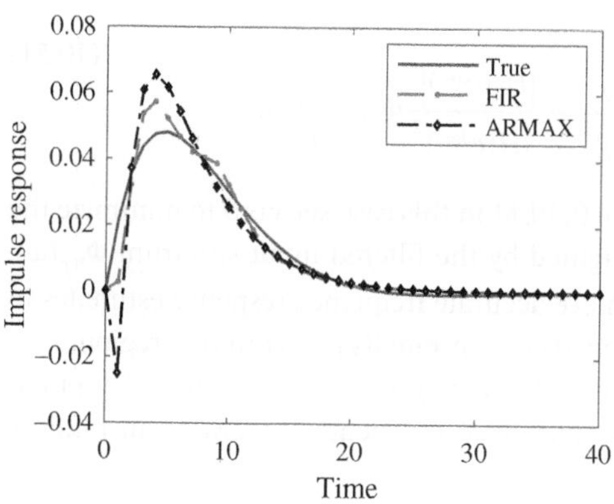

Figure 10.3 Identification of FIR and ARMAX models using a single realization of data from a one-step input applied at time $k = 100$.

the resulting model may still exhibit significant *bias* relative to the true system if the chosen model structure is insufficiently rich. For instance, an FIR model with too few coefficients may fail to capture the system dynamics accurately, even if the estimated coefficients are exact.

Understanding how the input signal influences the frequency-domain distribution of model bias is particularly important when identifying models for control applications. In such contexts, accurate modeling in certain frequency bands—such as the crossover frequency region—may be more critical than in others. This perspective becomes especially relevant when fitting low-order models to data.

In the PEM, the model parameters are chosen by minimizing the objective:

$$\min_{\theta} \frac{1}{N} \sum_{k=1}^{N} e_{\text{pred}}^2(k, \theta), \tag{10.47}$$

where $e_{\text{pred}}(k, \theta) = H^{-1}(q, \theta)\{y(k) - G(q, \theta)u(k)\}$. Suppose the true system is given by

$$y(k) = G_o(q)u(k) + H_o(q)\varepsilon(k). \tag{10.48}$$

Then the prediction error becomes

$$e_{\text{pred}}(k, \theta) = \frac{G_o(q) - G(q, \theta)}{H(q, \theta)} u(k) + \frac{H_o(q)}{H(q, \theta)} \varepsilon(k). \tag{10.49}$$

Applying Parseval's theorem, we obtain

$$\begin{aligned} \lim_{N\to\infty} \frac{1}{N} \sum_{k=1}^{N} e_{\text{pred}}^2(k, \theta) &= \int_{-\pi}^{\pi} \Phi_e(\omega) d\omega \\ &= \int_{-\pi}^{\pi} \left(\left|G_o(e^{j\omega}) - G(e^{j\omega}, \theta)\right|^2 \frac{\Phi_u(\omega)}{|H(e^{j\omega}, \theta)|^2} + \frac{\left|H_o(e^{j\omega})\right|^2}{|H(e^{j\omega}, \theta)|^2} \Phi_\varepsilon(\omega) \right) d\omega, \end{aligned} \tag{10.50}$$

where $\Phi_e(\omega)$ denotes the power spectral density of $e_{\text{pred}}(k)$.

To gain further insight, let us assume for the moment that the noise model contains no unknown parameters. In that case,

$$\begin{aligned} &\lim_{N\to\infty} \frac{1}{N} \sum_{k=1}^{N} e_{\text{pred}}^2(k, \theta) \\ &= \int_{-\pi}^{\pi} \left(\left|G_o(e^{j\omega}) - G(e^{j\omega}, \theta)\right|^2 \frac{\Phi_u(\omega)}{|H(e^{j\omega})|^2} + \frac{\left|H_o(e^{j\omega})\right|^2}{|H(e^{j\omega})|^2} \Phi_\varepsilon(\omega) \right) d\omega. \end{aligned} \tag{10.51}$$

Since the last term in the integrand is independent of θ, PEM in this case selects θ to minimize the L_2-norm of the modeling error $G_o(q) - G(q, \theta)$, weighted by the filtered input spectrum $\Phi_{u_f}(\omega)$, where $u_f(k) = H^{-1}(q)u(k)$. This implies that, to achieve accurate frequency response estimates in a particular frequency region, the filtered input u_f must concentrate its power in that region.

If accurate estimation is required across the entire frequency range, the preferred input is one with a flat power spectrum—for example, a sequence of independent, zero-mean random variables.

The frequency-domain bias distribution can also be shaped more flexibly by prefiltering the input–output data prior to applying PEM. This approach involves minimizing the filtered prediction error $e_{f\text{pred}} \triangleq L(q)e_{\text{pred}}$. In this case,

$$\lim_{N\to\infty} \frac{1}{N}\sum_{k=1}^{N} e_{f\text{pred}}^2(k,\theta)$$
$$= \int_{-\pi}^{\pi} \left(\left|G_o(e^{j\omega}) - G(e^{j\omega},\theta)\right|^2 \frac{\left|L(e^{j\omega})\right|^2 \Phi_u(\omega)}{|H(e^{j\omega})|^2} + \frac{\left|L(e^{j\omega})\right|^2 \left|H_o(e^{j\omega})\right|^2}{|H(e^{j\omega})|^2} \Phi_\varepsilon(\omega) \right) d\omega. \tag{10.52}$$

Therefore, by applying a prefilter $L(q)$ to the data, one can effectively influence the bias distribution in frequency. This is particularly useful when adjusting the input spectrum directly is impractical—for example, when it would require collecting a new dataset.

Finally, we note that the above discussion assumes the noise model contains no unknown parameters. When the noise model shares parameters with the process model—as in ARX or ARMAX structures—the noise spectrum $|H_o(e^{j\omega})|^2$ can influence the bias distribution. Nevertheless, the qualitative effects of the input spectrum and prefiltering remain essentially the same.

10.3.5 Parameter Estimation via Statistical Methods (*)

Mimimization of the prediction error serves as a logical and broadly applicable criterion for parameter estimation, even when the assumed model structure does not perfectly match the true system. In formulating the PEM approach, we did not assume an exact statistical description of the underlying plant.

However, when a statistical characterization of the plant is available, it becomes possible to estimate parameters in an optimal sense, according to a well-defined statistical criterion. Although such exact probabilistic models are rarely available in practice, studying these methods offers valuable insight into the behavior and limitations of more empirical approaches like PEM.

In this section, we present two of the most widely used statistical estimation methods.

10.3.5.1 Maximum Likelihood Estimation

In system identification, we aim to infer system parameters from measurements that are inherently noisy. MLE formalizes this by treating each measurement as a realization of a random variable governed by a specified probability distribution.

As an example, consider the model

$$y(k) = \phi^T(k)\theta + \varepsilon(k), \tag{10.53}$$

where $\varepsilon(k)$ is a Gaussian random variable with zero mean and variance r_ε. Then, the probability density function (PDF) of $y(k)$ is given by

$$dF(\zeta; y(k)) = \frac{1}{\sqrt{2\pi r_\varepsilon}} \exp\left\{-\frac{(\zeta - \phi^T(k)\theta)^2}{2r_\varepsilon}\right\}. \tag{10.54}$$

Here, ζ denotes a particular realization of the random variable $y(k)$ (see Appendix B for notation).

For a dataset comprising N observations, we can consider the joint PDF of the random vector $Y_N \triangleq (y(1), \ldots, y(N))$. Let this joint PDF be denoted by $dF(\zeta_N; Y_N)$, where ζ_N represents a realization of Y_N. Now suppose the actual observed data is given by $\hat{Y}_N = (\hat{y}(1), \ldots, \hat{y}(N))$. Substituting these observed values into the joint PDF yields a deterministic function of the parameter vector θ, known as the *likelihood function*. We denote this likelihood function corresponding to the observed data $\hat{Y}_N$ by $\ell(\theta|\hat{Y}_N)$. The basic idea behind MLE is to choose the parameter vector θ that makes the observed data most probable—that is, to maximize the likelihood function. Formally,

$$\hat{\theta}_N^{\mathrm{ML}} = \arg\left\{\max_{\theta} \ell(\theta|\hat{Y}_N)\right\}. \tag{10.55}$$

In general, deriving the likelihood function directly from a stochastic system model is challenging. A notable exception arises when the model can be expressed in a form where the observations are linear in the random variables.

Let us consider applying the maximum likelihood method to the following linear identification problem:

$$Y_N = \Phi_N\theta + \mathcal{E}_N. \tag{10.56}$$

Here, we assume that $\mathcal{E}_N$ is a zero-mean Gaussian random vector with covariance matrix $R_{\mathcal{E}}$. Under this assumption, the PDF of Y_N becomes

$$\begin{aligned} dF(\hat{Y}_N; Y_N) &= dF(\hat{Y}_N - \Phi_N\theta; \mathcal{E}_N) \\ &= \frac{1}{\sqrt{(2\pi)^N det(R_{\mathcal{E}})}} \exp\left\{-\frac{1}{2}(\hat{Y}_N - \Phi_N\theta)^T R_{\mathcal{E}}^{-1}(\hat{Y}_N - \Phi_N\theta)\right\}. \end{aligned} \tag{10.57}$$

The maximum likelihood estimate is then defined as

$$\begin{aligned} \hat{\theta}_N^{ML} &= \arg\{max_\theta dF(\hat{Y}_N; Y_N)\} \\ &= \arg\{max_\theta \log(dF(\hat{Y}_N; Y_N))\} \\ &= \arg\left\{max_\theta \left(-\frac{1}{2}(\hat{Y}_N - \Phi_N\theta)^T R_{\mathcal{E}}(\hat{Y}_N - \Phi_N\theta)\right)\right\} \\ &= \arg\left\{min_\theta \left(\frac{1}{2}(\hat{Y}_N - \Phi_N\theta)^T R_{\mathcal{E}}(\hat{Y}_N - \Phi_N\theta)\right)\right\}. \end{aligned} \tag{10.58}$$

This result shows that the maximum likelihood estimate coincides with a *weighted LS* estimate when the weighting matrix is the inverse of the OE covariance $R_{\mathcal{E}}$. In particular, if the OE is an i.i.d. Gaussian sequence with covariance matrix $r_\varepsilon I_N$, then the covariance is a scalar multiple of the identity matrix, and the unweighted LS estimate is also a maximum likelihood estimate.

10.3.5.2 Bayesian Estimation

Bayesian estimation takes a philosophically distinct approach to the parameter estimation problem. In this framework, uncertain parameters are treated as random variables characterized by a specified *prior* probability distribution. When observations are made, they inform and update our belief about the parameter values. This updated belief is expressed through the *posterior* probability distribution, which is the conditional distribution of the parameter vector given the observed data.

The parameter value at which the posterior PDF attains its maximum is called the *maximum a posteriori* (MAP) estimate. Alternatively, one may use the mean of the posterior distribution as the estimate, yielding the *minimum variance estimate.*

A central tool in Bayesian estimation is *Bayes' rule*. Let $dF(\hat{\theta} \mid \hat{Y}_N; \theta \mid Y_N)$ denote the conditional PDF of the parameter vector given the observations. Bayes' rule states

$$dF(\hat{\theta}|\hat{Y}_N;\theta|Y_N) = \frac{dF(\hat{Y}_N|\hat{\theta};Y_N|\theta)\cdot dF(\hat{\theta};\theta)}{dF(\hat{Y}_N;Y_N)}. \tag{10.59}$$

Here, $dF(\hat{Y}_N;Y_N)$ is the marginal PDF of the observations and is independent of θ; thus, it serves as a normalizing constant once the observations $\hat{Y}_N$ are fixed. Accordingly, the MAP estimate is given by

$$\hat{\theta}_N^{\text{MAP}} = \arg\left\{\max_{\hat{\theta}} dF(\hat{Y}_N|\hat{\theta};Y_N|\theta)\cdot dF(\hat{\theta};\theta)\right\}. \tag{10.60}$$

Consider again the identification problem

$$Y_N = \Phi_N\theta + \mathcal{E}_N, \tag{10.61}$$

where $\mathcal{E}_N$ is a Gaussian random vector with zero mean and covariance $R_{\mathcal{E}}$. In the Bayesian framework, the parameter vector θ is also modeled as a Gaussian random variable with prior mean $\hat{\theta}(0)$ and covariance $P(0)$. Hence, the prior distribution of θ is a normal distribution characterized by these parameters.

To evaluate the posterior PDF, we apply Bayes' rule:

$$dF(\hat{\theta}|\hat{Y}_N;\theta|Y_N) = [\text{constant}]\times dF_{\mathcal{N}}(\hat{Y}_N;Y_N)_{(\Phi_N\hat{\theta},R_{\mathcal{E}})}\cdot dF_{\mathcal{N}}(\hat{\theta},\theta)_{(\hat{\theta}(0),P(0))}, \tag{10.62}$$

where

$$dF_{\mathcal{N}}(\hat{x},x)_{(\overline{x},R)} = \frac{1}{\sqrt{(2\pi)^N\det(R)}}\exp\left\{-\frac{1}{2}(\hat{x}-\overline{x})^T R^{-1}(\hat{x}-\overline{x})\right\}. \tag{10.63}$$

The MAP estimate is obtained by maximizing the logarithm of the posterior PDF:

$$\begin{aligned}\hat{\theta}_N^{\text{MAP}} &= \arg\max_{\hat{\theta}}\left\{\left(-\frac{1}{2}(\hat{Y}_N-\Phi_N\hat{\theta})^T R_{\mathcal{E}}^{-1}(\hat{Y}_N-\Phi_N\hat{\theta})\right.\right.\\ &\qquad\left.\left.-\frac{1}{2}(\hat{\theta}-\hat{\theta}(0))^T P^{-1}(0)(\hat{\theta}-\hat{\theta}(0))\right)\right\}\\ &= \arg\min_{\hat{\theta}}\left\{\frac{1}{2}\left((\hat{Y}_N-\Phi_N\hat{\theta})^T R_{\mathcal{E}}^{-1}(\hat{Y}_N-\Phi_N\hat{\theta})\right.\right.\\ &\qquad\left.\left.+(\hat{\theta}-\hat{\theta}(0))^T P^{-1}(0)(\hat{\theta}-\hat{\theta}(0))\right)\right\}.\end{aligned} \tag{10.64}$$

Solving this LS problem yields

$$\hat{\theta}_N^{\text{MAP}} = \left(\Phi_N^T R_{\mathcal{E}}^{-1}\Phi_N + P^{-1}(0)\right)^{-1}\left(\Phi_N^T R_{\mathcal{E}}^{-1}\hat{Y}_N + P^{-1}(0)\hat{\theta}(0)\right). \tag{10.65}$$

Applying the matrix inversion lemma, this expression can be rewritten as

$$\hat{\theta}_N^{\text{MAP}} = \hat{\theta}(0) + P(0)\Phi_N^T\left(\Phi_N^T P(0)\Phi_N + R_{\mathcal{E}}\right)^{-1}\left(\hat{Y}_N - \Phi_N\hat{\theta}(0)\right). \tag{10.66}$$

This form highlights how the MAP estimate incorporates both prior knowledge and data-based information in a statistically optimal manner.

We make the following observations:

- From (10.64), we observe that if $P(0)$ is chosen to be nonsingular and the persistent excitation condition is satisfied, the first term will dominate the second as $N \to \infty$. Consequently, $\hat{\theta}_N^{\text{MAP}}$ converges to $\hat{\theta}_N^{LS}$. Hence, all the asymptotic properties of LS identification apply to this method as well.

- If $P(0)$ is chosen to be singular, the estimate of θ may be *biased*, since the null space of $P(0)$ corresponds to a subspace in which no parameter update occurs (i.e., zero update gain).
- Specifying the initial covariance matrix $P(0)$ to be other than ∞I is equivalent to penalizing deviations from the initial parameter guess using the weighting matrix $P^{-1}(0)$ in a LS framework. The standard LS solution is thus interpreted in the Bayesian framework as the MAP estimate under a uniform prior distribution (i.e., $P(0) = \infty I$).

Incorporating prior knowledge using this framework can help obtain smoother and more realistic impulse responses. In Section 2.5.5, we proposed using a diagonal weighting matrix to penalize large impulse response coefficients and thereby achieve a smoother step response. This is equivalent to specifying the initial parameter covariance matrix as a diagonal matrix (i.e., the inverse of the weighting matrix) within the Bayesian framework. This statistical interpretation provides a formal justification for the practice and a systematic way to select the weighting matrix (which may also be non-diagonal).

Equation (10.66) can also be written in the following recursive form:

$$\begin{aligned}\hat{\theta}(k) &= \hat{\theta}(k-1) + K(k)\left(y(k) - \phi^T(k)\hat{\theta}(k-1)\right)\\ K(k) &= \frac{P(k-1)\phi(k)}{r_\varepsilon + \phi^T(k)P(k-1)\phi(k)}\\ P(k) &= P(k-1) - \frac{P(k-1)\phi(k)\phi^T(k)P(k-1)}{r_\varepsilon + \phi^T(k)P(k-1)\phi(k)}.\end{aligned} \tag{10.67}$$

Here, $\hat{\theta}(k)$ represents $\hat{\theta}_k^{\text{MAP}}$ or $E\{\theta \mid Y_k\}$, and $P(k) = E\left\{(\theta - \hat{\theta}(k))(\theta - \hat{\theta}(k))^T \mid Y_k\right\}$. The derivation of this recursive formula is straightforward and is left as an exercise.

This framework can be generalized to handle time-varying parameters by adopting the following system model:

$$\begin{aligned}\theta(k) &= \theta(k-1) + \varepsilon_1(k)\\ y(k) &= \phi^T(k)\theta(k) + \varepsilon_2(k),\end{aligned} \tag{10.68}$$

where $\varepsilon_1(k)$ and $\varepsilon_2(k)$ are i.i.d. Gaussian sequences. Here, $\theta(k)$ is modeled as evolving according to a random walk. Alternatively, one may account for autocorrelation in $\varepsilon_1(k)$ and $\varepsilon_2(k)$ by augmenting the state vector. The recursive Bayesian estimator for such cases can be derived using Kalman filtering, which is left as an exercise to the reader.

We next demonstrate the application of the Bayesian approach to identifying impulse response coefficients through the following example.

Example 10.6: In practice, it may be more appropriate, prior to identification, to assume that the derivatives of the impulse response are zero-mean Gaussian random variables and specify the covariance of these derivatives. In other words, one may define:

$$E\left\{\frac{dh}{dt}_{t=i\cdot t_s}\right\} \approx E\left\{\frac{h_i - h_{i-1}}{t_s}\right\} = 0; \quad 1 \leq i \leq n. \tag{10.69}$$

$$E\left\{\left(\frac{dh}{dt}_{t=i\cdot t_s}\right)^2\right\} \approx E\left\{\left(\frac{h_i - h_{i-1}}{t_s}\right)^2\right\} = \frac{\sigma_i}{t_s^2}. \tag{10.70}$$

In this case, $P(0)$, the prior covariance matrix for θ, takes the following form:

$$P(0) = \begin{bmatrix} 1 & 0 & \cdots & \cdots & 0 \\ -1 & 1 & 0 & \cdots & 0 \\ 0 & \ddots & \ddots & \ddots & 0 \\ \vdots & \ddots & \ddots & \ddots & \vdots \\ 0 & \cdots & \cdots & -1 & 1 \end{bmatrix}^{-1} \begin{bmatrix} \sigma_1 & & & & \\ & \sigma_2 & & & \\ & & \ddots & & \\ & & & \ddots & \\ & & & & \sigma_n \end{bmatrix}^{T} \begin{bmatrix} 1 & 0 & \cdots & \cdots & 0 \\ -1 & 1 & 0 & \cdots & 0 \\ 0 & \ddots & \ddots & \ddots & 0 \\ \vdots & \ddots & \ddots & \ddots & \vdots \\ 0 & \cdots & \cdots & -1 & 1 \end{bmatrix}. \tag{10.71}$$

This formulation corresponds to penalizing the 2-norm of the difference between successive impulse response coefficients in the LS identification procedure. The same concept can be extended to incorporate prior estimates of time constants and other structural features of the system.

10.3.6 Other Methods (*)

Other parameter estimation techniques have been proposed in the literature. One method that stands out is the *instrumental variable (IV) method*, originally introduced by Young (1970). The central idea is that, for a model to be effective, the prediction error should exhibit little to no correlation with past data. A significant correlation would imply that the predictor has failed to extract all available information from past measurements. Several widely used choices for instrument variables are discussed in Ljung (1998).

In the IV method, a set of variables called instruments (denoted by the vector η) must first be defined. These instruments are typically constructed as transformations (either linear or nonlinear) of past data, i.e., $y(k-1), \ldots, y(0), u(k-1), \ldots, u(0)$. The parameter vector θ is then computed by solving the following equation:

$$\frac{1}{N}\sum_{k=1}^{N} \eta(k) e_{\text{pred}}(k, \theta) = 0. \tag{10.72}$$

Filtered prediction error e_{pred} may also be used in place of the raw error. The effectiveness of the IV method strongly depends on the choice of instruments. If $\eta(k)$ is chosen as $\phi(k)$, the resulting estimate reduces to the ordinary LS solution. It is also possible to select instruments that themselves depend on θ, which leads to a pseudo-linear regression formulation.

Other notable extensions of the LS approach include the so-called biased regression methods, in which the regression is constrained to a subspace of the full regressor space. Rather than being specified a priori, this subspace is constructed incrementally by successively adding linearly independent directions. In principal component regression (PCR), these directions are selected to maximize the variance of the regressors ϕ, whereas in partial least squares (PLS), they are chosen to maximize the covariance between ϕ and the output y. Both methods seek to reduce estimation variance at the expense of introducing some bias. In the Bayesian framework, this is analogous to adopting a singular initial covariance matrix, $P(0)$, where the singular directions are data-driven rather than imposed through prior knowledge. These techniques have been widely used in the chemometrics community for addressing multivariate calibration problems; see Lorber et al. (1987) for a comprehensive overview.

10.4 Nonparametric Identification

Nonparametric identification methods are particularly valuable when little prior knowledge about the system is available. These methods rely on model structures such as impulse response models and frequency response models, which do not assume a fixed, finite-dimensional parametric form.

In practice, however, the boundary between nonparametric and parametric identification is not always clear-cut. Even nonparametric approaches require certain assumptions—such as a FIR horizon or smoothness in the frequency domain—to ensure the estimation problem is well-posed. Conversely, in parametric identification, model order selection often relies on nonparametric tools such as residual analysis or cross-validation across models of varying complexity.

10.4.1 Impulse Response Identification

Identification of FIR models was introduced in Chapter 2. Impulse response identification is a widely used form of nonparametric identification. Before discussing FIR model identification more formally, let us revisit Example 2.7.

Example 10.7: *Identifying an FIR Model for the System of Example 10.4.*
We will repeat identification of an FIR model—as in Example 2.7—using the SysID Toolbox in MATLAB.

The FIR model coefficients are estimated using `impulseest`. Although the input–output dataset (`allData`) is the only required input for `impulseest`, additional options were used to control model complexity. Specifically, the FIR model order was limited to $n = 40$ to avoid overfitting, and a model delay of one sampling interval was imposed to enforce causality ($h_0 = 0$). Figure 10.4 compares the impulse response of the original system with that of the identified FIR model.

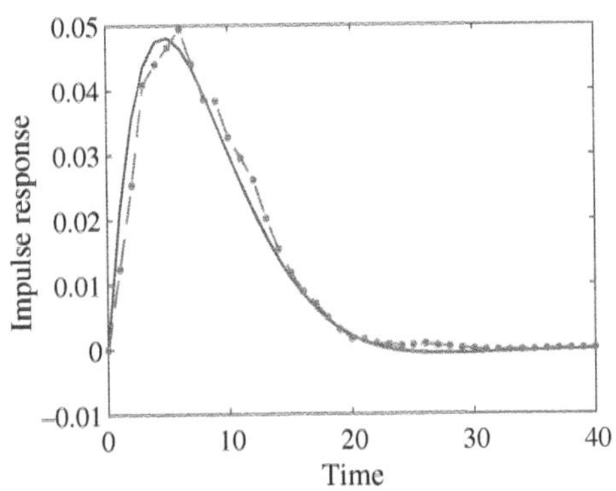

Figure 10.4 Identifying FIR model using the System Identification Toolbox. The impulse response coefficients of the identified model (dashed line with circles) closely match the true FIR coefficients (solid line).

Assume that the underlying system follows a convolution representation:

$$y(k) = \sum_{i=1}^{\infty} H_i u(k-i) + e_k. \tag{10.73}$$

Multiplying both sides by $u^T(k-\tau)$ yields

$$y(k)u^T(k-\tau) = \sum_{i=1}^{\infty} H_i u(k-i)u^T(k-\tau) + e(k)u^T(k-\tau). \tag{10.74}$$

Summing over $k = 1$ to N and normalizing gives

$$\left(\frac{1}{N}\sum_{k=1}^{N} y(k)u^T(k-\tau)\right) = \sum_{i=1}^{\infty} H_i \left(\frac{1}{N}\sum_{k=1}^{N} u(k-i)u^T(k-\tau)\right) + \left(\frac{1}{N}\sum_{k=1}^{N} e(k)u^T(k-\tau)\right). \tag{10.75}$$

Assuming quasi-stationary signals, taking the limit as $N \to \infty$ yields the cross-correlation form:

$$R_{yu}(\tau) = \sum_{i=1}^{\infty} H_i R_{uu}(\tau - i) + R_{eu}(\tau), \tag{10.76}$$

where the correlation functions are defined as

$$R_{yu}(\tau) = \lim_{N\to\infty} \frac{1}{N}\sum_{k=1}^{N} y(k)u^T(k-\tau), \tag{10.77}$$

$$R_{uu}(\tau) = \lim_{N\to\infty} \frac{1}{N}\sum_{k=1}^{N} u(k)u^T(k-\tau), \text{ and} \tag{10.78}$$

$$R_{eu}(\tau) = \lim_{N\to\infty} \frac{1}{N}\sum_{k=1}^{N} e(k)u^T(k-\tau). \tag{10.79}$$

The above relationship can also be derived using a statistical argument. Specifically, taking the expectation of (10.74), we obtain

$$E\{y(k)u^T(k-\tau)\} = \sum_{i=1}^{\infty} H_i E\{u(k-i)u^T(k-\tau)\} + E\{e(k)u^T(k-\tau)\}. \tag{10.80}$$

Assuming that $\{u(k)\}$ and $\{e(k)\}$ are stationary sequences, the terms R_{uu}, R_{yu}, and R_{eu} correspond to the expectations in the above equation.

Now, suppose that $u(k)$ is a zero-mean stationary sequence uncorrelated with $e(k)$, which is also assumed to be stationary. Under these assumptions, $R_{eu}(\tau) \to 0$ as $N \to \infty$. Further, assume that $H_i = 0$ for $i > n$. Such a value of n can be identified by examining $R_{yu}(\tau)$ under white noise excitation. When the input signal is white, $R_{uu}(i) = 0$ for $i \neq 0$. From this, it follows that $R_{yu}(\tau) = 0$ if $H_\tau = 0$. Thus, one can choose n such that $R_{yu}(\tau) \approx 0$ for $\tau > n$.

With these assumptions, we can express (10.76) as

$$\begin{bmatrix} R_{yu}(1) & R_{yu}(2) & \cdots & R_{yu}(n) \end{bmatrix} \approx \begin{bmatrix} H_1 & H_2 & \cdots & H_n \end{bmatrix} \times \begin{bmatrix} R_{uu}(0) & R_{uu}(1) & \cdots & R_{uu}(n-1) \\ R_{uu}(-1) & R_{uu}(0) & \cdots & R_{uu}(n-1) \\ \vdots & \ddots & \ddots & \vdots \\ R_{uu}(-n+1) & R_{uu}(-n+2) & \cdots & R_{uu}(0) \end{bmatrix}. \tag{10.81}$$

Taking the transpose and rearranging terms yields:

$$\begin{bmatrix} H_1^T \\ H_2^T \\ \vdots \\ H_n^T \end{bmatrix} \approx \begin{bmatrix} R_{uu}(0) & R_{uu}(1) & \cdots & R_{uu}(n-1) \\ R_{uu}^T(1) & R_{uu}(0) & \cdots & R_{uu}(n-2) \\ \vdots & \ddots & \ddots & \vdots \\ R_{uu}^T(n-1) & R_{uu}^T(n-2) & \cdots & R_{uu}(0) \end{bmatrix}^{-1} \begin{bmatrix} R_{yu}^T(1) \\ R_{yu}^T(2) \\ \vdots \\ R_{yu}^T(n) \end{bmatrix}. \tag{10.82}$$

Here, we have used the property $R_{uu}^T(i) = R_{uu}(-i)$.

In practice, $R_{uu}(i)$ and $R_{yu}(i)$ must be estimated from finite-length data. While such estimates can exhibit significant variance, the assumption that $H_i = 0$ for $i > n$ restricts the number of impulse response coefficients and therefore limits the required set of correlation estimates to $R_{uu}(i)$, $i = 0, \ldots, n-1$ and $R_{yu}(i)$, $i = 1, \ldots, n$. Under the condition of persistency of excitation, the variance of these estimates decays as $1/N$. However, the truncation introduces bias, leading to a trade-off between variance and bias governed by the choice of n. It is noteworthy that (10.82) gives virtually the same estimate as the LS identification when the number of data points is large. This can indeed be verified in Example 10.7.

10.4.2 Frequency Response Identification (*)

The dynamics of a general linear system can be described by its frequency response, which characterizes how the system responds to sinusoidal inputs of varying frequency. This response is defined by the amplitude ratio and phase angle at each frequency. Both quantities are naturally represented by a complex function of ω, whose modulus and argument correspond to the amplitude ratio and phase angle, respectively.

The frequency response function can be derived directly from the system's transfer function $G(q)$ by evaluating it on the unit circle, i.e., replacing q with $e^{j\omega}$. The amplitude ratio and phase angle at frequency ω are then given by

$$A.R.(\omega) = |G(e^{j\omega})| = \sqrt{\mathrm{Re}\{G(e^{j\omega})\}^2 + \mathrm{Im}\{G(e^{j\omega})\}^2}. \tag{10.83}$$

$$\phi(\omega) = \arg\{G(e^{j\omega})\} = \tan^{-1}\left[\frac{\mathrm{Im}\{G(e^{j\omega})\}}{\mathrm{Re}\{G(e^{j\omega})\}}\right]. \tag{10.84}$$

For systems with unit sampling time, $G(e^{j\omega})$ is evaluated over the interval $0 \le \omega \le \pi$, and fully characterizes the system's dynamics over that range.

One approach to system identification is to estimate $G(e^{j\omega})$ directly from input–output data without assuming a finite-dimensional parameterization. This is known as nonparametric identification, since the frequency response is defined over a continuous frequency range and not represented by a fixed set of parameters.

10.4.2.1 Frequency Response Computation

The most direct method for identifying the frequency response of a system is sine-wave testing, in which sinusoidal inputs of various frequencies are applied to the system, and the corresponding outputs are measured. While conceptually simple, this approach has limited practical utility due to two key drawbacks:

1. Generating pure sinusoidal perturbations over a wide frequency range can be technically challenging.
2. Each experiment yields information at only a single frequency, making the process time-consuming and inefficient.

A more practical alternative is to leverage Fourier analysis. From the z-domain input/output relationship for a linear system governed by $y(k) = G(q)u(k)$, it follows that

$$G(e^{j\omega}) = \frac{Y(\omega)}{U(\omega)}, \tag{10.85}$$

where

$$Y(\omega) = \sum_{k=1}^{\infty} y(k)e^{-j\omega k}. \tag{10.86}$$

$$U(\omega) = \sum_{k=1}^{\infty} u(k)e^{-j\omega k}. \tag{10.87}$$

Thus, the system's frequency response can be computed by dividing the Fourier transform of the output signal by that of the input signal.

In practice, however, several complications arise. The input and output signals are available only over a finite time horizon, and the output is typically contaminated by noise and disturbances. These factors introduce both bias and variance into the estimate of $G(e^{j\omega})$, requiring careful signal processing and averaging techniques to obtain reliable frequency response estimates.

Let us assume that the underlying system is given by

$$y(k) = G(q)u(k) + e(k), \tag{10.88}$$

where $e(k)$ is a zero-mean stationary sequence that accounts for the combined effects of noise and disturbances. Define the normalized discrete-time Fourier transforms (DTFTs) over a finite data record of length N as follows:

$$Y_N(\omega) \stackrel{\Delta}{=} \frac{1}{\sqrt{N}} \sum_{k=1}^{N} y(k)e^{-j\omega k}, \tag{10.89}$$

$$U_N(\omega) \stackrel{\Delta}{=} \frac{1}{\sqrt{N}} \sum_{k=1}^{N} u(k)e^{-j\omega k}, \text{ and} \tag{10.90}$$

$$E_N(\omega) \stackrel{\Delta}{=} \frac{1}{\sqrt{N}} \sum_{k=1}^{N} e(k)e^{-j\omega k}. \tag{10.91}$$

Then, the frequency-domain estimate of the system's transfer function can be written as

$$\hat{G}_N(\omega) \stackrel{\Delta}{=} \frac{Y_N(\omega)}{U_N(\omega)} = G(e^{j\omega}) + \frac{R_N(\omega)}{U_N(\omega)} + \frac{E_N(\omega)}{U_N(\omega)}. \tag{10.92}$$

where $R_N(\omega)$ denotes the residual error introduced by the finite-length data window. It can be shown that $|R_N(\omega)| = \frac{c_1}{\sqrt{N}}$ for some constant c_1.

Thus, $\hat{G}_N(\omega)$, computed from N data points as above, is an estimate of the true system frequency response $G(e^{j\omega})$. This estimate is commonly referred to as the *empirical transfer function estimate* (ETFE).

10.4.2.2 Statistical Properties of the ETFE

Taking the expectation of (10.92), we obtain:

$$E\{\hat{G}_N(\omega)\} = E\left\{G(e^{j\omega}) + \frac{R_N(\omega)}{U_N(\omega)} + \frac{E_N(\omega)}{U_N(\omega)}\right\} = G(e^{j\omega}) + \frac{R_N(\omega)}{U_N(\omega)}. \tag{10.93}$$

The variance of the estimate can be expressed as

$$E\left\{\left(\hat{G}_N(\omega) - G(e^{j\omega})\right)\left(\hat{G}_N(\omega) - G(e^{-j\omega})\right)\right\} = \frac{\Phi_e + \rho_N}{|U_N(\omega)|^2}, \tag{10.94}$$

where $\rho_N \leq \frac{c_2}{N}$ for some constant c_2.

The implications of the above results are as follows:

- The second term on the right-hand side of (10.93) decays as $\frac{1}{\sqrt{N}}$, indicating that $\hat{G}_N(\omega)$ is an *asymptotically unbiased estimate* of $G(e^{j\omega})$.
- If $u(k)$ is a periodic signal with period N, then $|U_N(\omega)|$ is nonzero only at m discrete frequency points $\omega = \frac{2\pi k}{m}$, $k \in [0, m-1]$. This implies that the ETFE is defined only at those frequencies. As $N \to \infty$, $|U_N(\omega)|$ increases at these frequencies, causing the variance in (10.94) to decay to zero.
- If $u(k)$ is a randomly generated signal, the number of frequency points at which the ETFE can be computed increases with N. However, $|U_N(\omega)|^2$ fluctuates around the power spectrum of $u(k)$ and does not grow unbounded with N. As a result, the variance in (10.94) does not decay to zero. This behavior is typical of nonparametric identification, where one is, in effect, estimating an infinite number of parameters.

A practical implication of the last comment is that the estimate can be very sensitive to noise in the data (no matter how many data points are used). Hence, some smoothing is needed. Some simple smoothing methods are:

- **Averaging over frequency bands:** Select a finite set of frequency points $\omega_1, \cdots, \omega_m$ in the interval $[0, \pi]$. Assume that $G(e^{j\omega})$ is approximately constant within the interval $\omega_i - \delta\omega \leq \omega \leq \omega_i + \delta\omega$ for each ω_i. The ETFE values $G_N(\omega)$ obtained within each window are then averaged using a weighting scheme based on the signal-to-noise ratio, such as $\frac{\Phi_e}{|U_N(\omega)|^2}$. With this assumption, the number of estimated frequency response values becomes finite and the variance of the estimate decays as $1/N$. However, this smoothing introduces bias due to the piecewise constant assumption on $G(e^{j\omega})$.
- **General smoothing using a weighting function:** A more flexible method is to apply a continuous weighting function $W_s(\zeta - \omega)$ to smooth the ETFE:

$$G_N^s(\omega) = \frac{\int_{-\pi}^{\pi} W_s(\zeta - \omega) G_N(\omega) \frac{|U_N(\zeta)|^2}{\Phi_e(\omega)} d\zeta}{\int_{-\pi}^{\pi} W_s(\zeta - \omega) \frac{|U_N(\zeta)|^2}{\Phi_e(\omega)} d\zeta}. \tag{10.95}$$

Here, W_s is a symmetric function centered at zero, typically parameterized by a window width that controls the trade-off between bias and variance. A wider window leads to lower variance but higher bias, and vice versa. As with the previous approach, the variance of the smoothed estimate decays as $1/N$ when the smoothing window is nonzero.

10.5 Subspace Identification

When identifying a multivariable system, it is advantageous to adopt a model structure that explicitly captures the correlations among output channels. Accounting for these correlations not only improves the data efficiency of estimating the deterministic system dynamics but also facilitates the development of a multivariable noise model—a key requirement in applications such as soft sensing and inferential control, where cross-channel interactions play a critical role. However, capturing such correlations requires a MIMO identification framework, rather than a SISO or MISO formulation.

Matrix polynomial model structures like MIMO ARMAX are generally not well-suited for this purpose, as they lead to numerically ill-conditioned, nonlinear estimation problems with many local minima. Furthermore, constructing a well-posed and identifiable model often demands significant prior knowledge, such as the system order and observability indices.

A more practical alternative is to identify a state-space model using a *subspace identification* method. Subspace identification has its roots in classical realization theory, notably from Ho and Kalman (1966) and Kung (1978). Among the various algorithms, we focus on *N4SID*, originally introduced by Van Overschee and De Moor (1994), which offers several advantages: it relies on non-iterative linear algebraic computations and requires minimal prior knowledge from the user. Other subspace algorithms—such as those in Larimore (1990) and Verhaegen (1994)—are based on similar principles but differ in how the data are processed. MATLAB's System Identification Toolbox includes implementations of several of these methods.

10.5.1 The Basic Method

Although subspace identification algorithms vary in their details, they share a common principle: A state sequence is first extracted—either explicitly or implicitly—by projecting input–output data. Once the state sequence is determined, the state-space system matrices are identified using this constructed sequence.

10.5.1.1 The Problem Formulation

Consider the standard linear stochastic state-space system:

$$\begin{aligned} x(k+1) &= Ax(k) + Bu(k) + \varepsilon_1(k) \\ y(k) &= Cx(k) + \varepsilon_2(k), \end{aligned} \tag{10.96}$$

where $\varepsilon_1(k)$ and $\varepsilon_2(k)$ are zero-mean white noise sequences. We assume the system is reachable, observable, and stationary.

The corresponding *steady-state Kalman filter* representation, which is the *input–output equivalent* to the above system, is given by

$$\begin{aligned} \hat{x}_\infty(k+1) &= A\hat{x}_\infty(k) + Bu(k) + K_\infty\varepsilon_\infty(k) \\ y(k) &= C\hat{x}_\infty(k) + \varepsilon_\infty(k), \end{aligned} \tag{10.97}$$

where $\hat{x}_\infty(k)$ denotes the steady-state Kalman filter estimate of the state, K_∞ is the steady-state Kalman gain, and $\varepsilon_\infty(k) = y(k) - C\hat{x}_\infty(k)$ is the innovation sequence, which is white by construction.

In subspace identification, we approximate this formulation using a finite-horizon, nonsteady-state Kalman filter:

$$\begin{aligned} \hat{x}_{i+1}(k+1) &= A\hat{x}_i(k) + Bu(k) + K_i\varepsilon_i(k) \\ y(k) &= C\hat{x}_i(k) + \varepsilon_i(k), \end{aligned} \tag{10.98}$$

where $\hat{x}_i(k)$ is the state estimate produced by running a Kalman filter initialized at time $k - i$ with $\hat{x}_0(k-i) = 0$ and a prescribed covariance (typically the open-loop steady-state covariance). The subscript i indicates that the filter has been run for i steps; i should be chosen larger than the system order n (i.e., $i > n$).

Equation (10.98) provides an approximation to Eq. (10.97). The deterministic components (A, B, C) are identical in both systems. Although the stochastic terms (K_i and $\text{Cov}\{\varepsilon_i(k)\}$) differ from their steady-state counterparts, they closely approximate K_∞ and $\text{Cov}\{\varepsilon_\infty(k)\}$ for sufficiently large i.

The objective of subspace identification is to recover the system matrices $(A, B, C, K_i, \text{Cov}\{\varepsilon_i(k)\})$—up to a similarity transformation—based on observed input–output data.

10.5.1.2 Multi-step Prediction Equation based on the Kalman Estimate

As a first step toward system identification, consider the multistep prediction of outputs using the Kalman state estimate. We can write

$$\begin{bmatrix} y(k) \\ y(k+1) \\ \vdots \\ \vdots \\ y(k+i-1) \end{bmatrix} = \begin{bmatrix} C \\ CA \\ CA^2 \\ \vdots \\ CA^{i-1} \end{bmatrix} \hat{x}_i(k) + \begin{bmatrix} 0 & 0 & \cdots & 0 & 0 \\ CB & 0 & \ddots & 0 & 0 \\ CAB & CB & \ddots & \vdots & \vdots \\ \vdots & \ddots & \ddots & \vdots & \vdots \\ CA^{i-2}B & \cdots & \cdots & CB & 0 \end{bmatrix} \begin{bmatrix} u(k) \\ u(k+1) \\ \vdots \\ u(k+i-1) \end{bmatrix} + \begin{bmatrix} e_i(k|k-1) \\ e_i(k+1|k-1) \\ \vdots \\ \vdots \\ e_i(k+i-1|k-1) \end{bmatrix}. \tag{10.99}$$

This expression represents the optimal predictions for $y(k), \ldots, y(k+i-1)$ based on the state estimate $\hat{x}_i(k)$ and the prediction errors $e_i(\cdot|k-1)$. For notational convenience, define

$$\mathcal{Y}_i^{0+}(k) = \mathcal{W}_i^o\hat{x}_i(k) + L_3\mathcal{U}_i^{0+}(k) + \mathcal{E}_i^{0+}(k|k-1), \tag{10.100}$$

where $\mathcal{Y}_i^{0+}(k)$ and $\mathcal{U}_i^{0+}(k)$ denote the stacked future outputs and inputs from time k to $k+i-1$, respectively.

We can easily verify by propagating the Kalman filter equation that $\hat{x}_i(k)$ is a linear function of $y(k-i), \ldots, y(k-1)$ and $u(k-i), \ldots, u(k-1)$. Defining

$$\mathcal{Y}_i^-(k) \triangleq \begin{bmatrix} y(k-i) \\ \vdots \\ y(k-1) \end{bmatrix}; \quad \mathcal{U}_i^-(k) \triangleq \begin{bmatrix} u(k-i) \\ \vdots \\ u(k-1) \end{bmatrix}, \tag{10.101}$$

we can express

$$\hat{x}_i(k) = \begin{bmatrix} H_1 & H_2 \end{bmatrix} \begin{bmatrix} \mathcal{Y}_i^-(k) \\ \mathcal{U}_i^-(k) \end{bmatrix}. \tag{10.102}$$

Here, H_1 and H_2 depend on the system matrices and the covariance structure used in the initialization of the nonstationary Kalman filter. From the structure of the Kalman filter, it can be shown that $\hat{x}_i(k)$ is a linear function of the past inputs and outputs, specifically $y(k-i), \dots, y(k-1)$ and $u(k-i), \dots, u(k-1)$.

Substituting (10.102) into (10.100), we obtain

$$\mathcal{Y}_i^{0+}(k) = \underbrace{\mathcal{W}_i^o \begin{bmatrix} H_1 & H_2 \end{bmatrix}}_{\begin{bmatrix} L_1 & L_2 \end{bmatrix}} \begin{bmatrix} \mathcal{Y}_i^-(k) \\ \mathcal{U}_i^-(k) \end{bmatrix} + L_3 \mathcal{U}_i^{0+}(k) + \mathcal{E}_i^{0+}(k|k-1). \tag{10.103}$$

A key observation is that, when the Kalman filter is initialized appropriately, the innovation sequence $\mathcal{E}_i^{0+}(k|k-1)$ is orthogonal (in the statistical sense) to the regressors $\mathcal{Y}_i^-(k)$, $\mathcal{U}_i^-(k)$, and $\mathcal{U}_i^{0+}(k)$, assuming $u(k)$ is a white noise sequence. This orthogonality ensures that LS regression yields an unbiased and consistent estimate for $[L_1; L_2]$ and L_3.

Remark 10.5: The white input assumption can be relaxed by modifying the Kalman filter initialization. Van Overschee and De Moor (1994) and Moor (1994) describe the necessary adjustments for correlated inputs and present a numerically robust algorithm using RQ factorization.

10.5.1.3 Obtaining Consistent Estimates of Predictor Matrices

Suppose we are given N input–output sample pairs $y(0), \dots, y(N)$ and $u(0), \dots, u(N)$. Using these, we construct the following data matrices:

$$\begin{aligned} U_{0|i-1} &\stackrel{\Delta}{=} \begin{bmatrix} \mathcal{U}^-(i) & \mathcal{U}^-(i+1) & \cdots & \cdots & \mathcal{U}^-(i+j-1) \end{bmatrix} \\ &= \begin{bmatrix} u(0) & u(1) & u(2) & \cdots & u(j-1) \\ u(1) & u(2) & u(3) & \cdots & u(j) \\ \vdots & \cdots & \cdots & \cdots & \vdots \\ u(i-1) & u(i) & u(i+1) & \cdots & u(i+j-2) \end{bmatrix}, \end{aligned} \tag{10.104}$$

$$\begin{aligned} U_{i|2i-1} &\stackrel{\Delta}{=} \begin{bmatrix} \mathcal{U}^{0+}(i) & \mathcal{U}^{0+}(i+1) & \cdots & \cdots & \mathcal{U}^{0+}(i+j-1) \end{bmatrix} \\ &= \begin{bmatrix} u(i) & u(i+1) & u(i+2) & \cdots & u(i+j-1) \\ u(i+1) & u(i+2) & u(i+3) & \cdots & u(i+j) \\ \vdots & \cdots & \cdots & \cdots & \vdots \\ u(2i-1) & u(2i) & u(2i+1) & \cdots & u(2i+j-2) \end{bmatrix}, \end{aligned} \tag{10.105}$$

$$Y_{0|i-1} \triangleq \left[\mathcal{Y}^-(i)\ \ \mathcal{Y}^-(i+1)\ \ \cdots\ \ \cdots\ \ \mathcal{Y}^-(i+j-1)\right] = \begin{bmatrix} y(0) & y(1) & y(2) & \cdots & y(j-1) \\ y(1) & y(2) & y(3) & \cdots & y(j) \\ \vdots & \cdots & \cdots & \cdots & \vdots \\ y(i-1) & y(i) & y(i+1) & \cdots & y(i+j-2) \end{bmatrix}, \text{ and} \tag{10.106}$$

$$Y_{i|2i-1} \triangleq \left[\mathcal{Y}^{0+}(i)\ \ \mathcal{Y}^{0+}(i+1)\ \ \cdots\ \ \cdots\ \ \mathcal{Y}^{0+}(i+j-1)\right] = \begin{bmatrix} y(i) & y(i+1) & y(i+2) & \cdots & y(i+j-1) \\ y(i+1) & y(i+2) & y(i+3) & \cdots & y(i+j) \\ \vdots & \cdots & \cdots & \cdots & \vdots \\ y(2i-1) & y(2i) & y(2i+1) & \cdots & y(2i+j-2) \end{bmatrix}. \tag{10.107}$$

Here, $j = N - 2i + 1$ is the number of columns in each Hankel-like matrix.

Using these matrices, the LS estimate of the predictor matrices $[L_1\ \ L_2\ \ L_3]$ is obtained as

$$\left[\hat{L}_1\ \ \hat{L}_2\ \ \hat{L}_3\right] = Y_{i|2i-1} \Big/ \begin{bmatrix} Y_{0|i-1} \\ U_{0|i-1} \\ U_{i|2i-1} \end{bmatrix}, \tag{10.108}$$

where the operator "/" denotes the LS solution, $A/B = AB^T(BB^T)^{-1}$. This formulation yields a consistent estimate of the predictor matrices under mild statistical assumptions.

10.5.1.4 Choosing the System Order and Obtaining the Kalman State Data Matrix for $\hat{x}_{i+1}$ – Theoretical Basis

Since $[L_1\ L_2] = \mathcal{W}_i^o \cdot [H_1\ H_2]$, and $i > n$, if the system is reachable and observable, the matrix $[L_1\ L_2]$ will have rank n. Thus, the system order n can be determined by examining the rank of this matrix. This implies that only n linear combinations of $\mathcal{Y}_i^-(k)$ and $\mathcal{U}_i^-(k)$ are needed to define the Kalman state. However, the choice of coordinate axes within this state space is arbitrary: the system matrices identified from input–output data are unique only up to a state coordinate transformation.

Suppose the singular value decomposition (SVD) of the matrix is

$$\left[L_1\ \ L_2\right] = \left[W_1\ \ W_2\right] \begin{bmatrix} \Sigma & 0 \\ 0 & 0 \end{bmatrix} \begin{bmatrix} V_1^T \\ V_2^T \end{bmatrix}. \tag{10.109}$$

Then, the extended observability matrix $\mathcal{W}_i^o$ can be chosen as $W_1 P$, where P is any nonsingular matrix. To obtain a balanced realization, one common choice is

$$\mathcal{W}_i^o = W_1 \Sigma^{1/2},$$

which leads to the following expression for the state estimate:

$$\hat{x}_i(k) = \left(\mathcal{W}_i^o\right)^\dagger \begin{bmatrix} L_1 & L_2 \end{bmatrix} \begin{bmatrix} \mathcal{Y}^-(k) \\ \mathcal{U}^-(k) \end{bmatrix}. \tag{10.110}$$

$$= \Sigma^{1/2} V_1^T \begin{bmatrix} \mathcal{Y}^-(k) \\ \mathcal{U}^-(k) \end{bmatrix}. \tag{10.111}$$

Using this result, we can easily construct the data matrix for the Kalman state estimates $\hat{x}_i(k)$ based on input–output measurements:

$$\begin{aligned} \hat{X}_i &= \begin{bmatrix} \hat{x}_i(i) & \cdots & \cdots & \hat{x}_i(N-i) \end{bmatrix} \\ &= \left(\mathcal{W}_i^o\right)^\dagger \begin{bmatrix} \hat{L}_1 & \hat{L}_2 \end{bmatrix} \begin{bmatrix} Y_{0|i-1} \\ U_{0|i-1} \end{bmatrix}. \end{aligned} \tag{10.112}$$

This matrix $\hat{X}_i$ will later be used to identify the state-space matrices via regression.

10.5.1.5 Choosing the System Order and Obtaining the Kalman State Data Matrix x_i – Practical Considerations

In practice, only a finite data set is available, and the rank of the estimated matrix $\begin{bmatrix} \hat{L}_1 & \hat{L}_2 \end{bmatrix}$, a LS approximation of $\begin{bmatrix} L_1 & L_2 \end{bmatrix}$, can be significantly higher than the true system order n due to identification noise and modeling errors.

A practical approach to select the system order and define the state space is to compute the SVD of the matrix:

$$\begin{bmatrix} \hat{L}_1 & \hat{L}_2 \end{bmatrix} \begin{bmatrix} Y_{0|i-1} \\ U_{0|i-1} \end{bmatrix}. \tag{10.113}$$

Based on this SVD, the extended observability matrix estimate $\hat{\mathcal{W}}_i^o$ can be defined in the same way as in the theoretical case. The system order is then determined by inspecting the singular values and identifying a significant drop between successive values. This step involves some subjectivity, reflecting a trade-off between bias and variance.

Once the system order is selected, the state data matrix $\hat{X}_i$ is constructed using the same formula as in Eq. (10.112). From a model reduction perspective, selecting a reduced state dimension in this manner effectively minimizes the squared sum of the prediction error vector $\mathcal{E}_i^{0+}(k|k-1)$ over $k = i+1, \ldots, N+1-i$. This criterion closely resembles the objective of Hankel-norm model reduction, with frequency weighting implicitly imposed by the input spectrum.

This procedure for determining the state is characteristic of the N4SID algorithm. Other subspace identification methods differ in how the system order and state coordinate system are chosen. For example, instead of using the product matrix above, one may apply SVD directly to the gain matrix:

$$\begin{bmatrix} \hat{L}_1 & \hat{L}_2 \end{bmatrix},$$

or the normalized matrix:

$$\text{Cov}^{-1/2}\{\mathcal{Y}_i^{0+}(k)\} \begin{bmatrix} \hat{L}_1 & \hat{L}_2 \end{bmatrix} \text{Cov}^{-1/2} \begin{Bmatrix} \mathcal{Y}_i^-(k) \\ \mathcal{U}_i^-(k) \end{Bmatrix},$$

to emphasize *correlation* rather than prediction accuracy for a given data set.

10.5.1.6 Obtaining the Kalman State Data Matrix for $\hat{x}_{i+1}$

We now construct the data matrix for $\hat{x}_{i+1}(k+1)$ using a similar approach as before. A key point is that we cannot obtain $\hat{x}_{i+1}(k+1)$ by simply time-shifting $\hat{x}_i(k)$. Doing so would yield $\hat{x}_i(k+1)$, which corresponds to a Kalman filter initialized at time $k-i+1$, rather than $k-i$. To maintain a consistent initialization for both estimates, we must follow the same procedure as before, but replace $\mathcal{Y}_i^-(k)$ and $\mathcal{U}_i^-(k)$ with:

$$\mathcal{Y}_{i+1}^-(k+1) \triangleq \begin{bmatrix} y(k-i) \\ \vdots \\ y(k-1) \\ y(k) \end{bmatrix}; \quad \mathcal{U}_{i+1}^-(k+1) \triangleq \begin{bmatrix} u(k-i) \\ \vdots \\ u(k-1) \\ u(k) \end{bmatrix}. \tag{10.114}$$

We also define the future input–output sequences (of length $i-1$) as

$$\mathcal{Y}_{i-1}^{0+}(k+1) \triangleq \begin{bmatrix} y(k+1) \\ \vdots \\ y(k+i-1) \end{bmatrix}; \quad \mathcal{U}_{i-1}^{0+}(k+1) \triangleq \begin{bmatrix} u(k+1) \\ \vdots \\ u(k+i-1) \end{bmatrix}. \tag{10.115}$$

Following the same structure, the Kalman state estimate is expressed as

$$\hat{x}_{i+1}(k+1) = \begin{bmatrix} \bar{H}_1 & \bar{H}_2 \end{bmatrix} \begin{bmatrix} \mathcal{Y}_{i+1}^-(k+1) \\ \mathcal{U}_{i+1}^-(k+1) \end{bmatrix}. \tag{10.116}$$

Substituting this into the multistep prediction equation gives

$$\mathcal{Y}_{i-1}^{0+}(k+1) = \underbrace{\mathcal{W}_{i-1}^o \begin{bmatrix} \bar{H}_1 & \bar{H}_2 \end{bmatrix}}_{\begin{bmatrix} \bar{L}_1 & \bar{L}_2 \end{bmatrix}} \begin{bmatrix} \mathcal{Y}_{i+1}^-(k+1) \\ \mathcal{U}_{i+1}^-(k+1) \end{bmatrix} + \bar{L}_3 \mathcal{U}_{i-1}^{0+}(k+1) + \mathcal{E}_{i-1}^{0+}(k+1|k). \tag{10.117}$$

As before, the predictor matrices can be consistently estimated via LS:

$$\begin{bmatrix} \hat{\bar{L}}_1 & \hat{\bar{L}}_2 & \hat{\bar{L}}_3 \end{bmatrix} = Y_{i+1|2i-1} \Big/ \begin{bmatrix} Y_{0|i} \\ U_{0|i} \\ U_{i+1|2i-1} \end{bmatrix}. \tag{10.118}$$

The data matrices $Y_{i+1|2i-1}$, $Y_{0|i}$, and others are defined analogously to those in Eq. (10.108).

The state vector $\hat{x}_{i+1}$, aligned with the same coordinate system used for $\hat{x}_i$, can be constructed as

$$\begin{aligned} \hat{X}_{i+1} &= \begin{bmatrix} \hat{x}_{i+1}(i+1) & \cdots & \cdots & \hat{x}_{i+1}(N-i+1) \end{bmatrix} \\ &= (\hat{\mathcal{W}}_{i-1}^o)^\dagger \begin{bmatrix} \hat{\bar{L}}_1 & \hat{\bar{L}}_2 \end{bmatrix} \begin{bmatrix} Y_{0|i} \\ U_{0|i} \end{bmatrix}. \end{aligned} \tag{10.119}$$

Here, $\hat{\mathcal{W}}_{i-1}^o$ is obtained by removing the last n_y rows from $\hat{\mathcal{W}}_i^o$, the extended observability matrix previously used to compute $\hat{x}_i$.

10.5.1.7 Obtaining the System Matrices

Once the data matrices for $\hat{x}_i(k)$ and $\hat{x}_{i+1}(k+1)$ are constructed, the system matrices A, B, and C can be estimated using linear LS. The Kalman gain K_i and the covariance matrix of $\varepsilon_i(k)$ can also be estimated from the residuals.

The system equation in matrix form, based on Eq. (10.98), is

$$\begin{bmatrix} \hat{X}_{i+1} \\ Y_{i|i} \end{bmatrix} = \begin{bmatrix} A & B \\ C & 0 \end{bmatrix} \begin{bmatrix} \hat{X}_i \\ U_{i|i} \end{bmatrix} + \begin{bmatrix} K \\ I \end{bmatrix} E_i, \tag{10.120}$$

where $$E_i = \begin{bmatrix} \varepsilon_i(i) & \varepsilon_i(i+1) & \varepsilon_i(i+2) & \cdots & \varepsilon_i(N-i) \end{bmatrix}. \tag{10.121}$$

The matrices A, B, and C are obtained via LS regression. From the output residual matrix $E_i(= Y_{i|i} - C\hat{X}_i)$, estimates for the innovation covariance matrix R and gain matrix K are computed as

$$\begin{aligned} \hat{R} &= \frac{1}{j} E_i E_i^T \\ \hat{K} &= \left(\frac{1}{j} (\hat{X}_{i+1} - A\hat{X}_i - BU_{i|i}) E_i^T \right) \hat{R}^{-1}, \end{aligned} \tag{10.122}$$

where $j = N - 2i + 1$.

Note: Equation (10.120) becomes considerably more complex when the white input assumption is relaxed. In such cases, consistent LS estimates require a modified Kalman filter initialization, where the initial estimate $\hat{x}_0(k-i)$ depends on the input history. Although the estimates obtained from the basic algorithm are biased under general input conditions, the bias diminishes as $i \to \infty$.

Example 10.8: *State-space Identification for the System in Example 10.7.*
MATLAB's System Identification Toolbox provides several options for state-space model estimation. The method discussed here is `n4sid`, which implements the subspace identification algorithm. The user may either specify a model order or allow MATLAB to automatically determine it.

To apply `n4sid` with a range of model orders, use the following command:

```
Gss=n4sid(allData,1:10);
```

Using this approach, the following model was identified:

$$\begin{aligned} x(k+1) &= \begin{bmatrix} 0.927 & -0.0804 \\ 0.277 & 0.711 \end{bmatrix} x(k) + \begin{bmatrix} -0.0174 \\ 0.0269 \end{bmatrix} u(k) + \begin{bmatrix} 0.00264 \\ -0.0113 \end{bmatrix} e(k) \\ y(k) &= \begin{bmatrix} -2.625 & -1.273 \end{bmatrix} x(k) + e(k). \end{aligned}$$

One can easily verify that the identified model is close to the discretized form of the true model (10.37). The impulse response of the model is very close to the ARMAX model (see Figure 10.2) and is skipped for brevity.

10.5.2 Analysis and Discussion

10.5.2.1 Properties and Advantages

The subspace identification method exhibits the following properties:

- The estimated deterministic system model is *asymptotically unbiased*, i.e., the bias vanishes as $N \to \infty$.
- The estimates of the covariance matrices are biased due to the use of a nonsteady-state Kalman filter in Eq. (10.98). This bias decreases as $i \to \infty$.

These results follow from the consistency properties of the LS estimator used in the algorithm.

Key advantages of subspace identification over the PEM include:

- Automatic model structure identification, avoiding the need to select model orders or determine structural parameters such as observability indices.
- Direct estimation of a stochastic state-space model without requiring nonlinear optimization, as is necessary for identifying ARMAX models.
- Built-in model reduction, which provides low-order models without additional postprocessing.

Example 10.9: *Identification of a 2 × 2 MIMO Model Using N4SID.*
Consider the following MIMO transfer function:

$$G = \begin{bmatrix} \frac{5.72}{60s+1}e^{-14s} & \frac{1.52}{25s+1}e^{-15s} \\ \frac{1.65}{30s+1}e^{-20s} & \frac{1.27}{6s+1} \end{bmatrix}. \tag{10.123}$$

Let the sampling interval be $t_s = 4$. Assume that the outputs are affected by a Gaussian white noise of variance $(0.1)^2$. We generate a 2-input PRBS signal (using `idinput`) to perform multi-input testing and store data using `iddata`. The MATLAB `n4sid` method is used to directly identify the MIMO model.

Figure 10.5 shows the order selection panel. The singular values give a convenient way to choose the model order. In contrast, identifying a parametric model such as `ARMAX` requires the user to specify the model order and delays for each input–output channel. The complete MATLAB code implementing the MIMO `n4sid` algorithm is available on GitHub: https://github.com/Lee-MPC-Book

To test the model, we generated another set of input signals. The responses of the two outputs are shown in Figure 10.6. The thin lines represent the experimental test data. Clearly, the identified model reliably represents the true system.

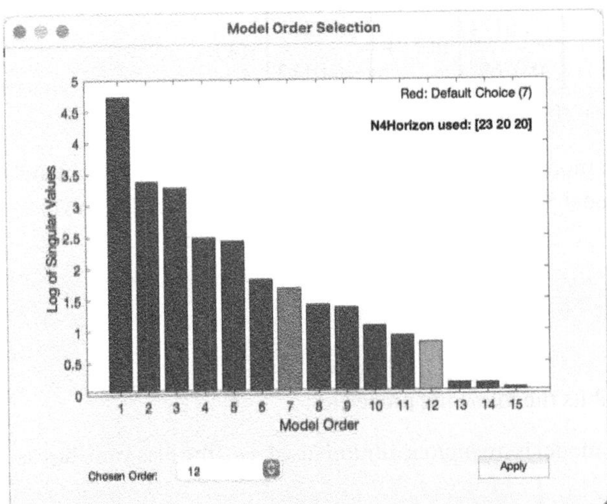

Figure 10.5 Selection of model order in `n4sid`.

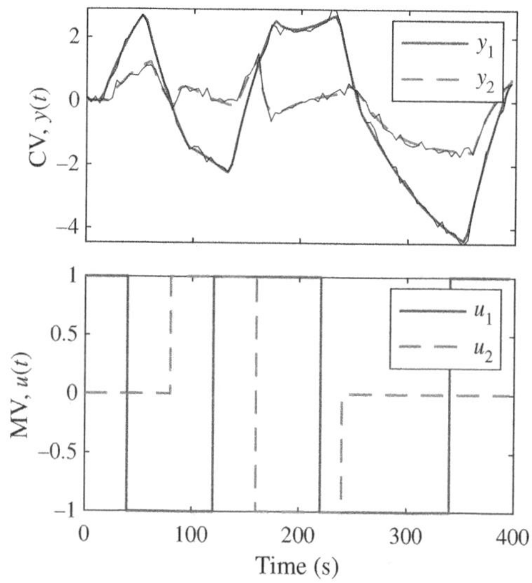

Figure 10.6 Comparison of the predicted output (thick lines) and test data (thin lines). The bottom panel represents the input signal.

10.5.2.2 Computational Issues

The subspace identification procedure described above consists entirely of linear algebra operations and does not require iterative optimization. As a result, its computational load is modest compared to that of the PEM.

While conceptually straightforward, the described implementation is not the most efficient or numerically stable. In practice, the method can be significantly improved using QR or RQ factorizations, which enhance both numerical stability and computational efficiency.

10.5.2.3 Other Issues

Despite its many advantages, subspace identification also has notable limitations. While the resulting model is asymptotically unbiased, little can be guaranteed about its quality when estimated from finite-length data, which is the reality in most practical applications.

Several nonideal factors, such as plant nonlinearity and data nonstationarity, can lead to correlation between the residual sequence $e(k+\ell|k)$, $\ell = 0, \ldots, i-1$ in Eq. (10.99) and the regressors. As a result, the matrices L_1 and L_2 estimated via LS—which are essential for determining system order and constructing Kalman state data—may exhibit both variance and bias. Variance becomes particularly problematic when the parameter i (the assumed upper bound of the state order) is chosen too large. While the variances in the estimated prediction matrices can sometimes be quantified, it remains difficult to trace their impact on overall model performance metrics such as prediction accuracy or frequency response fidelity.

One implication is that subspace methods typically require large data sets to be effective, which is expected given their minimal reliance on prior knowledge. Another is that subspace identification should not be seen as a replacement for PEM, but rather as a complement. For example, subspace methods are often used to provide high-quality initial estimates for PEM, which benefits from good parameter initialization and a suitable model structure.

A more critical issue arises when the input and output data are correlated—such as in closed-loop operation. In this case, the prediction errors $e(k+\ell|k)$ may become correlated with the input vector $\mathcal{U}_i^{0+}(k)$, breaking the orthogonality assumption and invalidating the consistency of the LS estimate. Several strategies have been proposed to mitigate this problem. One approach involves replacing $\mathcal{Y}^{0+}(k)$ with one-step-ahead predictions generated using a high-order ARX model, thereby removing the correlation. The bias associated with using closed-loop data in subspace identification has been addressed in detail by researchers; see Verhaegen (1993), Van Overschee and De Moor (1997), and Qin and Ljung (2003) for further discussion.

Lastly, it is important not to overlook the assumption that the stochastic component of the system is stationary. If nonstationarity arises primarily due to integrating disturbances (leading to slow mean shifts), the input–output data can be differenced prior to identification. Additional low-pass filtering may also be warranted to avoid excessive emphasis on high-frequency components—see the earlier discussion on bias distribution in the frequency domain.

10.6 Practice of System Identification: A User's Perspective

Thus far, we have explored various system identification methods and their theoretical foundations. However, successful system identification involves far more than simply understanding and applying these techniques. The practitioner must carefully design experiments to generate informative data, preprocess the data to mitigate noise and outliers, select an appropriate model structure, apply a suitable identification algorithm, and finally validate and refine the model to ensure its adequacy for model-based control applications. The overall identification process, including its key steps, is illustrated in Figure 10.7.

In this section, we adopt the user's perspective and provide a practical overview of these critical steps, highlighting key considerations and challenges encountered in real-world applications.

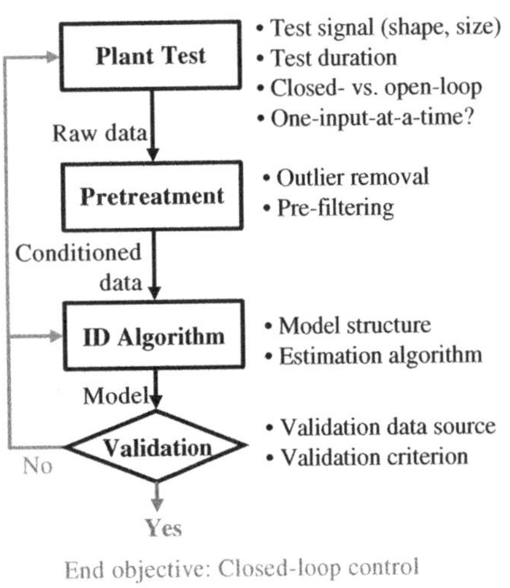

Figure 10.7 Overall process and key steps involved in system identification.

10.6.1 Experiment Design

In most practical scenarios, the data required for system identification must be obtained by conducting experiments on the actual process. These experiments are often expensive and time-consuming, as they require direct engagement with operational systems. In the process industries, it is estimated that up to 90% of the total cost and effort associated with commissioning a model-based controller is attributable to plant testing. This underscores the importance of careful experiment design and the need to optimize these tests for efficiency and informativeness.

Despite its significance, the field of experiment design for system identification remains relatively underdeveloped. As such, this discussion will focus on qualitative insights into key decisions, rather than a comprehensive treatment of formal design techniques.

Sample Time and Test Duration

The sampling interval for data collection is typically dictated by existing hardware constraints, such as data acquisition systems, actuators, and sensors. However, the *sampling theorem* (see Section 5.1.2) gives an upper bound on the sample period. At the same time, choosing an excessively small sample period for a slow system can lead to numerical challenges, such as clustering of system poles near (1, 0) in the z-domain, which may degrade numerical robustness and model quality. Longer experiments generally yield more informative data, but they also incur higher operational costs. Under certain assumptions, the variance of parameter estimates and its distribution can be analyzed in the context of LS estimation. These analyses provide useful insights into the expected information content of a given dataset and can guide the planning of experiment duration.

Test Signal:

The choice of input excitation is often the most critical design decision, as it has a major impact on model quality. However, the choice of excitation signal is often constrained by operational limitations. In practice, the potential impact of test perturbations on plant operations may restrict the amplitude, duration, or frequency content of the test signal. The power distribution across frequency components directly affects the bias and variance of the resulting model and, ultimately, the achievable closed-loop performance. Commonly used excitation signals in practice include sequences of steps or pulses, random signals, and PRBS signals. The latter two are often favored in theory due to their ability to deliver high power and relatively uniform spectral content within the Nyquist band. We will examine PRBS signals in greater detail in Section 10.6.2.

Single-input vs. Multiple-input Testing:

An important decision in experiment design is whether to vary inputs one at a time or simultaneously. The "one-input-at-a-time" approach is generally safer, as its effects on the process are more predictable and easier to manage. However, varying multiple inputs simultaneously provides more information within a given time frame. For highly interactive systems, it is often difficult to capture the necessary multivariable dynamics using single-input tests, as they emphasize SISO model accuracy, which may be inadequate for such systems. When multiple-input testing is employed, care must be taken to design the test signals such that both their autocorrelation and cross-correlation properties are optimized to ensure effective excitation and model identifiability.

Open-loop Testing vs. Closed-loop Testing:
The examples included in the chapter so far used open-loop testing, as illustrated in Figure 10.8a. Alternatively, experiments can be conducted in a closed-loop operation if a feedback controller is already present (Figure 10.8b). In principle, when a suitable controller is present, closed-loop testing more closely reflects the operating conditions under which the model will ultimately be used. If closed-loop testing is chosen, two important issues must be considered:

- First, naturally occurring disturbances alone may not provide sufficient excitation to generate data with adequate information content. The absence of an external excitation signal can lead to identification of the inverse of the controller rather than the true process dynamics.
- Second, the presence of feedback introduces a correlation between past outputs and future inputs in the data. This violates assumptions underlying many conventional identification algorithms, such as linear LS, and can result in biased parameter estimates (see the discussion on convergence in Section 10.3.2). To address this, a common strategy is to inject an external excitation—such as a dither signal (as seen in Figure 10.8c)—and identify the *closed-loop transfer function* between this signal and the system output. From this, an open-loop model of the plant can be recovered using appropriate signal processing techniques.

10.6.2 PRBS Signals

PRBS signals are preferred for system identification due to their ease of generation, implementation simplicity, and favorable spectral properties. Unlike truly random signals, PRBS signals exhibit consistent statistical characteristics, even with relatively short signal lengths. In this Section, we examine the properties and design of PRBS signals in more detail. An accessible overview can be found in Tan and Godfrey (2001), and its application to FIR model identification is further discussed by Sung and Lee (2003).

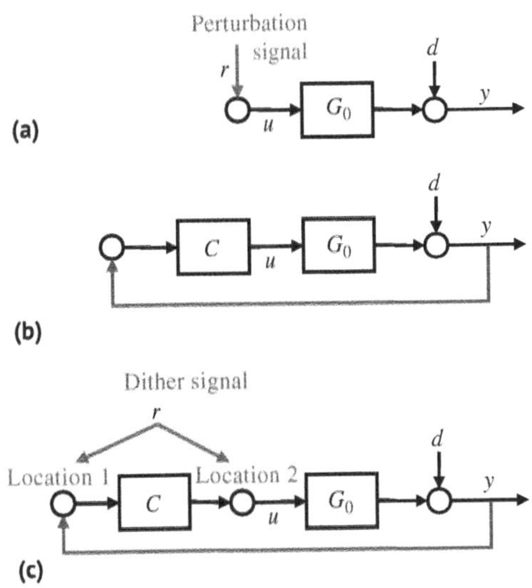

Figure 10.8 Illustration of three testing configurations: (a) Open-loop testing, (b) closed-loop testing, and (c) closed-loop testing with external dithering.

10.6.2.1 The Covariance and Power Spectrum of PRBS

A PRBS (denoted by η) is a periodic binary signal of period M that alternates between a high-level $\bar{b}$ and a low-level $\underline{b}$. Its covariance function $R_\eta(\tau)$ (assuming perfect periodicity) is given by

$$R_\eta(\tau) = \begin{cases} B^2 \text{ for } \tau = 0 \\ -\dfrac{B^2}{M} \text{ for } \tau \neq 0. \end{cases} \tag{10.124}$$

Here, $B = \bar{b} - \dfrac{\bar{b} + \underline{b}}{2}$ represents the deviation from the mean value. By choosing a sufficiently large M, the covariance function can be made to resemble that of white noise.

The corresponding power spectrum of the PRBS signal is given by

$$\begin{aligned}
\Phi_\eta(\omega) &= \sum_{\tau=-\infty}^{\infty} R(\tau)\exp(-j\omega\tau) \\
&\approx R(0) + R(M)\exp(-j\omega M) + R(2M)\exp(-j2\omega M) + \cdots \\
&\quad + R(-M)\exp(j\omega M) + R(-2M)\exp(j2\omega M) + \cdots \\
&= B^2 + 2B^2\cos(\omega M) + 2B^2\cos(2\omega M) + \cdots.
\end{aligned}$$

This power spectrum is zero except at discrete frequency points $\omega_i = i\frac{2\pi}{M}$. The power at each of these frequencies is equal but theoretically infinite. This makes sense because the integral of the power spectrum over $(0, 2\pi)$ must equal $R(0)$ (which is the signal power), and the spectrum is nonzero only at discrete points. For periodic signals, the discrete Fourier transform (DFT) of the covariance function is a convenient alternative:

$$\Phi(\omega_i) = \sum_{\tau=0}^{M-1} R(\tau)\exp(-j\omega_i\tau) \approx R(0) = B^2 \quad \forall\omega_i.$$

Thus, a PRBS distributes its power uniformly across the discrete frequency points within the Nyquist band.

One advantage of PRBS is that, being deterministic, its statistical properties hold even for relatively short signal lengths. For example, for a PRBS signal of length N that is an integer multiple of M,

$$\frac{1}{N}\sum_{i=1}^{N} \eta(k)\eta(k-\tau) = \begin{cases} B^2 \text{ for } \tau = 0 \\ -\dfrac{B^2}{M} \text{ for } \tau \neq 0. \end{cases} \tag{10.125}$$

For truly random signals, a much larger N would be needed for this relationship to hold approximately. Since in practice one is often limited to collecting short data records, this property of PRBS can be particularly advantageous.

10.6.2.2 Generation of a Single PRBS

PRBS signals are commonly generated using *shift registers*. Figure 10.9 illustrates a 4-bit shift register configuration. The symbol $\oplus$ denotes binary addition (modulo-2), defined as

$$\begin{aligned}
1 \oplus 1 &= 0 \\
0 \oplus 1 &= 1 \\
0 \oplus 0 &= 0.
\end{aligned}$$

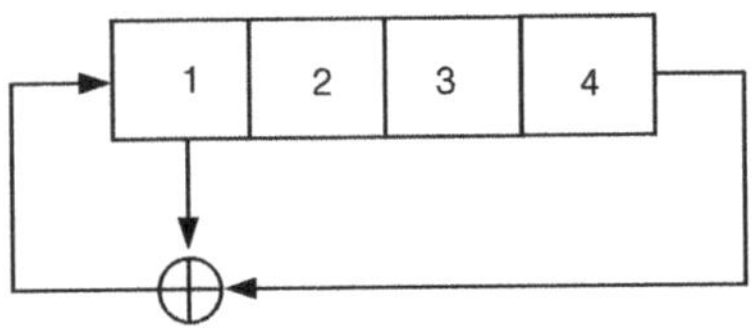

Figure 10.9 Generation of a PRBS signal using a linear feedback shift register (LFSR).

Using this method, a periodic binary sequence of length $2^{N_s} - 1$ can be generated, where N_s is the number of shift registers. Here, 0 represents the lower signal level $\underline{b}$ and 1 represents $\bar{b}$. Once a full-period sequence is obtained, it can be repeated as needed to extend the signal to a desired length.

The procedure for generating a PRBS using a shift register is

1. Initialize the shift register with a nonzero binary pattern (a combination of 0s and 1s). Different initial conditions produce different sequences.
2. Compute the binary sum (modulo-2) of selected register values, typically the first and last.
3. Shift all register values one position to the right.
4. Insert the computed binary sum into the first register.
5. Repeat steps 2–4 until $2^{N_s} - 1$ values have been generated. The time history of any register can be used as the PRBS sequence.

10.6.2.3 Generation of Multi-input PRBS Signals

When performing a multiple-input test—where several inputs are varied simultaneously—it is often desirable to design multiple PRBS signals that are mutually independent. There are several ways to achieve this.

One simple and robust way to do this is to design a PRBS of period $2M$. We represent this signal as $\eta \triangleq \begin{bmatrix} \eta_1^T & \eta_2^T \end{bmatrix}^T$, where η_1 and η_2 are the first and second halves of the PRBS, respectively. A two-dimensional PRBS of period $2M$ can be defined as

$$u_1 = \begin{bmatrix} \eta_1 \\ \eta_2 \end{bmatrix} \qquad u_2 = \begin{bmatrix} \eta_2 \\ \eta_1 \end{bmatrix}.$$

This idea can be naturally extended to higher-dimensional cases. For an n_u-input system, one can construct a PRBS of period $n_u \times M$ and define:

$$u_1 = \begin{bmatrix} \eta_1 \\ \eta_2 \\ \vdots \\ \eta_{n_u} \end{bmatrix} \qquad u_2 = \begin{bmatrix} \eta_2 \\ \vdots \\ \eta_{n_u} \\ \eta_1 \end{bmatrix} \qquad \cdots$$

All cross-variance terms become $-B^2/M$. Thus, the cross-correlation between input signals can be made negligibly small by choosing a large M.

It is recommended to select M based on the maximum settling time of the system (with respect to all inputs). This choice ensures that the number of identifiable parameters—$M \times n_u$—is appropriate for capturing the dynamic characteristics of the system.

10.6.2.4 Period and Number of Repetitions

The total duration of the experiment is given by $M \times n_p$, where M is the period of the PRBS signal and n_p is the number of repetitions. If M is chosen too small, the input power will not be distributed across a sufficient number of frequency points. Additionally, the quantity B^2/M may become significant, resulting in a covariance function that deviates from that of ideal white noise.

It is important to note that a PRBS signal is persistently exciting of order M, meaning it can be used to identify at most M parameters. Therefore, when identifying an FIR model, M should be chosen to be equal to or greater than the number of sampling steps corresponding to the system's settling time.

In some cases, it may be undesirable to spread the excitation power across too many frequency components—for example, when one is only interested in modeling a narrow frequency band. In such scenarios, it is advisable to limit the value of M, while increasing n_p to achieve the desired signal length and maintain sufficient excitation energy.

10.6.2.5 Bias and Variance of PRBS

The key parameters to be selected in PRBS generation are the upper value $\bar{b}$, the lower value $\underline{b}$, the period M, and the number of repeated periods n_p. By specifying $\bar{b}$ and $\underline{b}$, one determines both the mean value $((\bar{b}+\underline{b})/2)$ and the variance $\left((\bar{b}-\underline{b})^2/4\right)$ of the signal.

The mean value is typically chosen to match the steady-state input value of the process. The choice of variance requires careful consideration. A higher variance implies a stronger excitation signal, resulting in a higher signal-to-noise ratio, which generally improves the quality of parameter estimation. However, large amplitude signals may excite nonlinear behavior in the input–output dynamics, thereby violating the assumptions of linear model structures. In addition, aggressive input variation can disrupt ongoing plant operations or violate safety and operational constraints. Therefore, selecting an appropriate variance level involves a trade-off. One must balance the need for informative excitation with the risk of exciting nonlinearities or disturbing the process. Achieving this balance typically requires some familiarity with the system dynamics and operating constraints.

10.6.3 Data Preprocessing

Before applying an identification algorithm, the input–output data must be preprocessed. The first step is to perform basic cleaning tasks such as removing spikes and outliers. Next, the data should be detrended—that is, the bias removed and deviation variables computed. This can be done in several ways:

1. Use known steady-state values to compute deviation variables:

$$y(k) = y_p(k) - y_{ss} \qquad u(k) = u_p(k) - u_{ss}$$

   where $y_p(k)$ and $u_p(k)$ represent the raw output and input data, respectively, and y_{ss} and u_{ss} are the steady-state output and input values.
2. Use the average of the data to compute deviation variables:

$$y(k) = y_p(k) - y_{\mathrm{ref}} \qquad u(k) = u_p(k) - u_{\mathrm{ref}}$$

$$y_{\mathrm{ref}} = \frac{\sum_{k=1}^{N} y_p(k)}{N} \qquad u_{\mathrm{ref}} = \frac{\sum_{k=1}^{N} u_p(k)}{N}$$

3. Use the differenced data:

$$\Delta y(k) = y(k) - y(k-1) \qquad \Delta u(k) = u(k) - u(k-1)$$

The appropriate detrending method depends on the specific context. Option 1 requires the user to have knowledge of steady-state values. Option 2 may perform poorly if the dataset is not sufficiently long relative to the settling time. If the identification experiment spans a long period and the means of the input and output vary significantly (e.g., due to load disturbances), Option 3 (differencing) may yield the most robust results. Note that selecting Option 3 for detrending implicitly assumes that $H(q)$ contains an integrator.

After detrending, additional prefiltering of the data may be useful. Consider the case of a known or fixed structure of the noise model. Suppose that the model structure to be identified is

$$y(k) = G(q,\theta)u(k) + H_1(q)H_2(q,\theta)\varepsilon(k), \tag{10.126}$$

where $H_1(q)$ is a known component of the noise model and $H_2(q,\theta)$ is unknown. In this case, we can rewrite the model as

$$\underbrace{\frac{1}{H_1(q)}y(k)}_{y_f(k)} = G(q,\theta)\underbrace{\frac{1}{H_1(q)}u(k)}_{u_f(k)} + H_2(q,\theta)\varepsilon(k) \tag{10.127}$$

Thus, the input and output data can be prefiltered with $1/H_1(q)$ to recast the model into a standard form (e.g., ARX, ARMAX, OE).

Another reason for prefiltering is when one may wish to minimize a filtered prediction error, $L(q)e(i)$, in order to emphasize (or de-emphasize) certain frequency ranges. This can be accomplished by prefiltering the data with $L(q)$ before applying PEM.

10.6.4 Model Fitting and Validation

10.6.4.1 Getting a "Feel" for the Problem

It is advisable to begin system identification with simple and easy-to-implement methods. These allow the user to explore the basic characteristics of the system without being distracted by numerical artifacts or algorithmic complications, and aid the user to determine a suitable model order and process delays.

Identification of a FIR model is often the simplest, as it does not require much prior information and results in a linear LS problem. Likewise, PEM using an ARX model is also a linear LS problem. The subspace identification (e.g., N4SID) and frequency response identification (e.g., ETFE) are also relatively straightforward techniques. These methods can provide a quick "feel" for the nature and complexity of the identification problem.

Typically, one or more initial models are tried to get a feel for the problem. The performance of the identified model(s) is analyzed. If the models provide satisfactory predictions on the validation data, the identification problem is likely to be well-posed and relatively straightforward. However, if the results are inconsistent or unsatisfactory, the problem may be more complex and further analysis is warranted. Section 10.6.4.2 outlines these in more detail.

10.6.4.2 Model Assessment

A key step is to validate and analyze the identified model. Validation is often performed with an independent validation dataset. The following analyses can be performed using these models:

- Plot the step and frequency responses of the identified models. If multiple models are identified, compare the results for different model structures.
- Generate separate *validation data* and evaluate model predictions using this dataset. This allows assessment of model performance on unseen data and helps detect potential overfitting.
- Analyze the residuals from the model to assess whether they resemble white noise and whether there is significant cross-correlation with the input data. This residual analysis can help detect bias in the model. However, it may not reveal variance-related issues (e.g., overfitting), since models with more parameters typically produce smaller residuals on the training data.

The following are issues that typically arise with an identified model.

- **Bias Due to Insufficient Model Order:** If the model order is too low, the resulting model will be biased. For example, in the ARX structure, the noise is modeled as $\frac{1}{A(q)}\varepsilon(k)$. Thus, a poor representation of the noise via $1/A(q)$ leads to biased parameter estimates. Increasing the order of $A(q)$ may help reduce bias; however, higher-order models introduce more parameters and increase the parameter variance. Switching to an ARMAX structure can better account for the noise and reduce the required model order. However, near pole-zero cancellation and significant autocorrelation between the residuals may still be an issue. A similar principle applies to subspace identification: If the selected maximum model order is too low relative to the true system order, bias may result. Bias is generally not expected with FIR identification or ETFE. However, complex noise structures can still show slow convergence. If noise characteristics are known *a priori*, prefiltering the data may help mitigate this effect.
- **Large Parameter Variances:** The parameter variances can become large when the dataset is too short relative to the noise level and the number of estimated parameters. Although the model fit may appear satisfactory (i.e., low residual value during model fitting), the model may perform poorly on the validation data. This issue can also manifest itself as significant disagreement among models derived from different structures. FIR identification and ETFE are vulnerable to this issue due to their high variance. Parametric models—such as ARX—may also suffer if a high model order is used in an attempt to model noise. Subspace identification is similarly affected when both the upper bound and the chosen model order (e.g., based on SVD) are too high.
- **Feedback in the Data:** If the data contain feedback effects—such as from an existing control system—the resulting model may be biased. This bias often becomes evident through poor prediction performance on validation data. FIR identification, ETFE, and subspace identification are especially sensitive to this issue. The problem tends to be less severe for PEM with an ARX structure, provided that the model is sufficiently rich to account for the dynamics introduced by feedback.
- **Nonlinearity:** Severe nonlinearities in the system can significantly hinder identification, even when the goal is to obtain the best possible linear approximation. Rather than immediately turning to complex nonlinear models (e.g., neural networks), it is often preferable to consider the underlying physics of the system. In many cases, simple variable transformations—such as taking square roots or logarithms—can help linearize the behavior and improve model performance.

10.6.4.3 Further Improvements and Refinements

If the results obtained from the initial application of simple identification methods are unsatisfactory, one can attempt to improve them using more advanced techniques. Even if the initial results are acceptable, further refinement may still be desirable—especially when preparing the model for control design or long-term prediction.

At this stage, it is helpful to ask whether it is important to identify an accurate noise model. If an accurate noise model is not required, then simpler structures such as the OE model or the IV method may be used. The OE method is particularly well-suited when the structure of the noise model has already been specified *a priori*.

If an accurate noise model is desired, one needs to ask whether it is necessary to model the correlation among different outputs. If modeling the correlation among outputs is not necessary, a MISO ARMAX identification may be used. However, in certain cases—such as inferential control or when outputs are correlated—it is advisable to start with subspace identification and subsequently refine the model using PEM with an appropriate MIMO ARMAX structure.

Regardless of the approach, it is critical to validate the resulting models using a fresh dataset. We say that the model prediction is satisfactory when the prediction error is approximately white and exhibits minimal correlation with the input. If these conditions are not met, then the problem may lie in the input data itself (e.g., insufficient excitation), or the model may suffer from one or more of the issues discussed earlier (e.g., insufficient order, high variance, feedback bias, or nonlinearity).

When Is the Identified Noise Model Useful for Control?

Some identification techniques—such as subspace identification or ARMAX—produce explicit noise models. These noise models are only useful for control design if the noise characteristics observed during the identification experiment are representative of those expected during actual operation. If the system was operated under unusually stable conditions during data collection, the resulting noise model may not reflect the true disturbance environment of the process. In such cases, it is not advisable to use the identified noise model for control design; instead, one may need to design a more realistic noise model.

An inadequate noise model will bias the estimation of the deterministic component of the identified model as well. Thus, having an adequate noise model structure can still benefit the identification process. Some identification methods—such as OE—are relatively insensitive to poor noise modeling. Therefore, the modeling strategy should be chosen based on the importance of an accurate noise model, which is determined by the control objective and system characteristics.

10.6.5 Model Quality Assessment and an Integrated Framework

Ultimately, the quality of an identified model can only be fully judged by designing and implementing a model-based controller on the actual process. Nevertheless, it is desirable to have some—albeit approximate—means of assessing model quality and anticipating potential closed-loop performance prior to deployment.

In Section 10.6.4.2, we introduced two types of errors—bias and variance—that are used to quantify model quality. Bias, which reflects systematic modeling errors, is typically a result of inadequate model structure or insufficient input excitation and persists regardless of data volume. Variance results from data noise and decreases as more data are used. A trade-off between bias and variance is usually unavoidable; striking a balance between the two is central to effective model selection.

If sufficient data is available, bias can be estimated by fitting a higher-order model to the residuals to detect systematic patterns. In the absence of bias, variance can be quantified analytically in simple linear estimation settings (or approximately using linearization techniques).

The impact of model uncertainties on the achievable closed-loop performance can be assessed by expressing model uncertainty in the frequency domain and analyzing its effect using the tools of *robust control* theory. The integration of system identification with robust controller design has received considerable attention. Although promising theoretical developments have been made, a comprehensive and practical framework, particularly for general multivariable systems, has yet to be established.

An integrated framework for identification, model quality assessment, and controller design typically includes the following components:

- Quantitative assessment of model quality and its implications for control performance.
- Decisions on whether additional data collection is needed.
- Design of new experiments informed by current model quality.
- Iterative updates to the model as new data become available (i.e., iterative identification).
- Continuous adaptation of the model and controller (adaptive identification).

Although we do not yet possess all the tools required to fully realize such an integrated framework, keeping this holistic perspective in mind is valuable. It can guide ongoing research and development efforts aimed at bridging the gap between identification theory and practical control system design.

Exercises

1. Show that the pseudo-linear form described in Eq. (10.29):

$$\hat{y}(k|k-1) = B(q)u(k) + [1 - A(q)]\, y(k) + [C(q) - 1]\, [y(k) - \hat{y}(k|k-1)]$$

is the optimal one-step-ahead predictor for the ARMAX model.

2. Consider the following linear regression problem:

$$y = a_0 + a_1\varepsilon, \qquad a_0 = 2,\ a_1 = 0.1$$

Here, ε is a zero-mean Gaussian noise, which can be generated using `epsi=randn(1000,1);`. Compute the estimate of the parameter a_0 and the covariance of the noise term (note that $a_1\varepsilon$ is also a zero-mean Gaussian noise). Define and express the 95% confidence interval for the estimated parameters.

3. Consider the discrete-time state-space system:

$$x(k+1) = \begin{bmatrix} 0 & 1 & 0 \\ 0 & 0 & 1 \\ 0 & 0 & 0 \end{bmatrix} x(k) + \begin{bmatrix} 0 \\ 0 \\ 1 \end{bmatrix} u(k),$$

$$y(k) = \begin{bmatrix} 1 & 0 & 0 \\ 0 & 1 & 0 \end{bmatrix} x(k),$$

where $x(k) \in \mathbb{R}^3$, $u(k) \in \mathbb{R}$, and $y(k) = \begin{bmatrix} y_1(k) & y_2(k) \end{bmatrix}^T \in \mathbb{R}^2$.

(a) Derive the minimal ARX representation for each output separately. That is, find polynomials $a_{ij}(q^{-1})$ and $b_{ij}(q^{-1})$ such that:

$$\begin{bmatrix} y_1(k) \\ y_2(k) \end{bmatrix} = \begin{bmatrix} a_{11}(q^{-1}) & a_{12}(q^{-1}) \\ a_{21}(q^{-1}) & a_{22}(q^{-1}) \end{bmatrix} \begin{bmatrix} y_1(k) \\ y_2(k) \end{bmatrix} + \begin{bmatrix} b_{11}(q^{-1}) \\ b_{21}(q^{-1}) \end{bmatrix} u(k) + \begin{bmatrix} e_1(k) \\ e_2(k) \end{bmatrix}.$$

(b) Compute the *observability index* 0_1 for y_1 and 0_2 for y_2. Verify that $0_1 = 3$ and $0_2 = 2$.

(c) Suppose we identify a single SIMO ARX model for both outputs with a common autoregressive order n_a:
- Show that if $n_a = 2$ (chosen from 0_2), the regression vector for y_1 lacks information about x_3, making some parameters unidentifiable.
- Show that if $n_a = 3$ (chosen from 0_1), the regression vector for y_2 contains redundant terms, leading to linear dependence in the regressor matrix and *loss of identifiability*.

(d) Explain why the mismatch in observability indices for different outputs can cause rank deficiency in the regression matrix in MIMO ARX identification.

(e) Suggest a remedy for this identifiability issue (e.g., different model orders per output, or use of state-space subspace identification methods).

4. Use the PEM method to identify an ARMAX model for the system:

$$G(s) = \frac{0.5}{19.36s^2 + 7.04s + 1}.$$

Use the data generated in Example 10.4, where the input perturbation is chosen as a PRBS signal and the measurements include an additive white noise of covariance $(0.04)^2$.

Extend the MATLAB code of Example 10.2 (where we used linear regression to identify ARX model) to perform pseudo-linear regression for an ARMAX model. Verify the model by comparing the impulse response of the identified model with the true model.

5. Prove that for an ARX model with $n + m$ parameters, the use of a signal that is persistently exciting of order $n + m$ guarantees that the rank condition for consistency is satisfied.

6. Show that Eq. (10.64) can be solved using the given recursive formula.

7. Derive the Bayesian estimator for system (10.68) using Kalman filter theory.

8. Show that $\hat{x}_i(k)$ in subspace identification is a linear function of $y(k-1), \ldots, y(k-i)$ and $u(k-1), \ldots, u(k-i)$. Derive the explicit form, including expressions for H_1 and H_2.

9. Consider the following ARMAX model:

$$\begin{aligned} y(k) - 0.54y(k-1) + 0.14y(k-2) &= 0.2u(k-3) \\ &+ 0.1u(k-4) + \varepsilon(k) + 0.25\varepsilon(k-1) \end{aligned} \tag{10.128}$$

Generate test data using a PRBS input signal. Assume that $\varepsilon(k)$ is zero-mean white noise with variance 0.1^2. Note that the system order and input delay are not known *a priori*. Use the `SysID` toolbox in MATLAB to identify a suitable ARMAX model.

10. Since determining the input delay requires trial and error, a practical approach is to perform FIR identification first, then refine the model using ARMAX. Use the data from the previous exercise to identify a FIR model. Estimate the input delay based on the FIR coefficients, then reapply ARMAX identification using the estimated delay.

11. Apply subspace identification to the dataset used in Exercise 9. Compare the identified model with the ARMAX model from the previous exercises.

12. Consider a discrete-time, state-space model:

$$x(k+1) = \begin{bmatrix} 0.8 & 0.1 \\ 0 & 0.9 \end{bmatrix} x(k) + \begin{bmatrix} 0.1 \\ 0.05 \end{bmatrix} u(k) \tag{10.129}$$

$$y(k) = \begin{bmatrix} 1 & 0 \end{bmatrix} x(k) + v(k), \tag{10.130}$$

where $v(k)$ is a Gaussian measurement noise with $\sigma_v^2 = 0.01$.
Apply the *N4SID* method to estimate the system matrices A, B, and C, along with the noise model parameters of the innovation form, K and R_ε, using simulated input–output data as described in the chapter. Perform the following steps:

(a) *Data Generation*: Generate a PRBS input $u(k)$ of length $N = 511$ with amplitude ± 1. Simulate the system to obtain $y(k)$.

(b) *Data Matrix Construction*: Choose a user-defined *past horizon* i and *future horizon* j. By default, use $i = j = 15$. Construct the block Hankel matrices for past inputs $U_{0|i-1}$, past outputs $Y_{0|i-1}$, future inputs $U_{i|2i-1}$, and future outputs $Y_{i|2i-1}$.

(c) *Orthogonal Projection*: Obtain the LS estimate of the predictor matrices $[L_1 \ L_2 \ L_3]$ as

$$\begin{bmatrix} \hat{L}_1 & \hat{L}_2 & \hat{L}_3 \end{bmatrix} = Y_{i|2i-1} \Big/ \begin{bmatrix} Y_{0|i-1} \\ U_{0|i-1} \\ U_{i|2i-1} \end{bmatrix}, \tag{10.131}$$

where the LS solution, $A/B \triangleq AB^T(BB^T)^{-1}$.

(d) *SVD and Model Order Selection*: Perform SVD of the projected matrix to estimate the system order n_x. Compare the singular values to identify the correct system order.

(e) *System Matrices Recovery*: Apply the LS method to the estimated state sequence data matrices to determine the system matrices A, B, and C. Then, use the residuals to compute the Kalman gain K and the innovation covariance matrix R_ε.

(f) *Validation*: Simulate the identified model with the original PRBS input and compare the output $y_{\mathrm{id}}(k)$ with the true output $y(k)$. Also compute the fit percentage as

$$\mathrm{Fit} = 100\left(1 - \frac{\| y - y_{\mathrm{id}} \|_2}{\| y - \bar{y} \|_2}\right)\%. \tag{10.132}$$

Part III

Advanced MPC

11

Linear MPC: State-space Formulation

Innovation is taking two things that already exist and putting them together in a new way.

– Tom Freston

The basic industrial Model Predictive Control (MPC) algorithm discussed in Part I is based on finite impulse or step response models. While effective in many cases, these models have inherent limitations: they are often high-dimensional, offer limited flexibility in modeling noise and disturbances, and are unsuitable for open-loop unstable systems. Additionally, they are not well-suited for incorporating first-principles models (when available), handling time-varying dynamics or process drifts, or accommodating multi-rate sampling.

In contrast, the state-space modeling framework introduced in Part II offers a more general and flexible approach. Over the past few decades, most major research developments in MPC have adopted state-space models as the foundational representation. As discussed in Chapter 5, various modeling frameworks can be readily converted to state-space form—and often vice versa—enabling broader applicability and integration with estimation and control tools.

As noted in Part I, the origins of impulse response-based MPC can be traced back to early industrial implementations in the 1970s—such as Richalet's *Model-Based Heuristic Control* and Cutler's *Dynamic Matrix Control* (DMC). These pioneering works marked the first industrial use of explicit prediction of future trajectories in control design. However, the broader concepts of predictive modeling and dynamic optimization in control predate these implementations by at least two decades. Modern MPC also shares deep connections with linear quadratic control theory, as discussed in Chapters 7 and 8, and 9.

The standard state-space model structure used in classical linear quadratic control is given by

$$\begin{aligned} x(k+1) &= Ax(k) + Bu(k) + \varepsilon_1(k) \\ y(k) &= Cx(k) + \varepsilon_2(k), \end{aligned} \tag{11.1}$$

where $\varepsilon_1(k)$ and $\varepsilon_2(k)$ represent Gaussian white noise sequences with covariances R_1 and R_2, respectively. This model is adequate for regulatory control where the unknown inputs can be reasonably approximated by white noise. However, it offers limited guidance for practical control problems involving output regulations to set points under structured or correlated disturbances. To apply the optimal estimation and control tools introduced in Part II, practical system models must be transformed into the form of (11.1), with disturbances modeled using white noise terms.

In this chapter, we extend the discussion from Section 6.6 and demonstrate how to perform such transformations and develop output-feedback controller designs that assimilate and extend the original MPC formulations, such as DMC. We also address key implementation challenges and analyze the resulting state-space MPC frameworks.

11.1 Model Construction

The standard state-space model of the form (11.1) must be extended to accommodate various elements commonly encountered in practical MPC applications. The purpose of this section is to present a general framework for constructing an augmented state-space model that can incorporate structured disturbances and measurement noise. The desired form is

$$\begin{aligned} \bar{x}(k+1) &= \bar{A}\bar{x}(k) + \bar{B}u(k) + \varepsilon_1(k) \\ y_m(k) &= \bar{C}\bar{x}(k) + \varepsilon_2(k), \end{aligned} \tag{11.2}$$

which retains the same structure as (11.1). Here, $\bar{x}(k)$ denotes the augmented state vector, which will be defined in this section to ensure that the external input terms $\varepsilon_1(k)$ and $\varepsilon_2(k)$ represent white noise sequences. The matrices $(\bar{A}, \bar{B}, \bar{C})$ are constructed from the system matrices of the original model (11.1), modified as needed to meet practical modeling requirements.

When offset-free set point tracking is required in the presence of persistent disturbances (as in DMC), it is convenient to employ a *differenced model*. The corresponding augmented state-space representation is given by

$$\begin{aligned} \bar{x}(k+1) &= \bar{A}\bar{x}(k) + \bar{B}\Delta u(k) + \varepsilon_1(k) \\ y_m(k) &= \bar{C}\bar{x}(k) + \varepsilon_2(k), \end{aligned} \tag{11.3}$$

where the resulting augmented matrices and state are different than those in (11.2). For example, $\bar{A}$ for the differenced model inherently includes integrators introduced by the differencing operation. In (11.3), the manipulated input appears as its incremental change $\Delta u(k)$, enabling a quadratic penalty to be applied to Δu—rather than to u—in the usual way. The augmented system matrix $\bar{A}$ naturally incorporates integrators, ensuring offset-free behavior. The resulting controller is expressed in the so-called "velocity form," in which the incremental change of the manipulated input is computed instead of its absolute value.

The remainder of this section presents systematic methods for constructing augmented models of the forms in Eqs. (11.2) and (11.3), tailored for state-space-model-based MPC. We will discuss several alternative definitions of the augmented state $\bar{x}(k)$ and corresponding formulations of $\bar{A}$, $\bar{B}$, and $\bar{C}$ that guarantee offset-free tracking and robust control performance under various disturbance types. Insights from DMC (Part I) and LQ control theory (Part II) will be used to systematically develop these models for practical MPC design.

11.1.1 Model Structure for State-space MPC

Figure 11.1 illustrates the overall modeling structure assumed for state-space MPC. The linear state-space representation of this system is given by

$$\begin{aligned} x(k+1) &= Ax(k) + Bu(k) + B_d d(k) \\ y(k) &= Cx(k) + w_y(k) \end{aligned} \tag{11.4}$$

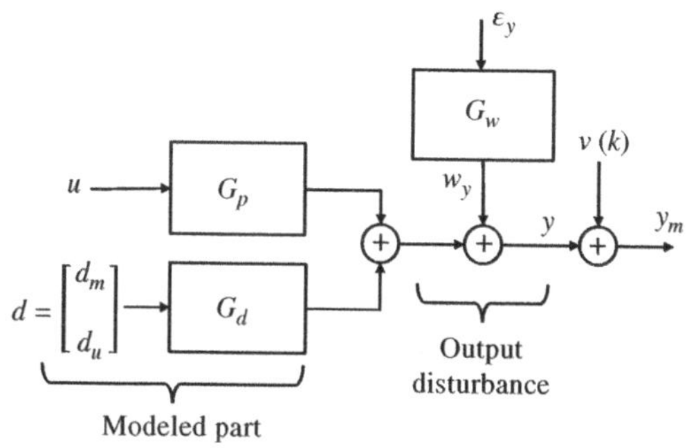

Figure 11.1 Overall modeling structure for state-space MPC. Here, $d(k)$ includes both measured ($d_m(k)$) and unmeasured ($d_u(k)$) disturbances. The effect of output disturbances (which are typically unmodeled) may be captured via $w_y(k)$; $v(k)$ represents measurement noise, and $y_m(k)$ are measured outputs.

Here, $u(k) \in \mathbb{R}^{n_u}$ denotes the manipulated inputs, $y(k) \in \mathbb{R}^{n_y}$ the outputs, and $d(k) \in \mathbb{R}^{n_d}$ the modeled disturbances. The term $w_y(k)$ represents additive output disturbances, accounting for the effects of unmodeled disturbances and model–plant mismatch. The disturbances $d(k)$ may be either measured (d_m) or unmeasured (d_u). Measured disturbances are relatively straightforward to handle—either by treating them as known system inputs, as in DMC, or more effectively, by including them as measured outputs. The latter approach will be discussed later in the context of state-estimation-based MPC formulations. In this chapter, unless otherwise stated, d will represent *unmeasured disturbances* (d_u). In cases where some or all modeled disturbances are measured, we will explicitly distinguish between measured and unmeasured disturbances using d_m and d_u, respectively.

We use the notation y_m to denote the measured values of the output (and possibly disturbance) variables. In a stochastic setting, measurement noise $v(k)$ affects the measured outputs $y_m(k)$ as follows:

$$y_m(k) = y(k) + v(k) \tag{11.5}$$

Here, $v(k)$ is assumed to be a zero-mean white noise sequence of covariance R_v. This assumption is reasonable because measurement noise is often well approximated as an independent and identically distributed (i.i.d.) sequence.

Remark 11.1: Recall from Part II that when only a subset of the outputs is measured, $y(k)$ consists of both measured and unmeasured outputs. When only some of the outputs are measured, we write the measurement equation as

$$y_m(k) = H_m y(k) + v(k). \tag{11.6}$$

When all outputs are measured, $H_m = I$ and the above reduces to Eq. (11.5).

11.1.1.1 Deterministic System Model for DMC

We begin by deriving a deterministic model that serves as the foundation for a basic state-space MPC algorithm analogous to the DMC approach introduced in Chapter 3. The overall schematic for industrial DMC was shown earlier in Figure 3.2. In DMC, measured disturbances were treated as known inputs, and the effects of unmeasured disturbances and model–plant mismatch were captured solely by the output disturbance $w_y(k)$.

Starting from the state-space model in (11.4), we take the difference between two successive time steps to obtain

$$\Delta x(k+1) = A\Delta x(k) + B\Delta u(k) + B_{d_m}\Delta d_m(k). \tag{11.7}$$

Let us define the unmeasured-disturbance-free output vector as $y' = Cx$ and introduce an augmented state vector:

$$\bar{x}(k) = \begin{bmatrix} \Delta x(k) \\ y'(k) \end{bmatrix}.$$

Since

$$y'(k+1) - y'(k) = C\Delta x(k+1) = CA\Delta x(k) + CB\Delta u(k) + CB_{d_m}\Delta d_m(k),$$

the resulting augmented state-space model is

$$\begin{aligned} \underbrace{\begin{bmatrix} \Delta x(k+1) \\ y'(k+1) \end{bmatrix}}_{\bar{x}(k+1)} &= \underbrace{\begin{bmatrix} A & 0 \\ CA & I \end{bmatrix}}_{\bar{A}} \underbrace{\begin{bmatrix} \Delta x(k) \\ y'(k) \end{bmatrix}}_{\bar{x}(k)} + \underbrace{\begin{bmatrix} B \\ CB \end{bmatrix}}_{\bar{B}_u} \Delta u(k) + \underbrace{\begin{bmatrix} B_{d_m} \\ CB_{d_m} \end{bmatrix}}_{\bar{B}_{d_m}} \Delta d_m(k) \\ y(k) &= \underbrace{\begin{bmatrix} 0 & I \end{bmatrix}}_{\bar{C}} \underbrace{\begin{bmatrix} \Delta x(k) \\ y'(k) \end{bmatrix}}_{\bar{x}(k)} + w_y(k). \end{aligned} \tag{11.8}$$

Note the similarity to the step response model introduced in Part I (Chapter 3); in particular, the input appears in differenced form, $\Delta u(k)$, consistent with the step response model formulation. As in DMC, when measured disturbances are present, an additional term involving $\Delta d_m(k)$ is included to provide feedforward compensation. In step response models (see Eq. 3.11), the effects of unmeasured disturbances and model–plant mismatch are represented by the additive output disturbance term $w_y(k)$.

$$w_y(k) \triangleq y_m(k) - y'(k). \tag{11.9}$$

11.1.1.2 Need for Further Disturbance Modeling

Early industrial MPC algorithms made simplifying assumptions about disturbance $d(k)$ and output error $w_y(k)$. Specifically, we assumed that the disturbance was measured and remains constant over the prediction horizon (i.e., $\Delta d_m(k+i|k) = 0,\ 1 \le i \le p$) and the output error was modeled as a constant bias, i.e., $w_y(k+i|k) = w_y(k)$ for all $i \ge 0$.

Although more general disturbance profiles (e.g., ramps) and their noisy measurements can, in principle, be incorporated, the step-response-based modeling framework employed in early algorithms is not well suited for generating optimal predictions in such cases. Since model–plant mismatch and unmeasured disturbances are collectively represented by a single output error term, $w_y(k)$, the framework provides only limited capability to capture more complex or structured disturbance dynamics.

In contrast, the state-space framework offers a more compact representation of system dynamics, accommodates open-loop unstable systems, handles multi-rate measurements and inferential control, and provides significant flexibility in modeling and handling various types of disturbances via stochastic noise processes.

The remainder of this section focuses on constructing state-space models that incorporate more general formulations of state and output disturbances. Stochastic processes play a central role here, as they are used to characterize disturbances, model uncertainty, and measurement noise. As discussed in Section 6.6, such disturbances can be modeled using linear systems driven by Gaussian white noise. We will now revisit some key results from that section, with a particular focus on constructing stochastic state-space models suitable for MPC. This modeling task is critical, as a well-designed controller depends on two things:

- A model that can accurately predict how disturbances affect process outputs.
- An associated estimator that improves prediction accuracy as new measurements become available.

We now proceed to construct such models under different levels of available process information and integrate them into an augmented state-space formulation consistent with (11.3).

11.1.2 Stochastic System Model with Output Disturbance Only

In many practical applications, the physical sources of disturbances are not explicitly modeled. When a disturbance model (G_d in Figure 11.1) is unavailable, the resulting model captures only the causal relationship between the manipulated inputs and the system outputs. Then, the cumulative effect of unmeasured disturbances, model-plant mismatch, and other uncertainties must be collectively captured by the output disturbance, $w_y(k)$. In this setting, disturbance modeling reduces to characterizing the statistical properties of $w_y(k)$.

11.1.2.1 Model with Simple Output Disturbances

Most disturbances encountered in industrial process systems exhibit persistent behavior (e.g., step-like changes). This was the underlying assumption in the basic DMC algorithm introduced in Part I. In a stochastic framework, this corresponds to modeling $w_y(k)$ as the integral of a white noise signal.

Augmenting the state with persistent output disturbances is a standard approach for introducing integrator dynamics into the output channels, while preserving the standard model structure for optimal estimation and control as given in Eq. (11.3):

$$
\begin{aligned}
\underbrace{\begin{bmatrix} x(k+1) \\ w_y(k+1) \end{bmatrix}}_{\bar{x}(k+1)} &= \underbrace{\begin{bmatrix} A & 0 \\ 0 & I \end{bmatrix}}_{\bar{A}} \underbrace{\begin{bmatrix} x(k) \\ w_y(k) \end{bmatrix}}_{\bar{x}(k)} + \underbrace{\begin{bmatrix} B \\ 0 \end{bmatrix}}_{\bar{B}} u(k) + \underbrace{\begin{bmatrix} 0 \\ I \end{bmatrix} \varepsilon_y(k)}_{\varepsilon_1(k)} \\
y_m(k) &= \underbrace{\begin{bmatrix} I & I \end{bmatrix}}_{\bar{C}} \begin{bmatrix} x(k) \\ w_y(k) \end{bmatrix} + \nu(k).
\end{aligned}
\tag{11.10}
$$

While this formulation effectively captures the persistent nature of output errors caused by disturbances, it introduces an augmented state $w_y(k)$ that is not reachable from the control input $u(k)$. Consequently, the overall system is non-stabilizable in the strict control-theoretic sense. In practice, however, this does not pose a problem: since $w_y(k)$ represents output bias terms arising from disturbances and modeling errors, it does not require active stabilization as long as it remains bounded.

A *differenced formulation* provides a convenient way to account for persistent output disturbances without introducing non-stabilizable modes. Recall that if $w_y(k)$ represents an integrated white-noise process exhibiting persistent behavior, its increment satisfies $\Delta w_y(k+1) = \varepsilon_y(k)$, where $\varepsilon_y(k)$ is white noise with zero mean and covariance of R_{ε_y}. The differenced form of the output equation can therefore be written as

$$y(k+1) - y(k) = C\Delta x(k+1) + \varepsilon_y(k),$$

from which the following augmented system model can be constructed:

$$\underbrace{\begin{bmatrix} \Delta x(k+1) \\ y(k+1) \end{bmatrix}}_{\tilde{x}(k+1)} = \underbrace{\begin{bmatrix} A & 0 \\ CA & I \end{bmatrix}}_{\tilde{A}} \underbrace{\begin{bmatrix} \Delta x(k) \\ y(k) \end{bmatrix}}_{\tilde{x}(k)} + \underbrace{\begin{bmatrix} B \\ CB \end{bmatrix}}_{\tilde{B}_u} \Delta u(k) + \underbrace{\begin{bmatrix} 0 \\ I \end{bmatrix} \varepsilon_y(k)}_{\varepsilon_1(k)} \tag{11.11}$$

$$y_m(k) = \underbrace{\begin{bmatrix} 0 & I \end{bmatrix}}_{\tilde{C}} \begin{bmatrix} \Delta x(k) \\ y(k) \end{bmatrix} + \underbrace{\nu(k)}_{\varepsilon_2(k)}.$$

The persistent output disturbance is captured through the output integrators driven by $\varepsilon_y(k)$, while $\nu(k)$ accounts for measurement noise. The process and output noise covariances in the augmented model are then given by

$$\bar{R}_1 = \begin{bmatrix} 0 \\ I \end{bmatrix} R_{\varepsilon_y} \begin{bmatrix} 0 & I \end{bmatrix} \qquad R_2 = R_\nu. \tag{11.12}$$

The covariances of R_{ε_y} and R_ν can be specified on the basis of process knowledge or treated as tuning parameters. This formulation has the advantage of maintaining stabilizability and providing a convenient structure for Kalman filtering and MPC design with integral action.

11.1.2.2 Model with General Output Disturbance

A more general model for output disturbances can be expressed as

$$w_y = G_w \frac{1}{1-q^{-1}} \varepsilon_y, \tag{11.13}$$

where $G_w(q)$ is a user-defined transfer function that characterizes the disturbance dynamics. This can be expressed in state-space form using the following equations:

$$x(k+1) = Ax(k) + Bu(k) \tag{11.14}$$

$$x_w(k+1) = A_w x_w(k) + B_w w(k) \tag{11.15}$$

$$w(k) = w_y(k-1) + \varepsilon_y(k) \tag{11.16}$$

$$y(k) = Cx(k) + C_w x_w(k). \tag{11.17}$$

Here, (A, B, C) represent the plant dynamics, and (A_w, B_w, C_w) represent the realization of the transfer function G_w.

As before, differencing the state and output equations yields:

$$\Delta x(k+1) = A\Delta x(k) + B\Delta u(k) \tag{11.18}$$

$$\Delta x_w(k+1) = A_w \Delta x_w(k) + B_w \varepsilon_y(k) \tag{11.19}$$

$$y(k+1) - y(k) = C\Delta x(k+1) + C_w \Delta x_w(k+1). \tag{11.20}$$

Defining the augmented state vector as

$$\bar{x}(k) \triangleq \begin{bmatrix} \Delta x(k) \\ \Delta x_w(k) \\ y(k) \end{bmatrix},$$

the overall system becomes

$$\bar{x}(k+1) = \begin{bmatrix} A & 0 & 0 \\ 0 & A_w & 0 \\ CA & C_w A_w & I \end{bmatrix} \bar{x}(k) + \begin{bmatrix} B \\ 0 \\ CB \end{bmatrix} \Delta u(k) + \begin{bmatrix} 0 \\ B_w \\ C_w B_w \end{bmatrix} \varepsilon_y(k) \tag{11.21}$$

$$y_m(k) = \begin{bmatrix} 0 & 0 & I \end{bmatrix} \bar{x}(k) + \nu(k). \tag{11.22}$$

This general formulation allows flexible and accurate modeling of structured output disturbances within the standard augmented MPC framework.

Remark 11.2: A step disturbance filtered through a first-order lag is a special case of the above: Here, each component of the output disturbance is modeled by an integrating process followed by a first-order lag filter:

$$w_y = \begin{bmatrix} \frac{1}{q-\alpha_1} & & \\ & \ddots & \\ & & \frac{1}{q-\alpha_{n_y}} \end{bmatrix} \frac{1}{1-q^{-1}} \varepsilon_y, \tag{11.23}$$

where $0 < \alpha_i \leq 1$ is the time constant of the filter for the ith output. Setting $\alpha_i = 1$ yields a double-integrated white noise signal. It is easy to see that

$$A_w = \begin{bmatrix} \alpha_1 & & \\ & \ddots & \\ & & \alpha_{n_y} \end{bmatrix} \qquad B_w = I \qquad C_w = I, \tag{11.24}$$

in the augmented model (11.22).

11.1.3 Stochastic System Model with State and Output Disturbances

The output disturbance introduced earlier can account for model bias and other uncertainties. In physical systems, certain variables act as load disturbances and enter the system through the state input term $d(k)$. Although the effects of such disturbances can, in linear systems, be lumped into the additive output error term w_y, their correlation structures, especially among different output channels, are often more naturally modeled at their sources. In addition to these modeled state disturbances, the output disturbance $w_y(k)$ may still be included to account for model–plant mismatch and other unmodeled effects.

11.1.3.1 Stochastic Model with State Disturbances

A common control problem considered in the LQ control literature is state regulation in the presence of disturbances and model uncertainty. The effect of all uncertainties is traditionally captured through a state (or input) disturbance model:

$$x(k+1) = Ax(k) + Bu(k) + B_d(k). \tag{11.25}$$

In process control, disturbances are commonly represented as integrated white-noise processes:

$$d(k+1) = d(k) + \varepsilon_d(k). \tag{11.26}$$

With $\bar{x}^T(k) \triangleq \begin{bmatrix} x^T(k) & d^T(k) \end{bmatrix}$, we can write the augmented model as

$$\underbrace{\begin{bmatrix} x(k+1) \\ d(k+1) \end{bmatrix}}_{\bar{x}(k+1)} = \underbrace{\begin{bmatrix} A & B_d \\ 0 & I \end{bmatrix}}_{\bar{A}} \underbrace{\begin{bmatrix} x(k) \\ d(k) \end{bmatrix}}_{\bar{x}(k)} + \underbrace{\begin{bmatrix} B \\ 0 \end{bmatrix}}_{\bar{B}} u(k) + \underbrace{\begin{bmatrix} 0 \\ I \end{bmatrix}}_{\overline{B_d}} \varepsilon_d(k)$$
$$y_m(k) = \underbrace{\begin{bmatrix} C & 0 \end{bmatrix}}_{\bar{C}} \bar{x}(k) + v(k). \tag{11.27}$$

Note that the state disturbance model was discussed in Chapter 8 for rejection of unmeasured disturbance under the linear quadratic regulation (LQR) framework. In this chapter, we examine model structures that could include both state and output disturbances to solve a large number of problems of practical interest. We also discuss how such models can be derived either from first principles or through system identification techniques.

11.1.3.2 General Model Form

The general model form that includes both state and output disturbances was previously given in Eq. (11.4):

$$x(k+1) = Ax(k) + Bu(k) + B_d d(k)$$
$$y(k) = Cx(k) + w_y(k).$$

If both d and w_y are integrating-type disturbances—such as randomly occurring step changes or integrated white noise processes—then,

$$w_y(k+1) = w_y(k) + \varepsilon_y(k)$$
$$d(k+1) = d(k) + \varepsilon_d(k).$$

Model augmentation with stochastic disturbances can be used to include these characteristics within the standard state-space model structure of Eq. (11.2) as follows:

$$\underbrace{\begin{bmatrix} x(k+1) \\ d(k+1) \\ w_y(k+1) \end{bmatrix}}_{\bar{x}(k+1)} = \underbrace{\begin{bmatrix} A & B_d & 0 \\ 0 & I & 0 \\ 0 & 0 & I \end{bmatrix}}_{\bar{A}} \underbrace{\begin{bmatrix} x(k) \\ d(k) \\ w_y(k) \end{bmatrix}}_{\bar{x}(k)} + \underbrace{\begin{bmatrix} B \\ 0 \\ 0 \end{bmatrix}}_{\bar{B}} u(k) + \underbrace{\begin{bmatrix} 0 & 0 \\ I & 0 \\ 0 & I \end{bmatrix} \begin{bmatrix} \varepsilon_d(k) \\ \varepsilon_y(k) \end{bmatrix}}_{\varepsilon_1(k)}$$
$$y_m(k) = \underbrace{\begin{bmatrix} C & 0 & I \end{bmatrix}}_{\bar{C}} \begin{bmatrix} x(k) \\ d(k) \\ w_y(k) \end{bmatrix} + v(k), \tag{11.28}$$

where $\varepsilon_d(k)$ and $\varepsilon_y(k)$—and hence $\varepsilon_1(k)$—are zero-mean Gaussian white noise sequences.

A major issue with this formulation is that the system becomes undetectable when the number of independent integrators exceeds the number of output measurements. This lack of detectability leads to divergence in the Riccati recursion during Kalman filter design. To address this, the number of integrators must be reduced to be equal to the number of outputs—typically through a linear projection—such that $(\bar{C}, \bar{A})$ form a detectable pair. Although this projection is theoretically feasible, it is cumbersome to implement in practice.

Using a *differenced system model* in Eq. (11.7) is an alternative to explicitly augmenting the state with integrating state and output disturbance variables. If all outputs are to be estimated and regulated without offset, the differenced disturbances $\Delta d(k)$ and $\Delta w_y(k)$ are treated as white noise processes. This approach leads to a more compact and well-posed formulation, as given in Eq. (11.3), and naturally accommodates both state and output disturbances.

$$\underbrace{\begin{bmatrix} \Delta x(k+1) \\ y(k+1) \end{bmatrix}}_{\bar{x}(k+1)} = \underbrace{\begin{bmatrix} A & 0 \\ CA & I \end{bmatrix}}_{\bar{A}} \underbrace{\begin{bmatrix} \Delta x(k) \\ y(k) \end{bmatrix}}_{\bar{x}(k)} + \underbrace{\begin{bmatrix} B \\ CB \end{bmatrix}}_{\bar{B}} \Delta u(k) + \underbrace{\begin{bmatrix} B_d & 0 \\ CB_d & I \end{bmatrix} \begin{bmatrix} \varepsilon_d(k) \\ \varepsilon_y(k) \end{bmatrix}}_{\varepsilon_1(k)}$$

$$y_m(k) = \underbrace{\begin{bmatrix} 0 & I \end{bmatrix}}_{\bar{C}} \begin{bmatrix} \Delta x(k) \\ y(k) \end{bmatrix} + \nu(k). \tag{11.29}$$

In this formulation, the number of integrators matches the number of outputs, ensuring that the system remains detectable (assuming the pair (C, A) is detectable). Consequently, a Kalman filter can be designed without additional modifications. As before, the noise covariances R_d, R_y, and R_ν may be specified based on process knowledge or used as tuning parameters.

11.1.3.3 Including Measured Disturbances in the Output

Modern process plants are increasingly integrated, necessitating flexibility in the treatment of measured disturbances. Thus far, we have assumed the disturbance signal $d(k)$ to be unmeasured and have not explicitly addressed the case of measured disturbances. In the DMC formulation, measured disturbances were treated as deterministic inputs. However, disturbance measurements are often noisy and therefore require separate filtering. To enable optimal filtering of measured disturbances together with output measurements, they can be incorporated within the above framework by including d_m (or Δd_m) as part of the measured output vector $y_m(k)$. In the following, we will describe ways to achieve this while preserving the standard model form.

Consider the following model:

$$x(k+1) = Ax(k) + Bu(k) + B_d d(k),$$

where $d(k) = \begin{bmatrix} d_m^T(k) & d_u^T(k) \end{bmatrix}^T$. In the augmented models (11.27) and (11.28), $d_m(k)$, the portion of the disturbance vector $d(k)$ that is measured, can be included in the output vector, since $d(k)$ is already part of the augmented state vector $\bar{x}$. The measurement equation can then be written as

$$y_m(k) = \underbrace{\begin{bmatrix} C & [0 \;\; 0] & I \\ 0 & [0 \;\; I] & 0 \end{bmatrix}}_{\bar{C}} \begin{bmatrix} x(k) \\ \begin{bmatrix} d_u(k) \\ d_m(k) \end{bmatrix} \\ w_y(k) \end{bmatrix} + \begin{bmatrix} \nu_y(k) \\ \nu_{d_m}(k) \end{bmatrix}. \tag{11.30}$$

This formulation enables optimal prediction under general disturbance and noise models, offering a systematic and robust means of incorporating measured disturbances into the estimation framework.

Including a measured disturbance signal as an output in the differenced model of (11.29) introduces a subtle complication. The issue arises because the measured disturbance $d_m(k)$ must influence the output instantaneously at time step k, whereas unmeasured disturbances affect the output y only after at least one sampling interval. To address this, the state vector can be augmented with the previous disturbance value $d_m(k-1)$, in addition to $d_m(k)$, resulting in:

$$\underbrace{\begin{bmatrix} \Delta x(k+1) \\ y(k+1) \\ d_m(k+1) \\ d_m(k) \end{bmatrix}}_{\tilde{x}(k+1)} = \underbrace{\begin{bmatrix} A & 0 & B_{d_m} & -B_{d_m} \\ CA & I & CB_{d_m} & -CB_{d_m} \\ 0 & 0 & I & 0 \\ 0 & 0 & I & 0 \end{bmatrix}}_{\tilde{A}} \underbrace{\begin{bmatrix} \Delta x(k) \\ y(k) \\ d_m(k) \\ d_m(k-1) \end{bmatrix}}_{\tilde{x}(k)} + \underbrace{\begin{bmatrix} B_u \\ CB_u \\ 0 \\ 0 \end{bmatrix}}_{\tilde{B}} \Delta u(k)$$
$$+ \underbrace{\begin{bmatrix} B_{d_u} & 0 \\ CB_{d_u} & 0 \\ 0 & I \\ 0 & 0 \end{bmatrix} \begin{bmatrix} \Delta d_u(k) \\ \Delta d_m(k+1) \end{bmatrix}}_{\varepsilon_1(k)} \tag{11.31}$$
$$y_m(k) = \underbrace{\begin{bmatrix} 0 & I & 0 & 0 \\ 0 & 0 & I & 0 \end{bmatrix}}_{\tilde{C}} \underbrace{\begin{bmatrix} \Delta x(k) \\ y(k) \\ d_m(k) \\ d_m(k-1) \end{bmatrix}}_{\tilde{x}(k)} + \underbrace{\begin{bmatrix} v_y(k) \\ v_{d_m}(k) \end{bmatrix}}_{\varepsilon_2(k)}.$$

This formulation preserves the standard state-space structure, allowing the direct application of the conventional Kalman filter for joint estimation of process states and measured disturbances.

11.1.3.4 Obtaining the Model from the First Principles

Physics-based modeling of a system often leads to a continuous-time nonlinear model of the following form:

$$\dot{x} = f(x, u, d)$$
$$y = g(x) + w_y.$$

The above model can be linearized and discretized to obtain a linear model of the form given by (11.4), where A, B, B_d and C are the Jacobian matrices resulting from the linearization. Then, depending on the characteristics of the disturbances d and w_y, appropriate system augmentation may be performed as before to arrive at the standard form of (11.3).

11.1.3.5 Obtaining the Model from System Identification

A discrete-time transfer function model of the form

$$y(k) = G(q)u(k) + H(q)\varepsilon(k), \tag{11.32}$$

is typically obtained through system identification using the prediction error method described in Chapter 10. In the standard formulation, H is chosen such that $H(0) = I$, implying that the instantaneous gain value is normalized.

If the realized state-space matrices corresponding to $\begin{bmatrix} G(q) & H(q) \end{bmatrix}$ are $(A, [B_G \; B_H], C, [0 \; I])$, then the system can be represented in state-space form as

$$\begin{aligned} x(k+1) &= Ax(k) + B_G u(k) + B_H \varepsilon(k) \\ y(k) &= Cx(k) + \varepsilon(k). \end{aligned}$$

Because $H(q)$ is identified with $H(0) = I$, the state and output noise terms are fully correlated as the above state-space realization shows. Although a Kalman filter formulation exists for the case of correlated state and output noises, the standard approach presented in Chapter 7 does not apply, as it assumes the process and measurement noises are independent. A simple remedy is to introduce a one-step delay in the noise term before performing the realization:

$$y(k) = G(q)u(k) + \underbrace{H(q)q^{-1}}_{\tilde{H}(q)}\varepsilon(k). \tag{11.33}$$

Note that introducing this additional delay in the noise model does not affect the statistical properties of the system or the optimality of the resulting prediction.

When all disturbances exhibit persistent behavior, it is more appropriate to model the residual term using an integrated noise process:

$$y(k) = G(q)u(k) + H(q)\frac{1}{1 - q^{-1}}\varepsilon(k).$$

Since $\left(1 - q^{-1}\right) y(k) = \Delta y(k)$, it is convenient to perform identification directly in differenced form:

$$\Delta y(k) = G(q)\Delta u(k) + H(q)\varepsilon(k). \tag{11.34}$$

As before, to avoid direct feedthrough of the noise term in the output, an additional delay can be introduced in the noise model, yielding

$$\Delta y(k) = G(q)\Delta u(k) + \underbrace{H(q)q^{-1}}_{\tilde{H}(q)}\varepsilon(k). \tag{11.35}$$

This model can then be realized and augmented into the standard state-space form as follows:

$$\begin{aligned} \underbrace{\begin{bmatrix} \Delta x(k+1) \\ y(k+1) \end{bmatrix}}_{\tilde{x}(k+1)} &= \underbrace{\begin{bmatrix} A & 0 \\ CA & I \end{bmatrix}}_{\tilde{A}} \underbrace{\begin{bmatrix} \Delta x(k) \\ y(k) \end{bmatrix}}_{\tilde{x}(k)} + \underbrace{\begin{bmatrix} B_G \\ CB_G \end{bmatrix}}_{\tilde{B}} \Delta u(k) + \underbrace{\begin{bmatrix} B_{\tilde{H}} \\ CB_{\tilde{H}} \end{bmatrix} \varepsilon(k)}_{\varepsilon_1(k)} \\ y_m(k) &= \underbrace{\begin{bmatrix} 0 & I \end{bmatrix}}_{\tilde{C}} \begin{bmatrix} \Delta x(k) \\ y(k) \end{bmatrix}. \end{aligned} \tag{11.36}$$

This representation aligns with the standard form of Eq. (11.7) and allows for Kalman filter design with integrated disturbance modeling.

11.1.4 Summary

Table 11.1 summarizes various model types used in model-based control and compares them based on three attributes: the model's origin (first principles or system identification), applicability to

Table 11.1 Types of models and their origins.

Model type	Origins of the model	Nonlinear	Unstable
ODE	Physics	L/NL	S/US
State space	Physics/identified	L/NL	S/US
Transfer function	Physics/identified	L	S/US
ARMAX	Identified	L*	S/US
Convolution	Identified	L†	S

* NARMAX models can handle nonlinearity.

† Volterra models are extensions of the FIR model to express nonlinearity.

nonlinear systems, and ability to handle unstable systems. FIR models are classified as convolution models. These model types can be readily converted into a state-space representation. The state-space structure also extends naturally to both nonlinear and unstable systems. Model conversions were discussed in Chapter 5, while this chapter focused on obtaining combined deterministic/stochastic state-space models in the standard form of Eq. (11.3) for both physics-based and identified models.

The next three sections focus on linear MPC. Extensions of linear MPC to nonlinear systems will be discussed in Section 11.5.

11.2 The Background: Deterministic State-space MPC

This section aims to demonstrate how MPC formulations for various deterministic control problems can be derived using a state-space model, building on the LQR framework discussed in Chapter 8.

11.2.1 State Regulation Problem

In Section 8.2, the infinite-horizon LQR problem was addressed by solving, at each time step k, a finite-horizon quadratic optimization problem. The goal was to determine an input trajectory that drives the linear system from the current state $x(k)$ to the origin while minimizing the corresponding finite-horizon cost:

$$\min_{v_0,\cdots,v_{p-1}} \left\{ \sum_{i=1}^{p-1} z_i^T Q z_i + \sum_{j=0}^{p-1} \left(v_i^T R v_i + \Delta v_i^T R_\delta \Delta v_i \right) + z_p^T Q_t z_p \right\} \tag{11.37}$$

with $z_0 = x(k)$, and subject to the system model:

$$z_{i+1} = A z_i + B v_i.$$

As in Part II, z_i and v_i are temporary variables introduced solely for the optimization performed at each time step. In the formulation at time step k, we adopt this compact notation in place of the more conventional $x(k+i|k)$ and $u(k+i|k)$ to facilitate exposition.

Compared to the standard LQR problem discussed in Part II, we introduce an additional input penalty term of the form $\Delta v_i^T R_\delta \Delta v_i$, which penalizes changes in the control input. This added generality is useful in certain applications—for example, when designing controllers with integral action. The MPC Toolbox supports both types of input penalties: those on the input itself and those on its increments.

Recall from an earlier chapter the p-step prediction equation:

$$\mathcal{Z} = \mathsf{S}^x z_0 + \mathsf{S}^u \mathcal{V}, \tag{11.38}$$

where

$$\mathcal{Z} = \begin{bmatrix} z_1 \\ z_2 \\ \vdots \\ z_p \end{bmatrix} \quad \mathcal{V} = \begin{bmatrix} \upsilon_0 \\ \upsilon_1 \\ \vdots \\ \upsilon_{p-1} \end{bmatrix}, \tag{11.39}$$

are stacked state and input vectors over the prediction horizon. The matrices S^x and S^u are defined as

$$\mathsf{S}^x = \begin{bmatrix} A \\ A^2 \\ \vdots \\ A^p \end{bmatrix} \quad \mathsf{S}^u = \begin{bmatrix} B & 0 & \cdots & 0 \\ AB & B & \cdots & 0 \\ \vdots & \vdots & \ddots & \ddots \\ A^{p-1}B & A^{p-2}B & \cdots & B \end{bmatrix}.$$

With $\upsilon_{-1} \stackrel{\Delta}{=} u(k-1)$, the differenced inputs are written as

$$\underbrace{\begin{bmatrix} \Delta\upsilon_0 \\ \Delta\upsilon_1 \\ \vdots \\ \Delta\upsilon_{p-1} \end{bmatrix}}_{\Delta\mathcal{V}} = \underbrace{\begin{bmatrix} I & 0 & \cdots & 0 \\ -I & I & \cdots & 0 \\ \vdots & \ddots & \ddots & \vdots \\ 0 & \cdots & -I & I \end{bmatrix}}_{\Delta_L} \underbrace{\begin{bmatrix} \upsilon_0 \\ \upsilon_1 \\ \vdots \\ \upsilon_{p-1} \end{bmatrix}}_{\mathcal{V}} - \underbrace{\begin{bmatrix} I \\ 0 \\ \vdots \\ 0 \end{bmatrix}}_{\mathcal{T}_p} \upsilon_{-1}. \tag{11.40}$$

Substituting this into the cost function yields a quadratic optimization problem in $\mathcal{V}$:

$$\mathcal{V} = \arg\min_{\mathcal{V}} \{\mathcal{V}^T \mathcal{H} \mathcal{V} + 2\left(z_0^T \tilde{G}_1^T - \upsilon_{-1}^T \tilde{G}_2^T\right) \mathcal{V}\} \tag{11.41}$$

$$\mathcal{H} = \mathsf{S}^{uT} \Lambda^x \mathsf{S}^u + \Lambda^u + \Delta_L^T (\Lambda^{\delta u}) \Delta_L$$
$$\tilde{G}_1^T = \mathsf{S}^{xT} \Lambda^x \mathsf{S}^u \qquad \tilde{G}_2^T = \mathcal{T}_p^T \Lambda^{\delta u} \Delta_L,$$

and Λ^x and Λ^u are block-diagonal matrices formed from the weighting matrices Q, Q_t, and R, corresponding to the state and input costs, respectively.

Solving this unconstrained quadratic program gives the optimal input sequence as a linear state-feedback:

$$\mathcal{V}^* = -\mathcal{H}^{-1}\left(\tilde{G}_1 z_0 - \tilde{G}_2 \upsilon_{-1}\right). \tag{11.42}$$

In the receding horizon implementation of MPC, only the first input υ_0^* from $\mathcal{V}^*$ is applied to the plant as $u(k)$. At the next time step $k+1$, the entire procedure is repeated based on the new state measurement. The following example illustrates the equivalence between unconstrained LQR and MPC under this framework.

Example 11.1: *Unconstrained MPC and Equivalence with LQR.*
Consider the system in Example 8.2:

$$x(k+1) = \begin{bmatrix} 1.41 & -0.4724 & 0 \\ 1 & 0 & 0 \\ 0 & 1 & 0 \end{bmatrix} x(k) + \begin{bmatrix} 1 \\ 0 \\ 0 \end{bmatrix}.$$

Recall that with horizon $p = 3$, $Q = I$, $R = 100$ and $R_\delta = 0$, the finite-horizon LQR control law was obtained as

$$u(k) = -\begin{bmatrix} 0.1922 & -0.0891 & 0 \end{bmatrix} x(k).$$

The values of S^x and S^u are calculated as described in this section, whereas $\Lambda^x = I_9$, $\Lambda^u = 100I_3$ and $\Lambda^{\delta u} = 0$. These are used to calculate the Hessian and gradient:

$$\mathcal{H} = \begin{bmatrix} 109.3 & 4.953 & 1.514 \\ 4.953 & 104 & 1.41 \\ 1.514 & 1.41 & 101 \end{bmatrix} \qquad \tilde{G} = \begin{bmatrix} 10.72 & -4.378 & 0 \\ 5.099 & -2.34 & 0 \\ 1.469 & -0.715 & 0 \end{bmatrix}.$$

Using these, the optimal control moves in the receding window are computed as

$$\begin{bmatrix} \upsilon_0 \\ \upsilon_1 \\ \upsilon_2 \end{bmatrix} = \begin{bmatrix} 0.0959 & -0.0390 & 0 \\ 0.0443 & -0.0206 & 0 \\ 0.0125 & -0.0062 & 0 \end{bmatrix} z_0.$$

At each step, only the solution corresponding to the current control move, i.e., $u(k) = \upsilon_0$, is implemented in MPC. Note that this is exactly the same as the feedback control law derived by iterating the Riccati equation in Example 8.2:

$$u(k) = \begin{bmatrix} 0.0959 & -0.0390 & 0 \end{bmatrix} x(k).$$

11.2.2 Constraints

One of the main advantages of MPC is its ability to handle both input and state constraints. These constraints are typically formulated as

$$\begin{aligned} u_{\min} &\le \upsilon_i \le u_{\max} \\ -\Delta u_{\max} &\le \Delta \upsilon_i \le \Delta u_{\max} \\ x_{\min} &\le z_i \le x_{\max}. \end{aligned}$$

The input constraints can be compactly expressed as

$$\begin{aligned} \mathcal{V} &\le \mathcal{I}_p u_{\max} \\ -\mathcal{V} &\le -\mathcal{I}_p u_{\min}. \end{aligned}$$

From (11.40), since $\Delta V = \Delta_L \mathcal{V} - \mathcal{T}_p \upsilon_{-1}$, the input rate constraint becomes

$$\begin{aligned} \Delta_L \mathcal{V} &\le \mathcal{I}_p \Delta u_{\max} + \mathcal{T}_p \upsilon_{-1} \\ -\Delta_L \mathcal{V} &\le \mathcal{I}_p \Delta u_{\max} - \mathcal{T}_p \upsilon_{-1}. \end{aligned}$$

Finally, the state constraints are similarly written as

$$\begin{aligned} \mathsf{S}^u\mathcal{V} &\leq \bar{\mathcal{I}}_p x_{\max} - \mathsf{S}^x z_0 \\ -\mathsf{S}^u\mathcal{V} &\leq -\bar{\mathcal{I}}_p x_{\min} + \mathsf{S}^x z_0. \end{aligned}$$

where $\mathcal{I}_p$ and $\bar{\mathcal{I}}_p$ defined as

$$\mathcal{I}_p = \left.\begin{bmatrix} I_{n_u} \\ \vdots \\ I_{n_u} \end{bmatrix}\right\} (p \text{ blocks}) \quad \bar{\mathcal{I}}_p = \left.\begin{bmatrix} I_{n_x} \\ \vdots \\ I_{n_x} \end{bmatrix}\right\} (p \text{ blocks}).$$

When constraints are present, the problem must be solved numerically using a quadratic programming (QP) solver (such as `quadprog` in MATLAB). The overall constrained optimization problem is

$$\min_{\mathcal{V}} \left\{ \frac{1}{2}\mathcal{V}^T\mathcal{H}\mathcal{V} + (\tilde{G}_1 z_0 - \tilde{G}_2 v_{-1})^T \mathcal{V} \right\}$$

$$\begin{bmatrix} I_{n_u \cdot p} \\ -I_{n_u \cdot p} \\ \Delta_L \\ -\Delta_L \\ \mathsf{S}^u \\ -\mathsf{S}^u \end{bmatrix} \mathcal{V} \leq \begin{bmatrix} \mathcal{I}_p u_{\max} \\ -\mathcal{I}_p u_{\min} \\ \mathcal{I}_p \Delta u_{\max} + \mathcal{T}_p v_{-1} \\ \mathcal{I}_p \Delta u_{\max} - \mathcal{T}_p v_{-1} \\ \bar{\mathcal{I}}_p x_{\max} - \mathsf{S}^x z_0 \\ -\bar{\mathcal{I}}_p x_{\min} + \mathsf{S}^x z_0 \end{bmatrix}, \tag{11.43}$$

Although we assumed state constraints of the form $x_{\min} \leq z_i \leq x_{\max}$ in the formulation above, any general linear constraint of the following form can be readily incorporated

$$C_z z_i + C_u u_i + C_{\Delta u} \Delta u_i - c_i \leq 0.$$

11.2.3 Offset-free Output Tracking and Regulation Problem

Traditional MPC techniques focused on tracking the set point trajectory, rather than the entire state. The state regulation problem formulation is inadequate to achieve offset-free set point tracking or rejecting load disturbances, both of which require the controller to incorporate integral action.

Integral action can be conveniently incorporated by reformulating the state-space model in terms of the input increment, $\Delta u(k)$. To achieve offset-free output tracking, the combined effects of unmeasured disturbances and model–plant mismatch are represented by the output disturbance term $w_y(k)$, which is typically modeled as a constant signal.

$$w_y(k + i|k) = w_y(k).$$

This disturbance can then be estimated and compensated by the controller.

Recall from Section 8.4.3 that differencing the model with the output disturbance and augmenting the state variables, we obtain

$$\begin{aligned} \bar{x}(k+1) &= \bar{A}\bar{x}(k) + \bar{B}\Delta u(k) \\ y(k) &= \bar{C}\bar{x}(k), \end{aligned}$$

where

$$\bar{x}(k) = \begin{bmatrix} \Delta x(k) \\ y(k) \end{bmatrix} \quad \bar{A} = \begin{bmatrix} A & 0 \\ CA & I \end{bmatrix} \quad \bar{B} = \begin{bmatrix} B \\ CB \end{bmatrix} \quad \bar{C} = \begin{bmatrix} 0 & I \end{bmatrix}.$$

The corresponding output regulation problem is given by

$$\min_{v_0,\cdots,v_{p-1}} \left\{ \sum_{i=1}^{p} \left[e_i^T (\Gamma_i^y)^T \Gamma_i^y e_i \right] + \sum_{i=0}^{p-1} \left[\Delta v_i^T (\Gamma_i^u)^T \Gamma_i^u \Delta v_i \right] + z_p^T Q_t z_p \right\} \tag{11.44}$$

subject to

$$\begin{aligned} z_{i+1} &= \bar{A} z_i + \bar{B} \Delta v_i \\ e_i &= \bar{C} z_i - r_i \\ \text{and} \quad z_0 &= \begin{bmatrix} \Delta x(k) \\ y(k) \end{bmatrix}. \end{aligned} \tag{11.45}$$

Γ_y and Γ_u are the output and input weighting matrices, respectively, consistent with the original DMC formulation. The terminal cost term $z_p^T Q_t z_p$ is included here to improve closed-loop stability and performance. Unlike in the state regulation case, we omit the penalty on v_i to avoid introducing steady-state offset during constant set point tracking. The optimization is performed subject to the usual input and output constraints.

Remark 11.3: Infinite-horizon LQ control and its desirable closed-loop properties—stability and predictable tuning behavior—were examined in Part II. We can select the terminal penalty matrix Q_t and the terminal constraint appropriately to realize infinite-horizon MPC. For *open-loop stable* systems, we can set $p = m + n$ (where n is the number of sampling steps corresponding to the settling time of the step response) and enforce the terminal error to be zero ($e_p = 0$). This is equivalent to the infinite-horizon case ($p = \infty$ with a finite m). The rationale is that, with $p = m + n$, the system response has essentially settled at the end of the prediction horizon, so $\Delta x \approx 0$ beyond it. Moreover, with $e_p = 0$, all integrating modes are zeroed. This strategy offers a simple and practical alternative to using a terminal constraint $z_p \in \hat{X}_{ma}$, which can be cumbersome to evaluate in practice.

The following example illustrates a disturbance rejection problem to demonstrate how the differenced model form can be used effectively within an MPC framework.

Example 11.2: *Set Point Tracking and Disturbance Rejection.*
Design an output feedback controller for the system in Example 11.1. The objective is to track a set point $r = 0.5$, while rejecting a step disturbance $w_y(k)$ of magnitude 0.1. The plant model is

$$x(k+1) = \begin{bmatrix} 1.41 & -0.4724 & 0 \\ 1 & 0 & 0 \\ 0 & 1 & 0 \end{bmatrix} x(k) + \begin{bmatrix} 1 \\ 0 \\ 0 \end{bmatrix} u(k)$$

$$y(k) = \begin{bmatrix} -0.1744 & 0.189 & 0.1545 \end{bmatrix} x(k) + w_y(k).$$

The model state can be augmented with the tracking error $e = y - r$ and written in a differenced form as follows:

$$\bar{x}(k+1) = \begin{bmatrix} 1.41 & -0.4724 & 0 & 0 \\ 1 & 0 & 0 & 0 \\ 0 & 1 & 0 & 0 \\ -0.0569 & 0.237 & 0 & 1 \end{bmatrix} \bar{x} + \begin{bmatrix} 1 \\ 0 \\ 0 \\ -0.1744 \end{bmatrix} \Delta u(k) + \begin{bmatrix} 0 \\ 0 \\ 0 \\ 1 \end{bmatrix} \Delta w_y(k),$$

$$e(k) = \begin{bmatrix} 0 & 0 & 0 & I \end{bmatrix} \bar{x}(k),$$

where

$$\bar{x}(k) = \begin{bmatrix} x(k) \\ e(k) \end{bmatrix}.$$

Omitting feedback correction: A controller can be designed using the augmented formulation shown above. This setup constitutes a valid state regulation problem, as the steady-state solution ($e = 0$, $\Delta u = 0$) satisfies the system dynamics. However, if the model is operated in open loop without feedback correction, the effect of the unmeasured disturbance on the output is not compensated. This behavior is illustrated by the dashed line in Figure 11.2. While the tracking error in the prediction model converges to zero, the actual output $y_m(k)$ remains offset from the set point by exactly 0.1, the magnitude of the additive disturbance.

Incorporating feedback correction: To compensate for the disturbance using measurement feedback, an observer can't be introduced. Since $\hat{y}(k|k-1) = \bar{C}\hat{\bar{x}}(k|k-1)$, the filter updates the state estimate as follows:

$$\hat{\bar{x}}(k|k-1) = \bar{A}\hat{\bar{x}}((k-1|k-1) + \bar{B}\Delta u(k-1)$$
$$\hat{\bar{x}}(k|k) = \hat{\bar{x}}(k|k-1) + K_f \left[y_m(k) - \bar{C}\hat{\bar{x}}(k|k-1) \right]$$

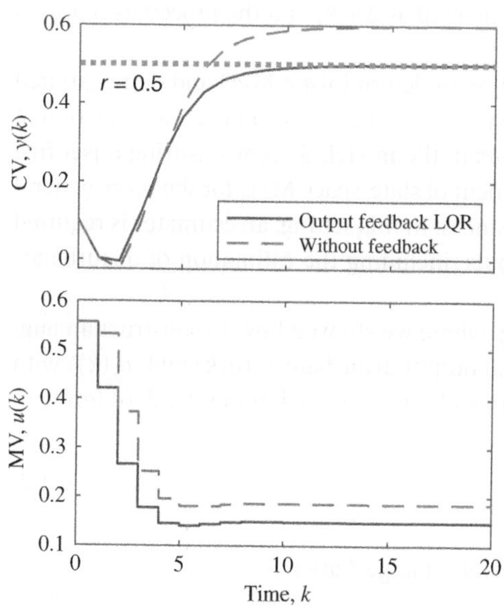

Figure 11.2 Output feedback control problem for set point tracking and step disturbance rejection. An estimator is required for state feedback implementation.

The unmeasured disturbance and model uncertainty are captured as output disturbance $w_y(k)$, as in DMC.

With this setup, the observer gain simplifies to $K_f = \begin{bmatrix} 0 & 0 & 0 & 1 \end{bmatrix}^T$. This choice reflects the fact that the disturbance enters only through the output (affecting $e(k)$), and no measurement noise is present. This filter does not adjust the internal states $\Delta x(k)$, but enforces a deadbeat correction on the augmented state $e(k)$. The solid curve in Figure 11.2 confirms that this estimator-based output feedback controller achieves perfect tracking of the set point $r = 0.5$, despite the presence of an unmeasured step disturbance in the output.

This example highlights how the differenced model formulation, as in DMC, naturally introduces integral action and enables offset-free tracking. While deterministic MPC with direct feedback correction of the error state is often adequate in practice, it may become inefficient for systems with disturbances exhibiting complex auto- and cross-correlation. Moreover, open-loop use of the model with just the output correction is unsuitable for unstable systems, as the state trajectories will diverge. In such cases, incorporating a full stochastic model and a state estimator within the MPC framework can yield significant performance improvements.

11.3 The Workhorse: MPC with State Estimation

For MPC with set point tracking, the minimization objective can be chosen as the following quadratic cost:

$$\min_{\Delta v_0,\cdots,\Delta v_{m-1}} \sum_{i=1}^{p} \|\Gamma^y(r_i - y_i)\|_2^2 + \sum_{j=0}^{m-1} \|\Gamma^u(\Delta v_j)\|_2^2, \tag{11.46}$$

where p is the prediction horizon, m is the control horizon, y_i is the i-step-ahead prediction of y and r_i denotes the set point for y_i. A control horizon of m implies that $\Delta v_{m+j} = 0$ for all $j \geq 0$. At each time step k, this quadratic optimization is solved with $z_0 = \hat{\bar{x}}(k|k)$, the estimate of the current augmented state. The resulting solution Δv_0 is implemented as $\Delta u(k)$ and the process is repeated in a receding horizon fashion.

In Section 11.2.3, we discussed the basic formulation of deterministic MPC and demonstrated its equivalence to LQR and DMC. In particular, we introduced a differenced model formulation of MPC, which guarantees the presence of n_y integrators in the model, thereby ensuring offset-free tracking of the outputs. We now turn to the development of state-space MPC for the more general case where the system is subject to stochastic disturbances. In this setting, an estimator is required to reconstruct the state $\hat{\bar{x}}(k|k)$ from measured outputs, including the estimation of disturbance effects that influence future outputs.

Section 11.1 was devoted to disturbance modeling, where we showed how to construct an augmented state-space model that incorporates state and output disturbances ($d(k)$ and $w_y(k)$) with persistent characteristics. The resulting augmented model was expressed in the standard form.

$$\begin{aligned} \bar{x}(k+1) &= \bar{A}\bar{x}(k) + \bar{B}u(k) + \varepsilon_1(k) \\ y_m(k) &= \bar{C}\bar{x}(k) + \varepsilon_2(k), \end{aligned} \tag{11.2}$$

or in one that uses the incremental manipulated variable change $\Delta u(k)$

$$\begin{aligned} \bar{x}(k+1) &= \bar{A}\bar{x}(k) + \bar{B}\Delta u(k) + \varepsilon_1(k) \\ y_m(k) &= \bar{C}\bar{x}(k) + \varepsilon_2(k), \end{aligned} \tag{11.3}$$

where $\varepsilon_1(k)$ and $\varepsilon_2(k)$ are mutually independent, zero-mean white noise sequences. Again, we emphasize that the definitions of the augmented state $\bar{x}$ and the state space matrices $(\bar{A}, \bar{B}, \bar{C})$ can vary based on the disturbance model employed and on whether differencing is used. These model forms provide a suitable foundation for applying optimal estimation techniques such as the Kalman filter. The state augmentation step is essential—omitting the disturbance states leads to biased estimates. With these elements in place, we now design an output feedback MPC. Since the MPC control law relies on the current state estimate $\hat{\bar{x}}(k|k)$, we begin by addressing the state estimation problem. Using the augmented model in Eq. (11.3), we will next show how the state estimate is computed and how the output prediction equation is constructed to yield a QP formulation for the output regulation problem. Although not explicitly shown, the same derivations can be performed using the augmented model in Eq. (11.2).

11.3.1 State Estimation Using Kalman Filter

The first step in MPC implementation is to estimate the current state vector, $\bar{x}(k)$ using the new measurement $y_m(k)$. We denote the estimate of $\bar{x}(k)$ as $\hat{\bar{x}}(k|k)$, and its i-step-ahead prediction as $\hat{\bar{x}}(k+i|k)$. While various estimation techniques may be applied, the Kalman filter is a natural choice for linear MPC when a stochastic state-space model is used. Moving horizon estimation (see Section 7.5) is another widely used alternative.

At each time step k, upon receiving the measurement $y_m(k)$, the state estimate is updated using the steady-state Kalman filter equations:

$$\hat{\bar{x}}(k|k-1) = \bar{A}\hat{\bar{x}}(k-1|k-1) + \bar{B}\Delta u(k-1) \tag{11.47}$$

$$\hat{\bar{x}}(k|k) = \hat{\bar{x}}(k|k-1) + K_f\left(y_m(k) - \bar{C}\hat{\bar{x}}(k|k-1)\right). \tag{11.48}$$

The steady-state Kalman gain K_f is computed as

$$K_f = P\bar{C}^T\left(\bar{C}P\bar{C}^T + R_2\right)^{-1}, \tag{11.49}$$

where P is the state estimate covariance, obtained as the solution to the following algebraic Riccati equation:

$$P = \bar{A}P\bar{A}^T + R_1 - \bar{A}P\bar{C}^T\left(\bar{C}P\bar{C}^T + R_2\right)\bar{C}P\bar{A}^T. \tag{11.50}$$

Using the steady-state Kalman gain simplifies implementation and is generally adequate for MPC. However, for finite-horizon problems, when a reliable estimate of the initial state covariance $P(0)$ is available, a time-varying (transient) Kalman filter may provide improved performance.

Stability of the Kalman filter is an important consideration in MPC design. As discussed in Chapter 7, the matrix P is referred to as a *stabilizing solution* if the observer poles—i.e., the eigenvalues of $\bar{A} - K_f\bar{C}\bar{A}$—lie strictly inside the unit disc. The Riccati equation admits a unique stabilizing solution when the following conditions are satisfied:

- The pair $(\bar{C}, \bar{A})$ is detectable.
- The pair $(\bar{A}, R_1^{1/2})$ is stabilizable.

The first condition is automatically met for the augmented model in Eq. (11.3) if the original pair (C, A) is detectable. The second condition ensures that all unstable modes of $\bar{A}$ are excited by the process noise $\varepsilon_1(k)$, which can be enforced through appropriate design of the state disturbance model.

11.3.2 Control Calculation Using State Estimate

The filtered state estimate $\hat{x}(k|k)$ serves as a basis to predict the future behavior of the system. In this section, we derive the output prediction equations, define the relevant constraints, and formulate the corresponding MPC optimization problem.

11.3.2.1 Future Predictions

Consider the system

$$\begin{aligned} z_{i+1} &= \bar{A}z_i + \bar{B}\Delta v_i \\ y_i &= \bar{C}z_i. \end{aligned} \tag{11.51}$$

The ℓ-step-ahead prediction of the output can be readily computed as

$$y_\ell = \bar{C}\bar{A}^\ell z_0 + (\bar{C}\bar{A}^{\ell-1}\bar{B}\Delta v_0 + \bar{C}\bar{A}^{\ell-2}\bar{B}\Delta v_1 + \quad \cdots \quad + \bar{C}\bar{B}\Delta v_{\ell-1}),$$

where $z_0 = \hat{x}(k|k)$.

Thus, the full p-step prediction equation can be compactly written as

$$\underbrace{\begin{bmatrix} y_1 \\ \vdots \\ y_p \end{bmatrix}}_{\mathcal{Y}} = \bar{S}^x z_0 + \bar{S}^u \underbrace{\begin{bmatrix} \Delta v_0 \\ \vdots \\ \Delta v_{m-1} \end{bmatrix}}_{\Delta\mathcal{V}}, \tag{11.52}$$

where

$$\bar{S}^x = \begin{bmatrix} \bar{C}\bar{A} \\ \bar{C}\bar{A}^2 \\ \vdots \\ \bar{C}\bar{A}^p \end{bmatrix} \quad \bar{S}^u = \begin{bmatrix} \bar{C}\bar{B} & 0 & 0 & \cdots & 0 \\ \bar{C}\bar{A}\bar{B} & \bar{C}\bar{B} & 0 & \cdots & 0 \\ \vdots & \vdots & \ddots & \ddots & \vdots \\ \bar{C}\bar{A}^{p-1}\bar{B} & \bar{C}\bar{A}^{p-2}\bar{B} & \cdots & \cdots & \bar{C}\bar{A}^{p-m}\bar{B} \end{bmatrix}. \tag{11.53}$$

Note that this prediction structure is identical in form to the one presented earlier, except that it is based on the augmented matrices $\bar{A}$, $\bar{B}$, and $\bar{C}$.

Remark 11.4: It should now be evident that, when state estimation is employed, there is no need to explicitly include the feedback correction term $w_y(k+i|k)$ in the prediction equation. The estimated state $z_0 = \hat{x}(k|k)$ already incorporates all necessary measurement updates for accurately predicting future output trajectories. An exception arises when the process and measurement noises are correlated ($R_{1,2} \neq 0$). In such cases, the Kalman filter applies an additional measurement correction when forming the one-step-ahead prediction from the filtered estimate $\hat{x}(k|k)$ (see Eq. [7.23]), and this correction must likewise be included in the multistep output prediction equation.

11.3.2.2 Objective Function

The prediction vector $\mathcal{Y}$ is obtained from (11.52). The future reference trajectory for $\mathcal{Y}$ is represented by the vector

$$\mathcal{R} = \begin{bmatrix} r_1 \\ \vdots \\ r_p \end{bmatrix}.$$

While it is common to have r_i constant within the horizon, the MPC formulation is flexible to accommodate any future reference trajectory through the vector $\mathcal{R}$.

The objective function (11.46) can be rewritten as

$$\arg\min_{\Delta\mathcal{V}} \{\|\Lambda_y(\mathcal{R} - \mathcal{Y})\|_2^2 + \|\Lambda_u \Delta\mathcal{V}\|_2^2\},$$

where $\Lambda_y = \text{blogdiag}[\underbrace{\Gamma^y, \cdots, \Gamma^y}_{p}]$ and $\Lambda_u = \text{blogdiag}[\underbrace{\Gamma^u, \cdots, \Gamma^u}_{m}]$.

Substituting $\mathcal{Y}$ from (11.52), we define the prediction error as $\mathcal{E} = \mathcal{R} - \bar{\mathcal{S}}^x z_0$, and the objective becomes

$$= \arg\min_{\Delta\mathcal{V}} \{[\mathcal{E} - \bar{\mathcal{S}}^u \Delta\mathcal{V}]^T \Lambda_y^T \Lambda_y [\mathcal{E} - \bar{\mathcal{S}}^u \Delta\mathcal{V}] + \Delta\mathcal{V}^T \Lambda_u^T \Lambda_u \Delta\mathcal{V}\}.$$

This can be expressed in standard QP form:

$$\begin{aligned} \Delta\mathcal{V} &= \arg\min_{\mathcal{V}} \{\Delta\mathcal{V}^T \mathcal{H} \Delta\mathcal{V} + 2\mathfrak{g}^T \Delta\mathcal{V}\} \\ \mathcal{H} &= \bar{\mathcal{S}}^{uT} \Lambda_y^T \Lambda_y \bar{\mathcal{S}}^u + \Lambda_u^T \Lambda_u \\ \mathfrak{g} &= \bar{\mathcal{S}}^{uT} \Lambda_y^T \Lambda_y (\mathcal{R} - \bar{\mathcal{S}}^x z_0). \end{aligned} \tag{11.54}$$

11.3.2.3 Constraints

The constraints in MPC are expressed in the standard linear inequality form:

- *Input rate constraints*, $|\Delta v_i| \le \Delta u_{\max}$, are expressed as

$$\begin{bmatrix} I_{n_u \cdot m} \\ -I_{n_u \cdot m} \end{bmatrix} \Delta\mathcal{V} \le \begin{bmatrix} \mathcal{I}_m \Delta u_{\max} \\ \mathcal{I}_m \Delta u_{\max} \end{bmatrix}. \tag{11.55}$$

- *Input magnitude constraints*, $u_{\min} \le v_i \le u_{\max}$, are expressed as

$$\begin{bmatrix} \mathcal{I}_{m\Sigma} \\ -\mathcal{I}_{m\Sigma} \end{bmatrix} \Delta\mathcal{V} \le \begin{bmatrix} \mathcal{I}_m \left(u_{\max} - u(k-1)\right) \\ -\mathcal{I}_m \left(u_{\min} - u(k-1)\right) \end{bmatrix}. \tag{11.56}$$

- *Output constraints*, $y_{\min} \le y_i \le y_{\max}$, are expressed as

$$\begin{bmatrix} \bar{\mathcal{S}}^u \\ -\bar{\mathcal{S}}^u \end{bmatrix} \Delta\mathcal{V} \le \begin{bmatrix} \tilde{\mathcal{I}}_p y_{\max} - \bar{\mathcal{S}}^x z_0 \\ -\tilde{\mathcal{I}}_p y_{\min} + \bar{\mathcal{S}}^x z_0 \end{bmatrix}. \tag{11.57}$$

Here, the matrix definitions are

$$\mathcal{I}_m = \left.\begin{bmatrix} I_{n_u} \\ \vdots \\ I_{n_u} \end{bmatrix}\right\} m \text{ blocks} \qquad \mathcal{I}_{m\Sigma} = \left.\begin{bmatrix} I_{n_u} & 0 & \cdots & 0 \\ I_{n_u} & I_{n_u} & \cdots & 0 \\ \vdots & \vdots & \ddots & \vdots \\ I_{n_u} & I_{n_u} & \cdots & I \end{bmatrix}\right\} m \text{ blocks.} \tag{11.58}$$

These matrix structures are used to appropriately enforce the constraints over the prediction and control horizon.

11.3.2.4 Control Implementation

The QP matrices, including the Hessian $\mathcal{H}$ and constraint matrices, are precomputed and stored. These remain constant unless the MPC formulation is modified (e.g., changes in prediction/control horizons, weighting matrices, or constraint bounds). At $k = 0$, the state estimates, $\hat{\tilde{x}}$, are initialized.

At each time step k, the following steps are performed:

1. Use the model (11.47) to compute the one-step-ahead prediction $\hat{\tilde{x}}(k|k-1)$.
2. Incorporate the latest measurement $y_m(k)$ to update the state estimate $\hat{\tilde{x}}(k|k)$ using the Kalman filter (11.48).
3. Set $z_0 = \hat{\tilde{x}}(k|k)$. Evaluate the gradient vector $\mathfrak{g}$ from Eq. (11.54) and the constraint vector (i.e., the right-hand sides of Eq. [11.55], Eq. [11.56], and Eq. [11.57]).
4. Solve the QP to obtain $\Delta\mathcal{V}^*$ and apply the first control move:

$$u(k) \leftarrow u(k-1) + \Delta v_0^\star.$$

5. Repeat the procedure at the next sampling instant.

Remark 11.5: Condensed vs. sparse QP formulation: In the formulation above, the prediction equations have been substituted into the objective function and constraints, thereby eliminating the equality constraints. As noted in Part I, this is referred to as the *condensed QP* formulation. While it reduces the number of decision variables, it produces a dense Hessian matrix without block-diagonal structure, limiting the applicability of sparse matrix techniques. An alternative is to retain the prediction equations as explicit equality constraints. In this *sparse QP* formulation, the optimization variables include both $\Delta\mathcal{V}$ and the predicted outputs $\mathcal{Y}$, increasing the problem dimensionality. However, the sparsity structure of the Hessian is preserved, enabling more efficient exploitation of sparsity.

The relative computational efficiency of the condensed and sparse QP formulations depends on the problem size and the solver used. For large-scale problems or applications requiring rapid computation (e.g., MPC implemented on embedded hardware), the sparse QP formulation is often more efficient—particularly when using primal-dual or dual interior-point solvers with warm starts. Moreover, the sparse formulation extends more naturally to nonlinear MPC, as discussed in Chapter 12, where the control problem involves online solution of nonlinear programs (NLPs).

11.3.3 MPC with Output Disturbance Only

We begin by considering the case of output feedback MPC, where the effect of disturbances is not explicitly modeled in the state-space dynamics. In this formulation, all model mismatches, disturbances, and uncertainties are lumped into an output disturbance term w_y. To ensure offset-free control—particularly important in process applications—this disturbance is typically modeled as an integrated white noise process.

The resulting augmented state-space model, previously derived in Eq. (11.11), is given by

$$\underbrace{\begin{bmatrix} \Delta x(k+1) \\ y(k+1) \end{bmatrix}}_{\tilde{x}(k+1)} = \underbrace{\begin{bmatrix} A & 0 \\ CA & I \end{bmatrix}}_{\tilde{A}} \underbrace{\begin{bmatrix} \Delta x(k) \\ y(k) \end{bmatrix}}_{\tilde{x}(k)} + \underbrace{\begin{bmatrix} B_u \\ CB_u \end{bmatrix}}_{\tilde{B}} \Delta u(k) + \begin{bmatrix} 0 \\ I \end{bmatrix} \varepsilon_y(k)$$
$$y_m(k) = \underbrace{\begin{bmatrix} 0 & I \end{bmatrix}}_{\tilde{C}} \tilde{x}(k) + v(k). \tag{11.59}$$

11.3.3.1 An Illustrative Example

Consider the following SISO system:

$$y(s) = \underbrace{\frac{1.7}{60.48s^2 + 15.6s + 1}e^{-3s}}_{G(s)} u(s). \tag{11.60}$$

Suppose that the actual plant is governed by

$$G(s) = \frac{1.36}{60.48s^2 + 15.6s + 1}e^{-3s}. \tag{11.61}$$

Assuming a sampling time of $t_s = 2$, we discretize the process transfer function (11.60) to obtain the following state-space form:

$$x(k+1) = \begin{bmatrix} 0 & 1 & 0 & 0 \\ 0 & 0 & 1 & 0 \\ 0 & 0 & 0 & 1 \\ 0 & 0 & -0.597 & 1.546 \end{bmatrix} x(k) + \begin{bmatrix} 0 \\ 0 \\ 0 \\ 1 \end{bmatrix} u(k) \tag{11.62}$$

$$y(k) = \begin{bmatrix} 0.009148 & 0.06531 & 0.0129 & 0 \end{bmatrix} x(k).$$

Due to the mismatch in the steady-state gain, the true plant differs from the above model only in

$$B_{\text{plant}} = \begin{bmatrix} 0 & 0 & 0 & 0.8 \end{bmatrix}^T.$$

Example 11.3: *Unconstrained MPC with Output Disturbance Model.*
Consider the problem of controlling the system (11.61) to a set point of 0.5, in absence of any disturbance. The model used by MPC is given in (11.60). The input constraints are

$$|\Delta u(k)| \leq 0.1, \qquad -1 \leq u(k) \leq 1.$$

An unconstrained MPC is solved, and the input moves are clipped to satisfy the above constraints.
Solution using output disturbance model: Since the disturbance is not modeled, the "model" used in MPC includes the state-space model (11.62). To ensure offset-free tracking, an integrated white noise sequence is added as an output disturbance, the model is differenced, and the augmented form (11.59) is used for the Kalman filter and MPC design.

The prediction and control horizons are set to $p = 20$ and $m = 5$, respectively. The weights are chosen as $\Gamma^y = 1$ and $\Gamma^u = 0.1$, leading to $\Lambda_y = I_p$ and $\Lambda_u = 0.1I_m$. The noise covariance is specified as $R_w = 0.01$, and no measurement noise is assumed, i.e., $R_v = 0$.

The unconstrained MPC problem is solved at each step and constraints are enforced through input clipping. In other words, the unconstrained input is clipped to satisfy magnitude and rate limits: $\Delta u(k) = \min(\Delta u(k), \Delta u_{\max})$ and $u(k) = \min(u(k), u_{\max})$ when $\Delta u(k)$ and $u(k)$ are positive, etc. Since the Hessian remains constant, it is precomputed and stored offline to reduce online computation.

Figure 11.3 shows the closed-loop responses for the set point tracking problem. The MPC based on the original (unaugmented) model will work only if the model is known perfectly, but does not provide offset-free control in presence of model–plant mismatch. The differenced formulation introduces integral action, and the persistent output disturbance ensures that the estimate $\hat{y}(k|k)$ remains unbiased despite the model–plant mismatch.

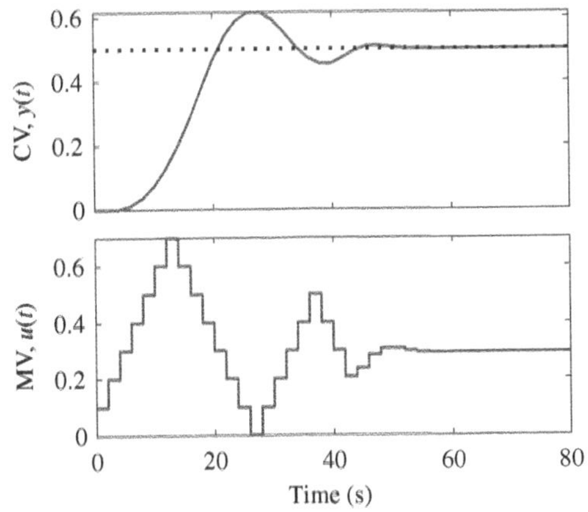

Figure 11.3 Closed-loop responses for set point change to 0.5 for MPC using an output disturbance model for an unconstrained MPC with input clipping.

Indeed, one can verify that the steady-state Kalman filter gain for this example is

$$K_{\infty} = \begin{bmatrix} 0 & 0 & 0 & 0 & 1 \end{bmatrix}^T.$$

This is expected: When $R_v = 0$, we obtain a "deadbeat" observer for the disturbance state.

Use of output disturbance model in the MPC toolbox: MATLAB offers the MPC Toolbox, which supports much of the functionality described in this chapter. Readers interested in implementation details are referred to Appendix F, which presents an example based on the Shell Oil Fractionator Control Problem.

Unlike the differenced formulation presented earlier, the MPC Toolbox follows a different approach. As described in Section 11.1, the model state itself is augmented with the output disturbance:

$$\bar{x}(k) \stackrel{\Delta}{=} \begin{bmatrix} x(k) \\ w_y(k) \end{bmatrix}.$$

This results in the following augmented model for unmeasured output disturbances:

$$\underbrace{\begin{bmatrix} x(k+1) \\ w_y(k+1) \end{bmatrix}}_{\bar{x}(k+1)} = \underbrace{\begin{bmatrix} A & 0 \\ 0 & I \end{bmatrix}}_{\bar{A}} \underbrace{\begin{bmatrix} x(k) \\ w_y(k) \end{bmatrix}}_{\bar{x}(k)} + \underbrace{\begin{bmatrix} B \\ 0 \end{bmatrix}}_{\bar{B}} u(k) + \underbrace{\begin{bmatrix} 0 \\ I \end{bmatrix}}_{\bar{B}_2} \varepsilon(k)$$
$$y_m(k) = \underbrace{\begin{bmatrix} C & I \end{bmatrix}}_{\bar{C}} \bar{x}(k). \tag{11.63}$$

The two approaches yield identical results.

Example 11.4: *Constrained Linear MPC with Output Disturbance Model.*
We solve the problem from the previous example using constrained MPC. All the parameters remain the same as before.

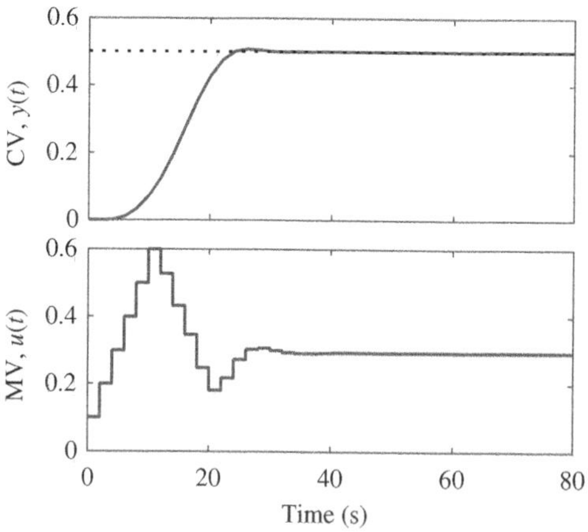

Figure 11.4 Closed-loop responses for $r = 0.5$ using constrained MPC.

The procedure closely follows that of the previous example. However, instead of solving an unconstrained optimization problem, the MATLAB command `quadprog` is used to solve a constrained QP problem. As before, the estimator is employed to compute the augmented state estimate $\hat{\bar{x}}(k|k)$. At each time step, the control sequence $\Delta\mathcal{V}$ is computed by solving the QP, and the first control input $\Delta u(k)$ is applied.

Figure 11.4 shows the input moves and the output response with the constrained MPC. A comparison with Figure 11.3 shows that the constrained MPC yields superior performance than the unconstrained MPC with input clipping.

The MATLAB code for unconstrained and constrained MPC is uploaded on GitHub at https://github.com/Lee-MPC-Book.

In the code, `A, B, and C` denote the original state-space matrices, while `Aaug, Baug, and Caug` correspond to the augmented matrices $\bar{A}, \bar{B}, \bar{C}$. The reader can adjust the model–plant mismatch. The `mpcType` flag can be used to choose unconstrained or constrained MPC.

11.3.3.2 The Kalman Filter

The Kalman gain obtained in the above example is $K_f = \begin{bmatrix} 0 & 0 & 0 & 0 & 1 \end{bmatrix}^T$. This implies that the differenced states $\Delta\hat{x}(k|k)$ are estimated using an open-loop observer, whereas the output $\hat{y}(k|k)$ is estimated with a deadbeat observer. Consequently, this closely resembles that of the DMC algorithm, where the correction term was computed as $w_y(k|k) = y_m(k) - Cx(k)$; state-estimation-based MPC produces the same result in the absence of measurement noise (recall that $R_v = 0$ was used in Example 11.4).

The Kalman filter for this system takes the form:

$$\hat{\bar{x}}(k+1) = \bar{A}\hat{\bar{x}}(k) + \bar{B}\Delta u(k) + \begin{bmatrix} 0 \\ K_f \end{bmatrix} \left(y_m(k) - \bar{C}\hat{\bar{x}}\right). \tag{11.64}$$

The Kalman gain K_f for this output disturbance model has a diagonal structure, $K_f = \text{diag}(f_1, \ldots, f_{n_y})$, where $f_i \in (0, 1]$ depends on the signal-to-noise ratio (SNR) of the ith output channel. If measurement noise is negligible, then $f_i = 1$, the filter applies no smoothing to the output update and the resulting controller effectively reduces to the original DMC formulation.

For the more general system given by Eq. (11.22), the Kalman filter takes the following form:

$$\hat{\bar{x}}(k+1) = +\begin{bmatrix} B \\ 0 \\ CB \end{bmatrix} \Delta u(k) \begin{bmatrix} A & 0 & 0 \\ 0 & A_\alpha & 0 \\ CA & A_\alpha & I \end{bmatrix} \hat{\bar{x}}(k)$$
$$+\begin{bmatrix} 0 \\ K_{f_b} \\ K_{f_a} \end{bmatrix} \left(y_m(k) - \begin{bmatrix} 0 & 0 & I \end{bmatrix} \hat{\bar{x}}(k) \right). \tag{11.65}$$

The Kalman gain in this structure adopts a parameterized form:

$$\begin{aligned} K_{f_b} &= \text{diag}\left[(f_b)_1, \cdots, (f_b)_{n_y} \right] \\ K_{f_a} &= \text{diag}\left[(f_a)_1, \cdots, (f_a)_{n_y} \right], \end{aligned} \tag{11.66}$$

where each element is defined as

$$(f_b)_i = \frac{(f_a)_i^2}{1 + \alpha_i - \alpha_i (f_a)_i} \quad \text{for } 1 \le i \le n_y, \tag{11.67}$$

and $(f_a)_i \in (0, 1]$ is a tuning parameter determined by the SNR of the ith output channel. Specifically, $(f_a)_i \to 1$ as the ratio $\sigma(\nu)/\sigma(\varepsilon) \to 0$, indicating negligible measurement noise.

The above parameterizations of the filter gain matrix provide a convenient and intuitive mechanism for tuning the estimator gain on a per-output basis, enabling tailored shaping of the closed-loop behavior. The structure is reminiscent of the internal model control (IMC) framework, where filter time constants are used to balance response speed against robustness.

A key implication of this discussion is that if a controlled output is unmeasured, this formulation cannot ensure offset-free tracking and disturbance rejection for that output, as it does not receive feedback correction from other measurements.

11.3.3.3 Disturbance Rejection Problem

Consider the system from Example 11.3 with an additive disturbance modeled as:

$$y(s) = \underbrace{\frac{1.7}{60.48s^2 + 15.6s + 1} e^{-3s}}_{G(s)} u(s) + \underbrace{\frac{0.5}{19.36s^2 + 7.04s + 1} e^{-8s}}_{G_d(s)} d(s). \tag{11.68}$$

For the true plant, the gains for the process and disturbance models are 1.36 and 0.6, respectively.

Example 11.5: *Disturbance Rejection with Output Disturbance Model.*
Consider the problem of tracking a desired set point while rejecting a unit step *unmeasured* disturbance, $d(k) = 1, k \ge 0$ for the system (11.68). The constraints are $|\Delta u(k)| \le 0.1$ and $-1 \le u(k) \le 1$.

We use the output disturbance model. The MPC parameters are the same as in Example 11.4. We first express the model in the standard form (11.3). The advantage of this is that the rest of the code for computing state estimates and control actions remains unchanged—aside from necessary adjustments to account for the augmented state vector dimensions.

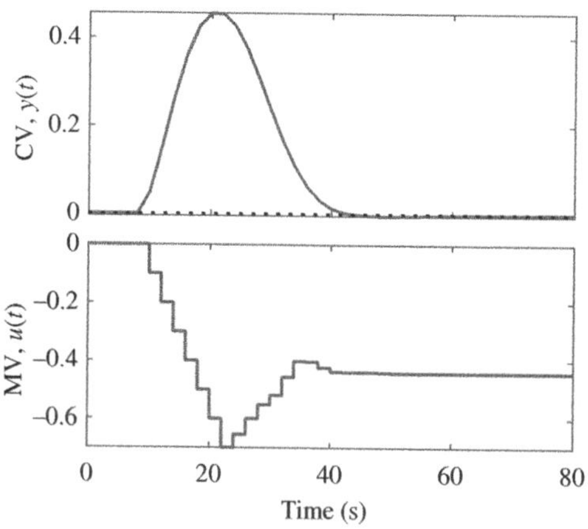

Figure 11.5 Closed-loop response for disturbance rejection using constrained MPC employing the output disturbance model.

Figure 11.5 illustrates the closed-loop performance of the system using the output disturbance model. Because the disturbance is unmeasured, the MPC can only compensate after its effect appears in the measured output. The feedback-based correction is therefore delayed by the process time delay, $\theta = 3$, as well as by the input constraints. Nevertheless, despite the lack of disturbance measurement, the inclusion of the output disturbance in the augmented model enables the MPC to achieve offset-free tracking.

11.3.4 MPC with State Disturbance Model

MPC with an output disturbance model is a valuable formulation, which is particularly useful when no information is available about state disturbances. In this case, all disturbances, model mismatch, and uncertainties are lumped into the output disturbance term $w_y(k)$, which is modeled as integrated white noise to ensure offset-free tracking.

An alternative scenario arises when the disturbance is unmeasured, but its effect on the system dynamics is known and can be modeled. This is common in practice, as many disturbances affecting process behavior are not directly measured, even though their influence on the system is well understood.

Consider the system model in the form of Eq. (11.4):

$$x(k+1) = Ax(k) + Bu(k) + B_d d(k)$$

The standard way to account for the characteristics of $d(k)$ is to augment the system state with the disturbance:

$$\bar{x}(k) \triangleq \begin{bmatrix} x(k) \\ d(k) \end{bmatrix},$$

and model $d(k)$ as an integrated white noise process. This allows the model to capture the persistent nature of disturbances and enables offset-free tracking of set point changes. The augmented model was presented in (11.27).

A key requirement for achieving offset-free control is the ability of the observer to accurately estimate the unmeasured disturbance. This necessitates:

- The augmented system $(\bar{C}, \bar{A})$ must be detectable. Since the augmentation introduces integrators into the model, detectability of the augmented pair $(\bar{C}, \bar{A})$ is guaranteed if and only if the original pair (C, A) is detectable and the following rank condition holds:

$$\text{rank}\left(\begin{bmatrix} I - A & -B_d \\ C & 0 \end{bmatrix}\right) = n + n_d.$$

- An important implication of the above rank condition is that the number of integrating n_d disturbances (e.g., $d(k)$) that can be added to the augmented state is fundamentally limited by the number of available measurements.
- In addition, all outputs must be excited independently by d to prevent bias in their estimates. As shown by Yu et al. (1994), this requirement translates to the condition that the matrix $C(I - A)^{-1}B_d$ must have a full row rank. This implies that the number of integrating disturbance states must equal the number of output variables.

Alternatively, *the differenced form of the model* can be used to enforce integral action on the outputs automatically:

$$\underbrace{\begin{bmatrix} \Delta x(k+1) \\ y(k+1) \end{bmatrix}}_{\bar{x}(k+1)} = \underbrace{\begin{bmatrix} A & 0 \\ CA & I \end{bmatrix}}_{\bar{A}} \underbrace{\begin{bmatrix} \Delta x(k) \\ y(k) \end{bmatrix}}_{\bar{x}(k)} + \underbrace{\begin{bmatrix} B_u \\ CB_u \end{bmatrix}}_{\bar{B}} \Delta u(k) + \underbrace{\begin{bmatrix} B_d \\ CB_d \end{bmatrix}}_{\bar{B}_d} \varepsilon_d(k) \tag{11.69}$$

$$y_m(k) = \underbrace{\begin{bmatrix} 0 & I \end{bmatrix}}_{\bar{C}} \bar{x}(k) + v(k). \tag{11.70}$$

Since disturbance $d(k)$ is modeled as an integrated white noise process, $\varepsilon_d(k) = \Delta d(k)$ is a zero-mean white noise sequence (with covariance R_{ε_d}). The overall augmented model takes the standard form of Eq. (11.3):

$$\begin{aligned} \bar{x}(k+1) &= \bar{A}\bar{x}(k) + \bar{B}\Delta u(k) + \varepsilon_1(k) \\ y_m(k) &= \bar{C}\bar{x}(k) + \varepsilon_2(k). \end{aligned}$$

The covariances of the two white noise sequences are given by

$$\bar{R}_1 = \begin{bmatrix} B_d \\ CB_d \end{bmatrix} R_{\varepsilon_d} \begin{bmatrix} B_d \\ CB_d \end{bmatrix}^T \qquad \bar{R}_2 = R_v.$$

This augmented system $(\bar{C}, \bar{A})$ is detectable if the original system (C, A) is detectable, regardless of the number of integrating disturbances considered.

Example 11.6: *Constrained MPC with State Disturbance Model.*
Let us revisit the control problem of Example 11.5 using a state disturbance model. The two scenarios—set point tracking with $r = 0.5$ and disturbance rejection with $d = 1.0$, with model–plant mismatch—remain the same as before.

Converting $G_d(s)$ to a discrete-time state-space form, with $t_s = 2$, yields:

$$x_d(k+1) = \underbrace{\begin{bmatrix} 0 & 1 & 0 & 0 & 0 & 0 \\ 0 & 0 & 1 & 0 & 0 & 0 \\ 0 & 0 & 0 & 1 & 0 & 0 \\ 0 & 0 & 0 & 0 & 1 & 0 \\ 0 & 0 & 0 & 0 & 0 & 1 \\ 0 & 0 & 0 & 0 & -0.4832 & 1.339 \end{bmatrix}}_{A_{x_d}} x_d(k) + \underbrace{\begin{bmatrix} 0 \\ 0 \\ 0 \\ 0 \\ 0 \\ 1 \end{bmatrix}}_{B_{x_d}} d(k)$$

$$y_d(k) = \underbrace{\begin{bmatrix} 0.03172 & 0.04045 & 0 & 0 & 0 & 0 \end{bmatrix}}_{C_{x_d}} x_d(k).$$

The true plant model differs only in

$$B_{x_d,\text{plant}} = \begin{bmatrix} 0 & 0 & 0 & 0 & 0 & 1.2 \end{bmatrix}^T.$$

Although a more compact representation is possible, we can easily stack G and G_d to obtain the following model:

$$\begin{aligned} x(k+1) &= \underbrace{\begin{bmatrix} A & 0 \\ 0 & A_{x_d} \end{bmatrix}}_{A_\text{model}} x(k) + \underbrace{\begin{bmatrix} B \\ 0 \end{bmatrix}}_{B_\text{model}} u(k) + \underbrace{\begin{bmatrix} 0 \\ B_{x_d} \end{bmatrix}}_{B_{d,\text{model}}} d(k) \\ y(k) &= \underbrace{\begin{bmatrix} C & C_{x_d} \end{bmatrix}}_{C_\text{model}} x(k). \end{aligned} \tag{11.71}$$

The standard approach of augmenting the state with the disturbance:

$$\bar{x}(k) \triangleq \begin{bmatrix} x(k) \\ d(k) \end{bmatrix}$$

can be directly applied to obtain the augmented model (11.27). This formulation is fully supported by the MPC Toolbox, allowing for straightforward implementation and result generation.

Alternatively, the following MATLAB code illustrates how to construct the augmented model in differenced form in Eq. (11.69).

```
% Augmentation for differenced form
Aaug = [Amodel,           zeros(nx,1);
        Cmodel*Amodel, 1];
Baug = [Bumodel;          Cmodel*Bumodel];
Caug = [zeros(1,nx),      1];

% Initialize estimator
xaug = zeros(nx+1,1);      % Current state estimate
yhat = xaug(end);          % Output estimate
```

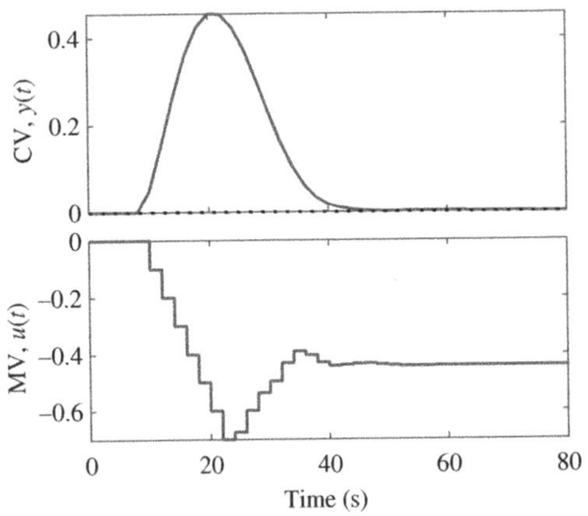

Figure 11.6 Closed-loop responses of constrained MPC employing the state disturbance model.

```
B2aug = [Bdmodel; Cmodel*Bdmodel];
R1aug = B2aug * 0.01 * B2aug';
R2aug = 0;
```

A key advantage of expressing the models in the standard form (Eq. 11.3) is that the rest of the code for computing state estimates and control actions remains unchanged—aside from necessary adjustments to account for the augmented state vector dimensions.
This differenced model-based formulation yields results consistent with those obtained with the MPC Toolbox. Figure 11.6 shows the closed-loop performance of the system using the state disturbance model. The results are comparable to those obtained in the previous example where integrating disturbances were directly added to the outputs.

This may be expected because the feedback correction is only possible after the effect of disturbance is visible in the measured output. In both the output and state disturbance models, this response is inherently delayed by the time delay of the process dynamics, $n_\theta = 3$.

The benefit of incorporating a state disturbance model becomes more pronounced when direct measurement of $d(k)$ or faster secondary measurements are available to enable earlier inference of the disturbance. Both allow the controller to take anticipatory corrective actions.

The solution to Example 11.6 using the measured disturbance as a feedforward signal can be obtained with the MPC Toolbox. Direct disturbance measurements can also be readily incorporated into y_m to facilitate feedforward compensation, since disturbance variables are already included in the state vector.

Example 11.7: *Constrained MPC with Measured Disturbance.*
Now assume that the disturbance is measured. The measured disturbance signal can be incorporated into the MPC design—either by using $\Delta d_m(k)$ directly as an input, as in DMC, or by including $d(k)$ as part of the measured output vector y_m. The latter approach enables optimal filtering when the disturbance measurements are affected by significant noise.

Consider the prediction model formulation in (11.31) applied to the process described by (11.68) with a sampling time of $t_s = 2$. No model–plant mismatch is assumed. We use this prediction model to design an MPC controller tuned as in Example 11.3 with the same input constraints:

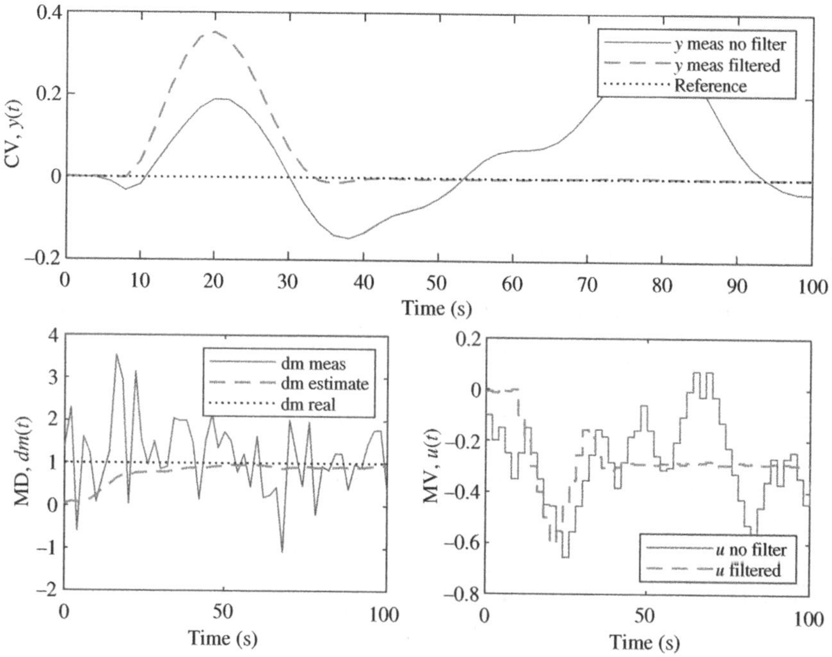

Figure 11.7 Closed-loop responses to a measured disturbance step change of 1.0 with measurement noise of variance = 1 and a Kalman filter d_m covariance $R_v = 10.0$.

$$|\Delta u(k)| \leq 0.1, \qquad -1 \leq u(k) \leq 1,$$

and a Kalman filter with process noise covariance of $R_{\Delta d_m} = 0.01$.

A disturbance step $d_m(k) = 1.0$ is introduced at time zero with measurements corrupted by zero-mean white noise of variance 1. Figure 11.7 shows the response to the unfiltered disturbance signal resulting when setting the Kalman filter noise covariances $v_{d_m} = 0$ - which is equivalent to using $\Delta d_m(k)$ as input to the system, as in DMC. This generates severe manipulated variable variability despite slow controller tuning and the low MV move rate limit. As a result, the output meanders around the reference target. When the d_m noise covariance of 10.0 is chosen, the measured disturbance estimate reaches the real value around time 40, with the output settling as fast as in the noise-free case.

11.3.5 MPC with Full Disturbance Model

The previous examples addressed MPC using only an output disturbance or a state disturbance model. Since we had only one measured output, only one integrating disturbance can be included, as was discussed in Section 11.3.4.

11.3.5.1 Motivation: Comparison of Output and State Disturbance Models

We addressed set point tracking and disturbance rejection for a SISO system with model–plant mismatch in Examples 11.5 and 11.6, using output and state disturbance models, respectively. Since the example involved only a single measured output, both formulations yielded qualitatively similar performance. In both cases, the MPC can only begin taking corrective action once the disturbance effect is observed in the output.

However, if additional outputs or secondary measurements are available that reveal the disturbance effect earlier, the state disturbance model can leverage this information to initiate corrective input adjustments before the disturbance significantly impacts the primary controlled variable. To fully exploit this advantage, the system must satisfy appropriate reachability and detectability conditions to ensure desirable closed-loop behavior. The remainder of this section explores these criteria through an extension of Example 11.4.

A historical note before proceeding: nearly three decades ago, Yu et al. (1994) compared several disturbance model structures in the context of the Shell oil fractionator control problem. This benchmark problem was introduced in Chapter 3 with step-response-based DMC and is revisited in Appendix F using the MPC Toolbox. In the early 2000s, Muske and Badgwell (2002) and Pannocchia and Rawlings (2003) provided insightful discussions on disturbance modeling strategies to ensure offset-free MPC. Later, Maeder et al. (2009) offered theoretical guarantees for offset-free linear MPC under various modeling assumptions. Interested readers are encouraged to consult these references for deeper insights.

Now, let us extend the original example by introducing a secondary measurement. The overall MIMO model may be written as

$$y(s) = \begin{bmatrix} \dfrac{1.7}{60.48s^2 + 15.6s + 1}e^{-3s} & \dfrac{0.5}{19.36s^2 + 7.04s + 1}e^{-8s} \\ \dfrac{0.3}{10.89s^2 + 4.62s + 1}e^{-s} & \dfrac{1}{36s^2 + 12s + 1} \end{bmatrix} \begin{bmatrix} u(s) \\ d(s) \end{bmatrix}. \tag{11.72}$$

As before, we consider one manipulated input $u(k)$, one primary controlled output $y_1(k)$, and an unmeasured disturbance $d(k)$. The key difference now is the availability of a secondary measurement $y_s(k) = y_2(k)$. Traditional control algorithms typically disregard this additional measurement. However, through appropriate disturbance modeling, MPC can utilize all available information for improved performance.

Example 11.8: *Comparison of Output and State Disturbance Models for SISO System with Secondary Measurement: Perfect Information Case.*
We revisit Example 11.4, but now include an additional secondary measurement as defined in Eq. (11.72). Assume there is no model–plant mismatch—i.e., the true plant model is identical to the one used in the MPC controller.

This example will be solved using the `MPC Toolbox`. Refer to Appendix F for a detailed tutorial on using the toolbox.

The transfer function (11.72) yields a two-input-two-output system, where the first input is the manipulated variable $u(k)$, and the second input is the unmeasured disturbance $d(k)$. Both outputs—$y \triangleq [y_1 \quad y_s]^T$—are measured, but only the first output is controlled. Consequently, we set the output weighting matrix as $\Gamma^y = \mathtt{diag}([1 \quad 0])$. As in Example 11.4, $\Gamma^u = 0.1$, and the horizons are set to $p = 20$ and $m = 5$.

All other parameters—sampling interval, simulation time, and constraints—are kept the same as in the previous examples. As before, we examine two scenarios: (1) Set point tracking with $r = 0.4$; (2) Rejection of a unit step disturbance. With this setup, we are ready to compare the performance of the two disturbance modeling approaches.

MPC-1: With output disturbance model: The output disturbance model used here is similar to the one described in Example 11.4. Since the system has two outputs, an integrating white noise disturbance is added to both channels. In the MPC Toolbox, the functions `setoutdist` and

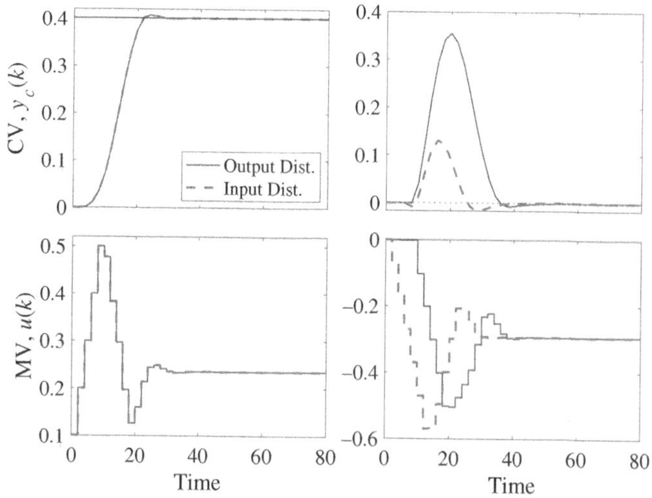

Figure 11.8 Comparison of MPC utilizing output (solid lines) and state (dashed lines) disturbance models for setpoint tracking and disturbance rejection, when the model is perfectly known.

`setindist` are used to specify output and state disturbance models, respectively. For example, an integrator model for two output channels can be defined and assigned as follows:

```
iwnTF = tf(1,[1 0]) * eye(2);
```

This model can then be set as the output disturbance model using the `setoutdist` command.

The solid lines in Figure 11.8 show the closed-loop responses of the system for both set point tracking and disturbance rejection scenarios using this formulation. The dynamic performance is consistent with what was observed in Example 11.4. Although an additional secondary measurement with faster dynamics is available, the output disturbance model cannot exploit this information, as it lacks the feedforward structure needed to do so.

MPC-2: With state disturbance model: Since there is only one unmeasured disturbance, as in the previous example, the implementation of the state disturbance model is the same as in Example 11.6. An integrator can be specified as the input disturbance model using the `setindist` command.

For the set point tracking case (left panels), the closed-loop behavior is identical to that obtained with the output disturbance model, as expected. However, the dashed lines in the right panels of Figure 11.8 reveal that, owing to disturbance estimation through y_2, the controller reacts immediately once the disturbance begins to influence the secondary measurement. This enables a faster corrective response by the MPC, illustrating the key benefit of incorporating an appropriate disturbance model.

The discussion above is by no means the final word on selecting a disturbance model for MPC. While it highlights the benefits of using a state or input disturbance model, the choice must be made judiciously, as illustrated in the next example. A careful reader may have noticed that Example 11.4 considered a case with model–plant mismatch, whereas the previous example assumed perfect model knowledge. We will return to this important distinction after presenting the next example.

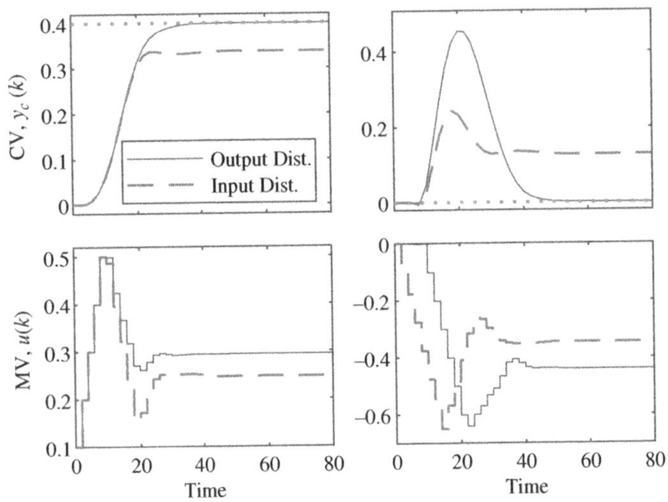

Figure 11.9 Comparison of MPC utilizing output disturbance (solid lines) and state disturbance (dashed lines) models in the presence of model–plant mismatch.

Example 11.9: *Effect of Model–Plant Mismatch.*
Repeat Example 11.8 in the presence of model–plant mismatch, where the model and the plant differ only in their steady-state gains. Specifically, the plant's steady-state relationship is given by

$$y_\infty = \begin{bmatrix} 1.36 & 0.6 \\ 0.27 & 1.1 \end{bmatrix} \begin{bmatrix} u_\infty \\ d_\infty \end{bmatrix}.$$

Model–plant mismatch can be introduced in the MPC Toolbox by using the options structure. The command `optMPM = mpcsimopt;` initializes the structure `optMPM` with default settings. The true plant model is then assigned to the field `optMPM.Model`. Please refer to Appendix F or the MPC Toolbox User Guide for further details on specifying simulation options.

Figure 11.9 shows the closed-loop performance under model–plant mismatch. The MPC with output disturbance model (solid lines) continues to yield offset-free behavior, whereas the MPC with state disturbance model (dashed lines) exhibits a steady-state offset in the controlled output. for both setpoint tracking and disturbance rejection. This illustrates the sensitivity of the state disturbance formulation to inaccuracies in the process model.

Since there are two outputs, two integrators are required to ensure offset-free tracking. The output disturbance model inherently provides this by modeling each output disturbance as an integrating white noise sequence. In contrast, the state disturbance model introduced only a single integrator—corresponding to the unmeasured disturbance $d(k)$—through an integrated white noise sequence. Consequently, offset-free tracking of arbitrary set points across both outputs is not guaranteed with this formulation. This example naturally motivates the use of a full disturbance model, in which a second integrator is added to one of the outputs in addition to the integrating state disturbance, thereby ensuring sufficient degrees of freedom for offset-free estimation and control of the output.

11.3.5.2 Model Formulation and MPC

When a disturbance model is available, incorporating both state and output disturbances can enhance the performance of MPC. For example, including an output disturbance model in addition to a state disturbance model may be necessary to achieve unbiased output estimates in the presence of model–plant mismatch. We therefore now consider the general case of a full disturbance model of (11.28).

Example 11.10: *Model–Plant Mismatch: Full Disturbance Model.*
We now repeat Example 11.9 using an appropriate disturbance model that includes both an unmeasured input disturbance and an output disturbance.

In the previous example, only a single integrator was added to model the unmeasured disturbance. As a result, a steady-state offset from the set point was observed. To eliminate this offset, an additional integrating disturbance must be introduced—this time as an additive output disturbance.

Integrating white noise for both output channels: An immediate approach would be to add integrated white noise disturbances to both output channels. However, with one unmeasured disturbance already modeled as an integrator, this would result in three integrators, while only two outputs are measured. The resulting augmented model is *not detectable*; in other words, three integrating disturbances cannot be estimated from just two measurements, leading to undetectability. The MPC Toolbox correctly raises an error in such a scenario.

Integrating white noise for the y_c channel only: An alternative approach is to introduce an additive integrating disturbance to only one of the output channels. Since offset-free estimation and control are required for y_c, the integrating disturbance should be added to the first output channel. The resulting closed-loop response is shown in Figure 11.10. This configuration is implemented using a custom output disturbance model,

$$G_w = \frac{1}{s}\begin{bmatrix} 1 & 0 \\ 0 & 0 \end{bmatrix},$$

which can be specified using the `setoutdist` command in the MPC Toolbox.

Thanks to the disturbance estimation facilitated by the unmeasured disturbance model, the controller is able to react as soon as the disturbance effect is detected in the secondary output. Furthermore, the inclusion of the second integrator ensures offset-free tracking of the controlled output.

MPC was originally designed for large-scale MIMO control. In such situations, one needs to make choices regarding the disturbance and control structures. Although we used simple examples, the overall observations summarized here provide key guidelines for designing state estimation-based MPC:

- The number of independent integrators can be at most equal to the number of measurements. They should be selected such that the augmented system remains detectable.
- The output disturbance model is convenient, as it directly assigns one integrator to each output channel.
- However, the output disturbance model does not account for correlations among the output channels.
- In contrast, state or full disturbance models enable cross-channel information exchange between the outputs.
- Introducing state disturbances that simultaneously excite both fast and slow output dynamics is often an effective way to capture such correlations and enhance controller performance.

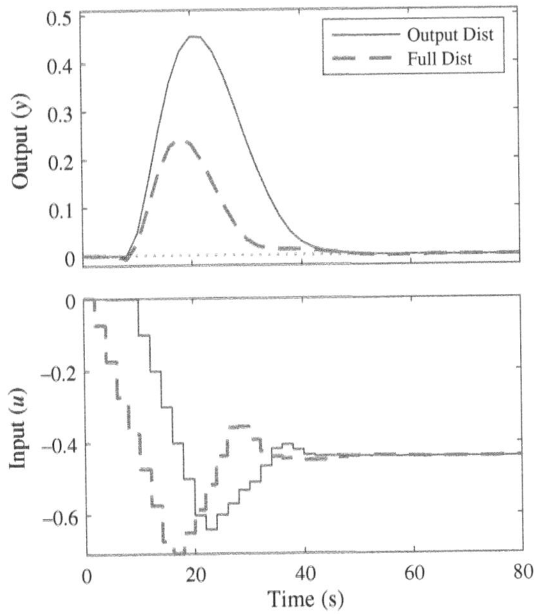

Figure 11.10 Comparison of MPC utilizing output disturbance (solid lines) and state disturbance (dashed lines) models in presence of model–plant mismatch.

11.3.6 Tracking a Set point Trajectory

The MPC objective function at time step k is typically defined as

$$\min_{\Delta\mathcal{V}} \sum_{i=1}^{p} |\Gamma^y(r_i - y_i)|_2^2 + |\Gamma^u(\Delta v_{i-1})|_2^2. \tag{11.73}$$

This optimization is subject to the model prediction equations and the usual constraints, and it is initialized with $z_0 = \hat{\bar{x}}(k|k)$. Note that the reference trajectory r_i may, in general, be time-varying.

In many applications—such as trajectory tracking for robots or autonomous vehicles—the controller must follow a prescribed reference path. In other scenarios, particularly when the system is initially far from the desired set point, it may be beneficial to guide the controller using a reference trajectory. This can help mitigate aggressive control actions and reduce the likelihood of constraint violations.

It is worth noting that, when deriving the quadratic objective function (11.54), we allowed for a general reference trajectory of the form:

$$\mathcal{R}(k+1|k) = \begin{bmatrix} r(k+1|k) \\ \vdots \\ r(k+p|k) \end{bmatrix}.$$

MPC Toolbox: If the reference trajectory is fixed and known *a priori*, it can be directly specified in the MPC Toolbox. For cases where the reference trajectory is generated or updated at each time step, the `mpcmove` function can be used to compute the control moves at every sampling instant. These control actions can then be applied to the simulated or real plant accordingly.

11.3.7 Constraint Softening

Constraint softening is often necessary to ensure that the MPC optimization problem remains feasible, particularly in the presence of conflicting objectives or disturbances. This is achieved by introducing slack variables and modifying the objective function and output constraints as follows:

$$\min_{\Delta \mathcal{V},\epsilon} \sum_{i=1}^{p} \|\Gamma^y(r_i - y_i)\|_2^2 + \|\Gamma^u(\Delta v_{i-1})\|_2^2 + \|\Gamma^\epsilon \epsilon\|_2^2, \tag{11.74}$$

subject to

$$y_{\min} - \epsilon \leq y_i \leq y_{\max} + \epsilon, \qquad \epsilon \geq 0. \tag{11.75}$$

Here, ϵ is a nonnegative slack variable included in the cost function and penalized with a weight Γ^ϵ, which is typically chosen to be large. This penalization allows for limited violations of output constraints when necessary, thereby ensuring that the controller remains feasible even under tight or conflicting conditions. In the formulation above, the 2-norm of $\Gamma^\epsilon \epsilon$ is used; however, the 1-norm is also commonly employed to enable *exact softening*, wherein ϵ remains zero unless the original (hard-constrained) problem is infeasible.

MPC Toolbox: The MPC Toolbox natively supports constraint softening. For manipulated variables, the fields `MV.MinECR`, `MV.MaxECR`, `MV.RateMinECR`, and `MV.RateMaxECR` control the extent to which lower, upper, and rate constraints can be relaxed. These fields are set to zero by default, corresponding to hard constraints. For output variables, the fields `OV.MinECR` and `OV.MaxECR` specify the relaxation parameters for the lower and upper output bounds. These values default to 1, allowing soft constraints. Smaller values make the constraints stricter, with 0 enforcing hard constraints.

11.4 Inferential Control

In certain applications, the primary control variable of interest is either not directly measured or is measured at a slower rate than other secondary measurements. This situation is common in process systems where a quality variable or concentration is the primary-controlled variable. In some cases, it can be completely unmeasured; in others, it can be measured infrequently or subject to a significant delay. In such cases, the primary variable must be inferred from secondary measurements. This *inferential estimation* was introduced in Chapter 7 (see Section 7.4.2).

11.4.1 Problem Formulation

Consider the scenario where the primary variable of interest is not directly measured, while a secondary variable $y_s(k)$ is measured frequently. The output vector $y(k)$ consists of both controlled $y_c(k)$ and secondary outputs $y_s(k)$—the latter referring to measured but uncontrolled variables. Recall that when the output vector contains both measured and unmeasured components, we write the measurement equation as

$$y_m(k) = H_m y(k) + \nu(k). \tag{11.6}$$

Similarly, controlled variables can be expressed as More generally, the other output relations are expressed as

$$y_c = H_c y(k), \tag{11.76}$$

where H_m and H_c are selection matrices: H_m and H_c each have one row per output of interest, with 1 in the column corresponding to the selected output and zeros elsewhere.

When persistent disturbances are present, the procedure outlined in Section 11.1 can be used to obtain an augmented model in the standard form:

$$\begin{aligned} \bar{x}(k+1) &= \bar{A}\bar{x}(k) + \bar{B}u(k) + \varepsilon_1(k) \\ y_m(k) &= H_m\bar{C}\bar{x}(k) + \nu(k). \end{aligned} \tag{11.77}$$

The augmented state $\bar{x}(k)$ should include persistent disturbance variables $d(k)$. However, this augmentation can render the system undetectable if the number of integrators in d exceeds the number of measurements. In such cases, two remedies are possible:

- Reduce the dimension of d to match the number of measurements n_{y_m} via an appropriate projection, or
- Adopt the differenced model formulation and augment the state with $y(k)$, which includes all output variables.

In the latter case, the resulting augmented system may remain undetectable, particularly when the number of integrators acting on y_c and y_s exceeds the dimension of y_m. However, the Riccati difference equation (RDE) can still converge under certain conditions. Specifically, if the pair (H_mC, A) is observable and the number of disturbance variables is sufficiently small to be estimated from y_m without bias, then a Kalman filter can be constructed to estimate the augmented state $\hat{\bar{x}}(k|k)$ as described earlier. The Kalman gain and estimation error covariance can be computed analogously to Eqs. (11.49) and (11.50), with $\bar{C}$ replaced by $H_m\bar{C}$.

Once the state estimate $\hat{\bar{x}}(k|k)$ is obtained, it is used to generate model predictions over the control horizon:

$$\mathcal{Y} = \bar{\mathcal{S}}^{\bar{x}} z_0 + \bar{\mathcal{S}}^{u}\Delta\mathcal{V}, \tag{11.78}$$

with $z_0 = \hat{\bar{x}}(k|k)$. Since not all outputs need to be regulated, we assign zero weights in the cost function to the outputs that do not require set point tracking.

Example 11.11: *Inferential Control in the Presence of Unmeasured Step Disturbance for the System of Example 7.10.*
Consider the true "plant" described by

$$\begin{aligned} x(k+1) &= \begin{bmatrix} 0.6346 & 0.02865 & 0.02325 & 0.2662 \\ 0.9891 & 0.04599 & 0.03975 & 0.6882 \\ 0.7102 & 0.03516 & 0.03416 & 0.9322 \\ 0.1060 & 0.00792 & 0.01213 & 0.6866 \end{bmatrix} x(k) \\ &\quad + \begin{bmatrix} 0.03277 \\ 0.2113 \\ 0.1856 \\ 0.01007 \end{bmatrix} u(k) + \begin{bmatrix} 0.01316 \\ 0.1228 \\ 0.01819 \\ 0.00249 \end{bmatrix} d_u(k) \\ y_m(k) &= \begin{bmatrix} 0 & -28.1 & -27.3 & 0 \end{bmatrix} x(k) + \nu(k) \\ y_c(k) &= \begin{bmatrix} 1 & 0 & 0 & 0 \end{bmatrix} x(k). \end{aligned} \tag{7.30}$$

The primary controlled variable $y_c(k)$ is not measured. The control objective is to drive $y_c(k)$ to a set point of $r = 0.1$. A step disturbance $d(k) = 0.5$ enters the system and is unmeasured.

Additionally, the MPC model does not perfectly match the true plant. The input and disturbance matrices used in the controller model are

$$B = \begin{bmatrix} 0.02949 \\ 0.1902 \\ 0.1670 \\ 0.009063 \end{bmatrix} \qquad B_d = \begin{bmatrix} 0.01448 \\ 0.1351 \\ 0.02001 \\ 0.002739 \end{bmatrix}. \tag{11.79}$$

Inferential control is necessary because $y_c(k)$ must be inferred from the measurement $y_m(k)$. According to the framework introduced above, we define

$$C = \begin{bmatrix} 1 & 0 & 0 & 0 \\ 0 & -28.1 & -27.3 & 0 \end{bmatrix}. \tag{11.80}$$

Since only the secondary output is measured, $H_m = \begin{bmatrix} 0 & 1 \end{bmatrix}$. This sets up the inferential estimation and MPC formulation for the given system.

Solution using the MPC toolbox: We begin by defining two state-space models. The first model, `Gtrue`, corresponds to the true plant as described in Eq. (7.30). This is a two-input, two-output system, where `Btrue` is constructed as $[B \quad B_d]$. The second model, `Gmodel`, represents the model used by the MPC controller and is created by replacing the B and B_d matrices with those given in Eq. (11.79).

The following command is used to specify the input/output structure of the MPC model: `Gmpc=setmpcsignals(Gmodel,'MV',1,'UD',2,'MO',2,'UO',1);`

This indicates that the second input channel is an unmeasured disturbance, and that the first output is unmeasured.

The constraints are set as $-2 \leq u(k) \leq 2$ and $|\Delta u(k)| \leq 0.2$. The prediction and control horizons are selected as $p = 10$ and $m = 5$, respectively. The weighting matrices are defined as

$$\Gamma^y = \begin{bmatrix} 1 & 0 \\ 0 & 0 \end{bmatrix} \qquad \Gamma^u = 0.1.$$

To simulate the true plant behavior during MPC execution, the plant model `Gtrue` is specified using the `mpcsimopt` command as MPC options structure.

Output Disturbance Model: Note that we have only one measurement available. Consequently, at most one nonstationary unmeasured disturbance can be estimated from the secondary measurement. This implies that a full disturbance model cannot be used, as it would render the augmented system undetectable. Indeed, an integrating output disturbance cannot be specified on the unmeasured output, since doing so will render the augmented model undetectable.

Alternatively, one may consider specifying an output disturbance affecting the measured (secondary) output:

$$\frac{1}{s}\begin{bmatrix} 1 & 0 \\ 0 & 0 \end{bmatrix},$$

as integrating white noise using the `setoutdist` command. However, as discussed earlier, this choice produces a deadbeat estimator for the secondary output while leaving the state estimates

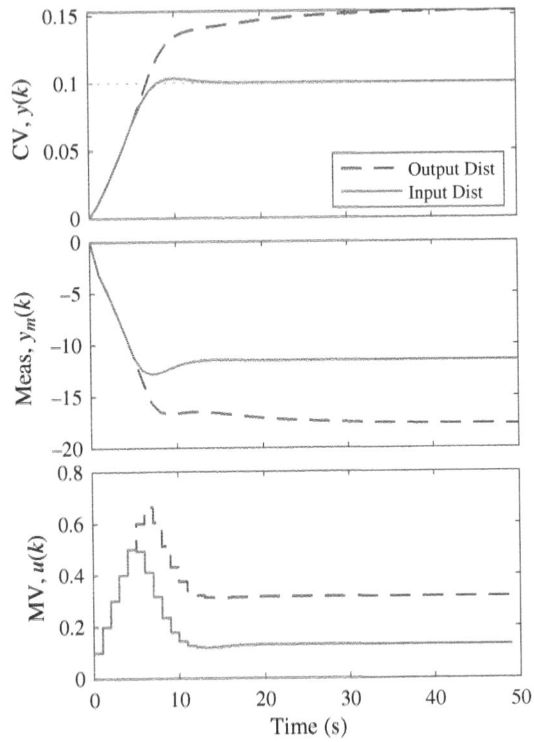

Figure 11.11 With unmeasured disturbance modeled as integrating white noise, MPC is able to control the primary variable at its set point. A very small offset is seen in the controlled variable.

unaffected by the filter. As a result, the estimate of the primary output remains significantly biased in steady state, leading to offset from the set point.

State Disturbance Model: Since there is only one unmeasured disturbance, it is appropriate to introduce an integrating white-noise process for it using the `setindist` command. The resulting augmented system is detectable, and the MPC achieves very good control performance, as illustrated in Figure 11.11. However, because there are two outputs, the available measurements are insufficient to guarantee offset-free estimation for both in the presence of model–plant mismatch. As a result, despite excellent transient performance, a small (almost imperceptible) steady-state offset remains in the controlled output. This offset becomes more pronounced as the model–plant mismatch increases.

11.4.2 Infrequent Primary Measurements

A more common scenario than the one discussed previously is when the primary control variable is measured only infrequently, while secondary variables are available at every sampling instance. Suppose the secondary measurements are taken at every sampling instant, but the primary measurement is available only at every m^{th} step. The measurement model is then given by

$$y_m(k) = \begin{cases} \begin{bmatrix} C_c \\ C_s \end{bmatrix} x(k) + \begin{bmatrix} v_c(k) \\ v_s(k) \end{bmatrix} & \text{when } k = m, 2m, 3m, \cdots \\ C_s x(k) + v_s(k) & \text{otherwise.} \end{cases} \tag{11.81}$$

If the pair $\left(\begin{bmatrix} C_c \\ C_s \end{bmatrix}, A\right)$ is detectable, a Kalman filter can be designed for this multi-rate measurement setup. Unlike the standard Kalman filter, the resulting filter does not converge to a time-invariant steady state. Instead, it converges to a periodic steady state with the covariance satisfying $P(k) = P(k+m)$ for all k.

In practice, it is convenient to initialize the filter with an initial guess for the covariance $P(0)$ and compute the time-varying filter recursively. Specifically, at each time step, the augmented state estimate, based on the augmented system matrices, is updated as follows:

$$\hat{\bar{x}}(k|k-1) = \bar{A}\hat{\bar{x}}(k-1|k-1) + \bar{B}u(k) \tag{11.82}$$

$$\hat{\bar{x}}(k|k) = \hat{\bar{x}}(k|k-1) + K_f(k)\Big(y_m(k) - \bar{C}(k)\hat{\bar{x}}(k|k-1)\Big), \tag{11.83}$$

where $K_f(k)$ is the time-varying Kalman gain calculated as

$$K_f(k) = P(k)\bar{C}(k)^T\left(\bar{C}(k)P(k)\bar{C}(k)^T + R_2(k)\right)^{-1},$$

using the covariance update equation

$$P(k) = \bar{A}P(k-1)\bar{A}^T + R_1 - \bar{A}P(k-1)\bar{C}(k-1)^T\left(\bar{C}(k-1)P(k-1)\bar{C}(k-1)^T + R_2\right)^{-1}.$$

The time-varying Kalman gain converges to a periodically varying sequence with period m as the error covariance $P(k)$ approaches a periodic steady-state solution of the same period. If implementing a time-varying Kalman filter is impractical, this periodic steady-state solution can be computed offline, allowing the corresponding sequence of Kalman gain matrices to be stored and used during online implementation.

This *multi-rate Kalman filter* enables accurate state estimation even when critical measurements are available only intermittently.

11.4.3 Handling Measurement Delays in Primary Measurements

Consider the case where the primary measurement is obtained with a delay of n_{θ_c} sample steps. This scenario commonly arises when the primary control variable—often related to quality or productivity—requires offline analysis, leading to delayed measurement availability. Specifically, the result of the sample taken at time k (sampled every m time steps) becomes available at time $k + n_{\theta_c}$. At these delayed time instances, both the primary and secondary measurements are available:

$$y_m(k) = \begin{bmatrix} C_c x\left(k - n_{\theta_c}\right) \\ C_s x(k) \end{bmatrix} + \begin{bmatrix} v_c(k) \\ v_s(k) \end{bmatrix} \qquad \text{when } k = i \cdot m + n_{\theta_c}. \tag{11.84}$$

At all other sampling instances, only the secondary measurements are available:

$$y_m(k) = C_s x(k) + v_s(k).$$

To incorporate delayed measurements into state estimation, various techniques have been proposed. A widely used approach—particularly suitable for integration with model-based control—augments the state vector with past values of the delayed measurement. Specifically, the augmented state includes the current state, the secondary output, and current and n_{θ_c} past values of the primary output:

$$\bar{x}(k) \triangleq \begin{bmatrix} x(k) \\ d(k) \\ y_c(k-1) \\ \vdots \\ y_c(k-n_{\theta_c}) \end{bmatrix}. \tag{11.85}$$

The state-space representation corresponding to this augmented state is given by

$$\underbrace{\begin{bmatrix} x(k+1) \\ d(k+1) \\ y_c(k) \\ \vdots \\ y_c(k+1-n_{\theta_c}) \end{bmatrix}}_{\bar{x}(k+1)} = \underbrace{\begin{bmatrix} A & B_d & 0 & \cdots & 0 \\ 0 & I & 0 & \cdots & 0 \\ C_c & 0 & 0 & \cdots & 0 \\ \vdots & \vdots & \ddots & \ddots & \vdots \\ 0 & 0 & 0\cdots & I & 0 \end{bmatrix}}_{\bar{A}} \underbrace{\begin{bmatrix} x(k) \\ d(k) \\ y_c(k-1) \\ \vdots \\ y_c(k-n_{\theta_c}) \end{bmatrix}}_{\bar{x}(k)} + \underbrace{\begin{bmatrix} B \\ 0 \\ 0 \\ \vdots \\ 0 \end{bmatrix}}_{\Gamma} u(k) + \begin{bmatrix} 0 \\ I \\ 0 \\ \vdots \\ 0 \end{bmatrix} \varepsilon(k), \tag{11.86}$$

with measurements at primary arrival times as

$$y_m(k) = \underbrace{\begin{bmatrix} 0 & 0 & \cdots & I \\ C_s & 0 & \cdots & 0 \end{bmatrix}}_{\bar{C}} \bar{x}(k) + \begin{bmatrix} v_c(k) \\ v_s(k) \end{bmatrix}, \tag{11.87}$$

and at all other time instances,

$$y_m(k) = \underbrace{\begin{bmatrix} C_s & 0 & \cdots 0 \end{bmatrix}}_{\bar{C}_s} \bar{x}(k) + v_s(k). \tag{11.88}$$

With this definition of the augmented state, the state estimates can be calculated as described in the previous section. Specifically,

$$\hat{\bar{x}}(k|k-1) = \bar{A}\hat{\bar{x}}(k-1|k-1) + \Gamma u(k) \tag{11.89}$$

$$\hat{\bar{x}}(k|k) = \hat{\bar{x}}(k|k-1) + K_f(k)\left(y_m(k) - \bar{C}(k)\hat{\bar{x}}(k|k-1)\right), \tag{11.90}$$

where $\bar{C}(k) = \bar{C}_s$ normally but becomes $\bar{C}(k) = \bar{C}$ whenever the delayed primary measurement result comes in.

The time-varying Kalman filter can be employed, which is calculated as

$$P(k) = \bar{A}P(k-1)\bar{A}^T + R_1 - \bar{A}K_f(k-1)\bar{C}(k-1)P(k-1)\bar{A}^T \tag{11.91}$$

$$K_f(k) = P(k)\overline{C}(k)^T\left(\overline{C}(k)P(k)\overline{C}(k)^T + R_2(k)\right)^{-1}, \tag{11.92}$$

where the covariance, R_1, of the state noise signal given by

$$R_1 = \mathtt{blockdiag}\left(R_{\epsilon_1}, 0_{n_y \times n_y}, \cdots, 0_{n_{yc} \times n_{yc}}\right), \tag{11.93}$$

and the covariance, $R_2(k)$, of the state noise signal is adjusted to reflect the noise covariance of the measurements—either y_s or y_s, y_c both—taken at time step k.

Remark 11.6: Delays in secondary measurements: When secondary measurements are also subject to a delay of θ_s sampling instances, it becomes necessary to retain past values of the secondary outputs for proper state estimation. A similar state augmentation approach as described earlier can be employed, with the augmented state defined as

$$\bar{x}(k) \triangleq \begin{bmatrix} x(k) \\ y_c(k) \\ \vdots \\ y_c(k-\theta_c) \\ y_s(k) \\ \vdots \\ y_s(k-\theta_s) \end{bmatrix}. \tag{11.94}$$

Remark 11.7: Rejecting persistent disturbances: Section 11.1 discussed how to formulate the overall model in standard state-space form under persistent noise sequences. To enable integral action for rejecting such disturbances, it is beneficial to use a differenced form of the model equations. In this case, the augmented state may be defined as

$$\bar{x}(k) \triangleq \begin{bmatrix} \Delta x(k) \\ y_c(k) \\ \vdots \\ y_c(k-\theta_c) \\ y_s(k) \end{bmatrix}. \tag{11.95}$$

The procedure outlined in this section can then be applied to accommodate multi-rate measurements and measurement delays.

11.5 Sequential Linearization-based MPC (for Nonlinear Systems)

The discussion so far has focused on linear systems. In this section, we extend the development of MPC to nonlinear systems. A fully nonlinear MPC approach—based on using a nonlinear model for prediction and solving the resulting nonlinear program (NLP)—will be presented in Chapter 12.

However, the practical implementation of such an approach is often hindered by significant computational demands and the complexity of solving large-scale NLPs. To address these challenges, various approximation techniques are commonly employed.

This section specifically focuses on sequential linearization, in which the nonlinear model is linearized at each sampling instant. The resulting linearized model is then used to formulate linear estimation and linear MPC problems.

Consider a general nonlinear discrete-time state-space model of the form:

$$x(k+1) = F\big(x(k), u(k), d(k)\big) \tag{11.96}$$

$$y(k) = g\big(x(k), d(k)\big) \tag{11.97}$$

$$y_m(k) = y(k) + \nu(k). \tag{11.98}$$

Here, $\nu(k)$ denotes measurement noise, assumed to be a Gaussian white noise sequence. The disturbance $d(k)$ is modeled using the following linear system:

$$x_w(k+1) = A_w x_w(k) + B_w \varepsilon(k) \tag{11.99}$$

$$d(k) = C_w x_w(k). \tag{11.100}$$

In many practical cases, the discrete-time function $F(\cdot)$ is not directly available. Instead, a first-principles continuous-time model is provided in the form:

$$\dot{x} = f(x, u, d). \tag{11.101}$$

To obtain the discrete-time dynamics, the ODE (11.101) is integrated over one sampling interval from kt_s to $(k+1)t_s$, using $x(k)$ as the initial condition and holding u and d constant at $u(k)$ and $d(k)$, respectively (i.e., zero-order hold). The resulting terminal state defines the function $F(x(k), u(k), d(k))$.

The remainder of this section mirrors the earlier development for linear MPC. We begin by discussing model construction, followed by the development of a state estimation scheme, the derivation of prediction equations over the prediction horizon, and finally, the formulation of the objective function and constraints used in solving the MPC problem.

11.5.1 Model Construction

11.5.1.1 Linearization

Before presenting the estimator and MPC algorithm, we first review the multivariate Taylor series expansion used to linearize a nonlinear function. Consider a vector-valued function $f(x, y, \dots)$, where the variables x, y, etc., are themselves vectors. The first-order Taylor expansion around a nominal point $(\hat{x}, \hat{y}, \dots)$ yields the following linear approximation:

$$f(x, y, \cdots) \approx f(\hat{x}, \hat{y}, \cdots) + f_{\hat{x}}(x - \hat{x}) + f_{\hat{y}}(y - \hat{y}) + \cdots;$$

$$\text{where } f_{\hat{x}} \triangleq \left.\frac{\partial f}{\partial x}\right|_{(\hat{x},\hat{y},\cdots)}$$

$$\text{and } f_{\hat{y}} \triangleq \left.\frac{\partial f}{\partial y}\right|_{(\hat{x},\hat{y},\cdots)}$$

are the Jacobian matrices evaluated at the expansion point.

As before, we use the surrogate variables z_i, v_i, and δ_i to represent the predicted future values $x(k+i|k)$, $u(k+i|k)$, and $d(k+i|k)$, respectively, at time step k. In sequential linearization-based

MPC, the key step is to linearize the nonlinear model (11.96) around the best available estimates at time step k: the projected state z_i^{proj} (to be defined later), the previous input $u(k-1)$, and the estimated disturbance $\delta_i = C_w A_w^{i-1} \hat{x}_w(k|k)$. The resulting linear approximation of the state update equation is

$$z_{i+1} \approx F(z_i^{\text{proj}}, u(k-1), \delta_i) + A_i\left(z_i - z_i^{\text{proj}}\right) + B_i\left(v_i - u(k-1)\right), \tag{11.102}$$

where the Jacobian matrices are defined as

$$A_i = \left.\frac{\partial F(x,u,d)}{\partial x}\right|_{z_i^{\text{proj}},u(k-1),\delta_i} \tag{11.103}$$

$$B_i = \left.\frac{\partial F(x,u,d)}{\partial u}\right|_{z_i^{\text{proj}},u(k-1),\delta_i}. \tag{11.104}$$

Note that the linearization does not include a first-order approximation with respect to δ_i, since the linear disturbance model already provides the best predicted value for δ_i, making further linearization unnecessary.

Remark 11.8: When $F(\cdot)$ is not explicitly available: A more common scenario arises when the system is described by a continuous-time model, so that the discrete-time function $F(\cdot)$ is only implicitly defined. In such cases, the Jacobian matrices for the continuous-time model (11.101) are given by

$$A_i^c = \left.\frac{\partial f(x,u,d)}{\partial x}\right|_{z_i^{\text{proj}},u(k-1),\delta_i}$$
$$B_i^c = \left.\frac{\partial f(x,u,d)}{\partial u}\right|_{z_i^{\text{proj}},u(k-1),\delta_i}.$$

The continuous-time linearized model can then be discretized using standard techniques (as described in Chapter 5) to obtain

$$A_i = \exp\left(A_i^c h\right)$$
$$B_i = \int_0^h \exp\left(A_i^c \tau\right) B_i^c \, d\tau.$$

A linear approximation of the output equation (11.97) can be obtained in a similar manner:

$$y_i \approx g(z_i^{\text{proj}}, \delta_i) + C_i(z_i - z_i^{\text{proj}}), \tag{11.105}$$

where the Jacobian matrices are defined as

$$C_i = \left.\frac{\partial g(z,\delta)}{\partial z}\right|_{z_i^{\text{proj}},u(k-1),\delta_i}. \tag{11.106}$$

11.5.1.2 State augmentation

Future predictions depend not only on the current process state $x(k)$ but also on the disturbance state $x_w(k)$, both of which must be estimated when new measurements become available. Following the approach discussed in Section 11.1, we define the augmented state as

$$\bar{x} = \begin{bmatrix} x(k) \\ x_w(k) \end{bmatrix}.$$

The nonlinear system dynamics (11.96) and the disturbance model (11.99) can be combined into an augmented form:

$$\begin{bmatrix} x(k+1) \\ x_w(k+1) \end{bmatrix} = \begin{bmatrix} F\left(x(k), u(k), C_w x_w(k)\right) \\ A_w x_w(k) \end{bmatrix} + \begin{bmatrix} 0 \\ B_w \end{bmatrix} \varepsilon(k). \tag{11.107}$$

An estimation scheme based on Kalman filter principles will be developed next to estimate the augmented state $\hat{\bar{x}}(k|k)$ using available measurements. Subsequently, multistep prediction equations will be derived to capture the effect of future input moves over the control horizon. The augmented nonlinear model will then be linearized to express system evolution as a linear function of the decision variables $u(k+i)$.

11.5.2 Extended Kalman Filter

The role of the estimator is to compute the state estimate $\hat{\bar{x}}(k|k)$ based on the previous estimate $\hat{\bar{x}}(k-1|k-1)$, the previous input $u(k-1)$, and the current measurement $y(k)$. The one-step prediction is given by

$$\hat{\bar{x}}(k|k-1) = \begin{bmatrix} F\left(\hat{x}(k-1|k-1), u(k-1), C_w \hat{x}_w(k-1|k-1)\right) \\ A_w \hat{x}_w(k-1|k-1) \end{bmatrix}. \tag{11.108}$$

This predicted state is then updated using the measurement $y(k)$ through the extended Kalman filter (EKF) update equation:

$$\hat{\bar{x}}(k|k) = \hat{\bar{x}}(k|k-1) + K_f(k)\left(y_m(k) - \bar{g}\left(\hat{\bar{x}}(k|k-1)\right)\right), \tag{11.109}$$

where the notation $\bar{g}\left(\hat{\bar{x}}(k|k-1)\right)$ is used to denote $g\left(\hat{x}(k|k-1), C_w \hat{x}_w(k|k-1)\right)$.

The Kalman gain $K_f(k)$ is derived by minimizing the estimation error covariance, where the error is defined as $\bar{x}_e(k) = \bar{x}(k) - \hat{\bar{x}}(k|k)$. To facilitate this, the nonlinear model (11.98) is linearized around the point $(\hat{\bar{x}}(k-1|k-1), u(k-1))$:

$$\begin{bmatrix} x(k) \\ x_w(k) \end{bmatrix} \approx \begin{bmatrix} F\left(\hat{x}(k-1|k-1), u(k-1), C_w \hat{x}_w(k-1|k-1)\right) \\ A_w \hat{x}_w(k-1|k-1) \end{bmatrix} +$$
$$\underbrace{\begin{bmatrix} \hat{A}_{k-1|k-1} & \hat{B}^d_{k-1|k-1} C_w \\ 0 & A_w \end{bmatrix}}_{\bar{A}_{k-1}} \begin{bmatrix} x(k-1) - \hat{x}(k-1|k-1) \\ x_w(k-1) - \hat{x}_w(k-1|k-1) \end{bmatrix} + \underbrace{\begin{bmatrix} 0 \\ B_w \end{bmatrix}}_{\bar{B}_w} \varepsilon(k) \tag{11.110}$$

In the above expressions, the Jacobians are defined as

$$\hat{A}_{k-1|k-1} = \left.\frac{\partial F}{\partial x}\right|_{\hat{\bar{x}}(k-1|k-1), u(k-1)} \tag{11.111}$$

$$\hat{B}^d_{k-1|k-1} = \left.\frac{\partial F}{\partial d}\right|_{\hat{\bar{x}}(k-1|k-1), u(k-1)}. \tag{11.112}$$

The linearized equation can be written compactly as

$$\bar{x}(k) = \hat{\bar{x}}(k|k-1) + \bar{A}_{k-1} \underbrace{\left(\bar{x}(k-1) - \hat{\bar{x}}(k-1|k-1)\right)}_{\bar{x}_e(k-1)} + \bar{B}_w \varepsilon(k). \tag{11.113}$$

With some abuse of notation, we rewrite the linearized output equation (11.105) in the compact form:

$$y_m(k) \approx \bar{g}\left(\hat{\bar{x}}(k|k-1)\right) + \underbrace{\begin{bmatrix} C_{k|k-1} & C^d_{k|k-1} \end{bmatrix}}_{\bar{C}_k} \left(\bar{x}(k) - \hat{\bar{x}}(k|k-1)\right) + v(k). \tag{11.105}$$

Substituting this into the EKF update Eq. (11.109), we obtain

$$\hat{\bar{x}}(k|k) = \hat{\bar{x}}(k|k-1) + K_f(k)\bar{C}_k\left(\bar{x}(k) - \hat{\bar{x}}(k|k-1)\right) + K_f(k)v(k). \tag{11.114}$$

Subtracting this from the linearized state Eq. (11.113) gives the error dynamics:

$$\begin{aligned} \bar{x}_e(k) = \bar{A}_{k-1}\bar{x}_e(k-1) + \bar{B}_w\varepsilon(k) \\ -K_f(k)\bar{C}_k\left\{\bar{x}(k) - \hat{\bar{x}}(k|k-1)\right\} - K_f(k)v(k). \end{aligned} \tag{11.115}$$

Substituting the expression for $\bar{x}(k) - \hat{\bar{x}}(k|k-1)$ from Eq. (11.113), we obtain

$$\bar{x}_e(k) = \left(I - K_f(k)\bar{C}_k\right)\bar{A}_{k-1}\bar{x}_e(k-1) + \left(I - K_f(k)\bar{C}_k\right)\bar{B}_w\varepsilon(k) + K_f(k)v(k). \tag{11.116}$$

This equation has the same structure as the estimation error dynamics derived in Eq. (7.10) of Chapter 7. Following the same reasoning, the Kalman filter gain that minimizes the estimation error covariance is given by

$$K_f(k) = P(k)\bar{C}_k^T\left(\bar{C}_k P(k)\bar{C}_k^T + R_2\right)^{-1}, \tag{11.117}$$

where the one-step-ahead estimation error covariance $P(k)$ is

$$P(k) = \bar{A}_{k-1}P(k-1)\bar{A}_{k-1}^T + R_1 - \bar{A}_{k-1}K_f(k-1)\bar{C}_{k-1}P(k-1)\bar{A}_{k-1}^T. \tag{11.118}$$

The process and measurement noise covariances are defined as

$$R_1 = \bar{B}_w R_\varepsilon \bar{B}_w^T \qquad R_2 = R_v.$$

Remark 11.9: Note that $\hat{A}_{k-1|k-1}$ is computed at time k using the most recent available information—specifically, the prior state estimate $\hat{\bar{x}}(k-1|k-1)$ and the applied input $u(k-1)$. This computation is performed before incorporating the current measurement $y_m(k)$ into the state estimate. As a result, the model used for prediction is linearized around $\hat{\bar{x}}(k-1|k-1)$ and $u(k-1)$.

This differs from A_0 introduced in Eq. (11.103), which is computed after the measurement $y(k)$ has been incorporated. Consequently, A_0 is linearized around the updated state estimate $z_0 = \hat{\bar{x}}(k|k)$ and the same previous input $u(k-1)$.

11.5.3 Multistep Prediction

The goal of MPC is to determine a sequence of future input moves, $\Delta v_0, \ldots, \Delta v_{k+m-1}$, that minimizes a quadratic objective function. The EKF described above provides the current state estimate $\hat{\bar{x}}(k|k)$. To make the MPC problem computationally tractable, the nonlinear model is linearized to obtain prediction equations that are linear functions of these decision variables.

11.5.3.1 Linearization of Future State Predictions

At the current time step k, the future input moves $\Delta u(k+i),\ i \geq 0$ are not yet known. To compute a nominal trajectory around which to linearize, the future state projections $z_1, \ldots, z_p$ are generated by assuming that the input remains constant at the most recent applied value: $u(k+\ell) = u(k-1)$ for $\ell = 0, \ldots, i$. The projected states are then computed recursively as

$$z_1^{\text{proj}} = F(z_0, u(k-1), \delta_0) \tag{11.119}$$

$$z_2^{\text{proj}} = F(F(z_0, u(k-1), \delta_0), u(k-1), \delta_1) \tag{11.120}$$

$$= F\left(z_1^{\text{proj}}, u(k-1), \delta_1\right) \tag{11.121}$$

$$\vdots \tag{11.122}$$

$$z_p^{\text{proj}} = F\left(z_{p-1}^{\text{proj}}, u(k-1), \delta_{p-1}\right), \tag{11.123}$$

and

$$\begin{aligned} \omega_{i+1} &= A_w \omega_i \\ \delta_i &= C_w \omega_i \end{aligned}, \tag{11.124}$$

initialized with

$$z_0 = \hat{x}(k|k) \text{ and } \omega_0 = \hat{x}_w(k|k).$$

If the system model is given in continuous time as an ODE (11.102), these projections are obtained by numerically integrating the ODE starting from the initial condition z_0, using a constant input $u = u(k-1)$ over the time interval $(0, p \cdot t_s]$, and a piecewise-constant disturbance input δ_i. The projected state z_i is then the value of the ODE solution at $i \cdot t_s$.

Consider the one-step state prediction:

$$\begin{aligned} z_1 &= F(z_0, v_0, \delta_0) \\ &\approx F(z_0, u(k-1), \delta_0) + B_0(v_0 - u(k-1)) \\ z_1 &\approx z_1^{\text{proj}} + B_0(v_0 - u(k-1)), \end{aligned} \tag{11.125}$$

where the approximation is linearized around $u(k-1)$, and z_1^{proj} is the nominal state projection assuming constant input $u(k-1)$.

Now consider the two-step prediction:

$$z_2 = F(z_1, v_1, \delta_1).$$

This expression depends explicitly on v_1 and implicitly on v_0 through the term z_1. To construct a linear approximation, we expand around the nominal trajectory using z_1^{proj} and $u(k-1)$:

$$z_2 \approx F\left(z_1^{\text{proj}}, u(k-1), \delta_1\right) + A_1\left(z_1 - z_1^{\text{proj}}\right) + B_1(v_1 - u(k-1)).$$

Substituting the expression for z_2 from Eq. (11.125):

$$z_2 \approx z_2^{\text{proj}} + \begin{bmatrix} A_1 B_0 & B_1 \end{bmatrix} \begin{bmatrix} v_0 - u(k-1) \\ v_1 - u(k-1) \end{bmatrix}. \tag{11.126}$$

The multistep state prediction can now be generalized as

$$z_i \approx z_i^{\text{proj}} + \left[\prod_{j=1}^{i-1} A_j B_0 \quad \prod_{j=2}^{i-1} A_j B_1 \quad \cdots \quad B_{i-1} \right] \times \begin{bmatrix} v_0 - u(k-1) \\ v_1 - u(k-1) \\ \vdots \\ v_{i-1} - u(k-1) \end{bmatrix}. \tag{11.127}$$

In this expression, the matrices A_j and B_j are the Jacobians defined as

$$A_j = \left. \frac{\partial F(x, u, d)}{\partial x} \right|_{z_j^{\text{proj}}, u(k-1), \delta_j} \tag{11.103}$$

$$B_j = \left. \frac{\partial F(x, u, d)}{\partial u} \right|_{z_j^{\text{proj}}, u(k-1), \delta_j} . \tag{11.104}$$

Note that the linearization is carried out around the nominal state trajectory z_j^{proj}, a constant input $u(k-1)$, and the predicted disturbance δ_j. All approximations assume that the input remains constant at $u(k-1)$ during the generation of the nominal trajectory.

11.5.3.2 Computation of Controlled Output Predictions

The predicted controlled output at time $k + i$ is given by

$$y_i = g\left(z_i, \delta_i\right). \tag{11.128}$$

To obtain a tractable expression, we linearize the function $g(\cdot)$ around the nominal state and disturbance trajectory, z_i^{proj} and δ_i:

$$y_i \approx g\left(z_i^{\text{proj}}, \delta_i\right) + C_i \left(z_i - z_i^{\text{proj}}\right), \tag{11.129}$$

where

$$C_i = \left. \frac{\partial g(x, d)}{\partial x} \right|_{z_i^{\text{proj}}, \delta_i} . \tag{11.106}$$

Note that the linear term in δ_i is omitted because its governing model is already linear.

Substituting the expression for $\hat{x}(k+i|k)$ from Eq. (11.127), we obtain the linearized output prediction:

$$y_i = g\left(z_i^{\text{proj}}, \delta_i\right) + C_i \left[\prod_{j=1}^{i-1} A_j B_0 \quad \prod_{j=2}^{i-1} A_j B_1 \quad \cdots \quad B_{i-1} \right] \times \begin{bmatrix} v_0 - u(k-1) \\ v_1 - u(k-1) \\ \vdots \\ v_{i-1} - u(k-1) \end{bmatrix}. \tag{11.130}$$

Here, the nominal trajectory $z_{k+i|k}^{\text{proj}}$ is computed using Eq. (11.123), and the matrices A_i, B_i, and C_i are defined as in Eqs. (11.103), (11.104), and (11.106), respectively.

Remark 11.10: In the preceding derivation, the nonlinear equations were linearized around a constant input trajectory given by $u(k-1)$. However, if an initial guess of the future input trajectory—denoted by $\tilde{\upsilon}_0, \tilde{\upsilon}_1, \ldots$—is available (e.g., from the optimal input sequence calculated at the previous sample time), the model can instead be linearized around this nominal trajectory for potentially improved accuracy.

Specifically, define the nominal state trajectory recursively as

$$\tilde{z}_{i+1}^{\text{proj}} = F\left(\tilde{z}_i^{\text{proj}}, \tilde{\upsilon}_i, \delta_i\right).$$

The corresponding linearized state-space matrices are

$$\begin{aligned}\tilde{A}_i &= \left.\frac{\partial F(x,u,d)}{\partial x}\right|_{\tilde{z}_i^{\text{proj}}, \tilde{\upsilon}_i, \delta_i}\\ \tilde{B}_i &= \left.\frac{\partial F(x,u,d)}{\partial u}\right|_{\tilde{z}_i^{\text{proj}}, \tilde{\upsilon}_i, \delta_i}.\\ \tilde{C}_i &= \left.\frac{\partial g(x,d)}{\partial d}\right|_{\tilde{z}_i^{\text{proj}}, \tilde{\upsilon}_i, \delta_i}\end{aligned}$$

With this formulation, the predicted output can be expressed as

$$y_i = g\left(\tilde{z}_i^{\text{proj}}, \delta_i\right) + \tilde{C}_i \begin{bmatrix} \prod_{j=1}^{i-1} \tilde{A}_j \tilde{B}_0 & \prod_{j=2}^{i-1} \tilde{A}_j \tilde{B}_1 & \cdots & \tilde{B}_{i-1} \end{bmatrix} \times \begin{bmatrix} \upsilon_0 - \tilde{\upsilon}_0 \\ \upsilon_1 - \tilde{\upsilon}_1 \\ \vdots \\ \upsilon_i - \tilde{\upsilon}_i \end{bmatrix}.$$

This approach enables linearization around the best available estimate of the time-varying input sequence, leading to improved prediction accuracy—particularly in strongly nonlinear systems.

Example 11.12: *Bilinear System.*
Consider a simple bilinear model:

$$x(k+1) = x(k)u(k). \tag{11.131}$$

For a horizon of 3, the system evolves as

$$\begin{aligned} z_3 &= z_2\upsilon_2 = (z_1\upsilon_1)\,\upsilon_2 \\ &= ((z_0\upsilon_0)\,\upsilon_1)\,\upsilon_2. \end{aligned}$$

Thus, the model prediction is

$$z_3 = \underbrace{z_0\upsilon_0\upsilon_1\upsilon_2}_{\mathfrak{f}(z_0,\upsilon_0,\upsilon_1,\upsilon_2)}. \tag{11.132}$$

We now linearize this prediction around z_0 and a nominal input trajectory held constant at $u(k-1)$. The partial derivatives are

$$\left.\frac{\partial \mathfrak{f}}{\partial z}\right|_{z_0,u(k-1)} = u^3(k-1)$$
$$\left.\frac{\partial \mathfrak{f}}{\partial v_0}\right|_{z_0,u(k-1)} = z_0u^2(k-1)$$
$$\left.\frac{\partial \mathfrak{f}}{\partial v_1}\right|_{z_0,u(k-1)} = z_0u^2(k-1)$$
$$\left.\frac{\partial \mathfrak{f}}{\partial v_2}\right|_{z_0,u(k-1)} = z_0u^2(k-1).$$

The resulting linear approximation is

$$\hat{x}(k+3|k) \approx \underbrace{z_0u^3(k-1)}_{\mathfrak{f}(z_0,u(k-1),u(k-1),u(k-1))} + \begin{bmatrix} z_0u^2(k-1) & z_0u^2(k-1) & z_0u^2(k-1) \end{bmatrix} \begin{bmatrix} v_0 - u(k-1) \\ v_1 - u(k-1) \\ v_2 - u(k-1) \end{bmatrix}.$$

This linearized expression corresponds to the actual nonlinear prediction in Eq. (11.132).

Next, we apply the recursive procedure that leads to Eq. (11.127). Assuming the input is held constant at $u(k-1)$, the state projections are

$$z_1^{\text{proj}} = z_0u(k-1)$$
$$z_2^{\text{proj}} = z_1^{\text{proj}}u(k-1) = z_0u^2(k-1)$$
$$z_3^{\text{proj}} = z_2^{\text{proj}}u(k-1) = z_0u^3(k-1).$$

The linearization of the model function $F(z, v)$ yields:

$$\frac{\partial F}{\partial z} = v \quad \text{and} \quad \frac{\partial F}{\partial v} = z.$$

From this, we obtain

$$A_i = u(k-1)$$
$$B_i = z_i^{\text{proj}} = z_0u^i(k-1).$$

Substituting into Eq. (11.127), we get

$$z_3 \approx \underbrace{z_0u^3(k-1)}_{z_3^{\text{proj}}} + \begin{bmatrix} u^2(k-1)\cdot z_0 & u(k-1)\cdot z_0u(k-1) & z_0u^2(k-1) \end{bmatrix} \begin{bmatrix} v_0 - u(k-1) \\ v_1 - u(k-1) \\ v_2 - u(k-1) \end{bmatrix}.$$

This matches the earlier linearization, confirming consistency between the recursive and direct approaches.

11.5.3.3 Simplification of Model Predictions

The strategy of computing nonlinear state updates and linearizing around the projected states $z^{\text{proj}}_{k+i|k}$ can be computationally demanding. This is because the linearized state-space matrices $A_{k+i|k}$, $B_{k+i|k}$, and $C_{k+i|k}$ must be recalculated at each step along the prediction horizon.

A simplification is to perform all linearizations around the current state estimate $\hat{x}(k|k)$, the most recent input $u(k-1)$, and the current disturbance estimate $\hat{x}_w(k|k)$—rather than around the full predicted trajectory. This reduces computational effort while still yielding reasonable prediction accuracy in many cases.

The output prediction equation (11.130) can be expressed as

$$y_i \approx g\left(z_i^{\text{proj}}, \delta_i\right) + \begin{bmatrix} C_0 A_0^{i-1} B_0 & C_k A_0^{i-2} B_0 & \cdots & C_0 B_0 \end{bmatrix} \begin{bmatrix} v_0 - u(k-1) \\ v_1 - u(k-1) \\ \vdots \\ v_{i-1} - u(k-1) \end{bmatrix}, \tag{11.133}$$

where

$$A_0 = \left.\frac{\partial F(x,u,d)}{\partial x}\right|_{z_0, u(k-1), \delta_0} \tag{11.103}$$

$$B_0 = \left.\frac{\partial F(x,u,d)}{\partial u}\right|_{z_0, u(k-1), \delta_0} \tag{11.104}$$

$$C_0 = \left.\frac{\partial g(x,d)}{\partial x}\right|_{z_0, u(k-1), \delta_0}. \tag{11.106}$$

This linearization assumes constant linearization points across the prediction horizon.

This simplification avoids the need for trajectory-dependent Jacobians while retaining a structure suitable for optimization.

Example 11.13: Consider the bilinear system from Example 11.12. We now derive a simplified approximation using the approach described in Eq. (11.133). Assume the output is directly equal to the state, i.e., $y(k) = x(k)$.

Since there is no disturbance state x_w in this example, all disturbance-related terms drop out. Furthermore, because the output function is linear ($g(z) = z$), we have $C_k = 1$, and the constant offset term in Eq. (11.133) vanishes. The prediction simplifies to:

$$y_3 \approx z_i^{\text{proj}} + \begin{bmatrix} A_k^2 B_k & A_k B_k & B_k \end{bmatrix} \begin{bmatrix} v_0 - u(k-1) \\ v_1 - u(k-1) \\ v_2 - u(k-1) \end{bmatrix}$$

$$= z_0 u^3(k-1) + \begin{bmatrix} u^2(k-1) z_0 & u(k-1) z_0 & z_0 \end{bmatrix} \begin{bmatrix} v_0 - u(k-1) \\ v_1 - u(k-1) \\ v_2 - u(k-1) \end{bmatrix}.$$

This expression approximates the earlier prediction obtained using time-varying system matrices over the horizon. While it is exact for the one-step-ahead case, it introduces errors in multistep predictions. However, such errors are typically acceptable in MPC, as the receding horizon framework continuously updates and corrects predictions at each time step.

11.5.4 Objective Function and Constraints

Recall that the standard quadratic objective function in MPC is given by

$$\Delta\mathcal{V} = \arg\min_{\Delta\mathcal{V}} \{||\left(\mathcal{R} - \mathcal{Y}\right)\Lambda_y||_2^2 + ||\Delta\mathcal{V}\Lambda_u||_2^2\},$$

where $\mathcal{R}$ is the p-step reference trajectory, $\mathcal{Y}$ is the predicted output vector, and $\Delta\mathcal{V}$ is the sequence of control increments over the control horizon.

The predicted output vector $\mathcal{Y}$ can be expressed using the simplified linearization from Eq. (11.133) as

$$\mathcal{Y} = \begin{bmatrix} g\left(z_1^{\text{proj}}, \delta_1\right) \\ g\left(z_2^{\text{proj}}, \delta_2\right) \\ \vdots \\ g\left(x_p^{\text{proj}}, \delta_p\right) \end{bmatrix}_{z_0=\hat{x}(k|k), v_i=u(k-1), \delta_0=\hat{d}(k|k)}$$

$$+ \begin{bmatrix} C_0B_0 & 0 & 0 & \cdots & 0 \\ C_0\left(A_0+I\right)B_0 & C_0B_0 & 0 & \cdots & 0 \\ \vdots & \vdots & \ddots & \ddots & \vdots \\ C_0\sum_{j=0}^{p-1} A_0^jB_0 & C_0\sum_{j=0}^{p-2} A_0^jB_0 & \cdots & \cdots & C_0\sum_{j=0}^{p-m} A_0^jB_0 \end{bmatrix} \begin{bmatrix} \Delta v_0 \\ \Delta v_1 \\ \vdots \\ \Delta v_{m-1} \end{bmatrix}. \quad (11.134)$$

The structure of the last term arises from the identity:

$$v_i - u(k-1) = \sum_{j=0}^{i} \Delta v_j,$$

which is substituted into the multistep output prediction expression.

For compact representation, the output prediction vector at time step k is written as

$$\mathcal{Y} = \mathcal{G}\left(z_0, u(k-1), \delta_0\right) + \mathcal{S}^u\Delta\mathcal{V}.$$

Here:

- $\mathcal{G}\left(z_0, u(k-1), \delta_0\right)$ denotes the vector of predicted outputs over the horizon, assuming the input remains fixed at the previous value $u(k-1)$. It is obtained by evaluating the output function $g(\cdot)$ at the projected states $z_1^{\text{proj}}, \dots, z_p^{\text{proj}}$, which are generated by forward integration of the nonlinear system over p steps starting from $z_0 = \hat{x}(k|k)$ with disturbance sequence $\delta_i = C_wA_w^i\hat{x}_w(k|k)$ and constant input $v_i = u(k-1)$.
- $\mathcal{S}^u$ is the dynamic matrix that relates input increments to output predictions. It is computed from the linearized system equations to ensure that the predicted outputs can be expressed in an affine form with respect to the optimization variables $\Delta\mathcal{V}$.

The objective function can then be rewritten as

$$\Delta\mathcal{V}^* = \arg\min_{\Delta\mathcal{V}} \{[\mathcal{E} - \mathcal{S}^u\Delta\mathcal{V}]^T \Lambda_y^T\Lambda_y[\mathcal{E} - \mathcal{S}^u\Delta\mathcal{V}] + \Delta\mathcal{V}^T\Lambda_u^T\Lambda_u\Delta\mathcal{V}\}$$
$$\mathcal{E} = \mathcal{R} - \mathcal{G}(z_0, u(k-1), \delta_0).$$

Following the same procedure used in deriving Eq. (11.54), the MPC optimization problem can be cast as a standard quadratic program:

$$\begin{aligned} \Delta\mathcal{V} &= \arg\min_{\Delta\mathcal{V}} \{\Delta\mathcal{V}^T\mathcal{H}\Delta\mathcal{V} + 2\mathfrak{g}^T\Delta\mathcal{V}\} \\ \mathcal{H} &= \mathcal{S}^{uT}\Lambda_y^T\Lambda_y\mathcal{S}^u + \Lambda_u^T\Lambda_u \\ \mathfrak{g} &= \mathcal{S}^{uT}\Lambda_y^T\Lambda_y\mathcal{E}. \end{aligned} \tag{11.135}$$

This optimization is subject to the following constraints:

$$\begin{bmatrix} I_{m.n_u} \\ -I_{m.n_u} \end{bmatrix} \Delta\mathcal{V} \leq \begin{bmatrix} \mathcal{I}_m\Delta u_{\max} \\ -\mathcal{I}_m\Delta u_{\max} \end{bmatrix} \tag{11.55}$$

$$\begin{bmatrix} \mathcal{I}_{m\Sigma} \\ -\mathcal{I}_{m\Sigma} \end{bmatrix} \Delta\mathcal{V} \leq \begin{bmatrix} \mathcal{I}_m\left(u_{\max} - u(k-1)\right) \\ -\mathcal{I}_m\left(u_{\min} - u(k-1)\right) \end{bmatrix} \tag{11.56}$$

$$\begin{bmatrix} \mathcal{S}^u \\ -\mathcal{S}^u \end{bmatrix} \Delta\mathcal{V} \leq \begin{bmatrix} \mathcal{I}_p y_{\max} - \mathcal{G}(z_0, u(k-1), \delta_0) \\ -\mathcal{I}_p y_{\min} + \mathcal{G}(z_0, u(k-1), \delta_0) \end{bmatrix}. \tag{11.136}$$

11.5.5 Implementation of Sequential Linearization-based MPC

Let us summarize the procedure for implementing the sequential linearization-based MPC algorithm. At the current time step k, the past state estimates $\hat{x}(k-1|k-1)$ and $\hat{x}_w(k-1|k-1)$ and the previous control input $u(k-1)$, are assumed to be available. Upon receiving the measurement $y_m(k)$, the state estimate is updated using the EKF.

The prediction model is then linearized, and the relevant matrices are constructed to express the MPC optimization problem in the standard QP form. After solving the QP, the optimal control move $u(k)$ is applied to the process. This procedure is repeated at each sampling instant, implementing the control strategy in a receding horizon framework.

The algorithm involves the following steps executed at each time step k:

1. **State estimation using the EKF**
   Compute the one-step-ahead predicted state:

$$\hat{\bar{x}}(k|k-1) = \begin{bmatrix} F\left(\hat{x}(k-1|k-1), u(k-1), C_w\hat{x}_w(k-1|k-1)\right) \\ A_w\hat{x}_w(k-1|k-1) \end{bmatrix}. \tag{11.108}$$

   Evaluate the Jacobians for use in the EKF update:

$$\hat{A}_{k-1|k-1} = \left.\frac{\partial F}{\partial x}\right|_{\hat{x}(k-1|k-1),u(k-1),C_w\hat{x}_w(k-1|k-1)} \qquad \hat{B}^d_{k-1|k-1} = \left.\frac{\partial F}{\partial d}\right|_{\hat{x}(k-1|k-1),u(k-1),C_w\hat{x}_w(k-1|k-1)}$$
$$C_{k|k-1} = \left.\frac{\partial g(x,d)}{\partial x}\right|_{\hat{x}(k|k-1),C_w\hat{x}_w(k|k-1)} \qquad C^d_{k|k-1} = \left.\frac{\partial g(x,d)}{\partial d}\right|_{\hat{x}(k|k-1),C_w\hat{x}_w(k|k-1)}.$$

Define the EKF matrices:

$$\overline{A}_{k-1} = \begin{bmatrix} \hat{A}_{k-1|k-1} & \hat{B}^d_{k-1|k-1} C_w \\ 0 & A_w \end{bmatrix}$$

$$\overline{B}_w = \begin{bmatrix} 0 \\ B_w \end{bmatrix}$$

$$\overline{C}_k = \begin{bmatrix} C_{k|k-1} & C^d_{k|k-1} \end{bmatrix}.$$

Compute the Kalman filter gain:

$$K_f(k) = P(k)\bar{C}_k^T \left(\bar{C}_k P(k) \bar{C}_k^T + R_2\right)^{-1} \tag{11.117}$$

$$P(k) = \bar{A}_{k-1} P(k-1) \bar{A}_{k-1}^T + R_1 - \bar{A}_{k-1} K(k-1) \bar{C}_{k-1} P(k-1) \bar{A}_{k-1}^T, \tag{11.118}$$

with noise covariances $R_1 = \bar{B}_w R_\varepsilon \bar{B}_w^T$ and $R_2 = R_v$. Use the measurement $y_m(k)$ to compute the updated state estimate:

$$\hat{\bar{x}}(k|k) = \hat{\bar{x}}(k|k-1) + K_f(k)\left(y_m(k) - \bar{g}\left(\hat{\bar{x}}(k|k-1)\right)\right). \tag{11.109}$$

2. **Future state and error projection**
Using the updated state estimate, compute future disturbance predictions

$$\begin{aligned} \delta_{i+1} &= C_w \hat{x}_w(k+i+1|k), \text{ where } \hat{x}_w(k+i+1|k) \\ &= A_w \hat{x}_w(k+i|k) \text{ initialized with } \hat{x}_w(k|k). \end{aligned}$$

Then, project future system states using the nonlinear model:

$$z_i^{\text{proj}} = F\left(z_{i-1}^{\text{proj}}, u(k-1), \delta_{i-1}\right) \quad \text{initialized with } z_0 = \hat{x}(k|k). \tag{11.123}$$

These calculations proceed recursively starting from $i = 1$ to $i = p$

Using the results, evaluate the projected output sequence $\mathcal{G}(z_0, u(k-1), \delta_0)$ and then the error sequence $\mathcal{E} = \mathcal{R} - \mathcal{G}(z_0, u(k-1), \delta_0)$.

3. **Calculation of linearization matrices**
Evaluate Jacobian matrices at the linearization point:

$$A_0 = \left.\frac{\partial F(x,u,d)}{\partial x}\right|_{z_0, u(k-1), \delta_0} \tag{11.103}$$

$$B_0 = \left.\frac{\partial F(x,u,d)}{\partial u}\right|_{z_0, u(k-1), \delta_0} \tag{11.104}$$

$$C_0 = \left.\frac{\partial g(x,d)}{\partial x}\right|_{z_0, u(k-1), \delta_0}. \tag{11.106}$$

Construct sensitivity matrix:

$$\mathcal{S}^u = \begin{bmatrix} C_0 B_0 & 0 & 0 & \cdots & 0 \\ C_0 (A_0 + I) B_0 & C_0 B_0 & 0 & \cdots & 0 \\ \vdots & \vdots & \ddots & \ddots & \vdots \\ C_0 \sum_{j=0}^{p-1} A_0^j B_0 & C_0 \sum_{j=0}^{p-2} A_0^j B_0 & \cdots & \cdots & C_0 \sum_{j=0}^{p-m} A_0^j B_0 \end{bmatrix}.$$

4. **Solving the quadratic optimization problem**
   Form the QP cost function matrices:

$$\begin{aligned} &\text{Hessian, } \mathcal{H} = \mathcal{S}^{uT}\Lambda_y^T\Lambda_y\mathcal{S}^u + \Lambda_u^T\Lambda_u \\ &\text{Gradient, } \mathfrak{g} = \mathcal{S}^{uT}\Lambda_y^T\Lambda_y\mathcal{E}. \end{aligned} \tag{11.135}$$

   Define constraint matrices as in Eqs. (11.55), (11.56), and (11.136). Solve the resulting QP using a suitable solver (e.g., MATLAB's `quadprog`) to obtain the optimal control move $\Delta\mathcal{V}$.
5. **Implement the input**
   From the QP solution $\Delta\mathcal{V}^*$, implement the first move Δv_0^* as $\Delta u(k)$.

Example 11.14: *MPC of a Nonlinear System.*
Lee and Ricker (1994) demonstrated the sequential linearization MPC using a paper-machine headbox control problem. We will use the same example and explain the procedure step-by-step. For the sake of brevity, this example and the complete MATLAB code (for this and other examples in this Chapter) are available at: https://github.com/Lee-MPC-Book

Exercises

1 Consider the discrete-time state-space model

$$x(k+1) = \begin{bmatrix} 0.90 & 0.10 & 0.5 & 2 & -2 \\ 0 & 0.95 & 1 & -3 & 0.4 \\ 0 & 0 & 1 & 0 & 0 \\ 0 & 0 & 0 & 1 & 0 \\ 0 & 0 & 0 & 0 & 1 \end{bmatrix} x(k) + \begin{bmatrix} 1 & 0 \\ 0 & 1 \\ 0 & 0 \\ 0 & 0 \\ 0 & 0 \end{bmatrix} u(k) + \begin{bmatrix} 0 & 0 & 0 \\ 0 & 0 & 0 \\ 1 & 0 & 0 \\ 0 & 1 & 0 \\ 0 & 0 & 1 \end{bmatrix} w(k)$$

$$y(k) = \begin{bmatrix} 1 & 0 & 1 & 0 & 0 \\ 0 & 1 & 0 & 1 & 0 \end{bmatrix} x(k) + v(k),$$

where $u \in \mathbb{R}^2$, $y \in \mathbb{R}^2$, w is the process noise, and v is the measurement noise. Let the state vector be

$$x = \begin{bmatrix} x_1 & x_2 & d_1 & d_2 & d_3 \end{bmatrix}^T,$$

where d_i $(i = 1, 2, 3)$ are disturbance states modeled as integrators.
In the above, w and v are modeled as Gaussian white noise with covariances:

$$R_w = \text{diag}(0, 0, 1, 1, 1) \qquad R_v = \text{diag}(0.5, 0.5).$$

(a) Verify that the system is not detectable, as $n_d > n_y$.
(b) Verify that the RDE started with $P(0) = I$ diverges.
(c) Instead of augmenting the state with d, use the differenced equation to augment them with y, reducing the number of integrators to 2.
(d) Verify that the system is detectable and RDE converges.
(e) Design a Kalman filter with the differenced model and plot its performance for 200 time steps.

2 Consider a discrete-time system with one control input $u(k)$, two outputs $y_s(k)$ and $y_c(k)$, and a single disturbance $d(k)$ that evolves as an integrator. Only $y_s(k)$ is measured; $y_c(k)$ must be inferred. The process dynamics are given by

$$x(k+1) = \begin{bmatrix} 0.90 & 0.20 \\ 0 & 0.95 \end{bmatrix} x(k) + \begin{bmatrix} 0.10 \\ 0.05 \end{bmatrix} u(k) + \begin{bmatrix} \varepsilon_1(k) \\ \varepsilon_2(k) \end{bmatrix}$$

$$\begin{bmatrix} y_s(k) \\ y_c(k) \end{bmatrix} = \begin{bmatrix} 1 & 0 \\ 0 & 1 \end{bmatrix} x(k) + \begin{bmatrix} 1 \\ 1 \end{bmatrix} d(k),$$

where

$$d(k+1) = d(k) + \varepsilon_3(k). \tag{11.137}$$

Since only the first output is measured,

$$y_m(k) = \begin{bmatrix} 1 & 0 \end{bmatrix} y(k) + v(k).$$

The covariance of process and measurement noise are $R_1 = \text{diag}(10^{-3}, 10^{-3}, 10^{-4})$ and $R_2 = 10^{-4}$.

(a) Define the augmented state $\bar{x}(k) \triangleq \begin{bmatrix} x(k) \\ d(k) \end{bmatrix} \in \mathbb{R}^3$. Obtain the augmented matrices $\bar{A}, \bar{B}, \bar{C}$ for the following:

$$\bar{x}(k+1) = \bar{A}\bar{x}(k) + \bar{B}u(k) + \varepsilon(k)$$
$$y_m(k) = \bar{C}\,\bar{x}(k) + v_1(k).$$

(b) Based on the augmented system, design the Kalman filter of the form

$$\hat{\bar{x}}(k|k) = \bar{A}\,\hat{\bar{x}}(k) + \bar{B}\,u(k) + K_f\big(y_s(k) - \bar{C}\,\hat{\bar{x}}(k)\big).$$

(c) Once the estimated augmented state is available, compute:

$$\hat{y}_c(k|k) = \begin{bmatrix} 0 & 1 & 1 \end{bmatrix} \hat{\bar{x}}(k|k).$$

Plot the actual y_c and the estimated $\hat{y}_c$ over 200 time steps, assuming the input $u(k) = 0$, $\forall k \geq 0$.

(d) Instead of augmenting the state with d, use the difference model to augment the state with both outputs:

$$\bar{x}(k+1) = \begin{bmatrix} \begin{bmatrix} A \end{bmatrix} & 0 & 0 \\ C_sA & 1 & 0 \\ C_cA & 0 & 1 \end{bmatrix} \bar{x}(k) + \begin{bmatrix} \begin{bmatrix} B \end{bmatrix} \\ C_sB \\ C_cB \end{bmatrix} \Delta u(k) + \begin{bmatrix} \begin{bmatrix} 1 & 0 \\ 0 & 1 \end{bmatrix} & 0 \\ \begin{bmatrix} 0 & 0 \end{bmatrix} & 1 \\ \begin{bmatrix} 0 & 0 \end{bmatrix} & 1 \end{bmatrix} \varepsilon(k)$$

$$y_m(k) = \begin{bmatrix} \begin{bmatrix} 0 & 0 \end{bmatrix} & 1 & 0 \end{bmatrix} \bar{x}(k) + v(k).$$

The above system is undetectable because the state contains two integrators while only one output is measured. Nevertheless, a Kalman filter can still be designed. Verify that the RDE converges using the same R_1 and R_2. Then, design the Kalman filter and compute the estimate of y_c as

$$\hat{y}_c(k|k) = \begin{bmatrix} \begin{bmatrix} 0 & 0 \end{bmatrix} & 0 & 1 \end{bmatrix} \hat{\tilde{x}}(k|k).$$

Compare the estimated y_c with that obtained from the previous approach.

3 In Eq. (11.30), we demonstrated how to modify the output equation of the augmented system model in Eq. (11.28) to account for measured disturbances. Similarly, to incorporate the measured disturbance d_m (a subset of d) into the differenced system model of Eq. (11.29), the state must be augmented as in Eq. (11.31). This ensures that the measured disturbances are properly integrated into the Kalman filter design and optimally utilized for state estimation and control.

An alternative formulation is given by

$$\underbrace{\begin{bmatrix} \Delta x(k+1) \\ y(k+1) \\ d_m(k) \end{bmatrix}}_{\tilde{x}(k+1)} = \underbrace{\begin{bmatrix} A & 0 & 0 \\ CA & I & 0 \\ 0 & 0 & I \end{bmatrix}}_{\tilde{A}} \underbrace{\begin{bmatrix} \Delta x(k) \\ y(k) \\ d_m(k-1) \end{bmatrix}}_{\tilde{x}(k)} + \underbrace{\begin{bmatrix} B \\ CB \\ 0 \end{bmatrix}}_{\tilde{B}} \Delta u(k)$$
$$+ \underbrace{\begin{bmatrix} B_d \\ CB_d \\ \begin{bmatrix} 0 & I \end{bmatrix} \end{bmatrix} \begin{bmatrix} \Delta d_u(k) \\ \Delta d_m(k) \end{bmatrix}}_{\varepsilon_1(k)} \tag{11.138}$$
$$y_m(k) = \underbrace{\begin{bmatrix} 0 & I & 0 \\ 0 & 0 & I \end{bmatrix}}_{\tilde{C}} \underbrace{\begin{bmatrix} \Delta x(k) \\ y(k) \\ d_m(k-1) \end{bmatrix}}_{\tilde{x}(k)} + \underbrace{\begin{bmatrix} \nu_y(k) \\ \Delta d_m(k) + \nu_{d_m}(k) \end{bmatrix}}_{\varepsilon_2(k)}.$$

In this formulation, the process noise term $\varepsilon_1(k)$ and the output noise term $\varepsilon_2(k)$ are correlated.

- Let R_{ε_d} denote the covariance of the white noise driving Δd, and let R_{ν_y} and $R_{\nu_{d_m}}$ be the covariances of the measurement noises. Derive expressions for the joint state/output noise covariance matrix:

$$E\left\{ \begin{bmatrix} \varepsilon_1(k) \\ \varepsilon_2(k) \end{bmatrix} \begin{bmatrix} \varepsilon_1^T(k) & \varepsilon_2^T(k) \end{bmatrix} \right\} = \begin{bmatrix} R_1 & R_{1,2} \\ R_{1,2}^T & R_2 \end{bmatrix}. \tag{11.139}$$

- The design of the Kalman filter for the case of correlated process and measurement noise was discussed in Section 7.3.2 of Chapter 7. Derive the corresponding Kalman filter for the above system.
- Note that the measured disturbance $d_m(k)$ appears explicitly in the one-step-ahead state prediction $\hat{\tilde{x}}(k+1|k)$—in contrast to the uncorrelated noise case. This dependence complicates the implementation of MPC, as $d_m(k)$ must be explicitly included in the prediction equations for future outputs. Derive the multistep output prediction equations for $y(k+1|k), \ldots, y(k+p|k)$.

4 Consider the nonlinear system:

$$\dot{x}_1 = x_2$$
$$\dot{x}_2 = \sin(x_1) + u,$$

where x_1 is the angle (rad), x_2 is angular rate (rad/s), and u is torque (normalized). The goal is to stabilize at the upright equilibrium $x^\star = [\pi,\ 0]^T$ with $u^\star = 0$.

(a) Discretize with forward Euler (with sampling time of $t_s = 0.05$):

$$x(k+1) = F(x(k), u(k)) = \begin{bmatrix} x_1(k) + t_s x_2(k) \\ x_2(k) + t_s(\sin(x_1(k)) + u(k)) \end{bmatrix}.$$

(b) At time step k, carry a nominal sequence $\{z_i^{\text{proj}}, v_i^{\text{proj}}\}_{i=0}^{p-1}$ (with $z_0^{\text{proj}} = x(k)$). In our case, z_i^{proj} represents the result of integrating the nonlinear model with input v_i^{proj}, which is chosen uniformly as the current input of $u(k-1)$. Linearize the dynamics at each stage:

$$A_i = \left.\frac{\partial F}{\partial x}\right|_{(z_i^{\text{proj}}, v_i^{\text{proj}})} = \begin{bmatrix} 1 & t_s \\ t_s\cos(z_{1,i}^{\text{proj}}) & 1 \end{bmatrix}, \quad B_i = \left.\frac{\partial F}{\partial u}\right|_{(z_i^{\text{proj}}, z_i^{\text{proj}})} = \begin{bmatrix} 0 \\ t_s \end{bmatrix}.$$

Define incremental variables $\delta z_i = z_i - z_i^{\text{proj}}$, $\delta v_i = v_i - v_i^{\text{proj}}$. The linearized, time-varying prediction is

$$z_{i+1} = f(z_i^{\text{proj}}, b_i^{\text{proj}}) + A_i\,\delta z_i + B_i\,\delta v_i$$

assuming $\{z_{i+1}^{\text{proj}}\}$ is generated by simulating f from $(z_0^{\text{proj}}, v_i^{\text{proj}})$.

(c) The control objective is to track the upright equilibrium using quadratic stage costs on the state deviation and control increments:

$$J = \sum_{i=0}^{p-1} \left(\| z_i - x^\star \|_Q^2 + \| \Delta v_i \|_R^2 \right) + \| z_p - x^\star \|_{Q_t}^2.$$

The constraints are

$$|v_i| \le 3, \quad |\Delta v_i| \le 0.5, \quad |(z_1)_i - \pi| \le \pi.$$

In incremental form, these constraints become affine inequalities in Δv.
The condensed QP is defined in terms of $\Delta \mathcal{V} = [\Delta v_0^T \ \cdots \ \Delta v_{p-1}^T]^T$ as

$$\min_{\Delta\mathcal{V}} \ \frac{1}{2}\Delta\mathcal{V}^T H \Delta\mathcal{V} + g^T \Delta\mathcal{V} \quad \text{s.t.} \quad C\Delta\mathcal{V} \le \mathrm{d},$$

where H, g, C, and d are constructed from A_i, B_i, the weighting matrices, and the bounds. Clearly derive the expressions for these matrices and compute their numerical values.

(d) Set the prediction horizon to $p = 20$, with $Q = \text{diag}(10, 1)$, $R = 0.1$, and the terminal cost Q_t. Simulate the performance of the successively linearized MPC starting from the initial state $x(0) = \text{diag}\{1, 1\}$.

(e) Compare the performance with that of a conventional LQR controller designed from the system matrices linearized at the equilibrium. The linearized model at the equilibrium is given by

$$A_\star = \begin{bmatrix} 1 & t_s \\ t_s\cos(\pi) & 1 \end{bmatrix} = \begin{bmatrix} 1 & t_s \\ -t_s & 1 \end{bmatrix}, \quad B_\star = \begin{bmatrix} 0 \\ t_s \end{bmatrix}.$$

12

Nonlinear Model Predictive Control

The problem with linear theory is that it is not nonlinear.

– John A. Adam

An increasing number of processes are operated over wider operating envelopes or follow dynamic operating recipes. This shift is driven by the adoption of advanced strategies such as real-time optimization, flexible operation, periodic modulation, and frequent grade transitions. In such settings, nonlinear models become essential for effective online optimization and control.

Nonlinear model predictive control (NMPC) enables accurate tracking of dynamic trajectories, making it well-suited for executing complex operating recipes. Furthermore, NMPC can optimize a defined economic performance index in real time while ensuring regulatory compliance. Beyond its control capabilities, NMPC provides a natural platform for integration with online optimization strategies, facilitating coordination with higher-level tasks such as scheduling and planning, as illustrated in Figure 12.1.

Earlier in Chapter 11, we briefly introduced how the standard linear state-space MPC formulation can be extended to utilize a nonlinear system model, if available (see Section 11.6). However, the nonlinear model was not used directly in the MPC optimization. Instead, it was linearized around the current state estimate to preserve the linear prediction structure, resulting in a quadratic programming (QP) problem for computing the control moves. This approach was paired with an EKF, which similarly applies successive linearization to extend the original Kalman filter to nonlinear systems.

Although this linearization-based strategy is computationally efficient, it introduces approximation errors—particularly when the current state estimate deviates significantly from the true state. In this chapter, we consider a more rigorous NMPC formulation that directly incorporates the full nonlinear system model into the optimization problem. By avoiding linearization in the prediction model, NMPC eliminates the associated approximation errors. Thanks to recent advances in large-scale optimization algorithms and modeling environments, such full-scale NMPC has become a viable option for real-time control of industrial processes.

The same nonlinear models can also be used for state and parameter estimation based on process measurements. Moving horizon estimator (MHE), introduced earlier for linear systems in Chapter 7, can be extended to nonlinear systems, resulting in a nonlinear estimation framework that—like NMPC—requires the solution of a nonlinear programming (NLP) problem at each sampling instant.

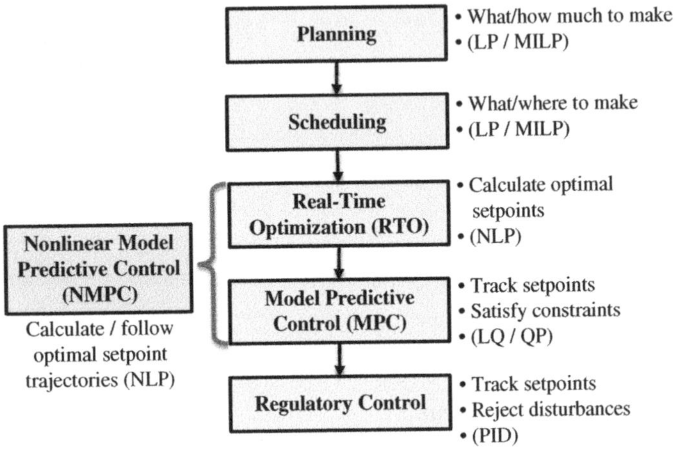

Figure 12.1 A typical corporate-wide decision hierarchy and the role of NMPC within this structure.

This chapter provides an introductory treatment of NMPC for readers new to the topic. We begin by presenting a general formulation of NMPC problems, followed by a discussion of key properties of the resulting optimization problems and the stability of the resulting closed-loop system. Various solution strategies are then introduced—starting with standard NLP algorithms and progressing to more recent methods that leverage sensitivity analysis to enable fast online updates informed by offline computations. A similar optimization-based modeling framework is then developed for MHE. The chapter concludes with numerical examples that demonstrate the practical application of these techniques.

12.1 NMPC Formulation

We consider the plant model described by a differential-algebraic equation (DAE) system in the following semi-explicit form:

$$\begin{aligned} \frac{dx}{dt} &= f(x(t), \xi(t), u(t), d(t)), \quad x(0) = z_0 \\ 0 &= g(x(t), \xi(t), u(t), d(t)). \end{aligned} \tag{12.1}$$

Here, $x(t) \in \mathbb{R}^{n_x}$, $\xi(t) \in \mathbb{R}^{n_\xi}$, $u(t) \in \mathbb{R}^{n_u}$, and $d(t) \in \mathbb{R}^{n_d}$ denote the differential states, algebraic states, manipulated inputs, and disturbance variables, respectively. The disturbance $d(t)$ is typically modeled using a separate disturbance model, which can be incorporated into the system for state estimation. As discussed in Chapter 11, an appropriate disturbance model is essential to achieve offset-free MPC performance.

We assume that the DAE system in Eq. (12.1) is of index-1, so that the algebraic variables can be explicitly solved as $\xi = \zeta(x, u, d)$. For digital implementation, we consider the inputs (u, d) to be held constant over each sampling interval, i.e., using zero-order hold (ZOH) at time t_k with values $(u(k), d(k))$. The DAE system is then integrated numerically over the interval $[t_k, t_{k+1}]$ to yield the following discrete-time expression:

$$x(k+1) = x(k) + \int_{t_k}^{t_{k+1}} f(x, \zeta\,(x, u(k), d(k)), u(k), d(k))dt. \tag{12.2}$$

where $t_k = k.t_s$ denotes the discrete sampling times indexed by $k \in \mathbb{I}_{0+}$. This integration yields the discrete-time nonlinear system:

$$x(k+1) = F\left(x(k), u(k), d(k)\right). \tag{12.3}$$

where the mapping F encapsulates one-step numerical integration of the underlying ODE system.

As discussed in Chapter 11, the sequential linearization-based MPC formulation relies on linearizing this nonlinear model at each sampling instant about the current state $x(k)$ and previous input $u(k-1)$. This yields a time-varying linear MPC problem that can be solved efficiently using QP techniques. However, the resulting control law is only an approximation of the true optimal control solution and may exhibit suboptimal performance due to the local linearization of the nonlinear dynamics.

Remark 12.1: To evaluate the integral in Eq.(12.2), a suitable numerical differential equation solver must be employed. Appendix G provides guidance on setting up and solving differential equations using MATLAB Simulink®. For non-stiff systems, the `ode45` solver—based on an adaptive step-size Runge–Kutta method—is typically sufficient. In contrast, stiff ODEs or DAEs require solvers such as `ode15s`, which implements a fully implicit backward differentiation formula (BDF) method (Kaisare (2017)).

In many practical cases, the effects of the manipulated inputs and disturbances are separable, allowing the model to be expressed in the following form:

$$x(k+1) = f_u\left(x(k), u(k)\right) + f_d\left(x(k), d(k)\right). \tag{12.4}$$

Here, the function $f_u : \mathbb{R}^{n_x+n_u} \to \mathbb{R}^{n_x}$, with $f_u(0,0) = 0$, represents the nominal system dynamics, while $f_d : \mathbb{R}^{n_x+n_d} \to \mathbb{R}^{n_x}$ captures the influence of disturbances and model uncertainties. We assume that both $f_u(\cdot,\cdot)$ and $f_d(\cdot,\cdot)$ are Lipschitz continuous with respect to their arguments, ensuring well-posedness of the resulting system.

In addition to the dynamic model (12.3), a key component of online dynamic optimization is the specification of the objective function. This is typically expressed in terms of a stage cost $\psi(x,u) : \mathbb{R}^{n_x+n_u} \to \mathbb{R}$ evaluated at each sampling instant t_k. The stage cost may take the form of a tracking objective, which penalizes deviations from a desired set point—as in conventional linear MPC—or an economic objective, which directly quantifies performance metrics such as cost, profit, yield, or energy consumption.

We assume that the states and controls are restricted to the domains $\mathbb{X}$ and $\mathbb{U}$, respectively. $\mathbb{X}$ is closed and contains the origin in its interior, while $\mathbb{U}$ is compact and contains the origin. These domains may be defined through a set of inequalities involving x and u, respectively.

Given the current state $x(k)$ (or its estimate $\hat{x}(k|k)$), we consider the standard NMPC formulation (see Rawlings et al. 2020), posed as a parametric NLP problem:

$$J_p(x(k)) \overset{\Delta}{=} \min_{z_l, v_l} \sum_{l=0}^{p-1} \psi(z_l, v_l) + \Psi(z_p) \tag{12.5a}$$

$$\text{s.t.} \quad z_0 = x(k) \tag{12.5b}$$

$$z_{l+1} = F(z_l, v_l, d_l) \quad l = 0, \ldots p-1 \tag{12.5c}$$

$$z_l \in \mathbb{X}, \quad v_l \in \mathbb{U} \tag{12.5d}$$

$$z_p \in \mathbb{X}_f. \tag{12.5e}$$

Here, z_l and d_l denote the predicted state and disturbance at step l in the prediction horizon, while v_l is the corresponding manipulated input. The optimization starts from $z_0 = x(k)$. The cost function generalizes the quadratic form used in linear quadratic regulator (LQR) and linear MPC, where both the stage cost $\psi(\cdot)$ and the terminal cost $\Psi(\cdot)$ are typically quadratic.

The optimization is performed over a finite prediction horizon of length p. For batch processes, this is commonly implemented as a shrinking-horizon problem, where the remaining batch time defines the horizon and appropriate terminal constraints are imposed at the endpoint. For continuous processes, the problem is often formulated over an infinite-horizon. As in previous chapters, to maintain computational tractability, we approximate the infinite-horizon by a receding horizon of length p, augmented with a terminal constraint $z_p \in \mathbb{X}_f$ (with $\mathbb{X}_f \subset \mathbb{X}$) and a terminal cost $\Psi(\cdot) : \mathbb{R}^{n_x} \to \mathbb{R}$.

The choice of terminal set and terminal cost plays a critical role in ensuring the feasibility and stability of the resulting closed-loop system. Numerous theories and methodologies have been developed to guide their selection, which we will explore in a later section. In general, the goal is to choose the terminal set and/or terminal cost so that the finite-horizon solution closely approximates the behavior of the infinite-horizon solution, which typically yields desirable closed-loop performance.

In some formulations, it is assumed that the prediction horizon p is sufficiently long and the terminal cost $\Psi(z_p)$ is appropriately large such that the condition $z_p \in \mathbb{X}_f$ is automatically satisfied by the optimizer, even without explicitly enforcing the terminal constraint. However, selecting such a horizon length p *a priori* can be challenging, and the resulting value may be too large to allow for real time solution of the optimization problem.

The dynamic optimization problem (12.5) yields an optimal trajectory $\left\{z^*_{0|k}, \ldots, z^*_{p|k}, v^*_{0|k}, \ldots, v^*_{p-1|k}\right\}$ at time k, with $z^*_{0|k} = x(k)$. The control action to be applied at the current time is given by $u(k) = v^*_{0|k}$. At the next time step, the plant evolves according to (12.3). The time index is then incremented, $k \leftarrow k + 1$, and the NMPC problem (12.5) is solved again over the shifted horizon, in a receding horizon fashion.

This recursive procedure defines a feedback control law of the form

$$u(k) = \kappa_p(x(k)), \text{ with } \kappa_p(\cdot) : \mathbb{R}^{n_x} \to \mathbb{R}^{n_u}, \tag{12.6}$$

which results in the following closed-loop system:

$$\begin{aligned} x(k+1) &= F\left(x(k), \kappa_p\left(x(k)\right), d(k)\right) \\ &= f_u\left(x(k), \kappa_p\left(x(k)\right)\right) + f_d\left(x(k), d(k)\right). \end{aligned} \tag{12.7}$$

This formulation assumes ideal conditions, neglecting the effects of computation time and any resulting delays in applying the control input.

12.2 Solution via NLP

To implement NMPC in practice, the nonlinear optimization problem (12.5) must be solved online at every sampling instant in a computationally efficient and reliable manner. Broadly speaking, there are two main approaches for solving such optimal control problems (Bryson and Ho, 1975):

- *Dynamic Programming (DP)*: This approach seeks to solve the Hamilton-Jacobi-Bellman (HJB) equation directly, resulting in a state-feedback law of the form $u^\star = \kappa_p(x)$, defined either explicitly or implicitly via the cost-to-go function.

- *Direct Solution of a Finite-dimensional NLP Problem*: In this approach, the optimal control problem is discretized over a finite prediction horizon and solved as a finite-dimensional NLP at each time step, yielding an open-loop optimal control sequence.

These two strategies mirror the corresponding methods discussed for linear systems in Chapters 8 and 11. Directly solving the HJB equation is rarely feasible because of the curse of dimensionality. While substantial advances have been made in NLP algorithms and solvers in recent years, NLPs remain computationally challenging and, in general, do not guarantee globally optimal or even feasible solutions. The efficiency and robustness of the online solution can be significantly influenced by how the NLP is formulated.

Accordingly, the remainder of this section focuses on the formulation and numerical solution of the NMPC problem as a finite-dimensional NLP, emphasizing strategies that improve computational performance and reliability.

12.2.1 Elements of NLP formulations

The NMPC formulation in (12.5) includes a stage-wise cost function, the dynamic model (12.3) as an equality constraint, and state and input constraints. While infinite-horizon formulations are desirable for ensuring stability and performance, only finite-horizon approximations are practically solvable using NLP.

There are multiple ways to incorporate the dynamic model into the NLP formulation. Broadly, two main approaches are employed:

(i) Sequential (or nested) approach, and
(ii) Simultaneous (or full-discretization) approach.

These two formulations are illustrated schematically in Figure 12.2.

In the sequential approach, model simulation and optimization are executed in a decoupled manner at each iteration. An initial value (IV) DAE solver—such as the BDF method—is embedded within the optimization loop to evaluate the system dynamics for each candidate input trajectory. This formulation results in a smaller optimization problem, as the dynamic model constraints are not included explicitly but are enforced through numerical integration. However, the sequential structure can be computationally demanding and may encounter difficulties with open-loop unstable systems, as the separation between simulation and optimization limits feedback between the two components.

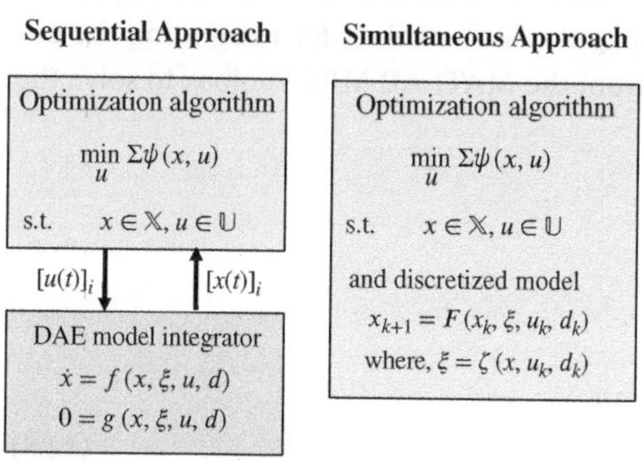

Figure 12.2 Structures of sequential and simultaneous approaches for dynamic optimization.

In contrast, the simultaneous approach discretizes the DAE system (12.1) across the prediction horizon and incorporates both state and input trajectories as decision variables in a single, large-scale NLP. This results in a sparse but high-dimensional optimization problem. With the advent of tools like automatic differentiation, this approach has gained renewed interest, as it enables efficient computation of first and even second derivatives, which are crucial for solving (12.5) effectively. Moreover, the simultaneous method is better suited to handle open-loop instabilities and can offer superior computational efficiency for large-scale systems. The effectiveness of the simultaneous approach depends on the availability of reliable and scalable NLP solvers—a field in which the optimization community has made significant advances in recent years.

12.2.2 NLP Solvers

NLP solvers applied to NMPC can be classified according to their optimization strategy:

- **Derivative-free Optimization Methods** (e.g., genetic algorithms, simulated annealing, particle swarm optimization, pattern search): These approaches do not rely on gradient information and are best suited for unconstrained or lightly constrained problems. They are relatively straightforward to implement—particularly within sequential approaches—but become computationally infeasible once the problem size exceeds roughly 100 variables. Consequently, they are rarely practical for large-scale NMPC applications.
- **Active-set Methods**, including successive quadratic programming (SQP) and generalized reduced gradient (GRG) algorithms: These approaches handle constraints and objective minimization simultaneously. They generally achieve superlinear convergence rates and are effective for medium-scale problems involving more than 100 decision variables.
- **Interior-point (barrier) Methods with Exact Hessians and Sparse Newton Solvers** (e.g., IPOPT): These solvers tackle the Karush-Kuhn-Tucker (KKT) conditions directly by exploiting second-order information. By leveraging the sparsity structure of large-scale NLPs, they achieve quadratic convergence and are particularly effective for NMPC problems with dimensions exceeding one million variables. A key advantage of barrier methods is their ability to compute sensitivity information for neighboring optimal solutions at essentially no additional computational cost, even in very large-scale settings. This sensitivity information can then be used to accelerate online computations.

For a more comprehensive discussion of NLP solvers and their application in dynamic optimization, see Betts (2001); Biegler (2010).

In Example 12.1, we consider the application of NMPC to a MIMO system. The sequential approach is convenient in this case, using the BDF-based `ode15s` solver for integrating the differential equations, combined with the `nmpc` object from the MATLAB MPC Toolbox to solve the dynamic optimization problem.

Example 12.1: *NMPC with Full State Feedback.*
Consider the following MIMO system:

$$\begin{aligned}
\dot{x}_1' &= -x_1 - x_1 x_3 + w_1 \\
\dot{x}_2' &= -x_2 + 0.5x_1 + d_1 \\
\dot{x}_3' &= -x_3 - \exp\left(-x_2^2\right) + u_1 \\
\dot{x}_4' &= -x_4 + x_1^2 + u_1 u_2.
\end{aligned} \tag{12.8}$$

Here, u denotes the manipulated variables (MV), while w and d represent unmeasured and measured disturbances, respectively. The objective of the NMPC controller is to regulate the following outputs to their desired set points while rejecting step disturbances:

$$\begin{aligned} y_1 &= x_2 + x_3 \\ y_2 &= -x_1 + x_4. \end{aligned} \tag{12.9}$$

The disturbance scenario is summarized in Table 12.1.

This problem will be solved using Simulink. The setup and solution procedure for the nonlinear model (12.8) is outlined in Appendix G. It is assumed that MATLAB's MPC Toolbox and Simulink are installed on the user's computer.

Full-state feedback is assumed, i.e., all state variables are measured, eliminating the need for state estimation. The controller computes optimal control actions directly from the measured states.

The MATLAB functions `System_StateFcn.m` and `System_OutputFcn.m` (see Appendix G) define the true plant model. These same functions are also used for prediction in NMPC, since there is no model-plant mismatch or disturbance modeling in this example. The MATLAB MPC Toolbox supports NMPC and follows similar procedures as for linear MPC (see Chapter 11 and Appendix F), with the primary difference being the specification of nonlinear state and output model functions, as illustrated below.

```
%% MPC parameters
% |nlmpc| command in MPC toolbox defines Nonlinear MPC. Set
% the control problem (e.g. Number of CV, MV, MD and UD).
nlobj = nlmpc(4, 2,'MV',[3,4],'MD',2,'UD',1);


% Set sampling time, and prediction and control horizons
% p should be chosen large enough for closed-loop stability
Ts = 1;
nlobj.Ts = Ts;
nlobj.PredictionHorizon = 20;
nlobj.ControlHorizon = 3;

%% Objective function
% Set weights on output tracking error and MV rate
nlobj.Weights.OutputVariables = [1, 1];
nlobj.Weights.ManipulatedVariablesRate = [0.1, 0.1];

%% Constraints
% Set constraints on MV rate
nlobj.MV(1).RateMin = -0.3;
nlobj.MV(1).RateMax = 0.3;
nlobj.MV(2).RateMin = -0.3;
nlobj.MV(2).RateMax = 0.3;

%% Predictive model
% Specify the nonlinear state and output functions for the
% predictive model used in NMPC. The model state/output
```

```
% function are separately defined from the "true'' system
nlobj.Model.StateFcn = 'PredModel_StateFcn';
nlobj.Model.OutputFcn = 'PredModel_OutputFcn';
% The above lines is where MPC and NMPC differ in MPC Toolbox

%% NMPC Check
% Best practice: Test your prediction model and any other
% custom functions before using them in a simulation.
% Use |validateFcns| command.

x0 = [0; 0; 0.5; 1];
u0 = [0; 0; 1.5; 0.8877];
validateFcns(nlobj,x0,u0(3:4),u0(2));
```

Now that the example has been set up, we proceed to simulate dynamic scenarios involving set point tracking and step changes in both measured and unmeasured disturbances. These scenarios are summarized in Table 12.1. Initially, the system performs set point tracking in the absence of disturbances. A measured disturbance $d_1 = 0.5$ is introduced at time $t \geq 50$, followed by an unmeasured disturbance $w_1 = 0.5$ applied during the interval $200 \leq t < 250$.

Example 12.2: (continued from Example 12.1)
After the `nlobj` object has been defined and verified, we proceed to set up the NMPC simulation. The dynamic scenario is specified according to Table 12.1.

With both the NMPC controller and simulation scenario defined, we now construct the closed-loop system in the Simulink environment. Figure 12.3 shows the flowsheet for the closed-loop implementation in Simulink. The NMPC block is used as the controller, and it references the previously defined NMPC object `nlobj`. Alternatively, the user may open and run the `NMPC_Example_Case1_Setup` script file to initialize the simulation setup. Note that the optimal control actions computed by the NMPC block replace the MV inputs to the plant model. This setup enables straightforward design and simulation of the closed-loop dynamic system.

Figure 12.4 presents the resulting closed-loop dynamics. The controlled outputs y_1 and y_2 successfully track their respective set points as specified in the simulation scenario. The NMPC controller also handles the measured disturbance d_1 introduced at $t = 50$ effectively, maintaining good

Table 12.1 Scenario for set point tracking and disturbance rejection in Example 12.1. The unmeasured disturbance w_1 is introduced at $t = 200$ and removed at $t = 250$.

Time	0	50	100	150	200	250
y_1^{sp}	0.5	0.5	0.5	0.3	0.3	0.3
y_2^{sp}	1.0	1.0	−0.5	−0.5	−0.5	−0.5
w_1	0	0	0	0	0.5	0
d_1	0	0.5	0.5	0.5	0.5	0.5

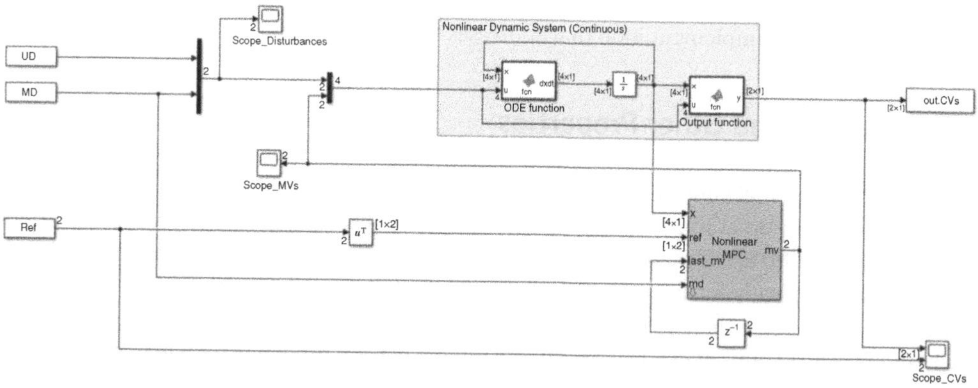

Figure 12.3 Flowsheet for closed-loop simulation in Simulink with full-state feedback.

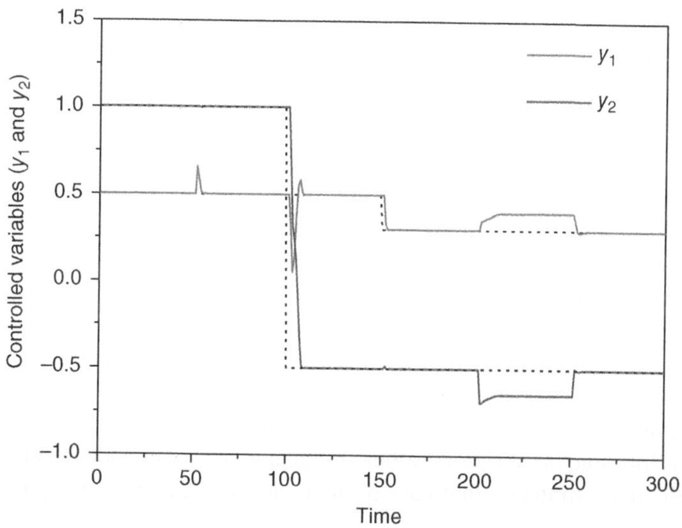

Figure 12.4 Closed-loop output responses under full-state feedback. The offset observed at $t = 200$ arises from the unmeasured disturbance w_1 and the lack of an integrator, which is necessary for offset-free control under model-plant mismatch.

performance. However, when the unmeasured disturbance w_1 is applied at $t = 200$, noticeable offsets emerge. Since the controller does not include a disturbance model or integral action, it is unable to reject the unmeasured disturbance, resulting in steady-state error. This highlights the importance of incorporating disturbance modeling or integral feedback when robustness to unmeasured disturbances is required.

Recall the importance of disturbance modeling and state-space model construction discussed in Section 11.2. The NMPC implementation in this example does not represent a complete NMPC framework, as it lacks both state estimation and feedback correction. This setup reflects a common oversight—namely, when all system states are directly measurable and unmeasured disturbances / model errors are absent, no integral action or disturbance rejection mechanism is needed. Such simplifications can lead to performance degradation in practical applications, where such assumptions rarely hold. A subsequent example will revisit this scenario with a more comprehensive and robust NMPC implementation that addresses these limitations.

12.3 Stability and Other Properties

MPC is a form of optimal control, for which a substantial body of literature exists concerning the characterization of optimal solutions and the stability of the resulting closed-loop system. The key distinction between MPC and classical optimal control lies in how the control law is defined. Classical approaches typically focus on deriving a parametric (precomputed) control law, while MPC defines the control law implicitly through an online optimization over a finite-horizon. This fundamental difference grants MPC the flexibility to handle control problems for which offline computation of optimal policies is infeasible. However, it also renders the direct application of classical analysis techniques impractical or, at the very least, nontrivial.

We encountered this challenge in the context of constrained LQR. There, we noted that while optimality does not always guarantee stability, under conditions such as stabilizability and detectability, infinite-horizon optimal controllers do yield stable closed-loop behavior, with the value function acting as a Lyapunov function. We further showed that infinite-horizon MPC can be approximated through the use of an appropriate terminal cost and terminal constraint set, thereby preserving stability.

The analysis of closed-loop stability for NMPC follows the same general principles as for constrained LQR. However, the nonlinear dynamics introduce additional challenges, particularly in the computation of the terminal cost (which represents the cost-to-go at the end of the horizon) and the terminal set (which must be an invariant subset of the feasible state space). In most practical cases, exact calculation of these quantities is not tractable. Instead, one must rely on suitable approximations—e.g., upper bounds on the cost-to-go function and conservative inner or outer approximations of invariant sets—to ensure stability.

A comprehensive review of the various methods developed in the literature to address these challenges is beyond the scope of this introductory chapter. Rather, our goal here is to provide insight into the key issues and outline the main strategies that have been employed to ensure stability in NMPC formulations.

12.3.1 Invariant Set and Output Admissible Set

We begin by revisiting two fundamental concepts critical to the determination of the terminal set $\mathbb{X}_f$: the notions of an invariant set and an output admissible set.

Definition 12.1: For a nonlinear system described by $x(k+1) = f(x(k))$, a compact set $\mathbb{O}$ containing the origin is said to be an *invariant set* if

$$\forall x \in \mathbb{O}, \quad f(x) \in \mathbb{O}. \tag{12.10}$$

In other words, any state trajectory that starts within $\mathbb{O}$ remains inside $\mathbb{O}$ for all future time steps.

To apply this concept in the MPC context, consider a local feedback controller $u(k) = \kappa_f(x(k))$ for the system $x(k+1) = f_u(x(k), u(k))$, which stabilizes the state to the origin within a compact region around the origin. For simplicity, we exclude the disturbance term $f_d(x(k), d(k))$, so that the origin $(x, u) = (0, 0)$ is an equilibrium point. This assumption can be relaxed by making appropriate assumptions on the disturbance (e.g., constant and bounded). Under the local controller, the closed-loop dynamics become

$$x(k+1) = f_u(x(k), \kappa_f(x(k))).$$

Suppose we define a terminal set $\mathbb{X}_f \subset \mathbb{X}$ such that:

1. For all $x \in \mathbb{X}_f$, the control action $\kappa_f(x)$ lies within the input constraint set $\mathbb{U}$, and
2. $\mathbb{X}_f$ is an invariant set for the closed-loop system above.

Then, any terminal state $x_p \in \mathbb{X}_f$ will remain within $\mathbb{X}_f$ under the local control law $\kappa_f(x(k))$, and both state and input constraints will be satisfied for all future times. Such a terminal set is called an *output admissible set*, and it plays an important role in ensuring the continued feasibility of the MPC optimization problem.

There are various approaches to selecting $\mathbb{X}_f$ to satisfy this property. The most conservative choice is a singleton set containing only the origin, which leads to the terminal equality constraint used in constrained LQR. Alternatively, one may define $\mathbb{X}_f$ as a small ball (e.g., a level set of a Lyapunov function) around the origin. However, overly small terminal sets impose tight constraints that may be difficult to satisfy for short horizons and can lead to suboptimal control actions that deviate from those of the infinite-horizon solution.

Therefore, it is desirable to choose $\mathbb{X}_f$ as large as possible while still maintaining output admissibility. The largest such set is called the *maximal output admissible set*, though computing it exactly is generally intractable for nonlinear systems due to its complex geometry. A practical approach is to approximate it with a simple shape, such as an ellipsoid defined by $x^T P x \leq \alpha$, and to enlarge it as much as possible while preserving admissibility.

A related concept is the *dual-mode control* strategy proposed by Mayne and Mischalska (1990). In this framework, the NMPC controller drives the state into a terminal region (an output admissible set), after which control is handed over to the local stabilizing controller $\kappa_f(x(k))$. To ensure that the terminal constraint is feasible, one may use a variable horizon approach, wherein the number of steps required to reach the terminal set is minimized. While this approach is popular in the literature, it deviates from the conventional fixed-horizon MPC formulation and will not be pursued further in this chapter.

It is also worth noting that for stabilizable systems, a sufficiently long prediction horizon may allow the terminal constraint to be satisfied implicitly, without enforcing it explicitly. However, determining such a horizon a priori is typically difficult and state-dependent. As a result, the MPC literature conventionally includes an explicit terminal constraint to ensure stability and recursive feasibility.

12.3.2 Cost-to-go and Terminal Penalty

The *cost-to-go* function, also referred to as the *value function*, represents the infinite-horizon accumulation of stage costs starting from a given initial state. For the dynamical system $x(k+1) = f_u(x(k), u(k))$ under a locally stabilizing feedback controller $u = \kappa_f(x)$ within a region $\mathbb{P}$, the cost-to-go function $V_\infty(x)$ is defined as follows:

$$V_\infty(x) = \sum_{l=0}^{\infty} \psi(z_l, \kappa_f(z_l)), \tag{12.11a}$$

$$\text{s.t. } z_{l+1} = f_u(z_l, \kappa_f(z_l)) \quad l \geq 0 \tag{12.11b}$$

$$z_0 = x, \tag{12.11c}$$

for all $x \in \mathbb{P}$.

In the linear case, the cost-to-go function can be computed analytically using a Lyapunov equation. Specifically, for a linear system with a stabilizing state feedback law $u(k) = -Lx(k)$, the cost-to-go becomes

$$V_\infty(x) = \sum_{l=0}^{\infty} z_l^T Q z_l + v_l^T R v_l \\ \text{s.t. } z_{l+1} = Az_l + Bv_l, \quad v_l = -Lz_l, \quad l \geq 0 \\ z_0 = x \tag{12.12}$$

which leads to the closed-form expression:

$$V_\infty(x) = x^T P x, \tag{12.13}$$

where P satisfies the discrete-time Lyapunov equation:

$$(A - \mathrm{BL})^T P(A - \mathrm{BL}) + Q + L^T \mathrm{RL} = P. \tag{12.14}$$

For nonlinear systems, however, the exact computation of $V_\infty(x)$ is typically intractable due to its potentially complex and non-convex structure. A more practical approach is to identify a simple, tractable function—such as a quadratic form $x^T Px$—that serves as an upper bound on the cost-to-go function. This upper bound should satisfy the following inequality:

$$\Psi(x) \geq \Psi(x^+) + \psi(x, \kappa_f(x)), \tag{12.15}$$

where $x^+ = f_u(x, \kappa_f(x))$.

A terminal penalty function $\Psi(x)$ satisfying this inequality, when combined with an appropriately chosen terminal set (as discussed previously), ensures the stability of the closed-loop system under the NMPC control law obtained by solving the finite-horizon problem (12.5) at each sampling instance.

12.3.3 Establishing Closed-loop Stability

We now establish the closed-loop stability of the MPC control law $u(k) = \kappa_p(x(k))$ defined by the nonlinear optimization problem (12.5), assuming that the terminal set is chosen as an output admissible set and the terminal penalty function satisfies the condition in (12.15) under a fictitious local control law $v_l = \kappa_f(z_l)$. This local law is not implemented in practice but is introduced purely to define the terminal set and terminal penalty.

It is important to note that the stability of the closed-loop system can only be discussed within the region of the state space where the optimization problem is feasible. This feasible region constitutes the domain of attraction, and it is assumed that the initial state lies within this domain.

Let $J_p(x(k))$ denote the optimal value of the finite-horizon optimization problem (12.5). Since both the stage cost $\psi(\cdot)$ and terminal cost $\Psi(\cdot)$ are positive definite, $J_p(x(k))$ is also positive definite. We aim to show that $J_p(x(k))$ decreases monotonically along the closed-loop trajectory, thereby serving as a Lyapunov function.

Consider the optimal solution to (12.5) at time k, denoted by

$$(z^*_{1|k}, \ldots, z^*_{p|k}, v^*_{0|k}, \cdot, v^*_{p-1|k}).$$

The control input applied to the plant is $u(k) = v^*_{0|k}$. At the next time step $k+1$, the following NMPC problem is solved:

$$J_p(x(k+1)) := \min_{z_l, v_l} \Psi(z_p) + \sum_{l=0}^{p-1} \psi(z_l, v_l), \tag{12.16a}$$

$$\text{s.t.} \quad z_{l+1} = f_u(z_l, v_l) \quad l = 0, \ldots p-1, \tag{12.16b}$$

$$z_0 = x(k+1), \tag{12.16c}$$

$$z_l \in \mathbb{X}, v_l \in \mathbb{U}, z_p \in \mathbb{X}_f. \tag{12.16d}$$

A feasible solution to this problem can be constructed by shifting the previously optimal input sequence forward by one step and appending the fictitious local control law at the end, i.e.,

$$v^*_{1|k}, \ldots, v^*_{p-1|k}, \kappa_f(z_{p|k}).$$

This is feasible because $z^*_{p|k} \in \mathbb{X}_f$, and $\mathbb{X}_f$ is invariant under the local control law κ_f by construction. Let the cost associated with this feasible (but not necessarily optimal) solution be denoted by $\hat{J}_p(x(k+1))$. Then,

$$J_p(x(k+1)) \leq \hat{J}_p(x(k+1)). \tag{12.17}$$

Now, evaluate the cost of the shifted sequence:

$$\begin{aligned}\hat{J}_p(x(k+1)) = {} & J_p(x(k)) - \psi(x(k), u(k)) \\ & -\Psi(z^*_{p|k}) + \psi(z^*_{p|k}, \kappa_f(z^*_{p|k})) + \Psi\Big(\underbrace{f_u\left(z^*_{p|k}, \kappa_f\left(z^*_{p|k}\right)\right)}_{(z^*_{p|k})^+}\Big).\end{aligned} \tag{12.18}$$

By the terminal penalty condition (12.15),

$$-\Psi(z^*_{p|k}) + \psi(z^*_{p|k}, \kappa_f(z^*_{p|k})) + \Psi\Big(\underbrace{f_u\left(z^*_{p|k}, \kappa_f\left(z^*_{p|k}\right)\right)}_{(z^*_{p|k})^+}\Big) \leq 0, \tag{12.19}$$

so we obtain

$$\hat{J}_p(x(k+1)) \leq J_p(x(k)) - \psi(x(k), u(k)). \tag{12.20}$$

Combining (12.17) and (12.20), we conclude

$$J_p(x(k+1)) + \psi(x(k), u(k)) \leq J_p(x(k)). \tag{12.21}$$

Since $\psi(x(k), u(k))$ is positive definite, the above inequality implies that $J_p(x(k))$ is a strictly decreasing sequence along the closed-loop trajectory. Furthermore, $J_p(x(k))$ is positive definite by construction, and it satisfies standard Lyapunov conditions. Properties such as properness and radial unboundedness also follow directly from its definition, completing the argument that $J_p(x(k))$ serves as a Lyapunov function, and hence the closed-loop system is asymptotically stable.

12.3.4 Implementation: Quasi-infinite-horizon MPC

We have established that the NMPC control law $u(k) = \kappa_p(x(k))$, defined by the optimization problem (12.5) with appropriately chosen terminal set and terminal penalty, ensures local asymptotic stability of the closed-loop system. However, this guarantee hinges critically on the proper selection of both the terminal set and the terminal penalty. In particular, if the terminal set is chosen too small, the terminal constraint may become infeasible for a wide range of initial conditions, significantly limiting the domain of attraction.

A key component in defining the terminal ingredients is the use of a fictitious local controller. This raises the natural question of how to choose this local controller in a systematic and effective manner. One widely adopted approach, proposed by Chen and Allgöwer (1998), is known as quasi-infinite-horizon MPC. In this framework, the local controller is designed based on the linearized system dynamics. Specifically, a stabilizing state feedback controller—such as one obtained from an LQR design—is derived from the Jacobian matrices evaluated at the operating point. Based on this linearized model, one can define a terminal region in the form of an ellipsoid, $x^T Px \leq \alpha$, by solving an appropriate Lyapunov equation. The matrix P and scalar α are selected so that this terminal region is an output admissible set, satisfying the required invariance and constraint satisfaction properties.

Moreover, the quadratic terminal penalty $\Psi(x) = x^T Px$ can be shown to upper-bound the infinite-horizon cost-to-go function and satisfy the condition in (12.15). While the details of the Lyapunov equation and the proof of these properties are mathematically involved, they are well documented in the literature (Chen and Allgöwer, 1998).

Another simplifying assumption made in our analysis is that the solution to the NLP problem becomes available instantaneously at each sampling instant. In practice, this is rarely the case. Depending on the complexity of the NLP and the available computational resources, the solution time may be non-negligible relative to the sampling period and must be accounted for in the control design.

To address this issue, Biegler and coworkers have proposed a method known as *Advanced Step MPC*, which allows partial computation of the solution in advance of the next sampling instant. Their work includes both analytical insights and practical implementation strategies for reducing the computational burden. Interested readers are referred to Zavala and Biegler (2009).

12.4 Nonlinear State Estimation

In the preceding sections, we considered NMPC under the assumption of full state feedback. However, implementing the NMPC strategy as presented requires knowledge of the full plant state vector $x(k)$, which is rarely available in practice. Therefore, it is essential to estimate the plant states using available measurements of $y(k)$. The measurement equation may itself be nonlinear:

$$y_m(k) = h\,(x(k)) + \nu(k). \tag{12.22}$$

State estimation techniques were introduced in Chapter 7 for linear systems. For nonlinear systems, several estimation methods are available. Among them, the EKF—along with its various enhancements—and MHE are the most widely used in practice. These two approaches will be discussed in the following subsections.

12.4.1 Extended Kalman Filter

The EKF, introduced in Section 11.6, is a widely used technique for state estimation in nonlinear systems. At each time step k, when the new measurement y_m becomes available, the EKF uses the previous state estimate $\hat{x}(k-1|k-1)$ and the input $u(k-1)$ to compute the current state estimate $\hat{x}(k|k)$. The method consists of a one-step prediction using the nonlinear model, followed by an update based on a linearized measurement model.

The prediction step is given by

$$\hat{x}(k|k-1) = F\Big(\hat{x}(k-1|k-1), u(k-1), d(k-1)\Big). \tag{12.23}$$

The filtered estimate is then calculated as

$$\hat{x}(k|k) = \hat{x}(k|k-1) + K_f(k)\left(y_m(k) - h\left(\hat{x}(k|k-1)\right)\right). \tag{12.24}$$

The Kalman gain K_f is calculated after linearizing the system dynamics and output equations around $\hat{x}(k-1|k-1)$ and $u(k-1)$. The details can be found in Section 11.6.

We now consider an example motivated by realistic scenarios in which only a subset of the output variables are measured, the measurements are affected by noise, and the system is influenced by disturbances. In such cases, NMPC and state estimation must be integrated to achieve robust performance, making state estimation an essential component of the overall control architecture.

Example 12.3: *Output Feedback NMPC.*
Consider the control problem of Example 12.1, now extended to the output feedback case. Since the full state vector is not assumed to be measurable, an observer is required to estimate the states from available output measurements.

To accommodate this, the system is modified to include a state estimation algorithm, enabling output-feedback control. Figure 12.5 shows the revised flowsheet of the closed-loop system. Instead of directly measuring the full state vector x, an estimator—specifically, an EKF—is used to generate the state estimate $\hat{x}$, which is then passed to the NMPC controller to compute the control inputs.

The EKF is implemented using the `Extended Kalman Filter` block in Simulink. Since this block requires a discrete-time state-space model, the continuous-time dynamics must be discretized. For this example, an explicit Euler integration scheme is sufficient:

$$\begin{aligned} &\int_{t_k}^{t_{k+1}} f(x, u, d)dt = x_N \\ &\qquad \text{where } x_0 = x(k), \text{ and} \\ &\qquad\qquad x_i = x_{i-1} + \delta t f\left(x_i, u(k), d(k)\right). \end{aligned} \tag{12.25}$$

Here, $\delta t = \dfrac{t_{k+1} - t_k}{N}$ denotes the Euler integration step size. For more complex systems or higher accuracy requirements, other numerical integration methods (e.g., `ode45`, `ode15s`) may be specified.

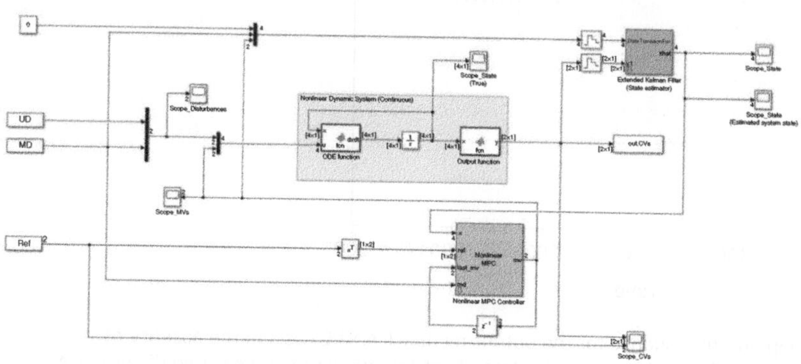

Figure 12.5 Simulink flowsheet of the closed-loop system using an EKF for state estimation in output-feedback NMPC.

Once the discrete-time model (12.25) is obtained, it must be incorporated consistently into both the EKF implementation and the NMPC predictive model definition (i.e., the nmpc object). This ensures that the controller and estimator operate on a consistent model representation.

```
%% Predictive model
% Specify the nonlinear state and output functions for the
% predictive model used in NMPC. The model state/output
% function are separately defined from the "true'' system
nlobj.Model.StateFcn = 'PredModel_StateFcn_discrete';
nlobj.Model.OutputFcn = 'PredModel_OutputFcn';

% Command below indicates that a discrete-time model is used
nlobj.Model.IsContinuousTime = false;
```

Figure 12.6 shows the closed-loop response of the NMPC with state estimates provided by the EKF. The behavior is comparable to the state-feedback NMPC results from Example 12.1. With accurate state estimation, the NMPC is able to predict and optimize future system behavior with only a minor degradation in performance. The controller effectively tracks set point changes and rejects the measured disturbance, confirming its capability in managing modeled disturbances. However, at $t = 200$, the introduction of an unmeasured disturbance results in a steady-state offset, revealing the lack of integral action and highlighting the limitation of the NMPC in compensating for unmodeled or unmeasured disturbances.

The results of the preceding example should not be surprising. They follow the familiar pattern observed earlier in Chapter 11: disturbance modeling is essential to reject the effects of unmeasured disturbances. MPC achieves offset-free control through a combination of the receding horizon principle and model correction. In this case, the critical element—model correction—is absent, leading to steady-state offsets when unmeasured disturbances or model-plant mismatch are present.

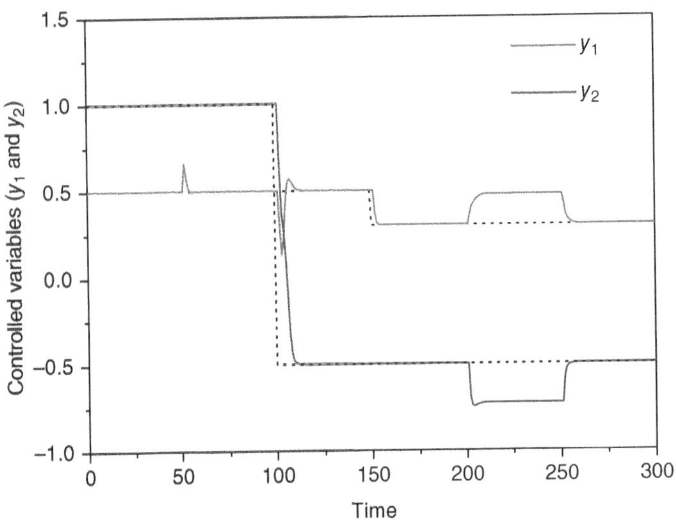

Figure 12.6 Closed-loop output response for the output-feedback NMPC case. In the absence of disturbance modeling, the NMPC exhibits a steady-state offset following the unmeasured disturbance introduced at $t = 200$, due to the lack of integral action.

As discussed in Section 11.2.1, a common approach to achieve offset-free control in the presence of persistent disturbances (e.g., step-type disturbances) is to include additive output disturbances and model them as integrated white-noise processes:

$$w_y(k+1) = w_y(k) + \varepsilon(k), \tag{12.26}$$

Note that one may use an input or a state disturbance instead. Because the disturbance is unmeasured, it must be estimated from the available output measurements. This requires augmenting the predictive model with an additional state:

$$\bar{x}(k) \triangleq \begin{bmatrix} x(k) \\ w_y(k) \end{bmatrix}, \tag{12.27}$$

and estimating the augmented state $\hat{\bar{x}}(k|k)$ using the measured outputs. For details on the estimation procedure for the augmented state, readers are referred to Section 11.6.

Example 12.4 illustrates how incorporating disturbance modeling can enable offset-free control performance in the presence of unmeasured disturbances.

Example 12.4: *NMPC: Effect of Disturbance Modeling.*

Since there are two measured outputs, we can introduce up to two disturbance variables modeled as integrating white-noise sequences:

$$\begin{bmatrix} x_{w1}(k+1) \\ x_{w2}(k+1) \end{bmatrix} = \begin{bmatrix} x_{w1}(k) \\ x_{w2}(k) \end{bmatrix} + \begin{bmatrix} \varepsilon_1(k) \\ \varepsilon_2(k) \end{bmatrix}. \tag{12.28}$$

It is convenient to model these as output load disturbances:

$$y(k) = \left[\begin{array}{cccc|cc} 0 & 1 & 1 & 0 & 1 & 0 \\ -1 & 0 & 0 & 1 & 0 & 1 \end{array}\right] \underbrace{\begin{bmatrix} x(k) \\ x_w(k) \end{bmatrix}}_{\bar{x}(k)}. \tag{12.29}$$

The MATLAB function defining the discrete predictive model should be modified accordingly to include the augmented state vector.

```
%% For Predictive model (state function, discretized)
function xk1 = PredModel_StateFcn_discrete(xk,uk)

    Ts = 1;
    Nstep = 10;
    dt = Ts/Nstep;

    xk1 = xk(1:4);     % Extract the state from augmented state
    uk1 = [uk(:);0];
    for ii = 1:Nstep
        xk1 = xk1 + dt*PredModel_StateFcn(xk,uk1);
    end

    xk1 = [xk1; xk(5:6)];  % Augment the disturbance state
end
```

The output model must also be modified to account for the augmented disturbance states. A possible MATLAB implementation for the output function is

```
%% For Predictive model (output function)
function y = PredModel_OutputFcn(xAug,u)
    y = zeros(2,1);
    y(1)  =   xAug(2)  + xAug(3)  + xAug(5);
    y(2)  =  -xAug(1)  + xAug(4)  + xAug(6);
end
```

The Simulink flowsheet remains the same as in the previous example (Figure 12.5). The key difference lies in the definition of the NMPC predictive model, which has been updated to include the disturbance states. To run the simulation, use either the Simulink model or the `NMPC_Example_Case3_Setup` m-file.

As shown in Figure 12.7, the NMPC now achieves satisfactory feedback control without steady-state offset. Figure 12.8 shows the dynamics of the estimated disturbance states, $\bar{x}(5)$ and $\bar{x}(6)$. These estimates remain near zero when the plant and model are well matched, including during measured disturbances and set point changes. However, when an unmeasured disturbance is introduced at $t = 200$, the disturbance state estimates rise sharply to compensate for the mismatch between model prediction and actual system behavior, thereby restoring offset-free control.

This example illustrates that appropriate disturbance modeling, combined with a suitable state estimator, is essential for achieving offset-free performance in NMPC. The augmented implementation presented here constitutes a complete and robust NMPC framework that effectively handles unmeasured disturbances.

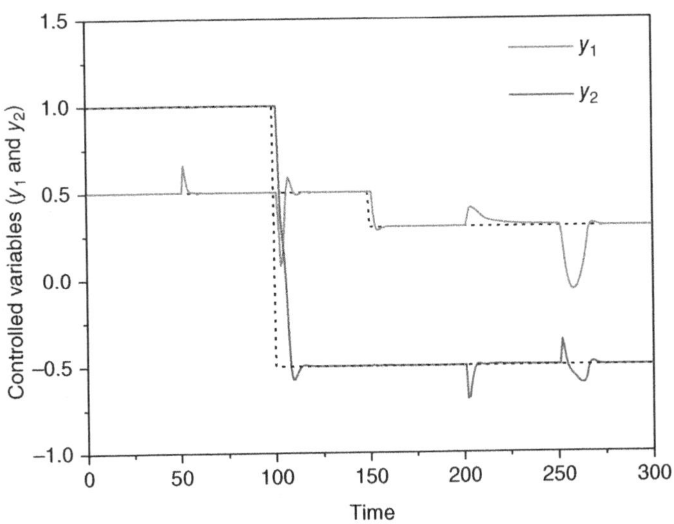

Figure 12.7 Closed-loop output response with disturbance modeling. The use of integrating disturbance states enables offset-free tracking even in the presence of unmeasured disturbances.

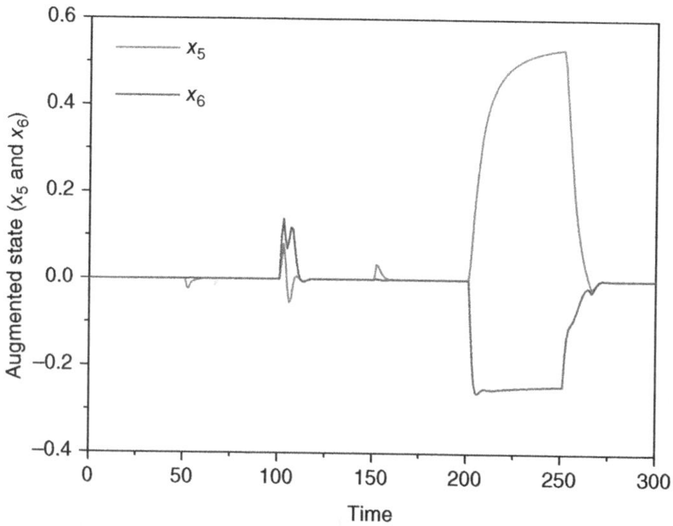

Figure 12.8 Dynamics of the disturbance states estimated using EKF. Note the rapid adjustment in response to the unmeasured disturbance introduced at $t = 200$.

12.4.2 MHE for Nonlinear State Estimation

While extensions of linear estimation techniques (e.g., EKF) can be applied to nonlinear systems, they do not fully exploit the information available in a nonlinear DAE model. MHE, introduced earlier in Chapter 7 for linear systems, provides a more general and powerful framework. It formulates state estimation as an optimization problem that uses the same nonlinear DAE model as NMPC and estimates states, parameters, and disturbances from available plant measurements over a moving time horizon. For this reason, MHE is often regarded as the estimation counterpart of NMPC.

Assume that the measurement vector $y(k)$ is related to the state through a nonlinear function:

$$y(k) = h(x(k)) + v(k), \tag{12.30}$$

where $v(k)$ is a white-noise sequence.

To estimate the current state, MHE uses a window of past measurements $y(k - M + 1), \dots, y(k)$ to infer the most likely sequence of past states $z_{-M+1|k}, \dots, z_{0|k}$, where $z_{l|k}$ denotes the estimate of $x(k + l)$ given data up to time k. These are referred to as "smoothed" estimates.

For simplicity, suppose that disturbances are modeled as integrated white noise:

$$d(k + 1) = d(k) + \varepsilon(k), \tag{12.31}$$

where $\varepsilon(k)$ is a zero-mean white-noise sequence. Then, the augmented system dynamics for estimation become:

$$\begin{bmatrix} x(k+1) \\ d(k+1) \end{bmatrix} = \begin{bmatrix} f_u(x(k), u(k)) + f_d(x(k), d(k)) \\ d(k) \end{bmatrix} + \begin{bmatrix} 0 \\ \varepsilon(k) \end{bmatrix}. \tag{12.32}$$

The MHE problem is then formulated as the following constrained optimization problem:

$$\min_{\substack{z_{-M+1},\delta_{-M+1},\\ \{\omega_l\}_{l=-M+1,\dots,0}}} \Phi_{-M}(z_{-M+1},\delta_{-M+1})$$

$$+\sum_{i=-M+1}^{0}(y(k+i)-h(z_i))^T R_2^{-1}(y(k+i)-h(z_i))$$

$$+\sum_{i=-M+1}^{0}\omega_i^T R_1^{-1}\omega_i \tag{12.33a}$$

$$\text{s.t. } z_{l+1}=f(z_l,u_l)+f_d(z_l,\delta_l)$$

$$\delta_{l+1}=\delta_l+\omega_l,\quad l\in\{-M+1,\dots,0\} \tag{12.33b}$$

$$z_l\in\mathbb{X},\quad l\in\{-M+1,\dots,0\} \tag{12.33c}$$

$$\delta_l\in\mathbb{D},\quad l\in\{-M+1,\dots,0\}. \tag{12.33d}$$

In this formulation, $\Phi_{-M}(z_{-M+1},\delta_{-M+1})$ is the so-called *arrival cost*, which encapsulates prior information before the start of the current estimation window at time $k-M+1$, as discussed in Chapter 7. The matrices R_2 and R_1 are the covariance matrices for the measurement noise ν and disturbance noise w, respectively.

Solving the MHE problem (12.33) yields the most likely sequence of states $z_{-M+1|k},\dots,z_{0|k}$ and disturbance estimates $\delta_{-M+1|k},\dots,\delta_{0|k}$, based on the current and past $M-1$ measurements. Theoretically, this corresponds to the joint maximum a posteriori (MAP) estimate of the states and disturbances, assuming Gaussian distributions for the process noise w, measurement noise ν, and the initial state z_{-M+1}.

An estimate of the current plant state and disturbance can be obtained from the MHE solution, specifically: $\hat{x}(k|k):=z_{0|k}$ and $\hat{d}(k|k):=\delta_{0|k}$. Note that only a finite data window of length M is used in the optimization. Consequently, the arrival cost $\Phi_{-M}(\cdot)$ is introduced to summarize all prior information before the current estimation window $[k-M+1,k]$.

For practical on-line implementation, this arrival cost must be updated recursively. However, exact recursive computation is generally intractable for nonlinear systems. Instead, a commonly used approximation can be derived from the EKF, expressed as

$$\begin{aligned}&\Phi_{-M}(z_{-M+1},\delta_{-M+1})\\&=\zeta^T(z_{-M+1},\delta_{-M+1})\{P(k-M+1|k-M)+\gamma I\}^{-1}\zeta(z_{-M+1},\delta_{-M+1}),\end{aligned} \tag{12.34}$$

where

$$\zeta(z_{-M+1},\delta_{-M+1})=\begin{bmatrix} z_{-M+1}-\hat{x}(k-M+1|k-M)\\ \delta_{-M+1}-\hat{d}(k-M+1|k-M)\end{bmatrix}. \tag{12.35}$$

Here, $P(k-M+1|k-M)$ is the state error covariance matrix propagated from the EKF's Riccati recursion. The estimates $\hat{x}(k-M+1|k-M)$ and $\hat{d}(k-M+1|k-M)$ are those stored M steps earlier. The term γI is added to accommodate the linearization error; a suitable choice of $\gamma>0$ ensures a conservative lower bound on the arrival cost and contributes to stable estimator behavior.

Based on the formulation in (12.33), the online implementation of a coupled MHE–NMPC scheme proceeds as follows. At time t_k, the MHE problem is solved first to yield the current state estimate $\hat{x}(k|k)$. This estimate is then used in the NMPC problem (12.5) as the initial state (invoking the *certainty equivalence principle*) to compute the control input $u(k)$, which is subsequently applied to the plant. At the next sampling instant t_{k+1}, both MHE and NMPC subproblems are updated, and the cycle repeats. The same discretization techniques and NLP solvers used in NMPC can also be applied to MHE for real-time implementation.

12.5 Case Study

We now consider a practical NMPC implementation using a data-driven model. The system under study is a continuous stirred-tank reactor (CSTR) with a cooling jacket, shown in Figure 12.9. An exothermic first-order reaction $A \rightarrow B$ occurs in the reactor. Due to the exothermic nature of the reaction, a cooling jacket is required to regulate the reactor temperature and maintain the desired outlet concentration of species A. The inlet and outlet flow rates are assumed to be equal, ensuring constant liquid holdup in the reactor.

The control objective is to maintain the outlet concentration C_A at a desired set point by manipulating the coolant inlet temperature $T_{C,\text{in}}$, in the presence of both measured and unmeasured disturbances (MD and UD). C_A is assumed to be measured for feedback control.

The dynamic behavior of the reactor is described by the following mass and energy balances:

$$\text{Mass balance:} \quad \frac{dC_A}{dt} = \frac{F}{V}\left(C_{A,in} - C_A\right) - r_A. \tag{12.36a}$$

$$\text{Energy balance:} \quad \frac{dT}{dt} = \frac{F}{V}\left(T_{in} - T\right) + \frac{-\Delta H}{\rho c_p} r_A - \frac{UA}{\rho c_p V}\left(T - T_{C,in}\right). \tag{12.36b}$$

$$\text{Reaction rate:} \quad r_A = k_0 e^{-E/(RT)} C_A. \tag{12.36c}$$

Here, we assume that the cooling jacket temperature T_c is the same as the inlet coolant temperature $T_{C,in}$ for simplicity. Without this assumption, an additional energy balance would be required to describe the cooling jacket temperature dynamics.

Details for building the Simulink model of the CSTR are provided in Appendix G.2. Once the system model is constructed, an NMPC controller is designed using the first-principles model.

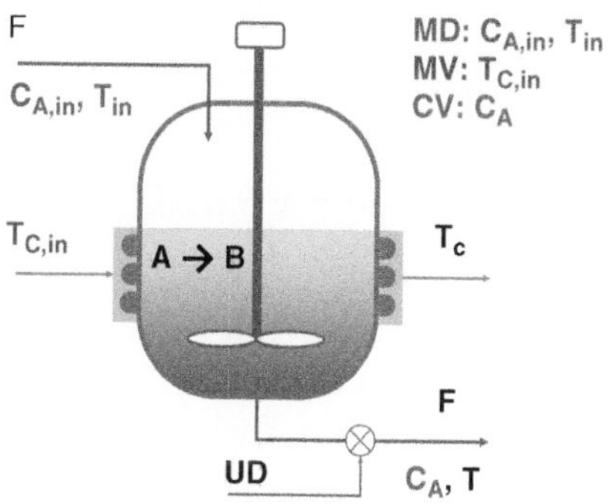

Figure 12.9 Case study: control of a CSTR.

To incorporate integral action, an integrating white-noise load disturbance is added, following the approach in Example 12.4:

$$x_w(k+1) = x_w(k) + \varepsilon(k)$$

And the augmented state and output

$$\bar{x}(k) = \begin{bmatrix} C_A \\ T \\ x_w \end{bmatrix}$$

$$\begin{aligned} y(k) &= C_A + x_w(k) \\ &= \begin{bmatrix} 1 & 0 & 1 \end{bmatrix} X(k). \end{aligned}$$

The closed-loop system model can now be assembled as described in the previous section. Please refer to the file `NMPC_Example_OutputControl_CSTR.slx` for implementation. The NMPC controller can be constructed using the following MATLAB code:

```
%% NMPC setup code
clear
clc

%% Dynamic model perturbation
Tp = 10;

x0 = [311.2639; 8.5698; 0];
u0 = [10; 298.15; 298.15];
%%

%% Nonlinear MPC design
%% MPC parameters
nlobj = nlmpc(3, 1,'MV',3,'MD',[1,2],'UD',4);

Ts = 0.5;
nlobj.Ts = Ts;
nlobj.PredictionHorizon = 20;
nlobj.ControlHorizon = [2, 2, 2, 14];

%% Scaling
nlobj.MV(1).ScaleFactor = 300;

nlobj.States(1).ScaleFactor = 300;
nlobj.States(2).ScaleFactor = 10;

%% Weights
nlobj.Weights.ManipulatedVariablesRate = 1;
nlobj.Weights.OutputVariables(1) = 1;
```

```
%% Constraints
nlobj.MV(1).RateMin = -1;
nlobj.MV(1).RateMax = 1;

nlobj.MV(1).Min = 273.15;
nlobj.MV(1).Max = 400;

%% Predictive model
% Specify the nonlinear state and output functions.
nlobj.Model.StateFcn = 'PredModel_StateFcn_discrete';
nlobj.Model.OutputFcn = 'PredModel_OutputFcn';
nlobj.Model.IsContinuousTime = false;

%% NMPC check with validateFcns function
x0 = [311.2639; 8.5698; 0];
u0 = [10; 298.15; 298.15];
validateFcns(nlobj,x0,u0(3),u0(1:2)');
```

Figure 12.10 presents the NMPC simulation results for several control scenarios of interest. Also included—though not discussed in detail—is the result obtained using a nonlinear ARX model identified via a neural network, as a surrogate for the first-principles model.

It is particularly instructive to examine the estimated augmented state shown in Figure 12.10b, which captures the effect of the load disturbance. The inclusion of this augmented disturbance state allows for offset-free tracking of the reference trajectory, even in the presence of model-plant mismatch and unmeasured disturbances. Notably, when an unmeasured disturbance is introduced at $t = 250$, the disturbance estimate adjusts rapidly, thereby preserving offset-free tracking performance.

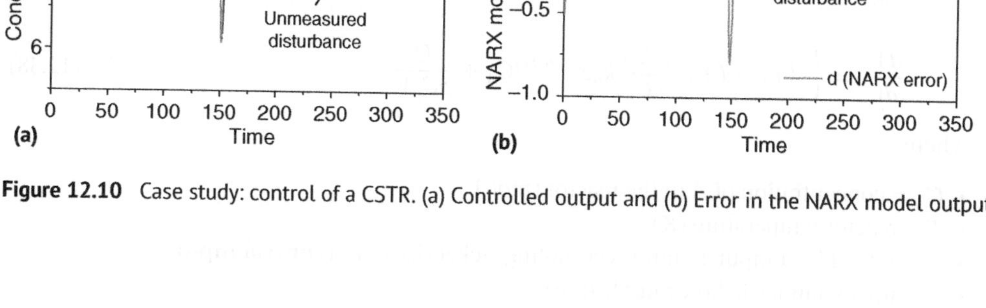

Figure 12.10 Case study: control of a CSTR. (a) Controlled output and (b) Error in the NARX model output.

12.6 Conclusions and Future Directions

NMPC has emerged as a powerful framework for nonlinear control and dynamic optimization. Its primary barrier to widespread industrial adoption has been the computational burden of solving large-scale NLP problems online, particularly given that most realistic process models are complex and high-dimensional. However, recent advances—such as full discretization methods using orthogonal collocation on finite elements—have led to robust and efficient strategies for solving such NLPs, moving NMPC closer to practical deployment in industrial settings.

Realizing these advances in practice requires comprehensive optimization frameworks that support: (1) high-level modeling of the underlying NLP; (2) seamless integration with efficient NLP solvers; and (3) coordinated execution of NMPC and MHE in a closed-loop control architecture.

Beyond computational developments, significant theoretical progress has been made in establishing stability guarantees for NMPC, at least in the nominal (i.e., disturbance-free) case. As in linear MPC, the infinite-horizon optimal control problem—known to yield desirable closed-loop properties—can be approximated through appropriate choices of terminal penalty and terminal constraints. This approximation preserves the essential stability guarantees without the computational complexity of a true infinite-horizon formulation.

Just as with linear MPC, NMPC relies on accurate state information, which is typically unavailable in industrial environments. Thus, robust state estimation is crucial. While nonlinear filtering has been studied extensively, a method that fully leverages the nonlinear model structure while maintaining robustness and accuracy has remained elusive. MHE fills this gap by formulating the estimation problem as an optimization over a moving time window. MHE is flexible enough to handle models of varying complexity, including large-scale nonlinear DAE systems. Like NMPC, MHE requires solving an NLP online, making it a natural estimation counterpart to NMPC.

Looking ahead, NMPC and MHE are poised to play a central role in the realization of integrated operational strategies that tightly couple real-time control, online optimization, and scheduling. Continued advances in computation, modeling platforms, and theoretical foundations will further solidify their position as key technologies for future process systems.

Exercises

1. A CSTR carries out an irreversible, exothermic, first-order reaction:

   $A \rightarrow B.$

   The nonlinear state-space model is given by

   $$\frac{\mathrm{d}C_A}{\mathrm{dt}} = \frac{F}{V}\left(C_{A_f} - C_A\right) - k_0 e^{-E/(RT)} C_A. \tag{12.37}$$

   $$\frac{\mathrm{dT}}{\mathrm{dt}} = \frac{F}{V}\left(T_f - T\right) + \frac{-\Delta H}{\rho C_p} k_0 e^{-E/(\mathrm{RT})} C_A + \frac{Q}{\rho C_p V}, \tag{12.38}$$

   where:

   - C_A = concentration of A in the reactor (mol/L)
   - T = reactor temperature (K)
   - Q = rate of heat input/removal via cooling jacket (kJ/min) **(control input)**
   - F = inlet volumetric flow rate (L/min)

- V = reactor volume (L)
- C_{A_f} = feed concentration (mol/L)
- T_f = feed temperature (K)
- k_0 = pre-exponential factor (1/min)
- E = activation energy (kJ/mol)
- R = gas constant (kJ/mol/K)
- ΔH = reaction enthalpy change (kJ/mol)
- ρ = density (kg/L)
- C_p = heat capacity (kJ/kg/K)

The nominal parameter values are

$$
\begin{aligned}
&F = 100\ \text{L/ min}, \quad V = 100\ \text{L}, \quad C_{A_f} = 1.0\ \text{mol/L}, \quad T_f = 350\ \text{K},\\
&k_0 = 7.2 \times 10^{10}\ \text{min}^{-1}, \quad E = 8.314 \times 10^3\ \text{kJ/mol}, \quad \Delta H = -4.78 \times 10^4\ \text{kJ/mol},\\
&\rho = 1.0\ \text{kg/L}, \quad C_p = 0.239\ \text{kJ/kg} \cdot K.
\end{aligned}
$$

The control objective is to regulate the reactor temperature T to a desired set point T_{ref} while ensuring that the concentration C_A remains within safe operating limits. The operational constraints are

$$
\begin{aligned}
-8 \times 10^5\ \text{kJ/min} \leq Q(t) \leq\ 8 \times 10^5\ \text{kJ/min}\\
330\ \text{K} \leq T(t) \leq 370\ \text{K}\\
0\ \text{mol/L} \leq C_A(t) \leq 1.0\ \text{mol/L}.
\end{aligned}
$$

Both T and C_A are assumed to be measured without error. The NMPC problem at each sampling instant k is

$$
\min_{Q(\cdot)} \sum_{i=0}^{p-1} \left\{ [\gamma_T\,(T(k+i|k) - T_{\text{ref}})]^2 + [\gamma_{C_A}\,(C_A(k+i|k) - C_{A,\text{ref}})]^2 + [\gamma_Q\,\Delta Q(k+i|k)]^2 \right\}, \tag{12.39}
$$

subject to

- Nonlinear CSTR dynamics (discretized model).
- State and input constraints above.
- Prediction horizon p, control horizon m.

The weights are given as $\gamma_T = 1.0, \gamma_{C_A} = 0.8, \gamma_Q = 10^{-2}$.

Perform the following steps:

(a) Discretize the CSTR nonlinear ODEs using an explicit Euler method with sampling time $t_s = 0.5$ min.

(b) Express the NMPC problem in a form suitable for solution by a NLP solver (e.g., IPOPT).

(c) Simulate the CSTR under NMPC starting from $T(0) = 345$ K and $C_A(0) = 0.8$ mol/L. Introduce a step change in T_{ref} from 345 to 360 K at $t = 10$ min. Compare the response with and without constraints.

(d) Discuss the effect of constraints on tracking performance and control effort. Compare the NMPC solution with a linearized MPC designed around the nominal steady state.

2. A Nonlinear CSTR System (Van de Vusse): Consider a CSTR with reactions A → B (desired), B → C (undesired), and 2A → D (undesired dimerization). Let the states be

$$x = \begin{bmatrix} C_A & C_B & T & T_{\text{jac}} \end{bmatrix}^T.$$

Concentrations are in mol/L, reactor and jacket temperatures in °C. The manipulated inputs are $u = \begin{bmatrix} \dot{F} & \dot{Q} \end{bmatrix}^T$, where $\dot{F} = \dot{V}/V_R$ (dilution rate,/h) and $\dot{Q}$ (net heat added to the jacket, kJ/h; negative for heat removal). The measured outputs are $y = \begin{bmatrix} C_B & T \end{bmatrix}^T$.

The nonlinear plant is described by the following first-principles model:

$$\dot{C}_A = \dot{F}\,(C_{A0} - C_A) - k_1(T)\,C_A - k_3(T)\,C_A^2, \tag{12.40}$$

$$\dot{C}_B = -\dot{F}\,C_B + k_1(T)\,C_A - k_2(T)\,C_B, \tag{12.41}$$

$$\dot{T} = \dot{F}\,(T_0 - T) - \frac{1}{\rho C_p}\left[k_1(T)C_A\,\Delta H_{\text{AB}} + k_2(T)C_B\,\Delta H_{\text{BC}} + k_3(T)C_A^2\,\Delta H_{\text{AD}}\right] + \frac{UA_R}{\rho C_p V_R}(T_{\text{jac}} - T) \tag{12.42}$$

$$\dot{T}_{\text{jac}} = \frac{1}{m_c C_{p,c}}[\dot{Q} + UA_R(T - T_{\text{jac}})], \tag{12.43}$$

$$k_i(T) = k_{i0}\,\exp\left(\frac{E_i}{T + 273.15}\right), \qquad i \in \{1, 2, 3\}. \tag{12.44}$$

Parameters and constants.

- Pre-exponential factors: $k_{10} = k_{20} = 1.287 \times 10^{12}$/h, $k_{30} = 9.043 \times 10^9$ [L/mol/h].
- Activation energies [K]: $E_1 = -9758.3$, $E_2 = -9758.3$, $E_3 = -8560$.
- Reaction enthalpies [kJ/mol]: $\Delta H_{\text{AB}} = 4.2$, $\Delta H_{\text{BC}} = -11.0$, $\Delta H_{\text{AD}} = -41.85$.
- Feed temperature: $T_0 = 104.9$ °C.
- Physical properties: $\rho = 0.9342$ [kg/L],/$C_p = 3.01$ [kJ/kg/K], $C_{p,c} = 2.0$ [kJ/kg/K].
- Heat-transfer and geometry: $U = 4032$ [kJ/m^2/K/h], $A_R = 0.215$ [m^2], $V_R = 10$ [L].
- Coolant mass: $m_c = 5.0$ [kg].

Initial conditions for x may be any accessible nominal steady state.

(a) With only C_B and T measurable in real time, the plant is partially observed; effective MPC therefore requires recovering the full input–state–output dynamics by system identification. A NARX surrogate can learn these dynamics from data, capturing nonlinearities and disturbance effects for multistep prediction. Let the nonlinear predictor be a neural NARX of the form

$$y(k+1) = f\big(y(k), y(k-1), \ldots, y(k-n), \quad u(k), u(k-1), \ldots, u(k-n)\big)$$
$$y = \begin{bmatrix} C_B & T \end{bmatrix}^T, \; u = \begin{bmatrix} \dot{F} & \dot{Q} \end{bmatrix}^T, \tag{12.45}$$

with a modest lag $n \in \{2, 3\}$ and $f(\cdot)$ implemented by a polynomial or MLP network.

Generate identification data under a closed-loop PID controller by varying output set points. For example, vary the C_B set point between 0.75 and 1.05 [mol/L] while keeping the T set point at 114 °C. Add sensor-consistent zero-mean Gaussian measurement noise and sample with $\Delta t \in [10, 30]$ s. At each time k, record $[C_B(k),\ T(k)]$ and $[\dot{F}(k-1),\ \dot{Q}(k-1)]$. Split the resulting dataset into train/validation/test *by time blocks* to prevent leakage, and train the NARX network.

(b) Use the trained neural NARX as a direct *output* predictor for the CSTR. Solve an output-tracking NMPC with move-rate penalties on inputs over horizons $p = m \in \{2, \ldots, 5\}$ and objective

$$\min_{\{\Delta u(k+i|k)\}} \quad \sum_{i=0}^{p-1} \|y(k+i|k) - r(k+i|k)\|_Q^2 + \sum_{i=0}^{m-1} \|\Delta u(k+i|k)\|_R^2. \tag{12.46}$$

Include box constraints and move-rate limits:

$$\dot{F}_{\min} \le \dot{F}(k+i|k) \le \dot{F}_{\max}, \qquad \dot{Q}_{\min} \le \dot{Q}(k+i|k) \le \dot{Q}_{\max}, \tag{12.47}$$

$$|\Delta \dot{F}(k+i|k)| \le 0.05, \quad |\Delta \dot{Q}(k+i|k)| \le 0.05. \tag{12.48}$$

Use, for example, $\dot{F} \in [3, 55]$ and $\dot{Q} \in [-3500, 2500]$ (engineering units). Track a piecewise-constant C_B set point while holding T near nominal.

Direct output prediction often exhibits steady-state offset. It is recommended to augment the NARX output with a drift/disturbance state w_y modeled as a random walk and estimate it within the NMPC formulation (e.g., by augmenting the predictor with $y^+ = f(\cdot) + w_y$, $w_y^+ = w_y + \varepsilon_k$). Optionally, add a terminal cost to enhance look-ahead and robustness.

(c) Both the NARX and the NMPC operate in discrete time under ZOH. Use the same scaling for y and u in training and in the controller (e.g., min–max to $[-1, 1]$); choose Q/R in that scaled space. Recommended durations: for identification, record ≥ 20 process time constants with persistently exciting inputs; for MPC, simulate ≥ 5 h per scenario with set point steps and disturbances.

Report the following.

i. Tracking/actuation metrics of NMPC (e.g., MSE of C_B and T; total variation of u).
ii. Constraint satisfaction (box bounds and move-rate limits).
iii. Closed-loop robustness under disturbances: vary k_{10} and/or T_0.
iv. Modeling/algorithmic choices: compare different horizon sizes p, m, and NARX lag orders n.

13

Repetitive MPC for Batch and Periodic Systems

A mistake repeated more than once is not a mistake anymore. It is a decision.

– Paulo Coelho

Certain process control problems are characterized by recurring signal patterns, such as periodic disturbances or setpoint trajectories. An important class of systems exhibiting repeated operating patterns is batch processes, which follow fixed recipes repeated across multiple runs. Furthermore, some continuous processes emulate batch operation by enforcing cyclic operating trajectories—examples include pressure swing adsorption (PSA) systems and simulated moving bed (SMB) chromatography.

A key feature of these systems is that nonlinear dynamics are often more pronounced, as operations span a wide range of conditions. Moreover, similar control error patterns tend to repeat across successive runs. For example, in tracking time-varying reference trajectories during a batch, feedback control may exhibit recurring errors due to factors such as time delays, inverse responses, or model mismatch. These same errors are likely to recur in the next run, given that the system dynamics remain unchanged. While such errors may appear inevitable from the perspective of feedback-only control, they can in fact be anticipated and corrected using appropriately designed feedforward actions. In other words, by learning from the errors of previous runs, one can apply corrective signals—either explicitly as feedforward inputs or as bias trajectory updates—to improve performance over time.

This core idea underlies a class of control strategies known as Iterative Learning Control (ILC) and Repetitive Control (RC), which were developed independently in the literature. We begin by offering a brief historical background on these methods before introducing their application within the context of Model Predictive Control (MPC).

13.1 Historical Background

The development of Iterative Learning Control (ILC) and Repetitive Control (RC) originated from different problem domains. ILC was initially proposed to improve performance over successive runs in systems such as robotic arms and servomechanisms. It focused on translating observed tracking errors into input corrections using model inversion techniques (Moore et al., 1992). In contrast, RC was developed to address periodic tracking errors in linear time-invariant (LTI) systems by leveraging the internal model principle (Hara et al., 1988). Longman (1997) later showed the conceptual similarity between the two and unified them under a common framework. The primary distinction lies in system behavior across runs: ILC assumes the state resets at the beginning of each run, as in batch processes, while RC assumes continuous state evolution across periods, as in cyclic processes.

Early ILC and RC designs were primarily developed in the frequency domain. For ILC, a learning filter structure was assumed, translating error trajectories into input bias corrections, with the filter designed to meet certain convergence criteria. RC, on the other hand, relied on the internal model principle to define a controller structure and used stability analysis—often via the small-gain theorem—for design. While these approaches were effective for unconstrained LTI systems, their applicability to chemical process control was limited due to the prevalence of constrained, multivariable, and nonlinear dynamics in such systems.

Beginning in the late 1990s, more general time-domain formulations emerged, enabling the use of optimal control techniques. Linear quadratic (LQ)-based formulations were proposed for both discrete-time (Tao et al., 1994) and continuous-time (Sogo and Adachi, 1994) systems. Subsequently, Amann et al. (1996) and Lee and coworkers (Lee et al., 2000, 2001) independently developed more comprehensive time-domain quadratic formulations that guaranteed convergence to the minimum achievable error. Lee and colleagues further extended this into a real-time MPC framework that combined feedback and learning, and demonstrated its effectiveness on experimental platforms (Lee et al., 1999; Yang et al., 2003). The methodology has also been extended to nonlinear systems (Pandit and Buchheit, 1995).

In this chapter, we explore how these ideas can be embedded within the state-space MPC framework to develop a controller that learns and compensates for repeating error patterns. The result is a *repetitive MPC* scheme that achieves perfect—or near-perfect—tracking over successive runs, even in the presence of constraints, nonlinearities, or structural challenges that limit the performance of traditional feedback control.

13.2 General Framework

13.2.1 Problem Formulation

The underlying system is modeled as:

$$\begin{aligned} x(k, t+1) &= A(t)x(k, t) + B(t)u(k, t) \\ y(k, t) &= C(t)x(k, t) + h(t), \qquad t \in \mathcal{I} \triangleq \{0, \ldots, N-1\}. \end{aligned} \tag{13.1}$$

Here, k denotes the run index, and t is the time index within a run. Each run consists of N sampling steps. The term $h(t)$ represents an *unknown but run-invariant* model mismatch, which may vary with t but remains fixed across runs. This term can capture effects such as persistent initial condition offsets, disturbances, or model biases—making $h(t)$ effectively N-periodic across runs. The system formulation allows for time-varying input and output dimensions, enabling treatment of more general scenarios such as multi-rate systems.

The state transition between successive runs can follow one of two forms:

- *Batch Process*: In a batch process, each run concludes with product withdrawal, followed by cleaning and recharging of the unit. When the same feedstock is reused, the state at the start of the next run is assumed to be reset to the same initial value:

$$x(k+1, 0) = x(k, 0). \tag{13.2}$$

- *Continuous Process*: For continuous systems operating in a periodic mode, the end state of one run becomes the initial state of the next:

$$x(k+1,0) = x(k,N). \tag{13.3}$$

In this case, the system evolves continuously from one run to the next, governed by its natural dynamics.

The periodically time-varying discrete-time system described above is general and flexible enough to cover a wide range of process control problems encountered in industry. Although most batch processes exhibit nonlinear dynamics over the course of a run, the operational patterns across runs are often repetitive. As a result, linearizing the system along a representative or nominal trajectory yields a linear time-varying model of the form given earlier. In some cases, the run length may vary—such as when the batch recipe is updated between runs. This variability complicates the assumption of periodicity, since the system is no longer strictly periodic.

The MPC design objective is to determine $u(k,t)$ such that, for a given reference trajectory $r(t)$, the deviation from the desired output trajectory is minimized:

$$J(x(0,0)) = \sum_{k=0}^{\infty}\sum_{t=0}^{N-1} \|y(k,t) - r(t)\|_{Q(t)}. \tag{13.4}$$

Here, the reference trajectory, disturbances, and system model may be periodically time-varying. This minimization is subject to the following constraints, for all k and $t = 0, \dots, N-1$:

$$y_{\min}(t) \le y(k,t) \le y_{\max}(t) \tag{13.5}$$

$$u_{\min}(t) \le u(k,t) \le u_{\max}(t) \tag{13.6}$$

$$\delta u_{\min}(t) \le \delta u(k,t) \le \delta u_{\max}(t), \tag{13.7}$$

where $Q(t)$ is a positive semi-definite weighting matrix. The input increment is defined as

$$\delta u(k,t) = u(k,t) - u(k,t-1)$$

so that the operator $\delta(\cdot)$ represents the change in the manipulated variable between consecutive time steps within the same run.

The closed-loop system is said to achieve *asymptotically perfect tracking* if:

$$Q^{1/2}(t)(y(k,t) - r(t)) \to 0 \text{ as } k \to \infty, \qquad \forall t \in \mathcal{I}. \tag{13.8}$$

Although the design objective primarily concerns asymptotic behavior, practical implementation also demands that convergence to the steady-state limit be achieved as rapidly as possible. The objective function in (13.4) accommodates this requirement. In practice, however, factors such as model-plant mismatch, measurement noise, and process disturbances constrain the achievable rate of convergence. The remainder of this chapter examines these practical considerations in the control of batch and periodic systems.

13.2.2 Limitations of Conventional Feedback Control for Periodic Processes

Conventional feedback controllers, such as PID or DMC, have limited effectiveness in handling time-varying reference trajectories and periodic disturbances. PID controllers, in particular, tend to produce significant transient errors when the system is subjected to dynamic reference trajectories or disturbances that repeat periodically. These errors often arise from system interactions, time delays, and other dynamic characteristics that prevent aggressive tuning. As a result, the same dynamic errors recur in each run, since the controller does not adapt or correct for them over time. Although the integral action in PID control is effective in eliminating constant offsets, it does not address time-varying errors, which are typical in periodic processes.

Model-based controllers such as DMC and its variants, though more advanced, are subject to similar limitations. Owing to its predictive formulation, DMC can effectively handle time-varying reference trajectories, yet it typically interprets feedback errors arising from unmeasured disturbances or model–plant mismatch as constant or ramp-type offsets. Consequently, when model inaccuracies or time-varying disturbances are significant, DMC may still exhibit persistent dynamic tracking errors.

For both batch and periodic processes, these limitations imply that identical error patterns will persist across repeated runs when conventional feedback controllers are used. There is no mechanism to carry the feedback error information from one run to the next, meaning that the controller lacks the ability to "learn" from previous executions. These deficiencies are illustrated in the following examples.

Example 13.1: *Tracking a Periodic Reference Trajectory using PID Control.*
Consider the problem of tracking a periodic reference trajectory (with period 10) for the following 2×2 continuous-time system:

$$G(s) = \frac{e^{-s}}{s^2 + 0.254s + 0.0071} \begin{bmatrix} 0.1406s + 0.0018 & 0.2664s + 0.0082 \\ 0.0015s + 0.0014 & -0.009s - 0.0001 \end{bmatrix}. \tag{13.9}$$

The system is discretized using a sampling time of 1. It exhibits strong interaction between loops, as indicated by the relative gain array (RGA):

$$\Lambda = \begin{bmatrix} 0.478 & 0.522 \\ 0.522 & 0.478 \end{bmatrix}. \tag{13.10}$$

We aim to track the periodic reference trajectory (dashed lines in Figure 13.1) using PID control.

A multiloop PID control scheme is first attempted. Based on the RGA analysis, y_1 is paired with u_2 and y_2 with u_1. The PID parameters for each loop are initially selected using the Cohen–Coon tuning method. To improve performance, delay compensators (assuming known delays) and decouplers are incorporated to mitigate loop interactions. The resulting closed-loop performance in tracking the periodic setpoint trajectories is shown in Figure 13.1.

As the figure illustrates, significant transient errors persist and repeat in every run. These dynamic errors are not eliminated by the controller, even with additional tuning around the initial Cohen–Coon parameters. The outcome highlights the limitations of conventional PID control in handling periodic reference trajectories in interactive systems.

Example 13.2: *Tracking a Periodic Reference Trajectory using MPC*
We now revisit Example 13.1 using MPC. A model–plant mismatch is introduced by underestimating the process time delay in the controller model as 0.1, whereas the true plant delay is 1.

An MPC controller is designed in the standard manner, with the model discretized using a sampling interval of $t_s = 1$. As shown in Figure 13.2, the MPC achieves slightly better performance than the PID controller, owing to its ability to explicitly incorporate time-varying reference trajectories into the control optimization. However, in the presence of model–plant mismatch, the limitations

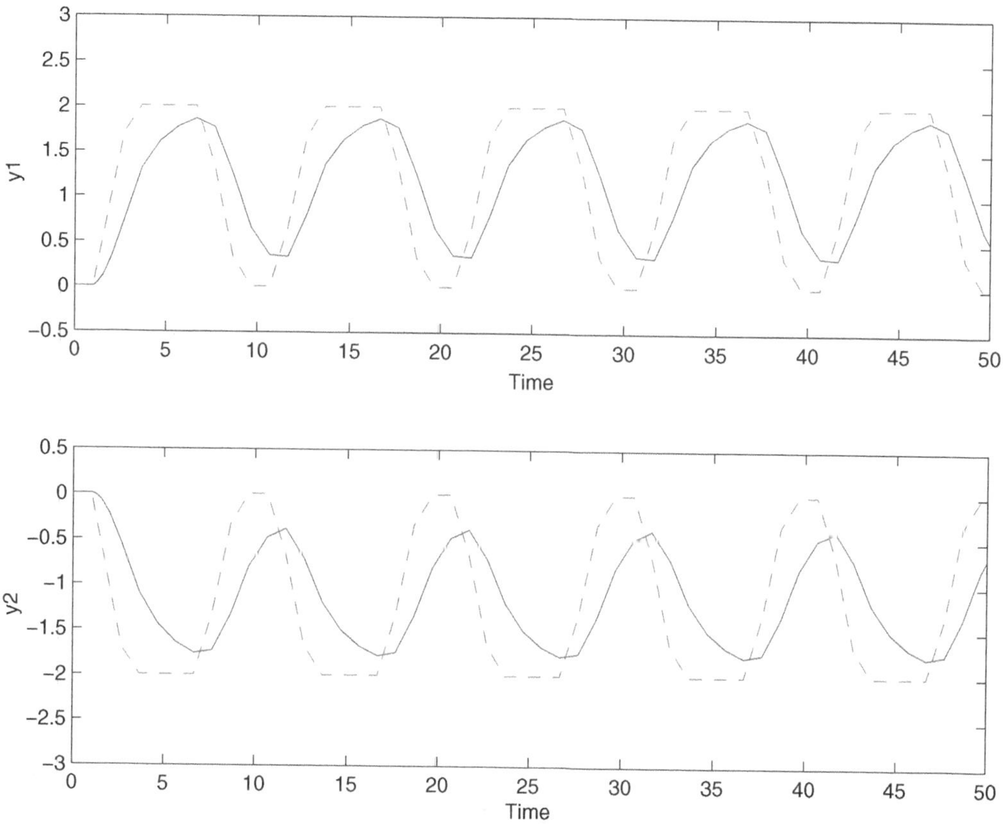

Figure 13.1 Tracking of periodic reference trajectories under PID control (dashed line: reference, solid line: output).

of MPC become apparent. The controller fails to eliminate the resulting periodic tracking errors, which reoccur consistently across successive cycles.

The mismatch between the model and the actual process produces a persistent periodic error that a conventional model-based controller cannot reject. This behavior is qualitatively similar to that observed with the PID controller, although the MPC exhibits somewhat improved transient response.

Recall from Chapter 11 that we introduced an integrator to eliminate bias in each controlled variable. Similarly, for periodic systems, a strategy is needed to compensate for the repeating error pattern caused by model mismatch.

13.3 Iterative Learning Model Predictive Control for Batch Systems

MPC possesses many features that make it attractive for process control applications, including those involving batch and periodic systems. However, a key limitation of conventional MPC is its inability to make systematic run-to-run improvements and to asymptotically reject dynamic tracking errors as operations are repeated. In other words, standard MPC lacks the mechanism to "learn" from past runs and correct for recurring dynamic errors.

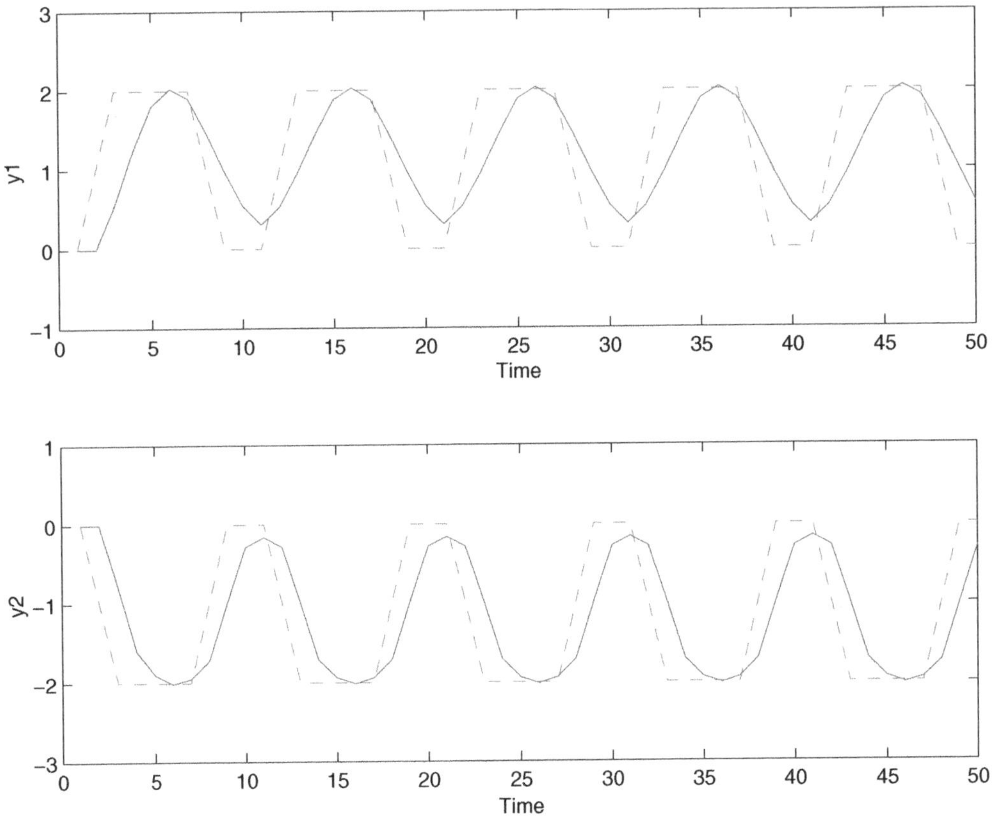

Figure 13.2 Tracking of periodic reference trajectories under DMC control with model/plant mismatch (dashed line: reference, solid line: output).

To overcome this limitation, the ideas from ILC and RC can be incorporated into the MPC framework. This integration begins with a modification of the model structure that explicitly links consecutive runs, thereby enabling the controller to exploit the repetitive nature of the system dynamics and improve performance over successive runs.

13.3.1 Notations for IL-MPC

We begin by defining some key mathematical objects that will be used repeatedly in the following developments.

Unless stated otherwise, bold lowercase letters denote "lifted" vectors—vectors that contain the sampled values of a variable over an entire run, i.e., from $t = 0$ to $t = N$. Specifically:

$$\mathbf{y}(k) \triangleq \begin{bmatrix} y(k,1) \\ \vdots \\ y(k,N) \end{bmatrix} \qquad \mathbf{r} \triangleq \begin{bmatrix} r(1) \\ \vdots \\ r(N) \end{bmatrix}$$
$$\mathbf{u}(k) \triangleq \begin{bmatrix} u(k,0) \\ \vdots \\ u(k,N-1) \end{bmatrix} \qquad \mathbf{h} \triangleq \begin{bmatrix} h(1) \\ \vdots \\ h(N) \end{bmatrix}. \tag{13.11}$$

Note that output is indexed from $t = 1$ to $t = N$ rather than from $t = 0$ to $t = N - 1$. This convention emphasizes that the terminal point of a batch ($t = N$) differs from the initial point of the next run ($t = 0$), unlike in continuous periodic systems. Since $y(k, N) = Cx(k, N) + h(N)$, this convention does not introduce any complications.

We use Δ to denote differencing with respect to the *run index*, both for lifted vectors and for individual variables:

$$\Delta \mathbf{u}(k) \triangleq \mathbf{u}(k) - \mathbf{u}(k-1) \tag{13.12}$$

$$\Delta u(k,t) \triangleq u(k,t) - u(k-1,t). \tag{13.13}$$

Recall that time differencing within a run was denoted by δ (see Eq. 13.7)

$$\delta u(k,t) = u(k,t) - u(k,t-1).$$

The following definitions of the observability matrix Π and the lifted input-output response matrix $\mathbf{G}$ will be useful for the developments that follow:

$$\Pi \triangleq \begin{bmatrix} C(1)A(0) \\ C(2)A(1)A(0) \\ \vdots \\ C(N)A(N-1)\cdots A(0) \end{bmatrix}$$
$$\mathbf{G} \triangleq \begin{bmatrix} C(1)B(0) & 0 & \cdots & 0 \\ C(2)A(1)B(0) & C(2)B(1) & \cdots & 0 \\ \vdots & \vdots & \ddots & \vdots \\ C(N)A(N-1)\cdots A(1)B(0) & \cdots & \cdots & C(N)B(N-1) \end{bmatrix}. \tag{13.14}$$

These matrices are defined for the time interval from $t = 1$ to $t = N$ and pertain to the time-varying linear system in (13.1).

13.3.2 "Run-to-run" IL-MPC Method for an Unconstrained System

For a batch system, the overall system dynamics can be expressed in terms of the run index using the concept of *lifting*:

$$\mathbf{y}(k) = \Pi x(k,0) + \mathbf{G}\mathbf{u}(k) + \mathbf{h}(k). \tag{13.15}$$

This representation is referred to as the *lifted form* of the original periodically time-varying system.

By differencing the above equation between two consecutive runs k and $k - 1$, we obtain:

$$\mathbf{y}(k) = \mathbf{y}(k-1) + \Pi\Delta x(k,0) + \mathbf{G}\Delta\mathbf{u}(k) + \Delta\mathbf{h}(k). \tag{13.16}$$

If both $x(k, 0)$ and $\mathbf{h}(k)$ are invariant across runs—i.e., the same error pattern repeats each run—the prediction simplifies to:

$$\mathbf{y}(k|k-1) = \mathbf{y}(k-1) + \mathbf{G}\Delta\mathbf{u}(k). \tag{13.17}$$

In practical situations where error patterns are not perfectly repetitive, a filtering term can be added to account for this deviation:

$$\mathbf{y}(k|k-1) = \mathbf{y}(k-1|k-2) + \mathbf{G}\Delta\mathbf{u}(k) + \mathbf{F}\big(\mathbf{y}(k-1) - \mathbf{y}(k-1|k-2)\big). \tag{13.18}$$

Here, $\mathbf{F}$ is a diagonal filter matrix, with diagonal elements between 0 (no transfer to the next run) and 1 (full transfer). This filtering mechanism enables partial or complete transfer of the $k-1$th run's error signal to the kth run, depending on how repeatable the errors are and the signal-to-noise characteristics.

A suitable objective for the control calculation is:

$$\min_{\Delta\mathbf{u}(k)} \|\mathbf{y}(k|k-1) - \mathbf{r}\|_{\mathbf{Q}}^2 + \|\Delta\mathbf{u}(k)\|_{\mathbf{R}}^2, \tag{13.19}$$

where $\mathbf{Q} = \text{blockdiag}\{Q(1), \ldots, Q(N)\}$ and $\mathbf{R} = \text{blockdiag}\{R(1), \ldots, R(N)\}$. This is a linear least-squares optimization problem, which admits an analytical solution:

$$\Delta\mathbf{u}(k) = \left(\mathbf{G}^T\mathbf{Q}\mathbf{G} + \mathbf{R}\right)^{-1}\{\mathbf{y}(k-1|k-2) + \mathbf{F}\left(\mathbf{y}(k-1) - \mathbf{y}(k-1|k-2)\right) - \mathbf{r}\}. \tag{13.20}$$

13.3.3 "Run-to-run" IL-MPC Method for a Constrained System

When constraints on inputs and outputs must be respected, the run-to-run IL-MPC formulation is modified as follows:

$$\min_{\Delta\mathbf{u}_k} \left\{\|\mathbf{y}(k|k-1) - \mathbf{r}\|_{\mathbf{Q}}^2 + \|\Delta\mathbf{u}_k\|_{\mathbf{R}}^2 + \|\varepsilon(k)\|_{\mathbf{S}}^2\right\}, \tag{13.21}$$

where $\varepsilon(k)$ is a slack variable introduced to soften the output constraints.

The constraints are expressed as linear inequalities:

$$\mathbf{B}\begin{bmatrix} \Delta\mathbf{u}(k) \\ \varepsilon(k) \end{bmatrix} \leq \mathbf{b}(k), \tag{13.22}$$

and the predicted output trajectory is given by

$$\mathbf{y}(k|k-1) = \mathbf{y}(k-1|k-2) + \mathbf{G}\Delta\mathbf{u}(k) + \mathbf{F}\left(\mathbf{y}(k-1) - \mathbf{y}(k-1|k-2)\right) \tag{13.18}$$

The matrices $\mathbf{B}$ and $\mathbf{b}(k)$ are constructed to represent the input and output constraints in terms of the decision variables. The slack variable vector $\varepsilon(k)$ allows the output constraints to be satisfied approximately when strict enforcement is infeasible.

The resulting optimization problem is a standard quadratic program (QP), for which reliable and efficient solvers are readily available.

13.3.4 Real-time-feedback IL-MPC Method for an Unconstrained System

While the run-to-run IL-MPC algorithm determines the control input trajectory for an entire run, it can also be adapted into a real-time feedback control algorithm to reject new disturbances that may arise during an ongoing batch. To do so, we reformulate the original model (13.1) to derive a periodically time-varying system structure that allows the tracking problem to be posed as a regulation problem.

We begin by defining the error trajectory

$$\mathbf{e}(k) = \mathbf{y}(k) - \mathbf{r} \tag{13.23}$$

and its time-dependent restriction

$$\mathbf{e}(k,t) \triangleq \mathbf{y}(k) - \mathbf{r} \text{ assuming } \Delta\mathbf{x}(k,j) = 0,\ j > t. \tag{13.24}$$

Here, $\mathbf{e}(k,t)$ represents the error trajectory for the kth run assuming that the state trajectory remains unchanged from the previous run for all $j > t$. Note that $\mathbf{e}(k,N) = \mathbf{e}(k)$.

By differencing the original state-space equations (13.1) for two successive runs, we obtain:

$$\begin{aligned} \Delta x(k,t+1) &= A(t)\Delta x(k,t) + B(t)\Delta u(k,t) \\ y(k,t+1) &= y(k-1,t+1) + C(t+1)\Delta x(k,t+1), \quad t = 0,\ \dots, N-1. \end{aligned} \tag{13.25}$$

From these, we derive expressions for $\mathbf{e}(k,t)$ and $\mathbf{e}(k,t+1)$

$$\mathbf{e}(k,t) = \mathbf{e}(k-1) + \begin{bmatrix} C(1)\Delta x(k,1) \\ \vdots \\ C(t)\Delta x(k,t) \\ 0 \\ \vdots \\ 0 \end{bmatrix} \tag{13.26}$$

$$\mathbf{e}(k,t+1) = \mathbf{e}(k-1) + \begin{bmatrix} C(1)\Delta x(k,1) \\ \vdots \\ C(t)\Delta x(k,t) \\ C(t+1)\Delta x(k,t+1) \\ \vdots \\ 0 \end{bmatrix}. \tag{13.27}$$

Subtracting the above gives

$$\mathbf{e}(k,t+1) = \mathbf{e}(k,t) + \mathbf{C}(t+1)\Delta x(k,t+1), \tag{13.28}$$

where

$$\mathbf{C}(t+1) \triangleq \begin{bmatrix} 0 \\ \vdots \\ 0 \\ C(t+1) \\ 0 \\ \vdots \\ 0 \end{bmatrix}$$

with $C(t+1)$ in the $(t+1)$th row and all other entries zero.

This leads to a periodically time-varying system description

$$\begin{bmatrix} \Delta x(k,t+1) \\ \mathbf{e}(k,t+1) \end{bmatrix} = \begin{bmatrix} A(t) & 0 \\ \mathbf{C}(t+1)A(t) & I \end{bmatrix} \begin{bmatrix} \Delta x(k,t) \\ \mathbf{e}(k,t) \end{bmatrix} + \begin{bmatrix} B(t) \\ C(t+1)B(t) \end{bmatrix} \Delta u(k,t) \tag{13.29}$$

$$e(k,t) = \begin{bmatrix} 0 & H(t) \end{bmatrix} \begin{bmatrix} \Delta x(k,t) \\ \mathbf{e}(k,t) \end{bmatrix}, \tag{13.30}$$

where $H(t)$ extracts the error vector at time t, denoted $e(k,t)$, from the lifted error vector $\mathbf{e}(k,t)$. This equation is initialized at the beginning of each run as follows

$$\begin{bmatrix} \Delta x(k,0) \\ \mathbf{e}(k,0) \end{bmatrix} = \begin{bmatrix} \mathbf{0} \\ \mathbf{e}(k-1) \end{bmatrix}. \tag{13.31}$$

For convenience of exposition, let us denote the above time-varying system as:

$$\begin{aligned} \bar{x}(k,t+1) &= \bar{A}(t)\bar{x}(k,t) + \bar{B}(t)\Delta u(k,t) \\ e(k,t) &= \bar{C}(t)\bar{x}(k,t), \quad t \in \mathcal{I}. \end{aligned} \tag{13.32}$$

The control objective in (13.19) can then be expressed, at time instant t, as:

$$\min_{\Delta u(k,t)} \left\{ \sum_{t=0}^{N-1} \left(\|\bar{x}(k,t)\|^2_{\bar{Q}(t)} + \|\Delta u(k,t)\|^2_{R(t)} \right) + \|\bar{x}(k,N)\|^2_{\bar{Q}_N} \right\}, \tag{13.33}$$

where

$$\bar{Q}(t) = \bar{C}^T(t)Q(t)\bar{C}(t). \tag{13.34}$$

The optimal solution is periodic and has the form:

$$\Delta u(k,t) = L(t)\bar{x}(k,t), \quad t \in \mathcal{I}. \tag{13.35}$$

The gain matrices $L(t)$ can be computed by solving a periodic Riccati difference equation backward in time, following the same procedure as for the standard LQ controller described in the chapter *Unconstrained LQ Control.*

13.3.5 Real-time-feedback-based IL-MPC for the Constrained System

For the constrained case, we take the receding horizon control approach and solve the following optimization problem at each sample time t of the kth batch:

$$\begin{aligned} &\min_{\Delta\mathcal{V},\varepsilon} \left\{ \|\mathbf{z}\|^2_{\mathbf{Q}_t} + \|\Delta\mathcal{V}\|^2_{\mathbf{R}_t} + \|\varepsilon\|^2_{\mathbf{S}_t} \right\} \\ &\text{subject to:} \quad \mathbf{B}\begin{bmatrix} \Delta\mathcal{V} \\ \varepsilon \end{bmatrix} \le \mathbf{b}. \end{aligned} \tag{13.36}$$

In this formulation

$$\Delta\mathcal{V} = \begin{bmatrix} \Delta u(k,t|k,t) \\ \vdots \\ \Delta u(k,t+m-1|k,t) \end{bmatrix}, \text{ and} \tag{13.37}$$

$$\mathbf{z} = \begin{bmatrix} e(k,t+1|k,t) \\ \vdots \\ e(k,t+p|k,t) \end{bmatrix} \text{ with } \Delta u(k,t+j|k,t) = 0 \text{ for } j \ge m, \tag{13.38}$$

where $e(k, t+i|k, t), i = 1, \ldots, p$ follow the system dynamics in (13.39):

$$\begin{aligned} \bar{x}(k, t+i|k, t) &= \bar{A}(t)\bar{x}(k, t+i-1|k, t) + \bar{B}(t)\Delta u(k, t+i-1|k, t) \\ e(k, t+i|k, t) &= \bar{C}(t)\bar{x}(k, t+i-1|k, t) \end{aligned} \tag{13.39}$$

with $\bar{x}(k, t|k, t)$ initialized with the updated current state $\bar{x}(k, t)$.

The predicted error trajectory $\mathbf{z}$ is computed over a p-step prediction horizon, assuming that control increments beyond $t + m - 1$ are zero. This trajectory can be expressed as a linear function of $\Delta\mathcal{V}$ using the system model. The matrices $\mathbf{B}$ and vectors $\mathbf{b}$ define the input and output constraints over the horizon. The slack variables ε are used to soften output constraints when necessary.

At each time t, the optimization problem minimizes tracking error and control effort over the prediction horizon while enforcing constraint admissibility. The values of p and m may be chosen independently to balance performance and computational cost.

As in standard MPC, this is a receding-horizon strategy: the optimization is solved at each sampling instant, but only the first control move $\Delta u(k, t)$ is implemented, and the remaining moves are discarded. In batch processes, both prediction and control horizons should be adjusted as the run progresses—shortened near the end—since prediction beyond the run endpoint is unnecessary.

13.4 Repetitive Model Predictive Control for Continuous Systems with Periodic Operations

We now consider continuous systems operating in a periodic manner. Unlike batch systems, where the state is reset at the beginning of each run, continuous systems retain continuity across runs—the terminal state of the current run becomes the initial state of the next. This results in a continuous run-to-run transition.

13.4.1 Notations for R-MPC

The following derivations parallel those for batch systems, with some modifications due to the continuous nature of the run-to-run transition. In particular, since the terminal state of one run directly becomes the initial state of the next, the indexing of the lifted variables differs slightly from the batch case.

As before, we use bold-face lowercase letters to denote lifted vectors that contain the sampled data of a variable over an entire run, now defined for $t = 0$ to $N - 1$:

$$\mathbf{y}(k) \triangleq \begin{bmatrix} y(k,0) \\ \vdots \\ y(k,N-1) \end{bmatrix} \qquad \mathbf{u}(k) \triangleq \begin{bmatrix} u(k,0) \\ \vdots \\ u(k,N-1) \end{bmatrix}$$

$$\mathbf{h} \triangleq \begin{bmatrix} h(0) \\ \vdots \\ h(N-1) \end{bmatrix} \qquad \mathbf{r} \triangleq \begin{bmatrix} r(0) \\ \vdots \\ r(N-1) \end{bmatrix}$$

We define Δ as the difference operator with respect to the run index:

$$\Delta\mathbf{u}(k) \triangleq \mathbf{u}(k) - \mathbf{u}(k-1) \quad \text{and} \quad \Delta u(k,t) \triangleq u(k,t) - u(k-1,t). \tag{13.40}$$

For the linear time-varying system in (13.1), the following matrices—state transition (Φ), controllability (Γ), observability (Π), and lifted impulse response ($\mathbf{G}$)—are defined over the interval $t = 0$ to $N - 1$:

$$
\begin{aligned}
\Phi &\triangleq A(N-1)A(N-2)\cdots A(0)\\
\Gamma &\triangleq \begin{bmatrix} \left(\prod_{i=N-1}^{1} A(i)\right) B(0) & \left(\prod_{i=N-1}^{2} A(i)\right) B(1) & \cdots & B(N-1) \end{bmatrix}\\
\Pi &\triangleq \begin{bmatrix} C(0) \\ C(1)A(0) \\ \vdots \\ C(N-1)A(N-2)\cdots A(0) \end{bmatrix}\\
\mathbf{G} &\triangleq \begin{bmatrix} 0 & 0 & \cdots & 0 & 0 \\ C(1)B(0) & 0 & \cdots & 0 & 0 \\ C(2)A(1)B(0) & C(2)B(1) & \cdots & 0 & 0 \\ \vdots & \vdots & \ddots & \vdots & \vdots \\ C(N-1)\left(\prod_{i=N-2}^{1} A(i)\right) B(0) & \cdots & \cdots & C(N-1)B(N-2) & 0 \end{bmatrix}
\end{aligned}
\tag{13.41}
$$

These lifted constructs are foundational for the formulation of repetitive MPC in continuous systems with periodic operation.

13.4.2 "Run-to-run" R-MPC Method for an Unconstrained System

Using the matrices defined earlier, the lifted model of the system is given by:

$$
\begin{aligned}
x(k+1,0) &= \Phi x(k,0) + \Gamma \mathbf{u}(k)\\
\mathbf{y}(k) &= \Pi x(k,0) + \mathbf{G}\mathbf{u}(k) + \mathbf{h}.
\end{aligned}
\tag{13.42}
$$

Note that the cost is defined over an infinite sequence of future runs because the control input in one run influences the initial condition of the next, due to the continuous run-to-run state transition.

In the absence of constraints, this is a standard infinite-horizon LQ problem that can be efficiently solved. To facilitate this, we rewrite the lifted model in terms of run-to-run increments:

$$
\begin{aligned}
\Delta x(k+1,0) &= \Phi \Delta x(k,0) + \Gamma \Delta \mathbf{u}(k)\\
\mathbf{y}(k) &= \mathbf{y}(k-1) + \Pi \Delta x(k,0) + \mathbf{G}\Delta \mathbf{u}(k).
\end{aligned}
\tag{13.43}
$$

To formulate the problem in a regulation framework, we augment the state with the tracking error:

$$
\begin{aligned}
\begin{bmatrix} \Delta x(k+1,0) \\ \mathbf{e}(k) \end{bmatrix} &= \underbrace{\begin{bmatrix} \Phi & 0 \\ \Pi & I \end{bmatrix}}_{\tilde{\Phi}} \begin{bmatrix} \Delta x(k,0) \\ \mathbf{e}(k-1) \end{bmatrix} + \underbrace{\begin{bmatrix} \Gamma \\ \mathbf{G} \end{bmatrix}}_{\tilde{\Gamma}} \Delta \mathbf{u}(k)\\
\mathbf{e}(k) &= \underbrace{\begin{bmatrix} \Pi & I \end{bmatrix}}_{\tilde{\Pi}} \begin{bmatrix} \Delta x(k,0) \\ \mathbf{e}(k-1) \end{bmatrix} + \mathbf{G}\Delta \mathbf{u}_k,
\end{aligned}
\tag{13.44}
$$

where $\mathbf{e}(k) \triangleq \mathbf{y}(k) - \mathbf{r}$.

Letting

$$\begin{aligned} \mathbf{z}(k+1) &= \bar{\Phi}\mathbf{z}(k) + \bar{\Gamma}\Delta\mathbf{u}(k) \\ \mathbf{e}(k) &= \bar{\Pi}\mathbf{z}(k) + \mathbf{G}\Delta\mathbf{u}(k) \end{aligned} \tag{13.45}$$

the optimal control problem becomes:

$$\min_{\Delta\mathbf{u}(k)} \sum_{i=1}^{\infty} \left(\left\| \begin{bmatrix} \mathbf{z}(k) \\ \Delta\mathbf{u}(k) \end{bmatrix} \right\|_{\mathbf{Q}_{\text{ex}}}^{2} \right) \tag{13.46}$$

with

$$\mathbf{Q}_{\text{ex}} = \begin{bmatrix} \bar{\Pi}^T\mathbf{Q}\bar{\Pi} & \bar{\Pi}^T\mathbf{Q}\mathbf{G} \\ \mathbf{G}^T\mathbf{Q}\bar{\Pi} & \mathbf{G}^T\mathbf{Q}\mathbf{G} + \mathbf{R} \end{bmatrix} \tag{13.47}$$

This is a classical LQ regulator problem. The optimal solution is obtained by solving the algebraic Riccati equation (ARE), as described in Chapter 8. The resulting control law takes the form:

$$\Delta\mathbf{u}(k) = \bar{\mathbf{L}}\mathbf{z}(k) = \bar{\mathbf{L}}_1\Delta x(k, 0) + \bar{\mathbf{L}}_2\mathbf{e}(k-1). \tag{13.48}$$

The control input for the entire run can be computed at its start using the above expression.

Example 13.3: *Application of Infinite-Horizon Unconstrained RMPC to the System in Example 13.1.*

To demonstrate the robustness of RMPC, we introduce model-plant mismatch by assuming the controller model has a delay of 0.1, whereas the true plant delay is 1.0.

An infinite-horizon unconstrained RMPC controller was designed by solving the equivalent linear-quadratic (LQ) problem, resulting in a state-feedback control law. Figure 13.3 illustrates the run-to-run improvement achieved over five runs when the RMPC controller was applied to the unconstrained system.

13.4.3 "Run-to-run" R-MPC Methods for the Constrained System

The algebraic Riccati equation approach used in the unconstrained case is not applicable when constraints are present. Instead, we solve the following finite-horizon optimization problem at each run k:

$$\min_{\Delta\mathbf{u}(i)} \left\{ \sum_{i=k}^{k+p-1} \|\mathbf{e}(i)\|_{\mathbf{Q}}^2 + \|\Delta\mathbf{u}(i)\|_{\mathbf{R}}^2 + \|\varepsilon(i)\|_{\mathbf{S}}^2 \right\} \tag{13.49}$$

subject to the constraint:

$$\mathbf{B} \begin{bmatrix} \Delta\mathbf{u}(i) \\ \varepsilon(i) \end{bmatrix} \leq \mathbf{b}(i), \, i = k, \, \dots, k+p-1 \tag{13.50}$$

Here, $\mathbf{b}(i)$ and $\mathbf{B}$ are defined to express input and output constraints in linear inequality form. The slack variable $\varepsilon(i)$ is introduced to soften the output constraints, allowing limited constraint violations when necessary. This formulation results in a quadratic programming problem, for which efficient solvers are readily available.

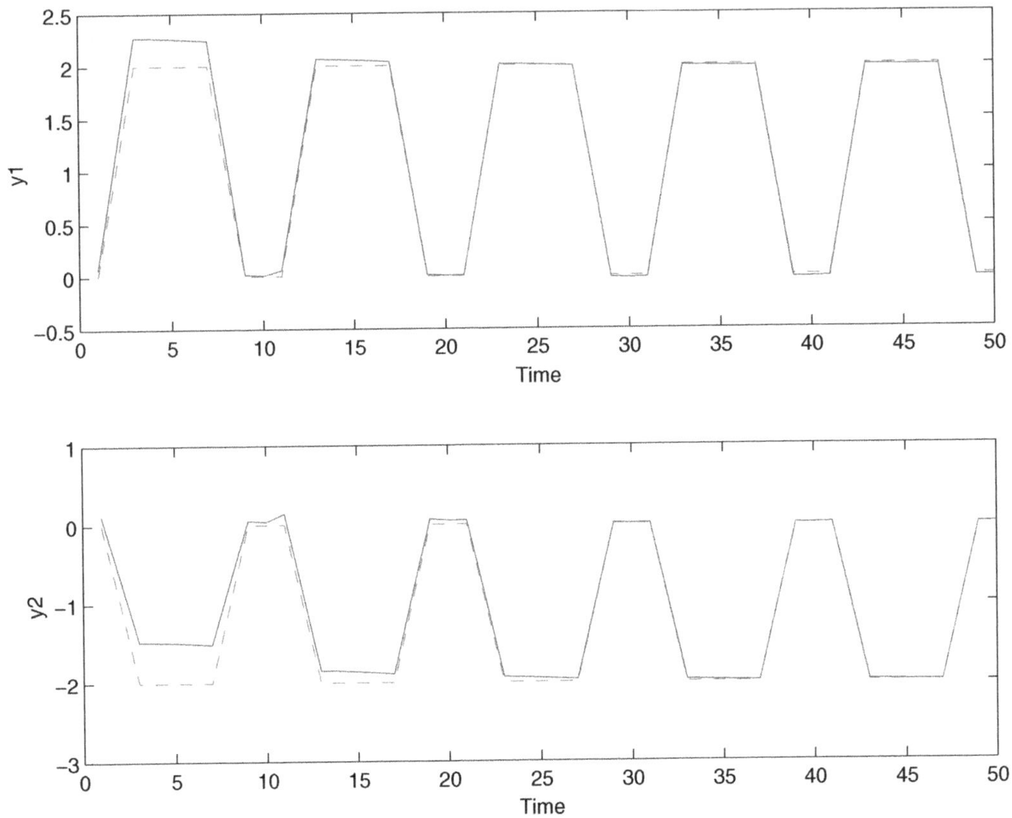

Figure 13.3 Performance of unconstrained RMPC applied to an unconstrained system with model-plant mismatch (dashed line: reference, solid line: output).

It is generally difficult to determine *a priori* the prediction horizon p required to ensure satisfactory performance. Moreover, achieving such performance may require a large value of p, leading to a high-dimensional optimization problem. An alternative approach is to design suitable terminal weights and terminal constraints so that, under mild assumptions, the resulting finite-horizon optimization approximates the behavior of the infinite-horizon problem.

To this end, consider solving the following finite-horizon optimization problem prior to each kth run to determine $\Delta\mathbf{u}(k)$:

$$\min_{\Delta\mathbf{u}(k)} \left\{ \sum_{i=k}^{k+m-1} \|\mathbf{e}(i)\|_{\mathbf{Q}}^2 + \|\Delta\mathbf{u}(i)\|_{\mathbf{R}}^2 + \|\varepsilon(i)\|_{\mathbf{S}}^2 \right\} + \|\mathbf{z}(k+m)\|_{\bar{\mathbf{M}}}^2 \tag{13.51}$$

subject to

$$\mathbf{B}\begin{bmatrix} \Delta\mathbf{u}(i) \\ \varepsilon_i \end{bmatrix} \leq \mathbf{b}(i), \quad i = k, \ldots, k+m-1 \tag{13.52}$$

and the system dynamics in (13.45), along with appropriately chosen terminal constraints on $\mathbf{z}(k+m)$.

The terminal penalty $||\mathbf{z}(k+m)||^2_{\bar{\mathbf{M}}}$ is introduced to capture the cost beyond the finite horizon, and the terminal constraint ensures recursive feasibility and stability. The constraints are enforced only over the horizon from run k to $k+m-1$.

Central to the above approach is the appropriate selection of the terminal weighting matrix $\bar{\mathbf{M}}$ and the terminal constraint. The following strategies, previously discussed in the context of conventional MPC, can be adapted here. We summarize the main ideas below without repeating the detailed derivations.

- *Strategy 1: Hard Terminal Constraint.*
  Choose $\bar{\mathbf{M}}$ to be sufficiently large, which is equivalent to enforcing the terminal equality constraint:

$$\mathbf{z}(k+m) = 0 \tag{13.53}$$

  This ensures that the augmented state returns to the origin at the end of the horizon.
- *Strategy 2: Autonomous Evolution Beyond the Horizon.*
  The above constraint may be overly restrictive, potentially leading to infeasibility or suboptimality. An alternative is to assume that no further input changes are made beyond the horizon, i.e., $\Delta\mathbf{u}(i) = 0$ for $i \geq k+m$. Under this assumption, the state evolves autonomously according to:

$$\mathbf{z}(i+1) = \bar{\Phi}\mathbf{z}(i) \tag{13.54}$$

  The terminal weight $\bar{\mathbf{M}}$ and terminal constraint should be chosen to approximate the infinite-horizon behavior under this autonomous system.
- *Strategy 3: Infinite-horizon Feedback Approximation.*
  Instead of assuming zero input changes, one can assume the input follows the infinite-horizon feedback policy $\Delta\mathbf{u}(k+i) = -\bar{\mathbf{L}}\mathbf{z}(k+i)$, where $\bar{\mathbf{L}}$ is the unconstrained infinite-horizon LQ gain matrix derived earlier. The resulting closed-loop system evolves as:

$$\mathbf{z}(i+1) = (\bar{\Phi} - \bar{\mathbf{L}}\Gamma)\,\mathbf{z}(i) \tag{13.55}$$

  Appropriate terminal weights and constraints can then be selected based on this feedback evolution.

Under reasonable assumptions, the closed-loop asymptotic stability of the augmented system can be established for all three strategies. This, in turn, guarantees *asymptotically perfect tracking*.

Example 13.4: *Comparison of Unconstrained and Constrained RMPC for a Plant with Input Constraints ($u_1 \leq 2$, $u_2 \leq 1$) Using the System in Example 13.3.*
We consider the same plant used in Example 13.3, now subject to input constraints $u_1 \leq 2$ and $u_2 \leq 1$. The constrained RMPC controller was implemented by solving a QP at the beginning of each run, using a prediction and control horizon of five runs.

As shown in Figure 13.4, the constrained RMPC still achieves run-to-run integration and converges to perfect tracking performance, despite the input limitations.

13.4.4 Real-time-feedback-based R-MPC Methods for the Unconstrained System

A real-time feedback control algorithm adjusts the input at every sampling time. We now derive the real-time R-MPC formulation for periodically time-varying systems, following the same approach as in real-time IL-MPC.

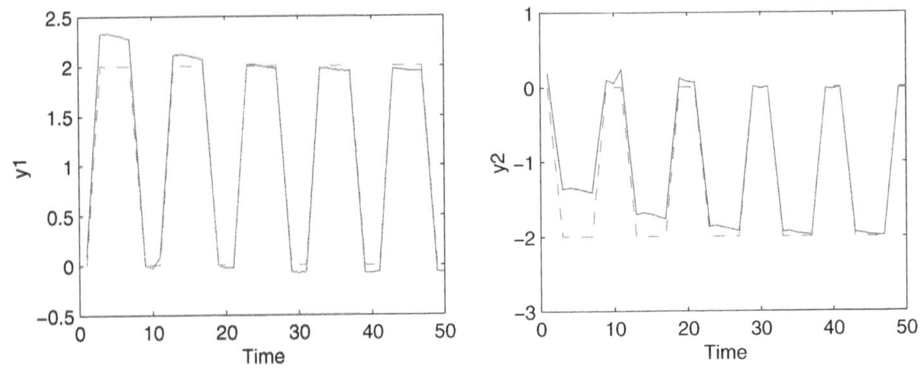

Figure 13.4 Performance of constrained "run-to-run" RMPC when applied to the constrained system with input constraints of $u_1 \leq 2$, $u_2 \leq 1$ (dashed line: reference, solid line: output).

The system evolution is described as:

$$\begin{bmatrix} \Delta x(k,t+1) \\ \mathbf{e}(k,t+1) \end{bmatrix} = \begin{bmatrix} A(t) & 0 \\ \mathbf{C}(t+1)A(t) & I \end{bmatrix} \begin{bmatrix} \Delta x(k,t) \\ \mathbf{e}(k,t) \end{bmatrix} + \begin{bmatrix} B(t) \\ C(t+1)B(t) \end{bmatrix} \Delta u(k,t)$$
$$e(k,t) = \begin{bmatrix} 0 & H(t) \end{bmatrix} \begin{bmatrix} \Delta x(k,t) \\ \mathbf{e}(k,t) \end{bmatrix}, \quad t \in \mathcal{I} \tag{13.56}$$

The key distinction from the batch case is the run-to-run transition, which is continuous

$$\begin{bmatrix} \Delta x(k+1,0) \\ \mathbf{e}(k+1,0) \end{bmatrix} = \begin{bmatrix} \Delta x(k,N) \\ \mathbf{e}(k,N) \end{bmatrix} \tag{13.57}$$

Let us represent the above time-varying system as

$$\bar{x}(k,t+1) = \bar{A}(t)\bar{x}(k,t) + \bar{B}(t)\Delta u(k,t)$$
$$e(k,t) = \bar{C}(t)\bar{x}(k,t), \quad t \in \mathcal{I} \tag{13.58}$$

with the transition between runs given by:

$$\bar{x}(k+1,0) = \bar{x}(k,N)$$
$$= \bar{A}(N-1)\bar{x}(k,N-1) + \bar{B}(N-1)\Delta u(k,N-1)$$

The associated cost function, analogous to (13.46), is

$$\min_{\Delta u(k,t)} \left\{ \sum_{k=1}^{\infty} \sum_{t=0}^{N-1} \left(\|\bar{x}(k,t)\|^2_{\bar{Q}(t)} + \|\Delta u(k,t)\|^2_{R(t)} \right) \right\} \tag{13.59}$$

where

$$\bar{Q}(t) = \bar{C}^T(t)Q(t)\bar{C}(t) \tag{13.60}$$

The optimal control law is periodic in t and takes the form:

$$\Delta u(k,t) = L(t)\bar{x}(k,t), t \in \mathcal{I} \tag{13.61}$$

The periodic solution can be obtained using the following procedure:

- **Step 1:** "Lift" the periodic system in (13.58), analogous to the lifting in (13.42), to obtain:

$$\begin{aligned} \bar{x}(k+1,0) &= \bar{\Phi}\bar{x}(k,0) + \bar{\Gamma}\Delta\mathbf{u}(k) \\ \mathbf{e}(k) &= \bar{\Pi}\bar{x}(k,0) + \mathbf{G}\Delta\mathbf{u}(k) \end{aligned} \tag{13.62}$$

Here, $\bar{\Phi}$, $\bar{\Gamma}$, and $\bar{\Pi}$ are the lifted matrices for the linear system defined by $(\bar{A}(t), \bar{B}(t), \bar{C}(t))$, constructed analogously to Φ, Γ and Π in (13.41).

- **Step 2:** Formulate the infinite-horizon quadratic cost as

$$\min_{\Delta\mathbf{u}(i)} \sum_{i=k+1}^{\infty} \left\{ \|\mathbf{e}(i)\|_{\mathbf{Q}}^2 + \|\Delta\mathbf{u}(i)\|_{\mathbf{R}}^2 = \left\| \begin{bmatrix} \bar{x}(i,0) \\ \Delta\mathbf{u}(i) \end{bmatrix} \right\|_{\mathbf{Q}_{ex}}^2 \right\} \tag{13.63}$$

$$\text{where, } Q_{ex} = \begin{bmatrix} \bar{\Pi}^T\mathbf{Q}\bar{\Pi} & \bar{\Pi}^T\mathbf{Q}\mathbf{G} \\ \mathbf{G}^T\mathbf{Q}\bar{\Pi} & \mathbf{G}^T\mathbf{Q}\mathbf{G} + \mathbf{R} \end{bmatrix} \tag{13.64}$$

This is equivalent to the optimization problem in (13.46). Solving the associated algebraic Riccati equation yields the optimal infinite-horizon cost:

$$\bar{x}(k+1,0)^T\bar{\mathbf{M}}_{LQ}\bar{x}(k+1,0) \tag{13.65}$$

where $\bar{\mathbf{M}}_{LQ}$ is the solution to the Riccati equation.

- **Step 3** Using $\bar{\mathbf{M}}_{LQ}$ from Step 2, we solve the finite-horizon problem for one period:

$$\begin{aligned} \min_{\Delta u(k,t)} \left\{ \sum_{t=0}^{N-1} \left(\|\bar{x}(k,t)\|_{\bar{Q}(t)}^2 + \|\Delta u(k,t)\|_{\bar{R}(t)}^2 \right) \right\} + \\ \bar{x}(k+1,0)^T\bar{\mathbf{M}}_{LQ}\bar{x}(k+1,0) \end{aligned} \tag{13.66}$$

The optimal control law over one period can then be computed by solving the associated Riccati difference equation backward in time. This yields:

$$\Delta u(k,t) = L(t)\bar{x}(k,t), \quad t \in \mathcal{I} \tag{13.67}$$

Note that the control law is independent of k, reflecting periodic steady-state behavior.

13.4.5 Real-time-feedback-based R-MPC Methods for the Constrained System

For the constrained case, an analytical solution to the infinite-horizon problem is not available. Therefore, as with IL-MPC, a suboptimal but practical strategy is to solve a QP online at each sampling time. Specifically, at each time step t of the kth batch, we propose solving the following optimization problem:

$$\begin{aligned} &\min_{\Delta\mathcal{V},\varepsilon} \left\{ \|\mathbf{z}\|_{\mathbf{Q}_t}^2 + \|\Delta\mathcal{V}\|_{\mathbf{R}_t}^2 + \|\varepsilon\|_{\mathbf{S}_t}^2 \right\} \\ &\text{such that } \mathbf{B} \begin{bmatrix} \Delta\mathcal{V} \\ \varepsilon \end{bmatrix} \le \mathbf{b} \end{aligned} \tag{13.68}$$

where

$$\Delta\mathcal{V} = \begin{bmatrix} \Delta u(k,t|k,t) \\ \vdots \\ \Delta u(k,t+m-1|k,t) \end{bmatrix} \tag{13.69}$$

$$\mathbf{z} = \begin{bmatrix} e(k,t+1|k,t) \\ \vdots \\ e(k,t+p|k,t) \end{bmatrix} \tag{13.70}$$

As before, $e(k,t+i|k,t)$ follow the system dynamics in (13.58):

$$\begin{aligned} \bar{x}(k,t+i|k,t) &= \bar{A}(t)\bar{x}(k,t+i-1|k,t) + \bar{B}(t)\Delta u(k,t+i-1|k,t) \\ e(k,t+i|k,t) &= \bar{C}(t)\bar{x}(k,t+i-1|k,t) \end{aligned} \tag{13.71}$$

with $\bar{x}(k,t|k,t)$ initialized with the updated current state $\bar{x}(k,t)$.

At each time t, this formulation considers the output behavior over a p-step prediction horizon and optimizes the control using m future input moves. The inequality constraint reflects both input and output constraints within the prediction window.

As in conventional MPC, this is a receding horizon optimization. Only the first control move $\Delta u(k,t)$ from the computed optimal sequence is implemented, and the remaining values are discarded.

Unlike batch systems, where the horizon shrinks as the end of the run approaches, here the prediction and control horizons should remain fixed. To handle cases where $t + p \geq N$, terms such as $e(k,t+p)$ should be interpreted as $e(k+1,t+p-N)$ to continue across runs.

To rigorously account for the influence of future terms beyond the prediction horizon and ensure stability, a terminal penalty on the final predicted state should be included. Following the same rationale used in Strategies 1–3 for run-to-run R-MPC, we set $p = m$ and include a penalty term of the form:

$$\| \bar{x}(k,t+m|k,t) \|_{\bar{\mathbf{M}}(t)}$$

where $\bar{\mathbf{M}}(t)$ is chosen such that:

$$\| \bar{x}(k,t+m)\|^2_{\bar{\mathbf{M}}(t)} = \sum_{j=t+m}^{\infty} \| e(k,j|k,t)\|^2_{Q(j)} + \| \Delta u(k,j|k,t)\|^2_{R(j)} \tag{13.72}$$

under the same assumptions made for the unconstrained case.

In addition, the following terminal constraint should be imposed:

$$\bar{x}(k,t+m|k,t) \in \mathcal{Z}(t), \tag{13.73}$$

where $\mathcal{Z}(t)$ is an output admissible set constructed to ensure constraint satisfaction beyond time $t+m$.

Example 13.5: *Constrained RMPC for the system in Example 13.3 with input rate constraints* $|\Delta u_i| \leq 0.1$.
We evaluate the performance of constrained RMPC for a system with input rate constraints $\Delta u_1, \Delta u_2 \leq 0.1$. At the beginning of each run, a QP was solved with prediction and control

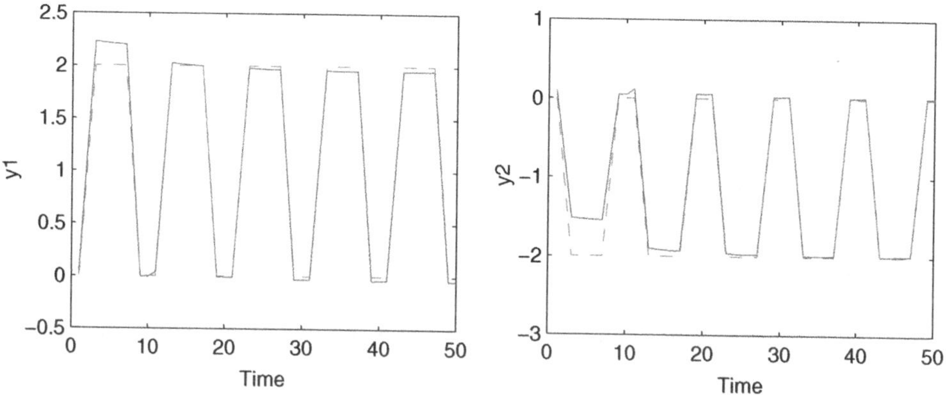

Figure 13.5 Performance of constrained real-time feedback RMPC when applied to the constrained (slew rate constraints of $\Delta u_1 \leq 0.1$, $\Delta u_2 \leq 0.1$) system (dashed line: reference, solid line: output).

horizons of five runs. To approximate the infinite-horizon solution, the terminal state was set to zero (Strategy1). Only the first control move was implemented, and the optimization was repeated at the start of the next run using updated state estimates based on the tracking error. The constrained RMPC computes optimal input profiles that satisfy the rate constraints. As shown in Figure 13.5, although some initial tracking errors are observed, the controller achieves near-perfect tracking within three runs.

13.5 Future Outlook

This chapter has focused exclusively on the deterministic formulation of repetitive MPC. However, the problem can also be approached within a stochastic framework. One motivation for doing so is that not all variations are repetitive across runs; for instance, measurement noise and transient disturbances may be specific to a given run and not carry over to future runs. The stochastic setting enables more flexible representations of disturbances and facilitates optimal filtering of feedback signals. This, in turn, allows for noise attenuation and online prediction of disturbances that have not appeared in prior runs.

Moreover, the current formulations treat model errors as additive disturbances, which may not be an accurate or sufficient representation. Developing more realistic descriptions of model uncertainty—such as bounded parameter variations or stochastic parameter distributions—along with the corresponding RC methodologies and supporting theory, represents an important direction for future research.

Another simplifying assumption made in this chapter is that the run length is fixed and system trajectories remain close to nominal profiles. In practice, however, run lengths often vary and may even be adjustable decision variables. Relaxing the fixed-run-length assumption would likely require the introduction of logical constraints, leading to mixed-integer optimization problems. A key conceptual challenge in this context is how to meaningfully embed information from previous runs when run lengths are no longer constant.

13.6 Exercises

1. A batch reactor operates for a fixed batch duration of $N = 20$ sampling intervals. The simplified discrete-time, linear state-space model is:

$$x(k, t+1) = \begin{bmatrix} 1 & 1 \\ 0 & 0.9 \end{bmatrix} x(k, t) + \begin{bmatrix} 0 \\ 0.1 \end{bmatrix} u(k, t), \tag{13.74}$$

$$y(k, t) = \begin{bmatrix} 1 & 0 \end{bmatrix} x(k, t) \tag{13.75}$$

Here k is the batch run index and $t(= 0, \cdots, N-1)$ is the time index within the run. In addition:

- $x_1(k, t)$ = product concentration (mol/L)
- $x_2(k, t)$ = rate of concentration change (mol/L per sample)
- $u(k, t)$ = heating rate (kJ/min) — *manipulated input*
- $y(k, t)$ = measured product concentration

The goal is to reach the *target concentration* $y_{\text{target}} = 1.0$ mol/L at the end of each batch ($t = N$) while respecting process constraints.
Constraints are given as:

$$\begin{aligned} 0\ \text{kJ/min} &\leq u(k, t) \leq 5\ \text{kJ/min} \\ 0\ \text{mol/L} &\leq y(k, t) \leq 1.2\ \text{mol/L} \end{aligned}$$

The iterative MPC cost over the batch horizon is:

$$J_N(k) = \sum_{t=0}^{N-1} \left[w_y (y(k, t) - r(t))^2 + w_u\, \Delta u(k, t)^2 \right] \tag{13.76}$$

where:

- $r(t)$ is the desired trajectory (e.g., a smooth ramp from 0 to y_{target})
- $\Delta u(k, t) = u(k, t) - u(k-1, t)$
- $w_y = 10$, $w_u = 0.1$

Before the start of batch run k, the state and input trajectory from run $k-1$ should be used to improve tracking in batch run k. Formulate and solve a standard constrained MPC problem over the full batch horizon $N = 20$, starting from $x(0) = [0\ 0]^T$. Then, run the iterative MPC for $k = 1, \ldots, 10$ batches. Plot $y(k, t)$ for each batch and compare with $r(t)$.
For each batch run k, report:

- Final concentration $y(k, N)$
- Total cost $J_N(k)$
- Maximum constraint violation (if any)

2. A single-input single-output (SISO) discrete-time plant is given by:

$$x(k+1) = \begin{bmatrix} 1 & 1 \\ 0 & 1 \end{bmatrix} x(k) + \begin{bmatrix} 0 \\ 1 \end{bmatrix} u(k), \tag{13.77}$$

$$y(k) = \begin{bmatrix} 1 & 0 \end{bmatrix} x(k) \tag{13.78}$$

The plant is operated in a *repetitive task* environment, where each run consists of $N = 10$ time steps, and the reference trajectory $r(t)$ is identical from run to run. The following are given:

- Initial state for the first run: $x(0) = \begin{bmatrix} 0 \\ 0 \end{bmatrix}$.
- Reference trajectory for each run:

$$r(t) = \sin\left(\frac{\pi t}{5}\right), \quad t = 0, 1, \ldots, 9 \tag{13.79}$$

- Input constraint: $|u(k)| \leq 2$
- Output constraint: $|y(k)| \leq 2$
- Cost function for a single run:

$$J_N = \sum_{t=0}^{N-1} \left[(y(t) - r(t))^2 + 0.1\, u(t)^2\right] \tag{13.80}$$

Formulate and solve a standard finite-horizon constrained MPC problem, ignoring repetitive improvements. Then, formulate the R-MPC and compare the performance with the conventional MPC. Run the R-MPC for 10 runs ($j = 1, \cdots, 10$), and plot

- Output $y(t)$ vs. reference $r(t)$ for each run
- Input $u(t)$ for each run
- Cost J_N per run

Compare the tracking error and total cost across the three runs. Discuss whether performance improved with repetition.

A

Review of Linear Transformation

In this chapter, we review the fundamentals of vector spaces and linear transformations. We begin with general concepts before focusing on the specific framework provided by linear algebra. To help bridge the gap between abstract theory and practical application, we will provide concrete examples wherever possible. However, we will not cover computational aspects of matrix algebra in this chapter. For those topics, readers are referred to standard references such as *Matrix Computations* by Golub and Van Loan (Golub and Van Loan, 1996).

A.1 Vector Space

A.1.1 Definition

A *vector space* is defined by an underlying scalar field (typically the real or complex numbers) and a set of elements called vectors, which are ordered collections of scalars from that field. For example, the n-dimensional real vector space, denoted $\mathbb{R}^n$, consists of vectors that are ordered groups of n real numbers. A typical vector in $\mathbb{R}^n$ can be represented as:

$$\mathbf{x} \in \mathbb{R}^n : \mathbf{x} = \begin{bmatrix} x_1 \\ \vdots \\ x_n \end{bmatrix}, \quad x_1, \ldots, x_n \in \Re. \tag{A.1}$$

Throughout this chapter, we adopt the following notation convention: boldface letters (e.g., $\mathbf{x}$) denote vectors, while regular font (e.g., x_1) is used for scalar components.

Geometrically, a vector in $\mathbb{R}^n$ can be viewed as a point in n-dimensional Euclidean space, where its components represent the coordinates of that point. This concept is illustrated in Figure A.1.

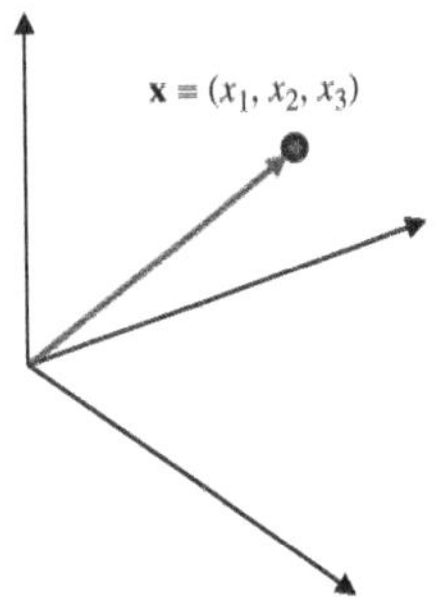

Figure A.1 Visualization of a vector in Euclidean space.

We define vector addition and scalar multiplication in $\mathbb{R}^n$ (or $\mathbb{C}^n$) as follows: Let $\mathbf{x}, \mathbf{y} \in \mathbb{R}^n$ (or $\mathbb{C}^n$) and $c \in \mathbb{R}$ (or $\mathbb{C}$). Then,

$$\mathbf{x} = \begin{bmatrix} x_1 \\ \vdots \\ x_n \end{bmatrix}; \quad \mathbf{y} = \begin{bmatrix} y_1 \\ \vdots \\ y_n \end{bmatrix}$$

$$\Downarrow$$

$$\mathbf{x} + \mathbf{y} = \begin{bmatrix} x_1 + y_1 \\ \vdots \\ x_n + y_n \end{bmatrix}; \quad c\mathbf{x} = \begin{bmatrix} cx_1 \\ \vdots \\ cx_n \end{bmatrix}.$$

With these operations, $\mathbb{R}^n$ (or $\mathbb{C}^n$) satisfies the properties of a vector space (or linear space). In particular, the following vector space axioms hold:

$$\mathbf{x} + \mathbf{y} \in \mathbb{R}^n \text{ (or } \mathbb{C}^n\text{)}, \quad c\mathbf{x} \in \mathbb{R}^n \text{ (or } \mathbb{C}^n\text{)}, \quad c(\mathbf{x} + \mathbf{y}) = c\mathbf{x} + c\mathbf{y}$$

$$(c_1 + c_2)\mathbf{x} = c_1\mathbf{x} + c_2\mathbf{x}, \quad (c_1c_2)\mathbf{x} = c_1(c_2\mathbf{x}).$$

These properties form the algebraic foundation of linear spaces and are central to the development of linear algebra.

A.1.2 Dimension of a Vector Space

Suppose $\mathbf{x_1}, \ldots, \mathbf{x_k} \in \mathbb{R}^n$. Then, the *span* of these vectors is defined as:

$$\text{Span}\{\mathbf{x_1}, \ldots, \mathbf{x_k}\} = \{c_1\mathbf{x_1} + \cdots + c_k\mathbf{x_k} : \quad c_1, \ldots, c_k \in \mathbb{R}\}.$$

The span of a set of vectors refers to the set of all linear combinations of those vectors. For example, if $\mathbf{x_1}, \mathbf{x_2}, \mathbf{x_3} \in \mathbb{R}^3$, then

$$\begin{aligned} \text{span}\{\mathbf{x_1}\} &= \text{a line (through the origin)} \\ \text{span}\{\mathbf{x_1}, \mathbf{x_2}\} &= \text{a plane (if the vectors are not collinear)} \\ \text{span}\{\mathbf{x_1}, \mathbf{x_2}, \mathbf{x_3}\} &= \text{3-dimensional space (if the vectors are not coplanar).} \end{aligned}$$

In general, a vector space is called *finite-dimensional* if there exists a finite set of vectors that spans the entire space. The minimum number of vectors required to span the space is called the *dimension* of the vector space.

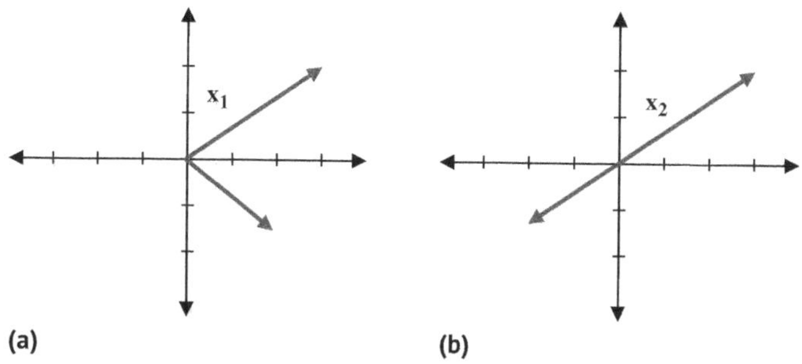

Figure A.2 Illustration of linear independence of vectors in two-dimensional space; $\mathbf{x}_1$ and $\mathbf{x}_2$ are: (a) linearly independent; (b) linearly dependent.

A.1.3 Linear Independence of Vectors

$\mathbf{x}_1, \ldots, \mathbf{x}_k \in \mathcal{X}$ are said to be *linearly independent* if the only solution to the equation

$$\sum_{i=1}^{k} \alpha_i \mathbf{x}_i = 0 \qquad \text{is } \alpha_i = 0, \forall i \in [1, k].$$

That is, $\mathbf{x}_1, \ldots, \mathbf{x}_k \in \mathcal{X}$ are linearly independent if

$$\sum_{i=1}^{k} \alpha_i \mathbf{x}_i = 0 \quad \Longrightarrow \quad \alpha_1 = \ldots, \alpha_k = 0 \tag{A.2}$$

This is illustrated pictorially in Figure A.2.

The following statements are true:

- If $\mathbf{x}_1, \cdots, \mathbf{x}_k \in \mathbb{R}^n$ are linearly independent, then for every vector $\mathbf{x} \in \text{span}\{\mathbf{x}_1, \ldots, \mathbf{x}_k\}$, there exists a unique set of scalars, $\alpha_1, \cdots, \alpha_k \in \Re$ such that

  $$\mathbf{x} = \alpha_1 \mathbf{x}_1 + \cdots + \alpha_k \mathbf{x}_k$$

- If $\mathbf{x}_1, \ldots, \mathbf{x}_k \in \mathbb{R}^n$ and $k > n$, then the set $\{\mathbf{x}_1, \ldots, \mathbf{x}_k\}$ is *linearly dependent.*

Readers may attempt to provide formal proofs for the above statements.

A.1.4 Basis

If $\mathbf{x}_1, \ldots, \mathbf{x}_n \in \mathbb{R}^n$ are linearly independent, then they form a *basis* of $\mathbb{R}^n$ because

$$\forall \mathbf{x} \in \mathbb{R}^n, \exists \text{ unique } \alpha_1, \ldots, \alpha_n \in \mathbb{R} \text{ such that } \mathbf{x} = \sum_{i=1}^{n} \alpha_i \mathbf{x}_i.$$

The vectors

$$\mathbf{e}_1 \stackrel{\Delta}{=} \begin{bmatrix} 1 \\ 0 \\ \vdots \\ 0 \end{bmatrix} \cdots \mathbf{e}_n \stackrel{\Delta}{=} \begin{bmatrix} 0 \\ 0 \\ \vdots \\ 1 \end{bmatrix}$$

are called the *natural basis* (or standard basis) of $\mathbb{R}^n$.

A.1.5 Subspace

Suppose $\mathcal{L} \subset \mathcal{X}$, where $\mathcal{X}$ is a linear (vector) space. If

$$\forall \mathbf{x}_1, \mathbf{x}_2 \in \mathcal{L} \qquad \text{and } \forall \alpha_1, \alpha_2, \qquad \alpha_1\mathbf{x}_1 + \alpha_2\mathbf{x}_2 \in \mathcal{L}$$

then, $\mathcal{L}$ is called a *subspace* of $\mathcal{X}$.

Accordingly,

- $\mathbb{R}^1 \subset \mathbb{R}^2$ is a subspace of $\mathbb{R}^2$.
- $\mathbb{R}^2 \subset \mathbb{R}^3$ is a subspace of $\mathbb{R}^3$.
- A line in $\mathbb{R}^2$ not passing through the origin is not a subspace of $\mathbb{R}^2$, since it violates the requirement of closure under scalar multiplication (e.g., the zero vector is not included).

A.1.6 Union, Intersection, Independence, and Internal Sum

$$\mathcal{R} + \mathcal{L} = \{\mathbf{r} + \mathbf{s} : \mathbf{r} \in \mathcal{R}, \mathbf{s} \in \mathcal{L}\}$$
$$\mathcal{R} \cap \mathcal{L} = \{\mathbf{x} : \mathbf{x} \in \mathcal{R} \text{ and } \mathbf{x} \in \mathcal{L}\}.$$

The subspaces $\mathcal{R}$ and $\mathcal{L}$ are said to be independent if and only if

$$\mathcal{R} \cap \mathcal{L} = \{\mathbf{0}\}.$$

If $\mathcal{R}$ and $\mathcal{L}$ are linearly independent, then we note their direct sum as:

$$\mathcal{R} \oplus \mathcal{L} = \mathcal{R} + \mathcal{L}$$

Here, the symbol $\oplus$ indicates that the sum is direct, i.e., the only common element between the subspaces is the zero vector. In this case, the dimension of the direct sum is:

$$d\{\mathcal{R} \oplus \mathcal{L}\} = d\{\mathcal{R}\} + d\{\mathcal{L}\}.$$

A.1.7 Change of Basis

Suppose $\mathbf{x} \in \mathcal{X}$, where $\mathcal{X}$ is a vector space, and let it be expressed in terms of a basis

$$\left\{ \mathbf{x}_i \stackrel{\Delta}{=} \begin{bmatrix} x_{1,i} \\ \vdots \\ x_{n,i} \end{bmatrix}, \; i = 1, \ldots, n \right\}.$$

Then, we can write

$$\mathbf{x} = \alpha_1\mathbf{x}_1 + \cdots + \alpha_n\mathbf{x}_n$$

Now, suppose $\mathbf{x}$ is also represented in terms of a different basis

$$\left\{ \hat{\mathbf{x}}_i \stackrel{\Delta}{=} \begin{bmatrix} \hat{x}_{1,i} \\ \vdots \\ \hat{x}_{n,i} \end{bmatrix}, \; i = 1, \ldots, n \right\},$$

$$\mathbf{x} = \hat{\alpha}_1\hat{\mathbf{x}}_1 + \cdots + \hat{\alpha}_n\hat{\mathbf{x}}_n.$$

How do we compute the coefficients $\hat{\alpha}_1, \ldots, \hat{\alpha}_n$ given $\alpha_1, \ldots, \alpha_n$? We can express this relationship algebraically as:

$$\begin{bmatrix} \hat{x}_{1,1} & \cdots & \hat{x}_{1,n} \\ \vdots & \ddots & \vdots \\ \hat{x}_{n,1} & \cdots & \hat{x}_{n,n} \end{bmatrix} \begin{bmatrix} \hat{\alpha}_1 \\ \vdots \\ \hat{\alpha}_n \end{bmatrix} = \begin{bmatrix} x_{1,1} & \cdots & x_{1,n} \\ \vdots & \ddots & \vdots \\ x_{n,1} & \cdots & x_{n,n} \end{bmatrix} \begin{bmatrix} \alpha_1 \\ \vdots \\ \alpha_n \end{bmatrix}.$$

Hence, the new coefficients are given by

$$\begin{bmatrix} \hat{\alpha}_1 \\ \vdots \\ \hat{\alpha}_n \end{bmatrix} = \underbrace{\begin{bmatrix} \hat{x}_{1,1} & \cdots & \hat{x}_{1,n} \\ \vdots & \ddots & \vdots \\ \hat{x}_{n,1} & \cdots & \hat{x}_{n,n} \end{bmatrix}^{-1} \begin{bmatrix} x_{1,1} & \cdots & x_{1,n} \\ \vdots & \ddots & \vdots \\ x_{n,1} & \cdots & x_{n,n} \end{bmatrix}}_{\substack{\text{matrix performing} \\ \text{the change of basis}}} \begin{bmatrix} \alpha_1 \\ \vdots \\ \alpha_n \end{bmatrix}$$

This matrix performs the *change of basis*, converting a vector's coordinate representation from the original basis to the new one.

Now, suppose we are converting from the natural (standard) basis to another basis. In this case, the original basis matrix is the identity matrix:

$$x_{i,j} = \delta_{i,j} \stackrel{\Delta}{=} \begin{cases} 0 \text{ for } i \neq j \\ 1 \text{ for } i = j. \end{cases}$$

Then, the change of basis simplifies to

$$\begin{bmatrix} \hat{\alpha}_1 \\ \vdots \\ \hat{\alpha}_n \end{bmatrix} = \begin{bmatrix} \hat{x}_{1,1} & \cdots & \hat{x}_{1,n} \\ \vdots & \ddots & \vdots \\ \hat{x}_{n,1} & \cdots & \hat{x}_{n,n} \end{bmatrix}^{-1} \begin{bmatrix} \alpha_1 \\ \vdots \\ \alpha_n \end{bmatrix}.$$

Note: In Euclidean space, this process may also be interpreted as a coordinate transformation rather than a change of basis, depending on context and application. This is illustrated in Figure A.3.

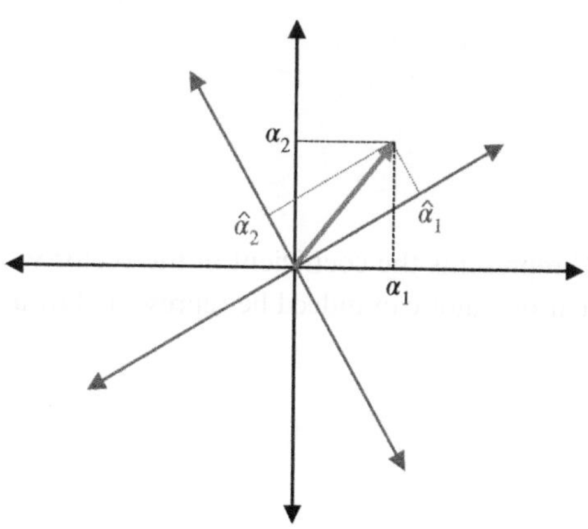

Figure A.3 Illustration of change of basis in two-dimensional Euclidean space.

A.2 Linear Operator

A.2.1 Definition

Let

$$\mathbf{C} : \mathcal{X} \rightarrow \mathcal{Y}$$
$$\mathcal{X} = \text{an } n\text{-dimensional vector space over some scalar field } \mathcal{F}$$
$$\mathcal{Y} = \text{an } m\text{-dimensional vector space over the same scalar field.}$$

Then map $\mathbf{C}$ is called a *linear operator* if

$$\mathbf{C}(\alpha_1 \mathbf{x}_1 + \alpha_2 \mathbf{x}_2) = \alpha_1 \mathbf{C}\mathbf{x}_1 + \alpha_2 \mathbf{C}\mathbf{x}_2,$$

where

$$\mathbf{x}_1, \mathbf{x}_2 \in \mathcal{X}; \quad \alpha_1, \alpha_2 \in \mathcal{F}.$$

A.2.2 Matrix Representation

Consider the set of all linear maps from $\mathcal{X}$ to $\mathcal{Y}$, denoted as $\mathcal{L}(\mathcal{X}, \mathcal{Y})$. This set can be structured as a vector space (of linear operators) by defining addition of two operators and multiplication by a scalar in a natural manner:

$$(\mathbf{C}_1 + \mathbf{C}_2)\,\mathbf{x} = \mathbf{C}_1\mathbf{x} + \mathbf{C}_2\mathbf{x}$$
$$(c\mathbf{C})\mathbf{x} = c(\mathbf{C}\mathbf{x}).$$

These operations are designed to ensure that $\mathcal{L}(\mathcal{X}, \mathcal{Y})$ satisfies the properties of a vector space. Readers should verify that these definitions indeed qualify $\mathcal{L}(\mathcal{X}, \mathcal{Y})$ as a linear space by checking the vector space axioms.

Let $\{\mathbf{x}_1, \ldots, \mathbf{x}_n\}$ and $\{\mathbf{y}_1, \ldots, \mathbf{y}_m\}$ be the basis sets for $\mathcal{X}$ and $\mathcal{Y}$ respectively. Then, for each basis vector $\mathbf{x}_i$ in $\mathcal{X}$, there exist unique $c_{\ell,i}$, $\ell = 1, \ldots, m$, $i = 1, \ldots, n$ such that

$$\mathbf{C}\mathbf{x}_i = c_{1,i}\mathbf{y}_1 + \cdots + c_{m,i}\mathbf{y}_m, \quad i = 1, \ldots, n.$$

Using the conventional framework of matrix algebra, we can express the operator $\mathbf{C}$ as a matrix multiplication:

$$\begin{bmatrix} y_1 \\ \vdots \\ y_m \end{bmatrix} = \underbrace{\begin{bmatrix} c_{11} & \cdots & c_{1n} \\ \vdots & \ddots & \vdots \\ c_{m1} & \cdots & c_{mn} \end{bmatrix}}_{C} \begin{bmatrix} x_1 \\ \vdots \\ x_n \end{bmatrix}.$$

In this formulation, each element $c_{\ell,i}$ of the matrix C represents the coefficient of the vector $\mathbf{y}_\ell$ in the linear combination forming $\mathbf{C}\mathbf{x}_i$. Thus, any linear operator can indeed be represented by a matrix corresponding to its action on the basis vectors of $\mathcal{X}$, transforming them into the vector space $\mathcal{Y}$.

Note: In this text, a linear operator will be denoted by a bold capital letter, while its matrix representation will be written using an italicized (nonbold) capital letter. This convention helps distinguish between the abstract operator and its representation with respect to a particular basis.

A.2.3 Change of Basis for Linear Operators

Let $\mathbf{C} : \mathcal{X} \to \mathcal{Y}$ be expressed as

$$\underbrace{\begin{bmatrix} \beta_1 \\ \vdots \\ \beta_m \end{bmatrix}}_{\mathbf{y}} = C \underbrace{\begin{bmatrix} \alpha_1 \\ \vdots \\ \alpha_n \end{bmatrix}}_{\mathbf{x}}.$$

Now, consider the following change of basis for both $\mathbf{x}$ and $\mathbf{y}$:

$$\mathbf{x} = \hat{\alpha}_1 \hat{\mathbf{x}}_1 + \cdots + \hat{\alpha}_n \hat{\mathbf{x}}_n$$
$$\mathbf{y} = \hat{\beta}_1 \hat{\mathbf{y}}_1 + \cdots + \hat{\beta}_m \hat{\mathbf{y}}_m.$$

Then, the transformation can be represented as:

$$\underbrace{\begin{bmatrix} \hat{\mathbf{y}}_1 & \cdots & \hat{\mathbf{y}}_m \end{bmatrix} \begin{bmatrix} \hat{\beta}_1 \\ \vdots \\ \hat{\beta}_m \end{bmatrix}}_{\mathbf{y}} = C \underbrace{\begin{bmatrix} \hat{\mathbf{x}}_1 & \cdots & \hat{\mathbf{x}}_n \end{bmatrix} \begin{bmatrix} \hat{\alpha}_1 \\ \vdots \\ \hat{\alpha}_n \end{bmatrix}}_{\mathbf{x}}$$

Hence, the new coefficients for the vector $\mathbf{y}$ in terms of the changed basis can be expressed as:

$$\begin{bmatrix} \hat{\beta}_1 \\ \vdots \\ \hat{\beta}_m \end{bmatrix} = \underbrace{\begin{bmatrix} \hat{\mathbf{y}}_1 & \cdots & \hat{\mathbf{y}}_m \end{bmatrix}^{-1} C \begin{bmatrix} \hat{\mathbf{x}}_1 & \cdots & \hat{\mathbf{x}}_n \end{bmatrix}}_{\hat{C}} \begin{bmatrix} \hat{\alpha}_1 \\ \vdots \\ \hat{\alpha}_n \end{bmatrix}$$

$\hat{C}$ is the matrix representation of the operator $\mathbf{C}$ when $\mathbf{x}$ and $\mathbf{y}$ are expressed in the bases $\{\hat{\mathbf{x}}_i, i = 1, \ldots, n\}$ and $\{\hat{\mathbf{y}}_j, j = 1, \ldots, m\}$, respectively.

Comment:

- If the operator $\mathbf{C}$ is represented by matrix multiplication with C in terms of the natural basis, then, using the bases $\{\hat{\mathbf{x}}_i, i = 1, \ldots, n\}$ and $\{\hat{\mathbf{y}}_i, i = 1, \ldots, m\}$, $\mathbf{C}$ is represented by $Y^{-1}CX$, where

$$X = \begin{bmatrix} \hat{\mathbf{x}}_1 & \cdots & \hat{\mathbf{x}}_n \end{bmatrix} \quad Y = \begin{bmatrix} \hat{\mathbf{y}}_1 & \cdots & \hat{\mathbf{y}}_m \end{bmatrix}.$$

- Note that we may express this transformation as:

$$\underbrace{\begin{bmatrix}\beta_1\\ \vdots\\ \beta_m\end{bmatrix}}_{\mathbf{y}} = C\underbrace{\begin{bmatrix}\alpha_1\\ \vdots\\ \alpha_n\end{bmatrix}}_{\mathbf{x}} = Y\hat{C}X^{-1}\underbrace{\begin{bmatrix}\alpha_1\\ \vdots\\ \alpha_n\end{bmatrix}}_{\mathbf{x}}$$

  Here, **C** can be interpreted as executing a series of three linear operations, providing a useful framework if one prefers viewing the operation as a coordinate transformation rather than a change of basis.
- **What is it good for?** Often, given a C, we seek matrices X and Y with certain properties so that the transformed matrix $\hat{C}$ is "nice" (e.g., diagonal). Finding such X and Y indicates the necessary change of basis in the domain and codomain spaces that simplifies the analysis of the operator.

  Example: High purity distillation column

$$\begin{bmatrix}\Delta x_D\\ \Delta x_B\end{bmatrix} = \begin{bmatrix}0.755 & -0.745\\ 0.745 & -0.755\end{bmatrix}\begin{bmatrix}\Delta L\\ \Delta V\end{bmatrix}$$

$$\begin{bmatrix}\Delta x_D\\ \Delta x_B\end{bmatrix} = \begin{bmatrix}1 & 1\\ 1 & -1\end{bmatrix}\begin{bmatrix}1.5 & 0\\ 0 & 0.01\end{bmatrix}\begin{bmatrix}1 & 1\\ -1 & 1\end{bmatrix}^{-1}\begin{bmatrix}\Delta L\\ \Delta V\end{bmatrix}.$$

- In representing a linear operator through matrix multiplication, it is crucial to specify which bases are being used. If no specific bases are mentioned, it is conventionally assumed that the standard basis (the original coordinate system) is utilized.

A.2.4 Null Space and Image Space

Consider $\mathbf{C} : \mathcal{X} \to \mathcal{Y}$. Define the kernel and image of **C** as follows:

$$\text{Ker}\mathbf{C} = \{\mathbf{x} : \mathbf{x} \in \mathcal{X} \text{ and } \mathbf{Cx} = \mathbf{0}\}$$
$$\text{Im}\mathbf{C} = \{\mathbf{Cx} : \mathbf{x} \in \mathcal{X}\}$$

Ker**C** and Im**C** are subspaces of $\mathcal{X}$ and $\mathcal{Y}$.

The following is known as the rank-nullity theorem, a fundamental result in linear algebra.

$$d\{\mathcal{X}\} = d\{\text{Ker}\mathbf{C}\} + d\{\text{Im}\mathbf{C}\}.$$

This theorem states that the dimension of the domain space $\mathcal{X}$ is partitioned between the null space (kernel) and the image (range) of the linear transformation **C**. In other words, the number of linearly independent directions in $\mathcal{X}$ is exactly the sum of those that are mapped to zero and those that contribute to the output space.

A.2.5 Inverse Operator

Consider a linear operator $\mathbf{C} : \mathcal{X} \to \mathcal{Y}$. For any $\mathbf{y} \in \text{Im}\mathbf{C}$, we define the preimage set

$$\mathbf{C}^{-1}\mathbf{y} = \{\mathbf{x} : \mathbf{x} \in \mathcal{X} \text{ and } \mathbf{Cx} = \mathbf{y}\}.$$

Note:

- $\mathbf{C}^{-1}\mathbf{y}$ is not a singleton if Ker$\mathbf{C} \neq \{\mathbf{0}\}$ as multiple elements in $\mathbb{C}$ may map to the same $\mathbf{y}$.
- $\mathbf{C}^{-1}\mathbf{y}$ is undefined if $\mathbf{y} \notin$ Im$\mathbf{C}$.

A mapping $\mathbf{C}_\ell^{-1}$ is called a *left inverse* of $\mathbf{C}$ if

$$\mathbf{C}_\ell^{-1}\mathbf{C}\mathbf{x} = \mathbf{x} \quad \forall \mathbf{x} \in \mathcal{X}$$

A mapping $\mathbf{C}_r^{-1}$ is called a *right inverse* of $\mathbf{C}$ if

$$\mathbf{C}\mathbf{C}_r^{-1}\mathbf{y} = \mathbf{y} \quad \forall \mathbf{y} \in \mathcal{Y}$$

A.2.6 Injection, Surjection, and Bijection

Let $\mathbf{C} : \mathcal{X} \to \mathcal{Y}$. Then

- $\mathbf{C}$ is a *surjection* if Im$\mathbf{C} = \mathcal{Y}$. In this case
  - A right inverse $\mathbf{C}_r^{-1}$ always exists.
  - A left inverse $\mathbf{C}_\ell^{-1}$ may or may not exist.
  - $\mathbf{C}$ corresponds to a *nonsingular* "fat" or "square" matrix (i.e., with more columns than rows or equal dimensions with full row rank).
- $\mathbf{C}$ is an *injection* if Ker$\mathbf{C} = \{\mathbf{0}\}$. In this case
  - A left inverse $\mathbf{C}_\ell^{-1}$ always exists.
  - A right inverse $\mathbf{C}_r^{-1}$ may or may not exist.
  - $\mathbf{C}$ corresponds to a *nonsingular* "tall" or "square" matrix (i.e., with more rows than columns or equal dimensions with full column rank).
- $\mathbf{C}$ is a *bijection* if it is both a surjection and an injection. In this case
  - The inverse $\mathbf{C}^{-1}$ exists and satisfies $\mathbf{C}^{-1} = \mathbf{C}_\ell^{-1} = \mathbf{C}_r^{-1}$.
  - $\mathbf{C}$ corresponds to a *nonsingular* "square" matrix (i.e., same number of rows and columns with full rank).

A.2.7 Inner Product Space

Let $\mathbf{x}, \mathbf{y} \in \mathcal{X}$ be represented (with respect to the same basis) as:

$$\mathbf{x} = \begin{bmatrix} \alpha_1 \\ \vdots \\ \alpha_n \end{bmatrix}, \quad \mathbf{y} = \begin{bmatrix} \beta_1 \\ \vdots \\ \beta_n \end{bmatrix}.$$

Then, their *inner product* $\langle \mathbf{x}, \mathbf{y} \rangle : \mathcal{X} \times \mathcal{X} \to \mathbb{R}$ is defined as:

$$\langle \mathbf{x}, \mathbf{y} \rangle = \alpha_1\beta_1 + \cdots + \alpha_n\beta_n.$$

Equivalently, this can be written in matrix notation as:

$$\langle \mathbf{x}, \mathbf{y} \rangle = \mathbf{x}^T\mathbf{y}.$$

With the definition, the vector space X becomes an *inner product space*.

A.2.8 Orthogonal Vectors and Orthonormal Basis

Two vectors $\mathbf{x}_1, \mathbf{x}_2 \in \mathcal{X}$ are said to be *orthogonal* if

$$\langle \mathbf{x}, \mathbf{y} \rangle = 0.$$

In Euclidean space, this condition implies that the two vectors form a 90° angle with each other. This geometric interpretation is illustrated in Figure A.4.

A vector $\mathbf{x} \in \mathcal{X}$ is *normalized* if

$$\langle \mathbf{x}, \mathbf{x} \rangle = 1.$$

A set of vectors $\mathbf{x}_1, \ldots, \mathbf{x}_n$ forms an *orthonormal basis* if

$$\langle \mathbf{x}_i, \mathbf{x}_j \rangle = \delta_{ij},$$

where δ_{ij} is the Kronecker delta. Note that the standard basis vectors form an *orthonormal basis.*

A.2.9 Change of Basis to Orthonormal Basis

Let $\mathbf{x} \in \mathcal{X}$ be represented in the standard basis as:

$$\mathbf{x} = \alpha_1 \mathbf{e}_1 + \cdots + \alpha_n \mathbf{e}_n$$

Now suppose $\mathbf{x}$ is instead represented with respect to another orthonormal basis $\{\mathbf{x}_1, \ldots, \mathbf{x}_n\}$:

$$\mathbf{x} = \hat{\alpha}_1 \mathbf{x}_1 + \cdots + \hat{\alpha}_n \mathbf{x}_n$$

Then, we can show that each coefficient $\hat{\alpha}_i$ is given by the inner product

$$\hat{\alpha}_i = \langle \mathbf{x}_i, \mathbf{x} \rangle = \mathbf{x}_i^T \mathbf{x}.$$

In matrix form, this becomes

$$\begin{bmatrix} \hat{\alpha}_1 \\ \vdots \\ \hat{\alpha}_n \end{bmatrix} = \underbrace{\begin{bmatrix} \mathbf{x}_1^T \\ \vdots \\ \mathbf{x}_n^T \end{bmatrix}}_{X^T} \begin{bmatrix} \alpha_1 \\ \vdots \\ \alpha_n \end{bmatrix}$$

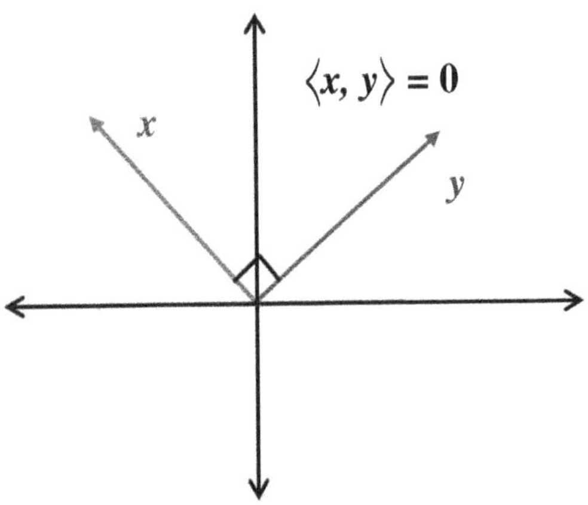

Figure A.4 Illustration of orthogonal vectors in a two-dimensional space.

Note that the matrix X^T represents the change of basis from the original coordinate system to the orthonormal basis $\{\mathbf{x}_1, \ldots, \mathbf{x}_n\}$, and since the basis is orthonormal, $X^T = X^{-1}$.

A.2.10 Orthogonal Matrix

A matrix $A \in \mathbb{R}^{n\times n}$ is called an *orthogonal matrix* if and only if

$$AA^T = I \text{ or } A^T = A^{-1}$$

Matrix A is orthogonal if and only if its columns (or rows) form an orthonormal basis for $\mathbb{R}^n$.

A change of basis from the standard (natural) basis to another orthonormal basis is represented by an orthogonal matrix. This transformation is pictorially illustrated in Figure A.5.

A.2.11 Projection, Orthogonal Projection

Suppose $\mathcal{X} = \mathcal{R} \oplus \mathcal{L}$. Then, every $\mathbf{x} \in \mathcal{X}$ can be uniquely expressed as

$$\mathbf{x} = \mathbf{r} + \mathbf{s}, \text{ where } \mathbf{r} \in \mathcal{R}, \mathbf{s} \in \mathcal{L}$$

A linear map $\mathbf{Q} : \mathcal{X} \to \mathcal{X}$ is called a *projection* onto $\mathcal{R}$ along $\mathcal{L}$ if

$$\mathbf{Q}\mathbf{x} = \mathbf{r}$$

The projection $\mathbf{Q}$ is called an *orthogonal projection* if $\mathcal{R} \perp \mathcal{L}$.

Assume that $\mathcal{R}$ and $\mathcal{L}$ are spanned by the basis vectors $\mathbf{x}_1, \ldots, \mathbf{x}_k$ and $\mathbf{x}_{k+1}, \ldots, \mathbf{x}_n$, respectively. For any $\mathbf{x} = \alpha_1\mathbf{e}_1 + \cdots + \alpha_n\mathbf{e}_n$, the projection can be written as

$$Q\mathbf{x} = \underbrace{\begin{bmatrix}\mathbf{x}_1 & \cdots & \mathbf{x}_k \,|\, \mathbf{0} & \cdots & \mathbf{0}\end{bmatrix}\begin{bmatrix}\mathbf{x}_1 & \cdots & \mathbf{x}_n\end{bmatrix}^{-1}}_{Q}\begin{bmatrix}\alpha_1 \\ \vdots \\ \alpha_n\end{bmatrix}.$$

Consider the specific case of orthogonal projection. Let $\mathbf{x}_1, \mathbf{x}_2$ be an orthonormal basis for the vector space $\mathcal{X}$. Any vector $\mathbf{x} \in \mathcal{X}$ can be expressed as

$$\mathbf{x} = \alpha_1\mathbf{x}_1 + \alpha_2\mathbf{x}_2.$$

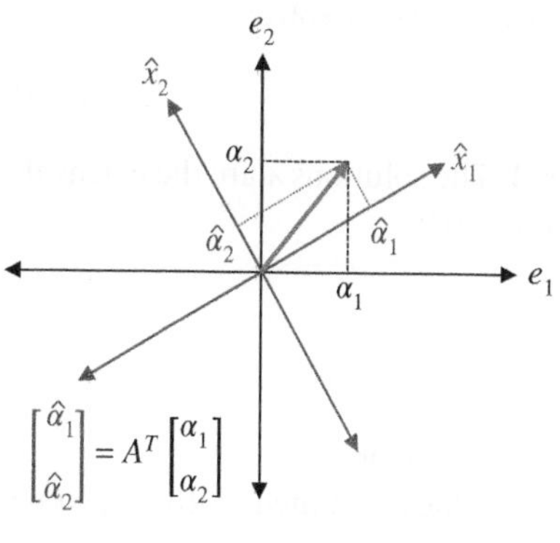

Figure A.5 Illustration of an orthogonal matrix as a change of basis from the natural basis to another orthonormal basis set.

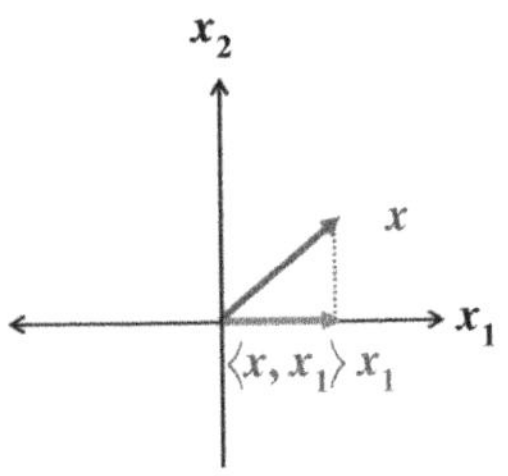

Figure A.6 Illustration of orthogonal projection onto subspaces defined by orthonormal basis vectors.

The orthogonal projection of $\mathbf{x}$ on the one-dimensional subspace spanned by $\mathbf{x}_1$ is given by

$$Qx = \underbrace{\begin{bmatrix} \mathbf{x}_1 & \mathbf{0} \end{bmatrix} \begin{bmatrix} \mathbf{x}_1^T \\ \mathbf{x}_2^T \end{bmatrix}}_{Q} \begin{bmatrix} \alpha_1 \\ \alpha_2 \end{bmatrix}.$$

This operation is illustrated in Figure A.6.

A.3 Matrix Algebra

A.3.1 Eigenvalues, Eigenvectors

Let $\mathbf{A} : \mathbb{C}^n \rightarrow \mathbb{C}^n$ be a linear operator. Then, there exists a scalar $\lambda \in \mathbb{C}$ and a nonzero vector $\mathbf{x} \in \mathbb{C}^n$ such that

$$\mathbf{Ax} = \lambda \mathbf{x}$$

where λ is the *eigenvalue* and $\mathbf{x}$ is the *eigenvector*.

The one-dimensional subspace spanned by an eigenvector, denoted span($\mathbf{x}$), is referred to as an *invariant subspace* (or *invariant manifold*) of the matrix $\mathbf{A}$. Figure A.7 provides a pictorial representation of an eigenvector and its associated invariant subspace.

A.3.2 Computing Eigenvalues and Eigenvectors

To find the eigenvalues and eigenvectors of a matrix $A \in \mathbb{R}^{n \times n}$ (or $\mathbb{C}^{n \times n}$), we solve:

$$(A - \lambda I)\mathbf{x} = \mathbf{0} \qquad \Longrightarrow \qquad \det(A - \lambda I) = \mathbf{0}. \tag{A.3}$$

Equation A.3 is called the *characteristic equation* of the matrix A. The solutions λ are the eigenvalues, and the corresponding eigenvectors $\mathbf{x}$ lie in the *kernel* of $(A - \lambda I)$:

$$\mathbf{x} \in \mathrm{Ker}\{A - \lambda I\}.$$

Useful Facts:

- A real matrix can have complex eigenvalues and complex eigenvectors.
- An $n \times n$ matrix has n (not necessarily distinct) eigenvalues, counted with algebraic multiplicity.
- If all n eigenvalues are distinct, then there exist n linearly independent eigenvectors.

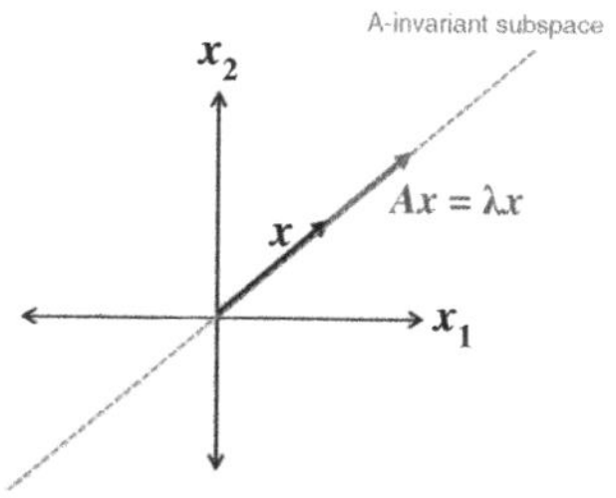

Figure A.7 Pictorial representation of an eigenvector and its corresponding invariant subspace.

- If an eigenvalue has multiplicity greater than one, the number of linearly independent eigenvectors corresponding to it may be less than its multiplicity.
- A real symmetric matrix has real eigenvalues and a complete set of n orthogonal (and real) eigenvectors.

Why is it useful?

Let $\mathbf{A} : \mathbb{C}^n \to \mathbb{C}^n$ be a linear operator with n linearly independent eigenvectors $\mathbf{x}_1, \dots, \mathbf{x}_n$. Then, any vector $\mathbf{x} \in \mathbb{C}^n$ can be expressed in two ways:

$$\begin{aligned}\mathbf{x} &= \alpha_1 \mathbf{e}_1 + \cdots + \alpha_n \mathbf{e}_n \\ &= \hat{\alpha}_1 \mathbf{x}_1 + \cdots + \hat{\alpha}_n \mathbf{x}_n\end{aligned}$$

where $\mathbf{e}_i$ is the standard basis and $\mathbf{x}_i$ is the eigenvector basis.

Applying $\mathbf{A}$ to $\mathbf{x}$ yields:

$$\mathbf{Ax} = \lambda_1 \hat{\alpha}_1 \mathbf{x}_1 + \cdots + \lambda_n \hat{\alpha}_n \mathbf{x}_n.$$

In matrix form, this operation can be expressed as:

$$A\mathbf{x} = \underbrace{\begin{bmatrix} \mathbf{x}_1 & \cdots & \mathbf{x}_n \end{bmatrix} \begin{bmatrix} \lambda_1 & & \\ & \ddots & \\ & & \lambda_n \end{bmatrix} \begin{bmatrix} \mathbf{x}_1 & \cdots & \mathbf{x}_n \end{bmatrix}^{-1}}_{A} \begin{bmatrix} \alpha_1 \\ \vdots \\ \alpha_n \end{bmatrix}.$$

Thus, changing the basis to the eigenvectors (in both the domain and codomain) transforms the matrix $\mathbf{A}$ into a diagonal form. This diagonalization greatly simplifies many computations, such as matrix powers, exponentials, and solutions to differential equations.

A.3.3 Jordan Decomposition and Its Applications

For $A \in \mathbb{C}^{n \times n}$ being a square matrix with n linearly independent eigenvectors. Then, A is *diagonalizable*, and we can express it as:

$$A = X \Lambda X^{-1} \quad \text{where } \Lambda = \text{diag}\{\lambda_1, \dots, \lambda_n\},$$

where

- X is the matrix whose columns are the eigenvectors $\mathbf{x}_1, \dots, \mathbf{x}_n$.
- $\Lambda = \text{diag}\{\lambda_1, \dots, \lambda_n\}$ is a diagonal matrix containing the corresponding eigenvalues.

This decomposition is known as the eigendecomposition or diagonalization of A. It reveals the change of basis that transforms A into a diagonal matrix, simplifying many computations.

For example, consider the continuous-time linear system:

$$\dot{\mathbf{x}} = A\mathbf{x}$$

Using the eigenvalue decomposition, the solution can be written as:

$$\mathbf{x}(t) = Xe^{\Lambda t}X^{-1}\mathbf{x}(0).$$

where $e^{\Lambda t}$ is a diagonal matrix with entries $e^{\Lambda_i t}$. This form makes it straightforward to analyze the system's behavior over time.

Stability Criterion

The stability of the system depends on the eigenvalues of A:

- The system is asymptotically stable if and only if $\text{Re}\{\lambda_i\} < 0$ for all i.
- If any eigenvalue has a nonnegative real part, the system is not asymptotically stable.

Discrete-time Systems

For the discrete-time system

$$\mathbf{x}(k+1) = A\mathbf{x}(k)$$

the solution is

$$\mathbf{x}(k) = A^k\mathbf{x}(0) = X\Lambda^k X^{-1}\mathbf{x}(0)$$

Λ^k is a diagonal matrix with entries λ_i^k. The discrete-time system is asymptotically stable if and only if $|\lambda_i| < 1$ for all i. This ensures that $\lambda_i^k \to 0$ as $k \to \infty$, leading to $\mathbf{x}(k) \to \mathbf{0}$.

Repeated Eigenvalue Case

An $n \times n$ matrix with repeated eigenvalues may not have n linearly independent eigenvectors. In such cases, full diagonalization via a *similarity transformation* is not possible. Instead, we use the *Jordan canonical form*, which generalizes diagonalization. For example, suppose $A \in \mathbb{C}^{7\times 7}$ has 4 distinct eigenvalues and one repeated eigenvalue with algebraic multiplicity 3 but fewer than 3 linearly independent eigenvectors. Then, the Jordan decomposition of A may take the form:

$$A = \begin{bmatrix} \mathbf{x}_1 & \cdots & \mathbf{x}_5 & \mathbf{x}_6 & \mathbf{x}_7 \end{bmatrix} \left[\begin{array}{cccc|ccc} \lambda_1 & 0 & 0 & 0 & 0 & 0 & 0 \\ 0 & \lambda_2 & 0 & 0 & 0 & 0 & 0 \\ 0 & 0 & \lambda_3 & 0 & 0 & 0 & 0 \\ 0 & 0 & 0 & \lambda_4 & 0 & 0 & 0 \\ \hline 0 & 0 & 0 & 0 & \lambda_5 & 1 & 0 \\ 0 & 0 & 0 & 0 & 0 & \lambda_6 & 1 \\ 0 & 0 & 0 & 0 & 0 & 0 & \lambda_7 \end{array}\right] \begin{bmatrix} \mathbf{x}_1 & \cdots & \mathbf{x}_5 & \mathbf{x}_6 & \mathbf{x}_7 \end{bmatrix}^{-1}$$

In this Jordan form, the lower-right 3×3 block is a Jordan block corresponding to the repeated eigenvalue λ with geometric multiplicity 1 and algebraic multiplicity 3. Among the corresponding basis vectors, only $\mathbf{x}_5$ is a true eigenvector. The vectors $\mathbf{x}_6$ and $\mathbf{x}_7$ are called generalized eigenvectors, satisfying:

$$(A - \lambda I)^2 \mathbf{x}_6 = \mathbf{0}, \quad (A - \lambda I)^3 \mathbf{x}_7 = \mathbf{0},$$

but $(A - \lambda I)\mathbf{x}_6 \neq \mathbf{0}$ and $(A - \lambda I)^2 \mathbf{x}_7 \neq \mathbf{0}$.

A.3.4 Singular Value Decomposition

The eigenvector-based Jordan decomposition has several limitations:

- It applies only to square matrices, corresponding to an *automorphism*.
- Full diagonalization is not always possible.
- Eigenvectors generally do not form an orthonormal basis.

Moreover, when the domain and codomain spaces differ, it may be unnecessary—or even impossible—to use the same basis for both. In such cases, the singular value decomposition (SVD) offers a more general and powerful alternative for analysis.

Let $\mathbf{A} : \mathbb{R}^n \to \mathbb{R}^n$ (or $\mathbb{C}^n \to \mathbb{C}^n$). Then, the SVD of $\mathbf{A}$ can be written as:

$$\begin{aligned} A &= U\Sigma V^T \\ &= \begin{bmatrix} \mathbf{u}_1 & \cdots & \mathbf{u}_n \end{bmatrix} \begin{bmatrix} \sigma_1 & & \\ & \ddots & \\ & & \sigma_n \end{bmatrix} \begin{bmatrix} \mathbf{v}_1^T \\ \vdots \\ \mathbf{v}_n^T \end{bmatrix} \end{aligned}$$

Here, U and V are orthogonal matrices (unitary if in $\mathbb{C}^n$), and:

$$\begin{aligned} \sigma_i &: \text{singular values} \\ \mathbf{u}_i &: \text{output (left) singular vectors} \\ \mathbf{v}_i &: \text{input (right) singular vectors} \end{aligned}$$

The largest singular value, σ_1, is called the *maximum singular value*, and the smallest, σ_n, is the *minimum singular value*. The ratio

$$\kappa = \sigma_1/\sigma_n$$

is known as the *condition number* of the matrix. If all singular values are nonzero, then $\mathbf{A}$ is nonsingular and thus invertible.

Why is this useful for analyzing a linear transformation? Consider the transformation

$$\mathbf{y} = A\mathbf{x} \Rightarrow U^T\mathbf{y} = \Sigma V^T\mathbf{x}.$$

This shows that by changing the basis of the domain and codomain spaces to orthonormal sets $\mathbf{v}_i, i = 1, \cdots, n$ and $\mathbf{u}_i, i = 1, \cdots, m$ respectively, the matrix representation of the linear map becomes diagonal.

According to the SVD, we have

$$\begin{aligned} A\mathbf{v}_1 &= \sigma_1\mathbf{u}_1 \\ &\vdots \\ A\mathbf{v}_n &= \sigma_n\mathbf{u}_n \end{aligned}$$

The remarkable feature of this decomposition is that both $\mathbf{v}_i$ and $\mathbf{u}_i$ are orthonormal bases. In other words, the transformation maps one orthonormal basis to another, scaled by the singular values. This orthogonality-preserving structure simplifies the analysis and reveals key properties such as the action of A along principal directions, amplification/attenuation factors (the σ_i), and numerical conditioning.

The input-output mapping associated with the SVD of a 2×2 matrix is illustrated in Figure A.8.

- **"Fat" Matrix:** Let $A \in \mathbb{R}^{m \times n}$ with $m < n$. Then the SVD takes the form

$$A = \begin{bmatrix} \mathbf{u}_1 & \cdots & \mathbf{u}_m \end{bmatrix} \left[\begin{array}{ccc|ccc} \sigma_1 & & & 0 & \cdots & 0 \\ & \ddots & & \vdots & \ddots & \vdots \\ & & \sigma_m & 0 & \cdots & 0 \end{array} \right] \begin{bmatrix} \mathbf{v}_1^T \\ \vdots \\ \mathbf{v}_m^T \\ \hline \mathbf{v}_{m+1}^T \\ \vdots \\ \mathbf{v}_n^T \end{bmatrix}.$$

  Thus, A has m nonzero singular values $\sigma_1, \ldots, \sigma_m$.
  If all singular values are nonzero, A has full row rank and admits a right inverse (i.e., there exists $A^\dagger$ such that $AA^\dagger = I$), though not a left inverse.
  The subspace spanned by $\mathbf{v}_{m+1}, \ldots, \mathbf{v}_n$ forms the kernel (null space) of A, since

  $$A\mathbf{v}_i = \mathbf{0} \text{ for } i > m.$$

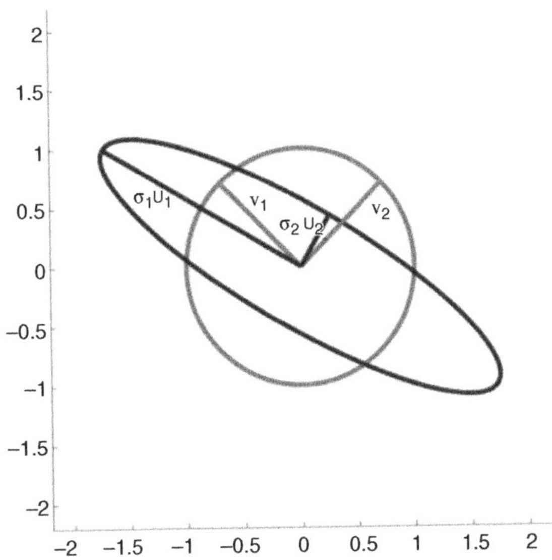

Figure A.8 Pictorial representation of the input-output mapping induced by the SVD of a 2×2 matrix.

- **"Tall" Matrix:** Let $A \in \mathbb{R}^{m\times n}$ with $m > n$. Then the SVD has the form

$$A = \left[\mathbf{u}_1 \quad \cdots \quad \mathbf{u}_n \;\middle|\; \mathbf{u}_{n+1} \quad \cdots \quad \mathbf{u}_m\right] \begin{bmatrix} \sigma_1 & & \\ & \ddots & \\ & & \sigma_n \\ \hline 0 & \cdots & 0 \\ \vdots & \ddots & \vdots \\ 0 & \cdots & 0 \end{bmatrix} \begin{bmatrix} \mathbf{v}_1^T \\ \vdots \\ \mathbf{v}_n^T \end{bmatrix}.$$

In this case, there are n singular values. If all singular values are nonzero, A has full column rank and admits a left inverse (i.e., there exists $A^\dagger$ such that $A^\dagger A = I$), though not a right inverse.

The subspace spanned by $\mathbf{u}_1, \ldots, \mathbf{u}_n$ is the image (or range) of A.

How do we compute the SVD?
For a "fat" matrix ($A \in \mathbb{R}^{m\times n}$ with $m < n$), observe

$$AA^T = \left[\mathbf{u}_1 \quad \cdots \quad \mathbf{u}_m\right] \begin{bmatrix} \sigma_1^2 & & \\ & \ddots & \\ & & \sigma_m^2 \end{bmatrix} \begin{bmatrix} \mathbf{u}_1^T \\ \vdots \\ \mathbf{u}_m^T \end{bmatrix}.$$

This shows that the eigenvalues and eigenvectors of AA^T yield the singular values (squared) and the output (left) singular vectors $\mathbf{u}_i$ of A.

Once Σ and U are known, the input (right) singular vectors can be computed via

$$\begin{bmatrix} \mathbf{v}_1^T \\ \vdots \\ \mathbf{v}_n^T \end{bmatrix} = \Sigma^{-1} \begin{bmatrix} \mathbf{u}_1^T \\ \vdots \\ \mathbf{u}_m^T \end{bmatrix} A.$$

Similarly, for a "tall" matrix, one can compute the eigenvalues and eigenvectors of A^TA to obtain the singular values and right singular vectors, and then recover the left singular vectors from

$$U = AV\Sigma^{-1}.$$

A.3.5 Cayley–Hamilton Theorem

Any square matrix ($A \in \mathbb{R}^{n\times n}$) satisfies its own characteristic equation $\det(\lambda I - A) = 0$: Let the characteristic polynomial be

$$g(\lambda) = \lambda^n + a_1\lambda^{n-1} + \cdots + a_n\lambda_n = 0.$$

Then the matrix substitution gives

$$g(A) = A^n + a_1A^{n-1} + \cdots + a_nI = 0.$$

This result implies that A^n (and all higher powers such as A^{n+1}, A^{n+2}, etc.) can be expressed as linear combinations of $I, A, \dots, A^{n-1}$. Because the Cayley–Hamilton theorem provides a degree-n polynomial identity that annihilates the matrix A, any higher power A^k for $k \geq n$ can be recursively reduced using the relation $A^n = -a_1 A^{n-1} - \cdots - a_n I$. This leads to a minimal generating set $\{I, A, \dots, A^{n-1}\}$ for all polynomial expressions in A.

A.3.6 Matrix Function

Suppose $f(\lambda)$ is defined by a power series

$$f(\lambda) = \sum_{i=0}^{\infty} c_i \lambda^i,$$

which is convergent for $|\lambda| < R$. Then, for a matrix A whose eigenvalues all satisfy $|\lambda_i| < R$, the *matrix function* $f(A)$ is defined via the same power series

$$f(A) = \sum_{i=0}^{\infty} c_i A^i$$

which also converges in the matrix sense.

Example: The exponential function is defined as:

$$\exp(\lambda) = 1 + \lambda + \frac{1}{2}\lambda^2 + \frac{1}{3!}\lambda^3 + \cdots\cdots$$

This extends to matrices as:

$$\exp(A) = 1 + A + \frac{1}{2}A^2 + \frac{1}{3!}A^3 + \cdots\cdots$$

From the Cayley–Hamilton Theorem, we know there exists scalars $\alpha_0, \dots, \alpha_{n-1}$, depending on A such that

$$f(A) = p_{n-1}(A) = \alpha_0 A^{n-1} + \cdots + \alpha_{n-1} I.$$

But how do we determine these coefficients? For this, we use the following important result:

If $f(A)$ is defined by a power series and λ_i is an eigenvalue of A with corresponding eigenvector $\mathbf{x}_i$, then

$$f(A)\mathbf{x}_i = f(\lambda_i)\mathbf{x}_i.$$

This implies that $p_{n-1}(\lambda)$ must interpolate the function $f(\lambda)$ at the eigenvalues of A, i.e.,

$$f(\lambda_i) = p_{n-1}(\lambda_i), \quad i = 1, \dots, n$$

Hence, the coefficients $\alpha_0, \dots, \alpha_n$ can be determined by solving this interpolation system.

How do we use $f(\lambda_i) = p_{n-1}(\lambda_i)$ to obtain $\alpha_0, \dots, \alpha_{n-1}$?

- **Case 1: Distinct Eigenvalues**
  If A has n distinct eigenvalues $\lambda_0, \dots, \lambda_n$, then the interpolation condition

  $$f(\lambda_i) = p_{n-1}(\lambda_i), \quad i = 1, \dots, n$$

  provides a system of n equations that can be solved directly to determine $\alpha_0, \dots, \alpha_{n-1}$.

- **Case 2: Repeated Eigenvalues (Multiplicity m)**
  If an eigenvalue λ_i has algebraic multiplicity m, then additional conditions must be imposed using derivatives:

$$\begin{aligned} f(\lambda_i) &= p_{n-1}(\lambda_i) \\ f^1(\lambda_i) &= p_{n-1}^1(\lambda_i) \\ &\vdots \\ f^{m-1}(\lambda_i) &= p_{n-1}^{m-1}(\lambda_i) \end{aligned}$$

  These derivative matching conditions ensure that the interpolating polynomial agrees with f not only in value but also in behavior at repeated eigenvalues.

Example:
Let

$$A = \begin{bmatrix} 0 & h \\ -h & 0 \end{bmatrix}, \quad \lambda_1 = jh, \lambda_2 = -jh.$$

We seek to compute the matrix exponential exp(A). Since $A^2 = -h^2 I$, powers of A alternate between A and I

$$\exp(A) = \sum_{i=1}^{\infty} \frac{1}{i!} A^i = \alpha_0 A + \alpha_1 I.$$

for some scalars α_0, α_1 to be determined. Consider the scalar exponential

$$\begin{aligned} e^{jh} &= \alpha_0 jh + \alpha_1, \\ e^{-jh} &= -\alpha_0 jh + \alpha_1. \end{aligned}$$

Solving this system yields

$$\alpha_0 = \frac{\sin h}{h}, \quad \alpha_1 = \cos h.$$

Therefore,

$$\exp(A) = \frac{\sin h}{h} \begin{bmatrix} 0 & h \\ -h & 0 \end{bmatrix} + \cos h \begin{bmatrix} 1 & 0 \\ 0 & 1 \end{bmatrix} = \begin{bmatrix} \cos h & \sin h \\ -\sin h & \cos h \end{bmatrix}.$$

A.3.7 Vector Norms

Let $\mathbf{x} \in \mathbb{R}^n$ (or $\mathbb{C}^n$). A function

$$\|\mathbf{x}\| : \mathcal{X} \to \mathbb{R}^{0+}$$

that satisfies certain conditions is called a *norm* of the vector $\mathbf{x}$.

Some commonly used norms include the following:

$$\begin{aligned} &\|\mathbf{x}\|_1 = |x_1| + \cdots + |x_n| \quad \text{(1-norm or Manhattan norm)} \\ &\|\mathbf{x}\|_2 = \sqrt{x_1^2 + \cdots + x_n^2} = \sqrt{\langle \mathbf{x}, \mathbf{x} \rangle} \quad \text{(Euclidean norm)} \\ &\|\mathbf{x}\|_\infty = \max_i |x_i| \quad \text{(Infinity norm).} \end{aligned}$$

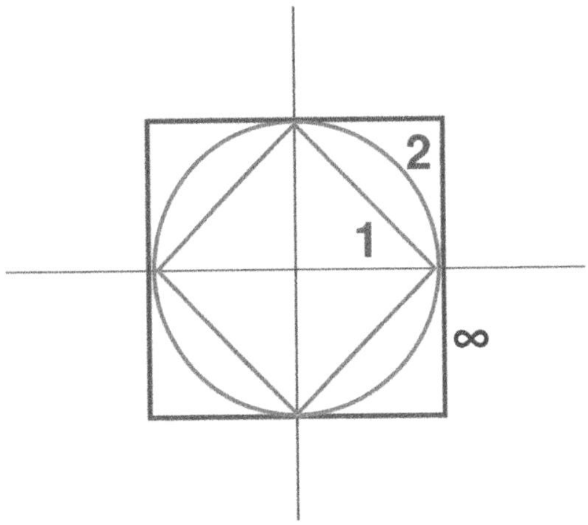

Figure A.9 Illustration of different vector norm types.

The norm of a vector represents its magnitude or size. The most commonly used norm is the 2-norm, which corresponds to the Euclidean distance from the origin. However, the appropriate choice of norm may vary depending on the specific application. Figure A.9 illustrates various types of vector norms. The square, diamond, and circle represent the set of all vectors with unit norm under different norm definitions.

If $\mathbf{x}_1, \ldots, \mathbf{x}_n$ is an orthonormal basis, then for $\mathbf{x} = \alpha_1\mathbf{x}_1 + \cdots + \alpha_n\mathbf{x}_n$, we have

$$\|\mathbf{x}\|_2 = \sqrt{\alpha_1^2 + \cdots + \alpha_n^2}$$

An orthogonal transformation (i.e., a change of basis via an orthonormal set) preserves the 2-norm. That is, for any orthogonal matrix Q, we have

$$\|Q\mathbf{x}\|_2 = \|\mathbf{x}\|_2.$$

In the 2-norm sense, orthogonality ensures that there is no "overlap" among the components, and each contributes independently to the total magnitude.

A.3.8 Matrix Norm

Similar to vector norms, matrix norms provide a way to measure the size of a matrix. They can be defined in different ways—such as the Frobenius norm, which captures the overall magnitude of all matrix elements, or induced norms, which quantify the maximum "gain" of the linear transformation represented by the matrix.

Let $A \in \mathbb{R}^{m\times n}$. Then:

- **Frobenius Norm:** The Frobenius norm is defined as:

$$\|A\|_F = \sqrt{\sum_{i=1}^{m}\sum_{j=1}^{n}(a_{i,j})^2} = \sqrt{\mathrm{trace}(A^T A)}.$$

  It can also be expressed as:

$$\|A\|_F = \sqrt{\|A\mathbf{e}_1\|_2^2 + \cdots + \|A\mathbf{e}_n\|_2^2}.$$

- **Induced Norm:** The induced (or operator) norm corresponding to a given vector norm $|\cdot|_i$ is defined by:

$$\|A\|_i = \max_{\substack{\mathbf{x} \in \mathbb{R}^n \\ \mathbf{x} \neq 0}} \frac{\|A\mathbf{x}\|_i}{\|\mathbf{x}\|_i} = \max_{\|\mathbf{x}\|_i=1} \|A\mathbf{x}\|_i.$$

Common examples include norms induced by the vector 1-norm, 2-norm, and ∞-norm.

You may try proving the following identities for the induced matrix norms:

- $\|A\|_{i1} = \max_j \sum_i |a|_{ij}$ (the maximum absolute column sum)
- $\|A\|_{i2} = \sigma_{\max}(A) = \sqrt{\lambda_{\max}(A^T A)}$ (i.e., the largest singular value of A)
- $\|A\|_{i\infty} = \max_i \sum_j |a|_{ij}$ (the maximum absolute row sum).

A.3.9 Positive (Negative) Definiteness and Semi-definiteness

Let $A \in \mathbb{R}^{n \times n}$. Then

Quadratic Form-based Definitions

- A is *positive definite* if $\mathbf{x}^T A\mathbf{x} > 0$ for all $\mathbf{x} \in \mathbb{R}^n$, $\mathbf{x} \neq \mathbf{0}$.
- A is *positive semi-definite* if $\mathbf{x}^T A\mathbf{x} \geq 0$ for all $\mathbf{x} \in \mathbb{R}^n$.
- A is *negative definite* if $\mathbf{x}^T A\mathbf{x} < 0$ for all $\mathbf{x} \in \mathbb{R}^n$, $\mathbf{x} \neq \mathbf{0}$.
- A is *negative semi-definite* if $\mathbf{x}^T A\mathbf{x} \leq 0$ for all $\mathbf{x} \in \mathbb{R}^n$.
- A is *indefinite* if it is neither positive semi-definite nor negative semi-definite.

Eigenvalue Characterization

The matrix A satisfies:

- *Positive definite* $\Leftrightarrow$ all eigenvalues are strictly positive.
- *Positive semi-definite* $\Leftrightarrow$ all eigenvalues are nonnegative.
- *negative definite* $\Leftrightarrow$ all eigenvalues are strictly negative.
- *negative semi-definite* $\Leftrightarrow$ all eigenvalues are nonpositive.

A.4 Exercises

A. *Linear independence:* Prove the followings:

- If $\mathbf{x}_1, \cdots, \mathbf{x}_k \in \mathbb{R}^n$ are linearly independent,

$$\forall \mathbf{x} \in \text{Span}\{\mathbf{x}_1, \cdots, \mathbf{x}_k\}, \quad \exists \text{ unique } \alpha_i \text{ s.t. } \mathbf{x} = \alpha_1 \mathbf{x}_1 + \cdots + \alpha_k \mathbf{x}_k$$

- $\mathbf{x}_1, \cdots, \mathbf{x}_k \in \mathbb{R}^n$ are *linearly dependent* if $k > n$.

B. *Change of Basis:* Express $\mathbf{x} = \begin{bmatrix} 1 \\ 3 \end{bmatrix}$ in terms of the basis $\begin{bmatrix} 1 \\ 1 \end{bmatrix}$ and $\begin{bmatrix} 1 \\ -1 \end{bmatrix}$.

C. *Matrix Representation of A Linear Operator:* Prove that any linear operator can be represented by multiplication with a matrix.

D. *Change of Basis for A Linear Operator:* Suppose a linear operator is represented by the matrix $\begin{bmatrix} 1 & 2 & 3 \\ 4 & 5 & 6 \\ 7 & 8 & 9 \end{bmatrix}$ in terms of the natural basis. How does the matrix representation change if we use $\left(\begin{bmatrix} 1 \\ 1 \\ 1 \end{bmatrix}, \begin{bmatrix} 1 \\ -1 \\ 1 \end{bmatrix}, \begin{bmatrix} 1 \\ 1 \\ -1 \end{bmatrix} \right)$ as the input basis and $\left(\begin{bmatrix} 1 \\ 2 \\ 1 \end{bmatrix}, \begin{bmatrix} 2 \\ 1 \\ 1 \end{bmatrix}, \begin{bmatrix} 1 \\ 1 \\ 2 \end{bmatrix} \right)$ for the output basis?

E. *Subspace:* Prove that null space and image space are subspaces of the domain and codomain spaces, respectively.

F. *Null Space and Image Space:* Consider the linear transformation represented by the matrix $\begin{bmatrix} 1 & 2 \\ 2 & 4 \end{bmatrix}$. Define the null space and the image space. Draw the picture of these subspaces on a two-dimensional space.

G. *Orthonormal Basis, Orthogonal Transformation, and Orthogonal Projection*

Consider vector $\mathbf{x} = \begin{bmatrix} 5 \\ 3 \end{bmatrix}$. Perform a change of basis to the orthonormal basis set of $\left(\begin{bmatrix} \frac{1}{\sqrt{2}} \\ \frac{1}{\sqrt{2}} \end{bmatrix}, \begin{bmatrix} \frac{1}{\sqrt{2}} \\ \frac{-1}{\sqrt{2}} \end{bmatrix} \right)$. What is the orthogonal matrix representing the change of basis?

What is the *orthogonal* projection of $\mathbf{x}$ on the one-dimensional space defined by $\begin{bmatrix} \frac{1}{\sqrt{2}} \\ \frac{1}{\sqrt{2}} \end{bmatrix}$?

Can you represent the orthogonal projection by a matrix multiplication?

H. *Eigenvalues, eigenvectors, Jordan decomposition:* Compute the eigenvalues and (generalized) eigenvectors of the matrix $\begin{bmatrix} 0 & 1 \\ -1 & 2 \end{bmatrix}$. What is its Jordan decomposition?

I. Prove that, if the n eigenvalues of an $n \times n$ matrix are distinct, the n eigenvectors are linearly independent.

J. For the system

$$x(k+1) = Ax(k)$$

derive the condition on A so that $x(k) \to 0$ as $k \to \infty \forall x(0)$. Use Jordan decomposition.

K. Perform the SVD on the following matrix:

$$\begin{bmatrix} 1 & 2 & 3 \\ 4 & 5 & 6 \\ 7 & 14 & 21 \end{bmatrix}.$$

Does it have an inverse? What is the null space and range (image) space?

L. Perform the SVD on the following matrix:

$$\begin{bmatrix} 1 & 1 & 1 \\ 2 & 2 & 1 \end{bmatrix}.$$

Express the set of all right inverses. What is the null space and range (image) space?

M. Prove the following statement using the definition of power series and properties of eigenvectors.

> If $f(A)$ is defined by a power series and λ_i is an eigenvalue of A with corresponding eigenvector $\mathbf{x}_i$, then:
>
> $$f(A)\mathbf{x}_i = f(\lambda_i)\mathbf{x}_i.$$

N. Evaluate $\ln A$ where $A = \begin{bmatrix} 1 & h \\ 0 & 1 \end{bmatrix}$.

O. Prove that, if $f(A)$ is a polynomial (possible of ∞ order) and λ_i and $\mathbf{x}_i$ are its eigenvalue and eigenvector,

$$f(A)\mathbf{x}_i = f(\lambda_i)\mathbf{x}_i.$$

P. Prove that orthogonal transformations are "2-norm preserving."

Q. Prove that

- $\|A\|_{i1} = \max_j \sum_i |a|_{ij}$ (maximum column sum)
- $\|A\|_{i2} = \sigma_{\max}(A) = \sqrt{\lambda_{\max}(A^T A)}$
- $\|A\|_{i\infty} = \max_i \sum_j |a|_{ij}$ (maximum row sum)

R. Prove that a matrix is positive definite if all the eigenvalues are positive.

S. Let $A \in \mathbb{R}^{m\times n}$ be a "tall" matrix ($m > n$). Parameterize the set of all left inverses in terms of the SVD parameters of A. Express $(A^T A)^{-1} A^T$ in terms of the SVD parameters and see that it is a left inverse. Now, consider

$$\mathbf{y} = A\mathbf{x}.$$

The system of equation is over-specified and there is no solution in general. Using the SVD, shows that

$$\arg\{\min_{\mathbf{x}} \|\mathbf{y} - \mathbf{Ax}\|_2\} = (A^T A)^{-1} A^T \mathbf{y}$$

$(A^T A)^{-1} A^T \mathbf{y}$ is called *least squares solution* of the linear equation.

T. Let $A \in \mathbb{R}^{m\times n}$ be a "fat" matrix $(m < n)$. Parameterize the set of all right inverses in terms of the SVD parameters of A. Express $A^T(AA^T)^{-1}$ in terms of the SVD parameters and see that it is a right inverse. Now, consider

$$\mathbf{y} = A\mathbf{x}.$$

The system of equation is under-specified and there are ∞ number of solutions in general. Parameterize the set of all solutions. Using the SVD, show that

$$\arg\{\min_{\mathbf{x}} \|\mathbf{x}\|_2 \text{ s.t. } \mathbf{y} = \mathbf{Ax}\} = A^T(AA^T)^{-1}\mathbf{y}$$

$A^T(AA^T)^{-1}\mathbf{y}$ is called *minimum norm solution* of the linear equation.

U. Let $A \in \mathbb{R}^{m\times n}$ be a "tall" matrix $(m > n)$. Parameterize the set of all left inverses in terms of the SVD parameters of A. Express $(A^T A)^{-1} A^T$ in terms of the SVD parameters and see that it is a left inverse. Now, consider

$$\mathbf{y} = A\mathbf{x}.$$

The system of equation is over-specified and there is no solution in general.
Using the SVD, shows that

$$\arg\{\min_{\mathbf{x}} \|\mathbf{y} - \mathbf{Ax}\|_2\} = (A^T A)^{-1} A^T \mathbf{y}$$

$(A^T A)^{-1} A^T \mathbf{y}$ is called *least squares solution* of the linear equation.

B

Random Variables and Stochastic Processes

In this chapter, we briefly review random variables and stochastic processes, which form the basis for modeling disturbances and noise encountered in practical control problems. We begin with the fundamentals of probability distributions and then move on to stochastic processes—collections of random variables indexed by a parameter set. We discuss how these processes are characterized and present simple mathematical descriptions of commonly encountered processes.

B.1 Random Variables

B.1.1 What Is a Random Variable?

We often encounter physical phenomena that exhibit randomness, characterized by the following:

- The outcome of any single occurrence (or trial) cannot be predicted with certainty.
- However, the probability of any subset of possible outcomes is well-defined.

A *random variable* is a formal way to describe such phenomena. Importantly, a random variable is not defined by a specific number or value. Instead, it is defined by the set of possible outcomes and the probability distribution assigned to those outcomes. The result of a particular trial is referred to as a *realization* of the random variable.

For example, consider the outcomes of rolling a die. Let x represent the outcome—not the result of a specific roll, but the abstract variable representing possible outcomes. Then, x is defined by the set $\{1, 2, 3, 4, 5, 6\}$ and the probability assigned to each value (e.g., $\Pr(x = i) = 1/6$, where Pr denotes the probability). When we say $x = 1$ or $x = 2$, we are actually referring to a realization of the random variable x.

Random variables are typically categorized as either discrete or continuous:

- A *discrete random variable* takes values in a finite or countably infinite set. The die roll is a canonical example.
- A *continuous random variable* takes values in an uncountably infinite set, such as the real numbers. Examples include the composition or temperature at a point in a distillation column.

B.1.2 Statistics

Statistics is the application of probability theory to real-world problems. Broadly speaking, statistics addresses two fundamental tasks:

- **Prediction Based on a Known Probabilistic Model:**
Given a probabilistic model, the goal is to predict the outcome of future trial(s). For example, one might choose a prediction $\hat{x}$ that minimizes the expected squared error:

$$\hat{x} = \arg\min_{\hat{x}} E\left\{(x - \hat{x})^2\right\}$$

- **Inference: Estimating or Improving the Probabilistic Model from Data:**
Often, the model contains unknown parameters (e.g., θ). Given observed data X generated according to the underlying model, the task is to construct an estimate $\hat{\theta}(X)$ of the true parameter θ.

$$\hat{\theta}(X) = \arg\min_{\hat{\theta}} E\left\{|\theta - \hat{\theta}|_2^2\right\}$$

A related task is hypothesis testing, where the goal is to determine whether a given hypothesis about the system is supported by the data—quantified in terms of probabilities of error or confidence levels.

In practice, these two processes are often combined: as new data become available, the probabilistic model is continually refined and used to make updated predictions.

B.1.3 Basic Probability Concepts

B.1.3.1 Probability Distribution and Density: Scalar Case

A random variable is characterized by a function that describes the probability of various outcomes, rather than assigning a fixed value. Let d be a *continuous* random variable ($d \in \mathbf{R}$). Then one of the following functions is used to define d:

- **Cumulative Distribution Function (CDF)**
The CDF of the random variable d is defined as

$$F(\zeta; d) = \Pr\{d \leq \zeta\}. \tag{B.1}$$

The function $F(\zeta; d)$ is nondecreasing in ζ and asymptotically approaches 1 as ζ tends toward the upper bound of the distribution.
- **Probability Density Function (PDF)**
The PDF $\mathcal{P}(\zeta; d)$ is defined as the derivative of the CDF:

$$\mathcal{P}(\zeta; d) = \frac{dF(\zeta; d)}{d\zeta}. \tag{B.2}$$

This function satisfies the following normalization condition:

$$\int_{-\infty}^{\infty} \mathcal{P}(\zeta; d)\, d\zeta = 1. \tag{B.3}$$

Furthermore, the probability that d lies within the interval $(a, b]$ is given by

$$\int_{a}^{b} \mathcal{P}(\zeta; d)\, d\zeta = F(b; d) - F(a; d) = \Pr(a < d \leq b) \tag{B.4}$$

Example: Gaussian (normal) distribution:

$$\mathcal{P}(\zeta; d) = \frac{1}{\sqrt{2\pi\sigma^2}} \exp\left\{-\frac{1}{2}\left(\frac{\zeta - m}{\sigma}\right)^2\right\}. \tag{B.5}$$

This distribution is fully determined by two parameters: the mean m and the standard deviation σ.

B.1.3.2 Probability Distribution and Density: Vector Case

Consider a *continuous* random vector ($d \in \mathbb{R}^n$):

$$\mathbf{d} = \begin{bmatrix} d_1 & \cdots & d_n \end{bmatrix}.$$

In this case, we must describe not only the distribution of each individual component but also their joint behavior and correlations.

- **Joint CDF**
  The *joint CDF* of the random vector d is defined as

$$F(\zeta_1, \ldots, \zeta_n; d_1, \ldots, d_n) = Pr\{d_1 \le \zeta_1, \ldots, d_n \le \zeta_n\}. \tag{B.6}$$

  The domain of F is $\mathbb{R}^n$. For example, when $n = 2$, F corresponds to a surface over the 2D plane. As all $\zeta_i \to \infty$, the function approaches 1:

$$F(\zeta_1, \ldots, \zeta_n; d_1, \ldots, d_n) \to 1 \text{ as } \zeta_1, \ldots, \zeta_n \to \infty.$$

- **Joint and Marginal PDFs**
  The *joint PDF* $\mathcal{P}(\zeta_1, \cdots, \zeta_n; d_1, \cdots, d_n)$ is defined as the mixed partial derivative of the joint CDF:

$$\mathcal{P}(\zeta_1, \ldots, \zeta_n; d_1, \ldots, d_n) = \frac{\partial^n F(\zeta; d)}{\partial \zeta_1 \ldots \partial \zeta_n}. \tag{B.7}$$

For notational simplicity, we denote the joint density function compactly as $\mathcal{P}(\zeta; d)$, where $\zeta = [\zeta_1, \ldots, \zeta_n]^T$ and $d = [d_1, \ldots, d_n]^T$. Then,

$$\begin{aligned} &\int_{a_1}^{b_1} \cdots \int_{a_n}^{b_n} \mathcal{P}(\zeta_1, \ldots, \zeta_n; d_1, \ldots, d_n)\ d\zeta_1 \ldots d\zeta_n \\ &\quad = \Pr\{a_1 < d_1 \le b_1, \ldots, a_n < d_n \le b_n\}. \end{aligned} \tag{B.8}$$

Naturally, the joint density is normalized:

$$\int_{-\infty}^{\infty} \cdots \int_{-\infty}^{\infty} \mathcal{P}(\zeta_1, \cdots, \zeta_n; d_1, \cdots, d_n) d\zeta_1 \cdots d\zeta_n = 1. \tag{B.9}$$

The *marginal density* of an individual component (e.g., d_1) is obtained by integrating the joint density over all other variables:

$$\mathcal{P}(\zeta_1; d_1) = \int_{-\infty}^{\infty} \cdots \int_{-\infty}^{\infty} \mathcal{P}(\zeta_1, \cdots, \zeta_n; d_1, \cdots, d_n)\ d\zeta_2 \cdots d\zeta_n. \tag{B.10}$$

The joint probability density represents the likelihood that several random variables simultaneously take certain values, while the marginal density reflects the likelihood that a single component achieves a value independent of the values of the others.

In general, the joint density does not equal the product of marginals:

$$\mathcal{P}(\zeta_1, \cdots, \zeta_n; d_1, \cdots, d_n) \ne \mathcal{P}(\zeta_1; d_1) \cdots \mathcal{P}(\zeta_n; d_n). \tag{B.11}$$

Equality holds only if $d_1, \ldots, d_n$ are mutually independent:

$$\mathcal{P}(\zeta_1, \cdots, \zeta_n; d_1, \cdots, d_n) = \mathcal{P}(\zeta_1; d_1) \cdots \mathcal{P}(\zeta_n; d_n). \tag{B.12}$$

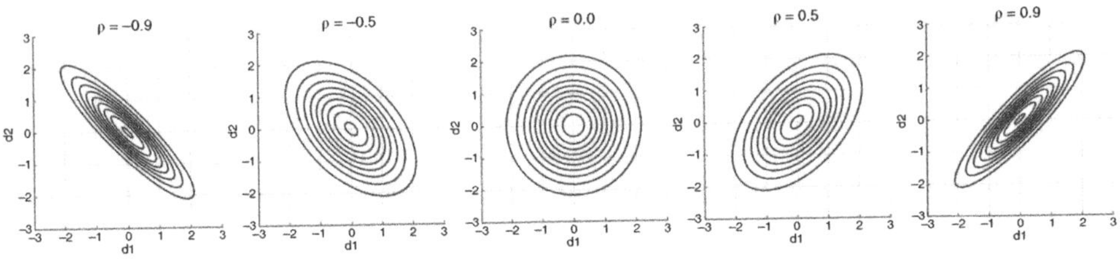

Figure B.1 Contour plots of 2D Gaussian densities for various correlation values ρ.

Example: Jointly Gaussian random variables:

Suppose the random vector $\mathbf{d} = \begin{bmatrix} d_1 & d_2 \end{bmatrix}^T$ is jointly Gaussian. The joint density takes the form:

$$\mathcal{P}(\zeta_1, \zeta_2; d_1, d_2) = \frac{1}{2\pi\sigma_1\sigma_2(1-\rho^2)^{1/2}} \exp\left\{-\frac{1}{2(1-\rho^2)}\left[\left(\frac{\zeta_1 - m_1}{\sigma_1}\right)^2 - 2\rho\frac{(\zeta_1 - m_1)(\zeta_2 - m_2)}{\sigma_1\sigma_2} + \left(\frac{\zeta_2 - m_2}{\sigma_2}\right)^2\right]\right\} \tag{B.13}$$

This density is defined by five parameters: the means m_1, m_2, standard deviations σ_1, σ_2, and the correlation coefficient ρ.

- $\rho = \pm 1$ implies perfect correlation,
- $\rho = 0$ implies no correlation.

Level plots of two-dimensional Gaussian densities for different values of the correlation parameter ρ are shown in Figure B.1.

The marginal densities can be obtained by integrating over the other variable:

$$\mathcal{P}(\zeta_1; d_1) = \int_{-\infty}^{\infty} \mathcal{P}(\zeta_1, \zeta_2; d_1, d_2)\ d\zeta_2 \tag{B.14}$$

$$= \frac{1}{\sqrt{2\pi\sigma_1^2}} \exp\left\{-\frac{1}{2}\left(\frac{\zeta_1 - m_1}{\sigma_1}\right)^2\right\}. \tag{B.15}$$

$$\mathcal{P}(\zeta_2; d_2) = \int_{-\infty}^{\infty} \mathcal{P}(\zeta_1, \zeta_2; d_1, d_2)\ d\zeta_1 \tag{B.16}$$

$$= \frac{1}{\sqrt{2\pi\sigma_2^2}} \exp\left\{-\frac{1}{2}\left(\frac{\zeta_2 - m_2}{\sigma_2}\right)^2\right\}. \tag{B.17}$$

Thus, (m_1, σ_1) and (m_2, σ_2) define the marginal distributions of d_1 and d_2, respectively. Note that unless $\rho = 0$, the joint distribution does not factorize:

$$\mathcal{P}(\zeta_1, \zeta_2; d_1, d_2) \neq \mathcal{P}(\zeta_1; d_1)\mathcal{P}(\zeta_2; d_2) \tag{B.18}$$

For a general n-dimensional Gaussian random vector $d = \begin{bmatrix} d_1 & \cdots & d_n \end{bmatrix}^T$, the density is given by

$$\mathcal{P}(\zeta; d) \triangleq \mathcal{P}(\zeta_1, \cdots, \zeta_n; d_1, \cdots, d_n) \tag{B.19}$$

$$= \frac{1}{(2\pi)^{\frac{n}{2}} |P_d|^{1/2}} \exp\left\{ -\frac{1}{2}(\zeta - \bar{d})^T P_d^{-1} (\zeta - \bar{d}) \right\}, \tag{B.20}$$

where $\bar{d} \in \mathbb{R}^n$ is the mean vector and $P_d \in \mathbb{R}^{n\times n}$ is the covariance matrix. The interpretation and properties of these parameters will be discussed in Section B.1.3.3.

B.1.3.3 Expectation of a Random Variable: Scalar Case

Random variables are completely characterized by their distribution functions or density functions. However, in general, these functions are nonparametric. Hence, random variables are often characterized by their moments up to a finite order; in particular, use of the first two moments is quite common.

- **Expectation of Random Variable Function**

  Any function of d is a random variable. Its expectation is computed as follows:

$$E\{f(d)\} \triangleq \int_{-\infty}^{\infty} f(\zeta)\mathcal{P}(\zeta; d)\ d\zeta. \tag{B.21}$$

- **Mean**

$$\bar{d} \triangleq E\{d\} = \int_{-\infty}^{\infty} \zeta \mathcal{P}(\zeta; d)\ d\zeta. \tag{B.22}$$

  The above is called mean or expectation of d.

- **Variance**

$$\mathrm{Var}\{d\} \triangleq E\{(d - \bar{d})^2\} = \int_{-\infty}^{\infty} (\zeta - \bar{d})^2 \mathcal{P}(\zeta; d)\ d\zeta. \tag{B.23}$$

  The variance quantifies the extent of d deviating from its mean.

Example: Gaussian variable:

For Gaussian variable with density

$$\mathcal{P}(\zeta; d) = \frac{1}{\sqrt{2\pi\sigma^2}} \exp\left\{ -\frac{1}{2}\left(\frac{\zeta - m}{\sigma}\right)^2 \right\}, \tag{B.24}$$

it is easy to verify that

$$\bar{d} \triangleq E\{d\} = \int_{-\infty}^{\infty} \zeta \frac{1}{\sqrt{2\pi\sigma^2}} \exp\left\{ -\frac{1}{2}\left(\frac{\zeta - m}{\sigma}\right)^2 \right\}\ d\zeta = m. \tag{B.25}$$

$$\mathrm{Var}\{d\} \triangleq E\{(d - \bar{d})^2\} = \int_{-\infty}^{\infty} (\zeta - m)^2 \frac{1}{\sqrt{2\pi\sigma^2}} \exp\left\{ -\frac{1}{2}\left(\frac{\zeta - m}{\sigma}\right)^2 \right\}\ d\zeta = \sigma^2. \tag{B.26}$$

Hence, the parameters m and σ^2 in the normal distribution correspond to the mean and variance, respectively, of the Gaussian random variable.

B.1.3.4 Expectation of a Random Variable: Vector Case

The concepts of *mean* and *variance* can be naturally extended to vector-valued random variables. Let $\mathbf{d} \in \mathbb{R}^n$ be a continuous random vector.

The mean (or expected value) of the ℓ-th component of $\mathbf{d}$ is defined as

$$\bar{d}_\ell = E\{d_\ell\} = \int_{-\infty}^{\infty} \zeta_\ell \mathcal{P}(\zeta_\ell; d_\ell)\, d\zeta_\ell = \int_{-\infty}^{\infty} \cdots \int_{-\infty}^{\infty} \zeta_\ell \mathcal{P}(\zeta_1, \ldots, \zeta_n; d_1, \ldots, d_n)\, d\zeta_1 \cdots d\zeta_n. \tag{B.27}$$

The variance of d_ℓ is given by

$$\mathrm{Var}\{d_\ell\} = E\{(d_\ell - \bar{d}_\ell)^2\} = \int_{-\infty}^{\infty} (\zeta_\ell - \bar{d}_\ell)^2 \mathcal{P}(\zeta_\ell; d_\ell)\, d\zeta_\ell \tag{B.28}$$

$$= \int_{-\infty}^{\infty} \cdots \int_{-\infty}^{\infty} (\zeta_\ell - \bar{d}_\ell)^2 \mathcal{P}(\zeta_1, \ldots, \zeta_n; d_1, \ldots, d_n)\, d\zeta_1 \cdots d\zeta_n. \tag{B.29}$$

In the vector case, it is also important to quantify the relationships between different components through their covariance:

$$\mathrm{Cov}\{d_\ell, d_m\} = E\{(d_\ell - \bar{d}_\ell)(d_m - \bar{d}_m)\} \tag{B.30}$$

$$= \int_{-\infty}^{\infty} \cdots \int_{-\infty}^{\infty} (\zeta_\ell - \bar{d}_\ell)(\zeta_m - \bar{d}_m) \mathcal{P}(\zeta_1, \ldots, \zeta_n; d_1, \ldots, d_n)\, d\zeta_1 \cdots d\zeta_n. \tag{B.31}$$

Note that variance is a special case of covariance:

$$\mathrm{Cov}\{d_\ell, d_\ell\} = \mathrm{Var}\{d_\ell\}. \tag{B.32}$$

The correlation coefficient between d_ℓ and d_m is defined as

$$\rho = \frac{\mathrm{Cov}\{d_\ell, d_m\}}{\sqrt{\mathrm{Var}\{d_\ell\}\mathrm{Var}\{d_m\}}}. \tag{B.33}$$

- $\rho = \pm 1$ indicates perfect (linear) correlation.
- $\rho = 0$ indicates no correlation.

To capture all pairwise variances and covariances between the components of $\mathbf{d}$, we define the covariance matrix:

$$\mathrm{Cov}\{\mathbf{d}\} = \mathbb{E}\left\{(\mathbf{d} - \bar{\mathbf{d}})(\mathbf{d} - \bar{\mathbf{d}})^T\right\} \tag{B.34}$$

$$= \int_{-\infty}^{\infty} \cdots \int_{-\infty}^{\infty} (\zeta - \bar{\mathbf{d}})(\zeta - \bar{\mathbf{d}})^T \mathcal{P}(\zeta_1, \ldots, \zeta_n; d_1, \ldots, d_n)\, d\zeta_1 \cdots d\zeta_n. \tag{B.35}$$

The (i, j)-th entry of $\mathrm{Cov}\{\mathbf{d}\}$ is $\mathrm{Cov}\{d_i, d_j\}$. The diagonal entries are the variances of the components. The matrix is symmetric:

$$\mathrm{Cov}\{d_i, d_j\} = \mathrm{Cov}\{d_j, d_i\}. \tag{B.36}$$

For two random vectors $x \in \mathbb{R}^n$ and $y \in \mathbb{R}^m$, the cross-covariance matrix is defined as

$$\mathrm{Cov}\{x, y\} = E\left\{(x - \bar{x})(y - \bar{y})^T\right\}. \tag{B.37}$$

This matrix has dimensions $n \times m$, and satisfies the identity:

$$\mathrm{Cov}\{x, y\} = (\mathrm{Cov}\{y, x\})^T. \tag{B.38}$$

Example: Gaussian variables—two-dimensional case:

Let $\mathbf{d} = \begin{bmatrix} d_1 & d_2 \end{bmatrix}$ be a jointly Gaussian random vector with the following joint PDF:

$$\mathcal{P}(\zeta; \bar{\mathbf{d}}) = \frac{1}{2\pi\sigma_1\sigma_2(1-\rho^2)^{1/2}} \exp\left\{-\frac{1}{2(1-\rho^2)}\left[\left(\frac{\zeta_1 - m_1}{\sigma_1}\right)^2 -2\rho\frac{(\zeta_1 - m_1)(\zeta_2 - m_2)}{\sigma_1\sigma_2} + \left(\frac{\zeta_2 - m_2}{\sigma_2}\right)^2\right]\right\}. \tag{B.39}$$

Then, the expected value (mean vector) of $\mathbf{d}$ is

$$\begin{aligned} E\{\bar{\mathbf{d}}\} &= \int_{-\infty}^{\infty}\int_{-\infty}^{\infty} \begin{bmatrix} \zeta_1 \\ \zeta_2 \end{bmatrix} \mathcal{P}(\zeta; \bar{\mathbf{d}})\ d\zeta_1 d\zeta_2 \\ &= \begin{bmatrix} m_2 \\ m_2 \end{bmatrix}. \end{aligned} \tag{B.40}$$

Similarly, the covariance matrix of $\mathbf{d}$ is

$$\begin{aligned} \text{Cov}\{\bar{\mathbf{d}}\} &= \int_{-\infty}^{\infty}\int_{-\infty}^{\infty} \begin{bmatrix} \zeta_1 - m_1 \\ \zeta_2 - m_2 \end{bmatrix} \begin{bmatrix} (\zeta_1 - m_1) & (\zeta_2 - m_2) \end{bmatrix} \mathcal{P}(\zeta; \bar{\mathbf{d}})\ d\zeta_1 d\zeta_2 \\ &= \begin{bmatrix} \sigma_1^2 & \sigma_1\sigma_2\rho \\ \sigma_1\sigma_2\rho & \sigma_2^2 \end{bmatrix}. \end{aligned} \tag{B.41}$$

This covariance matrix encodes the variances of d_1 and d_2 along the diagonal and their correlation through the off-diagonal terms.

Example: Gaussian variables—n-dimensional case:

Let $\mathbf{d} = [d_1 \ \cdots \ d_n]^T$ be an n-dimensional Gaussian random vector. Its PDF is given by

$$\mathcal{P}(\zeta; \mathbf{d}) = \frac{1}{(2\pi)^{\frac{n}{2}} |P_{\mathbf{d}}|^{1/2}} \exp\left\{-\frac{1}{2}(\zeta - \bar{\mathbf{d}})^T P_{\mathbf{d}}^{-1} (\zeta - \bar{\mathbf{d}})\right\}. \tag{B.42}$$

Then, it can be shown that:

$$E\{\mathbf{d}\} = \int_{-\infty}^{\infty} \cdots \int_{-\infty}^{\infty} \zeta \mathcal{P}(\zeta; \bar{\mathbf{d}})\ d\zeta_1, \cdots, d\zeta_n = \bar{\mathbf{d}}. \tag{B.43}$$

$$\text{Cov}\{\mathbf{d}\} = \int_{-\infty}^{\infty} \cdots \int_{-\infty}^{\infty} (\zeta - \bar{\mathbf{d}})(\zeta - \bar{\mathbf{d}})^T \mathcal{P}(\zeta; \bar{\mathbf{d}})\ d\zeta_1, \cdots, d\zeta_n = P_{\mathbf{d}}. \tag{B.44}$$

Hence, the parameters $\bar{d}$ and P_d that define the multivariate Gaussian density correspond to the mean vector and the covariance matrix, respectively.

Exercise: Verify that, with

$$\bar{\mathbf{d}} = \begin{bmatrix} m_1 \\ m_2 \end{bmatrix}; \ P_{\mathbf{d}} = \begin{bmatrix} \sigma_1^2 & \sigma_1\sigma_2\rho \\ \sigma_1\sigma_2\rho & \sigma_2^2 \end{bmatrix}, \tag{B.45}$$

the expression for the multivariate Gaussian density reduces to the two-dimensional case presented earlier.

The covariance matrix P_d encodes information about both the spread (i.e., the variance) of each component of the random vector $\mathbf{d}$ and their correlations. Specifically,

$$\mathrm{Var}\{d_\ell\} = [\mathrm{Cov}\{\mathbf{d}\}]_{\ell,\ell}. \tag{B.46}$$

$$\rho\{d_\ell, d_m\} = \frac{[\mathrm{Cov}\{\mathbf{d}\}]_{\ell,m}}{\sqrt{[\mathrm{Cov}\{\mathbf{d}\}]_{\ell,\ell}\,[\mathrm{Cov}\{\mathbf{d}\}]_{m,m}}}, \tag{B.47}$$

where $[\cdot]_{i,j}$ denotes the $(i, j)_{\text{th}}$ entry of the matrix. However, when n is large, it becomes difficult to intuitively understand how the components interact and to visualize the overall "shape" of the density. In such cases, singular value decomposition (SVD) provides a powerful visualization tool. Since the covariance matrix $P_\mathbf{d}$ is symmetric and positive semi-definite, it admits an orthogonal eigendecomposition (equivalent to SVD in this context):

$$P_\mathbf{d} = E\left\{(\mathbf{d} - \bar{\mathbf{d}})(\mathbf{d} - \bar{\mathbf{d}})^T\right\}. \tag{B.48}$$

$$= \begin{bmatrix} v_1 & \cdots & v_n \end{bmatrix} \begin{bmatrix} \sigma_1 & & \\ & \ddots & \\ & & \sigma_n \end{bmatrix} \begin{bmatrix} v_1^T \\ \vdots \\ v_n^T \end{bmatrix}. \tag{B.49}$$

Pre- and post-multiplying by V^T and V, we get

$$E\left\{V^T(\mathbf{d} - \bar{\mathbf{d}})(\mathbf{d} - \bar{\mathbf{d}})^T V\right\} = \begin{bmatrix} \sigma_1 & & \\ & \ddots & \\ & & \sigma_n \end{bmatrix}. \tag{B.50}$$

Define a rotated random vector $\mathbf{d}^* = V^T\mathbf{d}$. Then $\mathbf{d}^*$ is the coordinate representation of $\mathbf{d}$ in the orthonormal basis $\{v_1, \ldots, v_n\}$, and

$$E\left\{(\mathbf{d}^* - \bar{\mathbf{d}}^*)(\mathbf{d} - \bar{\mathbf{d}}^*)^T\right\} = \begin{bmatrix} \sigma_1 & & \\ & \ddots & \\ & & \sigma_n \end{bmatrix}. \tag{B.51}$$

This diagonal covariance matrix implies that the components of $\mathbf{d}^*$ are uncorrelated. Thus, the eigenvectors $v_1, \ldots, v_n$ define a rotated coordinate system in which the multivariate Gaussian is axis-aligned, and the variances along those axes are given by $\sigma_1^2, \ldots, \sigma_n^2$.

Exercise: Suppose $\mathbf{d} \in \mathbb{R}^2$ is a zero-mean Gaussian random vector with covariance matrix:

$$P_\mathbf{d} = \begin{bmatrix} \frac{\sqrt{2}}{2} & \frac{\sqrt{2}}{2} \\ \frac{\sqrt{2}}{2} & -\frac{\sqrt{2}}{2} \end{bmatrix} \begin{bmatrix} 10 & 0 \\ 0 & 0.1 \end{bmatrix} \begin{bmatrix} \frac{\sqrt{2}}{2} & \frac{\sqrt{2}}{2} \\ \frac{\sqrt{2}}{2} & -\frac{\sqrt{2}}{2} \end{bmatrix}. \tag{B.52}$$

From the decomposition above, the eigenvectors of P_d are

$$v_1 = \begin{bmatrix} \frac{\sqrt{2}}{2} \\ \frac{\sqrt{2}}{2} \end{bmatrix}, \quad v_2 = \begin{bmatrix} \frac{\sqrt{2}}{2} \\ -\frac{\sqrt{2}}{2} \end{bmatrix}. \tag{B.53}$$

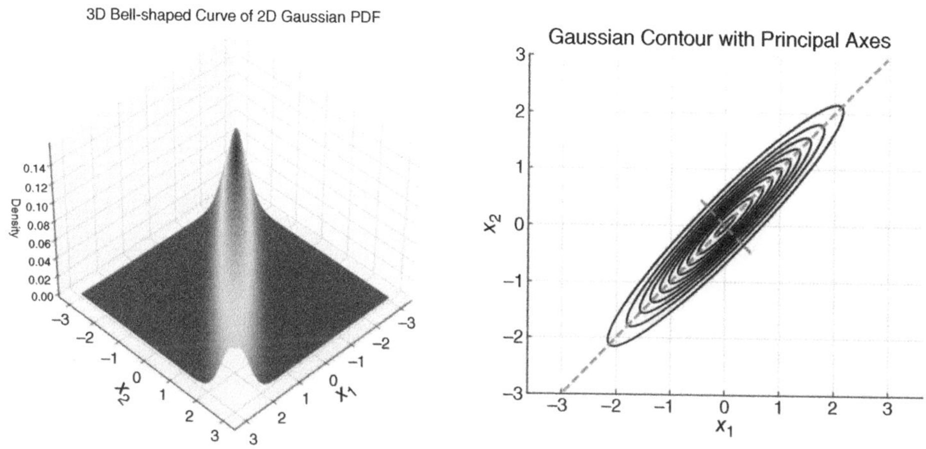

Figure B.2 Visualization of a Gaussian probability density function and its Elliptical level sets.

The PDF and its corresponding elliptical level sets are illustrated in Figure B.2.
Questions:

1. Can you visualize the overall shape of the Gaussian density function?
2. What is the variance of $\mathbf{d}$ along the direction (1, 1) (i.e., along v_1)?
3. What is the variance along the direction (1, −1) (i.e., along v_2)?
4. Based on the shape and orientation of the density, what would the conditional distribution of d_1 given $d_2 = \beta$ look like?
5. Plot the joint, marginal, and conditional densities to confirm your analysis.

B.1.3.5 Conditional Probability Density: Scalar Case

When two random variables are statistically related, the probability density of one variable changes when the other is observed to take on a specific value.

> The probability density of a random variable conditioned on the value(s) of other variable(s) is called the conditional probability density.

This concept plays a central role in stochastic estimation, where we aim to infer unknown variables based on the observed values of others.

Let x and y be continuous random variables with joint probability density $\mathcal{P}(\zeta, \beta; x, y)$. The conditional probability density of x given that $y = \beta$, denoted $\mathcal{P}(\zeta \mid \beta; x \mid y)$, is formally defined as

$$\mathcal{P}(\zeta|\beta; x|y) = \frac{\lim\limits_{\epsilon \to 0} \displaystyle\int_{\beta-\epsilon}^{\beta+\epsilon} \mathcal{P}(\zeta, \beta^*; x, y)\, d\beta^*}{\underbrace{\displaystyle\int_{-\infty}^{\infty} \int_{\beta-\epsilon}^{\beta+\epsilon} \mathcal{P}(\zeta, \beta^*; x, y)\, d\beta^* d\zeta}_{\text{normalization factor}}} \tag{B.54}$$

$$= \frac{\mathcal{P}(\zeta, \beta; x, y)}{\displaystyle\int_{-\infty}^{\infty} \mathcal{P}(\zeta, \beta; x, y) d\zeta} \tag{B.55}$$

$$= \frac{\mathcal{P}(\zeta, \beta; x, y)}{\mathcal{P}(\beta, y)}. \tag{B.56}$$

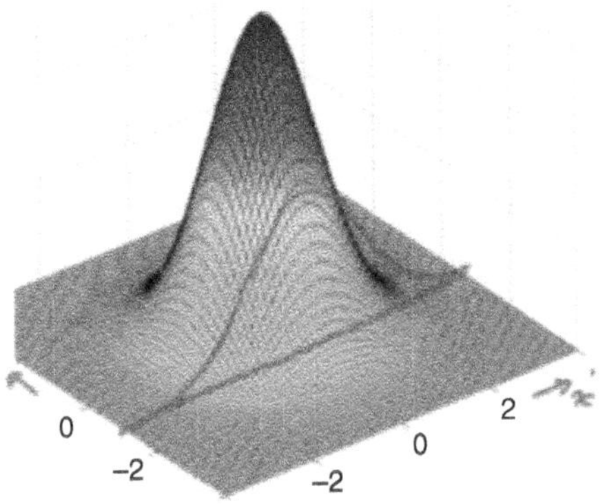

Figure B.3 Illustration of conditional density of x given $y = -1$.

Figure B.3 illustrates how the conditional density $\mathcal{P}(\zeta|\beta; x|y)$ is derived from the joint density $\mathcal{P}(\zeta, \beta; x, y)$.

Key observations:

- The conditional density is computed as

$$\begin{pmatrix}\text{Conditional Density} \\ \text{of } x \text{ given } y\end{pmatrix} = \frac{\text{Joint Density of } x \text{ and } y}{\text{Marginal Density of } y}. \tag{B.57}$$

  This relationship is conceptually intuitive and foundational in probability theory.

- As a valid PDF, the conditional density integrates to 1:

$$\int_{-\infty}^{\infty} \mathcal{P}(\zeta|\beta; x|y)\ d\zeta = 1. \tag{B.58}$$

- The conditional density equals the marginal density only when x and y are statistically independent:

$$\mathcal{P}(\zeta|\beta; x|y) = \mathcal{P}(\zeta, x). \tag{B.59}$$

  if and only if

$$\mathcal{P}(\zeta, \beta; x, y) = \mathcal{P}(\zeta, x)\mathcal{P}(\beta, y). \tag{B.60}$$

We are particularly interested in the conditional density because, in many practical situations, some random variables are measurable while others are not. For a given trial, if variable x is unmeasured but y is observed (i.e., $y = \beta$), then the conditional density $\mathcal{P}(\zeta|\beta; x|y)$ provides the basis for estimating x from the available information.

Summary of Key PDFs:

- Joint Probability Density: $\mathcal{P}(\zeta, \beta; x, y)$

$$\int_{a_2}^{b_2}\int_{a_1}^{b_1} \mathcal{P}(\zeta, \beta; x, y)\, d\zeta d\beta = \Pr\{a_1 < x \leq b_1 \text{ and } a_2 < y \leq b_2\}. \tag{B.61}$$

- Marginal Probability Density of x: $\mathcal{P}(\zeta, x)$

$$\mathcal{P}(\zeta, x) = \int_{-\infty}^{\infty} \mathcal{P}(\zeta, \beta; x, y)\, d\beta. \tag{B.62}$$

- Marginal Probability Density of y: $\mathcal{P}(\beta, y)$

$$\mathcal{P}(\beta, y) = \int_{-\infty}^{\infty} \mathcal{P}(\zeta, \beta; x, y)\, d\zeta. \tag{B.63}$$

- Conditional Probability Density of x given y: $\mathcal{P}(\zeta|\beta; x|y)$

$$\mathcal{P}(\zeta|\beta; x|y) = \frac{\mathcal{P}(\zeta, \beta; x, y)}{\mathcal{P}(\beta, y)}. \tag{B.64}$$

- Conditional Probability Density of y given x: $\mathcal{P}(\beta|\zeta; y|x)$:

$$\mathcal{P}(\beta|\zeta; y|x) = \frac{\mathcal{P}(\zeta, \beta; x, y)}{\mathcal{P}(\zeta, x)}. \tag{B.65}$$

Bayes' Rule: Combining Eqs. (B.63) and (B.64) gives what is known as Baye's Rule:

$$\mathcal{P}(\zeta|\beta; x|y) = \frac{\mathcal{P}(\beta|\zeta; y|x)\mathcal{P}(\zeta, x)}{\mathcal{P}(\beta, y)}. \tag{B.66}$$

This expresses the conditional density of x given y in terms of the conditional density of y given x. This relation is widely used in inference, estimation, and learning problems, especially when direct modeling of $\mathcal{P}(\zeta|\beta; x|y)$ is difficult, but $\mathcal{P}(\beta|\zeta; y|x)$ is tractable.

Conditional Expectation and Variance

Given the conditional density $\mathcal{P}(\zeta|\beta; x|y)$, we define

- **Conditional Expectation:**

$$E\{x|y\} \triangleq \int_{-\infty}^{\infty} \zeta \mathcal{P}(\zeta|\beta; x|y)\, d\zeta. \tag{B.67}$$

- **Conditional Variance:**

$$\mathrm{Var}\{x|y\} \triangleq E\{(\zeta - E\{x|y\})^2\} \tag{B.68}$$

$$= \int_{-\infty}^{\infty} (\zeta - E\{x|y\})^2 \mathcal{P}(\zeta|\beta; x|y)\, d\zeta. \tag{B.69}$$

Example: Jointly normally distributed (Gaussian) variables:

Suppose that x and y are jointly normally distributed with the following parameters: means $\bar{x}$, $\bar{y}$, standard deviations σ_x, σ_y, and correlation coefficient ρ:

$$\mathcal{P}(\zeta, \beta; x, y) = \frac{1}{2\pi\sigma_x\sigma_y(1-\rho^2)^{1/2}} \tag{B.70}$$

$$\times \exp\left\{-\frac{1}{2(1-\rho^2)}\left[\left(\frac{\zeta - \bar{x}}{\sigma_x}\right)^2 - 2\rho\frac{(\zeta - \bar{x})(\beta - \bar{y})}{\sigma_x\sigma_y} + \left(\frac{\beta - \bar{y}}{\sigma_y}\right)^2\right]\right\}.$$

Some algebraic manipulations of the above yield:

$$\mathcal{P}(\zeta,\beta;x,y) = \underbrace{\frac{1}{\sqrt{2\pi\sigma_y^2}} \exp\left\{-\frac{1}{2}\left(\frac{\beta-\bar{y}}{\sigma_y}\right)^2\right\}}_{\text{marginal density of } y} \tag{B.71}$$

$$\times \underbrace{\frac{1}{\sqrt{2\pi\sigma_x^2(1-\rho^2)}} \exp\left\{-\frac{1}{2}\left(\frac{\zeta-\bar{x}-\rho\frac{\sigma_x}{\sigma_y}(\beta-\bar{y})}{\sigma_x\sqrt{1-\rho^2}}\right)^2\right\}}_{\text{conditional density of } x}$$

$$= \underbrace{\frac{1}{\sqrt{2\pi\sigma_x^2}} \exp\left\{-\frac{1}{2}\left(\frac{\zeta-\bar{x}}{\sigma_x}\right)^2\right\}}_{\text{marginal density of } x} \tag{B.72}$$

$$\times \underbrace{\frac{1}{\sqrt{2\pi\sigma_y^2(1-\rho^2)}} \exp\left\{-\frac{1}{2}\left(\frac{\beta-\bar{y}-\rho\frac{\sigma_y}{\sigma_x}(\zeta-\bar{x})}{\sigma_y\sqrt{1-\rho^2}}\right)^2\right\}}_{\text{conditional density of } y}.$$

From this, we obtain the conditional densities:

$$\mathcal{P}(\zeta|\beta;x|y) = \frac{1}{\sqrt{2\pi\sigma_x^2(1-\rho^2)}} \exp\left\{-\frac{1}{2}\left(\frac{\zeta-\bar{x}-\rho\frac{\sigma_x}{\sigma_y}(\beta-\bar{y})}{\sigma_x\sqrt{1-\rho^2}}\right)^2\right\}. \tag{B.73}$$

$$\mathcal{P}(\beta|\zeta;y|x) = \frac{1}{\sqrt{2\pi\sigma_y^2(1-\rho^2)}} \exp\left\{-\frac{1}{2}\left(\frac{\beta-\bar{y}-\rho\frac{\sigma_y}{\sigma_x}(\zeta-\bar{x})}{\sigma_y\sqrt{1-\rho^2}}\right)^2\right\}. \tag{B.74}$$

Thus, the conditional distributions are also Gaussian. Specifically, $\mathcal{P}(\zeta|\beta;x|y)$ is Gaussian with mean of $\bar{x}+\rho\frac{\sigma_x}{\sigma_y}(\beta-\bar{y})$ and variance of $\sigma_x^2(1-\rho^2)$. The conditional expectation becomes

$$E\{x|y\} = \bar{x} + \rho\frac{\sigma_x}{\sigma_y}(\beta-\bar{y}) \tag{B.75}$$

$$= \bar{x} + \frac{\rho\sigma_x\sigma_y}{\sigma_y^2}(\beta-\bar{y}) \tag{B.76}$$

$$= E\{x\} + \mathrm{Cov}\{x,y\}\mathrm{Var}^{-1}\{y\}(\beta-\bar{y}). \tag{B.77}$$

The conditional variance of x given $y = \beta$ is

$$E\{(x - E\{x|y\})^2|y\} = \sigma_x^2(1-\rho^2) \tag{B.77}$$

$$= \sigma_x^2 - \frac{\sigma_x^2\sigma_y^2\rho^2}{\sigma_y^2} \tag{B.78}$$

$$= \sigma_x^2 - (\sigma_x\sigma_y\rho)\frac{1}{\sigma_y^2}(\sigma_x\sigma_y\rho) \tag{B.79}$$

$$= \mathrm{Var}\{x\} - \mathrm{Cov}\{x,y\}\mathrm{Var}^{-1}\{y\}\mathrm{Cov}\{y,x\}. \tag{B.80}$$

As $\rho \to 1$, the conditional variance $\mathrm{Var}\{x|y\} \to 0$, meaning the conditional distribution becomes sharply peaked, i.e., deterministic. This reflects the intuition that if two variables are perfectly correlated, one fully determines the other.

B.1.3.6 Conditional Probability Density: Vector Case

The concept of conditional probability density extends naturally to the vector case. Let $x \in \mathbb{R}^n$ and $y \in \mathbb{R}^m$ be random vectors; the conditional density of x given $y = [\beta_1, \cdots, \beta_m]^T$ is defined as

$$\mathcal{P}(\zeta_1,\cdots,\zeta_n|\beta_1,\cdots,\beta_m; x_1,\cdots,x_n|y_1,\cdots,y_m)$$
$$= \frac{\mathcal{P}(\zeta_1,\cdots,\zeta_n,\beta_1,\cdots,\beta_m; x_1,\cdots,x_n,y_1,\cdots,y_m)}{\mathcal{P}(\beta_1,\cdots,\beta_m; y_1,\cdots,y_m)}. \tag{B.81}$$

Bayes' Rule in the multivariate case is given by

$$\mathcal{P}(\zeta_1,\cdots,\zeta_n|\beta_1,\cdots,\beta_m; x_1,\cdots,x_n|y_1,\cdots,y_m) \tag{B.82}$$
$$= \frac{\mathcal{P}(\beta_1,\cdots,\beta_m|\zeta_1,\cdots,\zeta_n; y_1,\cdots,y_m|x_1,\cdots,x_n)\mathcal{P}(\zeta_1,\cdots,\zeta_n; x_1,\cdots,x_n)}{\mathcal{P}(\beta_1,\cdots,\beta_m; y_1,\cdots,y_m)}.$$

The conditional expectation and conditional covariance of x given $y = \beta$ are defined as

$$E\{x|y\} = \int_{-\infty}^{\infty}\cdots\int_{-\infty}^{\infty}\begin{bmatrix}\zeta_1\\ \vdots\\ \zeta_n\end{bmatrix}\mathcal{P}(\zeta|\beta; x|y)\; d\zeta_1\cdots d\zeta_n. \tag{B.83}$$

$$\mathrm{Cov}\{x|y\} = \int_{-\infty}^{\infty}\cdots\int_{-\infty}^{\infty}\begin{bmatrix}\zeta_1 - E\{x_1|y\}\\ \vdots\\ \zeta_n - E\{x_n|y\}\end{bmatrix}\begin{bmatrix}\zeta_1 - E\{x_1|y\}\\ \vdots\\ \zeta_n - E\{x_n|y\}\end{bmatrix}^T\mathcal{P}(\zeta|\beta; x|y)\; d\zeta_1\cdots d\zeta_n. \tag{B.84}$$

Example: Jointly Gaussian random vectors:

Let $x \in \mathbb{R}^n$ and $y \in \mathbb{R}^m$ be jointly normally distributed random vectors. Define the concatenated vector:

$$z = \begin{bmatrix}x\\ y\end{bmatrix}. \tag{B.85}$$

Then the joint distribution of z is

$$\mathcal{P}(\zeta,\beta; x,y) = \frac{1}{(2\pi)^{\frac{n+m}{2}}|P_z|^{1/2}}\exp\left\{-\frac{1}{2}(\eta - \bar{z})^T P_z^{-1}(\eta - \bar{z})\right\}, \tag{B.86}$$

where

$$\bar{z} = \begin{bmatrix} \bar{x} \\ \bar{y} \end{bmatrix}; \quad \eta = \begin{bmatrix} \zeta \\ \beta \end{bmatrix} \tag{B.87}$$

$$P_z = \begin{bmatrix} \text{Cov}(x) & \text{Cov}(x, y) \\ \text{Cov}(y, x) & \text{Cov}(y) \end{bmatrix}. \tag{B.88}$$

Then the conditional mean and covariance are given by:

$$E\{x|y\} = \bar{x} + \text{Cov}(x, y)\text{Cov}^{-1}(y)(\beta - \bar{y}). \tag{B.89}$$

$$E\{y|x\} = \bar{y} + \text{Cov}(y, x)\text{Cov}^{-1}(x)(\zeta - \bar{x}), \tag{B.90}$$

and

$$\text{Cov}\{x|y\} \triangleq E\left\{(\zeta - E\{x|y\})(\zeta - E\{x|y\})^T\right\} \tag{B.91}$$

$$= \text{Cov}\{x\} - \text{Cov}\{x, y\}\text{Cov}^{-1}\{y\}\text{Cov}\{y, x\}. \tag{B.92}$$

$$\text{Cov}\{y|x\} \triangleq E\left\{(\beta - E\{y|x\})(\beta - E\{y|x\})^T\right\} \tag{B.93}$$

$$= \text{Cov}\{y\} - \text{Cov}\{y, x\}\text{Cov}^{-1}\{x\}\text{Cov}\{x, y\}. \tag{B.94}$$

These results are foundational in linear estimation theory (e.g., Kalman filtering), where predictions and updates are based on conditioning over Gaussian priors.

B.1.4 Statistics

B.1.4.1 Prediction

The first fundamental problem in statistics is the prediction of the outcome of a future trial based on a known probabilistic model.

> Given a PDF $\mathcal{P}(x)$ for a random variable x, predict the outcome of x for a new trial.

Note that, if $\mathcal{P}(x)$ is not a point distribution (i.e., a Dirac delta function), the exact outcome of x cannot be predicted deterministically. Instead, one must define a prediction criterion to identify an optimal estimate.

Two common criteria are

- **Maximum Likelihood Estimate (MLE):** Choose the value of x that maximizes the density:

  $$\hat{x}_{\text{MLE}} = \arg\left[\max_x \mathcal{P}(x)\right].$$

- **Minimum Mean Squared Error (MMSE) Estimate:** Choose $\hat{x}$ to minimize the expected squared error:

  $$\hat{x}_{\text{MMSE}} = \arg\left[\min_{\hat{x}} E\left\{\|x - \hat{x}\|_2^2\right\}\right].$$

  It can be shown that $\hat{x}_{\text{MMSE}} = E\{x\}$.

Exercise: Prove that the MMSE estimate is the mean $E\{x\}$.

If additional information is available—such as a related variable y measured in the same trial—then a more accurate estimate is given by the conditional expectation:

$$\hat{x} = E\{x|y\}.$$

B.1.4.2 Sample Mean and Covariance: Estimating a Probabilistic Model

The second fundamental problem of statistics is to infer the probabilistic model from observed data. The simplest formulation is

> Given observations $x(1), \cdots x(N)$ from N independent trials of a random variable x, estimate the PDF of x.

Often, to make the problem well-posed, a parametric family (e.g., Gaussian) is assumed. Under the assumption that x follows a multivariate normal distribution, the sample mean and covariance are commonly used as estimators:

- **Sample Mean:**

$$\hat{\bar{x}}_N = \frac{1}{N}\sum_{i=1}^{N} x(i).$$

- **Sample Covariance:**

$$\hat{R}_{xN} = \frac{1}{N-1}\sum_{i=1}^{N}\left(x(i) - \hat{\bar{x}}_N\right)\left(x(i) - \hat{\bar{x}}_N\right)^T.$$

These are *consistent estimators*, meaning they are unbiased and converge to the true mean $\bar{x}$ and R_x (i.e., they converge to true values as $N \to \infty$).

A more general case arises in the context of parameterized models, such as

$$y = f(\theta, u) + x,$$

where

- y is an observed random vector,
- x is an additive random disturbance,
- u is a *known* deterministic input (which may vary across trials),
- θ is an *unknown* deterministic parameter (constant across trials).

Then,

> Given input-output data $\{u(i), y(i)\}_{i=1}^{N}$, estimate the noise statistics (e.g., $\bar{x}$, R_x) and the unknown parameter θ.

This is a classic regression problem, which will be addressed in detail in Section B.2.

B.2 Stochastic Process

A stochastic process is a collection of random variables indexed by a parameter set, which may be continuous or discrete. Figure B.4 provides a conceptual illustration of a stochastic process $X(k)$. Stochastic processes can be classified based on whether their index set is continuous or discrete, and whether the associated random variables are continuous or discrete (see Figure B.5). In this discussion, we focus on processes with a discrete index set and continuous-valued random variables, as our interest lies in discrete-time systems. Accordingly, a stochastic process in our context is defined as a sequence of random variables evolving over discrete time steps.

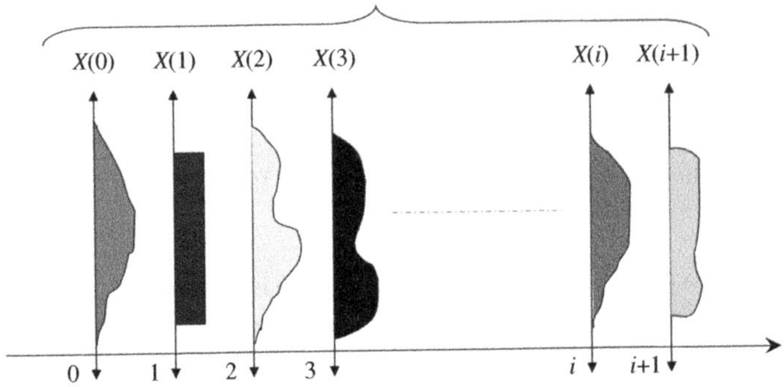

Figure B.4 Conceptual illustration of a stochastic process $X(k)$.

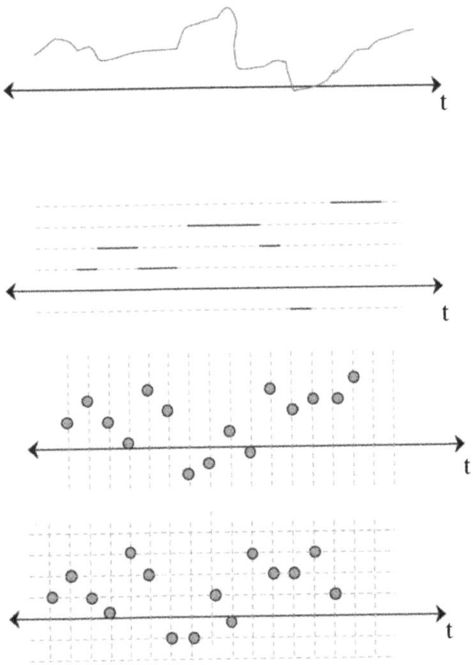

Figure B.5 Classification of stochastic processes based on index set and variable type.

B.2.1 Basic Probability Concepts

B.2.1.1 Distribution Function

Let $x(k)$ be a stochastic sequence. Then, for any finite collection of time indices $k_1, \ldots, k_\ell$, the tuple $(x(k_1), \ldots, x(k_\ell))$ constitutes an ℓ-dimensional random variable. As such, one can define the corresponding finite-dimensional distribution and density functions. For example, the joint CDF is given by

$$F(\lambda_1, \ldots, \lambda_\ell; x(k_1), \ldots, x(k_\ell)) = \Pr(x(k_1) \leq \lambda_1, \ldots, x(k_\ell) \leq \lambda_\ell). \tag{B.95}$$

The associated joint density function can be defined analogously, provided it exists.

These definitions naturally extend to vector-valued stochastic sequences, where each $x(k_i)$ and λ_i is a vector. In such cases, the inequalities and integrals are interpreted component-wise over the appropriate multidimensional domains.

B.2.1.2 Mean and Covariance

The mean of the stochastic variable $x(k)$ is defined as

$$\bar{x}(k) = E\{x(k)\} = \int_{-\infty}^{\infty} \lambda dF(\lambda; x(k)). \tag{B.96}$$

The covariance between $x(k_1)$ and $x(k_2)$ is given by

$$\begin{aligned} R_x(k_1, k_2) &= E\left\{[x(k_1) - \bar{x}(k_1)][x(k_2) - \bar{x}(k_2)]^T\right\} \\ &= \int_{-\infty}^{\infty}\int_{-\infty}^{\infty} [\lambda_1 - \bar{x}(k_1)][\lambda_2 - \bar{x}(k_2)]^T dF(\lambda_1, \lambda_2; x(k_1), x(k_2)). \end{aligned} \tag{B.97}$$

The cross-covariance between two stochastic processes $x(k)$ and $y(k)$ is defined as

$$\begin{aligned} R_{xy}(k_1, k_2) &= E\left\{[x(k_1) - \bar{x}(k_1)][y(k_2) - \bar{y}(k_2)]^T\right\} \\ &= \int_{-\infty}^{\infty}\int_{-\infty}^{\infty} [\lambda_1 - \bar{x}(k_1)][\lambda_2 - \bar{y}(k_2)]^T dF(\lambda_1, \lambda_2; x(k_1), y(k_2)). \end{aligned} \tag{B.98}$$

A Gaussian process is a stochastic process for which every finite-dimensional distribution is multivariate normal. Such a process is fully characterized by its mean function and covariance function.

B.2.1.3 Stationary Stochastic Processes

In this book, we define a *stationary stochastic process* as one whose finite-dimensional distribution functions are invariant under time shifts. A weaker notion, known as *weak stationarity* (or wide-sense stationarity), requires only that the first and second moments be time-invariant. Specifically, a stochastic process $x(k)$ is weakly stationary if

$$\begin{aligned} E\{x(k)\} &= \bar{x}, \quad \forall k \\ E\{[x(k) - \bar{x}][x(k-\tau) - \bar{x}]^T\} &= R_x(\tau), \quad \forall k, \end{aligned} \tag{B.99}$$

i.e., the mean is constant and the covariance depends only on the time lag τ, not on the absolute time index. For Gaussian processes, weak stationarity implies stationarity in the strict sense, as the distribution is completely characterized by the first two moments. In the scalar case, $R(0)$ corresponds to the variance of the signal, and the normalized covariance $\frac{R(\tau)}{R(0)}$ indicates the degree of time correlation. A value of 1 implies perfect correlation, while 0 indicates no correlation.

Many real-world signals contain both deterministic and stochastic components. It is often useful to treat such signals within a unified framework by defining empirical estimates of the mean and covariance:

$$\begin{aligned} \bar{x} &= \lim_{N\to\infty} \frac{1}{N} \sum_{k=1}^{N} x(k), \\ R_x(\tau) &= \lim_{N\to\infty} \frac{1}{N} \sum_{k=1}^{N} [x(k) - \bar{x}][x(k-\tau) - \bar{x}]^T. \end{aligned} \tag{B.100}$$

These definitions average out both deterministic trends and stochastic fluctuations. Signals for which the above limits exist are referred to as *quasi-stationary*. In the case of a purely stochastic stationary process, any particular realization should satisfy these relationships in the limit, consistent with the theoretical definitions of $\bar{x}$ and $R_x(\tau)$.

B.2.1.4 Spectra of Stationary Stochastic Processes

Throughout this chapter, we assume that continuous time has been rescaled such that each discrete time step corresponds to one unit of continuous time. If the actual sampling interval is $t_s \neq 1$, then frequency values in the discrete-time domain must be scaled by a factor of $1/t_s$ to correctly represent continuous-time frequencies.

The spectral density $\Phi_x(\omega)$ of a stationary stochastic process $x(k)$ is defined as the Fourier transform of its autocovariance function:

$$\Phi_x(\omega) = \frac{1}{2\pi} \sum_{\tau=-\infty}^{\infty} R_x(\tau) e^{-j\tau\omega}. \tag{B.101}$$

The spectral density describes how the power of the signal is distributed across different frequencies. Specifically, the power of $x(k)$ in a frequency band (ω_1, ω_2) is given by

$$2 \cdot \int_{\omega=\omega_1}^{\omega=\omega_2} \Phi_x(\omega) d\omega.$$

Peaks in the spectral density indicate the presence of dominant periodic components in the signal at the corresponding frequencies.

The autocovariance function $R_x(\tau)$ can be recovered from the spectrum via the inverse Fourier transform:

$$R_x(\tau) = \int_{-\pi}^{\pi} \Phi_x(\omega) e^{j\omega\tau} d\omega. \tag{B.102}$$

In particular, setting $\tau = 0$ yields:

$$R_x(0) = E\{x(k)x(k)^T\} = \int_{-\pi}^{\pi} \Phi_x(\omega) d\omega. \tag{B.103}$$

This result states that the total area under the spectral density curve equals the variance of the signal and is commonly referred to as *Parseval's relation*.

Example: Figure B.6 illustrates the time-series realizations, covariance functions, and corresponding spectral densities for the following stationary processes:

- **A.** White noise
- **B.** Mixture of two sinusoidal signals
- **C.** White noise passed through a low-pass filter.

These examples highlight key differences in temporal structure and frequency content across the three signal types. In particular, white noise exhibits no time correlation and a flat spectrum, while sinusoidal signals show strong periodicity and concentrated spectral peaks. Filtered white noise retains its stochastic nature but exhibits smoother temporal variations and a low-pass spectral profile.

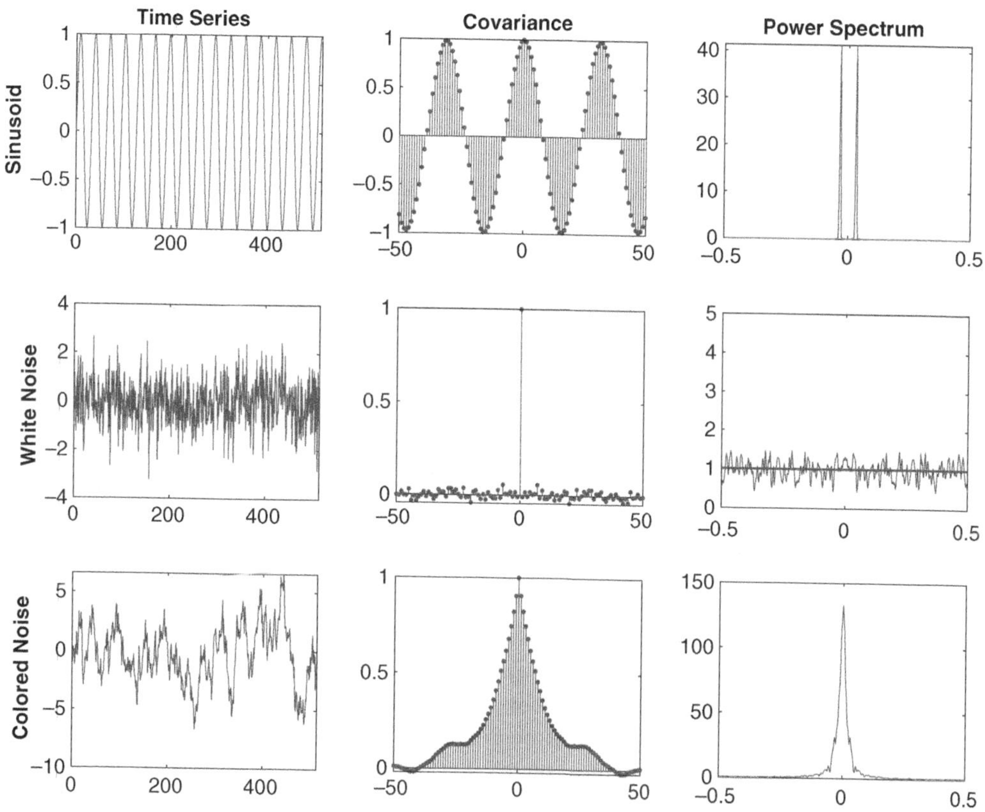

Figure B.6 Covariance and spectra of different signal types.

B.2.1.5 Discrete Time White Noise

A particular class of stochastic processes, known as white noise, will be used extensively in this book. A sequence $x(k)$ is said to be *white* if

$$\mathcal{P}(x(k) \mid x(\ell)) = \mathcal{P}(x(k)) \quad \text{for all } \ell < k, \tag{B.104}$$

i.e., the elements of the sequence are mutually independent across time. In such cases, knowledge of $x(\ell)$ provides no information about $x(k)$ for any $\ell < k$.

If the white noise process is also stationary, it satisfies the following properties:

$$\begin{aligned} E\{x(k)\} &= \overline{x}, \quad \forall k \\ E\left\{(x(k) - \overline{x})(x(k-\tau) - \overline{x})^T\right\} &= \begin{cases} R_x, & \text{if } \tau = 0 \\ 0, & \text{if } \tau \neq 0. \end{cases} \end{aligned} \tag{B.105}$$

Thus, the covariance structure of white noise is fully characterized by a single matrix R_x, corresponding to the variance at lag zero.

The power spectral density of white noise is constant across all frequencies. Using the definition in Eq. (B.101), we have

$$\Phi_x(\omega) = \frac{1}{2\pi} R_x, \tag{B.106}$$

indicating that white noise distributes its power uniformly over the entire frequency domain. The term "white noise" is inspired by the analogy with white light, which contains all frequencies of the visible spectrum with equal intensity.

B.2.1.6 Colored Noise

A stochastic process generated by passing white noise through a linear dynamic system is referred to as *colored noise*.

Any stationary stochastic process with a given mean and covariance function can be generated by filtering white noise through an appropriately chosen dynamical system. To illustrate this, consider the model:

$$d(k) = H(q)\varepsilon(k) + \bar{d}, \tag{B.107}$$

where $\varepsilon(k)$ is a white noise process with identity covariance, $\bar{d}$ is the constant mean, and $H(q)$ is a stable and stably invertible transfer function (or transfer matrix).

Using standard spectral analysis, one can show that the spectral density of $d(k)$ is given by

$$\Phi_d(\omega) = \frac{1}{2\pi} H(e^{j\omega})H^T(e^{-j\omega}). \tag{B.108}$$

According to the *spectral factorization theorem*, there exists a stable and stably invertible filter $H(q)$ that satisfies Eq. (B.108). That is, any stationary process with a valid spectral density can be synthesized using a linear filter driven by white noise.

This result is particularly valuable for modeling disturbances or stochastic inputs whose covariance functions are known or prescribed. Moreover, since a stationary Gaussian process is fully characterized by its mean and covariance, such a process can be synthesized by filtering a zero-mean Gaussian white sequence through a suitable dynamic system and adding a constant bias if the mean is nonzero.

B.2.1.7 Integrated White Noise and Nonstationary Processes

Some stochastic processes exhibit random mean shifts, where both the magnitude and the direction of the shifts are random. A fundamental model that captures such behavior is the integrated white noise $\varepsilon_{\text{int}}(k)$ (also known as a random walk), defined as

$$\varepsilon_{\text{int}}(k) = \varepsilon_{\text{int}}(k-1) + \varepsilon(k),$$

where $\varepsilon(k)$ is a white noise sequence. This process is nonstationary, as its mean and variance evolve over time. It is widely used to model systems with accumulative random effects, such as financial prices or drifting sensor biases. Figure B.7 displays realizations of this process for different distributions of $\varepsilon(k)$. Despite the stochasticity in step sizes, the cumulative nature of the process results in persistent trends.

More generally, many real-world signals exhibit a combination of stationary fluctuations and randomly occurring level shifts, i.e.,

$$y(k) = \varepsilon_{\text{int}}(k) + H(q^{-1})\varepsilon(k).$$

Figure B.8 shows combinations of integrated white noise and stationary noise in different proportions.

Alternatively, the combined effects can be equivalently represented as an integrated white noise process passed through a different shaping filter $\tilde{H}(q^{-1})$:

$$y(k) = \tilde{H}(q^{-1})\varepsilon_{\text{int}}(k).$$

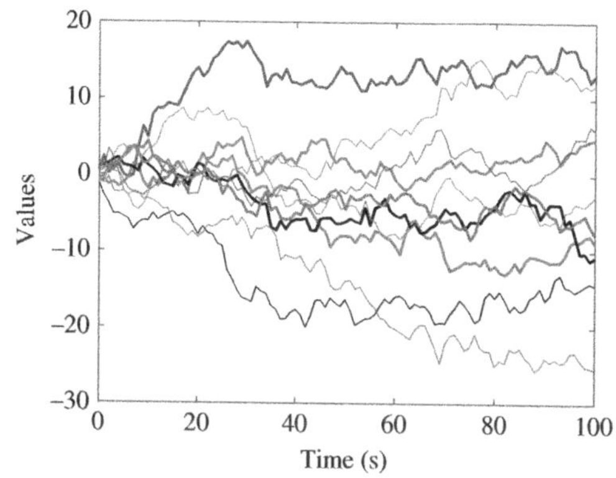

Figure B.7 Realizations of integrated white noise process.

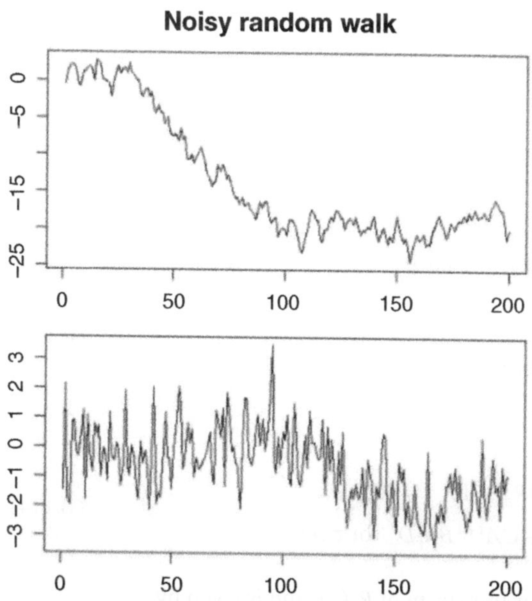

Figure B.8 Combined effects of integrated white noise and stationary noise of varying proportions. The top plot has a higher proportion of integrated white noise than the bottom plot.

This formulation is particularly useful when the integrated component dominates long-term behavior. It captures both stationary and non-stationary dynamics.

While the process $y(k)$ is nonstationary, the first-order difference,

$$\Delta y(k) = y(k) - y(k-1),$$

is stationary. This property is essential in time-series modeling (e.g., ARIMA models), where differencing is used to transform nonstationary data into stationary form.

B.2.1.8 Stochastic Difference Equation

A general framework for modeling stochastic processes is the *stochastic difference equation*, given by

$$\begin{aligned} x(k+1) &= Ax(k) + B\varepsilon(k) \\ y(k) &= Cx(k) + D\varepsilon(k), \end{aligned} \tag{B.109}$$

where $\varepsilon(k)$ is a white noise sequence with zero mean and covariance R_ε.

The mean and covariance of the state $x(k)$ evolve according to

$$\begin{aligned} E\{x(k)\} &= AE\{x(k-1)\} = A^k E\{x(0)\} \\ E\{x(k)x^T(k)\} &= AE\{x(k-1)x^T(k-1)\}A^T + BR_\varepsilon B^T. \end{aligned} \tag{B.110}$$

If all eigenvalues of A lie strictly inside the unit disk (i.e., A is stable), then as $k \to \infty$

$$\begin{aligned} \lim_{k\to\infty} E\{x(k)\} &= 0 \\ \lim_{k\to\infty} E\{x(k)x^T(k)\} &= R_x, \end{aligned} \tag{B.111}$$

where R_x is the unique solution to the *discrete-time Lyapunov equation*:

$$R_x = AR_xA^T + BR_\varepsilon B^T. \tag{B.112}$$

Since the output is given by $y(k) = Cx(k) + D\varepsilon(k)$, we have

$$\begin{aligned} E\{y(k)\} &= CE\{x(k)\} + DE\{\varepsilon(k)\} = 0 \\ E\{y(k)y^T(k)\} &= CE\{x(k)x^T(k)\}C^T + DE\{\varepsilon(k)\varepsilon^T(k)\}D^T = CR_xC^T + DR_\varepsilon D^T. \end{aligned} \tag{B.113}$$

The autocovariance function of $y(k)$ is defined as

$$R_y(\tau) \triangleq E\{y(k+\tau)y^T(k)\} = \begin{cases} CR_xC^T + DR_\varepsilon D^T \text{ for } \tau = 0 \\ CA^\tau R_xC^T + CA^{\tau-1}BR_\varepsilon D^T \text{ for } \tau > 0. \end{cases} \tag{B.114}$$

The power spectral density of $y(k)$ is the Fourier transform of $R_y(\tau)$, and is given by

$$\Phi_y(\omega) = \frac{1}{2\pi}\left(C(e^{j\omega}I - A)^{-1}B + D\right) R_\varepsilon \left(C(e^{j\omega}I - A)^{-1}B + D\right)^T. \tag{B.115}$$

If the matrix A contains eigenvalues on or outside the unit circle, the process $x(k)$—and thus $y(k)$—is nonstationary due to unbounded growth in the state covariance (as seen in Eq. [B.110]). This nonstationarity is often intentional, particularly when modeling mean-shifting behavior, such as in random walks.

A common approach is to include integrators in the system and focus on modeling the differenced output:

$$\begin{aligned} x(k+1) &= Ax(k) + B\varepsilon(k) \\ \Delta y(k) &= Cx(k) + D\varepsilon(k). \end{aligned} \tag{B.116}$$

In this case, even though $y(k)$ is nonstationary, the differenced signal $\Delta y(k)$ is stationary as long as A is stable.

C

Model Reduction

Model reduction refers to the process of transforming a high-order model into a lower-order one that preserves the essential input–output behavior of the original system. Lower-order models are generally preferred because they reduce computational complexity and simplify both controller design and implementation.

In many cases, the modeling approach used—such as step response identification—naturally produces high-order models, which may not even be minimal. For systems with many inputs and outputs, this can result in models containing hundreds or even thousands of states. The same is often true for first-principles models derived from detailed physical equations. These scenarios motivate the use of model reduction techniques that produce compact models without sacrificing the dynamic fidelity needed for controller design.

Model reduction goes beyond the task of eliminating redundant states to produce a minimal realization. It typically involves a trade-off: sacrificing some accuracy in the input–output behavior to achieve a significantly reduced model order. This is often done by transforming the system into a coordinate basis where the states can be ranked by importance, then discarding those deemed negligible. Because the relative importance of different dynamic behaviors depends on the application, no single model reduction method is universally optimal.

In this section, we focus on a widely used and effective technique called *balanced truncation*, which has proven especially useful in control applications. It also offers significant numerical advantages when applied to finite impulse response (FIR)-based models.

C.1 Model Reduction Problem

Let us consider a system described by the discrete-time state-space model (6.1), with state dimension n. The goal of model reduction is to obtain a lower-order system of order $r < n$ that closely approximates the input–output behavior of the original system.

It is important to note that the state-space representation of a given input–output behavior is not unique; a change of state coordinates does not affect the system's external behavior. For the purpose of model reduction, we aim to find a realization in which the states are defined and ordered according to their relative importance in describing the system's input–output dynamics.

To achieve this, we introduce a state coordinate transformation:

$$\tilde{x}(k) = Tx(k),$$

where T is a nonsingular matrix. In the transformed coordinates, the state-space equations become

$$\begin{aligned} \tilde{x}(k+1) &= \tilde{A}\tilde{x}(k) + \tilde{B}v(k) \\ y(k) &= \tilde{C}\tilde{x}(k), \end{aligned} \tag{C.1}$$

where the transformed matrices are defined as

$$\tilde{A} = TAT^{-1}, \tilde{B} = TB, \tilde{C} = CT^{-1}.$$

If the transformation T is selected such that the transformed states are ranked by their contribution to the input–output behavior, then model reduction can be accomplished by truncating the last $n - r$ state variables. This yields a reduced-order model that retains only the most dynamically significant modes of the original system.

$$\begin{aligned} \tilde{x}_r(k+1) &= \tilde{A}_r\tilde{x}_r(k) + \tilde{B}_r v(k) \\ y(k) &= \tilde{C}_r\tilde{x}_r(k), \end{aligned} \tag{C.2}$$

where $\tilde{x}_r(k) = \begin{bmatrix} I_r & 0 \end{bmatrix} \tilde{x}(k)$, and

$$\begin{aligned} \tilde{A}_r &= \begin{bmatrix} I_r & 0 \end{bmatrix} \tilde{A} \begin{bmatrix} I_r \\ 0 \end{bmatrix} \\ \tilde{B}_r &= \begin{bmatrix} I_r & 0 \end{bmatrix} \tilde{B} \\ \tilde{C}_r &= \tilde{C} \begin{bmatrix} I_r \\ 0 \end{bmatrix}. \end{aligned} \tag{C.3}$$

The remainder of this section discusses how to construct a suitable coordinate transformation T that facilitates such a reduction.

C.2 Hankel Matrix and Hankel Singular Values

Let us define the controllability and observability matrices as

$$\mathcal{W}^c = \begin{bmatrix} B & AB & A^2B & \cdots \end{bmatrix} \tag{C.4}$$

$$\mathcal{W}^o = \begin{bmatrix} C^T & (CA)^T & (CA^2)^T & \cdots \end{bmatrix}^T. \tag{C.5}$$

Recall that the truncated versions of these matrices, $\mathcal{W}_n^c$ and $\mathcal{W}_n^o$, were previously introduced to assess reachability and observability, which explains their naming conventions.

The Hankel matrix $\mathcal{H}_G$ is then defined as

$$\mathcal{H}_G \triangleq \mathcal{W}^o \cdot \mathcal{W}^c. \tag{C.6}$$

The subscript G indicates that this is the Hankel matrix associated with the discrete-time system operator $G = C(zI - A)^{-1}B$. The Hankel matrix $\mathcal{H}_G$ is a doubly infinite matrix that maps past inputs to current and future outputs. This structure becomes evident by noting the following relationships:

$$x(0) = \mathcal{W}^c \begin{bmatrix} u(-1) \\ u(-2) \\ u(-3) \\ \vdots \end{bmatrix}, \tag{C.7}$$

and

$$\begin{bmatrix} y(0) \\ y(1) \\ y(2) \\ \vdots \end{bmatrix} = \mathcal{W}^o x(0). \tag{C.8}$$

Here, $\mathcal{W}^c$ maps the past inputs $u(-\infty, 0)$ to the current state $x(0)$, which is then mapped to the current/future outputs $y[0, \infty)$ through $\mathcal{W}^o$.

Assuming the original system is controllable and observable, both the controllability matrix $\mathcal{W}^c$ and the observability matrix $\mathcal{W}^o$ have full rank n, and therefore $\mathcal{H}_G$ has rank n. In this case, $\mathcal{H}_G$ has exactly n nonzero singular values, which are defined as

$$\begin{aligned} \sigma_i(\mathcal{H}_G) &= \sigma_i(\mathcal{W}^o\mathcal{W}^c) \\ &\triangleq \left[\lambda_i((\mathcal{W}^c)^T(\mathcal{W}^o)^T\mathcal{W}^o\mathcal{W}^c)\right]^{1/2} \\ &= \left[\lambda_i(\mathcal{W}^c(\mathcal{W}^c)^T(\mathcal{W}^o)^T\mathcal{W}^o)\right]^{1/2}. \end{aligned} \tag{C.9}$$

Here $\lambda_i(\cdot)$ denotes the ith eigenvalue. The quantities $\sigma_i(\mathcal{H}_G)$'s are called *Hankel singular values* of G, and they represent the relative importance of the system modes in shaping the input–output behavior. The largest singular value, denoted $\bar{\sigma}(\mathcal{H}_G)$, is known as the *Hankel norm* of G.

We define the following two matrices:

$$P \triangleq \mathcal{W}^c(\mathcal{W}^c)^T = \sum_{j=0}^{\infty} A^j BB^T A^{Tj}. \tag{C.10}$$

$$Q \triangleq (\mathcal{W}^o)^T\mathcal{W}^o = \sum_{j=0}^{\infty} A^{Tj} C^T C A^j. \tag{C.11}$$

Here, P and Q are known as the *controllability* and *observability Gramians*, respectively. These Gramians satisfy the following discrete-time Lyapunov equations:

$$APA^T - P + BB^T = 0. \tag{C.12}$$

$$A^TQA - Q + C^TC = 0. \tag{C.13}$$

These equations can be verified by substituting the definitions of P and Q into the respective expressions. Matrix equations of this form are referred to as Lyapunov equations, and their numerical properties have been extensively studied due to their fundamental role in control theory and applications.

The ith Hankel singular value of a system is given by $[\lambda_i(PQ)]^{1/2}$, where P and Q are the positive-definite solutions to the Lyapunov equations (C.12) and (C.13), respectively. Computing the Hankel singular values therefore requires solving this pair of Lyapunov equations. Importantly, since the Hankel matrix $\mathcal{H}_G$, and consequently its singular values, depend solely on the system's input–output behavior, they are invariant under state coordinate transformations.

A natural goal in model reduction is to find a state-space transformation such that the truncated model—obtained via the projection in (C.2)—produces a Hankel matrix that approximates the original $\mathcal{H}_G$ as closely as possible. It has been shown that there exists a reduced-order model G_r of order r such that:

$$\bar{\sigma}\left(\mathcal{H}_G - \mathcal{H}_{G_r}\right) = \sigma_{r+1}(\mathcal{H}_G). \tag{C.14}$$

This result establishes that the error in the optimal reduced Hankel matrix, measured in the Hankel norm, is precisely the Hankel singular value of the original system.

Although this optimal reduced-order model exists, the procedure to construct such a system is quite involved. A seemingly straightforward approach would be to compute the singular value decomposition (SVD) of $\mathcal{H}_G$ and truncate it to retain only the top r principal components. However, the resulting rank-r matrix may no longer possess the Hankel structure required to correspond to a realizable system with valid state-space matrices (A_r, B_r, C_r).

C.3 Balanced Realization and Truncation

While it is true that

$$\min_{G_r} \bar{\sigma}\left(\mathcal{H}_G - \mathcal{H}_{G_r}\right) = \sigma_{r+1}(\mathcal{H}_G), \tag{C.15}$$

computing such an optimal reduced model G_r can be numerically challenging. In this section, we introduce a model reduction technique known as *balanced truncation*, which provides a close approximation to the optimal Hankel-norm solution while being computationally efficient and numerically robust.

In a *balanced realization*, the controllability and observability Gramians are both equal to a diagonal matrix composed of the Hankel singular values, ordered by magnitude:

$$P = Q = \Sigma_{\mathcal{H}_G} \triangleq \begin{bmatrix} \sigma_1(\mathcal{H}_G) & & \\ & \ddots & \\ & & \sigma_n(\mathcal{H}_G) \end{bmatrix}. \tag{C.16}$$

In this form, each state variable contributes a distinct and quantifiable amount to both the reachability and observability of the system. The last state is the least controllable and observable, the second-to-last is the next least, and so on.

Before presenting the procedure for constructing a balanced realization, let us recall how the Gramians transform under a change of coordinates. If $\tilde{x} = Tx$ is a state transformation, then the Gramians in the new coordinates are

$$\tilde{P} = TPT^T. \tag{C.17}$$

$$\tilde{Q} = (T^{-1})^T Q T^{-1}. \tag{C.18}$$

The first step in obtaining a balanced realization is to factor the positive-definite matrix Q as follows:

$$Q = R^T R. \tag{C.19}$$

This factorization is known as the *Cholesky factorization*. Applying the state coordinate transformation $\tilde{x}_1 = Rx$ results in

$$P_1 \stackrel{\Delta}{=} RPR^T. \tag{C.20}$$

$$Q_1 \stackrel{\Delta}{=} (R^{-1})^T Q R^{-1} = I. \tag{C.21}$$

Next, we perform a SVD of the symmetric positive-definite matrix P_1:

$$P_1 = U_1 \Sigma^2_{\mathcal{H}_G} U_1^T. \tag{C.22}$$

Then, applying a second state coordinate transformation $\tilde{x}_2 = \Sigma^{-1/2}_{\mathcal{H}_G} U_1^T \tilde{x}_1$ yields balanced controllability and observability Gramians, as shown in below equation:

$$P_2 = \Sigma^{-1/2}_{\mathcal{H}_G} U_1^T \left[U_1 \Sigma^2_{\mathcal{H}_G} U_1^T \right] U_1 \Sigma^{-1/2}_{\mathcal{H}_G} = \Sigma_{\mathcal{H}_G}. \tag{C.23}$$

$$Q_2 = \Sigma^{1/2}_{\mathcal{H}_G} U_1^T U_1 \Sigma^{1/2}_{\mathcal{H}_G} = \Sigma_{\mathcal{H}_G}. \tag{C.24}$$

In summary, the complete coordinate transformation

$$\tilde{x} = \Sigma^{-1/2}_{\mathcal{H}_G} U_1^T R x, \tag{C.25}$$

produces a balanced realization. Numerically, computing this transformation involves solving a pair of Lyapunov equations, followed by Cholesky factorization and SVD of a symmetric positive-definite matrix.

The model reduction technique that combines the balanced realization with the truncation of type (C.2) is called *balanced truncation*. Originally proposed as a method for eliminating states that are practically unreachable and/or unobservable, this technique has since been shown to offer a close approximation to the optimal Hankel-norm reduction. While it does not produce the exact optimal Hankel-norm approximation, it performs very well in practice and provides a guaranteed bound on the approximation error:

$$\bar{\sigma}\left(\mathcal{H}_G - \mathcal{H}_{G_r}\right) \leq 2 \sum_{j=r+1}^{n} \sigma_j(\mathcal{H}_G). \tag{C.26}$$

Similarly, balanced truncation provides an upper bound on the L_∞ norm of the truncation error:

$$\|G(q) - G_r(q)\|_\infty \stackrel{\Delta}{=} \sup_\omega \bar{\sigma}\left(G(e^{j\omega}) - G_r(e^{j\omega})\right) \leq 2 \sum_{j=r+1}^{n} \sigma_j(\mathcal{H}_G). \tag{C.27}$$

C.4 Application to FIR Models

Recall that the observable canonical realization of a FIR model results in

$$A = \left.\begin{bmatrix} 0 & I & 0 & \dots & \dots & 0 & 0 \\ 0 & 0 & I & 0 & \dots & 0 & 0 \\ \vdots & & & & & & \vdots \\ 0 & 0 & \dots & \dots & \dots & 0 & I \\ 0 & 0 & \dots & \dots & \dots & 0 & 0 \end{bmatrix}\right\} n \times n_v \tag{C.28}$$

$$B = \begin{bmatrix} H_1^T & H_2^T & \cdots & H_n^T \end{bmatrix}^T. \tag{C.29}$$

$$C = \begin{bmatrix} I & 0 & \cdots & 0 \end{bmatrix}. \tag{C.30}$$

To obtain a balanced realization for such a system, we first compute the controllability and observability Gramians by solving the Lyapunov equations:

$$APA^T - P + BB^T = 0. \tag{C.12}$$

$$A^TQA - Q + C^TC = 0. \tag{C.13}$$

Owing to the special structure of the A, B, and C matrices for FIR models, we can derive explicit solutions. In particular, the observability Gramian Q is simply the identity matrix. The controllability Gramian P can be computed recursively as follows:

$$P_{i,n} = (BB^T)_{i,n} \qquad 1 \le i \le n. \tag{C.31}$$

$$P_{i,j} = P_{i+1,j+1} + (BB^T)_{i,j} \qquad 1 \le i \le j \le n-1. \tag{C.32}$$

$P_{i,j} = P_{j,i}$.. with summary

Since P is symmetric and positive-definite, it admits a SVD of the form:

$$P = U\Sigma_{\mathcal{H}G}^2 U^T. \tag{C.33}$$

A balanced realization $(\tilde{A}, \tilde{B}, \tilde{C})$ can then be obtained via the coordinate transformation $\tilde{x} = \Sigma_{\mathcal{H}G}^{-1/2} U^T x$.

In summary, by exploiting the structured realization of FIR models, we bypass the computationally intensive steps of solving Lyapunov equations and performing Cholesky factorization when constructing a balanced realization. Once the balanced model is obtained, model order reduction can be performed by examining the Hankel singular values and truncating the subspace associated with negligible singular values. If $\sigma_{r+1} \ll \sigma_r$, then the rth-order model obtained through balanced truncation will closely approximate the input–output behavior of the original system.

Example C.1: *Model Reduction of a FIR System.*
A FIR model was developed in Eq. (5.69) of Example 5.8 for the system:

$$g = \frac{2.69(-6s+1)}{(20s+1)(5s+1)} e^{-1.5s}.$$

With $t_s = 3$, a total of $n = 40$ states were retained to construct a state-space model. In this example, we apply balanced truncation to reduce the model order.

First, the controllability and observability Gramians P and Q were generated using the procedure outlined earlier for FIR models. Alternatively, one could solve the Lyapunov equations (C.12) and (C.13) and confirm that the resulting matrices are identical. As expected, Q is the $n \times n$ identity matrix, and P is given by

$$P = \begin{bmatrix} 0.5474 & 0.4957 & 0.4132 & \dots \\ 0.4957 & 0.5170 & 0.4858 & \dots \\ 0.4132 & 0.4858 & 0.5137 & \dots \\ \vdots & \vdots & \vdots & \ddots \end{bmatrix}.$$

Next, we performed a SVD of P to obtain

$$\Sigma_{\mathcal{H}_G} = \sqrt{\Sigma_P} = \text{diag}\left[1.8800 \quad 0.6663 \quad 0.0945 \quad 0.0097 \quad 0.0069 \quad \dots\right].$$

Figure C.1 presents a bar chart of the Hankel singular values. While the state-space model in Example 5.8 used three states, the singular value spectrum suggests that a second-order in Figure C.2.

The original model was

$$g_z = \frac{-0.1744z^2 + 0.189z + 0.1545}{z^3 - 1.41z^2 + 0.4724z},$$

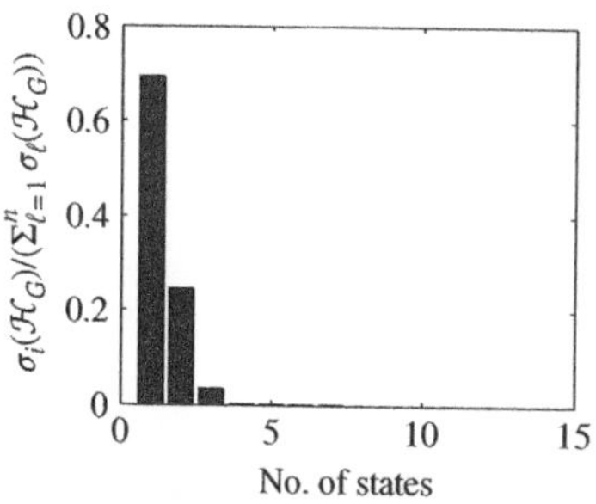

Figure C.1 Hankel singular values, $\sigma_{i(\mathcal{H}_G)} / \left(\sum_{\ell=1}^{n} \sigma_\ell(\mathcal{H}_G) \right)$.

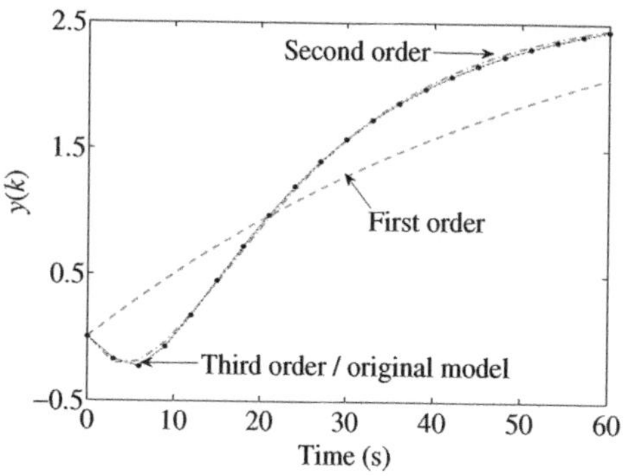

Figure C.2 Comparison of step responses of various reduced-order models with the original model.

whereas the reduced-order models in transfer function form are

$$G_{r=1} = \frac{0.1591}{z - 0.9512} \qquad G_{r=2} = \frac{-0.2087z + 0.3393}{z^2 - 1.544z + 0.5937}$$

$$G_{r=3} = \frac{-0.1744z^2 + 0.1891z + 0.1542}{z^3 - 1.41z^2 + 0.473z - 8.7 \times 10^{-5}}.$$

D

Optimality of Kalman Filter and LQG Controller for Linear Gaussian Systems

The Kalman filter (KF) and linear quadratic Gaussian (LQG) controller were introduced in the main text. In this appendix, we revisit their derivations to rigorously establish their optimality for linear systems with Gaussian noise using probabilistic arguments. We also show the equivalence between the KF and the moving horizon estimator (MHE) in the case of linear, unconstrained systems, providing a theoretical foundation for MHE in its simplest form. While this case is limited, MHE's appeal lies in its natural extension to constrained and nonlinear settings. As the analysis draws on advanced concepts from stochastic processes, it is intended for readers with a more technical background. To preserve the flow of Chapters 7 and 8, this material is presented here in the appendix.

D.1 KF as the Bayesian Estimator for Gaussian Systems

In Chapter 7, we introduced a linear observer structure and formulated the state estimation problem as a parametric optimization task—minimizing the expected estimation error variance with respect to the observer gain. However, the KF can also be derived from a fully probabilistic perspective: as a Bayesian estimator that recursively computes the conditional density of the state $x(k)$. This probabilistic approach reveals the KF as the optimal estimator for linear systems with Gaussian noise. For further discussion on least squares estimation as a form of maximum a posteriori (MAP) estimation in Markov systems, refer to Jazwinski (2007).

Assume that $\varepsilon_1(k)$ and $\varepsilon_2(k)$ are zero-mean Gaussian noise sequences. If the initial state $x(0)$ is also Gaussian, then the state and output sequences, $x(k)$ and $y(k)$, are jointly Gaussian for all k. This allows us to formulate the state estimation problem as computing the conditional expectation $E\{x(k) \mid Y(k)\}$, where the observation history is defined as

$$Y(k) = [y^T(1),\ \ldots, y^T(k)]^T.$$

We denote the conditional expectation $E\{x(i) \mid Y(j)\}$ by $\hat{x}(i|j)$. The KF recursively computes this estimate in two steps:

- **Model Update:** *Compute* $E\{x(k)|Y(k-1)\}$ *given* $E\{x(k-1)|Y(k-1)\}$ *and* $u(k-1)$.

Since the system evolves as $x(k) = Ax(k-1) + Bu(k-1) + \varepsilon_1(k-1)$ and $\varepsilon_1(k-1)$ is independent of $Y(k-1)$ and has zero mean, we obtain

$$\begin{aligned}\hat{x}(k|k-1) &= E\{Ax(k-1) + Bu(k-1) + \varepsilon_1(k-1) \mid Y(k-1)\}\\ &= AE\{x(k-1) \mid Y(k-1)\} + Bu(k-1).\end{aligned} \tag{D.1}$$

Hence, we obtain

$$\hat{x}(k|k-1) = A\hat{x}(k-1|k-1) + Bu(k-1). \tag{D.2}$$

The prediction error is

$$x(k) - \hat{x}(k|k-1) = A\,(x(k-1) - \hat{x}(k-1|k-1)) + \varepsilon_1(k-1). \tag{D.3}$$

The corresponding covariance is

$$\begin{aligned}P(k) &= E\left\{(x(k) - \hat{x}(k|k-1))\,(x(k) - \hat{x}(k|k-1))^T\right\}\\ &= A\bar{P}(k-1)A^T + R_1.\end{aligned} \tag{D.4}$$

Because the conditional distribution of $x(k)$ given $Y(k-1)$ is Gaussian, it is fully specified by its mean $\hat{x}(k|k-1)$ and covariance $P(k)$.

- **Measurement Update:** *Compute $E\{x(k)|Y(k)\}$ given $E\{x(k)|Y(k-1)\}$, $P(k)$ and $y(k)$.*

  Prior to measurement $y(k)$, (k) has the Gaussian density of mean $\hat{x}(k|k-1)$ and covariance $P(k)$. Given

  $$y(k) = Cx(k) + \varepsilon_2(k),$$

  and the independence of $\varepsilon_2(k)$, the pair $(x(k), y(k))$ is jointly Gaussian with the following joint covariance

  $$E\left\{\begin{bmatrix} x(k) - \hat{x}(k|k-1) \\ y(k) - y(k|k-1) \end{bmatrix}\begin{bmatrix} x(k) - \hat{x}(k|k-1) \\ y(k) - y(k|k-1) \end{bmatrix}^T\right\} = \begin{bmatrix} P(k) & P(k)C^T \\ CP(k) & CP(k)C^T + R_2 \end{bmatrix}. \tag{D.5}$$

  From the known formula for the conditional distribution of jointly Gaussian variables:

  $$E\{x|y\} = E\{x\} + R_{xy}R_y^{-1}(y - E\{y\}). \tag{D.6}$$

  $$\mathrm{Cov}\{x|y\} = R_x - R_{xy}R_y^{-1}R_{yx}, \tag{D.7}$$

  we obtain the Kalman update equations:

  $$\begin{aligned}\hat{x}(k|k) &= E\{x(k)|y(k)\}\\ &= \hat{x}(k|k-1) + P(k)C^T\left(CP(k)C^T + R_2\right)^{-1}(y(k) - C\hat{x}(k|k-1)).\end{aligned} \tag{D.8}$$

  $$\begin{aligned}\bar{P}(k) &= \mathrm{Cov}\{x(k)|y(k)\}\\ &= P(k) - P(k)C^T\left(CP(k)C^T + R_2\right)^{-1}CP(k).\end{aligned} \tag{D.9}$$

In summary, for Gaussian systems, the conditional mean and covariance of the state $x(k)$ can be computed recursively using the following equations:

$$\hat{x}(k|k-1) = A\hat{x}(k-1|k-1) + Bu(k-1). \tag{D.10}$$

$$\hat{x}(k|k) = \hat{x}(k|k-1) + \underbrace{P(k)C^T\left(CP(k)C^T + R_2\right)^{-1}}_{\bar{K}(k)}(y(k) - C\hat{x}(k|k-1)). \tag{D.11}$$

The associated error covariance updates are given by

$$P(k) = A\bar{P}(k-1)A^T + R_1. \tag{D.12}$$

$$\bar{P}(k) = P(k) - P(k)C^T\left(CP(k)C^T + R_2\right)^{-1}CP(k). \tag{D.13}$$

These equations define a linear estimator whose gain, $\bar{K}(k)$, is determined by the KF.

D.2 Moving Horizon Estimation: Recursive Solution to the Unconstrained Linear Problem

In this section, we derive the so-called *arrival cost* to obtain a recursive formulation of the moving horizon estimator (MHE) and demonstrate its equivalence to the KF in the unconstrained, linear case. The derivation follows the principles of dynamic programming.

For simplicity, we omit the terms involving the known deterministic input u, as their exclusion does not impact the core analysis. The remainder of the section is dedicated to deriving the recursive expression given in Eq. (7.59).

D.2.1 Dynamic Programming and Arrival Cost

We now explore the possibility of computing the estimate $\hat{x}(k+1|k)$ recursively by solving the least squares problem in (7.54) as follows:

$$\begin{aligned} J_k = \min_{z_1, z_2, \ldots, z_{k+1}} & \left[(z_1 - \hat{x}(1|0))^T Q_0(1)(z_1 - \hat{x}(1|0))\right. \\ & + \sum_{i=1}^{k} \{(z_{i+1} - Az_i)^T Q_1 (z_{i+1} - Az_i) + (y(i) - Cz_i)^T Q_2 (y(i) - Cz_i)\}. \end{aligned} \tag{D.14}$$

Definition D.1: The function $\Phi_{K_f(\zeta)} \triangleq J_k$ with the constraint $z_{k+1} = \zeta$ is called the *k-step arrival cost* for ζ. It represents the minimum value of the cost functional J_k under the additional constraint that the state at time $k+1$ is fixed at ζ.

Based on the definition above, the arrival cost $\Phi_{K_f(z_{k+1})}$ can be expressed as

$$\begin{aligned} \Phi_{K_f(z_{k+1})} = \min_{z_1, \cdots, z_k} & \left[(z_1 - \hat{x}(1|0))^T Q_0(1)(z_1 - \hat{x}(1|0))\right. \\ & + \sum_{i=1}^{k} \{(z_{i+1} - Az_i)^T Q_1 (z_{i+1} - Az_i) + (y(i) - Cz_i)^T Q_2 (y(i) - Cz_i)\}. \end{aligned} \tag{D.15}$$

This formulation is particularly useful for constructing a recursive solution to J_k through stage-wise dynamic programming. We now outline the basic idea behind this approach.

Consider the problem of computing the one-step arrival cost:

$$\begin{aligned} \Phi_1(z_2) = \min_{z_1} & \left[(z_1 - \hat{x}(1|0))^T Q_0(1)(z_1 - \hat{x}(1|0)) + (z_2 - Az_1)^T Q_1 (z_2 - Az_1)\right. \\ & \left. + (y(1) - Cz_1)^T Q_2 (y(1) - Cz_1)\right]. \end{aligned} \tag{D.16}$$

Here, $\Phi_1(z_2)$ represents the arrival cost associated with reaching state z_2, i.e., the minimum cost incurred up to time step $k = 1$ that results in a state value of z_2.

Given this definition, we can now rewrite Eq. (D.14) in a recursive form:

$$\min_{z_2,\cdots,z_{k+1}} \left[\Phi_1(z_2) + \sum_{i=2}^{k} \{(z_{i+1} - Az_i)^T Q_1 (z_{i+1} - Az_i) + (y(i) - Cz_i)^T Q_2 (y(i) - Cz_i)\} \right]. \tag{D.17}$$

This reformulation is valid because, *for a fixed value of* z_2, the remaining terms in the objective function do not depend on z_1.

At the next stage, we define the two-step arrival cost recursively:

$$\Phi_2(z_3) = \min_{z_2} \left[\Phi_1(z_2) + (z_3 - Az_2)^T Q_1 (z_3 - Az_2) + (y(2) - Cz_2)^T Q_2 (y(2) - Cz_2) \right]. \tag{D.18}$$

By repeating this process recursively, we obtain the general update rule for the arrival cost at time step k:

$$\Phi_k(z_{k+1}) = \min_{z_k} \left[\Phi_{k-1}(z_k) + (z_{k+1} - Az_k)^T Q_1 (z_{k+1} - Az_k) + (y(k) - Cz_k)^T Q_2 (y(k) - Cz_k) \right]. \tag{D.19}$$

Finally, we solve

$$J_k = \min_{z_{k+1}} \Phi_k(z_{k+1}) \tag{D.20}$$

to obtain the optimal estimate $\hat{x}(k+1|k)$.

The key question at this point is whether a recursive expression for $\Phi_i(z_{i+1})$ can be derived at each stage. If so, we can compute $\hat{x}(i+1|i)$ in a recursive manner. As we will show next, the answer is indeed affirmative.

D.2.2 Recursive Calculation of the Arrival Cost and One-step-ahead Prediction

To derive a recursive expression for the arrival cost and obtain the one-step-ahead state estimate, it is helpful to introduce the following lemma:

Lemma D.1: *Let* x^* *be the solution to the least squares problem.*

$$\min_x x^T H x - 2g^T x. \tag{D.21}$$

Then, the optimal solution and associated cost can be expressed as

$$x^* = H^{-1} g, \tag{D.22}$$

and

$$\begin{aligned} x^T H x - 2g^T x &= (x - x^*)^T H (x - x^*) - g^T H^{-1} g \\ &= -g^T H^{-1} g \ with\ x = x* \end{aligned} \tag{D.23}$$

(D.22) can be obtained by setting the derivative of the objective function with respect to x to zero. The first equation of (D.23) can be proved by substituting (D.22) for x^* in the right-hand-side of the equation and showing that it reduces to the left-hand side. The second equation follows immediately by setting $x = x^*$.

We now show that a closed-form expression for the arrival cost can be derived at the first stage. Specifically, we need to solve the following optimization problem:

$$\Phi_1(z_2) = \min_{z_1} \left[(z_1 - \hat{x}(1|0))^T Q_0(1)(z_1 - \hat{x}(1|0))\right.$$

$$+ (z_2 - Az_1)^T Q_1(z_2 - Az_1) \tag{D.24}$$

$$\left. + (y(1) - Cz_1)^T Q_2(y(1) - Cz_1)\right]. \tag{D.25}$$

To proceed, we rewrite the objective in the standard least squares form so that we can apply Lemma D.1 to obtain a closed-form solution.

$$\begin{aligned}\Phi_1(z_2) = \min_{z_1} &\left[z_1^T(Q_0(1) + C^T Q_2 C + A^T Q_1 A)z_1\right.\\ &- 2z_1^T(Q_0(1)\hat{x}(1|0) + C^T Q_2 y(1) + A^T Q_1 z_2) + z_2^T Q_1 z_2\\ &\left.+ \text{terms involving only } y(1) \text{ and } \hat{x}(1|0)\right].\end{aligned} \tag{D.26}$$

Using Lemma D.1, we can now minimize with respect to z_1 to obtain:

$$\begin{aligned}\Phi_1(z_2) = &- (Q_0(1)\hat{x}(1|0) + C^T Q_2 y(1) + A^T Q_1 z_2)^T (Q_0(1) + C^T Q_2 C + A^T Q_1 A)^{-1}\\ &\times (Q_0(1)\hat{x}(1|0) + C^T Q_2 y(1) + A^T Q_1 z_2) + z_2^T Q_1 z_2\\ &+ \text{terms involving only } y(1) \text{ and } \hat{x}(1|0)].\end{aligned} \tag{D.27}$$

We now reorganize $\Phi_1(z_2)$ into quadratic form in z_2:

$$\begin{aligned}\Phi_1(z_2) = &\, z_2^T \left(-Q_1 A (Q_0(1) + C^T Q_2 C + A^T Q_1 A)^{-1} A^T Q_1 + Q_1\right) z_2\\ &- 2z_2^T \left(Q_1 A (Q_0(1) + C^T Q_2 C + A^T Q_1 A)^{-1} \{Q_0(1)\hat{x}(1|0) + C^T Q_2 y(1)\}\right)\\ &+ \text{terms independent of } z_2.\end{aligned} \tag{D.28}$$

Let the one-step-ahead estimate be defined as

$$\hat{x}(2|1) = \arg\left\{\min_{z_2} \Phi_1(z_2)\right\}. \tag{D.29}$$

This definition is consistent with the notation used throughout. Applying Lemma D.1, we can now derive a recursive expression for both the optimal prediction and the associated arrival cost.

The one-step-ahead estimate is given by

$$\begin{aligned}\hat{x}(2|1) = &\left(-Q_1 A (Q_0(1) + C^T Q_2 C + A^T Q_1 A)^{-1} A^T Q_1 + Q_1\right)^{-1} Q_1 A\\ &\times (Q_0(1) + C^T Q_2 C + A^T Q_1 A)^{-1} \left(Q_0(1)\hat{x}(1|0) + C^T Q_2 y(1)\right).\end{aligned} \tag{D.30}$$

The corresponding arrival cost can be written in the quadratic form:

$$\begin{aligned}\Phi_1(z_2) = &(z_2 - \hat{x}(2|1))^T \underbrace{\left(-Q_1 A (Q_0(1) + C^T Q_2 C + A^T Q_1 A)^{-1} A^T Q_1 + Q_1\right)}_{Q_0(2)} (z_2 - \hat{x}(2|1))\\ &+ \Phi_1(\hat{x}(2|1)) + \text{other terms independent of } z_2.\end{aligned} \tag{D.31}$$

At the next stage, the arrival cost becomes

$$\begin{aligned}\Phi_2(z_3) = \min_{z_2} \big[&(z_2 - \hat{x}(2|1))^T Q_0(2)(z_2 - \hat{x}(2|1)) \\ &+ (z_3 - Az_2)^T Q_1(z_3 - Az_2) \\ &+ (y(2) - Cz_2)^T Q_2(y(2) - Cz_2) \\ &+ \Phi_1(\hat{x}(2|1))\big] + \text{other terms independent of } z_3. \end{aligned} \tag{D.32}$$

Except for the constant term $\Phi_1(\hat{x}(2|1))$, this is in the same form as the previous stage, $\Phi_1(z_2)$, and thus the same recursive formula can be applied. Continuing this procedure yields the following general recursive expression:

$$\begin{aligned}\Phi_j(z_{j+1}) = (z_{j+1} - \hat{x}(j+1|j))^T Q_0(j+1)(z_{j+1} - \hat{x}(j+1|j)) + \Phi_j(\hat{x}(j+1|j)) \\ + \text{terms independent of } z_{j+1} \qquad \text{for } j = 1, \ldots, k, \end{aligned} \tag{D.33}$$

where the predicted estimate is given by

$$\begin{aligned}\hat{x}(j+1|j) = &\left\{-Q_1 A(Q_0(j) + C^T Q_2 C + A^T Q_1 A)^{-1} A^T Q_1 + Q_1\right\}^{-1} \\ &\times Q_1 A(Q_0(j) + C^T Q_2 C + A^T Q_1 A)^{-1} \left(Q_0(j)\hat{x}(j|j-1) + C^T Q_2 y(j)\right), \end{aligned} \tag{D.34}$$

and the updated weighting matrix is

$$Q_0(j+1) = -Q_1 A(Q_0(j) + C^T Q_2 C + A^T Q_1 A)^{-1} A^T Q_1 + Q_1. \tag{D.35}$$

Remarks:

- Equations (D.34)–(D.35) provide a recursive procedure for computing the one-step-ahead estimate $\hat{x}(j+1|j)$.
- The expression for the arrival cost $\Phi_j(z_{j+1})$ includes the constant term $\Phi_j(\hat{x}(j+1|j))$. Although this term is propagated to the next stage in the expression for $\Phi_{j+1}(z_{j+2})$, it does not influence the optimal estimate $\hat{x}(j+2|j+1)$ and can therefore be omitted from the recursion.

D.2.3 Equivalence with the KF

Now we show that (D.34) and (D.35) can be written as (7.58)–(7.59), thereby establishing their equivalence with the KF formulation.

To facilitate this equivalence, we make use of the following result, commonly referred to as the *Matrix Inversion Lemma*.

Lemma D.2: *Let $\mathcal{A}$, $\mathcal{C}$, and $\mathcal{C}^{-1} + \mathcal{D}\mathcal{A}^{-1}\mathcal{B}$ be nonsingular square matrices. Then*

$$(\mathcal{A} + \mathcal{B}\mathcal{C}\mathcal{D})^{-1} = \mathcal{A}^{-1} - \mathcal{A}^{-1}\mathcal{B}(\mathcal{C}^{-1} + \mathcal{D}\mathcal{A}^{-1}\mathcal{B})^{-1}\mathcal{D}\mathcal{A}^{-1}. \tag{D.36}$$

The lemma can be readily verified by multiplying both sides of the equation by $(\mathcal{A} + \mathcal{B}\mathcal{C}\mathcal{D})$ and confirming that the right-hand side simplifies to the identity matrix. The proof is straightforward and is left as an exercise for the reader.

We begin by showing the equivalence between (D.35) and (7.59). To do so, we first invert both sides of Eq. (D.35) to obtain

$$Q_0^{-1}(j+1) = \left\{-Q_1 A(Q_0(j) + C^T Q_2 C + A^T Q_1 A)^{-1} A^T Q_1 + Q_1\right\}^{-1}. \tag{D.37}$$

Next, we apply the *Matrix Inversion Lemma* to the right side of the equation. Let us define:

$$\begin{aligned}
\mathcal{A} &= Q_1 \\
\mathcal{B} &= Q_1 A \\
\mathcal{C} &= -(Q_0(j) + C^T Q_2 C + A^T Q_1 A)^{-1} \\
\mathcal{D} &= A^T Q_1.
\end{aligned} \tag{D.38}$$

to obtain

$$\begin{aligned}
Q_0^{-1}(j+1) &= \mathcal{A}^{-1} - \mathcal{A}^{-1}\mathcal{B}(\mathcal{C}^{-1} + \mathcal{D}\mathcal{A}^{-1}\mathcal{B})^{-1}\mathcal{D}\mathcal{A}^{-1} \\
&= Q_1^{-1} - Q_1^{-1} Q_1 A\left(-Q_0(j) - C^T Q_2 C - A^T Q_1 A + A^T Q_1 Q_1^{-1} Q_1 A\right)^{-1} A^T \\
&= Q_1^{-1} + A\left(Q_0(j) + C^T Q_2 C\right)^{-1} A^T.
\end{aligned} \tag{D.39}$$

Applying the *Matrix Inversion Lemma* once again to the term in parenthesis yields:

$$Q_0^{-1}(j+1) = Q_1^{-1} + A\left(Q_0^{-1}(j) - Q_0^{-1}(j) C^T (Q_2^{-1} + C Q_0^{-1}(j) C^T)^{-1} C Q_0^{-1}(j)\right) A^T,$$

which is (7.59).

We now show the equivalence between (D.34) and (7.58).

$$\begin{aligned}
\hat{x}(j+1|j) &= \left\{-Q_1 A(Q_0(j) + C^T Q_2 C + A^T Q_1 A)^{-1} A^T Q_1 + Q_1\right\}^{-1} \\
&\quad \times Q_1 A(Q_0(j) + C^T Q_2 C + A^T Q_1 A)^{-1}.
\end{aligned} \tag{D.40}$$

$$(Q_0(j)\hat{x}(j|j-1) + C^T Q_2 y(j)). \tag{D.41}$$

Using the earlier result:

$$\left\{-Q_1 A(Q_0(j) + C^T Q_2 C + A^T Q_1 A)^{-1} A^T Q_1 + Q_1\right\}^{-1} = Q_1^{-1} + A\left(Q_0(j) + C^T Q_2 C\right)^{-1} A^T,$$

we substitute to get:

$$\begin{aligned}
\hat{x}(j+1|j) &= \left\{Q_1^{-1} + A\left(Q_0(j) + C^T Q_2 C\right)^{-1} A^T\right\} Q_1 A(Q_0(j) + C^T Q_2 C + A^T Q_1 A)^{-1} \\
&\quad \times (Q_0(j)\hat{x}(j|j-1) + C^T Q_2 y(j)) \\
&= \left\{A + A\left(Q_0(j) + C^T Q_2 C\right)^{-1} A^T Q_1 A\right\} (Q_0(j) + C^T Q_2 C + A^T Q_1 A)^{-1} \\
&\quad \times (Q_0(j)\hat{x}(j|j-1) + C^T Q_2 y(j)) \\
&= A(Q_0(j) + C^T Q_2 C)^{-1} \left\{(Q_0(j) + C^T Q_2 C) + A^T Q_1 A\right\} (Q_0(j) + C^T Q_2 C + A^T Q_1 A)^{-1} \\
&\quad \times (Q_0(j)\hat{x}(j|j-1) + C^T Q_2 y(j)) \\
&= A(Q_0(j) + C^T Q_2 C)^{-1} \left(Q_0(j)\hat{x}(j|j-1) + C^T Q_2 y(j)\right).
\end{aligned} \tag{D.42}$$

Now, observe the following two simplifications, using the Matrix Inversion Lemma:

$$\begin{aligned}
&A(Q_0(j) + C^T Q_2 C)^{-1} Q_0(j)\hat{x}(j|j-1) \\
&= A\left\{Q_0^{-1}(j) - Q_0^{-1}(j) C^T (Q_2^{-1} + C Q_0^{-1}(j) C^T)^{-1} C Q_0^{-1}(j)\right\} Q_0(j)\hat{x}(j|j-1) \\
&= \{A - A Q_0^{-1}(j) C^T (Q_2^{-1} + C Q_0^{-1}(j) C^T)^{-1} C\}\hat{x}(j|j-1).
\end{aligned} \tag{D.43}$$

and

$$\begin{aligned}
&A(Q_0(j) + C^TQ_2C)^{-1}C^TQ_2y(j) \\
&= A\left\{Q_0^{-1}(j) - Q_0^{-1}(j)C^T(Q_2^{-1} + CQ_0^{-1}(j)C^T)^{-1}CQ_0^{-1}(j)\right\}C^TQ_2y(j) \\
&= AQ_0^{-1}(j)C^T\left\{I - (Q_2^{-1} + CQ_0^{-1}(j)C^T)^{-1}CQ_0^{-1}(j)C^T\right\}Q_2y(j) \\
&= AQ_0^{-1}(j)C^T(Q_2^{-1} + CQ_0^{-1}(j)C^T)^{-1}\left\{(Q_2^{-1} + CQ_0^{-1}(j)C^T) - CQ_0^{-1}(j)C^T\right\}Q_2y(j) \\
&= AQ_0^{-1}(j)C^T(Q_2^{-1} + CQ_0^{-1}(j)C^T)^{-1}y(j).
\end{aligned} \tag{D.44}$$

Combining both results, we recover the expression for $\hat{x}(j+1|j)$ as given in Eq. (7.58), completing the proof of equivalence.

D.3 Stochastic State Feedback Problems

The primary results in the main results in Chapter 8 were developed under a deterministic framework. However, in practical applications, measurements are often corrupted by random noise, and unmeasured disturbances or model uncertainties are typically present. Therefore, it is important to revisit the linear quadratic (LQ) control problem in the context of a stochastic system.

We consider the following stochastic state-space model:

$$x(k+1) = Ax(k) + Bu(k) + \varepsilon_1(k), \tag{D.45}$$

where ε_1 is a zero-mean white noise sequence with covariance R_1. For the analysis that follows, we assume full state measurement, i.e., the state vector $x(k)$ is available for feedback.

D.3.1 Open-loop Optimal Solution via Least Squares

We now consider the following *open-loop* quadratic optimal control problem for the stochastic system, assuming the initial condition $x(0)$:

$$J_p^o(x(0)) = \min_{u(0),\ldots,u(p-1)} E\left\{\sum_{k=0}^{p-1}\left[x^T(k)Qx(k) + u^T(k)Ru(k)\right] + x^T(p)Q_tx(p)\right\}. \tag{D.46}$$

This is an *open-loop* optimal control problem, meaning that all inputs are determined at time $k = 0$, without the benefit of future feedback measurements. As before, the system trajectory over the horizon can be expressed as

$$\begin{bmatrix} x(0) \\ x(1) \\ \vdots \\ \vdots \\ x(p) \end{bmatrix} = \begin{bmatrix} I \\ A \\ \vdots \\ \vdots \\ A^p \end{bmatrix} x(0) + \begin{bmatrix} 0 & \dots & \dots & 0 \\ B & 0 & \dots & 0 \\ AB & \ddots & \ddots & \vdots \\ \vdots & \ddots & \ddots & \vdots \\ A^{p-1}B & \dots & \dots & B \end{bmatrix} \begin{bmatrix} u(0) \\ \vdots \\ \vdots \\ u(p-1) \end{bmatrix}$$
$$+ \begin{bmatrix} 0 & \dots & \dots & 0 \\ I & 0 & \dots & 0 \\ A & \ddots & \ddots & \vdots \\ \vdots & \ddots & \ddots & \vdots \\ A^{p-1} & \dots & \dots & I \end{bmatrix} \begin{bmatrix} \varepsilon_1(0) \\ \vdots \\ \vdots \\ \varepsilon_1(p-1) \end{bmatrix}, \tag{D.47}$$

which will be denoted as

$$\mathcal{X} = \mathcal{S}^x x(0) + \mathcal{S}^u \mathcal{U} + \mathcal{S}^\varepsilon \mathcal{E} \tag{D.48}$$

Then,

$$\begin{aligned} V_p &= E\left\{\mathcal{X}^T \bar{Q} \mathcal{X} + \mathcal{U}^T \bar{R} \mathcal{U}\right\} \\ &= E\left\{\left(\mathcal{S}^x x(0) + \mathcal{S}^u \mathcal{U} + \mathcal{S}^\varepsilon \mathcal{E}\right)^T \bar{Q}\left(\mathcal{S}^x x(0) + \mathcal{S}^u \mathcal{U} + \mathcal{S}^\varepsilon \mathcal{E}\right)\right\} + \mathcal{U}^T \bar{R} \mathcal{U} \\ &= (\mathcal{S}^x x(0) + \mathcal{S}^u \mathcal{U})^T \bar{Q}\,(\mathcal{S}^x x(0) + \mathcal{S}^u \mathcal{U}) + \mathcal{U}^T \bar{R} \mathcal{U} + E\left\{\mathcal{E}^T \mathcal{S}^{\varepsilon T} \bar{Q} \mathcal{S}^\varepsilon \mathcal{E}\right\}. \end{aligned} \tag{D.49}$$

Using the property that $E\{x^T y\} = E\left\{\text{trace}\{y x^T\}\right\}$ in evaluating the last term, we obtain

$$V_p = (\mathcal{S}^x x(0) + \mathcal{S}^u \mathcal{U})^T \bar{Q}\,(\mathcal{S}^x x(0) + \mathcal{S}^u \mathcal{U}) + \mathcal{U}^T \bar{R} \mathcal{U} + \bar{Q}\,\mathcal{S}^\varepsilon\,\bar{R}_1\,\mathcal{S}^{\varepsilon T}, \tag{D.50}$$

where $\bar{R} = \text{blockdiag}\,\{R, \ \dots, R\}$. We note that:

- The first two terms are identical to the deterministic cost function in Eq. (8.9).
- The third term is independent of the control sequence $\mathcal{U}$ and thus does not affect the optimization.

Therefore, the optimal control for the stochastic case is the same as in the deterministic case:

$$\mathcal{U} = -\left(\mathcal{S}^{uT} \bar{Q} \mathcal{S}^u + \bar{R}\right)^{-1} \mathcal{S}^{uT} \bar{Q} \mathcal{S}^x x(0). \tag{D.51}$$

The corresponding optimal cost is

$$J_p^o(z_0) = x^T(0)\left[\mathcal{S}^{xT} \mathcal{S}^x - \mathcal{S}^{xT} \bar{Q} \mathcal{S}^u \left(\mathcal{S}^{uT} \bar{Q} \mathcal{S}^u + \bar{R}\right)^{-1} \mathcal{S}^{uT} \bar{Q} \mathcal{S}^x\right] x(0) + \bar{Q}\,\mathcal{S}^\varepsilon\,\bar{R}_1\,\mathcal{S}^{\varepsilon T}. \tag{D.52}$$

Note that the above represents only an *open-loop* optimal solution. In stochastic systems, there is an important distinction between open-loop and closed-loop (feedback) optimal control. In general, open-loop control results in significantly higher expected cost because it does not adapt to future measurements. To derive the optimal feedback policy, it is necessary to account for the impact of future observations on future control actions. This leads naturally to a stochastic dynamic programming formulation, which we now consider.

D.3.2 Optimal Feedback Policy via Dynamic Programming

We begin by considering the one-step-ahead problem posed at time $p-1$:

$$J_1(x(p-1)) = \min_{u(p-1)} E\{x^T(p-1)Qx(p-1) + u^T(p-1)Ru(p-1) \\ + x^T(p)S(p)x(p)\}. \tag{D.53}$$

$$x(p) = Ax(p-1) + Bu(p-1) + \varepsilon(p-1). \tag{D.54}$$

$$S(p) = Q_t. \tag{D.55}$$

Substituting the system equation (D.54) into the objective function (D.53), we obtain

$$J_1(x(p-1)) = \min_{u(p-1)} \big((Ax(p-1) + Bu(p-1))^T S(p)(Ax(p-1) + Bu(p-1)) \\ + x^T(p-1)Qx(p-1) + u^T(p-1)Ru(p-1) \\ + E\{\varepsilon_1^T(p-1)S(p)\varepsilon_1(p-1)\}\big). \tag{D.56}$$

The first three terms are identical to those in the deterministic case. The final term evaluates to a constant of trace$\{S(p)R_1\}$, and does not affect the optimal control policy. Thus, the optimal control input is the same as in the deterministic case:

$$u(p-1) = -\underbrace{(B^T S(p)B + R)^{-1} B^T S(p)A}_{L(p-1)} x(p-1). \tag{D.57}$$

The corresponding optimal cost-to-go is

$$J_1(x(p-1)) = x^T(p-1)S(p-1)x(p-1) + \text{trace}\{S(p)R_1\}, \tag{D.58}$$

where

$$S(p-1) = A^T S(p)A + Q - A^T S(p)B\left(B^T S(p)B + R\right)^{-1} B^T S(p)A. \tag{D.59}$$

Proceeding to the next stage, we solve the two-step-ahead problem at time $p-2$:

$$J_2(x(p-2)) = \min_{u(p-2)} E\left\{x^T(p-1)S(p-1)x(p-1) + x^T(p-2)Qx(p-2) \\ + u^T(p-2)Ru(p-2)\right\} + \text{trace}\{S(p)R_1\}. \tag{D.60}$$

This is structurally identical to the previous stage, so we again obtain

$$J_2(x(p-2)) = x^T(p-2)S(p-2)x(p-2) + \sum_{j=p-1}^{p} \text{trace}\{S(j)R_1\}. \tag{D.61}$$

$$u(p-2) = -\underbrace{(B^T S(p-1)B + R)^{-1} B^T S(p-1)A}_{L(p-2)} x(p-2), \tag{D.62}$$

where

$$S(p-2) = A^T S(p-1)A + Q - A^T S(p-1)B(B^T S(p-1)B + R)^{-1} B^T S(p-1)A. \tag{D.63}$$

By induction, we obtain the optimal input sequence for the full horizon problem in (D.46):

$$u(k) = -(B^T S(k+1)B + R)^{-1} B^T S(k+1)Ax(k), \qquad k = p-1, \dots, 0, \tag{D.64}$$

where

$$S(k) = A^T S(k+1)A + Q - A^T S(k+1)B(B^T S(k+1)B + R)^{-1} B^T S(k+1)A, \tag{D.65}$$

with terminal condition $S(p) = Q_t$. This feedback policy is identical to the one derived for the deterministic case. However, the optimal cost in the stochastic case includes an additional term due to process noise:

$$J_p(x(0)) = x^T(0)S(0)x(0) + \sum_{j=1}^{p} \text{trace}\{S(j)R_1\}. \tag{D.66}$$

This cost is strictly lower than the open-loop optimal cost $J_p^o(z_0)$, highlighting the performance advantage of feedback in stochastic control.

D.3.3 Open-loop Optimal Feedback Control vs. Optimal Feedback Control

In the deterministic setting, dynamic programming and multistep prediction-based least squares are simply alternative methods for solving the same control problem. However, in the general stochastic case, these two approaches address conceptually different problems.

The dynamic programming approach yields the optimal feedback control policy, explicitly accounting for the stochastic nature of future disturbances and measurements. In contrast, the multistep prediction-based least squares approach computes the optimal open-loop control trajectory for a given initial condition $x(0)$, without adjusting future inputs based on future measurements.

A common strategy is to resolve the open-loop optimal control problem at every time step, using updated state estimates. This approach is referred to as *open-loop optimal feedback control* (OLOFC). While it results in a feedback law, OLOFC is generally not optimal in the stochastic setting, as it fails to consider how future measurements will influence future input decisions.

Specifically, at time k, OLOFC solves

$$\min_{u(k),\cdots,u(k+p-1)} E\left\{\sum_{i=k}^{k+p-1} \left[x^T(i)Qx(i) + u^T(i)Ru(i)\right] + x^T(k+p)Q_t x(k+p)\right\}. \tag{D.67}$$

This formulation assumes that future control actions $u(k+1), \dots, u(k+p-1)$ are fixed at time k, and therefore ignores the potential value of future measurements $x(k+1), \dots, x(k+p)$ in shaping those future decisions. In contrast, dynamic programming solves the problem recursively, taking into account the influence of future observations:

$$\begin{aligned}
J_p(x(k)) &= \min_{u(k)} E\{x^T(k)Qx(k) + u^T(k)Ru(k) + J_{p-1}(x(k+1))\} \\
J_{p-1}(x(k+1)) &= \min_{u(k+1)} E\{x^T(k+1)Qx(k+1) + \\
&\qquad u^T(k+1)Ru(k+1) + J_{p-2}(x(k+2))\} \\
&\vdots \quad \vdots \\
J_1(x(k+p-1)) &= \min_{u(k+p-1)} E\{x^T(k+p-1)Qx(k+p-1) + \\
&\qquad u^T(k+p-1)Ru(k+p-1) + x^T(k+p)Q_t x(k+p)\}.
\end{aligned}$$

It is important to note that in general, the minimization operator and the expectation operator do not commute. This means that optimizing a deterministic input trajectory (as OLOFC does) is not equivalent to optimizing a feedback policy under uncertainty.

That said, in the special case of the LQ problem, the optimal feedback control law for the stochastic system turns out to be identical to that of the deterministic case. In other words, for linear dynamics with quadratic costs and Gaussian disturbances, OLOFC coincides with the optimal feedback control policy. This is a fortunate but exceptional property of the LQ problem and should not be generalized.

Moreover, the dynamic programming framework extends naturally to the infinite-horizon case ($p = \infty$). In the limit, the optimal policy converges to the same infinite-horizon linear quadratic regulator (LQR) solution derived for the deterministic setting in Eqs. (8.32 and 8.33).

D.4 Stochastic Output Feedback Problems

We now consider the case where the state is only partially observed through noise-corrupted measurements. Instead of having direct access to the full state vector $x(k)$, we observe an output $y(k)$ that provides only partial information about the state:

$$\begin{aligned} x(k+1) &= Ax(k) + Bu(k) + \varepsilon_1(k) \\ y(k) &= Cx(k) + \varepsilon_2(k). \end{aligned} \tag{D.68}$$

The following assumptions are made:

- The initial state $x(0)$ is a Gaussian random variable with mean $\hat{x}_0$ and covariance R_0.
- The process noise $\varepsilon_1(k)$ and $\varepsilon_2(k)$ are zero-mean, white Gaussian sequences with covariances R_1 and R_2, respectively.

D.4.1 Optimal Output Feedback Controller

We now consider the problem of deriving the optimal *output feedback* control policy based on the quadratic performance index:

$$V_p = E\left\{\sum_{i=0}^{p-1} \left[x^T(i)Qx(i) + u^T(i)Ru(i)\right] + x^T(p)Q_t x(p)\right\}. \tag{D.69}$$

The optimal solution turns out to be a combination of the optimal state feedback controller (i.e., LQR) and the optimal state estimator (i.e., KF). Specifically, the control policy is given by

$$\begin{aligned} \hat{x}(i+1|i+1) &= A\hat{x}(i|i) + Bu(i) + K_f(i+1)\left[y(i+1) - C(A\hat{x}(i|i) + Bu(i))\right], \\ u(i) &= -L(i)\hat{x}(i|i), \qquad i = p-1, \cdots, 0. \end{aligned} \tag{D.70}$$

The state feedback gain $L(i)$ is defined as

$$\begin{aligned} L(i) &= (B^T S(i+1)B + R)^{-1} B^T S(i+1)A \\ S(i) &= A^T S(i+1)A + Q - A^T S(i+1)B(B^T S(i+1)B + R)^{-1} B^T S(i+1)A \\ &\quad \text{with the initialization of } S(p) = Q_t. \end{aligned} \tag{D.71}$$

The Kalman gain $K_f(i+1)$ and the error covariance update are given by

$$\begin{aligned} K_f(i+1) &= P(i+1|i)C^T(CP(i+1|i)C^T + R_2)^{-1} \\ P(i+1|i) &= AP(i|i-1)A^T + R_1 - AP(i|i-1)C^T(CP(i|i-1)C^T + R_2)^{-1}CP(i|i-1)A^T \\ &\quad \text{with the initialization of } P(1|0) = AR_0A^T + R_1. \end{aligned} \tag{D.72}$$

The estimator (KF) is initialized with the prior mean and the covariance:

$$\hat{x}(0|0) = \hat{x}_0, \quad P(0|0) = R_0.$$

D.4.2 Derivation via Dynamic Programming

We begin with the one-step-ahead optimal output feedback problem posed at time $p-1$:

$$J_1(\mathcal{I}_{p-1}) = \min_{u(p-1)} E\{x^T(p)S(p)x(p) + x^T(p-1)Qx(p-1) + u^T(p-1)Ru(p-1) \mid \mathcal{I}_{p-1}\}. \tag{D.73}$$

Here, $\mathcal{I}_{p-1}$ represents all available information at time $p-1$, including prior statistics and measurements (e.g., $\hat{x}_0, R_0, y(1), \dots, y(p-1)$). Since the conditional mean and covariance of $x(p-1)$ summarize all relevant information, we may take:

$$\mathcal{I}_{p-1} = \{\hat{x}(p-1|p-1), P(p-1|p-1)\}. \tag{D.74}$$

These quantities are computed using a KF initialized with $x(0|0) = \hat{x}_0$ and $P_{0|0} = R_0$. Note that

$$\hat{x}(p|p-1) = A\hat{x}(p-1|p-1) + Bu(p-1) \tag{D.75}$$

Substituting this into the objective function yields:

$$\begin{aligned} J_1(\mathcal{I}_{p-1}) = \min_{u(p-1)} &(A\hat{x}(p-1|p-1) + Bu(p-1))^T S(p)(A\hat{x}(p-1|p-1) + Bu(p-1)) \\ &+ \hat{x}^T(p-1|p-1)Q\hat{x}(p-1|p-1) + u^T(p-1)Ru(p-1) \\ &+ E\left\{\hat{x}_e^T(p)S(p)\hat{x}_e(p) + x_e^T(p-1)Qx_e(p-1)\right\} \mid \mathcal{I}_{p-1}\}. \end{aligned} \tag{D.76}$$

where $\hat{x}_e(p) = x(p) - \hat{x}(p|p-1)$ and $x_e(p-1) = x(p-1) - \hat{x}(p-1|p-1)$, both zero-mean and uncorrelated with the corresponding estimates.

The first three terms match those of the deterministic case, with $x(p-1)$ replaced by its conditional mean. The final two terms reduce to constants: $\hat{x}(p-1|p-1)$. The last two terms, trace$\{S(p)P(p|p-1)\}$ and trace$\{QP(p-1|p-1)\}$, which do not affect the minimization. Therefore, the optimal control is

$$u(p-1) = \underbrace{-(B^TS(p)B + R)^{-1}B^TS(p)A}_{L(p-1)}\hat{x}(p-1|p-1). \tag{D.77}$$

and the optimal cost is

$$\begin{aligned} J_1(\mathcal{I}_{p-1}) = \hat{x}^T(p-1|p-1)S(p-1)\hat{x}(p-1|p-1) + \text{trace}\{QP(p-1|p-1)\} \\ + \text{trace}\{S(p)P(p|p-1)\}, \end{aligned} \tag{D.78}$$

with:

$$S(p-1) = A^T S(p)A + Q - A^T S(p)B(B^T S(p)B + R)^{-1}B^T S(p)A. \tag{D.79}$$

We now move to the two-step-ahead problem posed at time $p-2$:

$$\begin{aligned} J_2(\mathcal{I}_{p-2}) &= \min_{u(p-2)} E\left\{J_1(\mathcal{I}_{p-1}) + x^T(p-2)Qx(p-2) + u^T(p-2)Ru(p-2) \mid \mathcal{I}_{p-2}\right\} \\ &= \min_{u(p-2)} E\left\{\hat{x}^T(p-1|p-1)S(p-1)\hat{x}(p-1|p-1) + x^T(p-2)Qx(p-2)\right. \\ &\quad + u^T(p-2)Ru(p-2) + \text{trace}\{QP(p-1|p-1)\} \\ &\quad \left. + \text{trace}\{S(p)(AP(p-1|p-1)A^T + R_1)\} \mid \mathcal{I}_{p-2}\right\}. \end{aligned} \tag{D.80}$$

In writing the last term, we used the fact that $P(p|p-1) = AP(p-1|p-1)A^T + R_1$. Now, note that

$$\hat{x}(p-1|p-1) = A\hat{x}(p-2|p-2) + Bu(p-2) + K_f(p-1)\underbrace{(y(p-1) - Cx(p-1|p-2))}_{e(p-1)} \tag{D.81}$$

Note that the innovation term $e(p-1)$ has the property that $E\{e(p-1) \mid \mathcal{I}_{p-2}\} = 0$ and it is uncorrelated with $\hat{x}(p-2|p-2)$ or $u(p-2)$. Substituting the above and evaluating the expectation yields

$$\begin{aligned} &J_2(\mathcal{I}_{p-2}) \\ &= \min_{u(p-2)} \left(A\hat{x}^T(p-2|p-2) + Bu(p-2)\right)^T S(p-1)\left(A\hat{x}^T(p-2|p-2) + Bu(p-2)\right) \\ &+ \hat{x}^T(p-2|p-2)Q\hat{x}(p-2|p-2) + u^T(p-2)Ru(p-2) \\ &+ E\{e^T(p-1)K_f^T(p-1)S(p-1)K_f(p-1)e(p-1) + x_e^T(p-2)Qx_e(p-2) \mid \mathcal{I}_{p-2}\} \\ &+ \text{trace}\{QP(p-1|p-1)\} + \text{trace}\{S(p)(AP(p-1|p-1)A^T + R_1)\}. \end{aligned}$$

Evaluating the first term in the expectation, it can be shown that

$$\begin{aligned} &E\{e^T(p-1)K_f^T(p-1)S(p-1)K_f(p-1)e(p-1) \mid \mathcal{I}_{p-2}\} \\ &= \text{trace}\{S(p-1)K_f(p-1)E\{e(p-1)e^T(p-1) \mid \mathcal{I}_{p-2}\}K_f^T(p-1)\} \\ &= \text{trace}\{S(p-1)(AP(p-2|p-2)A^T + R_1 - P(p-1|p-1))\}. \end{aligned} \tag{D.82}$$

The proof for the above step is left as an exercise. Substituting this into the previous expression,

$$\begin{aligned} J_2(\mathcal{I}_{p-2}) &= \min\nolimits_{u(p-2)}(A\hat{x}^T(p-2|p-2) + Bu(p-1))^T S(p-1)(A\hat{x}^T(p-2|p-2) + Bu(p-1)) \\ &\quad + \hat{x}^T(p-2|p-2)Q\hat{x}(p-2|p-2) + u^T(p-2)Ru(p-2) \\ &\quad + \text{trace}\{S(p-1)(AP(p-2|p-2)A^T + R_1)\} + \text{trace}\{QP(p-2|p-2)\} \\ &\quad - \text{trace}\{S(p-1)P(p-1|p-1)\} \\ &\quad + \text{trace}\{S(p)(AP(p-1|p-1)A^T + R_1)\} + \text{trace}\{QP(p-1|p-1)\}. \end{aligned}$$

The above is in the same form as $J_1(\mathcal{I}_{p-1})$ except for the constant terms, which do not affect the solution. As with the previous stage, the solution is

$$u(p-2) = \underbrace{-(B^T S(p-1)B + R)^{-1}B^T S(p-1)A}_{L(p-2)}\hat{x}(p-2|p-2). \tag{D.83}$$

And the associated cost is

$$
\begin{aligned}
J_2(\mathcal{I}_{p-2}) &= \hat{x}^T(p-2|p-2)S(p-2)\hat{x}(p-2|p-2) \\
&\quad + \left[\sum_{j=p-1}^{p} \text{trace}\{S(j)(AP(j-1|j-1)A^T + R_1)\} + \text{trace}\{QP(j-1|j-1)\}\right] \\
&\quad - \text{trace}\{S(p-1)P(p-1|p-1)\},
\end{aligned}
\tag{D.84}
$$

with the usual Riccati recursion:

$$
S(p-2) = A^T S(p-1)A + Q - A^T S(p-1)B(B^T S(p-1)B + R)^{-1}B^T S(p-1)A, \tag{D.85}
$$

and

$$
\begin{aligned}
P(p-1|p-1) &= P(p-1|p-2) \\
&\qquad -P(p-1|p-2)C^T(CP(p-1|p-2)C^T + R_2)^{-1}CP(p-1|p-2) \\
P(p-1|p-2) &= AP(p-2|p-2)A^T + R_1.
\end{aligned}
\tag{D.86}
$$

By induction, the previously stated optimal control law and cost expression follow. The optimal output feedback cost is

$$
\begin{aligned}
J_p(\mathcal{I}_0) &= \hat{x}^T(0|0)S(0)\hat{x}(0|0) \\
&\quad + \sum_{j=1}^{p} \left[\text{trace}\{S(j)(AP(j-1|j-1)A^T + R_1)\} + \text{trace}\{QP(j-1|j-1)\}\right] \\
&\quad - \sum_{\ell=1}^{p-1} \text{trace}\{S(\ell)P(\ell|\ell)\}.
\end{aligned}
\tag{D.87}
$$

D.4.3 Extension to the Infinite-Horizon Case: LQG Controller and Separation Principle

The previous finite-horizon results extend naturally to the infinite-horizon setting. For this case, the performance index is defined as

$$
V_\infty = \lim_{p\to\infty} E\left\{\frac{1}{p}\sum_{k=0}^{p} \left[x^T(k)Qx(k) + u^T(k)Ru(k)\right]\right\}. \tag{D.88}
$$

Dividing by p ensures that the cost remains bounded. Minimizing V_∞ corresponds to minimizing the steady-state value of $E\{x^TQx + u^TRu\}$.

Taking the limit of the finite-horizon solution yields the optimal output feedback control law, which consists of

- a KF (the optimal state estimator)
- and an infinite-horizon LQR (the optimal state-feedback regulator for the deterministic case):

$$
\begin{aligned}
\hat{x}(k|k) &= (A - BL_\infty)\hat{x}(k-1|k-1) + K_f(k)\left(y(k) - (A - BL_\infty)\hat{x}(k-1|k-1)\right)) \\
u(k) &= -L_\infty \hat{x}(k|k).
\end{aligned}
\tag{D.89}
$$

As $k \to \infty$, the Kalman gain $K_f(k)$ typically converges to a steady-state value K_∞ yielding the linear time-invariant LQG controller:

$$
\begin{aligned}
\hat{x}(k|k) &= (A - BL_\infty)\hat{x}(k-1|k-1) + K_\infty\{y(k) - (A - BL_\infty)\hat{x}(k-1|k-1)\} \\
u(k) &= -L_\infty \hat{x}(k|k),
\end{aligned}
\tag{D.90}
$$

where the steady-state Kalman gain is given by

$$K_\infty = P_\infty C^T \left(C P_\infty C^T + R_2\right)^{-1}. \tag{D.91}$$

This is the celebrated LQG controller.

In conclusion, the LQG solution demonstrates the separation principle: for linear systems with quadratic cost and Gaussian noise, the optimal output feedback controller can be constructed by separately designing the optimal state estimator (KF) and the optimal state feedback controller (LQR).

D.4.4 Analysis

The system

$$\begin{aligned} x(k+1) &= Ax(k) + Bu(k) + \varepsilon_1(k) \\ y(k) &= Cx(k) + \varepsilon_2(k), \end{aligned} \tag{D.92}$$

with controller

$$\begin{aligned} \hat{x}(k+1|k+1) &= A\hat{x}(k|k) + Bu(k) + K_f\left[y(k+1) - C\left(A\hat{x}(k|k) + Bu(k)\right)\right] \\ u(k) &= -L\hat{x}(k|k), \end{aligned} \tag{D.93}$$

results in the coupled system:

$$\begin{bmatrix} x(k+1) \\ x_e(k+1) \end{bmatrix} = \underbrace{\begin{bmatrix} A - BL & -BL \\ 0 & A - K_f CA \end{bmatrix}}_{\Phi} \begin{bmatrix} x(k) \\ x_e(k) \end{bmatrix} + \underbrace{\begin{bmatrix} I \\ I - K_f C \end{bmatrix}}_{\Gamma_1} \varepsilon_1(k) + \underbrace{\begin{bmatrix} 0 \\ K_f \end{bmatrix}}_{\Gamma_2} \varepsilon_2(k). \tag{D.94}$$

This system is one-way coupled: the evolution of the state $x(k)$ does not influence the estimation error $\tilde{x}_e(k)$. The eigenvalues of the overall closed-loop transition matrix Φ are simply the union of the eigenvalues of $(A - BL)$ and $(A - K_f CA)$. Therefore, if both sets of eigenvalues lie strictly inside the unit disk, the state and estimation error processes converge to zero-mean stationary distributions with bounded variances.

In particular, the joint second-order statistics satisfy:

$$E\left\{\begin{bmatrix} x(k) \\ \tilde{x}_e(k) \end{bmatrix}\begin{bmatrix} x(k) \\ \tilde{x}_e(k) \end{bmatrix}^T\right\} = \Sigma, \tag{D.95}$$

where Σ is the unique positive semi-definite solution to the discrete-time Lyapunov equation:

$$\Phi\Sigma\Phi^T + \Gamma_1 R_1 \Gamma_1^T + \Gamma_2 R_2 \Gamma_2^T = \Sigma. \tag{D.96}$$

Because the LQG controller combines the LQR and KF—both of which are stabilizing under mild assumptions—the resulting closed-loop system is stable under the same conditions.

E

Internal Model Control Basics

In this chapter, we explore the control design methodology of Internal Model Control (IMC) and compare it with DMC and other related controllers. This is achieved by examining MPC/DMC from a transfer function and block-diagram perspective, thereby complementing the algebraic, equation-based approach adopted throughout this book.

A comprehensive treatment of IMC in its most general form lies beyond the scope of this book. For our purposes, it is sufficient to present the key concepts for linear, stable, single-input–single-output (SISO) systems, with emphasis on practical implementation rather than rigorous theoretical analysis. Readers interested in the theoretical foundations and broader developments of IMC are referred to García (1982), García and Morari (1982), García and Morari (1985), and Morari and Zafiriou (1989).

E.1 Discrete-time Systems

The basics of modeling discrete sampled-data systems used in computer control were covered in Chapter 5. For the purposes of this chapter, we assume discrete-time processes to be represented by the following input/output difference equation of order n:

$$\begin{aligned} y(k) = &-\alpha_1 y(k-1) - \alpha_2 y(k-2) - \cdots - \alpha_n y(k-n) \\ &+ \beta_1 u(k-1-\tau) + \beta_2 u(k-2-\tau) + \cdots + \beta_n u(k-n-\tau) \end{aligned} \tag{E.1}$$

which is an equivalent representation of Eq. (5.10) with $m = n - 1$, explicitly accounting for the response delay (dead time) τ. This formulation can represent any discrete-time model by appropriately setting certain coefficients α_i and/or β_i to zero. For instance, a truncated finite impulse response (FIR) model of order n is obtained by setting all $\alpha_i = 0$.

It is also important to note that, in the discrete-time setting, the earliest an input change can affect the measured output is at the next sampling interval. Consequently, discrete-time difference equations inherently include a one-sample time delay.

The z-transform of (E.1) is

$$y(z) = z^{-\tau} \frac{\beta_1 z^{-1} + \beta_2 z^{-2} + \cdots + \beta_n z^{-n}}{1 + \alpha_1 z^{-1} + \alpha_2 z^{-2} + \cdots + \alpha_n z^{-n}} u(z) \tag{E.2}$$

which is expressed in terms of the backward shift operator z^{-1}. Multiplying both the numerator and denominator by z^n yields the following generic process transfer function:

$$G(z) = z^{-\tau}\frac{N(z)}{D(z)} = z^{-\tau}\frac{\beta_1 z^{n-1} + \beta_2 z^{n-2} + \cdots + \beta_n}{z^n + \alpha_1 z^{n-1} + \alpha_2 z^{n-2} + \cdots + \alpha_n}. \tag{E.3}$$

The difference between the orders of the denominator and numerator polynomials, $D(z)$ and $N(z)$, reflects the inherent one-sample delay in the discrete-time model. The steady-state (DC) gain for a stable system is given by

$$G(1) = \frac{\beta_1 + \beta_2 + \cdots + \beta_n}{1 + \alpha_1 + \alpha_2 + \cdots + \alpha_n}. \tag{E.4}$$

A convenient form of the transfer function in Eq. (E.3), which explicitly reveals the system's poles and zeros, is

$$G(z) = K\frac{(z-\zeta_1)(z-\zeta_2)\cdots(z-\zeta_{n-1})}{(z-\pi_1)(z-\pi_2)\cdots(z-\pi_n)}z^{-\tau}, \tag{E.5}$$

where ζ_i and π_i denote the system zeros and poles, respectively, and K is a constant so that the steady-state gain equals $G(1)$ in (E.4). The following example illustrates these concepts.

Example E.1: *Z-transform of a Multieffect Evaporator Model.*
In Example 2.5 we introduced this continuous-time model relating the effect of steam flowrate on the product concentration of a multieffect evaporator system:

$$y_{pc}(s) = \frac{2.69(-6s+1)e^{-1.5s}}{(20s+1)(5s+1)}u_{sf}(s). \tag{E.6}$$

The process is second order with time constants of 20 and 5 minutes and a delay of 1.5 minutes. Sampling this continuous system every 30 seconds yields the following discrete-time transfer function in the z domain:

$$y_{pc}(z) = \frac{-0.0772z + 0.08392}{z^2 - 1.88z + 0.8825}z^{-3}\,u_{sf}(z),$$

or its equivalent zero-pole form:

$$y_{pc}(z) = \frac{-0.0772\,(z - 1.0870)}{(z-0.9732)(z-0.9068)}z^{-3}\,u_{sf}(z).$$

Some observations are in order:

- The sampled system poles (0.9732 and 0.9068) indicate that the transfer function is stable.
- The sampled system zero at 1.0870 lies outside the unit circle and is therefore unstable. Note that the continuous-time transfer function in Eq. (E.6) has a zero at $s = +1/6$, which also corresponds to an unstable (non-minimum phase) zero and results in an inverse response. The chosen sampling interval of 30 seconds is sufficiently short to capture this inverse response behavior in the discrete-time system as well.
- In general, the presence of an inverse response in a discrete-time system indicates the existence of unstable zeros. However, the converse is not necessarily true. For the evaporator process, the discrete-time model exhibits no visible inverse response when the sampling interval exceeds 9 minutes, even though the zero remains unstable. The zero becomes stable only when the sampling period is increased beyond 18 minutes.

- The dead time term $z^{-\tau}$ (e.g., z^{-3} in this example) can be interpreted as representing τ zeros at infinity, which are considered unstable in the discrete-time domain.

From here on we assume every transfer function and variable is in z-transform. Therefore, unless necessary for clarity, (z) will be dropped in the nomenclature.

E.2 The IMC Loop Structure and Properties

In Figure 4.2, we showed how the basic IMC and conventional feedback structures are interchangeable. Any conventional feedback controller C can be modified to yield the IMC structure. And any IMC controller can be put in conventional form as follows:

$$C = \frac{C_{\text{IMC}}}{1 - C_{\text{IMC}}\tilde{P}_u}. \tag{E.7}$$

Though fully interchangeable, in this section we show how C_{IMC} is much easier to design than C and how the IMC structure addresses robustness as a design objective in an explicit manner.

The most complete embodiment of the IMC structure is depicted in Figure E.1. Its elements are

- r is the reference signal to be followed by the plant output y. It is usually expressed as a dynamic trajectory to a setpoint.
- d is the measured/feedforward disturbance. Transfer function P_d represents its real plant dynamics impacting the output, while $\tilde{P}_d$ is the available model of those dynamics. The feedforward disturbance compensator C_{FF} is designed to cancel or minimize the effect of d on the controlled output.
- w_y captures the combined unmeasured effects of measurement noise and unmeasured disturbances on output y.
- y_m is the measured output containing disturbance effects and corrupted by noise.

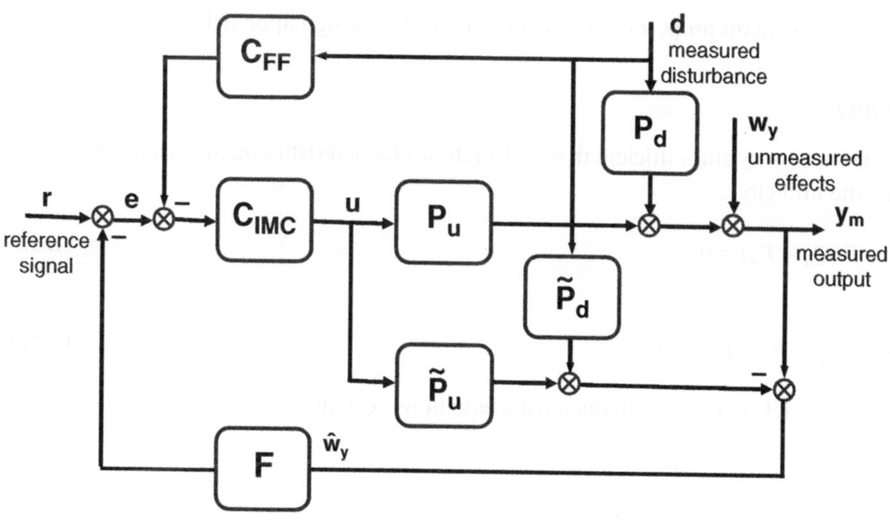

Figure E.1 The complete internal model control (IMC) loop structure.

- P_u is the real plant transfer function relating the manipulated variable u to the output. $\tilde{P}_u$ is its available model, which together with $\tilde{P}_d$ are denoted as the "internal model" of the plant.
- F is a low-pass unit gain dynamic filter acting on the IMC feedback signal $\hat{w}_y$. This is the same bias signal introduced in Eq. (1.4) to correct the model-predicted output with the measured output feedback signal.
- C_{IMC} is the IMC controller acting on the feedback error e between the reference signal and the filtered signal $\hat{w}_y$.

Control systems are usually designed to meet two objectives: reference signal tracking (servo) and disturbance rejection (regulatory). Let's derive the closed-loop transfer functions that define those objectives for the IMC loop.

The IMC feedback signal is

$$\hat{w}_y = (P_u - \tilde{P}_u)\, u + (P_d - \tilde{P}_d)\, d + w_y. \tag{E.8}$$

Note that it contains the combined effect of model mismatch, unmeasured noise and disturbance effects. When $\tilde{P}_u = P_u$ this signal only contains the external disturbance and noise effects. And since it does not contain information on the manipulated variable moves, the IMC controller becomes a *de facto* feedforward controller. To simplify the notation, let's combine the external effects on the output under the term:

$$\hat{w}_y = (P_d - \tilde{P}_d)\, d + w_y. \tag{E.9}$$

Using straightforward block diagram manipulations, we obtain the following closed-loop transfer functions for the manipulated variable and the measured output in response to the reference signal and disturbances:

$$u = \frac{C_{\text{IMC}}}{1 + C_{\text{IMC}} F\,(P_u - \tilde{P}_u)} \left[r - C_{FF} d - F\hat{w}_y\right] \tag{E.10}$$

$$y_m = \frac{P_u C_{\text{IMC}}}{1 + C_{\text{IMC}} F\,(P_u - \tilde{P}_u)} \left[r - C_{FF} d - F\hat{w}_y\right] + P_d d + w_y. \tag{E.11}$$

We now introduce the main properties of IMC that support its design approach.

E.2.1 Stability

For stability, it is necessary and sufficient that both of these characteristic equations have their roots strictly inside the unit circle:

$$\frac{1}{C_{\text{IMC}}} + F\,(P_u - \tilde{P}_u) = 0 \tag{E.12}$$

$$\frac{1}{C_{\text{IMC}} P_u} + \frac{F}{P_u}\,(P_u - \tilde{P}_u) = 0. \tag{E.13}$$

In the case of no model error, these characteristic equations become:

$$\frac{1}{C_{\text{IMC}}} = 0 \tag{E.14}$$

$$\frac{1}{C_{\text{IMC}} P_u} = 0. \tag{E.15}$$

Thus, stability of both controller and plant is sufficient for loop stability. This means that their poles have to be inside the unit circle. This is better understood by examining the transfer functions for the manipulated variable and the measured output when $\tilde{P}_u = P_u$ and $\tilde{P}_d = P_d$:

$$u = C_{\text{IMC}}\left[r - C_{FF}d - Fw_y\right] \tag{E.16}$$

$$y_m = P_u C_{\text{IMC}}\left[r - C_{FF}d - Fw_y\right] + P_d d + w_y. \tag{E.17}$$

The stability result follows for F and C_{FF} stable.

These equations also clarify why an IMC controller, in its direct form, is unable to stabilize open-loop unstable processes. Achieving closed-loop stability would require the zeros of the model-based control transfer function C_{IMC} to *exactly* cancel the unstable poles of the actual plant—an unrealistic expectation, since the true process dynamics can never be known with perfect accuracy. Moreover, even minor numerical errors or model mismatches would render the closed loop unstable. For unstable systems, an IMC controller can still be implemented but only after first stabilizing the plant with a conventional controller. Nonetheless, because the vast majority of industrial processes are open-loop stable, this limitation seldom presents a practical concern.

E.2.2 The Perfect Controller

The IMC construct inherently allows assessing the best achievable control performance by any controller. We define the "perfect controller" as the controller that achieves the smallest possible reference signal tracking error in the face of disturbances. Using the transfer function in Eq. (E.11), one can obtain the reference signal tracking error $r - y_m$ as a function of loop external signals r, d, and w_y:

$$\begin{aligned} r - y_m &= \frac{(1 - P_u C_{\text{IMC}}) + C_{\text{IMC}}F\,(P_u - \tilde{P}_u)}{1 + C_{\text{IMC}}F\,(P_u - \tilde{P}_u)}\,r \\ &- \frac{(P_d - P_u C_{\text{IMC}} C_{FF}) + C_{\text{IMC}}F\,(P_u - \tilde{P}_u)\,P_d - P_u C_{\text{IMC}}F\,(P_d - \tilde{P}_d)}{1 + C_{\text{IMC}}F\,(P_u - \tilde{P}_u)}\,d \\ &- \frac{(1 - P_u C_{\text{IMC}}F) + C_{\text{IMC}}F\,(P_u - \tilde{P}_u)}{1 + C_{\text{IMC}}F\,(P_u - \tilde{P}_u)}\,w_y. \end{aligned} \tag{E.18}$$

In the perfect model case, this tracking error equation reduces to

$$r - y_m = (1 - P_u C_{\text{IMC}})\,r \; - (P_d - P_u C_{\text{IMC}} C_{FF})\,d \; - (1 - P_u C_{\text{IMC}}F)\,w_y. \tag{E.19}$$

This *suggests* the following rules for designing the "perfect controller":

Servo: yields perfect reference signal r tracking. The next two follow as a result.

Measured disturbance regulation: Selection of $C_{FF} = \tilde{P}_d$ yields perfect measured disturbance rejection.

Unmeasured effects regulation: F rejects or attenuates the dynamics of w_y on the tracking error.

However, since C_{IMC} must be realizable and stable, the complete inversion of the discrete-time transfer function model is never implemented. To start with, the inherent sampling delay prevents any controller from achieving perfect control inside one time interval, and even longer when there are process dead times. Then, there are the limitations on stability. A good illustration is the evaporator process in the earlier example, which contains one unstable zero. Therefore, the suggested design rules above need to be modified.

There are many ways to obtain a "stable inverse" of a plant model. We introduce here a factorization method and show how DMC produces stable inverses. Consider the following model factorization:

$$\tilde{P}_u = \tilde{P}_u^+ \tilde{P}_u^- \tag{E.20}$$

where $\tilde{P}_u^+$ has unit steady-state gain and contains all time delays (whose inversion would make the controller unrealizable) and unstable zeros (whose inversion would render the controller unstable). These features are commonly referred to as the non-minimum-phase (NMP) characteristics of the process. Such a factorization is not unique; however, for the purposes of this discussion, we adopt this most convenient form:

$$\tilde{P}_u^+ = z^{-\tau-1} \prod_{i=1}^{p} \left(\frac{z-\zeta_i}{z-\hat{\zeta}_i} \right) \left(\frac{1-\hat{\zeta}_i}{1-\zeta_i} \right) \tag{E.21}$$

where $\tau + 1$ are the total units of delay, ζ_i are the p zeros of $\tilde{P}_u$ and $\hat{\zeta}_i$ their images inside the unit circle:

$$\hat{\zeta}_i = \begin{cases} \zeta_i & if \quad |\zeta_i| \leq 1 \\ 1/\zeta_i & if \quad |\zeta_i| > 1. \end{cases}$$

Once $\tilde{P}_u^+$ is factored out, we can design the stable IMC controller as

$$C_{\text{IMC}} = \frac{1}{\tilde{P}_u^-} \tag{E.22}$$

For the case of no model error, substituting this controller and the feedforward compensator selection $C_{FF} = P_d$ into the tracking error equation (E.19) we obtain

$$r - y_m = \left(1 - P_u^+\right) r - \left(1 - P_u^+\right) P_d d - \left(1 - P_u^+ F\right) w_y, \tag{E.23}$$

corresponding to the following manipulated variable and measured output responses:

$$u = \frac{1}{P_{\bar{u}}} \left(r - P_d d - F w_y\right) \tag{E.24}$$

$$y_m = P_u^+ r + \left(1 - P_u^+\right) P_d d + \left(1 - P_u^+ F\right) w_y. \tag{E.25}$$

Note that these intrinsic "non-invertible" dynamics limit the loop's response from achieving perfect servo and regulatory performance, even with a perfect model. As we now show, these limitations apply to any feedback controller.

For $\tau = 0$, Eq. (E.21) is the closed-loop response of y_m to a unit step change in r of the controller obtained when solving the infinite horizon problem in Eq. (9.9), in the limit as R approaches zero and $Q = C^T C$ (Kwakernaak and Sivan [1972]). Note that C is the matrix in state space model (6.1), and thus the objective function minimizes output variability while maintaining loop stability (property of the infinite horizon optimal control formulation).

Moreover, since unmeasured disturbances impact y_m the same way as r (per Eq. [E.25]), it is also optimal for feedback regulatory control. In case of delays, optimality is obtained after τ time intervals.

Because this controller produces the lowest achievable sum of squared errors extending to infinity under minimal constraints on the manipulated variable, it constitutes the ideal stable optimal controller. But most importantly, this result demonstrates that the NMP characteristics of a system cannot be eliminated by any control strategy.

It should also be clear that this is the closest achievable stable controller to the inverse of the plant. Therefore, since it is obtained by setting the MPC/DMC parameters $p = m = \infty$ and $\Gamma^u = 0$, any stable MPC controller produced by a different parameter selection is considered an "approximate stable inverse" of the plant.

These final two properties hold for any stable process and model, regardless of modeling accuracy.

E.2.3 Zero Offset

An important property of IMC is that it yields zero offset for any stable process and model as long as the following gain requirements are met:

$$\begin{aligned} C_{\text{IMC}}(1) &= \tilde{P}_u^{-1}(1) \\ C_{FF}(1) &= \tilde{P}_d(1) \\ F(1) &= 1. \end{aligned} \tag{E.26}$$

This can be proven by setting $z = 1$ in Eq. (E.18) and substituting the above. Of course, selecting $C_{FF} = \tilde{P}_d$ and F with unit gain satisfy the last two conditions. The final step is then to select a stable controller with its gain equal to the inverse of the plant model gain, which the controller in Eq. (E.22) exhibits.

E.2.4 Robustness

As with all controllers, an IMC loop can also be tuned to achieve stability in the presence of model error. Consider the characteristic equation in (E.12):

$$1 + C_{\text{IMC}}F\,(P_u - \tilde{P}_u) = 0.$$

For a given plant-model mismatch, both C_{IMC} and F can be selected such that its roots lie inside the unit circle. A popular choice is the first order filter:

$$F(z) = \frac{1-\alpha}{1-\alpha z^{-1}}$$

where $0 < \alpha < 1$. For a prespecified C_{IMC} and model error $P_u - \tilde{P}_u$ it can be proven that a sufficiently high value of α exists so that the roots of the characteristic equation lie inside the unit circle (García [1982]). Of course, the larger the filter pole, the slower the loop performance.

This provides a simple illustration of how the IMC structure offers a systematic means to balance performance and robustness. Analytical tools for evaluating these trade-offs are presented in Section E.2.8. More accurate knowledge of model error characteristics (for example, errors arising from nonlinearities) or disturbance properties (e.g., a ramp disturbance) can be used to design more sophisticated filters that deliver improved performance (Morari and Zafiriou, 1989). However, in cases where significant performance degradation occurs due to process changes, the most effective remedy is to reidentify or update the process model, ensuring it is revalidated with new operating data.

E.2.5 IMC Feedforward Compensator Design

Selection of the feedforward compensator as $C_{FF} = \tilde{P}_d$ was *suggested* above after conjecturing that the perfect inverse of the plant model could be used as the IMC controller. But since we've established that NMP process characteristics cannot be inverted, we must revisit its design. Here we present a simple design approach for systems with dead times and no unstable zeros, which is an improvement over feedback, though not optimal. A rigorous design approach of an optimal IMC feedforward compensator can be found in Morari and Zafiriou (1989).

The measured disturbance contribution to the tracking error in the perfect model case (E.19) using a feedforward compensator is

$$e_{ff} = -(P_d - P_u C_{IMC} C_{FF})d. \tag{E.27}$$

Using the factorization in (E.20) and the controller in (E.22) we obtain

$$e_{ff} = -(P_d - P_u^+ C_{FF})d. \tag{E.28}$$

In order to perfectly cancel the impact of d the compensator can be designed as

$$C_{FF} = \frac{P_d}{P_u^+}. \tag{E.29}$$

Note that $C_{FF}(1) = P_d(1)$ and therefore meets the requirement for zero offset. Also, since we assume the system does not have unstable zeros P_u^+ only contains the process dead time.

Assuming P_d and P_u have dead times of τ_d and τ_u, respectively, we can factor them out along with the inherent dead time to obtain the compensator:

$$C_{FF} = z^{-(\tau_d - \tau_u)} \frac{N_d(z)}{D_d(z)}, \tag{E.30}$$

where $N_d(z)$ and $D_d(z)$ are the disturbance transfer function numerator and denominator after factorization, and thus of equal order per Eq. (E.3).

This compensator is stable, given that P_d is stable. But for realizability, the net dead time of $(\tau_d - \tau_u)$ must be nonnegative, so a reasonable final design for the compensator is

$$C_{FF} = z^{-\tau_{ff}} \frac{N_d(z)}{D_d(z)} \tag{E.31}$$

where $\tau_{ff} = \max(0, \tau_d - \tau_u)$. The feedforward contribution to the tracking error now becomes

$$e_{ff} = -\left(1 - z^{-\tau_{ff} + (\tau_d - \tau_u)}\right) P_d(z) d(z). \tag{E.32}$$

Note that for $\tau_d \geq \tau_u$ the tracking error becomes zero regardless of the dynamics of the measured disturbance d or of its transfer function P_d, thus achieving optimal rejection of the disturbance by the compensator. This is physically intuitive. The compensator measures a change in the disturbance d right away, but its effect is not felt on the output for another $\tau_d + 1$ time intervals. Since the canceling effect of the plant controller via the IMC feedback starts $\tau_u + 1$ intervals later, the compensator needs to start sending its signal $\tau_d - \tau_u$ intervals after the disturbance is measured.

For $\tau_d < \tau_u$,

$$e_{ff} = -\left(1 - z^{-(\tau_u - \tau_d)}\right) P_d(z) d(z),$$

which means the controller can only start closing the tracking error after $\tau_u - \tau_d$ time intervals. If d is not fedforward, the tracking error due to feedback action is

$$e_{fb} = -\left(1 - z^{-\tau_u - 1}\right) P_d(z)d(z).$$

So, even when $\tau_d = 0$ feedback compensation is always one interval slower than feedforward due to the inherent time delay. Therefore, since e_{fb} can never be completely canceled, feedforward compensation is always better for any combination of τ_d and τ_u.

It is possible to completely cancel the disturbance effect after $\tau_u - \tau_d$ intervals. Such compensator would be optimal since it brings the response error to zero at the fastest possible time. MPC/DMC controllers tuned for deadbeat action produce such compensation as we illustrate in the following example.

Example E.2: *Heavy Oil Fractionator IMC Feedforward Compensation.*
The heavy oil fractionator first order plus dead time models in Section (3.8) provide many examples to demonstrate the concepts above. Since no transfer function contains unstable zeros, the benefit of feedforward disturbance compensation can be quickly assessed by looking at τ_{FF} for combinations of u_i and d_i for each output y_i. For example, for top end point (TEP) control, both disturbances can be completely canceled with feedforward compensation, except for the side draw (SD) pairing. And even then, the feedforward action is only delayed by one minute.

Here we illustrate the performance of C_{FF} for the case of $\tau_d < \tau_u$ where perfect disturbance cancellation is not physically possible. Consider controlling the side draw temperature (SDT) with the top draw (TD). This pairing exhibits the high-frequency dynamic response needed to reject the fast effect of both disturbances as compared to using the side draw (SD). The bottom reflux duty (BRD) is usually reserved for optimization. Both disturbances can be fed forward; however, $\tau_u - \tau_d = 5$ for both, which delays the compensation five minutes.

Here we demonstrate the concepts for the upper reflux duty (URD). The discrete-time transfer functions for a sampling time of one minute are

$$\mathrm{SDT}(z) = \frac{0.3058}{z - 0.926} z^{-5}\,\mathrm{TD}(z) + \frac{0.5154}{z - 0.6065}\,\mathrm{URD}(z). \tag{E.33}$$

We do this by comparing the feedforward and feedback behavior of unconstrained IMC and MPC tuned for output deadbeat response (or "perfect control" in the IMC parlance) in the case of no model errors. This selection allows us to better assess the inherent disturbance rejection capabilities of both techniques.

We leave it to the reader to derive the IMC feedforward compensator in (E.31):

$$C_{FF}(z) = \frac{0.5154z}{z - 0.6065}. \tag{E.34}$$

MPC is designed with $p = m$ sufficiently large and no input weights. The estimator is the simple DMC BIAS correction also used in IMC.

Figure E.2 shows the unconstrained closed-loop responses to a step in URD of 0.5 units at time zero. Feedback response of both controllers is identical, as should be expected, given both use the same BIAS factor to estimate the disturbance effect. They reject the disturbance step in 15 minutes. The IMC feedforward compensator in Eq. (E.34) speeds up the response by five minutes. MPC produces the minimum squared errors response since it brings the output to zero in the shortest

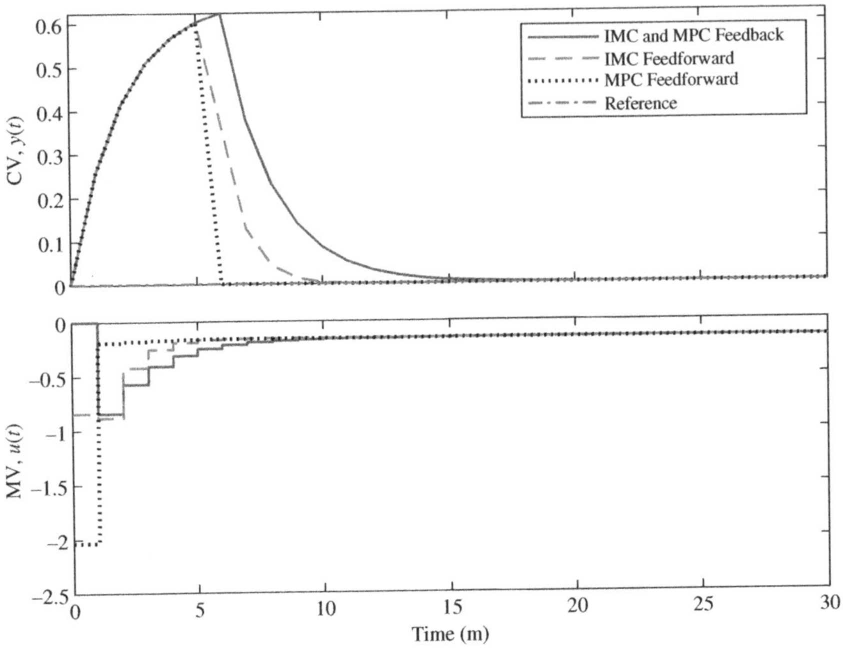

Figure E.2 Heavy oil fractionator control of the SDT with the TD and disturbance URD. Comparison of feedback vs. feedforward compensation of unconstrained IMC and MPC controllers tuned for output deadbeat response. A disturbance of 0.5 in URD is introduced at time zero.

physically possible time of six minutes. Following the method in Morari and Zafiriou (1989) one can obtain the compensator that produces the optimal response identical to that of MPC:

$$C_{\mathrm{FF}}(z) = 1.2446\,\frac{z - 0.5859}{z - 0.6065}.$$

Although not shown, constraining the inputs to ± 0.5 slows down all responses with MPC being the fastest. This will be illustrated in Section E.2.6.

E.2.6 Saturation Constraints

A beneficial feature of the IMC structure is its straightforward handling of manipulated variable saturation at their physical constraint limits. For controllers with integral action implemented under the conventional feedback structure, saturation can lead to "reset windup" due to continued integration of the target offset. This may cause severe overshoots and/or long settling times.

The IMC structure handles this by feeding the internal model the same constrained manipulated variable signal sent to the plant. This ensures that, when $P_u = \tilde{P}_u$, the IMC structure remains open-loop and thus stable for a stable plant and model. Such open-loop stability is preserved even when the controller, plant, and model are nonlinear, stable, and there is no model error (Morari and Zafiriou [1989]).

Since MPC enjoys the same feedback structure, its stability is assured under input constraints for stable loop elements. But its performance is superior to just clipping unconstrained move calculations since the QP produces moves that optimally handle both current *and* predicted violations. We illustrate this with an example.

Example E.3: *Heavy Oil Fractionator Saturation Constraints.*
As in Example E.2, we select the heavy oil fractionator pairing of the SDT controlled with the TD. We also use the IMC and MPC controllers tuned for output deadbeat control to better isolate the impact of constraints on control performance. The reader can verify that for this first-order system with time delays, an output deadbeat controller reaches and stays at the setpoint after $\tau_u + 1 = 6$ intervals in two moves. For an SDT setpoint change of 2.0 units, the controller produces two TD moves of 6.540 and −6.056, settling at 0.484.

In Figure E.3 we compare the responses of constrained IMC, MPC, and proportional–integral–derivative (PID) controllers to an SDT setpoint change of 2.0 units with the TD limited to ±0.5 units. The maximum constraint does not prevent the controllers from reaching 0.484 and meeting the SDT target. But it is close enough to influence the closed-loop dynamics.

Notice that the limit of 0.5 prevents the deadbeat-tuned IMC from implementing its computed first move of +6.540. Once at the limit, it produces one additional move to settle at 0.484, with the SDT reaching the setpoint in about 80 min with a sum of squared setpoint errors (SSE) of 47.9.

Deadbeat-tuned MPC (with $n = p = 100$) behaves quite differently since it is calculating $m = 100$ future feasible moves every minute. The TD stays at the upper limit for 45 minutes to compensate for the clipped first move before settling at 0.484. The SSE = 46.5 is the lowest of all the cases. Even

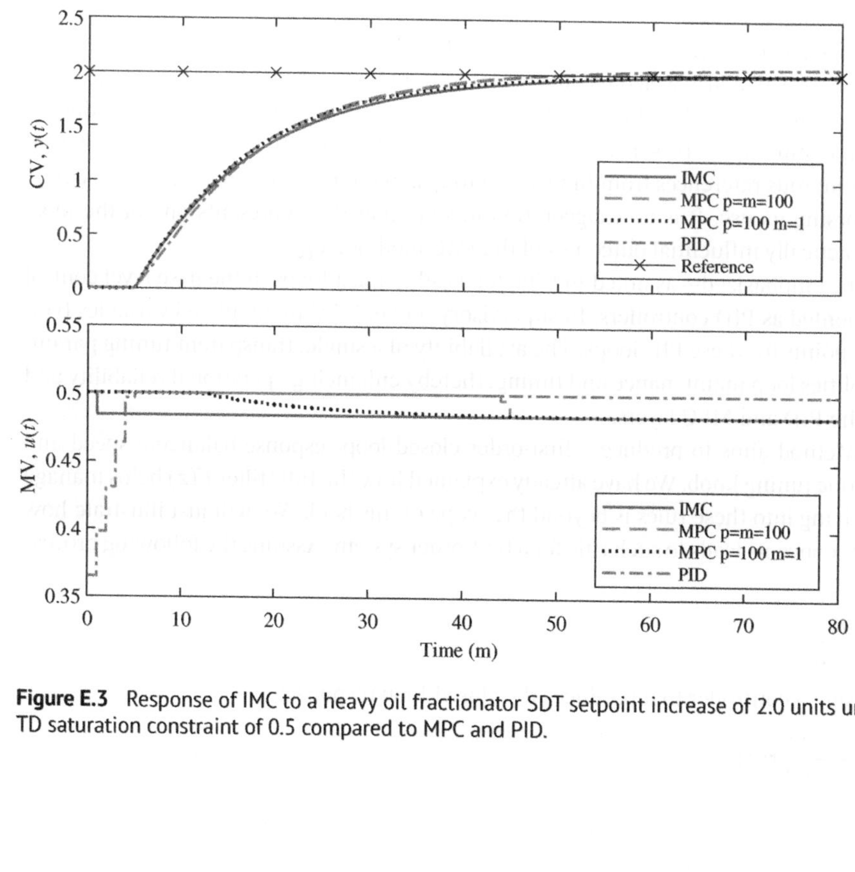

Figure E.3 Response of IMC to a heavy oil fractionator SDT setpoint increase of 2.0 units under a maximum TD saturation constraint of 0.5 compared to MPC and PID.

when tuned using $m = 1$ to constrain just the present move like IMC does, its SSE = 46.8 is still superior because of the optimal constraint handling.

The response of a well-tuned PI controller ($P = 0.1824$ and $I = 0.01643$ in Simulink Discrete PID block) in the conventional structure illustrates the impact of windup. Though trends beyond minute 80 are not shown, the SDT overshoots the setpoint from minutes 50 to 170, settling at 2.0 in minute 200. This behavior is effectively handled by PID anti-windup features offered in commercial DCS equipment. IMC addresses it in a simpler way with minimal suboptimality and computational effort vs. MPC. Of course, as covered extensively in the book, MPC pays off handsomely for multivariable systems under multiple constraints, since it uses every degree of freedom to optimally avert saturation.

E.2.7 IMC Tuning Method for PID Controllers

Invented in 1911 by Elmer Sperry for automatic ship steering in the US Navy, the PID controller has been the workhorse control tool for more than a century across every industry. Åström and Hagglund (1995) stated that 95% of all process control loops were PID. That number has been sustained as demand has continued to grow.

Applying the IMC design methodology to first- and second-order process models naturally yields conventional PID-plus-lag controllers. This result was first reported by Rivera et al. (1986) for continuous-time systems and later refined by Morari and Zafiriou (1989). Both studies provided tables mapping model parameters of first- and second-order dynamics to the corresponding PID-plus-lag controller parameters as functions of a single tuning parameter, λ, representing the desired closed-loop time constant. A comparable table for discrete-time systems was subsequently presented by Prett and García (1988).

Given the extensive body of research on PID tuning developed over the years, this single-knob approach has been a particularly welcome advancement in industry. A 2006 overview of the state of the art in PID control noted that the so-called "Lambda Method" is "the most widely adopted tuning method in commercial PID software packages" (Li et al., 2006). Indeed, a cursory internet search reveals numerous references from industrial groups, major control vendors, and academic researchers endorsing its use. It is no exaggeration to state that this represents one of the most impactful and practically influential outcomes of the IMC methodology.

It should also be emphasized—as noted in Chapter 1—that most loops in the base level control layer are implemented as PID controllers. In supervisory mode, MPC manipulated variables typically serve as set points for these PID loops. The availability of a single, transparent tuning parameter greatly simplifies loop maintenance and tuning, thereby enhancing operational reliability and uptime of both the PID and MPC layers.

The Lambda Method aims to produce a first-order closed-loop response balancing speed and robustness with one tuning knob. We have already explained how the IMC filter $F(z)$ helps manage that tradeoff. Delving into these rules is beyond the scope of this book. We will just illustrate how applying the IMC concepts leads to such rule for a first-order system. Assume the following process model:

$$P_u(z) = \frac{K}{z - \pi}.$$

The aim is to design a controller that produces the closed-loop response:

$$y_m(z) = \left(\frac{1 - \delta}{z - \delta}\right) r(z),$$

where $\delta = e^{-t_s/\lambda}$ is the discrete-time closed-loop response pole corresponding to the desired continuous closed-loop time constant λ, for a sampling time of t_s.

The reader can verify that, for this system—having no time delays or unstable zeros—the resulting IMC controller is

$$C_{\text{IMC}}(z) = \left(\frac{1-\delta}{K}\right)\left(\frac{z-\pi}{z-\delta}\right).$$

Note that this is the IMC controller in Eq. (E.22) multiplied by a first-order lag, which is equivalent to moving the IMC feedback signal filter block $F(z)$ ahead of the controller.

Finally, we obtain the equivalent conventional controller using Eq. (E.7):

$$C(z) = \left(\frac{1-\delta}{K}\right)\left(\frac{z-\pi}{z-1}\right),$$

which produces the following PI controller in discrete-time difference equation form:

$$\Delta u(k) = K_c[e(k) - \tau_I e(k-1)].$$

The proportional gain $K_c = (1-\delta)/K$ is a function of the closed-loop response pole tuning knob δ, while the integral time τ_I is the system pole π. So only the gain has to be adjusted online to achieve the "Lambda" single-parameter tuning effect.

The most common industrial approach for deriving the Lambda tuning rules begins by approximating the process with a first- or second-order model with dead time, typically obtained from a step test. The time delay is then represented by an equivalent pole–zero transfer function, leading to the model forms summarized in the standard tuning tables. The work of Skogestad (2004) provides an excellent exposition on developing such low-order process models for PID parameterization. This so-called SIMC ("Skogestad IMC") method is one of the standard PID tuning options available in the MATLAB™ Control System Designer application.

The success of this tuning method is the most compelling evidence of the transparency of the IMC methodology, essentially bridging the *implicit* model control approaches of the past with the *explicit* model-based approaches of today.

E.2.8 Analysis Tools

We have now demonstrated how the IMC design methodology yields controllers that address many of the challenges of advanced SISO problems. Yet its most significant feature is that it naturally provides a framework for assessing robustness to model uncertainty. In-depth coverage of this topic is beyond the scope of this book. The interested reader is referred to Morari and Zafiriou (1989), which is the most comprehensive treatise on this subject.

In this section we illustrate using the analysis tool for assessing closed-loop stability under model error over a family of plants that do not match the nominal model. A common way to define such a family is to use a frequency domain multiplicative uncertainty bound, on the magnitude of the relative model error:

$$\left|\frac{P_u(e^{i\omega t_s}) - \tilde{P}_u(e^{i\omega t_s})}{\tilde{P}_u(e^{i\omega t_s})}\right| \le |\ell_m(\omega)|, \quad \text{for } 0 \le \omega \le \pi/t_s. \tag{E.35}$$

Note that $z = e^{i\omega t_s}$ is substituted in the transfer functions to obtain their frequency domain representation. Closed-loop stability for all plants within the family defined above is guaranteed when the following criterion is satisfied:

$$\left|\tilde{\eta}(e^{i\omega t_s})\right| < \left|\frac{1}{\ell_m(e^{i\omega t_s})}\right| \quad \text{for } 0 \le \omega \le \pi/t_s. \tag{E.36}$$

$\tilde{\eta}(\omega)$ is the complementary sensitivity function that relates the reference signal $r(z)$ to the output $y(z)$ in the absence of model error. This criterion can be evaluated by plotting $\tilde{\eta}(z)$ and the inverse uncertainty bound $1/\ell_m(z)$ together on a Bode plot.

Example E.4: *Heavy Oil Fractionator Robustness to Time Delay Errors.*
Consider the heavy oil fractionator pairing of SDT control with TD in Example (E.2):

$$\mathrm{SDT}(z) = \frac{0.3058}{z - 0.926} z^{-5}\,\mathrm{TD}(z). \tag{E.37}$$

Intuitively, one would expect that any tightly tuned controller would be extremely sensitive to model errors. Indeed, using the IMC characteristic equation, one can verify that the IMC deadbeat controller is unstable even for plant dead time errors of ± 1 minute. As we discussed in Section E.2.4, the first-order unit gain filter $F(z)$ can be used in IMC to stabilize a system under model uncertainty.

To illustrate the use of this tool, consider a plant with the same model as in Eq. (E.37), except with a dead time of six minutes, and use (E.35) to generate $1/\ell_m(z)$. The reader can verify that the deadbeat controller tuning violates the robust stability condition (E.36) at high frequencies. As shown in Figure E.4, applying an IMC filter with pole $\alpha = 0.4$ attenuates $\tilde{\eta}(e^{i\omega t_s})$ at high frequencies, satisfying criterion (E.36) and thereby ensuring closed-loop stability. However, this choice results in severe manipulated variable oscillations and slow settling, as shown in Figure E.5. Increasing the filter pole to $\alpha = 0.8$ improves the response and provides a larger stability margin against model uncertainty.

We also include the plot of the sensitivity function, $1 - \tilde{\eta}(z)$, which characterizes the relationship between unmeasured disturbances and the output $y(z)$. Because the sensitivity and complementary sensitivity functions sum to one, they together quantify the fundamental trade-off between

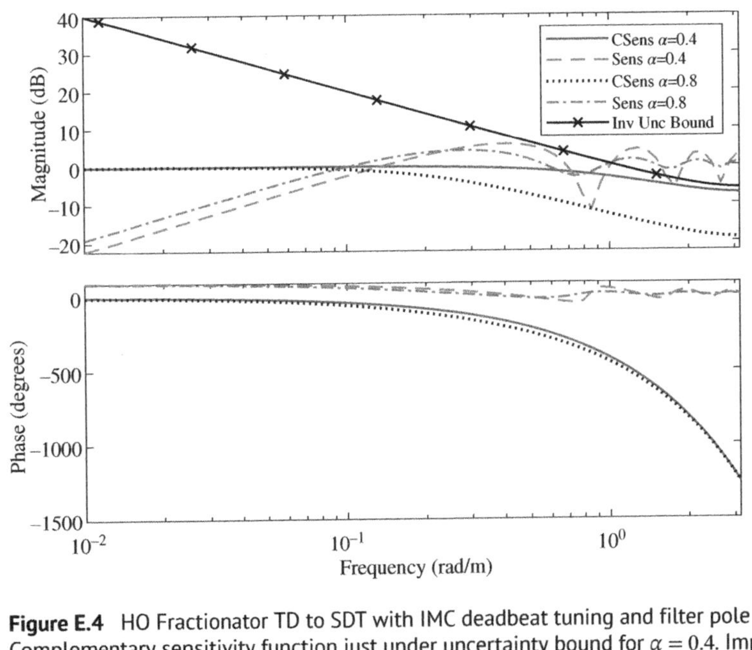

Figure E.4 HO Fractionator TD to SDT with IMC deadbeat tuning and filter pole $\alpha = 0.4$ and 0.8. Complementary sensitivity function just under uncertainty bound for $\alpha = 0.4$. Improved stability margin and reduced disturbance amplification (per the sensitivity function) is obtained for $\alpha = 0.8$.

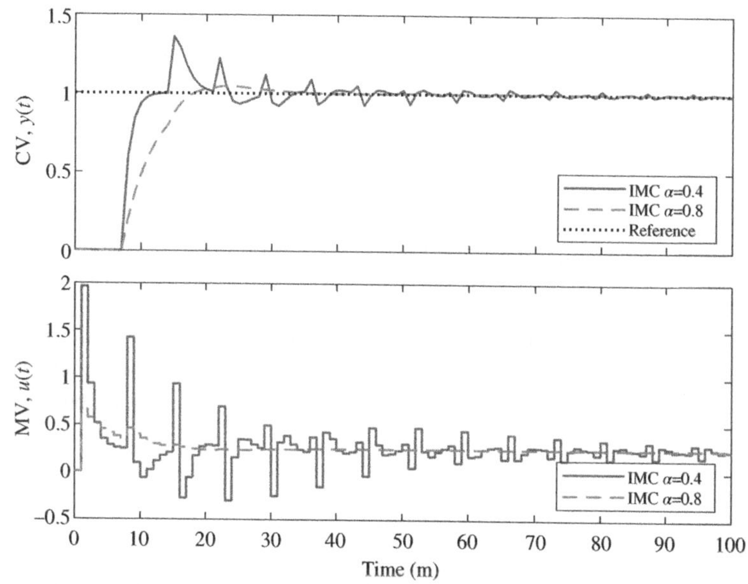

Figure E.5 IMC simulations corresponding to the IMC filter pole selections in Figure E.4 for plant delay $\tau_u = 6$.

regulatory performance and robustness. For example, employing a slower IMC filter to enhance robustness reduces the system's sensitivity to model error, but at the cost of a longer settling time. This trade-off motivates the use of a higher-order filter to better balance robustness and dynamic performance.

Although presented here within the IMC theoretical framework, the existing tools for robust stability and robust performance can be used on any controller. Their application to state-space MPC has been thoroughly examined in Lee and Yu (1994), including analyses of how tuning parameters (e.g., filter gain parameters) affect robustness. In that work, a coupled system state-space representation of the closed-loop dynamics as in Eq. (D.94) was used to derive the relevant transfer functions for robustness analysis. This formulation established an explicit correspondence between the IMC filter parameter(s) and the Kalman gains, derived under specific disturbance–noise models (e.g., output step or ramp disturbances with white measurement noise).

An example of such closed-loop MPC state-space representation based on the augmented model formulation in Eq. (11.11) after adding measured disturbances is as follows:

$$\begin{bmatrix} \bar{x}(k+1) \\ \bar{x}_e(k+1) \end{bmatrix} = \begin{bmatrix} \bar{A} - \bar{B}L_x & \bar{B}L_x \\ 0 & \bar{A} - K_f\bar{C} \end{bmatrix} \begin{bmatrix} \bar{x}(k) \\ \bar{x}_e(k) \end{bmatrix} + \begin{bmatrix} \bar{B}_d - \bar{B}L_d & \bar{G} & 0 & \bar{B}L_r \\ 0 & \bar{G} & K_f & 0 \end{bmatrix} \begin{bmatrix} \Delta d_m(k) \\ \varepsilon_y(k) \\ \nu_y(k) \\ r(k) \end{bmatrix}$$

$$y_m(k) = \begin{bmatrix} \bar{C} & 0 \end{bmatrix} \begin{bmatrix} \bar{x}(k) \\ \bar{x}_e(k) \end{bmatrix} + \nu_y(k), \tag{E.38}$$

where:

- $\bar{A}, \bar{B}, \bar{B}_d, \bar{G}$, and $\bar{C}$ are the matrices of the augmented $\bar{x}(k)$ state-space model,
- $\bar{x}_e(k) = \bar{x}(k) - \hat{\bar{x}}(k|k)$ is the estimation error,

- $\Delta d_m(k)$ is the measured disturbance change,
- $\varepsilon_y(k)$ and $\nu_y(k)$ are the unmeasured disturbance effects and measurement noise on output $y(k)$,
- $r(k)$ is the reference signal,
- L_x, L_d, and L_r are the unconstrained MPC control law gains on $\hat{\hat{x}}(k|k)$, $\Delta d_m(k)$, and $r(k)$, respectively, obtained from Eq. (11.54) when model (11.51) includes measured disturbances and
- K_f is the Kalman filter gain in the estimator for $\hat{\hat{x}}(k|k)$.

Using the MATLAB™ function `clsys=ss(A,B,C,D)` on this model, one can obtain the closed-loop state space relationship of the ith-loop input by simply selecting `clsys(i)`. In addition to using in simulations, one can generate frequency response Bode plots with the function `bodeplot(clsys(i))` to assess how controller and estimator tuning selections impact performance and robustness.

Exercises

1. Use the evaporator z-transform model in Example E.1 to derive the IMC controller that produces the closed-loop minimum squared error response to a step in Eq. (E.21).

2. Apply the IMC PID tuning rules to the evaporator model. Approximate this inverse response model as a first- or second-order dead time model. Suggest reviewing the methodology in Skogestad (2004). Compare with the PID tuning produced by MATLAB™ Control System Designer. Simulate PID, IMC, and MPC closed-loop responses to compare ease of tuning.

3. Program the closed-loop MPC state-space representation (E.38) for the evaporator model in MATLAB™. Set the deadbeat controller tuning $p = m =$ large and Γ^u small but ensuring the system is stable. Set the measurement noise covariance in the Kalman calculation to zero to generate the DMC-like gain. Use the $r(z)$ to $y(z)$ state-space function to calculate the closed-loop poles and zeros using the MATLAB™ functions `zero(clsys(4))` and `pole(clsys(4))`. Progressively reduce Γ^u and observe how the poles and zeros approach those of Eq. (E.21) as you reach the numerical zero limit. (Hint: need to manually remove pole/zero cancellations and numerically small roots).

4. Produce Bode plots of the evaporator complementary sensitivity and sensitivity functions to assess how MPC tuning parameters shape the magnitude plots to achieve improved robustness to model errors and disturbance rejection at different frequencies. Verify with simulations.

5. Produce Bode plots of the evaporator closed-loop responses to inputs $\Delta d_m(k)$, $\varepsilon_y(k)$, and $\nu_y(k)$ and test how different tuning parameters impact their shape. In particular, observe how the signals are amplified or attenuated. Verify with simulations.

6. Design the IMC controller of Eq. (E.22) for the shower process in Eq. (1.6) of Example 1.1. Obtain the equivalent conventional controller using Eq. (E.7) and show that it is exactly the MPC control law in Eq. (1.10). Now design the controller using a dead time of 4 seconds vs. the true delay of 6 seconds. Apply the robust stability analysis in Section E.2.8 to show that such controller is unstable.

F

MPC Toolbox Tutorial: Shell Oil Fractionator

F.1 Problem Description

F.1.1 Background

The Shell Oil Fractionator Control problem was previously addressed in Chapter 3. In this appendix, we revisit the same problem using the MPC Toolbox. Several simulation scenarios, adapted from Yu et al. (1994), are used for comparison. Interested readers are encouraged to consult that work for further details. Additionally, the MPC Toolbox User's Guide (Bemporad et al., 2021) provides a helpful introduction and is recommended reading before tackling this implementation.

F.1.2 Model Definition

Recall the schematic of Shell Oil Fractionator in Figure F.1, which shows the manipulated variables u_i, unmeasured disturbances d_i and output variables y_i. The MIMO model for this system is given by the matrix of transfer functions shown in Table F.1. Please refer to Section 3.8 for additional details.

Five different cases were considered in Section 3.8. We will consider the first two cases in this Appendix: (i) When there is no model-plant mismatch; and (ii) Model-plant mismatch as per Case-2, wherein the plant gains are as follows:

	u_1	u_2	u_3	d_1	d_2
y_1	4.05 − 2.11	1.77 − 0.39	5.88 − 0.59	1.2 + 0.12	1.44 + 0.16
y_2	5.39 − 3.29	5.72 − 0.57	6.9 − 0.89	1.52 + 0.13	1.83 + 0.13
y_7	4.38 − 3.11	4.42 − 0.73	7.2 − 1.33	1.14 + 0.18	1.26 + 0.18

The control objective is to reject the effect of step disturbances in d_1 and d_2 and ensure that the controlled variables y_1 and y_2 are kept at their setpoints and y_7 is kept above the minimum level of −0.5, preferably u_3 is minimized, and the following constraints are satisfied:

- $-0.5 \leq u_i \leq 0.5$
- $|\Delta u_i| \leq 0.05$
- $-0.5 \leq y_1, y_2 \leq 0.5$ and $-0.5 \leq y_7$
- We choose the sampling interval as $t_s = 5$.

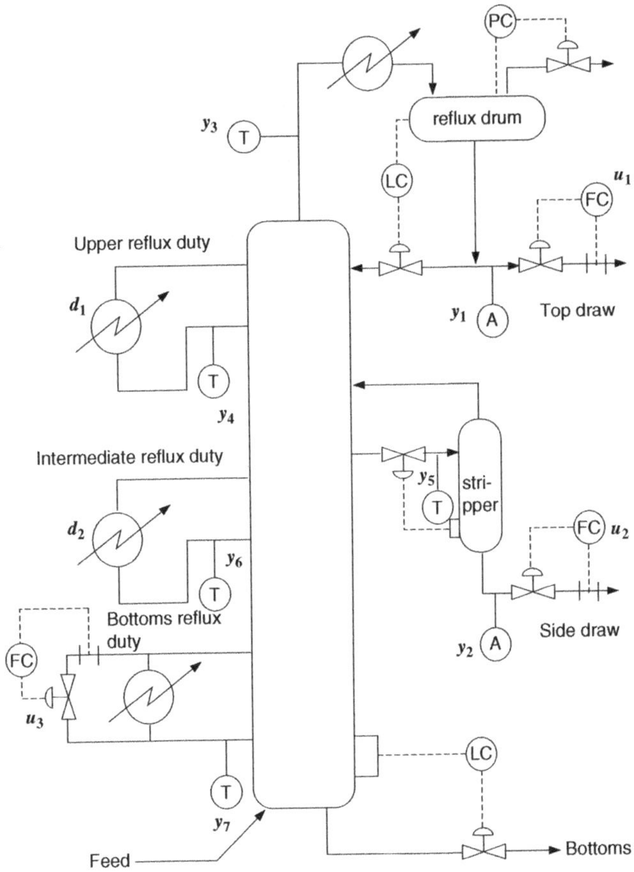

Figure F.1 Schematic of the control structure of the Shell Oil Fractionator. Recall that u, y and d represent input, output and disturbance variables, respectively.

Table F.1 Transfer functions for various inputs and outputs for the Shell Oil Fractionator. These transfer functions represent the model used in MPC. The true plant model differs with respect to the steady state gains.

	TD u_1	**SD u_2**	**BRD u_3**	**URD d_1**	**IRD d_2**
TEP y_1	$\frac{4.05e^{-27s}}{50s+1}$	$\frac{1.77e^{-28s}}{60s+1}$	$\frac{5.88e^{-27s}}{50s+1}$	$\frac{1.2^{-27s}}{45s+1}$	$\frac{1.44e^{-27s}}{40s+1}$
SEP y_2	$\frac{5.39e^{-18s}}{50s+1}$	$\frac{5.72e^{-14s}}{60s+1}$	$\frac{6.9e^{-15s}}{40s+1}$	$\frac{1.52e^{-15s}}{25s+1}$	$\frac{1.83e^{-15s}}{20s+1}$
TT y_3	$\frac{3.66e^{-2s}}{9s+1}$	$\frac{1.65e^{-20s}}{30s+1}$	$\frac{5.53e^{-2s}}{40s+1}$	$\frac{1.16}{11s+1}$	$\frac{1.27}{6s+1}$
URT y_4	$\frac{5.92e^{-11s}}{12s+1}$	$\frac{2.54e^{-12s}}{27s+1}$	$\frac{8.1e^{-2s}}{20s+1}$	$\frac{1.73}{5s+1}$	$\frac{1.79}{19s+1}$
SDT y_5	$\frac{4.13e^{-5s}}{8s+1}$	$\frac{2.38e^{-7s}}{19s+1}$	$\frac{6.23e^{-2s}}{10s+1}$	$\frac{1.31}{2s+1}$	$\frac{1.26}{22s+1}$
IRT y_6	$\frac{4.06e^{-8s}}{13s+1}$	$\frac{4.18e^{-4s}}{33s+1}$	$\frac{6.53e^{-1s}}{9s+1}$	$\frac{1.19}{19s+1}$	$\frac{1.17}{24s+1}$
BRT y_7	$\frac{4.38e^{-20s}}{33s+1}$	$\frac{4.42e^{-22s}}{44s+1}$	$\frac{7.2}{19s+1}$	$\frac{1.14}{27s+1}$	$\frac{1.26}{32s+1}$

F.1.3 Overview of Our Approach

Section F.3 will focus on solving the Shell Oil Fractionator Control problem using three different control structures, each incorporating disturbance modeling in a distinct way: (a) output disturbance, (b) unmeasured disturbance, and (c) both, represented as integrating white noise sequences. The techniques discussed in Chapter 11 will be instrumental for tuning the controller and achieving offset-free setpoint tracking in the presence of unmeasured disturbances (UD) and model-plant mismatch (MPM).

Before addressing these more complex scenarios, Section F.2 introduces a simpler case involving setpoint tracking without any UD or MPM. This example, implemented using the MPC Designer GUI, serves as a tutorial to familiarize the reader with the MPC Toolbox environment.

F.2 Tutorial: Using the MPC GUI

F.2.1 Simplified Problem Definition

Consider the following three-input, three-output system. For now, we neglect any UD:

$$\begin{bmatrix} y_1 \\ y_2 \\ y_7 \end{bmatrix} = \begin{bmatrix} \frac{4.05e^{-27s}}{50s+1} & \frac{1.77e^{-28s}}{60s+1} & \frac{5.88e^{-27s}}{50s+1} \\ \frac{5.39e^{-18s}}{50s+1} & \frac{5.72e^{-14s}}{60s+1} & \frac{6.9e^{-15s}}{40s+1} \\ \frac{4.38e^{-20s}}{33s+1} & \frac{4.42e^{-22s}}{44s+1} & \frac{7.2}{19s+1} \end{bmatrix} \begin{bmatrix} u_1 \\ u_2 \\ u_3 \end{bmatrix} \tag{F.1}$$

The constraints are the same as those in the original problem

$$-0.5 \le u_i \le 0.5, \quad |\Delta u_i| \le 0.05, \quad -0.5 \le y_1, y_2 \le 0.5, \quad -0.5 \le y_7$$

We consider a setpoint tracking problem with targets: $r_1 = 0.2$, $r_2 = 0.2$.

F.2.2 Using the MPC GUI

The first step in using the MPC GUI is to define the system model as specified in Eq. (F.1), and then launch the MPC Designer using the command `mpcDesigner`. The corresponding MATLAB code is shown below:

```
g11=tf(4.05, [50 1],'IODelay',27);
g12=tf(1.77, [60 1],'IODelay',28);
g13=tf(5.88, [50 1],'IODelay',27);
g21=tf(5.39, [50 1],'IODelay',18);
g22=tf(5.72, [60 1],'IODelay',14);
g23=tf(6.90, [40 1],'IODelay',15);
g71=tf(4.38, [33 1],'IODelay',20);
g72=tf(4.42, [44 1],'IODelay',22);
g73=tf(7.20, [19 1]);
```

```
% Defining the system and starting MPC GUI
Gm=[g11, g12, g13; g21, g22, g23; g71, g72, g73];
Gm.timeUnit='minutes';
mpcDesigner
```

Now that the MIMO transfer function model `Gm` has been created, we can proceed with setting up the MPC controller using `mpcDesigner`.

After launching the `mpcDesigner` GUI, the default view opens to the `MPC Designer` tab. Click the "`MPC Structure`" button within this tab. As shown in Figure F.2, this window allows the user to select the plant model for the controller. Choose the MIMO transfer function model `Gm`, which defines the input-output relationship as $Y(s) = G_m U(s)$. In this setup, all three outputs, $[y_1 \quad y_2 \quad y_7]^T$, are measured , and the inputs, $[u_1 \quad u_2 \quad u_3]^T$, are manipulated variables (MVs). This tutorial considers the case with no disturbances.

When the model is imported, `mpcDesigner` automatically generates and runs a default control scenario. Please disregard this default setup and follow the steps outlined here to properly configure the controller and simulate the intended control scenario.

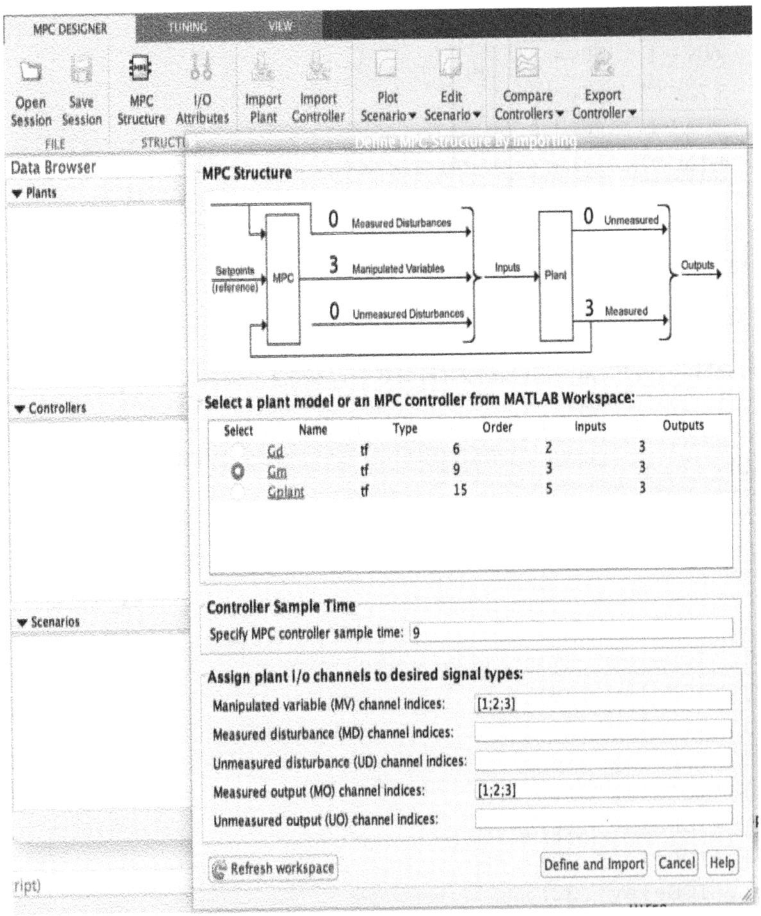

Figure F.2 Choosing the model using the MPC Designer tab.

We begin by setting up the control scenario using the "`Edit Scenario`" button. The control objective is to optimally drive the outputs y_1 and y_2 to their respective setpoints, $r_1 = 0.2$ and $r_2 = 0.2$. Note that y_7 does not have a setpoint but is still subject to process constraints. This information can be specified in the window shown in Figure F.3, accessible through the "`Edit Scenario`" option—either from the top menu bar or the data browser on the left.

The simulation duration is set to 300 minutes, corresponding to 60 sampling intervals. No disturbances are included in this scenario.

At this point, `mpcDesigner` automatically solves the problem—setpoint tracking to $r(k) = [0.2 \quad 0.2 \quad 0]^T$—using default tuning parameters and treating it as an unconstrained MPC problem. We will complete the full controller configuration in the next steps.

Following the setup in Section 3.8, set the sampling time to $t_s = 5$ minutes, the prediction horizon to $p = 20$, and the control horizon to $m = 10$ under the `Tuning` tab (see Figure F.4). The rationale for choosing $p = 20$—as discussed in Chapter 3—is to ensure the horizon is long enough to capture the desired closed-loop dynamics, especially given the process delay of six sampling intervals. Accordingly, the control horizon should be less than $p - 6$, and we select $m = 10$ as a suitable value.

The final steps involve setting the MPC constraints and tuning the controller weights. Constraints are defined using the `Constraints` button in the `TUNING` tab, as illustrated in Figure F.5. This

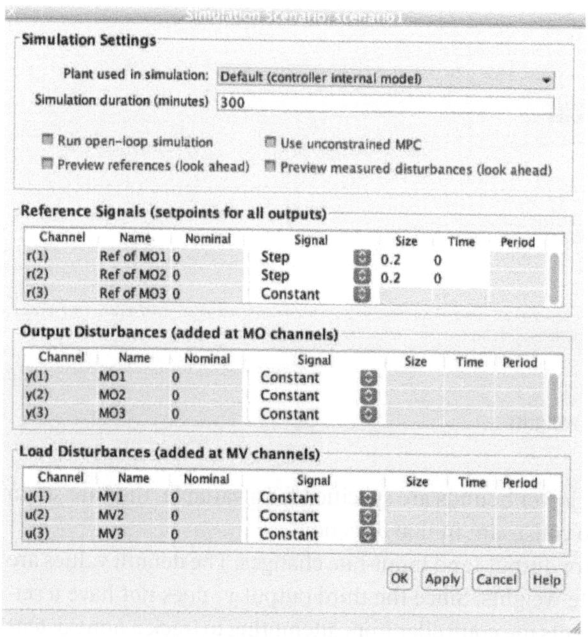

Figure F.3 Simulation scenario screen and the settings used for the setpoint tracking problem.

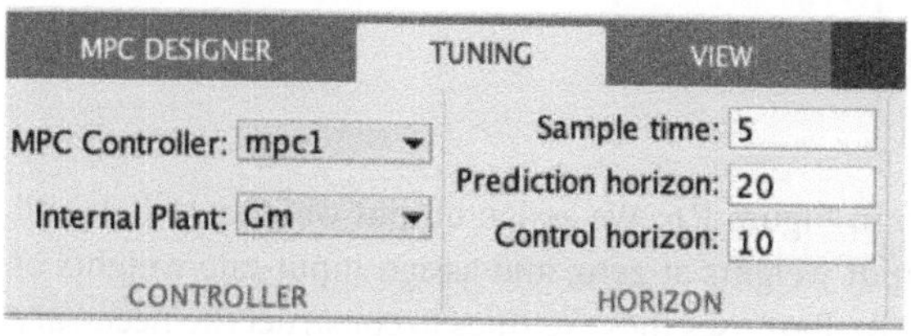

Figure F.4 Choosing sampling time and horizons.

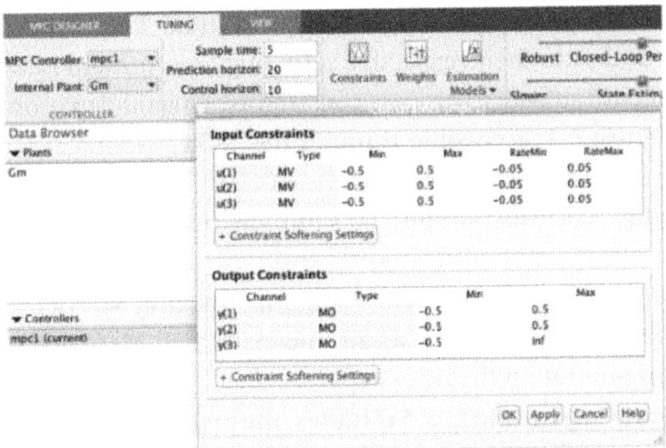

Figure F.5 Setting the constraints. Since there is no upper-limit on y_7, it is set as `inf`.

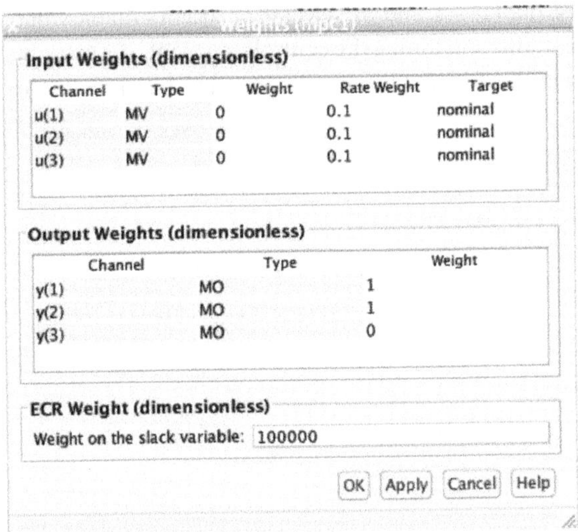

Figure F.6 Setting the input-rate and output weights.

interface is fairly intuitive. If no upper or lower bounds are specified for a variable, they are set to `inf` and `-inf`, respectively, indicating no constraint in that direction.

Next, we configure the tuning weights for outputs and input-rate changes. The default values are 1 for output weights and 0.1 for input-rate weights. Since the third output y_7 does not have a setpoint, we assign it a weight of 0. This prevents the controller from attempting to track a nonexistent setpoint. The resulting tuning weight matrices are:

$$Q = \begin{bmatrix} 1 & 0 & 0 \\ 0 & 1 & 0 \\ 0 & 0 & 0 \end{bmatrix} \qquad R = \begin{bmatrix} 0.1 & 0 & 0 \\ 0 & 0.1 & 0 \\ 0 & 0 & 0.1 \end{bmatrix}$$

These values are set via the interface shown in Figure F.6: We assign output weights of 1, 1, and 0 to the three output channels, leave the input weights at zero, and assign input-rate weights of 0.1 for each manipulated variable. The MPC Toolbox uses these settings to construct the necessary matrices for solving the optimization problem at each control interval.

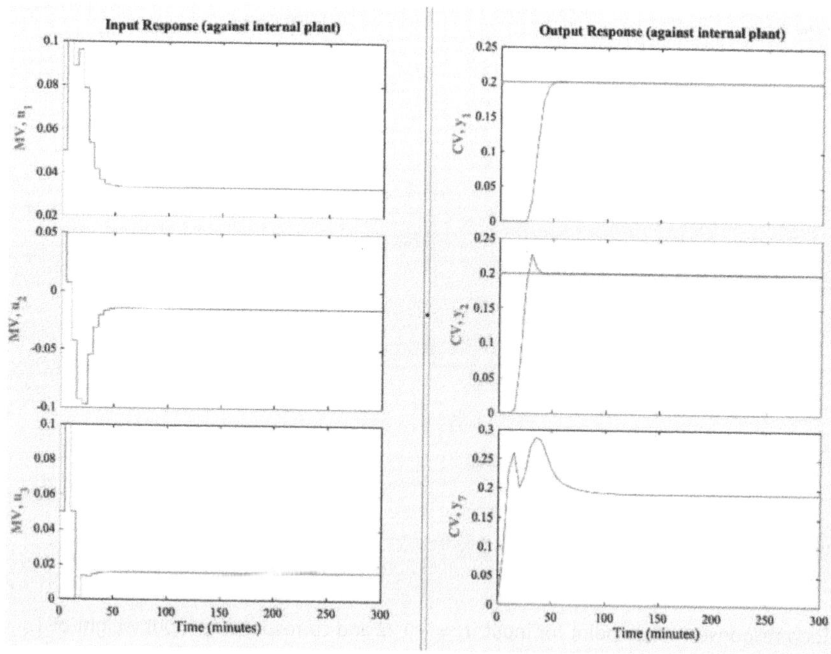

Figure F.7 Closed-loop response of the system for setpoint tracking.

This completes the controller tuning. Figure F.7 shows the closed-loop performance of the MPC controller for the given setpoint tracking problem.

Another control objective specified in the Shell Oil Fractionator problem is to maintain the input u_3 at a desired value. In Chapter 3, this was accomplished by redefining u_3 as an additional output variable. In contrast, the state-space MPC framework allows for direct specification of both input and input-rate weights. Accordingly, we can achieve this objective by assigning a weight on the input deviation for u_3.

Note that the system has three inputs and three outputs. In general, it is not feasible to assign a setpoint to an input in addition to tracking all output setpoints. However, since the output y_7 does not have a setpoint requirement, the available degree of freedom can be used to regulate u_3. In effect, we now have three control targets: the outputs y_1 and y_2 are to be tracked at their respective setpoints, and the input u_3 is to be maintained at a specified value.

Following the approach of Yu et al. (1994), we select the desired input target as $u_{3,\text{target}} = -0.22$, and define the input weight matrix as:

$$R^u = \begin{bmatrix} 0 & 0 & 0 \\ 0 & 0 & 0 \\ 0 & 0 & 1 \end{bmatrix}$$

This configuration can be easily implemented in the MPC Designer GUI by assigning the input target and corresponding input weight for the third input channel, as shown in Figure F.6.

Figure F.8 shows the resulting closed-loop response when the controller is configured to track outputs y_1 and y_2 at their setpoints of $[0.2 \quad 0.2]^T$, while simultaneously regulating u_3 to -0.22.

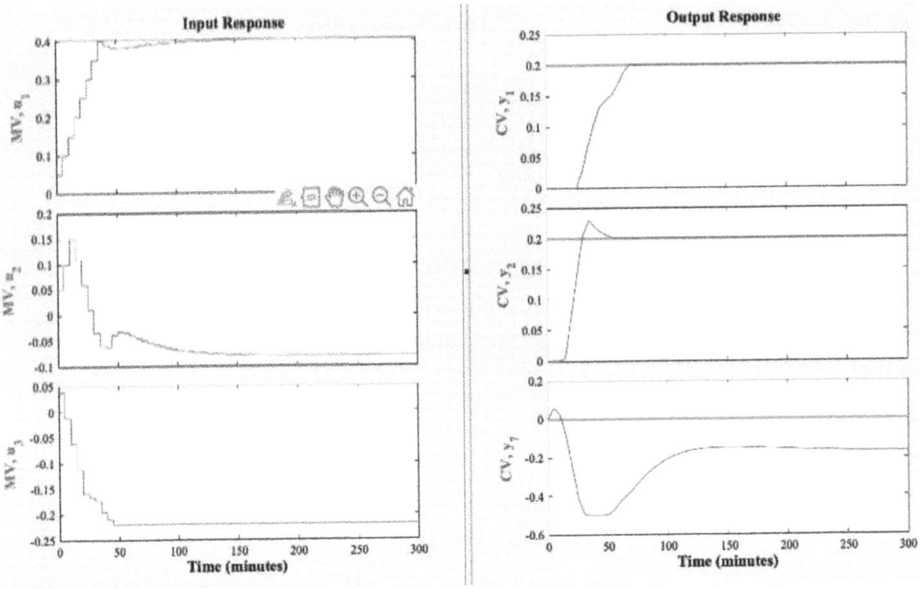

Figure F.8 Closed-loop response with setpoint for input $u_3 = -0.22$ and corresponding input weight of 1.

Remark F.1: Input weighting should be used with care. It is not standard practice in conventional MPC and may lead to steady-state bias. This is because enforcing an input to remain near a desired value can limit the degrees of freedom available for offset-free control.

Remark F.2: Economic objectives involving input usage—such as maintaining inputs above, below, or at certain values—are more appropriately handled in the optimization layer, either via a real-time optimization (RTO) layer or through economic MPC (Ellis et al., 2014).

F.3 Solving the Shell Oil Control Problem Using the MPC Toolbox

F.3.1 Comparison of Control Structures

Figure F.9 illustrates three different control structures, inspired by the work of Yu et al. (1994). In Design-A, output disturbances are modeled; in Design-B, input disturbances are considered; and in Design-C, both input and output disturbances are included.

Design-A represents a conventional structure where output disturbances are modeled as integrating white noise (IWN) sequences. This ensures offset-free tracking of the controlled outputs, even in the presence of unmeasured disturbances and MPM. As discussed in Chapters 3 and 11, this setup is equivalent to the step-response MPC formulation. Following the nomenclature of Yu et al. (1994), we refer to this configuration as Design-A.

Design-B models the two unmeasured disturbances as IWNs at the input side. This appears to be a natural approach, as the primary control goal is to mitigate the impact of these disturbances. While effective in the absence of model-plant mismatch (Case-A), this design fails to deliver offset-free tracking in the presence of mismatch (Case-B). This is because only two integrating disturbances are modeled, whereas three are required to offset errors in all controlled variables.

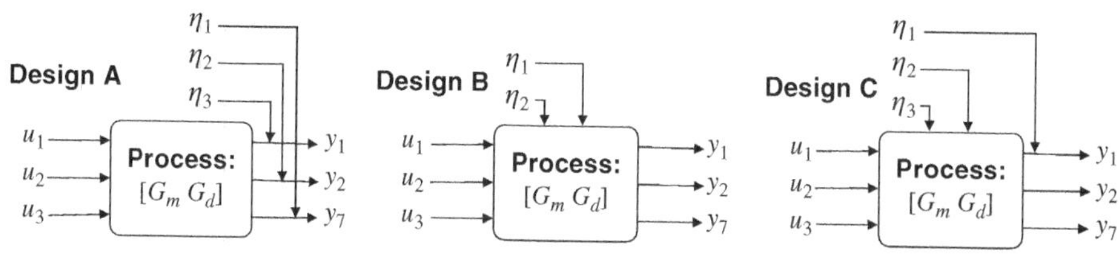

Figure F.9 Control structures for state-space MPC with state estimation. *Source*: Adapted from Yu et al. (1994).

Design-C addresses this shortcoming by incorporating one output disturbance model along with the two input disturbances. This hybrid configuration enables offset-free tracking through the output disturbance model, while leveraging feedforward compensation via the input disturbance models. As we will show, Design-C achieves robust performance even in the presence of model-plant mismatch.

The performance of all three designs—under both Case-A (no MPM) and Case-B (with MPM)—will be examined in the remainder of this section. Although these designs can be implemented using the MPC Designer GUI, we will use command-line tools from the MPC Toolbox for this analysis.

Unlike the DMC algorithm, the state-space MPC framework allows the incorporation of additional measurements to enhance control performance. For instance, including y_3 and y_5 as measured outputs can improve disturbance rejection, since the unmeasured disturbances influence these outputs without delay. Interested readers are encouraged to explore the remaining test cases from Section 3.8 using state-space MPC. The control structures discussed are all straightforward to implement using the methodology presented in the following sections.

F.3.1.1 Design-A: Output Disturbance Model

Step-1: Model definition: Consistent with the case study in Section 3.8, we begin by defining the control structure shown in Figure F.9 Design-A. The following code defines the nominal plant and disturbance models:

```
%% Defining Plant and Model
denom_Sys={[50, 1], [60, 1], [50, 1];
    [50, 1], [60, 1], [40, 1];
    [33, 1], [44, 1], [19, 1]};
delay_Sys=[27, 28, 27;  18, 14, 15; 20, 22, 0];
num_Sys_M={4.05, 1.77, 5.88;  5.39, 5.72, 6.9;   4.38, 4.42, 7.2};

denom_Dist={[45, 1], [40, 1]; [25, 1], [20, 1]; [27, 1], [32, 1]};
delay_Dist=[27, 27; 15, 15; 0, 0];
num_Dist_M={1.20, 1.44;  1.52, 1.83;  1.14, 1.26};

% Defining Model
Gm=tf(num_Sys_M,denom_Sys,'IODelay',delay_Sys,'timeUnit','min');
Gd=tf(num_Dist_M,denom_Dist,'IODelay',delay_Dist,'timeUnit','min');
Gmodel=[Gm Gd];
```

In Case-1, where the true model is exactly known (i.e., no MPM), the above definition is complete. However, in other cases, MPM is considered. In the Shell Oil Fractionator problem, only the numerators of the transfer functions differ between model and true plant. Hence, the numerators `num_Sys_P` and `num_Dist_P` need to be redefined and the true plant is defined as follows:

```
% Defining Plant (for MPM case)
num_Sys_P={1.94, 1.38, 5.29;  2.10, 5.15, 6.01;  1.27, 3.69, 5.87};
num_Dist_P={1.32, 1.60;  1.65, 1.96;  1.32, 1.44};
Gp=tf(num_Sys_P,denom_Sys,'IODelay',delay_Sys,'timeUnit','min');
Gd_p=tf(num_Dist_P,denom_Dist,'IODelay',delay_Dist,'timeUnit','min');
Gplant=[Gp Gd_p];
```

At this point, the two models—$G = \begin{bmatrix} G & G_d \end{bmatrix}$—are stored in the variables `Gmodel` and `Gplant`.

Step-2: Defining simulation scenario: We now define key simulation parameters including the sampling time, total simulation duration, setpoints, and disturbance magnitudes, as used throughout the example:

```
%% Scenarios for simulation
ts=5;                    % Sampling time (minutes)
Tsim=500;                % Total simulation time (minutes)
r=[0, 0, 0];             % Set-Points
d=[0.5, 0.5];            % Unmeasured disturbances
```

Step-3: Controller Tuning Parameters: The tuning parameters for the MPC controller include the prediction and control horizons:

```
%% Controller tuning parameters
p=20;
m=10;
```

Output, input, and input rate weights are specified via the `Weights` structure:

```
% Weights
Weights=struct('OV',[1, 1, 0],...        % Output weights
    'MV',[0, 0, 0],...                   % Input rate weights
    'MVRate',[1.5, 0.15, 1.5]);          % Input rate weights
```

Constraints on outputs and manipulated variables are defined as follows:

```
% Output constraints
OV=struct('Min',{-0.5, -0.5, -0.5}, ...
    'Max',{0.5, 0.5, inf});

% Input and input rate constraints
MV=struct('Min',{-0.5, -0.5, -0.5}, ...
    'Max',{0.5, 0.5, 0.5}, ...
    'RateMin',{-0.05, -0.05, -0.05}, ...
    'RateMax',{0.05, 0.05, 0.05});
```

Step-4: Defining MPC Controller Object: This step creates the MPC object with the specified structure and assigns the disturbance models.

```
%% MPC Design-A
% Define MPC Controller Object
Gmodel_MPC=setmpcsignals(Gmodel,'MV',[1,2,3],...
    'MO',[1,2,3],'UD',[4,5]);
mpc_DesignA=mpc(Gmodel_MPC,ts,p,m,Weights,MV,OV);

% Design-A: Input disturbance: None; Output Disturbance: IWN
setindist(mpc_DesignA,'model',ss([],[],[],[0;0]));
setoutdist(mpc_DesignA,'model',tf(1,[1 0])*eye(3));
```

This relatively compact block of code encapsulates a significant amount of information used by the MPC Toolbox to configure the controller. Before proceeding further, we strongly encourage readers to consult the official documentation for these MPC Toolbox functions through the Help Browser. This can be accessed by typing `doc setmpcsignals` in the MATLAB command window.

For quick reference, a brief explanation of each line is provided below:

- `setmpcsignals` assigns the signal types for the inputs and outputs in the model. Inputs can be manipulated variables (MV), measured disturbances (MD), or unmeasured disturbances (UD); outputs can be measured outputs (MO) or unmeasured outputs (UO). In this case, the process model G is a 3×3 transfer function, and the disturbance model is 3×2. When combined, `Gmodel` becomes a 3×5 model, where the first three inputs are MVs and the last two are UDs. All outputs are measured. This signal assignment is specified in the first command and stored in `Gmodel_MPC`.
- `mpc` creates a model predictive controller object using the model `Gmodel_MPC`. It accepts several input arguments. The initial ones—sampling time, prediction horizon, control horizon, and weights—are straightforward. The last three optional arguments are structures: `MV`, `OV`, and `DV`, which define properties of the manipulated variables, output variables, and disturbance variables, respectively:
  - `MV` is a structure used to specify various properties of the manipulated variables. The MV and MV-rate constraints defined in Step-3 are passed through this structure. Constraint softening, if needed, can be implemented using the fields `MV.MinECR`, `MV.MaxECR`, `MV.RateMinECR`, and `MV.RateMaxECR`. If an MV has a target value, it can be assigned using `MV.Target`. Optional fields such as `Name`, `Units`, and `ScaleFactor` can also be specified for each MV channel.
  - `OV` defines the properties of the output variables. Similar to `MV`, it allows specification of output constraints, constraint softening, and optional fields like names, units, and scaling factors.
  - `DV` is used to define optional metadata for disturbance variables, such as names, units, and scale factors.

  In this example, we do not utilize the optional fields `Name`, `Units`, `ScaleFactor`, or the constraint softening (ECR) options.
- `setindist` and `setoutdist` assign the input and output disturbance models for the controller. In Design-A, since no input disturbance model is used, it is set to an empty model. For the output disturbance model, we use integrators—one for each output—by specifying

`tf(1,[1 0])*eye(3)`. This corresponds to the continuous-time model $\frac{1}{s}I_{3\times3}$, which models each output disturbance as an integrating white noise process.

Step-5: Simulation: With the MPC controller defined and scenario parameters specified, we can perform closed-loop simulation using the `sim` command. This function takes the MPC object, simulation time (in number of samples), and setpoint vector as inputs. The disturbances and model-plant mismatch (if any) are provided via an MPC simulation options object created using `mpcsimopt`:

```
%% Simulating Case-1: No MPM
opt=mpcsimopt;
opt.UnmeasuredDisturbance=d;
[y1,t1,u1]=sim(mpc_DesignA,Tsim/ts,r,opt);

subplot(2,1,1); plot(t1,y1);
ylabel('CVs (y_1,y_2,y_7)');
[tt,uu]=stairs(t,u);
subplot(2,1,2); plot(tt,uu);
```

The second line above sets the values of the unmeasured disturbance variables as per vector `d`.

The performance of MPC with Design-A (output disturbance model) in rejecting unmeasured disturbances is shown in Figure F.10, assuming perfect model knowledge (i.e., no model-plant mismatch). The controller reacts only after the disturbances have affected the measured outputs. Consequently, it is not possible to avoid constraint violations entirely—most notably, y_2 significantly exceeds its upper bound. When constraints cannot be met, MPC automatically softens the affected

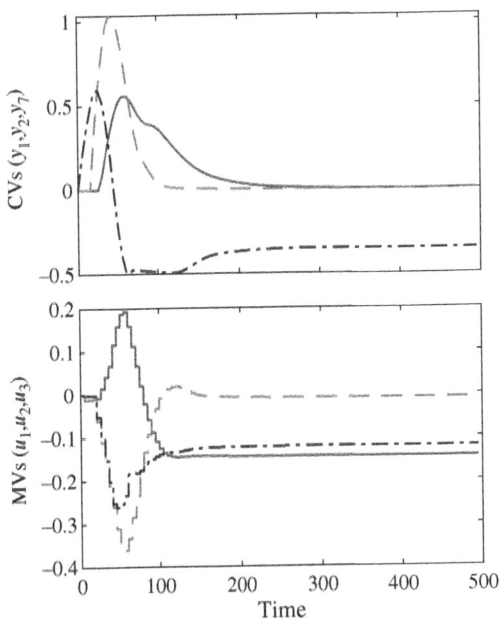

Figure F.10 Closed-loop simulation results for MPC Design-A. These results are for the case where the plant model is known perfectly (Case-1: No MPM).

constraints and solves a modified optimization problem. The system eventually regains control and settles at the desired setpoint. Modeling the disturbances as integrating white noise is essential for achieving offset-free tracking, as discussed in Chapter 11. A clear time lag is observed before the MPC begins to respond—this delay corresponds to the transport lags in the measurements y_1 and y_2, during which the inputs remain flat.

Next, we consider Case-2, which involves rejection of the same unmeasured disturbances in the presence of MPM. A different set of transfer functions representing the true plant is communicated to the MPC Toolbox via the MPC simulation options structure. The true plant model, already defined in Step-1 and stored as `Gplant`, is specified to the simulator using the `mpcsimopt` command. Recall that this structure was also used previously to provide disturbance values during simulation. To incorporate MPM, we set the `model` property of the options structure is set to the true plant, G_{plant}:

```
%% Case-2: With MPM
G_True=setmpcsignals(Gplant,'MV',[1,2,3],'MO',[1,2,3],'UD',[4,5]);
opt=mpcsimopt;                    % Reset MPC Options
opt.UnmeasuredDisturbance=d;
opt.Model=G_True;                 % Specify true plant model

[y2,t2,u2]=sim(mpc_DesignA,Tsim/ts,r,opt);
```

Since the MPC object and tuning parameters remain unchanged from Case-1, only Step-5 (simulation) is modified as shown above. The `opt.Model` field is set to `G_True`, which represents the true plant.

Figure F.11 shows the resulting closed-loop simulation with the same MPC controller as in Case-1, but now under MPM ($G_{\text{model}} \neq G_{\text{plant}}$). Note that in Design-A, unmeasured disturbance

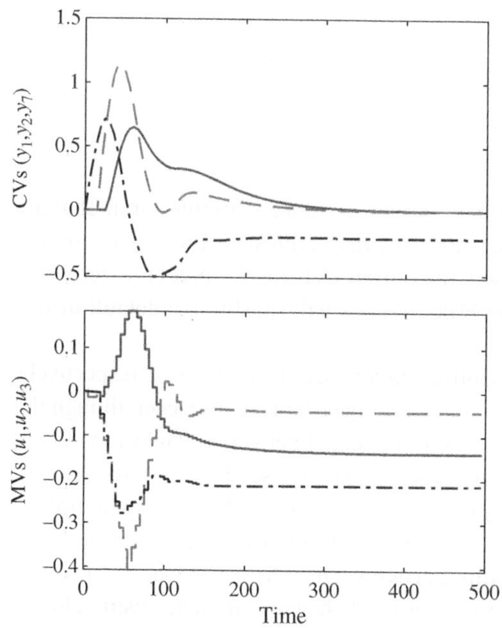

Figure F.11 Closed-loop simulation results for MPC Design-A for the case of model-plant mismatch (Case-2). The G_{model} representing the model used in MPC design is same as the previous case, whereas the true plant is represented by G_{plant}.

models are not explicitly used. Although MPM leads to larger output errors compared to the previous case, the qualitative behavior remains similar. In particular, the MPC controller—augmented with output disturbance models as integrating white noise sequences—still achieves offset-free tracking in steady state, despite the mismatch.

F.3.1.2 Design-B: Input Disturbance Model

We now turn our attention to Design-B, in which the effect of unmeasured disturbances is explicitly modeled. Since these disturbances are not measured, they must be estimated by the controller. They are modeled as integrating white noise (IWN), with their impact on the system described through G_d. The structure of the MPC model for Design-B is illustrated in Figure F.9 Design-B. The model definitions for `Gmodel` and `Gplant` remain unchanged, as do the simulation scenarios—Case-1 (no MPM) and Case-2 (with MPM)—and all controller tuning parameters (horizons, weights, and constraints). The key difference lies in Step-4, where the controller is configured to match the Design-B architecture.

In this design, the two unmeasured disturbances are modeled as integrating white noise, while no output disturbances are considered. In other words, the input disturbance model is:

$$G_{int} = \frac{1}{s}\begin{bmatrix} 1 & 0 \\ 0 & 1 \end{bmatrix}$$

The corresponding MATLAB code to configure the MPC object for Design-B is shown below:

```
%% MPC Design-B
% Define MPC Controller Object
Gmodel_MPC=setmpcsignals(Gmodel,'MV',[1,2,3],...
    'MO',[1,2,3],'UD',[4,5]);
mpc_DesignB=mpc(Gmodel_MPC,ts,p,m,Weights,MV,OV);

% Design-B: Input dist.: IWN (2 channels); Output dist.: None
iwnTF=tf(1,[1 0])*eye(2);                   % Two integrators
outDist=ss([],[],[],[0;0;0]);               % No output dist
setindist(mpc_DesignB,'model',iwnTF);
setoutdist(mpc_DesignB,'model',outDist);
```

The first two lines define the MPC object—identical to Design-A—since all tuning parameters are the same. The next two lines specify the disturbance models: `iwnTF` represents the two integrating disturbance channels, while `outDist` explicitly sets the output disturbance model to zero. It is important to provide this explicitly; otherwise, the MPC Toolbox will substitute a default disturbance model.

Figures F.12 and F.13 show the closed-loop responses under Case-1 and Case-2, respectively. Modeling unmeasured disturbances as IWN leads to significantly faster response, even though the disturbances are not directly measured. In Case-1, where the model perfectly matches the plant, the controller achieves fast and accurate tracking with no steady-state offset. However, in Case-2, a steady-state offset appears. This occurs because only two of the three output channels are indirectly corrected through the modeled input disturbances. To guarantee offset-free tracking in the presence of MPM, the number of independent integrating disturbances must match the number of controlled outputs. Since only two IWN channels are used here, offset-free behavior cannot be ensured for all outputs.

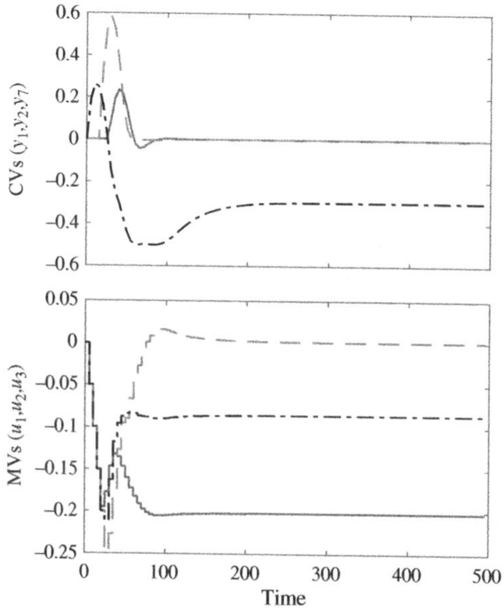

Figure F.12 Closed-loop simulation results for MPC Design-B, rejecting the effect of unmeasured disturbances when the model is known perfectly.

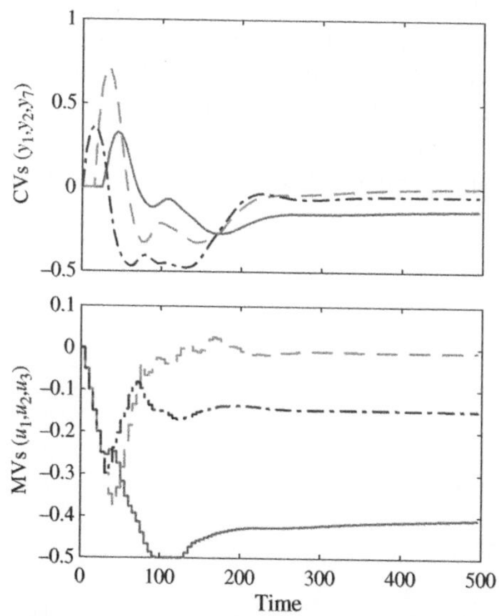

Figure F.13 Closed-loop simulation results for MPC Design-B in presence of model-plant mismatch (Case-2), with G_{model} used in MPC design and the true plant represented by G_{plant}. Note the output y_1 has a steady state offset.

F.3.1.3 Input and Output Disturbance Model

The case study thus far has illustrated key concepts from Chapter 11 in the context of the Shell Oil Fractionator problem. To ensure bias-free control with integral action on the controlled outputs, disturbances were modeled as integrating white noise (IWN).

The simplest configuration, Design-A, introduced additive IWN sequences to each output channel. While this guarantees offset-free tracking, the assumption of independent IWN sequences for all outputs leads to sluggish control action. Design-B improved transient performance by modeling unmeasured load disturbances as IWN. However, since only two integrating-type disturbances were used, they were insufficient to influence all three controlled outputs, resulting in steady-state offset under model-plant mismatch.

Design-C addresses this shortcoming by augmenting the two unmeasured input disturbances with an additional output disturbance, modeled as IWN. One might consider naively adding IWN to all three output channels in addition to the input disturbances (as in Design-A and Design-B combined), but the MPC Toolbox does not permit this: the resulting augmented system would not be detectable. Therefore, we add an IWN disturbance only to the first output channel:

$$G_{iwn} = \frac{1}{s}\begin{bmatrix} 1 \\ 0 \\ 0 \end{bmatrix}$$

The corresponding MATLAB configuration for Design-C is as follows:

```
% Design-C: Input dist.: IWN (2 channels); Output dist.: IWN (1 channel)
iwnTF=tf(1,[1 0])*eye(2);                  % Two integrators
outDist=tf(1,[1 0])*[1;0;0];               % Output dist in y1
```

For brevity, only the results for Case-2 (with MPM) are shown in Figure F.14. The response is faster, and all outputs settle at their respective setpoints. Compared to Design-B, one additional independent disturbance was added. Since there are three measured outputs, the system can support at most three independent IWN disturbances. This constraint arises because IWN models introduce unstable poles that must be estimated using a Kalman filter. To ensure observability (detectability) of the augmented system, the number of independent IWN disturbances must not exceed the number of measurements.

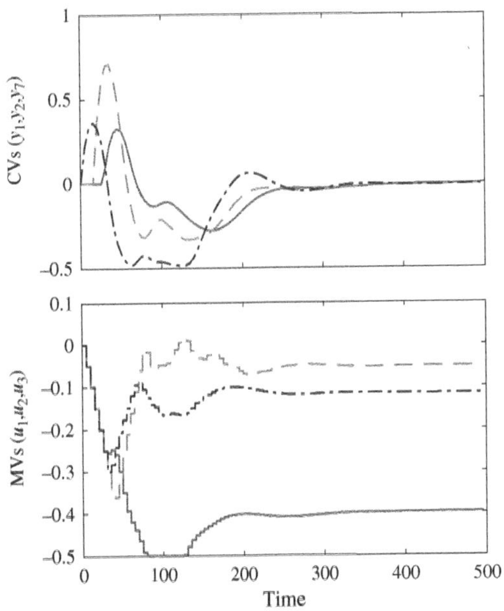

Figure F.14 MPC Design-C shows good performance for Case-2 (with MPM).

F.3.2 Specifying Target for MV

An additional control objective in the Shell problem was to reduce the magnitude of MV u_3, which correlates directly with heat recovery. Although there are three inputs, only two outputs (y_1 and y_2) are assigned setpoints; the third, y_7, is subject to constraints but not setpoint tracking. Based on analysis by Yu et al. (1994), a suitable target value for u_3 is −0.22.

In traditional MPC (e.g., DMC), this objective requires including MVs in the output model via direct feedthrough. This is not necessary with state-space MPC in the MPC Toolbox, which allows assigning an input target and input weight directly. The input target is specified via the `target` field of the `MV` structure within the `mpc` object. Input weights have already been used via the `Weights` property; in previous designs, we set them to `[0, 0, 0]`. Compared to Design-C, only the following changes in Step-3 (controller tuning parameters) are required:

```
% Weights
Weights=struct('OV',[1, 1, 0],...           % Output weights
    'MV',[0, 0, 0.1],...                    % Input weights
    'MVRate',[0.5, 0.1, 0.5]);              % Input rate weights

% Input and input rate constraints
MV=struct('Min',{-0.5, -0.5, -0.5}, ...
    'Max',{0.5, 0.5, 0.5}, ...
    'RateMin',{-0.05, -0.05, -0.05}, ...
    'RateMax',{0.05, 0.05, 0.05},...
    'Target',{0, 0, -0.22});
```

The changes from Design-C are the inclusion of a non-zero input weight $\Gamma^{u_3} = 0.1$ and the specification of an input target for u_3. The result of implementing this setpoint on MV is shown in Figure F.15. The "economic advantage" of keeping u_3 at −0.22 comes with a slightly sluggish response of the CVs to reach the setpoint.

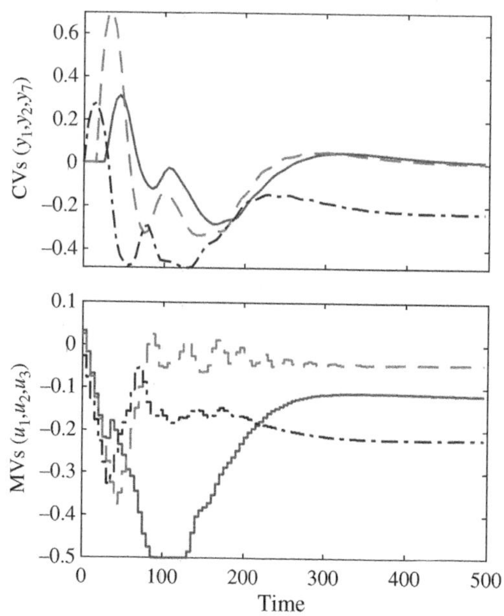

Figure F.15 Case-2 with target for $u_3 = -0.22$.

G

A Brief Tutorial on Simulink

G.1 A MIMO System Example

We present a brief tutorial on using MATLAB Simulink® for setting up a nonlinear model and getting ready for closed-loop control simulations. Consider the following model from Chapter 12:

$$\begin{aligned}
\dot{x}_1' &= -x_1 - x_1 x_3 + d_1 \\
\dot{x}_2' &= -x_2 + 0.5x_1 + d_2 \\
\dot{x}_3' &= -x_3 - \exp\left(-x_2^2\right) + u_1 \\
\dot{x}_4' &= -x_4 + x_1^2 + u_1 u_2,
\end{aligned} \tag{G.1}$$

where u represents the manipulated variables, d_1 is an unmeasured disturbance and d_2 is a measured disturbance. The two outputs defined as:

$$\begin{aligned}
y_1 &= x_2 + x_3 \\
y_2 &= -x_1 + x_4.
\end{aligned} \tag{G.2}$$

The first step in solving a closed-loop control problem is to build the dynamic system in the simulation environment. An advantage of using Simulink for the dynamic simulation is its clear and straightforward flow-sheeting capability. Figure G.1a shows the overall block-diagram used in Simulink for solving the dynamic model. We have shown three "Scope" for plotting the manipulated, disturbance and output variables. The central set of blocks are for dynamic simulation of the above system.

The "MATLAB function" blocks in Simulink are used for specifying state function and output function. The first function (labelled as "ODE function"), which is used for specifying the dynamic model (G.1), has four states and takes four inputs, $\begin{bmatrix} d_1 & d_2 & u_1 & u_2 \end{bmatrix}^T$. The dynamical equations are defined in the MATLAB function below:

```
function dxdt = System_StateFunction(x,u)
dxdt = zeros(4,1);
dxdt(1) = -x(1) - x(3)*x(1) + u(1);
dxdt(2) = -x(2) + 1/2*x(1)  + u(2);
dxdt(3) = -x(3) - exp(-x(2)^2) + u(3);
dxdt(4) = -x(4) + x(1)^2 + u(3)*u(4);
end
```

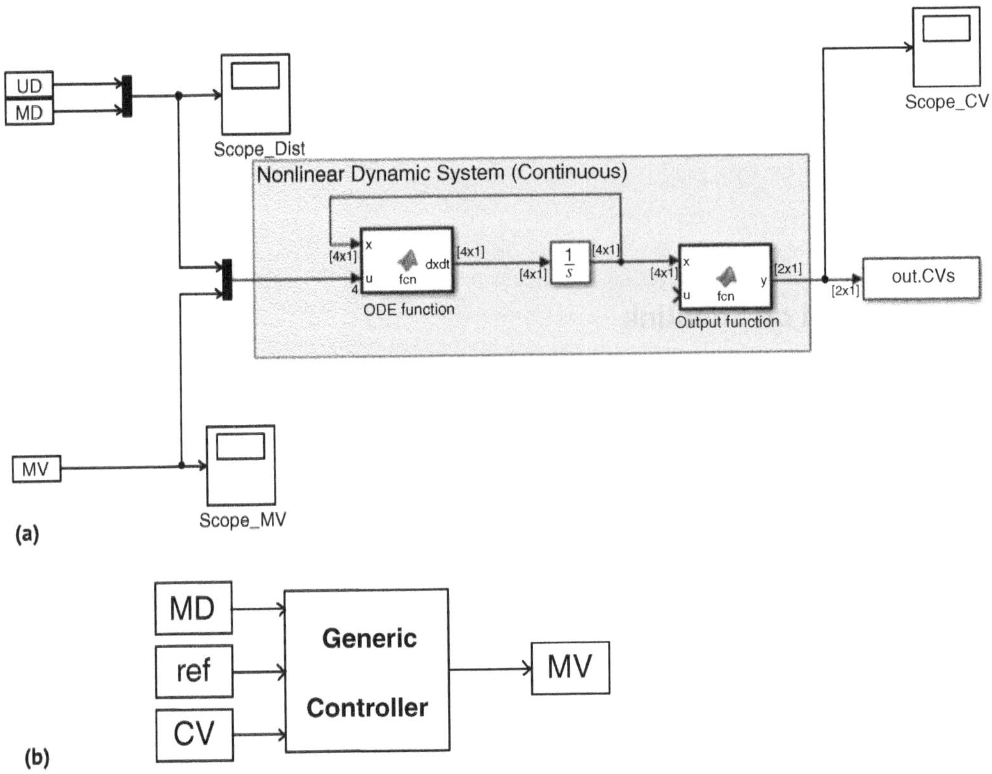

Figure G.1 (a) Flowsheet of a dynamic process in Simulink; and (b) Block diagram representation of a controller.

The integrator block is included for integrating the differential equations (G.1). The static "Output function" computes the outputs as per Eq. (G.2) is specified below:

```
function y = System_OutputFunction(x,u)
y = zeros(2,1);
y(1) = x(2) + x(3);
y(2) = x(4) - x(1);
end
```

This completes the specification of the nonlinear model. An interested user can connect the four inputs to respective Simulink Source blocks for performing open-loop simulations. Simulating the system with various input and disturbance changes allows users to gain familiarity with the system dynamics and provides insights into how the system responds under different conditions.

With the nonlinear system prepared, it is time to proceed to the controller design step. Figure G.1b shows a schematic of a typical controller. The controller uses the current values of controlled variable, reference and measured disturbances (and past input values, if relevant) and computes the corresponding manipulated variables. Any controller of choice may be used here, including the nonlinear MPC that is discussed in Chapter 12.

G.2 Simulink Model for a CSTR

The model for a CSTR is briefly explained here. The overall model equations are given by

$$\begin{aligned}
&\text{Mass balance: } \frac{dC_A}{dt} = \frac{F}{V}\left(C_{A,in} - C_A\right) - r_A \\
&\text{Energy balance: } \frac{dT}{dt} = \frac{F}{V}(T_{in} - T) + \frac{-\Delta H}{\rho c_p} r_A - \frac{UA}{\rho c_p V}\left(T - T_{C,in}\right) \qquad \text{(G.3)} \\
&\text{Reaction rate: } r_A = k_0 e^{-E/(RT)} C_A.
\end{aligned}$$

The system parameters used for this example are listed in Table G.1.

Please use `Dynamic_System.slx` file or use following MATLAB code below to build the target system model. The model framework for building a Simulink model is similar to the one in the previous section. The system state and output functions for the CSTR are given below.

```
%% State function
function dxdt = System_StateFcn(x,u)

%% States
T = x(1);
CA = x(2);

%% Inputs
CA_in = u(1);
T_in = u(2);
Tc_in = u(3);

%% parameters
FoV = 1;                % F over V
UAoRhoCP = 0.3;         % Heat transfer coefficient
HroRhoCp = -11.92;      % Heat of reaction
k0 = 3.49e7;            % Reaction kinetics
EoR = 5963.6;           % Activation energy (E over R)

%% State equations
dxdt = zeros(2,1);
```

Table G.1 System parameters for the NMPC case study of control of CSTR.

Variables	Values	Variables	Values
$\frac{F}{V}$	1	$\frac{UA}{\rho c_p V}$	-11.92
$\frac{-\Delta H}{\rho c_p}$	0.3	$\frac{-E}{RT}$	5963.6
k_0	3.49e7		

```
k = k0*exp(-EoR/T);
r = k*CA;

dxdt(1) = FoV*(T_in-T) + UAoRhoCP*(Tc_in-T) - HroRhoCp*r;
dxdt(2) = FoV*(CA_in-CA) - r;

end

%% Output function
function y = System_OutputFcn(x,u)
    y(1,1) = x(1);
    y(2,1) = x(2);
end
```

Figure G.2 shows the Simulink block diagram for the CSTR model. The MATLAB code-based and block-diagram-based approaches are shown. If you use `Dynamic_system.slx` file, please confirm that the above MATLAB code is identical to the CSTR model built in Simulink. Since both implementations are the same, you can use any of them for the target system model.

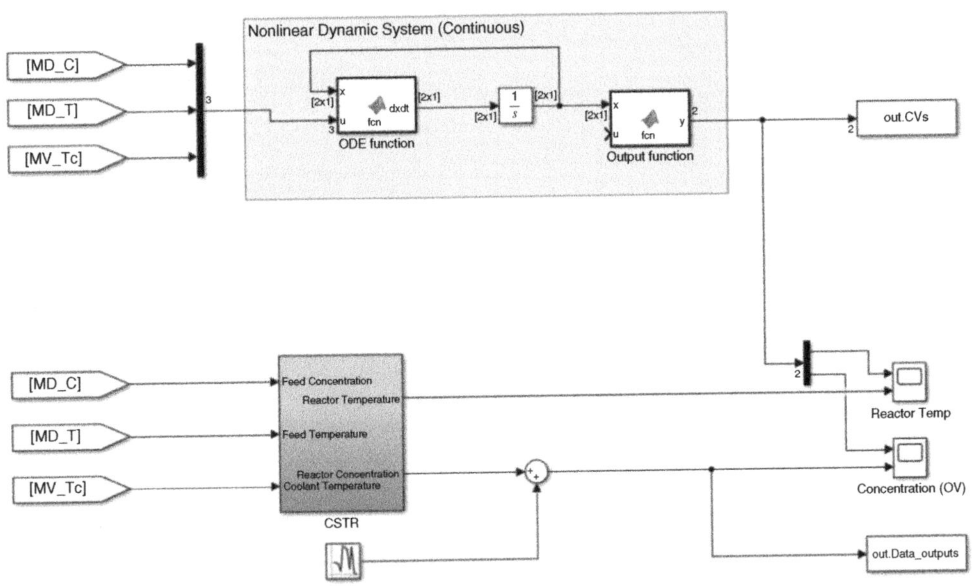

Figure G.2 Flowsheet of a dynamic system (CSTR) in Simulink: (top) MATLAB code based Simulink model; and (bottom) Block diagram based Simulink model.

Bibliography

ABB Ltd. Industrial software that increases plant profitability. 2025. https://new.abb.com/industrial-software/process-performance/apc-advanced-process-control

N. Amann, D. H. Owens, and E. Rogers. Iterative learning control for discrete-time systems with exponential rate of convergence. *IEE Proceedings-Control Theory and Applications*,143:217–224, 1996.

R. Amrit, W. Canney, P. Carrette, R. Linn, A. Martinez, A. Singh, T. Skrovanek, J. Valiquette, J. Williamson, and J. Zhou. Platform for advanced control and estimation (pace): Shell's and yokogawa's next generation advanced process control technology. *IFAC 9th International Symposium on Advanced Control of Chemical Processes*, 46:73–87, 2015.

B. D. O. Anderson and J. B. More, editors. *Optimal Control: Linear Quadratic Methods*. Prentice-Hall, 1996.

Anon. Computer runs refinery unit. *Business Week*, 1959.

K. J. Åström and T. Hagglund. *PID Controllers,* 2nd edition. Instrument Society of America, 1995.

K. J. Åström and B. Wittenmark, editors. *Computer-Controlled Systems: Theory and Design*. Prentice-Hall, 3rd edition, 1990.

AVEVA Group Ltd. AVEVA process optimization: Reactor models for refineries, ammonia and petrochemical plants, 2025.https://www.aveva.com/en/solutions/operations/value-chain-optimization/romeo-reactor-models/

A. Bemporad and M. Morari. Control of systems integrating logic, dynamics, and constraints. *Automatica*, 35(3):407–427, 1999.

A. Bemporad, M. Morari, V. Dua, and E. N. Pistikopoulos. The explicit linear quadratic regulator for constrained systems. *Automatica*, 38(1):3–20, 2002.

A. Bemporad. N. L. Ricker. and M. Morari. *MATLAB Model Predictive Control Toolbox User's Guide.* The Mathworks Inc., 2021.

J. Betts. *Practical Methods for Optimal Control Using Nonlinear Programming*. Philadelphia, PA, SIAM, 2001.

L. T. Biegler. Efficient solution of dynamic optimization and NMPC problems. In F. Allgöwer and A. Zheng, Editors, *Nonlinear Model Predictive Control, volume 26 of Progress in Systems and Control Theory,* Basel, Birkhäuser, 2000.

L. T. Biegler. *Nonlinear Programming: Concepts, Algorithms, and Applications to Chemical Processes.* Philadelphia, PA, SIAM, 2010.

F. Borrelli, A. Bemporad, M. Fodor, and D. Hrovat. An MPC/hybrid system approach to traction control. *IEEE Transactions on Control System Technology*, 14:541–552, 2006.

F. Borrelli, A. Bemporad, and M. Morari. *Predictive Control for Linear and Hybrid Systems.* Cambridge, Cambridge University Press, 2017.

A. E. Bryson and Y. C. Ho. *Applied Optimal Control.* Hemisphere, 1975.

E. Camacho and C. Bordons. *Model Predictive Control.* Berlin, Springer, 1998.

P. J. Campo and M. Morari. Robust model predictive control. In *Proceedings of the 1987 American Control Conference,* pages 1021–1026, 1987.

G. Cesconetto. Aam aircraft: Levels of autonomous flying. Technical report, CAAM, 2022. https://canadianaam.com/2022/06/07/aam-aircraft-levels-of-autonomous-flying/

H. Chen and F. Allgöwer. A quasi-infinite horizon nonlinear model predictive control scheme with guaranteed stability. *Automatica*, 34:1205–1217, 1998.

D. Chmielewski and V. Manousiouthakis. On constrained infinite-time linear quadratic optimal control. *Systems and Control Letters*, 29:121–129, 1996.

D. W. Clarke, C. Mohtadi, and P. S. Tuffs. Generalized predictive control-I. the basic algorithm. *Automatica*, 23:137–148, 1987a.

D. W. Clarke, C. Mohtadi, and P. S. Tuffs. Generalized predictive control–II. Extensions and interpretations. *Automatica*, 23:149–160, 1987b.

D. Correa. Advanced process control (APC) market to reach usd 5.3 billion, globally, by 2033 at 9 SE: Sensors and Controls, Report Code: A34949, 2025. https://www.prnewswire.com/news-releases/advanced-process-control-apc-market-to-reach-5-3-billion-globally-by-2033-at-9-cagr-allied-market-research-302391371.html.

C. R. Cutler and R. B. Hawkins. Application of a large predictive multivariable controller to a hydrocracker second stage reactor. In *Proceedings of American Control Conference,* Atlanta, GA, pages 284–291, 1988

C. R. Cutler and B. L. Ramaker. Dynamic matrix control – a computer control algorithm. In *AIChE National Meeting*, Houston, TX, 1979.

C. R. Cutler and B. L. Ramaker. Dynamic matrix control – a computer control algorithm. In *Proceedings of the Joint Automatic Control Conference,* San Francisco, CA, 1980.

S. Di Cairano, A. Bemporad, I. V. Kolmanovsky, and D. Hrovat. Model predictive control of magnetically actuated mass spring dampers for automotive applications. *International Journal of Control*, 80(11):1701–1716, 2007.

W. B. Dunbar, M. B. Milam, R. Franz, and R. M. Murray. Model predictive control of a thrust-vectored flight control experiment. In *Proceedings of the 15th IFAC World Congress,* Barcelona, Spain, 2002.

M. Ellis, H. Durand, and P. D. Christofides. A tutorial review of economic model predictive control methods. *Journal of Process Control*, 24 (8):1156–1178, 2014.

C. E. García. *Studies in Optimizing and Regulatory Control of Chemical Process Systems.* PhD thesis, University of Wisconsin-Madison, 1982.

C. E. García. Quadratic dynamic matrix of nonlinear processes. In *AIChE Annual Meeting*, San Francisco, CA, 1984.

C. E. García and M. Morari. Internal model control 1. a unifying review and some new results. *Industrial & Engineering Chemistry Process Design and Development*, 21:308–322,1982.

C. E. García and M. Morari. Internal model control 3. multivariable control law computation and tuning guidelines. *Industrial & Engineering Chemistry Process Design and Development*, 24:484–494, 1985.

C. E. García and A.M. Morshedi. Quadratic programming solution of dynamic matrix control (qdmc). *Chemical Engineering Communications*, 46:73–87, 1986.

C. E. García, D. M. Prett, and M. Morari. Model predictive control: Theory and practice – a survey. *Automatica*, 25(3):335–348, 1989.

E. G. Gilbert and K. T. Tan. Linear systems with state and control constraints: The theory and application of maximal output admissable sets. *IEEE Transactions on Automatic Control*, 36(9):1008–1020, 1991.

N. Giorgetti, A. Bemporad, H. E. Tseng, and D. Hrovat. Hybrid model predictive control application towards optimal semiactive suspension. In *Proceedings of IEEE International Symposium on Industrial* Electronics, Dubrovnik, Croatia, 2005.

N. Giorgetti, G. Ripaccioli, A. Bemporad, I. V. Kolmanovsky, and D. Hrovat. Hybrid model predictive control of direct injection stratified charge engines. *IEEE/ASME Transactions on Mechatronics*, 11(5):499–506, 2006.

A. Gopalakrishnan, N. S. Kaisare, and S. Narasimhan. Incorporating delayed and infrequent measurements in extended kalman filter based nonlinear state estimation. *Journal of Process Control*, 21:119–129, 2011.

A. Halanay. Quelques questions de la theorie de la stabilite pour les systemes aux differences finies, *Archive for Rational Mechanics and Analysis*, 12:150–154, 1963.

S. Hara, Y. Yamamoto, T. Omata, and N. Nakano. Repetitive control system: A new type servo system for periodic exogeneous signals. *IEEE Transactions on Automatic Control*,33: 659–668, 1988.

B. L. Ho and R. E. Kalman. Effective construction of linear state-variable models from input-output functions. *Regelungstechnik*, 14:545–548, 1966.

A. H. Jazwinski. *Stochastic Processes and Filtering Theory*. Mineola, New York, Dover Publications, Inc., 2007.

N. S. Kaisare. *Computational Techniques for Process Simulation and Analysis Using MATLAB*. 1st edition, Boca Raton, FL, USA, CRC Press, 2017.

R. E. Kalman. A new approach to linear filtering and prediction problems. *Transactions of the ASME. Series D, Journal of Basic Engineering*, 82:35–45, 1960.

H. Kawamura. Fundamentals of advanced process control. Technical report, Yokogawa Corporation, 2015.

S. S. Keerthi and E. G. Gilbert. Optimal infinite-horizon feedback laws for a general class of constrained discrete-time systems: stability and moving-horizon approximations. *Journal of Optimization Theory and Applications*, 57:265–293, 1988.

F. H. Kishi. On line computer control techniques and their application to re-entry aerospace vehicle control. *Advances in Control System*, 1:245–357, 1964.

D. L. Kleinman. Stabilizing a discrete, constant, linear system with application to iterative methods for solving the Riccati equation. *IEEE Transactions on Automatic Control*, AC-19:252–254, 1974.

M. V. Kothare, V. Balakrishnan, and M. Morari. Robust constrained model predictive controlusing linear matrix inequalities. *Automatica*, 32(10):1361–1379, 1996.

S. Kouro, P. Cortes, R. Vargas, U. Ammann, and J. Rodriguez. Model predictive control - a simple and powerful method to control power converters. *IEEE Transactions on Industrial Electronics*, 56(6):1826–1838, 2009.

S. Y. Kung. A new identification and model reduction algorithm via singular value decomposition. In *Proceedings of the 12th Asilomar Conference on Circuits, Systems and Computers*, pages 705–714, 1978.

H. Kwakernaak and R. Sivan, editors. *Linear Optimal Control Systems*. New York, John Wiley & Sons, 1972.

W. Kwon and A. Pearson. A modified quadratic cost problem and feedback stabilization of a linear system. *IEEE Transactions on Automatic Control*, 22(5):838–842, 1977.

A. Lagerberg and B. Egardt. Model predictive control of automotive powertrains with backlash. In *Proceedings of the 16th IFAC World Congress*, Prague, Czech Republic, 2005.

W. E. Larimore. Canonical variate analysis in identification, filtering and adaptive control. In *Proceedings of the 29th IEEE Conf. on Decision and Control*, pages 596–604, 1990.

K. Larson. Remembering digital pioneer charles r. cutler. 2020, https://www.controlglobal.com/home/article/11296813/remembering-digital-pioneer-charles-r-cutler

K. S. Lee, J. H. Lee, I. Chin, and H. J. Lee. Model predictive control technique combined with iterative learning for batch processes. *AIChE Journal*, 45:2175–2187, 1999.

J. H. Lee, K. S. Lee, and W. C. Kim. Model-based iterative learning control with a quadratic criterion for time-varying linear systems. *Automatica*, 36:641–657, 2000.

E. B. Lee and L. Markus. *Foundations of Optimal Control Theory*. New York, NY, Wiley, 1967

J. H. Lee, M. Morari, and C. E. García. State space interpretation of model predictive control. *Automatica*, 30:707–717, 1994.

J. H. Lee, S. Natarajan, and K. S. Lee. A model-based predictive control approach to repetitive control of continuous processes with periodic operations. *Journal of Process Control*, 11:195–207, 2001.

J. H. Lee and N. L. Ricker. Extended Kalman filter based nonlinear model predictive control. *Industrial & Engineering Chemistry Research*, 33:1530–1541, 1994.

W. Lee and V. W. Weekman. Advanced control practice in the chemical process industry: A view from industry. *AIChE Journal*, 22:27–38, 1976.

J. H. Lee and Z. H. Yu. Tuning of model predictive controllers for robust performance. *Computers & Chemical Engineering*, 18: 15–37, 1994.

J. H. Lee and Z. Yu. Worst-case formulation of model predictive control for systems with bounded parameters. *Automatica*, 33(5):763–781, 1997.

Y. Li, K. H. Ang, and G. C. Y. Chong. Patents, software, and hardware for PID control: An overview and analysis of the current art. *IEEE Control Systems Magazine*, pages 42–54, 2006.

L. Ljung, editor. *System Identification: Theory for the User*. 2nd edition, Prentice-Hall, 1998.

R. W. Longman. Design methodologies in learning and repetitive control. In *Proceedings of the 2nd Asian Control Conference*, 1997.

A. Lorber, L. E. Wangen, and B. R. Kowalski. A theoretical foundation for the pls algorithm. *Journal of Chemometrics*, 1:19–31, 1987.

K. Lundberg. 25 years ago. *IEEE Control Systems Magazine*, 30:15–16, 2010.

U. Maeder, F. Boorrelli, and M. Morari. Linear offset-free model predictive control. *Automatica*, 45:2214–2222, 2009.

G. D. Martin. Idcom hierarchical control of an oil refinery reactor. In *Proceedings of American Control Conference*, San Diego, CA, pages 675–678, 1984.

D. Q Mayne and H. Mischalska. Receding horizon control of nonlinear systems. *IEEE Transactions on Automatic Control*, 35:814–824, 1990.

D. Q. Mayne, J. B. Rawlings, C. V. Rao, and P. O. M. Scokaert. Constrained model predictive control: Stability and optimality. *Automatica*, 36:789–814, 2000.

H. Michalska and D. Mayne. Robust receding horizon control of constrained nonlinear systems. *IEEE Transactions on Automatic Control*, 38(11):1623–1633, 1993.

H. Michalska and D. Q. Mayne. Moving horizon observers and observer-based control. *IEEE Transactions on Automatic Control*, 40:995–1006, 1995.

K. L. Moore, M. Dahleh, and S. Bhattacharyya. Iterative learning control: A survey and new results. *Journal of Intelligent & Robotic Systems*, 9:563–594, 1992.

M. Morari and J. H. Lee. Model predictive control: the good, the bad, and the ugly. In Y. Arkun and W. Ray, editors, *Proceedings of 4th International Conference on Chemical Process Control – CPC IV*, pages 419–444, South Padre Island, TX, 1991.

M. Morari and E. Zafiriou. *Robust Process Control*. Englewood Cliffs, NJ, Prentice Hall, 1989

K. Muske and T. Badgwell. Disturbance modeling for offset-free linear model predictive control. *Journal of Process Control*, 12:617–632, 2002.

K. R. Muske and J. B. Rawlings. Model predictive control with linear models. *AIChE Journal*, 39(2):262–287, 1993.

M. Pandit and K. H. Buchheit. Optimizing iterative learning control of cyclic production with application to extruders. *IEEE Transactions on Control Systems Technology*, 7:382–389,1995

G. Pannocchia and J. B. Rawlings. Disturbance models for offset-free model predictive control. *AIChE Journal*, 49:426–437, 2003.

D. M. Prett and C. E. García. *Fundamental Process Control*. Butterworth, 1988.

D. M. Prett, C. E. García, and B. L. Ramaker. *The Shell Process Control Workshop. Solutions to the Shell Standard Control Problem*. Butterworth, Stoneham, MA, 1990.

D. M. Prett and M. Morari. *The Shell Process Control Workshop. Process Control Research: Industrial and Academic Perspectives*. Butterworth, Stoneham, MA, 1987.

D. M Prett, B. L. Ramaker, and R. C. Cutler. The dynamic matrix control method, US patent 4,349,869, 1982.

A. I. Propoi. Use of linear programming methods for synthesizing sampled-data automatic systems. *Automn. Remote Control*, 24(7):837–844, 1963.

S. J. Qin and T. A. Badgwell. A survey of industrial model predictive control technology. *Control Engineering Practice*, 11:733–764, 2003.

S. J. Qin and L. Ljung. Closed loop subspace identification with innovation estimation. In *IFAC Proceedings Volumes, volume 36*, pages 861–866, 2003.

C. V. Rao, J. B. Rawlings, and J. H. Lee. Constrained linear state estimation–a moving horizon approach. *Automatica*, 37:1619–1628, 2001.

C. V. Rao, J. B. Rawlings, and D. Q. Mayne. Constrained state estimation for nonlinear-discrete-time systems: stability and moving horizon approximations. *IEEE Transactions on Automatic Control*, 48:246–258, 2003.

J. B. Rawlings, D. Q. Mayne, and M. M. Diehl. *Model Predictive Control: Theory, Computation and Design*. Nob Hill Publishing, LLC., 2020.

J. B. Rawlings and K. R. Muske. The stability of constrained receding horizon control. *IEEE Transactions on Automatic Control*, 38:1512–1516, 1993.

J. Richalet. Industrial applications of model based predictive control. *Automatica*, 29:1251–1274, 1993.

J. Richalet, A. Rault, J. L. Testud, and J. Papon. Model predictive heuristic control: Applications to industrial processes. *Automatica*, 14(5):413–428, 1978.

S. Richter, S. Mariethoz, and M. Morari. High-speed on-line MPC based on a fast gradient method applied to power converter control. In *Proceedings of American Control Conference*, Baltimore, MD, pages 4737–4743, 2010.

N. L. Ricker. Model predictive control: State of the art. In Y. Arkun and W. Ray, editors, *Proccedings of 4th International Conference on Chemical Process Control – CPC IV* , pages 271–296, South Padre Island, TX, 1991.

D. E. Rivera, M. Morari, and S. Skogestad. Internal model control 4. pid controller design. *Industrial & Engineering Chemistry Process Design and Development*, 25: 252–265, 1986.

D. G. Robertson and J. H. Lee. On the use of constraints in least squares estimation and control. *Automatica*, 38:1113–1123, 2002.

D. G. Robertson, J. H. Lee, and J. B. Rawlings. A moving horizon-based approach for least-squares estimation. *AIChE Journal*, 42(8):2209–2224, 1996.

Rockwell Automation. Advanced process control: Continuously optimize production performance by reducing process variability with advanced process control, 2025.https://kalypso.com/services/data-science/advanced-process-control

R. Rouhani and R. K. Mehra. Model algorithmic control (mac);basic theoretical properties. *Automatica*, 18(4):401–414, 1982.

B. Saerens, M. Diehl, J. Swevers, and E. Van den Bulck. Model predictive control of automotive powertrains - first experimental results. In *Proceedings of the 47th IEEE Conference on Decision and Control*, Cancun, Mexico, page ThC18.2, 2008.

P. O. M. Scokaert and J. B. Rawlings. Constrained linear quadratic regulation. *IEEE Transactions on Automatic Control*, 43(8):1163–1169, 1998.

P. O. M. Scokaert, J. B. Rawlings, and E. S. Meadows. Discrete-time stability with perturbations: Applications to model predictive control. *Automatica*, 33(3):463–470, 1997.

S. Skogestad. Simple analytic rules for model reduction and pid controller tuning. *Modelling, Identification, and Control*, 25:85–120, 2004.

Smart Buildings Academy (SBA). The ultimate guide to building automation systems, 2019. https://guides.smartbuildingsacademy.com/building-automation-system

T. Soderstrom and P. Stoica, editors. *System Identification*. 1st edition, Prentice-Hall, 1989.

T. Sogo and N. Adachi. A gradient-type learning control algorithm for linear systems. In *Proceedings of the 1st Asian Control Conference*, 1994.

T. M. Stout and T. J. Williams. Pioneering work in the field of computer process control. *IEEE Annals of the History of Computing*, 17(1):6–18, 1995.

S. W. Sung and J. H. Lee. Pseudo-random binary sequence design for finite impulse response identification. *Control Engineering Practice*, 11:935–947, 2003.

Synopsis Editorial Staff. The 6 levels of vehicle autonomy explained, 2025. https://www.synopsys.com/blogs/chip-design/autonomous-driving-levels.html#pagesize=10&pageno=undefined

A. H. Tan and K. R. Godfrey. The generation of binary and near-binary pseudo-random signals: An overview. In *Proceedings of the 18th IEEE Instrumentation and Measurement Technology Conference*, pages 766–771, 2001.

K. M. Tao, R. L. Kosut, and G. Aral. Learning feedforward control. In *Proceedings of the American Control Conference*, pages 2575–2579, 1994.

P. Van Overschee and B. De Moor. N4SID: Subspace algorithms for the identification of combined deterministic-stochastic systems. *Automatica*, 30:75–93, 1994.

P. Van Overschee and B. De Moor. Closed loop subspace system identification. In *Proceedings of the 36th IEEE Conference on Decision and Control*, pages 1848–1853, 1997.

M. Verhaegen. Application of a subspace model identification technique to identify LTI systems operating in closed loop. *Automatica*, 29:1027–1040, 1993.

M. Verhaegen. Identification of the deterministic part of MIMO state space models given in innovations form from input-output data. *Automatica*, 30(1):61–74, 1994.

D. R. Yang, K. S. Lee, H. J. Ahn, and J. H. Lee. Experimental application of a quadratic optimal iterative learning control method for control of wafer temperature uniformity in rapid thermal processing. *IEEE Transactions on Semiconductor Manufacturing*, 16:36–44, 2003.

Yokogawa Corporation. Platform for advanced process control and estimation brochure, 2015. https://web-material3.yokogawa.com/BU36J06D01-01EN.pdf

P. C. Young. An instrumental variable method for real-time identification of a noisy process. *Automatica*, 6:271–287, 1970.

Z. Yu, W. Li, J. H. Lee, and M. Morari. State estimation based model predictive control applied to Shell control process: A case study. *Chemical Engineering Science*, 18:15–37, 1994.

F. Zanini, C. N. Jones, D. Atienza, and G. De Micheli. Multicore thermal management using approximate explicit model predictive control. In *Proceedings of IEEE ISCAS*, Paris, France, pages 3321–3324, 2010.

V. Zavala and L. Biegler. The advanced step NMPC controller: Optimality, stability and robustness. *Automatica*, 45:86–93, 2009.

Index

Note: Page numbers in *italics* and **bold** refers to figures and tables respectively.

d

i

j

k

l

o

p

q

r

s

Printed and bound by CPI Group (UK) Ltd, Croydon, CR0 4YY

21/09/2026

14965823-0004